Unified Logic:

How to Divide by Zero,
Solve the Liar's Paradox,
and Understand the Nature of Truth

Jesse Bollinger

Revised: 2019-07-26

Authorship Info, Assets, Contributors, and Tools Used:

Copyright © 2018 by Jesse Bollinger. All rights reserved.

The cover art was based on a stock image by an artist who goes by the alias "Megin". After buying non-exclusive royalty-free rights to use the image, the image was then edited to create the final cover.

The decision to write this book dates back to roughly 2014. The exact time that each specific idea was originally conceived varies greatly though. Some ideas included in the book were conceived before 2014 and some were conceived after.

This book was typeset using LaTeX via TeXstudio and MiKTeX. See page 787 for more info on which packages were used. Information on the authors of each LaTeX package may be found by searching the internet.

The author's personal website may be found at: **JesseBollinger.com**

Copyright Warning, Legal Disclaimer, and Contact Info:

No part of this publication may be reproduced, distributed, or transmitted in any form or by any means, including photocopying, recording, or other electronic or mechanical methods, without the prior written permission of the author, except where permitted by fair use under copyright law. To request an exception, contact the author.

Contact info for the author is available at the end of the book, starting on page 803.

Just to be extra safe, some additional legal constraints are specified in the preface in the form of a liability disclaimer, starting on page 15. By reading the book or following any of its suggestions you are implicitly agreeing to the terms of that disclaimer. The author is not responsible for any consequences of your own choices.

Book Identification Info:

Hardcover ISBN:	978-1-7325366-0-9
Paperback ISBN:	978-1-7325366-1-6
eBook ISBN:	978-1-7325366-2-3

The first finished version of this book was compiled on 2018-07-01 (year-month-day). It was published and printed shortly thereafter. The first publicly available version however is the 2018-07-13 revision. The only copies predating that revision are pre-published author's proof copies. If you would like to see what version of the book this copy is then simply look at the revision date at the bottom of the title page of the book.

Contents

Preface		**9**
1	**What is truth?**	**19**
	1.1 What do people say truth is?	19
	1.2 What is a better definition of truth?	20
	1.3 How do we define truth for objects?	23
	1.4 How do we resolve statements?	30
	1.5 What's wrong with statement-based truth?	32
	1.6 What's the price of equivocation?	40
	1.7 How do we solve the liar's paradox?	45
	1.8 How do we solve the ouroboros paradox?	47
	1.9 What is the deeper meaning of truth?	55
	1.10 What's next?	67
2	**What is logic?**	**69**
	2.1 What makes us believe something?	69
	2.2 How did the formal study of logic begin?	75
	2.3 What is classical logic?	82
	2.4 What is set theory?	123
3	**What is transformative logic?**	**169**
	3.1 The surprising strength of mindless forms	169
	3.2 Transformative implication basics	179
	3.3 Transience delimiters	193
	3.4 Selective transformative implication (part 1)	199
	3.5 Tracing notation	208
	3.6 Selective transformative implication (part 2)	211
	3.7 Transformative languages and interpretation injectors	218
	3.8 Well-formed formulas and rule set adherence	228
	3.9 Quotation delimiters	234
	3.10 Miscellaneous fundamentals	241
	3.11 Transformative derivation and integration	256

3.12	Rules about rules (meta implication)	261
3.13	Inclusion directives	277
3.14	Intention marks	279
3.15	The power of conceptual clarity and proper perspective	287

4 What is non-classical logic? 291
4.1	Why do we care?	291
4.2	Informal logic	300
4.3	Predicate logic	301
4.4	Multi-valued logic	313
4.5	Relevance logic	349
4.6	Modal logic	352
4.7	Paraconsistent logic	357
4.8	Constructive logic	362
4.9	Connexive logic	366

5 What is unified logic? 373
5.1	How is it related to the other logics?	373
5.2	Introduction to primitive unified logic	375
5.3	Unified implication and its relation to language	386
5.4	Existential statements and their relation to language	406
5.5	Reasoning with primitive unified logic	426
5.6	Unified logic's relationship to the other branches of logic	463
	5.6.1 Set theory	463
	5.6.2 Informal logic	464
	5.6.3 Predicate logic	465
	5.6.4 Multi-valued logic	469
	5.6.5 Relevance logic	472
	5.6.6 Modal logic	475
	5.6.7 Paraconsistent logic	480
	5.6.8 Constructive logic	484
	5.6.9 Connexive logic	488
	5.6.10 Classical logic	491
5.7	Introduction to relational unified logic	494
5.8	The relational essence of what a thing is	513
5.9	Relational operations	519
5.10	Blueprint notation	545
5.11	Considering the broader context	567

6 A rant-filled introduction to informal logic 591
6.1	The value of informal logic	591
	6.1.1 An underappreciated field of study	591
	6.1.2 An introduction to logical fallacies	592
6.2	A guided tour of some informal fallacies	593
	6.2.1 Non-sequitur	593

	6.2.2	Red herring	594
	6.2.3	Appeal to emotion	594
	6.2.4	Appeal to authority	596
	6.2.5	Appeal to popularity	596
	6.2.6	Appeal to tradition	596
	6.2.7	Appeal to force	597
	6.2.8	Appeal to indignation	597
	6.2.9	Appeal to consequences	597
	6.2.10	Appeal to nature	598
	6.2.11	Appeal to hypocrisy	598
	6.2.12	Strawman	599
	6.2.13	Ad hominem	599
	6.2.14	No true group member	600
	6.2.15	Special pleading	600
	6.2.16	Slippery slope	601
	6.2.17	Cherry picking	601
	6.2.18	Anecdotal evidence	602
	6.2.19	Circular reasoning	602
	6.2.20	Correlation-causation fallacy	602
	6.2.21	Reification fallacy	602
	6.2.22	False dichotomy	603
	6.2.23	False balance	603
	6.2.24	Shifting the burden of proof	607
	6.2.25	Hasty generalization	607
	6.2.26	Faulty analogy	608
	6.2.27	The fallacy fallacy	609
	6.2.28	Fitting fallacy	609
	6.2.29	Proof by intimidation	611
6.3	The consequences of poor reasoning		612
	6.3.1	A society rife with ignorance	612
	6.3.2	The poison of anti-intellectualism	614
	6.3.3	An introduction to cognitive biases	627
6.4	A guided tour of some cognitive biases		629
	6.4.1	Confirmation bias	629
	6.4.2	Survivorship bias	630
	6.4.3	Observer-expectancy effect	632
	6.4.4	Pareidolia	634
	6.4.5	Subjective validation	635
	6.4.6	Denomination effect	641
	6.4.7	Dunning-Kruger effect and impostor syndrome (together: the DKI spectrum)	643
	6.4.8	Actor-observer bias	644
	6.4.9	Halo effect	645
	6.4.10	Ostrich effect	647
	6.4.11	Identifiable victim effect	648

	6.4.12	Zero-sum bias	650
	6.4.13	Statistical history bias	654
	6.4.14	Contrast effect	661
	6.4.15	Fluency heuristic	664
	6.4.16	Bikeshedding	669
	6.4.17	Attentional bias	669
	6.4.18	Anchoring bias	670
	6.4.19	Framing effect	671
	6.4.20	Hindsight bias	671
	6.4.21	Outcome bias	672
	6.4.22	Choice-supportive bias	673
	6.4.23	Sunk cost fallacy	674
	6.4.24	Optimism bias	677
	6.4.25	Golden hammer effect	678
	6.4.26	Spotlight effect	679
	6.4.27	False consensus bias	681
	6.4.28	Conformity	682
	6.4.29	In-group bias	685
	6.4.30	Reactance bias	687
	6.4.31	Moral luck	689
	6.4.32	Rosy retrospection and declinism	695
	6.4.33	Cultural calibration bias	699
	6.4.34	Tunnel bias	713
	6.4.35	Bias blind spot	717
	6.4.36	The curse of knowledge	718
6.5	The journey towards a more rational tomorrow		724

7 Extras — 727

- 7.1 Introduction — 727
- 7.2 Logic, math, and programming — 727
  - 7.2.1 A much more intuitive unit for working with angles, hidden in plain sight — 727
  - 7.2.2 π is actually not the best circle constant (not my idea, but I support it) — 730
  - 7.2.3 A new term for the study of mathematics, with less of a dry connotation — 731
  - 7.2.4 Two terms designed to disambiguate the meaning of "intuition", so that rationally justified intuitions are not so easily sneered at or dismissed — 733
  - 7.2.5 Mathematicians should borrow the concept of namespaces from computer science — 736
  - 7.2.6 An axis of rotation is *not* the conceptually correct basis for a rotation — 737
  - 7.2.7 Mathematicians need to try harder to create descriptive names — 737
  - 7.2.8 A more rational standard format for dates and times — 738

	7.2.9	Logical punctuation is much better than traditional American punctuation	738
	7.2.10	Natural selection forces sometimes *increase* complexity in collaborative documents	739
	7.2.11	Mathematics and computer science will gradually become inseparable	740
	7.2.12	The tendency of some software to drift around randomly, without ever actually solving the most important underlying issues	740
	7.2.13	Set-based file systems are much more conceptually flexible than hierarchical file systems	742
	7.2.14	A simplified alternative to traditional software versioning	743
7.3	Society		746
	7.3.1	A possible solution for out-of-control medical prices, one which will cost almost nothing and carry almost no risks	746
	7.3.2	The pillars of human prosperity and poverty	749
	7.3.3	Some obvious but currently underappreciated ways of greatly improving the fundamental structure and stability of democracy	750
	7.3.4	Ban the use of "national security" as an excuse for reducing the freedoms and rights of innocent people	754
	7.3.5	Expanding the foundational principles of a properly separated and sovereign government	755
	7.3.6	A few simple ideas for making the power dynamics between businesses and consumers slightly more symmetrical	757
	7.3.7	A theory on the evolutionary benefits of depression and why it probably exists, and therefore also a possible idea for how to naturally prevent, treat, or cure it	758
	7.3.8	The underappreciated value of self-study and automated education resources	761
	7.3.9	Diversity and decentralization of control are essential to good education	764
	7.3.10	Prisons should focus less on punishment and more on rehabilitation and personal growth	765
7.4	Productivity, creativity, and economics		765
	7.4.1	Time pools: a method for ensuring productivity even for people who hate schedules	765
	7.4.2	Crucial components of achieving high productivity while simultaneously enhancing perception of quality	767
	7.4.3	The cult of minimalism: when simplification is taken too far	771
	7.4.4	Value parasitism: an underappreciated source of economic instability	773
	7.4.5	The counterproductive nature of worrying about things you can't control	777
	7.4.6	Creativity at will: how to be creative even when you don't feel inspired	779
7.5	Addendums, reference, and technical		784

- 7.5.1 Some existing literature 784
- 7.5.2 Sources / citations / references / bibliography / etc 785
- 7.5.3 Some recommendations for good math resources 786
- 7.5.4 LaTeX typesetting code (packages and commands) 787
- 7.5.5 Contact info 803
- 7.5.6 Other books by me 804

Preface

Welcome to my book, *Unified Logic*. There are quite a few fun and unusual ideas we will be talking about in here. As you will soon see, the breadth of subjects we will be covering is quite wide and diverse. This book covers a mix of both standard and non-standard perspectives within the study of logic, as well as a mix of both formal and informal material. Many of the included ideas are original research by me. Even for the strangest of the ideas though, great care is always taken to keep such ideas well-defined and rigorous to the best of my knowledge. In fact, in some cases I actually found that the standard treatment was too vague and too poorly defined to be usable, so I had to build up some new systems of my own from scratch in some cases in order to fill in the gaps in the required foundation for the discussion.

The book also contains a bunch of other random tangential ideas that seemed interesting, which I figured might be fun to include as extras. Some of these extras are virtually irrelevant to the main content really, but I felt like they added value regardless. Clarity of thought, freedom of expression, and imaginative thinking were among the highest priorities in the writing style I chose to use for this book. I've never liked obscurantism in writing, especially of the suffocatingly opaque and convoluted academic kind. I think that it is crucial to provide people with as many "aha" moments and deep insights as possible, and I hope that you think so as well. A formal writing style could not always be completely avoided though, so the book is a mix of both formal and informal writing, but I still always try to soften the technical content to make it more easily digestible and to make sure the ideas are understood clearly.

This is a book about both broadening the scope of what logic is capable of and clarifying the basic underlying principles and assumptions of logic by drawing some interesting connections between a wide-reaching range of many different parts of the field. Logic is so much more than just applying some instinctive reasoning to whatever problems you encounter in life. There are so many different possible flavors of how exactly logic should work. Which ways are the best and which are the worst? What should the underlying philosophy of logic really be? These are some of the big questions we will be seeking answers to in this book. There are actually many different possible foundational systems for logic, and we will be surveying quite a few of them. The deeper underlying nature of logic is not at all a cut and dry issue.

You'll probably be surprised by how many hidden nuances there actually are in how logic works. There are many different options available for how a system of logic could behave. It isn't immediately clear which ones of those options are best. This is something we'll be investigating quite a bit in this book. However, the sheer size and diversity of the multitude of different possibilities for how systems of logic can behave will make comprehensive coverage impractical. As such, sometimes you will notice that I only cover the broad strokes and core motivations of many of the theories in the literature. That's because there was just too much material to learn and also because so much of it yielded very rapidly diminishing returns the deeper I tried to investigate it. Much of the details were irrelevant to the purpose of this book, so I didn't bother with them.

The reader's experience matters a lot to me. This book was written with clarity and empathy for the reader in mind. I try to always provide the underlying motivations and reasons for concepts whenever I can. I try to guide you and gently hand-hold you towards understanding, regardless of how much extra time and effort and extra pages it takes me to do so. That's part of why this book is as long as it is. I tried to go above and beyond what is normally done to ensure clarity and genuine understanding. Whereas many books on logic often give you nothing but the mechanical details, without any insight into where any of the concepts even came from, I have strove to do the opposite here. There's a big difference between merely knowing how to do something and actually understanding it. Genuine understanding takes more time to achieve, but is vastly more empowering.

Probably at least a quarter of the book (mostly chapters 2 and 4) is devoted to simply acquiring a deeper conceptual understanding of the nature of existing systems of logic and how they differ from each other. However, much of the rest of the book is my own original research. The line between standard material and my own work is not always clear in the text though, so be careful. *Unified Logic* is at its core a book about trying to solve longstanding problems in the foundations of logic, elusively enigmatic paradoxes and flaws that have apparently defied resolution for (in some cases) literally thousands of years.

It is possible that this book may have finally solved some of those. In a sense, the goal in this book is sort of to try to create "one logic to rule them all"[1], a unifying system of logic spanning a broad range of multiple different branches of logic, one that seeks to satisfy as many ideal properties of logic as it can, while still somehow keeping the resulting system very easy to use, intuitive, and flexible. The system I have created for this (i.e. unified logic) seems to come fairly close to this goal in at least several respects.

Besides unified logic though, I have also created another system of logic called transformative logic that also fulfills some very useful and interesting roles. Transformative logic is necessary as an intermediate logic for talking about other kinds of systems of logic in a more general purpose way, including for unified logic and many other non-classical systems of logic. Don't worry if you don't know what any of this means though. You'll see what I mean later.

In addition to all that though, besides the formal logic stuff, we will also be dis-

---

[1] Feel free to imagine some evil Sauron-themed *Lord of the Rings* music here... Sadly though, there will be no discussion of rings in this book. You'll need to buy a book on abstract algebra if you want that. ☺

cussing informal logic some, including logical fallacies and cognitive biases, plus various other random things unrelated to the main trust of the book (e.g. random bits of useful or interesting insights spread across multiple different subjects). Look at the table of contents if you want more detailed info. Most of this bonus content will come at the end of the book, in chapters 6 (starting on page 591) and 7 (starting on page 727), separated from the main part of the book. These last two chapters are almost completely independent from the other parts of the book, and can be read at any time really, if you want to. Chapters 6 and 7 have no real prerequisites from any of the other parts of the book, whereas all the other chapters are meant to be read sequentially without skipping anything.

This book has been written in an intentionally uninhibited, irreverent, and free-ranging style. The goal throughout writing this book was always to simply create as much potential value as possible. Inevitably, once the dust settles, some of the ideas will stick better than others. To create an abundance of interesting ideas though, one must naturally allow an abundance of different pathways of thoughts to be followed. Thus, I purposefully allowed the writing style to sometimes get temporarily derailed from the main discussion onto wild tangents.

In fact, a vast multitude of such random tangents exist inside the book. Wherever there was any opportunity to follow an interesting or valuable thread of thought, even if it sent us far afield of what we were originally talking about, I often took it. This admittedly created some pacing issues and verbosity in some parts of the book, but overall I still think that it was very worth it because it enabled me to discover a much wider range of interesting ideas by allowing myself to follow my thoughts wherever they led.

The text has been carefully edited multiple times and I have tried hard to adhere to a strict standard of logical consistency, correctness, and rigor. However, like any highly exploratory text of this nature, it is of course possible that errors have crept in. Such is the nature of science and research though, so this is to be expected and is entirely normal. Seldom does every single idea a person conceives of end up sticking. If even a fraction of the multitude of ideas I've included in this book end up sticking then that would still be a victory for science and would still be a cause for celebration. Every little bit helps, especially in the long run, and I'd be happy to have contributed to that in any way.

If the full title of this book sounds a bit nuts, well, then you probably won't be disappointed. It's pretty nuts, in a sense, but in a good way, in a way generally supported by rigorous reasoning and thought experiments. Don't leap to judgment too fast. Like all good science, the material in this book has predictive power and expressive ability beyond merely what its surface-level goals and claims are. Despite being the person who wrote the book, not even I was able to predict many of the things that the system ended up being capable of doing. It was quite a pleasant surprise really.

One could even argue that some of the intermediate results and new avenues of exploration that this book tangentially opened up are actually more important and more valuable than the goals that originally inspired it are. This unexpected predictive power is quite unlike what tends to happen in the work of nutjobs and cranks, whose material in contrast generally utterly lacks predictive power or is of a hand-wavy and meaningless quality. Nonetheless, I do not presume that all of my ideas are correct. I merely hope that my work ends up making at least some beneficial impact on the literature, the state of knowledge, and the clarity of understanding of concepts in the field of logic. That would be enough for me.

It is also important to be aware that the system of logic that we will be gradually building up in this book (not counting any review of pre-existing material from the known literature) often will rest on different axioms (i.e. different foundational assumptions) than some existing systems do, and thus that some of the most basic rules of how to think about things will sometimes be different. Don't make assumptions too quickly. You'll need to give the system a bit of time so that you can grasp the full picture, otherwise you may not really understand the point I'm making. The exact nuances of how a system of logic is designed to work matter quite a lot, as you will soon see as we progress through the book.

Anyway though, that's enough of talking about the book in loosely general terms. Let's now take some time to get a better sense of the specifics of the book, i.e. what's in it and why you should care. There's actually a pretty weird and broadly varying mix of different things in here. The book grew organically like a plant. Lots of random ideas and unexpected little bits of serendipitous value occurred to me along the way. I just went with the flow mostly.

Many of the ideas I came up with during the writing of this book would almost certainly never have occurred to me if I hadn't been willing to let my ideas flow wherever they naturally led. Without that relaxed "go with the flow" kind of approach, this book probably wouldn't even exist. Or, at least, if it did still exist then it would contain far fewer valuable ideas. That's just how causality works though. Chains of events often get tangled up in far-reaching ways. Chaos theory and emergent phenomena are also another example of that kind of thing. Everything tends to quickly become interconnected and interdependent. This same force that makes things difficult to predict is also what drives much of the natural spontaneous emergence of complexity and beauty in the world, as well as much of human creativity. That's causality for you.

Putting all that aside though, really there are far too many ideas in this book to list them all here, so I will just mention a few random interesting landmarks instead. I guess you could say that this will be sort of like an appetizer, like the preview of a movie, where they only flash a few incomplete glimpses of a few of the scenes so that the viewer can use that basic impression as a basis for deciding if the movie is the kind of thing they might be interested in. It won't give a comprehensive sense of the book though. It's just a sample. Anyway though, let's begin. Here are some of the things in the book:

- A new proposed solution to the liar's paradox, a legendary paradox of logic that has eluded humanity for more than roughly 2500 years since its inception… The solution to this longstanding enigma requires deep insight into the fundamental nature of truth, insight that (as you will see) provides illuminating new perspectives on far more than just the liar's paradox itself.

- Numerous easily understood thought experiments and "aha moments" designed to instill you with a more genuine and principled understanding of the structure of logical thought... This includes both the old material (i.e. pre-existing literature) covered by the book and the new material (i.e. my own original research). This will empower you with more than merely some memorized arsenal of dry instructions. The goal is to give you a proper understanding of the underlying essence of the concepts in play, so that you can truly attain a substantive and freely expressive grasp of how to use them.

- An overview of many different paradoxes and conceptual flaws that afflict some of the existing systems of logic that currently dominate the field... For example, we will explore the shortcomings of classical logic, as well as of some other less known systems of logic as well. For instance, we will investigate the famous paradox of entailment and the closely related principle of explosion to uncover the real underlying nature of them. We will thereby eventually come to more fully grasp exactly why classical logic's definition of implication (i.e. material implication) feels so strange and indeed actually *is* fundamentally broken as a model of implication when interpreted properly, contrary to popular belief.

- Multiple ideas for alternative terminology and alternative notation for some existing concepts, designed to clarify those concepts and to make them easier to understand and to work with... A lot of the existing terms and systems in the literature can be very obtuse, opaque, and difficult to understand. On a few occasions, when it doesn't take us too far off track from our main goal in this book, I have taken the liberty of suggesting new alternatives for these. For example, in one part of the book I suggest a new (and vastly more descriptively clear and appropriate) term for "complex numbers". That's just a random example though.

- This book is much more about logic than about math. It is designed to be useful and interesting both for laypersons and for experts. The book hand-holds the reader through almost everything. You don't need to know much logic or math to read it. Basic knowledge of algebra (e.g. knowing what variables and equations are), combined with strong attention to detail, is probably sufficient. It also often isn't important to fully understand many parts of the book to still nonetheless be able to continue reading regardless. It depends though. Reading everything from front to back is probably the safest choice if you are patient enough. However, the last two chapters are completely independent though, and typically require zero knowledge of logic or math, so less technically inclined readers are free to skip to those parts at any time.

- Although this book is titled "*Unified Logic*", its content is actually more of an even split between two new systems of logic: transformative logic and unified logic. Transformative logic is arguably actually even more general than unified logic, even though unified logic inspired the name of the book and was the original impetus for writing it. Transformative logic provides a wonderfully flexible and diverse way of expressing arbitrary logical systems, one with quite a few unexpected capabilities. If breaking down the barriers in what is possible for a system of logic to express is something that interests you, then transformative logic will probably interest you, perhaps even more so than unified logic itself. Transformative logic has a lot of highly generalized and useful reasoning mechanisms and notations.

- A wide-reaching conceptual overview of the various different non-classical logics and what their underlying motivations and guiding principles are... Most people who have received education or training in formal logic are only familiar with classical logic. Many aren't even aware that other alternative approaches to logic exist, each with their own distinct personalities and quirks. This book will open your eyes to that wider world and will teach you that the world of logic is actually far more diverse and far more vibrant than what most people assume. We won't have time for in-depth coverage of the non-classical branches generally, but I will still cover enough to give you a good sense for what the main motivating ideas involved in each are. The exact amount of coverage varies by each branch though.

- We will discover new forms of logical implication that have been hidden in plain sight all along but were somehow missed by earlier logicians. These new forms of implication provide a highly compelling alternative to classical logic's material implication, and will form part of the primary foundation upon which we will build unified logic. Furthermore, we will use the knowledge gained from uncovering this secret to bridge the gap between formal logic (e.g. classical logic, math, etc) and natural language (e.g. English, German, etc) in an interesting new way. In fact, we will see that a large subset of natural language can actually be immediately formalized entirely into terms of logical expressions in a direct and natural correspondence, a property of which I am not aware of any other system of formal logic possessing so transparently.

- We will explore a vast and nuanced web of underappreciated subtle connections between many different concepts in logic. We will free ourselves from one-dimensional black-and-white thinking and will learn to see the world in a much more colorful and vibrant light. Previously rigid and brittle concepts will be made to bend in interesting new ways, thereby giving us a better sense of where the real boundaries of justified reasoning really are. Hidden complementary relationships will be revealed, contrived artificial barriers will be broken, and strange new avenues of thought will be illuminated.

- In addition to a new proposed solution to the liar's paradox, this book also contains a new proposed solution to Russell's paradox (yet another world-famous paradox of logic), one that seems to remove the need to use a laboriously verbose system of awkward axioms (e.g. ZFC) for set theory. This allows you to have a system of set theory which is both "naive" (i.e. intuitive) and rigorous at the same time, more or less, and thus feels much more satisfying, natural, and elegant. This proposed solution also has other unexpected consequences and removes additional limitations beyond just Russell's paradox itself. It even provides some insights into the nature of what it really means for something to exist as a distinct and meaningful entity within any hypothetical logical world, in a way that enables some bizarre new operations to be performed.

- Among the many bizarre new operations that unified logic makes possible, many are so weird and unexpected as to seem almost unbelievable perhaps (depending on your personality). This even includes crazy things like division by zero, and that's not even the weirdest example, not even close. This all sounds totally nuts of course. And yet, as crazy as it sounds, the system works. The results of these operations appear to be rigorously well-defined within unified logic and are freely usable in arbitrary expressions, just like any other normal objects in logic or mathematics would be, although admittedly the results are also sometimes computationally intractable, depending on each specific case.

- Want to learn how to protect your mind from manipulation by others? Want to be able to more consistently spot the telltale signs of poor reasoning and misinformation, both from others and from within yourself? The chapter on informal logic (chapter 6, starting on page 591) will help you with that. You might be surprised just how ridiculously common incorrect and counterproductive patterns of thought really are in our society. The study of informal fallacies and cognitive biases will open your eyes to that, and will also make you able to deflect bad arguments and manipulation attempts by others more often and more easily. I'll also occasionally give you a few useful bits of life advice too, randomly mixed in with this.

- Lots of miscellaneous useful ideas, located in the Extras chapter (chapter 7, starting on page 727)... For example, have you ever been frustrated that sometimes you just can't seem to find enough "inspiration" to get creative work done? Well, if so, then the "creativity at will" extra (see page 779) would interest you. It will teach you how to force yourself to always be able to be at least somewhat creative, even when you don't feel inspired, using some easy foolproof techniques. For another example, I came up with an idea for how to potentially significantly help reduce out-of-control healthcare prices, one which would cost almost nothing and carry almost no risks for society. Those are just two examples though. There are many more. For instance, do you hate rigid schedules but still want to be productive? I have a possible solution for that too.

In addition to this bulleted list above, which provides a small sample of what kind of content to expect in this book, remember that you can also look at the table of contents (which starts on page 3) for additional overview information. By the way, the reason why many of the chapter and section titles were phrased as questions was to emphasize the importance of asking questions over making premature assumptions, but sometimes this convention was dropped in cases where it became too cumbersome.

Also, notice that my contact info is available at the end of the book (see page 803) in case you would like to contact me for whatever reason. You can also find some of the typesetting code that I used to create this book near there (see page 787), in case you would like to use some of the same features and symbols that this book uses inside your own documents. There is also some info about the sources I used and some pre-existing literature near the end of the book.

Since this book does contain some life advice and similar items, I must include a legal disclaimer for my own safety against opportunistic legal claims. I am not fond of this kind of tedious and unfriendly text, but nonetheless it is better to be safe than sorry. I need to protect myself against the often capricious and draconian nature of our current legal system. My apologies in advance. You are probably accustomed to these kinds of things by now though, especially if you've read a lot of books or used a lot of software or other products, etc. Anyway though, here's the **legal disclaimer**:

> The information provided within this book is for informational, educational, and entertainment purposes only. All content in this book is purely the author's own personal thoughts and opinions, even when presented as if fact, and is not being provided to you in any kind of professional capacity. There are no representations or warranties, express or implied, about the completeness, accuracy,

reliability, suitability, efficacy, or claimed effects of any of the content contained in this book for any purpose. You should not assume that any of the information contained in this book is correct. It may contain unintentional errors, flaws, and omissions. Regardless of whether the information is correct or not though, any use of the information is at your own risk.

On rare occasions, this book may make statements related to physical or mental health. However, this book is not meant to be used, nor should it be used, to diagnose or treat any medical condition. The author is not a medical professional of any kind. For diagnosis or treatment of any medical problem, whether physical or mental, consult your physician, psychologist, psychiatrist, or other qualified professional. The publisher and author are not responsible for any physical or mental health needs that may require medical supervision and are not liable for any damages or negative consequences from any treatment, action, application, or preparation to any person reading or following the information in this book. Always consult your doctor for your individual needs.

The author does not assume and hereby disclaims any liability to any party for any loss, damage, or disruption caused by errors or omissions, whether such errors or omissions result from accident, negligence, or any other cause. Any names of people (besides obviously intentional citations), businesses or places, events or incidents, are fictitious. Any resemblance to actual persons, living or dead, or actual events is purely coincidental. This content disclaimer is not the only legal notice or contract that applies to this book. The separate copyright notice also applies. All rights are reserved and significant portions of the book may only be reproduced with the author's permission, except where permitted by fair use. The author retains full copyright.

By choosing to read this book, you agree that you are 100% responsible for all of your own choices and that under no conceivable circumstance whatsoever can the author ever be held liable or responsible for the consequences. The author of this book assumes absolutely zero liability for any advice or other content contained in this book. Your choices are your own, not the author's.

Even for cases not explicitly covered by the text of this disclaimer, by choosing to use any of the information, opinions, or advice contained in this book in any way you thereby instantly and irrevocably waive all rights to any claims against the author in connection with this book. Furthermore, you agree that if any part of this disclaimer is struck down or rendered unenforceable or in any other way invalid then the other parts will still apply, such that the author's liability is always minimized as much as possible under the law.

Also, in the time since the first publicly released version of this book (the 2018-07-13 revision) was made available, I have become more aware of what kinds of people seem potentially more likely to react negatively to the book and to not enjoy themselves. It is not my intent to give anyone a bad experience, nor do I want anyone to waste any of their money. As such, in the interest of better serving my readers, I have decided to collect together a list of traits that seem to maybe be highly correlated with negative reactions to the book. Often, some of the people who react negatively to the book seem to have at least one of these traits. Ask yourself if any of these apply to you (but feel free to only read the boldface text if you want to save some time):

This book may offend you if...

1. **You equate non-standard ideas with arrogance.** Being acquainted with the main ideas and standard literature of a field of study does indeed generally provide you with a very useful and empowering foundation. However, it would be naive to treat that as some kind of gospel. There's a difference between merely absorbing an existing body of knowledge and actually being in touch with the underlying spirit of discovery. Someone whose knowledge is only surface-deep may treat dissent as heresy. In reality though, irreverence is the heart of innovation. Progress happens through experimentation, not censorship.

2. **You are highly religious and easily offended.** On a few occasions, religious beliefs are used as examples of invalid reasoning, especially with respect to their often untestable nature and their tendency towards suppressing dissenting ideas. Such cases constitute only maybe 1% of the book, and could easily be ignored, but I've noticed that highly religious people seem to be among the most likely people to react negatively to this book.

3. **You give status symbols more weight than evidence.** Science requires a willingness to give logic and evidence more weight than traditions or social structures. Nonetheless, there are unfortunately quite a few people even in science who think primarily based on authority fallacies instead of real merit. This often manifests as gatekeeping behavior, where the evidence is ignored if an idea isn't mainstream enough or didn't come from a sufficiently high-ranking authority figure. If you have such authoritarian tendencies, then you may be too biased to understand this book.

4. **You are quick to grab a pitchfork and slow to express empathy.** There are many people, especially recently, who seem disproportionately preoccupied with finding excuses to attack and condemn other people without first giving adequate time to question their own assumptions or to empathize with the other human being on the receiving end. Such social posturing provides an easy way for many people to make themselves feel superior, and is closely related to bullying in terms of its underlying motive. Such behavior is toxic. Personally though, I believe in cultivating a spirit of generosity and being slow to assume things about other people, and I hope you do as well.

5. **You become jealous easily and it distorts your perspective.** I've noticed that there's always some people who will become increasingly likely to become hostile or dismissive the more the richness and diversity of experience in someone else's life becomes apparent. When someone is jealous of someone else, they will often make any possible excuse to label the target of their envy as a "jerk" or as "arrogant" etc, when it is really the jealous person's own emotions speaking. I'm a creative and diverse kind of person, so I'm often the target of such feelings.

6. **You want a reflowable ebook.** An ebook can be either reflowable or non-reflowable. Reflowable ebooks allow the text to be moved and resized to dynamically fit a screen. Non-reflowable ebooks in contrast have a rigid predetermined format for each page. Books that contain special content often require a non-reflowable format in order to work correctly. This book is one such example. Yet, despite the fact that the book is clearly non-reflowable and despite displaying multiple warnings about this, some people keep accusing the book of being "defective" just for not being reflowable. The book is not defective though. These people are simply not aware that non-reflowable ebooks exist.

More broadly speaking though, and perhaps most importantly, this book was designed to be exploratory in nature and should be interpreted with that same kind of liberated spirit and adventurous pragmatism. The book is meant to be consumed with a healthy "separating the wheat from chaff" kind of attitude. If you're getting bent out of shape about it then you're reading it wrong, in a way I didn't intend it. Highly creative ideas tend to require a willingness to sometimes just roll the dice and see what it lands on.

Innovation often requires some degree of assertiveness and boldness. If I had been too uptight and fearful about how I wrote the book then it very likely never would have been created and thus none of the corresponding value would have ever been created either. Thus, you will see that I am often bold in my writing style. I do this because I know that regardless of how I frame the ideas a critical eye will still nonetheless be applied and so ultimately everything will still work out one way or the other.

Natural selection pressures will ultimately decide what ideas live on and what ideas die off, just as it should be. Thus, an aggressively creative and uninhibited approach seems like the approach most likely to maximize value. That's the kind of philosophy I applied to writing this book. The dust will settle however it will, hopefully with at least some of the ideas having made a positive impact. This is the kind of easygoing attitude that I hope all readers would adopt, but I sometimes forget that some people are quite a bit more high-strung and hypersensitive than that.

Anyway though, that basically sums up the preface. We're now ready to move on to the main content of the book. Oh, and remember that everything except for the last two chapters is designed to be read sequentially, without skipping around much. This book is exploratory. It is not a reference book. There's lots of fun and useful stuff ahead for us. I hope you have as much fun reading it as I had writing it, if not more. It's been a wonderful journey of discovery for me and now it's time to finally share that with the world. I hope to see you at the conclusion at the end of the book. Enough talk though. Let's begin.

Chapter 1

What is truth?

1.1 What do people say truth is?

Humankind has always felt irresistibly compelled to pursue the truth. Why do we feel drawn to the truth so strongly and what exactly is it?

We spend most of our lives trying to ensure our personal well-being and to fulfill our interests. Every now and then though, we look at the world around us and wonder what the underlying truth behind our observations is. What is it that we are really seeking when we go looking for the truth?

What do dictionaries say truth is? Perhaps we might find insight into the nature of truth by looking up the term. Let's check out a few definitions and see. When I type "truth" into a search engine on my web browser the first definition that pops up is "the quality or state of being true" which is mostly unhelpful except insofar as it sort of explains the grammar of the word "truth" relative to the word "true".

Exploring the search results further though, I do see a few somewhat more helpful definitions such as "that which is true or in accordance with fact or reality", "the true or actual state of a matter", "conformity with fact or reality", "a proven or verified principle or statement", and so on. I could list even more examples of published definitions of truth, but doing so wouldn't really add any substantive value to our discussion, so I won't. It would just be more of the same. We already have enough.

These definitions do feel somehow dissatisfying though, don't they? Why is that exactly? What is it that distinguishes dictionary entries that fall flat from those that give real insight? A little bit of thought reveals the reason why these definitions feel somehow empty: All of these definitions of truth listed above have at least some element of circularity or tautology.

Each of these definitions either directly uses the word "true" or else uses a word which is a synonym of truth in some sense or which itself would depend upon a definition of truth in order to be defined. For example, the words "fact", "proven", and "verified" are all nearly synonymous with truth and would depend on some notion of truth in order to themselves be defined.

However, there are a few parts of the definitions above which provide some substantive non-circular information. Take for example the "conformity with fact or reality" definition. It suggests that something is true by virtue of corresponding to reality in some way, but what exactly is the "something" which is doing the corresponding?

Another one of the definitions gives us a hint for determining what the "something" that does the corresponding might be. The definition that says that truth is "a proven or verified principle or statement" suggests that perhaps the thing doing the corresponding is a statement. A statement is essentially any piece of information communicated through some kind of physical medium (e.g. text, speech, etc) that makes an assertion.

Combining these two definitions together we might then suppose that truth is "statements that correspond with reality". If we take this to be the definition of truth, then it also implies that truth is somehow "owned" by statements or is a property of statements and is inherently bound to them.

This is indeed more or less the definition of truth that most people who study logic subscribe to, provided that you interpret the word "reality" loosely, as referring to any arbitrary but still consistent logical system, rather than necessarily as referring only to the physical world.

Currently, in modern logic, truth is considered to be a property of statements and not of objects. For example, if I asked a logician "Is it true that multiplying an integer by 2 always results in an even number?" then the logician would say yes, provided they are sufficiently competent in math. If on the other hand I asked a logician "Is the spoon true?" then they would probably look at me quizzically or laugh or assume they misheard me.

The question "Is the spoon true?" would be considered to be nonsensical gibberish by logicians. Just because you can string a sequence of words together doesn't make it a meaningful or well-defined expression. If truth is strictly a property of statements, then a spoon can't be true, because a spoon is not a statement — a spoon is an object.

This all sounds perfectly reasonable of course. It seems illogical to say that a spoon is true, doesn't it? Our common sense and intuition immediately suggest to us that it would of course be absurd to assert such a strange notion as a spoon being true. This all seems to confirm our tentative definition of truth as "statements that correspond with reality".

You might wonder why we're even talking about all of this. Isn't it all so incredibly obvious that it hardly merits even being said? Isn't it obvious that an object has no truth value, since it doesn't state a fact about the world? It sure seems like it.

We are smart enough to be able to understand the nature of this distinction between statements and objects immediately. We can trust our common sense and intuition here, since this is such a simple and easy to understand case. Can't we?

Think again. We cannot.

1.2 What is a better definition of truth?

Let me pose a thought experiment that will show you that our supposedly "obvious" assumptions here, far from being enlightening, actually have us swimming in dark waters, unaware of the magnitude of the error we have just made.

Suppose we have two universes. Let's call the first universe S and the second universe O (the letter O, not the number 0). Suppose that both universes have essentially the same physical laws and content as ours, except as otherwise specified. Suppose also that the two universes are identical to each other in all respects, except as otherwise specified.

Let's first consider universe S. Some of the sentient beings living in universe S are capable of forming logical statements about truth or falsehood, which may then be evaluated. The S in "universe S" stands for "statements", which is why we call it universe S.

For example, consider a meandering path through a forest located in universe S. Like most paths through most forests, this path is surrounded variously by shrubs, wildlife, trees, rocks, streams, and so on. Anyway though, suppose that one day a sentient being (e.g. a human) is walking down the path. We will call this person Aristotle in honor of the ancient logician of the same name.

As Aristotle walks down the path, he can look up through the forest canopy and see that the sky is clear and blue and therefore infer that it is still daytime, since the sky would be black if it was night or perhaps red or orange if it was dawn or dusk. As he realizes this he says "It is daytime now." and his statement is true.

Likewise, Aristotle can take a detour off the path and walk up to the edge of a stream, where he can see that there are fish swimming quietly through the water and insects buzzing nosily above. He can then declare "There are fish in this stream and bugs in the air. Also, I can hear more sound coming from the bugs than from the fish." and yet again he would be correct.

Aristotle can say "This rock here is metamorphic." or "That fox got scared when it saw me and ran off to its den." or "These berries are poisonous when consumed by humans." and all of these statements could be true, i.e. could correspond to reality.

However, on the other hand, consider universe O. This other universe is identical to universe S in all respects except for that logical statements about truth or falsehood do not exist anywhere in it. In that respect, it has the opposite property of universe S. Universe O is a world of objects only, where there is no such thing as a statement in any form whatsoever. The O in "universe O" stands for "objects", which is why we call it universe O.

Here is the crux of the thought experiment though: Aren't all the statements that Aristotle made during his walk true in both universes, regardless of whether he said them or not? If truth were indeed strictly a property of statements, and if objects could not have truth values (as is the current implicit assumption), then wouldn't it follow that universe O has no truths?

Under our current definition of truth, there would be nothing in universe O that could ever hold a truth value. We have now arrived at a dilemma. On the one hand, it seems unreasonable to think that the mere absence of statements in universe O would somehow make it lack any form of truth. On the other hand though, the only way universe O could ever have any truth values would be if objects have truth values.

This directly contradicts our intuitive common sense notion that truth is "statements corresponding with reality" and that asking questions like "Is the spoon true?" is somehow nonsensical. What is true of universe S must also be true of universe O, with regards to all of the aspects in which they are identical.

However, in order for universe O to have truth values we have no choice but to deny the idea that truth is defined merely as "statements corresponding with reality", otherwise we would be contradicting ourselves. Therefore truth must be a property of objects and not of statements.

Notice that this is the exact opposite of our original intuitive notion of truth. What we were previously assuming was obviously the correct thing to attribute truth values to (i.e. statements) is in fact the wrong answer, and what we were previously assuming was obviously an absurd thing to attribute truth values to (i.e. objects) is in fact the only possible choice that doesn't result in contradictions in our notions of what reality is.

Are you shocked? If you are, then let this be a lesson in how reliable "common sense" really is, which is to say it is hardly reliable at all. Common sense is little more than merely lazy thinking and blind preconceptions. One of the most apt descriptions of the real nature of so called "common sense" (and also one of my favorite quotes) is one attributed to Einstein: "Common sense is the collection of prejudices acquired by age eighteen."[1]

One counterargument to this idea of objects having truth values though that I imagine might occur to some readers is to draw a comparison to the classic Buddhist thought experiment where one is asked "If a tree falls in a forest and no one is around to hear it, does it make a sound?". The idea is to make us question the relationship between reality and the mind. It also arguably conflates the difference between hearing a sound and producing a sound, but that's not the important point for our purposes.

The question compels us to consider the possibility that reality is nothing more than a figment of our minds. By suggesting that perhaps reality only exists within our own minds, one could thereby argue that because our minds are not present in universe O, then the objects within it and their associated truth values do not truly exist. This idea that reality is a figment of the mind is known broadly as philosophical idealism.

The opposite of philosophical idealism is philosophical realism, which instead roughly asserts that it is best to assume that reality exists independently of us and that we are merely observers of it. Closely related to realism is philosophical naturalism which is the idea that only natural laws and natural forces operate within the universe, as opposed to any supernatural forces or any other entities external to the universe itself.

Philosophical realism and philosophical naturalism are highly compatible with science and the systematic pursuit of truth. Philosophical idealism is not. No amount of effort could ever make philosophical idealism scientifically sound or useful. In order for a claim to be considered scientific it must be falsifiable, which is to say that it must be possible to prove or disprove the claim in principle. Philosophical idealism claims that reality is a figment of the mind, but there is no way one could ever test this claim.

---

[1] I searched the internet and it appears that mostly likely the quote is actually a paraphrasing of Einstein's opinion on the matter rather than a direct quote, and it was published in a book written by someone else. However, it still appears to capture something Einstein did indeed express, which is good enough for me. Ultimately it's the substance behind the quote that really matters, not so much who said it. For more information, including citations of historical documents, see *quoteinvestigator.com/2014/04/29/common-sense*

In contrast, realism keeps us focused on only the testable aspects of reality. The pursuit of untestable claims is fundamentally a futile endeavor and always will be.

In the pursuit of knowledge it is important to not get sidetracked by claims that are inherently impossible to resolve or ultimately meaningless. Some philosophers are often fond of endlessly debating epistemology[2] and arguing that we in some sense "can't really know anything" because of a lack of certainty about any higher realms of existence that we are unaware of. For example, we might all be part of a computer simulation, etc.

Philosophers often consider such discussions to be "deep" and somehow of profound significance. Far from being "deep" or profoundly significant however, such discussions are generally simply worthless and are merely an exercise in futility, except for perhaps insofar as such discussions exercise a person's mind (however slightly) by at least making them think a bit more. Be that as it may, unfalsifiable claims are claims that cannot be tested, and therefore no amount of discussion of them will ever yield any answer nor any practical use.

This is my response to the philosophical idealism counterargument that might have occurred to some readers when I argued my case that truth has to be a property of objects rather than of statements. The idea that reality is nothing but an illusion in our minds is simply not productive. Not only does it not predict anything, but it adds on extraneous untestable assumptions to our worldview that have no merit and no practical use.

To reiterate the point, consider the following apt words of David Hume: "If we take in our hand any volume; of divinity or school metaphysics, for instance; let us ask: Does it contain any abstract reasoning concerning quantity or number? No. Does it contain any experimental reasoning concerning matter of fact and existence? No. Commit it then to the flames: for it can contain nothing but sophistry and illusion."

Anyway, having now addressed that particular objection, let's return to the main line of thought. As I was saying, in order for universe O to have truth values, we have no choice but to deny the idea that truth is defined as "statements corresponding with reality", otherwise we would be contradicting ourselves. Therefore truth must be a property of objects and not of statements.

That means that questions like "Is the spoon true?" will need to have well-defined truth values. But how are we supposed to give a spoon — or any other object for that matter — a truth value exactly? Doesn't that not make any sense? Are we screwed either way, regardless of whether we attribute truth to statements or to objects?

1.3 How do we define truth for objects?

Luckily we are not. As it happens, there is a thought experiment that will shine some light on this rather perplexing situation.

Let's see if we can find any distinguishing features of objects that would give us a hint as to how to go about assigning truth values to them. In other words, let's see if

---

[2] Epistemology is "the study of the nature of knowledge". More cynically though, in practice it often ends up just being "the study of how to waste time asking pointless existential questions instead of doing real science".

there are any patterns or commonalities among all the different kinds of objects we can conceive of which somehow seem to fit some definition of truth.

Consider the following list of objects, for example:

Object	Truth Value
spoons	?
rocks	?
dry water	?
bread	?
spiders	?
rain	?
unicorns	?
castles	?
bright darkness	?
nuclear reactors	?
flat mountains	?
electromagnetism	?
written language	?
the living dead	?
hot ice	?
web browsers	?

Suppose I asked you to scan through each item in this list in turn and assign a truth value to each. To which ones would you assign true and to which ones would you assign false? Reading over the list, can you see some kind of special distinction between the objects which would allow us to separate the items out into two distinct classes of objects based on some feature they share in common? Take a moment to think about it before continuing on.

Ignore any urge you may have to think of assigning truth values to objects as being inherently nonsensical, for the time being, and just indulge me anyway. If you had no choice but to assign these objects truth values anyway, then which objects would you choose to be true or false, in such a way as to make the most meaningful and reasonable allocation of truth values possible (ignoring pedantry)? Once you've seen the distinction, it should be easy.

Had enough time to think? Here's the answer: The way to assign a truth value to any given object in the list is to simply assign the truth value based on whether that object exists or not. To any object which is known to exist, assign the value true. To any object which is either self contradictory or empirically not evident, assign the value false.

Here is what the resulting table ends up looking like:

Object	Truth Value
spoons	true
rocks	true

dry water	false
bread	true
spiders	true
rain	true
unicorns	false
castles	true
bright darkness	false
nuclear reactors	true
flat mountains	false
electromagnetism	true
written language	true
the living dead	false
hot ice	false
web browsers	true

Simply put, in a system of thought in which truth values are a property of objects (as opposed to of statements), whether or not something is true hinges on whether or not that thing exists. What exactly does it mean to exist though? Does software on a computer count as existing? What about a fictional entity such as an elf or a dragon?

To answer that question, you have to understand that when I say "existence" I am speaking broadly about any entity that exists in any logically coherent system, and which is in the implicit universe of discourse we are currently considering.

What is a "universe of discourse" exactly? A universe of discourse is a kind of conceptual world to which we have confined our thoughts or considerations. For example, if the universe of discourse is physical reality then things that do not exist within physical reality are outside the universe of discourse and hence are considered false, even if they exist in some sense in some other universe of discourse (e.g. in a fictional or virtual world) as "true" entities.

What about things like software? Does software exist in the sense of fulfilling truth? Yes, indeed it does. At least, it does unless you've defined your universe of discourse to intentionally exclude software and other virtual existences. For example, suppose someone is playing a video game on their computer. If the player uses their wizard army to rain fireballs down upon a village of unsuspecting goblins, then as far as the game world is concerned wizards and goblins and fireballs all really do exist. The fact that these things don't exist in conventional reality doesn't stop them from existing virtually in a simulated self-consistent logical system such as a game.

What about entities in works of fiction, outside the context of tangible simulations such as games? Take for example the elves from the novelist J.R.R. Tolkien's well-known series *The Lord of the Rings*. The key to figuring out in what sense a fictional entity "exists" is to make a careful distinction between the idea of it existing in reality and the idea of it existing in fiction.

Given any fictional entity X, for example, you could make a point of creating two new distinct words that allow you to disambiguate which one you are talking about. One possible format, for example, is to name one of them X-in-reality and the other X-in-fiction. So take for example Tolkien's elves. Do elves-in-reality exist? Certainly not, as far as any evidence we have is concerned. On the other hand, do elves-in-fiction exist? They sure do. I can pick up a Tolkien book and point out passages describing them, hence in the context of that particular fictional universe of discourse they really do exist.

If you were interpreting "unicorns" in the table of object truth values above as referring to the existence of unicorns in a purely fictional environment then you would be justified in arguing that in that fictional context they are true, i.e. fictionally existent. In the case of filling out that table, I considered the universe of discourse to be physical reality, since that was the most reasonable default. It is possible to choose other universes of discourse though.

So in summary, whenever I talk about "existence" in this book I mean to say that it exists in some specific logically consistent system. That system could be reality (the physical world) or it could be something virtual (like software). It could be dynamic (like a game) or static (like a novel).

However, that does not mean that anything goes. It merely means you have to interpret "existence" carefully based on the context. Leave your preconceptions at the door, but at the same time don't think that the broadness of this criteria gives you some kind of excuse to believe in unjustified claims.

Notice also the interesting fact that even virtual existences and fictional existences are still ultimately tied to the physical world. Virtual existences exist in a computer, which is a physical object. Likewise, fictional existences exist in a book, which is also a physical object.

Thus, in some sense one can argue that reality is indeed the only true plane of existence. I thought it merited discussion regardless though, so that you would not fall prey to any popular misconceptions or philosophical traps[3] about what the notion of existence can reasonably encompass. Everything that can be said to truly exist must have something tethering it to the physical world, even if indirectly.

Having now sufficiently discussed the subtleties involved in correctly defining what it means to be true and what it means to exist, it is now time to give an official formal definition of the basic truth terms.

When we first began this chapter, we suggested the possibility that truth might be defined as "statements that correspond with reality" and then demonstrated that such a definition was somewhat circular and would ultimately create a dilemma. We solved this dilemma by carefully resisting our urge to rely on our intuition and by using thought experiments to uncover a plausible alternative.

As such, we are now able to suggest a definition of truth that has the wonderful property of being defined purely in terms of more primitive concepts[4] and which is completely non-circular. This may be the first definition of truth to ever fully satisfy

---

[3] I vaguely recall once reading about a philosopher who got so tangled up by the question of whether or not entities that reside inside the mind "exist" that he concluded that there must be something wrong with logic itself, rather than with his reasoning. Of course, the real problem was that his pretentiousness kept him from realizing he was simply being incredibly sloppy. Poor reasoning often goes hand in hand with so called "philosophical depth" unfortunately.

[4] Objects and existence are more primitive concepts than the concept of truth because all statements about truth depend upon referring to objects or existence in some way, and while some people might question the nature of truth few would honestly question the mere existence or non-existence of objects.

these criteria, as far as I am aware, but like all human beings my knowledge is inherently limited and error-prone. Anyway, here is the definition:

Definition 1. *An object is said to be true if it exists within the universe of discourse being considered. Otherwise, if it does not exist, then it is said to be false. Truth is a property of objects, not of statements. However, statements are still necessary for referring to objects and discussing them. To distinguish this definition from other definitions of truth we refer to this definition as* **existence-based truth**.

Definition 2. *For reference purposes, truth that is defined as being "statements that correspond with reality" (i.e. the traditional definition) will henceforth be referred to in this book as* **statement-based truth**. *In statement-based truth, statements own the truth values to which they refer. This is in contrast to existence-based truth, where the referent objects own their own truth values and statements merely serve as a pragmatic communication tool.*

OK... So we've defined what true and false mean and we have stated that truth is a property of objects. However, one may now wonder how we are supposed to handle statements properly if those statements no longer have truth values in the sense that they used to under statement-based truth.

Well, the key to handling statements in an existence-based truth system is to realize that statements actually have two different kinds of truth values associated with them rather than merely one. This is in stark contrast to popular opinion and intuition which say that every statement is simply true or false and that hence there is only one truth value associated with any given statement taken as a whole.

The necessary insight is to realize that statements are objects too, no less than the referent objects are. We now feel compelled to ask: What exactly is the substance of a statement made up of as an object? What will that ultimately imply about how to properly handle its truth value?

Statements can be communicated via a vast plethora of different forms. Statements can be communicated through letters on a piece of paper (text), through pressure waves transmitted through the air (speech), through raw electrical pulses (telegraph), through electromagnetic radiation (radio waves), through digital network data (IP packets), through gestures and body language (sign language), and through ancient stone carvings (runes, cuneiform, and hieroglyphs) just to name a few.

The matter and energy that is transmitted through or stored within a medium of communication, and which can be subsequently interpreted as a representation of some concept known in advance by the receiver, forms a set of symbols.

Symbols do not have to be text in order to be valid symbols. A symbol just has to be any arbitrary object which stands for or implies a meaning of some kind. As objects, symbols include only the matter and energy that was essential for the transmission or storage of the meaning. In short, symbols are the physical substance of communication, in whatever form it takes.

Symbols are one of the two types of objects associated with a statement. The other type is the referent objects, i.e. the collection of objects to which the statement refers. By interpreting the symbols of a statement to discover what objects the statement is talking about, and also what it is saying about them, we are able to derive the actual meaning of the statement. This underlying intention or meaning of a statement is referred to as its semantics.

We now know enough to officially define the two different types of objects associated with any given statement. Here are the definitions:

Definition 3. *The raw matter and energy through which a statement is communicated, over any arbitrary physical medium capable of conveying it, will henceforth be referred to as the **symbolic set** of the statement. The symbolic set refers only to the raw substance of the statement and should not be considered inherently meaningful.*

Definition 4. *The set of all objects to which a statement refers once properly interpreted will henceforth be referred to as the **semantic set** of the statement. The semantic set captures as many objects as necessary to capture the full meaning of the statement, including not just physical objects but also any conceptual objects and relationships that are relevant. The semantic set of a statement and the meaning of a statement are synonymous. The semantic set fully expresses a statement's entire meaning. Just like the symbolic set, it is simply an inert object. Once constructed, the semantic set is no longer open to reinterpretation or symbol parsing, although the resulting relationships can still be analyzed of course etc.*

When a statement has been decomposed successfully into a symbolic set and semantic set, then that statement's meaning has been fully captured and the original statement no longer has any value for consideration. Attempting to think in terms of the original statement again after already decomposing the statement into a symbolic set and a semantic set will often only result in confusion and fallacious thinking.

Don't do it. You will see why later, once we get to that part. Put another way though, a statement stops being able to "talk" once it has been converted into its constituent symbolic and semantic sets. This will become important for how certain cases are resolved. It is sometimes necessary to make this distinction in order to ensure that things will be interpreted correctly.

With all this talk of sets though, it would probably be wise to give you some more information on what a set is. You may already know what a set is or may have inferred it from the context, but let's define it and its relation to our notions of objects and existence here anyway:

Definition 5. *A **set** is an arbitrary collection of objects. The order in which the objects are listed is irrelevant and duplicates don't matter. If a set includes at least one object (i.e. if it is non-empty), then we will say that the set exists and is therefore true. Whereas, if a set does not include any objects (i.e. if it is empty), then we will say that the set does not exist and is therefore false. There is more to the definition of a set than just this, but this partial definition will suffice for now.*

As you can see, the symbolic set and the semantic set both have well-defined truth values according to our notion of existence-based truth. Each set could conceivably be empty or non-empty, so there are several possible outcomes. Let us now consider then, to ensure that we properly understand it, what some examples of true and false sets of each of these respective kinds might be.

Consider the notion of a symbolic set. What is a good example of a true symbolic set? I'll give you a hint: You shouldn't have to look any farther than the sentence you are reading at this very moment in order to find one. I am referring of course to the previous sentence itself, which is a symbolic set that exists[5] and is therefore symbolically true.

Notice however that saying an object is symbolically true is not the same as saying it is semantically true. In this specific case both sets are true, but I was referring specifically only to the existence of the symbolic one when I made the inference that the symbolic set was true. It is critical that you fully grasp this distinction in order to be able to understand the system of truth we will be developing.

In fact, this book is of course riddled from top to bottom with a vast number of statements, as are the overwhelming majority of books. As I mentioned earlier, there are numerous possible mediums of communication so there are likewise numerous other examples of true symbolic sets.

In fact, it is not merely that tons of statements exist, but rather we can say something even stronger and more interesting: You literally cannot even find a physical example of a false symbolic set. It is impossible. Finding such a set would contradict the set's very own definition of being a set that does not exist, hence it cannot possibly be found.

You cannot observe a statement that does not exist. A symbolic set is simply the physical substance by which a statement is communicated, and thus if it was empty then it would be the same as being a non-existent statement. It would be like searching for messages in an empty notebook or a blank audio tape.

However, that does not mean that the notion of a false symbolic set is without meaning. While we may not ever be able to witness a false symbolic set in the flesh and blood in the real world, we can describe what one would conceivably "look like".

For example, consider the hypothetical statement "She stroked the [X]at." such that the "[X]" in the statement is simultaneously strictly equal to the letter "c" while also being strictly equal to the letter "b". Hence the hypothetical statement purports to be both "She stroked the cat." and "She stroked the bat." at the same time. A single letter cannot be two different letters at the same time, and therefore this hypothetical statement's existence is self-contradictory. Self-contradictory objects cannot exist, therefore this statement cannot exist, and therefore the symbolic set of this statement must be false.

Now, having covered examples of symbolic sets, let's consider examples of semantic sets. Consider the statement "All numbers which are multiples of 6 are also multiples of 3." and ask yourself if it is true. Indeed, the statement is true. It's semantic set is true, which is the same as saying that its semantic set exists. However, what exactly is the semantic set in this case?

By recalling our definitions, we can see that the semantic set here is the collection of all objects that meet the criteria specified in the statement, plus whatever conceptual relationships hold between them. In this particular case then, the semantic set includes the set of all multiples of 6 that are also multiples of 3. As we can see, $6 = 2 \cdot 3$ and therefore any number which is a multiple of 6 must also be a multiple of 3. Since we

---

[5] The fact that you are looking at it proves that it exists.

found a non-empty set that satisfies all the criteria implied by the statement, then we know that the statement is therefore semantically true.

On the other hand, consider the statement "The number x is less than 7 and greater than 12." and ask yourself if it is true. What is its semantic set? In other words, what are the concrete objects to which it is referring? There aren't any. The semantic set is empty. There is no such thing as an object which is a number and is less than 7 and greater than 12. Even without knowing anything about x besides its name, we can still infer that it could never satisfy the criteria.

We have now explored an example of each of the possible truth sets associated with any given statement: one true symbolic set, one false symbolic set, one true semantic set, and one false semantic set. However, as a point of additional nuance, notice that if the symbolic set is empty then the statement cannot have a non-empty semantic set. A statement that does not even physically exist cannot successfully refer to anything and hence cannot have a non-empty semantic set.

1.4 How do we resolve statements?

There is yet another possible form that a statement can take that we have not yet covered, but which is necessary for fully understanding how to handle statement semantics. Some statements are gibberish, and we should know how to handle them. Let us define what we mean by gibberish exactly, so that there is no ambiguity:

Definition 6. *Whenever a statement consists of symbols that do not appear to have any coherent meaning, such that it is impossible for an interpreter to determine what objects and relations the statement refers to (if anything), then the semantic set of the statement is said to be* **gibberish relative to the interpreter**. *When the choice of interpreter is implicitly understood in the context, then we may refer to such a set simply as a* **gibberish semantic set**. *All gibberish semantic sets are false, which is to say that the semantics of the statement are meaningless and hence specify no objects and hence cannot possibly express any truths.*

There are many different ways that a statement might be gibberish. The statement could be malformed in some way, such as if the symbols are arranged in a nonsensical way or bear no relationship to each other. The symbols could just be drawings that were never meant to mean anything to begin with. Use your imagination.

For example, consider the alleged statement "☺∞☐☒☉♪" interpreted according to your own vocabulary. The sequence of symbols is gibberish and hence it is semantically false, which is to say that it lacks any meaning. However, hypothetically if some person somewhere has made up a language in which this sequence has meaning then relative to them it may not be gibberish.

To be very clear, it is important to realize that this fact that gibberish is relative to the interpreter does not in even the slightest way imply that different interpreters get to have their own arbitrary definitions of truth. Truth is not relative to any individual, society, species, culture, or worldview. Truth is a property of reality itself or of some arbitrary self-consistent logical system.

The idea that truth could depend on who is interpreting statements about a system is essentially putting the cart before the horse. The truth does not depend on the state of our minds. Rather, the state of our minds depends on the truth[6]. The idea that truth is relative to individual minds (or whatever else) is a backwards way of thinking. The truth controls you. You do not control it.

Anyway, there is another aspect of how statements work that we should explore. Namely, we should explore how exactly the mechanism where statements refer to other objects and statements works. What rules might apply? Can a symbolic set talk about other sets? How does semantic reference work exactly?

With regards to the question of whether fully resolved symbolic sets are capable of talking about other sets (including other objects or statements) the answer is (in a sense) actually no. As you may recall, a symbolic set is nothing but the raw substance of a message transmitted through a physical medium of some kind. It is not inherently meaningful and it cannot talk about anything. It is simply an object.

What about semantic sets? Obviously, the whole purpose of the semantics of a statement is to be able to refer to other objects. The semantic set itself however is essentially inert. When our minds receive a statement, we use the pre-existing patterns of interpretation already there to attempt to construct a matching logical object of some kind. If the semantics don't make sense or are contradictory or the referent objects don't exist, then we will fail to construct the semantic set.

Most statements you will encounter talk about external objects. However, some statements talk about other statements. The case where a statement refers to external objects is rather straightforward in that you simply directly analyze the referent objects. On the other hand, the case where statements can talk about other statements merits more careful thought and a precise understanding of how semantics actually work. How do such cases work?

Well, in particular: Saying "the semantic set of statement A" is the same thing as saying "the set of objects and relationships that statement A *refers* to". This in turn implies that any time that we make a statement about the semantic set of another statement we are actually making an assertion about *what that other statement refers to*. Hence, we need to stay aware of the referential nature of semantic sets when thinking about them, otherwise we are liable to make mistakes.

Suppose for example we have the statement "The semantic set of this statement is the number 5.". This statement is literally an assertion that the sequence of symbols of which it is composed collectively has the same meaning as (or equivalently, refers to the same thing as) the number 5. In other words, we are asserting that "The semantic set of this statement is the number 5." and "5" have the same meaning.

For another example, suppose we have two statements named A and B. Suppose statement A says that statement B's semantic set is empty. This is the same thing as saying that B does not refer to any tangible objects or concepts that actually exist, or equivalently that statement B has no meaning. Statement A could be saying that state-

---

[6]This is independent of the fact that the knowledge that exists within our minds may or may not accurately reflect reality. The errors in our minds are also a part of reality. To put it another way: Delusional thoughts cannot be true, but they can still be truly delusional. Notice the distinction.

ment *B* is gibberish, but it could also simply be saying that statement *B* is false and hence that statement *B*'s referent objects are not ultimately real. If you think about it carefully, meaningless concepts and false concepts are actually closely related.

Anyway though, having now explored the concept of existence-based truth enough to broach the subject, we are now ready to further expand our understanding of what it is that makes statement-based truth so problematic compared to existence-based truth.

1.5 What's wrong with statement-based truth?

We already demonstrated a problem with statement-based truth during the universe S and universe O thought experiment where we showed that making statements (but not objects) own truth values leads to the absurd notion that truth does not apply to universe O. However, the case for statement-based truth is even worse than that. There are more problems than merely the fact that truth should exist independently of statements.

To be specific: Believe it or not, the concept of statement-based truth also *contains a logical fallacy*. Take a moment to realize the potential weight of what I'm saying. Statement-based truth is arguably what virtually all modern logic is based on and is part of the mathematical consensus. I just claimed that within this framework *the definition of truth* contains a logical fallacy.

Our modern logic is of course very effective at solving many different things, is extremely useful, and is truly a monument to the brilliance of the minds of the mathematical, logical, and scientific communities. Nothing I'm saying denies any of that. Logic, as it stands today, is incredibly powerful and generally accurate.

Regardless of what I say here, there is no doubt that the bulk of mathematics would still remain intact, because so much of it is so clearly unambiguously correct and thoroughly proven. However, modern logic also has many strange anomalies and paradoxes in it. The logical fallacy we are about to discuss appears to perhaps be the source of a substantial subset of these anomalies that occur in modern logic.

It is quite understandable that people would have missed the presence of this fallacy. After all, not only is it subtle but it is also hidden within our basic intuitive concept of truth. Who would ever think to even consider the possibility that such a thing could contain a fallacy? How crazy would that be? It seems absurdly unlikely and utterly insane.

I couldn't believe all this stuff either at first. I thought I must have been out of my mind, but the more I thought about it and explored the consequences the more I had to admit that it seemed to make sense. Thus, here I am writing this book. Anyway, ready to hear it? Here we go:

Statement-based truth is an *equivocation fallacy*.

OK... So what exactly is an equivocation fallacy though? Well, we have actually already seen one earlier in this chapter. Specifically I am referring to the part where we were talking about what the nature of existence is and what its scope is and I mentioned that in order to reason about whether Tolkien's fictional elves "existed" or not it would be necessary to precisely distinguish between the concept of elves-in-reality on the one hand and the concept of elves-in-fiction on the other. Conflating these two concepts and treating them like the same thing would be an example of an equivocation fallacy.

In other words, the definition of an equivocation fallacy is this:

Definition 7. *Equivocation is when multiple different meanings of a single term are conflated with each other and treated as if they are equivalent, but the term is still used as if it only means one thing, thereby creating ambiguity and other distortions of meaning.*

*An **equivocation fallacy** is when, during a logical argument, a term is equivocated, thereby introducing conceptual errors into the argument and possibly invalidating it. For example, an argument might shift back and forth between different meanings of the term, causing unrelated results to be treated as if they are about the same concept when they are not.*

Having now defined equivocation precisely, the next natural question to ask is: How exactly is statement-based truth an equivocation fallacy? What is it about the structure of the definition that causes an equivocation fallacy to be created? Let's look again at what our original definition for statement-based truth was. It was this:

> For reference purposes, truth that is defined as being "statements that correspond with reality" (i.e. the traditional definition) will henceforth be referred to in this book as **statement-based truth**. In statement-based truth, statements own the truth values to which they refer. This is in contrast to existence-based truth, where the referent objects own their own truth values and statements merely serve as a pragmatic communication tool.

So, can you see where the equivocation is coming from? In particular, look at the part where it says "statements own the truth values to which they refer". Now that we have developed a more extensive vocabulary for existence-based truth we can give a more precise and more illuminating description of what this actually means.

Here is a more precise definition of statement-based truth, described from the perspective of existence-based truth:

Definition 8. ***Statement-based truth*** *is a definition of truth where the truth value of the semantic set of a statement is assigned to the symbolic set of the statement, thereby conflating the two (i.e. equivocating them). In other words, in statement-based truth we are essentially trying to imbue the literal symbols through which the statement's message is communicated with the actual physical or logical reality (i.e. the meaning) to which it refers. This results in statement-based truth sometimes producing invalid results in cases where this distinction matters.*

The more terms we define for things, the easier it is to reason about those things, to communicate with other people, and to explore new possibilities systematically. Therefore, let's also give an official name to the equivocation fallacy in statement-based truth:

Definition 9. *When the symbolic set and the semantic set of a statement are conflated with each other, we will refer to this as **symbolic-semantic equivocation**. Statement-based truth is the prototypical example of it.*

You are probably wondering at this point what an example of symbolic-semantic equivocation causing invalid results would be. We will get to that soon and I will provide a concrete example. However, before that we need to better understand how semantic reference works.

All statements have both a symbolic set and a semantic set. If either set is not coherently defined then that set should be treated as being empty (or equivalently, in existence-based truth, as being false). Through interpreting a statement, we construct a semantic set based on that statement's symbolic set. Both symbolic sets and semantic sets are inert and are not capable of "doing" anything interpretive on their own. It is only through our own interpretation of the symbolic set as a source for deriving a corresponding semantic set that the sets are assigned.

When a statement A refers to a statement B, the semantic set[7] of A may refer to a part of or the entirety of either the symbolic set or the semantic set of B (or both). The way it does this is by aliasing its own semantic set to stand for whatever part or whole of either the symbolic set or semantic set of B is referred to, and then combining it with any other criteria it specifies for the referent to satisfy.

The various aliases of the semantic sets are resolved in order to identify which aliases refer to the same objects. After this, each semantic set specification is evaluated to see if the objects it refers to exist and satisfy the criteria it specifies. This results in a fully constructed semantic set. If the resulting objects exist as specified according to the criteria, then the statement is semantically true. Otherwise, if the criteria are collectively contradictory or if the object simply does not exist, then the statement is semantically false.

Let's consider a concrete example. Suppose we have a statement A which says "M is a monkey." and suppose that this is indeed true in our universe of discourse. Suppose we also have a statement B which says "The object to which statement A refers is a primate.". Let us now walk through the resolution process and what all the variables are, to better understand how the semantics work.

Before we do though, let me very briefly explain the notation I'm going to use. The notation I will be using here is a slightly informal version of standard set notation. There are two especially common ways to describe sets in standard notation. One way is to use explicit lists of zero or more objects, such as $S = \{a, b, c, d\}$, which means that S is a set containing a, b, c, and d.

The other way is to use set builder notation, which expresses the conditions that each member of the set must satisfy in order to be included. For example,

$$S = \{x : x \text{ is even and } x \geq 10\}$$

is the set of all even numbers that are greater than or equal to 10, i.e. $\{10, 12, 14, 16, \ldots\}$.

The colon ":" reads as "such that" or "where", thus this example reads as "S is the set of x such that x is even and x is greater than or equal to 10". The variable x is

---

[7] The meaning (the semantic set) of a statement is entirely separate from the symbols (the symbolic set) through which the statement is communicated. Don't conflate the two. The symbolic set is used by the interpreter to construct the semantic set and it is solely the semantic set (not the symbolic set) which represents the result of this process. The symbolic set itself does not have any meaning independent of interpretation.

called a dummy variable or a bound variable and is just a placeholder to help specify conditions for objects. You could use any other variable name and the effect would be the same.

Anyway though, back to our example about the monkey. The following statements represent what statement A, which says "M is a monkey.", is expressing:

$$A_{sym} = \text{"}M \text{ is a monkey."}$$
$$A_{sem} = \{x : x \text{ is } M \text{ and } x \text{ is a monkey}\}$$

Likewise, the following statements represent what statement B is expressing:

$$B_{sym} = \text{"The object to which statement } A \text{ refers is a primate."}$$
$$B_{sem} = \{x : x \text{ is } M \text{ and } x \text{ is a primate}\}$$

Note that A_{sym} and B_{sym} are both intended to represent symbolic sets, and likewise A_{sem} and B_{sem} are both intended to represent semantic sets. In other words, we will use subscripts of "sym" or "sem" to indicate whether a set is intended to be symbolic or semantic in origin. Also, if my direct use of quoted text as sets here for the symbolic sets confuses you, then it is important to realize that a quotation can always be thought of as just being an ordered set in disguise[8]. As such, it really is fair to treat quotations as sets, as a notational convenience.

Anyway though, we can see that the semantic sets of both statements A and B above refer to the same object M, and thus we evaluate all the criteria that both sets specify together, to see if any matching objects actually do exist. Although it was not explicitly specified, we know from real life that all monkeys are primates. We also specified as a given that we know that M is a monkey. Thus, we can infer that statement A and B are both indeed semantically true, i.e. (in an existence-based truth system) that the objects they refer to really do exist.

Next, let's consider a case where the statements are in conflict with each other. Suppose that statement A is "8 is an even number." and statement B is "Statement A is false.". Interpreting these statements to derive the underlying constituent parts results in the following list of sets:

$$A_{sym} = \text{"8 is an even number."}$$
$$A_{sem} = \{n : n = 8 \text{ and } n \text{ is even}\}$$
$$B_{sym} = \text{"Statement A is false."}$$
$$B_{sem} = \{n : n = 8 \text{ and } n \text{ is not even}\}$$

When the semantic sets of this list are evaluated they become:

$$A_{sem} = \{8\}$$
$$B_{sem} = \{\} = \emptyset$$

---

[8] For example, the text "Jesse" could alternatively be conceived of as actually being an indexed set $\{(1, J), (2, e), (3, s), (4, s), (5, e)\}$, to impose order upon it, and it would still effectively mean the same thing.

The symbol ∅ and the empty curly brackets { } represent an empty set, i.e. a collection of objects with nothing in it. We have been talking about sets occasionally without going too much into them. We will talk about sets more during the next chapter though. Also, be aware that the way we will be treating sets in this book is mostly the same as in the standard approach, but it is not identical. Thus, even if you already know some set theory, it may not be wise to skip our discussion of it. You'll see what I mean as we progress through the book.

Anyway though, let's return to our main line of discussion again. The set B_{sem} above is empty and therefore the corresponding statement B is semantically false. In contrast though, the set A_{sem} is not empty even after applying all the criteria and therefore the corresponding statement A is semantically true.

These past two examples of resolving semantics were pretty normal. You'll encounter a lot of similar ones probably. However, let's see what happens for some of the more pathological cases. In particular, next let's explore some statements that contain circular references.

Suppose that statement A is "The meaning of this statement is the same as statement B.". Likewise, suppose that statement B is "The meaning of this statement is the same as statement A.". We now have two statements that are each deferring their own meaning to the other statement. How are we to make sense of this situation? Is there any way to make sense of it at all?

Luckily, yes. As long as we think carefully we can figure out what is going on here. Firstly, we need to understand what it is these statements actually mean. In particular, what does "the meaning of this statement" refer to with regards to any given statement we might consider.

As you may recall, the semantic set of a statement *is* the meaning of the statement. Saying "the semantic set of a statement" is essentially just a more formal way of saying "the meaning of a statement", in that the semantic set includes all the objects and relationships that the statement refers to and hence does indeed fully capture the statement's entire meaning. Knowing this, we can now see that statement A is equivalent to "The semantic set of this statement is the same as the semantic set of statement B.". Likewise, we can now also see that statement B is equivalent to "The semantic set of this statement is the same as the semantic set of statement A.". Thus the two statements A and B result in the following sets:

A_{sym} = "The meaning of this statement is the same as statement B"
$A_{sem} = B_{sem}$
B_{sym} = "The meaning of this statement is the same as statement A"
$B_{sem} = A_{sem}$

As you can see, we have inferred that A_{sem} and B_{sem} are the same. In other words, taken together, statements A and B simply state that they have the same meaning. That is all well and good, but *what* is it that they mean? Are the statements true? Are they false? Do we not have enough information to decide?

It may surprise you, but in the absence of any other statements that might resolve the meaning of these two equivalent semantic sets we can still resolve their truth values. To do so though, we are going to go ahead and assume that these two statements together constitute our entire universe of discourse and that there aren't any other statements sitting around somewhere that could somehow give them additional meaning.

So, how are we supposed to figure this out? We actually covered a closely related subject earlier on that gives a hint about what we should do here. In particular, I am speaking of the definition of a gibberish semantic set. Here it is again for reference:

> Whenever a statement consists of symbols that do not appear to have any coherent meaning, such that it is impossible for an interpreter to determine what objects and relations the statement refers to (if anything), then the semantic set of the statement is said to be **gibberish relative to the interpreter**. When the choice of interpreter is implicitly understood in the context, then we may refer to such a set simply as a **gibberish semantic set**. All gibberish semantic sets are false, which is to say that the semantics of the statement are meaningless and hence specify no objects and hence cannot possibly express any truths.

The criteria for it being "impossible for an interpreter to determine what objects and relations the statement refers to (if anything)" is similar to our current situation. However, the sets we are considering here arguably do not quite qualify as gibberish sets. The case we have, where $A_{sem} = B_{sem}$ but neither defines itself, is an example of a similar but slightly different logical phenomenon. In particular, in this case it is an example of an undefined term, which has been placed in an otherwise semantically coherent statement. Here is a corresponding definition:

Definition 10. *When a term occurs within a statement, or within some set of statements considered together, and it is impossible to resolve what concrete logical object the term refers to, then that term is called an **undefined term**. In the context of sets, such an undefined term can also be referred to as an **undefined set**. In contrast to mere gibberish, an undefined set may still have a degree of referential integrity; it simply does not ever resolve to anything tangible. Hence, statements with semantic sets that are undefined sets can sometimes act like a middle ground between meaningless gibberish statements on the one hand and concrete well-formed statements on the other.*

However, we still haven't answered the question of what the value of an undefined set should be. To our intuition, it may perhaps seem at first that there shouldn't be any way to resolve this question, but there nonetheless actually is a way out of this dilemma. The key is to realize that when we consider a set of statements we are generally implicitly considering our universe of discourse to be closed. This implies that undefined terms within the system will ultimately have no hope of ever being given concrete meanings, at least insofar as we don't allow any new changes to be made to the system, since any such modified system would technically be a different system.

In other words, undefined sets are meaningless. But what does it mean to be meaningless exactly? Well, as you may recall, the meaning of a statement is the same thing as the semantic set of that statement. What then does it mean for something to be meaningless? Simple: If a statement is meaningless then that implies that its semantic set is empty. Having an empty semantic set is the same thing as being semantically false. Hence any statement whose semantic set is undefined is essentially false.

Therefore we can now resolve the earlier case under consideration where statement *A* was "The meaning of this statement is the same as statement *B*." and statement *B* was "The meaning of this statement is the same as statement *A*.". Here is the resolution:

$$A_{sem} = B_{sem} = \{\ \} = \emptyset$$

Therefore, in other words, both of these statements are semantically false. They are talking about nothing, and I mean that literally. They are talking about an empty set, which is conceptually the same thing as nothing[9]. Indeed this framework fits well with how people often talk about such things in natural language. If one encounters something that sounds absurd or contradictory[10], one might then exclaim "That has no meaning." and this would be pretty much synonymous with saying "That is a false concept.", i.e. that it was lacking in any true substance.

Let's give an official name to this notion that undefined sets are equivalent to empty sets:

Definition 11. *Within the confines of a particular universe of discourse, any set that is left undefined has the same value as an empty set. An undefined set refers to nothing, which is the same thing as saying it refers to an empty set. This principle will be referred to as the **definition of undefined**.*

For example, within any logical system, all variable names or terms that are never defined are each equivalent to an empty set. Likewise, any other set which never resolves to a concrete logical object is also equivalent to an empty set.

We have just considered an example where two statements refer to each other in a cyclic way. Next, let's consider an example of a single statement referring to itself. Just as two statements *A* and *B* each referring to each other is cyclic, a single statement referring to itself is also cyclic.

We already mentioned a self referential statement once before. It was this one: "The semantic set of this statement is the number 5.". Let's now call this statement *A* and resolve its semantics. Here are the resulting sets:

$$A_{sym} = \text{"The semantic set of this statement is the number 5."}$$
$$A_{sem} = \{5\}$$

As you can see, we are essentially taking statement *A* at its word and letting it assign its own semantic set to 5. However, this might strike us as being kind of a weird thing to do. After all, should the statement really be allowed to assign the sequence of letters in A_{sym} to mean 5? Also, how can we use the meaning of the words to construct its intention while simultaneously changing it to mean 5?

---

[9] If you don't agree that an empty set is conceptually the same thing as nothing, then just humor me anyway for now. I talk about this detail (i.e. about the nature of the empty set) in much greater depth later in the book.

[10] Such as, for example, a number that is simultaneously less than 7 and greater than 12. As you may recall, I have used this example once before.

The reason why we are confused on this point is because we are thinking of the relationship between A_{sym} and A_{sem} as being tighter than it really is. This is mostly due to the way it is worded. We can perhaps rephrase statement A as something more like this: "This statement refers to the number 5, but doesn't bother to say anything else about it.".

Incidentally, statement A is semantically true. You might be wondering though, what does this statement have anything to do with truth? Why can it be true? The key is to remember that statements don't own truth, but rather, only the objects to which they refer do.

Truth in our framework is all about objects. Statements are a way of constructing references to such objects by specifying a set of relationships a matching object must satisfy. Not all statements say anything substantive about what they refer to. But that is not a problem, because in existence-based truth it is only the actual concrete logical objects that already exist which have truth values. Statements are just a way of referring to them.

The truth remains the truth regardless of whether we say anything about it. Our earlier thought experiment involving universe S and universe O illustrated that point especially well. Statement A is semantically true simply because the number 5 exists as a valid concept. This is a subtle but important point about how existence-based truth works compared to how statement-based truth works.

Next, let's look at a more complicated example of circular references. This time, let's consider what would happen if we had a group of four statements, each referring to the next in a circular way, but with one of them also containing some extra stuff too, and see how that would work out. Consider the following statements, expressed as symbolic sets:

A_{sym} = "$-4 < 3$ and this statement means the same thing as statement B."
B_{sym} = "This statement means the same thing as statement C."
C_{sym} = "This statement means the same thing as statement D."
D_{sym} = "This statement means the same thing as statement A."

We can use these statements to construct the following corresponding semantic sets:

$$A_{sem} = \{(x, y) : x = -4 \text{ and } y = 3 \text{ and } x < y\}$$
$$A_{sem} = B_{sem}$$
$$B_{sem} = C_{sem}$$
$$C_{sem} = D_{sem}$$
$$D_{sem} = A_{sem}$$

As we can easily see, $A_{sem} = B_{sem} = C_{sem} = D_{sem}$ and $A_{sem} \neq \{\ \}$, and therefore all of these sets are non-empty (a.k.a. existent), and therefore all of these four statements are semantically true. This is in contrast to our earlier example of cyclic statements in which the set was left undefined and hence became false. The $-4 < 3$ part of statement A is what makes the difference. It grounds the entire group of statements in reality, so that the statements are no longer talking about nothing.

OK... So, we've been exploring some examples of various different semantic scenarios and resolving their meaning. It has taken a while to cover them. Do you remember what we were talking about when we originally went off on this tangent though? The answer: We were talking about statement-based truth and symbolic-semantic equivocation and I said that I would give you an example of where the fallacy results in invalid conclusions. That time has finally come.

1.6 What's the price of equivocation?

The example of symbolic-semantic equivocation resulting in invalid conclusions that I am referring to here, that we are about to be exploring, is none other than the legendary liar's paradox itself, an ancient and immensely famous paradox of logic. The earliest known examples of the liar's paradox in history appear to be two variants known as the Epimenides paradox and the Eubulides paradox.

The Epimenides paradox is not actually a real example of the paradox, but is sometimes cited in popular culture as if it is one. It was supposedly stated by a Greek philosopher named Epimenides sometime around 600 BCE (i.e. roughly 2600 years ago). According to the story, Epimenides was a Cretan and said "All Cretans are liars." which is supposed to be a paradox. However, it is not, because if Epimenides found anybody else who was Cretan and not a liar then that would falsify the statement.

The Eubulides paradox on the other hand is a genuine example of the liar's paradox. The Eubulides paradox was supposedly given in approximately 400 BCE (i.e. roughly 2400 years ago) and poses the question "A man says that he is lying. Is what he says true or false?". This phrase was later paraphrased and streamlined until it mutated into the most common modern day variant of the liar's paradox: "This statement is false.". The variants "This statement is a lie." and "I am lying." are also common.

There are also other related paradoxes that take a different approach or have a more complicated structure. One particularly common variant uses two statements that each refer to each other instead of merely one statement that refers to itself. I could not find the source or the name for this particular two-statement variant (if any exists), so I'm going to instead just give it a name myself for ease of reference.

Definition 12. *When a statement of the form "The next statement is false." is followed immediately by a statement of the form "The previous statement is true." we will refer to this as the **ouroboros paradox**. There are also other equivalent variants, such as one that uses below/above instead of next/previous for example or uses named variables or changes the wording in other superficial ways. However, we will refer to all such variants equally as the ouroboros paradox. At heart though, the ouroboros paradox is essentially just a different form of the liar's paradox, although admittedly a somewhat more structurally complicated one.*

The ouroboros is a famous mythological symbol of a serpent or dragon eating its own tail. Therefore it has the connotation of a circular relationship, and also of two distinct yet still connected things involved in the process, namely the head and the tail. Hence "the ouroboros paradox" seems like a good name for it.

Anyway though, let's analyze the liar's paradox using the conventional intuitive approach (i.e. statement-based truth) and see what happens.

The liar's paradox: "This statement is false."

1. Any given statement is always either true or false. Therefore either the liar's paradox is true or the liar's paradox is false. Hence there are two cases we need to consider (one for true and one for false).

2. **Case 1**: If the liar's paradox is true, then "This statement is false." is true which implies that it is false. Therefore in this case the liar's paradox would be both true and false.

3. **Case 2**: If the liar's paradox is false, then "This statement is false." is false which implies that it is true. Therefore in this case the liar's paradox would be both true and false.

4. In both cases the liar's paradox seems to evaluate as being both true and false.

5. However, it makes no sense for something to be both true and false at the same time. Therefore the liar's paradox seems impossible to resolve and hence appears to be a genuine paradox.

Before I demonstrate how to resolve the liar's paradox using an existence-based truth system, let's first briefly review several already existing theories on how to solve the liar's paradox, in order to give proper respect to the existing literature. For each proposed solution, I will concisely explain why it seems to not be a real solution.

One of the most popular proposed solutions is that of Alfred Tarski: the so called "meta-language" solution. The idea of the meta-language solution is that a statement A can only refer to another statement B when statement A exists in a "higher-level meta-language" than statement B. The lower-level language is called the "object language" and the higher-level one is called the "meta language".

The object language can refer to objects, but not to its own statements. Therefore, to refer to and talk about the statements of the object language, we supposedly need the meta-language, which treats those statements like objects within its own domain. If we accept Tarski's idea, then the implication is that when we talk about statements we are in effect working in a different language than when we are only talking about concrete objects.

Thus, in this way, Tarski's solution prevents the liar's paradox from ever being stated in the object language. This seems effective at first, but it has problems. One problem with it is that it kind of just sweeps the problem under the rug instead of addressing it directly. The liar's paradox is a statement referring to itself in its *own* language, and Tarski's solution just arbitrarily declares that this is not allowed.

A second problem with Tarski's solution is that it does not actually prevent liar-like statements from being made. Suppose we have some arbitrary well-defined concrete statement in the object language (at language level 0). Suppose in the meta-language at language level 1 we state that the statement at level 0 is true. Suppose furthermore that in the meta-language at language level 2 we state that the statement at level 1 is false. The statement at level 0 is now both true and false according to the statements. It is true at level 1 and false at level 2.

Essentially, Tarski's solution is vulnerable to an infinite regress of meta-languages. The idea of the truth being different at different language levels also doesn't actually seem to make much sense. What is that even supposed to mean? It seems like it merely reformulates the liar's paradox in a more convoluted and opaque way. The existence of a statement that is both true and false at different "language levels" seems no less paradoxical than the original liar's paradox was, if you really think about it.

The third problem with Tarski's solution is that it blatantly ignores the fact that both natural and formal languages *can* and *do* refer to themselves frequently. Programming languages are an example of a purely formal language which can be self referential. Tarski's solution says that you need to use meta-languages in order to talk about lower-level languages, but this completely ignores the fact that such meta-languages are un-natural and seldom occur in natural or formal languages, and that the liar's paradox itself can be construed as a counterexample to the claim that you need meta-languages in order to address the possibility that a statement could successfully talk about itself.

A second proposed solution for the liar's paradox is that of Arthur Prior. The idea is that every statement implicitly asserts itself to be true, in addition to whatever else it says, and thus that the liar's paradox is merely a self contradiction rather than a real paradox. In other words, Prior says that "This statement is false." is logically equivalent to "This statement is false and this statement is true." which is a contradiction and is therefore simply false.

This solution at first seems elegant and promising. However, like Tarski's solution, it has problems. One problem is that if you accept the argument that it is false, then doesn't that perhaps still end up implying that it must instead be true when you reconsider the statement once more? Another problem with Prior's solution is that the liar's paradox may in fact be a counterexample to the claim that every statement asserts itself to be true, which if true would invalidate the entire premise of the argument.

However, the fact that most clearly condemns Prior's argument to failure is that, even if we assume his argument is correct, it would still be incapable of solving the ouroboros paradox, which is a variant of the liar's paradox. As you may recall, the ouroboros paradox is the statement "The next statement is false." followed immediately by the statement "The previous statement is true.".

When we apply Prior's implicit truth assertion argument to these two statements we get the following:

1. The next statement is false and this statement is true.

2. The previous statement is true and this statement is true.

When we evaluate these two statements they just revert back to their original forms immediately, instead of forming the direct contradiction that Prior would use to make his argument. There is no difference between the original version and the implicit truth assertion version of this paradox. Thus Prior's argument can't possibly be the real solution to the fundamental problem underlying the liar's paradox.

A third proposed solution to the liar's paradox is to accept dialetheism. Dialetheism is the belief that contradictions exist in reality, i.e. that there are cases where both a statement A and it's negation $\neg A$ are true within the same universe of discourse. For example, a dialetheist might accept the idea that it is possible for "My shirt is wet." and "My shirt is not wet." to be true simultaneously.

The word dialetheism is derived from the Greek prefix "di-" meaning "two" or "double" and "aletheia" which means roughly "truth" or "disclosure". Dialetheism is an example of a truth-glut theory, which means it accepts the existence of statements that are both true and false at the same time.

There are also related (but perhaps less widely studied) theories which accept truth-gaps, which is the notion that things exist which are neither true nor false. In addition, Buddhist and Indian logic has a notion (called the Catuskoti) that all four combinations are possible: true, false, both true and false, and neither true nor false.

Indeed, even outside of Buddhist and Indian logic, in more widely accepted formal logics, multi-valued logic and paraconsistent logic sometimes employ some form of one or more of these unusual truth values, depending on the specific variant of logic being used. This is tangential information but I thought I'd mention it. Don't worry if you don't understand it right now.

Anyway though, the main problem with dialetheism is that it just doesn't really make much sense. Accepting it as true greatly weakens your ability to draw valid conclusions with logic. Also, most cases where people think dialetheism is necessary can actually still be solved with conventional logic as long as one is sufficiently careful.

Arguments in favor of dialetheism usually rely on exploiting ambiguous terms and imprecise wording. For example, suppose we have an antique lamp that is made partly of brass and partly of silver. Suppose we then said "The lamp is made with brass.". A dialetheist might argue that the statement is both true and false, because it both is and is not made with brass, by virtue of silver not being brass.

In reality though, the real problem is that the statement is being treated in an ambiguous way. The statement can be interpreted to mean two different things. In other words, the dialetheist is equivocating. The statement could be interpreted as either "The lamp is made partly of brass." or as "The lamp is made entirely of brass.".

If the statement means "The lamp is made partly of brass." then it is simply true, and clearly is not also false. The fact that it also is made partly of silver and silver is not brass is completely irrelevant to the truth value of it being made partly of brass. On the other hand, if the statement means "The lamp is made entirely of brass." then it is simply false, and clearly is not also true.

Thus the statement is either true or false (depending on which interpretation you pick), but never both. It is only by equivocating the different possible meanings of the statement that it becomes possible to confuse oneself into thinking that a statement can be both true and false at the same time.

Another similar example is to imagine a scenario where an athlete is in the process of crossing over the finish line of a race. One of the athlete's legs is over the line and the other is behind it, with his body dead center above the line. We then ask: Has the athlete crossed the line?

At first it may seem like one would be equally justified in saying the athlete has crossed the line as in saying that the athlete has not. However, this again is caused entirely by the ambiguous and underspecified nature of the question. There are several different ways to define what "crossing over the line" means.

One way of defining "crossing over the line" is to say that the moment *any* part of the athlete's body passes over the line then it counts as crossing the line. Another way is to say that the moment *every* part of the athlete's body passes the line then it counts as crossing the line. Yet another way of defining it would be to say that the moment when the athlete's *center of mass* passes over the line is what counts as crossing the line, as a balance between the two extremes.

Another problem with dialetheism is that it effectively changes the meaning of negating a statement. To a dialetheist, denying a claim still allows the possibility of it being true. That doesn't seem like a denial at all, now does it? One might argue that dialetheists aren't even talking about negation anymore and hence are not actually even addressing the question of contradiction at all. It is like some sort of logical sleight of hand.

Suppose a nuclear physicist named John Doe is monitoring a fusion reactor and one device at his workstation says that the reactor core is about to meltdown whereas another device says that it is not. If John Doe[11] is a true dialetheist then he might think himself justified in believing that both are potentially true, and would therefore perhaps begin performing the procedures for handling both outcomes simultaneously, even though this makes absolutely no sense and the corresponding handling procedures would likely conflict and be incompatible with each other.

Philosophers and logicians sometimes call themselves dialetheists, but pay close attention to how they actually live their lives. If they truly did believe in dialetheism, then they would probably often have a hard time even functioning and making simple decisions in many situations in their daily life, due to their inability to draw sane deductive conclusions in so many different contexts. In other words, even when people say they believe that true contradictions exist, their actions typically tell quite a different story.

As for how dialetheism is applied to the liar's paradox, basically you argue that the paradox is not really a problem after all and just accept the contradiction as fact. You just treat the liar's paradox as simply legitimately being both true and false at the same time. However, as we have seen, such a choice comes at a high price and does not seem to actually make much sense.

In addition to these three pre-existing proposals for solving the liar's paradox (namely: Tarski's meta-languages, Prior's implicit truth assertions, and dialetheism) there are supposedly potentially some other ones too. However, as far as I can tell they don't seem to merit me researching them any further. Some of them are probably variants of concepts similar to the three proposals we have already discussed here. Others simply seem too miscellaneous or contrived to bother with probably. I don't really know much about them. Further investigation seemed fruitless so I gave up on looking into them

---

[11] "John Doe" is a name sometimes used by crime investigators and morgues for a mysterious dead man who they've discovered but can't identify. In literature class they would say that my choice of his name here is an example of *foreshadowing*...

anymore. I'm not really sure, but in any case though, I believe we've probably covered the most important ones.

1.7 How do we solve the liar's paradox?

It is now time to apply our theory of existence-based truth to the liar's paradox. We will see that, unlike the previous proposed solutions, this one seems to work for real. We will now see through the fog of our unjustified intuitions about truth and at long last see the liar's paradox for what it really is: an illusion.

Consider the paradox once more:

The liar's paradox: "This statement is false."

Now, think back to our original definition of existence-based truth. How is the truth or falsehood of a statement defined in our system? As you may recall, we said that in an existence-based truth system the statements don't own the truth values to which they refer, but rather only the respective objects themselves do. Truth is a property of objects, not of statements.

Nonetheless statements are themselves another kind of object. Statements are necessary for speaking about objects and for reasoning about them, and in this respect we found that statements are really just blueprints for constructing two concrete sets: the symbolic set and the semantic set.

So let's read the liar's paradox in light of this perspective. Immediately a problem jumps out at us. The liar's paradox is *ambiguous* from this perspective. The statement says of itself that it is false, but in an existence-based truth system every statement has *two* different "truth values" associated with it, which are those of the symbolic and the semantic sets respectively.

The liar's paradox is not one statement. It is actually two different statements *pretending* to be one. It is an example of a symbolic-semantic equivocation fallacy. Here is the real form of the liar's paradox:

The symbolic liar's paradox: "This statement's symbolic set is false."

The semantic liar's paradox: "This statement's semantic set is false."

For the symbolic liar's paradox: saying that a statement's symbolic set is false, is the same thing as saying its symbolic set is empty, which is the same thing as saying that the literal symbols[12] of which the statement is composed do not exist.

For the semantic liar's paradox: Saying that a statement's semantic set is false, is the same thing as saying its semantic set is empty, which is the same thing as saying that the statement refers to nothing, or equivalently that the statement has no meaning. Any statement that is meaningless has an empty semantic set. Empty sets also happen to represent falsehood in our existence-based truth system, whereas non-empty sets represent truth.

---

[12] As you may recall, the "symbols" of a statement are comprised of the substance via which the statement is communicated through some medium. In this case the symbols are the text of the statement on the page, but it could be anything else.

The way we phrased the symbolic liar's paradox and the semantic liar's paradox after we split them off as separate cases from the original liar's paradox above was technical and a bit opaque at first glance. Let's rewrite them again with a more natural phrasing though, so that their intuitive meaning is clearer:

The symbolic liar's paradox: "This statement does not exist."

The semantic liar's paradox: "This statement means nothing."[13]

My how the tables have turned... The liar's paradox sure is starting to look a lot less intimidating now. Wouldn't you agree?

In fact, I daresay these new forms are utterly wimpy and trite. There is no longer any compulsion to have an epistemological panic attack at the mere sight of the liar's paradox, now that we can see it for what it really is.

We've got the liar's paradox on the run. All we have to do now is pick off the stragglers and declare victory. Nock your arrows ladies and gentlemen, and take aim at the symbolic liar's paradox. We'll pick it off first. It's practically got a bullseye painted on its back.

Let SymLP represent the symbolic liar's paradox statement. Applying symbolic and semantic resolution to SymLP we get the following sets:

$$\text{SymLP}_{sym} = \text{"This statement does not exist."}$$
$$\text{SymLP}_{sem} = \{x : x = \text{SymLP}_{sym} \text{ and } x = \emptyset\}$$

In other words, the semantic set is trying to find an object which is both equal to the symbolic set of SymLP and empty at the same time. It will obviously fail. The symbolic liar's paradox is a truly laughable statement. It attempts to assert that its own text does not exist. The moment you see the statement it has already been proven false, because you couldn't see it to begin with if it didn't already exist.

The semantic liar's paradox is not much better, if at all. It's only real defense is to possibly confuse us about how semantic truth works in an existence-based truth system. It is arguably even weaker than the symbolic liar's paradox.

Let SemLP represent the semantic liar's paradox. Here are the sets that it resolves to:

$$\text{SemLP}_{sym} = \text{"This statement means nothing."}$$
$$\text{SemLP}_{sem} = \emptyset$$

In this case, we don't even have to think about it. The semantic liar's paradox simply declares itself to be meaningless, which is the same as saying that its semantic set is empty, i.e that it refers to nothing. Furthermore, if you remember, an empty set is the same thing as falsehood in our framework. Hence any statement that has an empty semantic set is semantically false.

The danger in interpreting these results though is that the reader might be tempted to try to say that "This statement means nothing." is a true statement because we've

---

[13] Alternatively: "This statement refers to nothing."

46

shown that it is indeed meaningless or false. The reader might then be tempted to argue that this is contradictory and implies that the semantic liar's paradox is both true and false. However, that would be an incorrect interpretation of what's going on.

By asserting that "This statement means nothing." is true because its semantic set is empty, you are simply committing another symbolic-semantic equivocation fallacy. You are thinking in a statement-based truth framework again, rather than in an existence-based one.

In an existence-based truth system, our statements only have meaning and truth insofar as they refer to logical objects. The objects that the symbolic set and the semantic set refer to are together the *entirety* of all truth associated with a statement. The statement does not exist independently of the symbolic and semantic sets. The semantic set is just a inert set. It can't assert itself to be true or false. Neither can the symbolic set. Once the references are resolved, that's the end of the line.

We talk about statements a lot, but it's important to remember they are just a tool we use to communicate sets and the relationships between them. The assertion made by any statement is completely consumed by the resolution process and no longer exists in any meaningful sense afterwards. Taken together, the symbolic and semantic sets *are* the statement and it has no identity separate from them. Sets are just sets. They are inert objects. They can't make statements about themselves, no more than a rock could ever make a statement about itself.

We have now explored both interpretations of the liar's paradox and what their implications are. In both cases the liar's paradox ended up being semantically false. A death knell has sounded for the liar's paradox and it is time to deliver the final judgment: **Properly understood and disambiguated, the liar's paradox is false regardless of which of the two logically distinct ways you choose to interpret it.**

It's almost time to celebrate. However, we're not quite finished yet. We may have defeated the classical liar's paradox, but some of the other variants still remain standing. In particular: The ouroboros paradox and its related cousins should be resolved also.

1.8 How do we solve the ouroboros paradox?

To review: The ouroboros paradox consists of the statement "The next statement is false." followed immediately by the statement "The previous statement is true.". We can convert this into the following equivalent statements:

Statement A: "Statement *B* is false."

Statement B: "Statement *A* is true."

Just as with the liar's paradox, these statements are ambiguous because it is not clear whether they are talking about the symbolic or the semantic set of each respective statement. Let's consider the case where they are both talking about the symbolic set first, and call it the symbolic ouroboros paradox.

The symbolic ouroboros paradox looks like this:

Statement A: "Statement *B* does not exist."

Statement B: "Statement A does exist."

Let A and B represent the symbolic ouroboros paradox statements. Symbolic and semantic resolution results in the following sets:

$$A_{sym} = \text{"Statement } B \text{ does not exist."}$$
$$A_{sem} = \{x : x = B_{sym} \text{ and } x = \varnothing\}$$
$$B_{sym} = \text{"Statement } A \text{ does exist."}$$
$$B_{sem} = \{x : x = A_{sym} \text{ and } x \neq \varnothing\}$$

Further processing of the semantic sets results in:

$$A_{sem} = \varnothing$$
$$B_{sem} = \{\text{"Statement } B \text{ does not exist."}\}$$

Notice that this time statement A is semantically false and statement B is semantically true. This is different from the kind of result we got from the classical liar's paradox, where both cases were false, but that's ok. The statements are still fully resolved and there is no paradox. The truth values of the pairs of statements don't have to be the same. They just have to be well-defined.

Now let's consider the semantic ouroboros paradox instead. It looks like this:

Statement A: "The semantic set of statement B is empty."

Statement B: "The semantic set of statement A is not empty."

Let A and B represent the semantic ouroboros paradox statements. Symbolic and semantic resolution results in the following sets:

$$A_{sym} = \text{"The semantic set of statement } B \text{ is empty."}$$
$$A_{sem} = \{x : x = B_{sem} \text{ and } x = \varnothing\}$$
$$B_{sym} = \text{"The semantic set of statement A is not empty."}$$
$$B_{sem} = \{x : x = A_{sem} \text{ and } x \neq \varnothing\}$$

This time understanding the situation is a bit more complicated. The key to fully resolving the semantic sets here is to realize that A_{sem} and B_{sem} refer to the same set. Statement A's semantic set is a reference to statement B's semantic set, but statement B's semantic set is also a reference to statement A's semantic set, hence they are the same set.

Let's make it easier to understand. We know that A and B refer to the same set, whatever it ends up being. Thus we can simplify our thought process by just handling a single combined set. Since $A_{sem} = B_{sem}$ then the resulting truth value will apply to both sets equally. Let's call the combined set AB_{sem}.

Essentially, when you remove the extraneous information, it has the following form:

$$AB_{sem} = \{x : x = \varnothing \text{ and } x \neq \varnothing\}$$

Now it is much easier to see what the value will be. It isn't possible to find a set where the set is both empty and not empty. Therefore, when fully evaluated, the set becomes:

$$AB_{sem} = \emptyset$$

Combining this with the fact I mentioned earlier, that $A_{sem} = B_{sem} = AB_{sem}$, we now know what the values of the semantic sets must be: $A_{sem} = B_{sem} = \emptyset$. Therefore, both of the statements in the semantic ouroboros paradox are semantically false. This is analogous to the result we got earlier for the classical liar's paradox, so ultimately we got what we were looking for.

What about the other possible forms of the liar's paradox though? Can we resolve those also? A genuine solution to the liar's paradox should be able to resolve all possible forms of it. Only then can we be sure that the liar's paradox has truly been defeated.

Fortunately, this argument generalizes to any arbitrary number of statements that refer to each other in a circular way and make simple true/false claims about each other's truth values. Let's define a term for this generalized form:

Definition 13. *Suppose we have N statements, each of which refer to some statement other than itself among the N statements and simply assert that it is true or false. Suppose also that the references amongst these statements form a circular relationship. Suppose also that at least one statement asserts that another statement is false. We will refer to this as the **generalized ouroboros paradox**.*

Notice that I did not say that it was necessary for at least one statement to assert that another statement is true, but rather I only specified that at least one statement must assert that another statement is false. Also notice that I did not say there had to be an even or odd number of statements either. Neither of these things will matter.

You might encounter logicians who believe that the parity[14] of the number of statements is important in determining whether a sequence of statements is an example of the paradox. However, those logicians seem to be wrong.

Consider, for example, an instance of the generalized ouroboros paradox where we have two statements each asserting that the other is false. In other words we have these statements:

Statement A: "Statement *B* is false."

Statement B: "Statement *A* is false."

Some logicians might think that you have to have an odd number of statements asserting falsehood in order to have an instance of the paradox and may cite these two statements above as not being an example of the paradox. The reasoning would go like this:

---

[14]The parity of a number is its property of being either even or odd. For example, the number 16 has even parity and the number 9 has odd parity.

> Suppose that statement *A* is true, therefore statement *B* is false, therefore statement *A* is true, therefore statement *B* is false, therefore statement *A* is true, therefore statement *B* is false…

The logician would then argue that statement *A* always has the value true and statement *B* always has the value false, regardless of how long you continue this sequence, and therefore that this pair of statements is *not* an example of the paradox.

However, such a conclusion would actually be premature and misleading. What the logician failed to account for is that which statement you choose to evaluate first changes what truth values the statements appear to stabilize on.

Look at what happens if we instead start by assuming that statement *B* is true:

> Suppose that statement *B* is true, therefore statement *A* is false, therefore statement *B* is true, therefore statement *A* is false, therefore statement *B* is true, therefore statement *A* is false…

Notice that this is the exact opposite of the previous conclusion that we got when we started by evaluating *A* first. Furthermore, even if we tried treating both statements as true simultaneously or false simultaneously, we would still get nonsense. If both statements were true, then they would both be false. Likewise, if both statements were false, then they would both be true. In other words, the paradox would still be with us regardless of the parity of the number of statements.

This same pattern of misleading stabilization of truth values depending upon which statement you evaluate first applies just as well to any even number of statements that assert other statements to be false. For example, consider the following four statements:

Statement A: "Statement *B* is false."

Statement B: "Statement *C* is false."

Statement C: "Statement *D* is false."

Statement D: "Statement *A* is false."

If we choose to start evaluation with statement *A* then this is the result:

> Suppose *A* is true, thus *B* is false, thus *C* is true, thus *D* is false, thus *A* true, thus *B* is false, thus *C* is true, thus *D* is false…

As before, it has stabilized to a certain pattern. Let's see what happens if we choose to start with statement *B* instead though:

> Suppose *B* is true, thus *C* is false, thus *D* is true, thus *A* is false, thus *B* is true, thus *C* is false, thus *D* is true, thus *A* is false…

As you can see, it is now stabilizing to the opposite set of truth values. It is clear that the same phenomenon will occur with any even number of such statements.

Sequences of odd numbers of such statements actually also repeat a pattern depending on which statement you start on, its just that the pattern they repeat oscillates between two different states. For example, consider these three statements:

Statement A: "Statement B is false."

Statement B: "Statement C is false."

Statement C: "Statement A is false."

If we start the evaluation on statement A then it will go like this:

> Suppose **A is true**, thus B is false, thus C is true, thus A is false, thus B is true, thus C is false, thus **A is true**, thus B is false, thus C is true, thus A is false, thus B is true, thus C is false...

I put the phrase "A is true" in bold for you so that it would be easier to see where the cycle restarts. During every cycle the process passes through each of the statements and alternatingly gives each statement a value of true or false.

If we start evaluation with statement B being assumed true instead of statement A, then the cycle just starts at the point where B is true in the above cycle and continues onward from that point.

Also, the sequence of statements does not have to be all false in order for the cycle to occur. All it takes is a single assertion of falsehood in order for the paradox to appear. For example, consider the following four statements:

Statement A: "Statement B is false."

Statement B: "Statement C is true."

Statement C: "Statement D is true."

Statement D: "Statement A is true."

Starting with statement A, the sequence of inferences will go like this:

> Suppose **A is true**, thus B is false, thus C is false, thus D is false, thus A is false, thus B is true, thus C is true, thus D is true, thus **A is true**, thus B is false, thus C is false, thus D is false...

As you can see, we are getting the same kind of cyclic pattern of oscillating values.

Let's explore one more example. Consider the same four statements, but with statement B negated so that there are now an even number of assertions of falsehood mixed in with assertions of truth. Suppose it looked like this:

Statement A: "Statement B is false."

51

Statement B: "Statement C is false."

Statement C: "Statement D is true."

Statement D: "Statement A is true."

Suppose we then evaluate the statements starting at statement A, as before. Here is the result:

> Suppose A is true, thus B is false, thus C is true, thus D is true, thus A is true, thus B is false, thus C is true, thus D is true…

As with our previous examples that had even numbers of statements that asserted other statements were false, the pattern appears to stabilize. However, as before, watch what happens when we choose a different starting point:

> Suppose B is true, thus C is false, thus D is false, thus A is false, thus B is true, thus C is false, thus D is false, thus A is false…

Therefore, as you can see, the paradox appears here too. From looking at all these different examples of the generalized ouroboros paradox, we can see that all it takes for the paradox to appear is for at least one of the statements to assert that another statement is false.

As we traverse the statements, passing from one to the next, if we encounter a negative statement then we will begin propagating that negation forward to the rest of the statements in the cycle. However, since the relationship between the statements is circular, we will eventually propagate the negation back to the same statement the negation came from, thus negating it again.

In other words, within any circular set of statements any statement that says that another statement is false will end up inadvertently also asserting itself to be false, once the negation propagates back to itself. Thus paradoxical self contradiction is inevitable in such cases.

However, this big complicated mess of misleading reasoning only occurs when you are thinking in terms of statement-based truth. When you instead think in terms of existence-based truth, then the situation becomes far simpler and can be resolved easily.

What is it that makes the situation so much simpler for existence-based truth? Well, to talk about that more easily lets define a new term. We've encountered this concept earlier before when we were resolving the original (non-generalized) ouroboros paradox. Here's the definition:

Definition 14. *Suppose we have two statements A and B. In existence-based truth, if statement A's semantic set refers to the entirety of statement B's semantic set then statement A's semantic set becomes indistinguishable from statement B's semantic set. Anything that either statement says about its own semantic set is therefore also a constraint on the same set that the other statement refers to. We will call this process of statements losing their distinct semantic identity* **semantic collapse**.

Notice that in the generalized ouroboros paradox each statement always refers to the entirety of the semantic set of the next statement in the sequence. Therefore, in any instance of the generalized ouroboros paradox, all of the statements will undergo semantic collapse and will thus all refer to the same semantic set. Any constraints each statement specifies will therefore also apply to the others.

As with the liar's paradox, you can interpret the generalized ouroboros paradox two different ways: symbolically and semantically. We've mostly been emphasizing the semantic case because that's usually the intended meaning. Regardless though, let's now consider both possible interpretations (symbolic and semantic) of the generalized ouroboros paradox.

The symbolic generalized ouroboros paradox is not very interesting. The mere fact that we are looking at a set of statements in text proves that all of those statements exist and are therefore considered symbolically true. Thus it will always be easy to resolve which of those statements refer to true or false sets. Statements that assert that another statement exists will be true, whereas statements that assert that another statement does not exist will be false, since the raw uninterpreted symbols of all of the statements clearly do exist in all cases.

Each constituent statement has its own independent truth value. This is fine. As I said before, the statements don't have to have the same truth value, but rather they just need to be resolved in a coherent way. This is all that is needed to resolve the symbolic variant of the generalized ouroboros paradox.

It is merely a matter of checking whether each statement another statement refers to exists or not. Granted, the symbolic generalized ouroboros paradox is almost certainly not the intended interpretation of the paradox, but nonetheless it is still wise to know how to resolve it for the sake of completeness.

The semantic generalized ouroboros paradox is the more interesting interpretation though, and is almost certainly the interpretation that people intend to focus on and care the most about when they discuss the generalized ouroboros paradox. There are two possible cases for the semantic generalized ouroboros paradox:

Case 1: All of the statements assert that some other statement is false.

Case 2: At least one of the statements asserts that some other statement is false and at least one of the statements asserts that some other statement is true.

In the first case, since all the statements refer to the same semantic set and simply assert that it is false, the semantic general ouroboros paradox effectively collapses into a single statement that asserts itself to be semantically false, i.e. that asserts its own semantic set to be empty. However, we have already met precisely this statement before and have already resolved it: it is simply the semantic liar's paradox yet again. It is false. Equivalently, in an existence-based truth system, we can say it means nothing. Or, put yet another way, we can say it refers to nothing and hence cannot express anything true.

In the second case, we have all the statements again referring to the same semantic set. However, this time we have at least one statement which is trying to assert that the semantic set must be empty, while another statement is trying to assert that the semantic set must be non-empty[15]. That means that regardless of what else might be going on, the semantic set is guaranteed to have self contradictory constraints imposed upon its contents. It must be both empty and non-empty to satisfy the constraints, but that is impossible. No sets which are both empty and non-empty exist, and hence the semantic set must be empty, i.e. false.

Therefore, in all possible cases, the semantic generalized ouroboros paradox is false, just like the semantic liar's paradox. We have our answer.

Anyway though, it appears that we have now truly solved the liar's paradox once and for all. It is logically impossible to form any of these variants of the liar's paradox in an existence-based truth system. Since we have shown that we have eliminated not only the classical liar's paradox, but also its many variants, it is now officially time to celebrate.

For approximately 2500 years, logicians have struggled to defeat the liar's paradox, hoping that through the process of doing so they would uncover some great insight into the fundamental nature of truth. I believe that we have found that insight in existence-based truth. Today the paradox finally crumbles to dust, bringing long sought peace to a deeply perplexing question. We are now closer to understanding the real nature of truth.

Before we move on though, let's tie up one last loose end. We covered the cases where at least one statement in a circular relationship asserts that another statement is false, but what about the case where all of the statements assert that the other statements are true? Let's assume we're talking about the semantic interpretation of the statements, rather than the symbolic interpretation, since the semantic interpretation is the most likely intent and we already know how to deal with the symbolic interpretation of circular statements rather easily.

In the case where all of the statements assert that the others are true, what the collapsed semantic set resolves to will depend on whether or not any of the statements have anything tangible they refer to that exists.

In the case where none of the statements do anything but refer to other statements, but none of them are grounded in anything concrete, then the definition of undefined will come into effect and the statements will all be semantically false.

On the other hand, if at least one of the statements refers to a logically consistent object that really does exist, and if none of the other statements assert any contradictory constraints upon that object, then the set of circular statements can be true.

Consider this example:

Statement A: Statement *B* is true.

Statement B: Statement *A* is true.

The semantic set that both of these statements refer to never actually resolves to any concrete logical object, and therefore this set of statements is semantically false.

In contrast, consider this next example:

---

[15] Remember: In existence-based truth, an empty set is the same thing as falsehood and a non-empty set is the same thing as truth.

Statement A: All organisms evolve and statement *B* is true.

Statement B: Statement *C* is true.

Statement C: Statement *A* is true.

The set of all organisms certainly exists, and moreover all organisms clearly do evolve[16]. Hence statement *A* does refer to something tangible that fits the specified criteria and is thus not undefined. Therefore, because all of these statements will undergo semantic collapse, all of them will end up being semantically true.

Anyway though, that basically wraps up our discussion of the liar's paradox and the related subject matter. It sure has been interesting. We've learned a lot from these thought experiments.

1.9 What is the deeper meaning of truth?

However, there is still more to understanding what truth is than just what we've discussed so far. We've gained some strong insights into the nature of truth, but there is yet more to learn. As it happens, I've actually been giving you an oversimplified view of truth, so as to not overload you with too many new concepts at once and to avoid confusion.

What we've been discussing so far is just the tip of the iceberg. There is a vast depth of additional insight into the nature of truth awaiting us and it is time to dive in. I've only been telling you part of the story. Truth is far more than just a binary true or false value.

To make an analogy: It is like we have been sailing on a vast ocean of the mind, seeking to catch a glimpse of the truth, which we are compelled by fascination to pursue. When we first begin our journey the air is full of fog and we can hardly see a thing. This fog represents cognitive biases, logical fallacies, poor reasoning, and preconceptions. Through careful thought and objectivity we are eventually able to navigate our way out of the fog.

Once the fog is lifted we can see something glimmering and glowing deep under the waves: the truth. However, we are still just hovering at the surface level and the water is distorting our view. If we want to stand face to face with the truth, and see it for what it really is, we will have to dive deep into the water. To truly understand the nature of truth we have to understand all of its aspects, rather than merely some of its aspects. Only then will it truly come into focus and we will at last acquire the clarity we seek.

When people look at the universe with awe and wonder and think to themselves "What is the truth of all this?" they are not merely wondering if the universe will toss

---

[16] To be clear, some species evolve very slowly because their natural selection process has become very stable, but they still do evolve nonetheless. As for why evolution itself is true, there is so much overwhelming evidence that I daresay it would be fair to say that evolution is even more strongly confirmed than the laws of gravity are. This of course doesn't stop bigoted anti-intellectuals, uneducated people, clergy, and the brainwashed masses from spreading misinformation to the contrary. Read some biology books. You wouldn't trust someone who never studied electricity to be your electrician; likewise, you would be a fool to trust what anyone who hasn't studied real biology well has to say about it.

them back a value of true or false. If someone asked "What is the meaning of life?" and the universe replied back saying "true" then this would be a completely unsatisfying answer that would explain nothing.

So, there's more to truth than merely being true or false, but how do we express this? If truth is not a binary true or false value then what exactly is it? Let's see if we can work out a way to capture this notion in our system of thought. Concrete examples are usually helpful in these kinds of situations.

For example, let's consider a prism. A prism is a crystalline volume which has the ability to refract light[17]. Most typically, when one thinks of a prism one thinks of a crystalline transparent object with triangular bases and rectangular sides[18] which can generate a rainbow by splitting light that shines into it at the right trajectory.

Now, as you may recall, in existence-based truth all objects have truth values. Any object that exists is considered true. Prisms certainly exist, hence our prism must be true. Saying that something is true in existence-based truth is the same as simply saying that it exists as specified. However, merely acknowledging the existence of the prism hardly seems to capture its essence. One would think that the truth would be more enlightening than that. How do we capture that missing part?

The key is to realize that a fully formed notion of truth should capture everything about the prism, not merely whether or not it exists. In other words, the truth value we seek is one that captures all of the substance of the object and all the relationships which apply to it. So, for example, the truth value should capture the fact that prisms can refract light, among other things.

What is the logical object that is capable of expressing all of this information about the prism? Whatever it is, it is what the prism's truth value should really be. The answer is actually staring us right in the face: in a sense, the prism is its own truth value. Only the prism itself, or at least a complete logical representation of it, could ever hope to fully encompass all of the underlying truth about it.

In fact, every object that exists in the universe is its own unique truth value. True and false are just *properties* of truth values, not actual truth values in themselves. Only the objects themselves are the real truth values. We previously treated true and false as the only truth values, but that was only to prevent your mind from being overloaded, so that we could learn gradually.

To drive the point home: the stars in the sky, the plants in a garden, the rocks in a quarry, the rivers and oceans of the world... All of them are distinct truth values in and of themselves. So is the chair I am sitting in as I write this. So is whatever book or device you are reading this book from. Objects which don't exist can't be anything but false, but objects which do exist all have their own distinct truth values.

Notice how this changes what it means to know the truth of something. If truth was just a binary true or false, then knowing it wouldn't necessarily mean that much. However, if truth is the essence of an object itself, then knowing that object's truth

---

[17] Refraction is the bending of light, i.e. a change in the direction in which a beam of light is traveling as it passes through a volume. Refraction should not be confused with reflection. However, refraction and reflection both change the direction of light.

[18] Naturally, if its bases are triangular, it will have three rectangular sides. Look up a picture of a prism on the internet for clarity if you don't know what I'm talking about.

value is the same thing as fundamentally knowing everything there is to know about that object.

In other words, when we think of truth values as being the objects themselves, combined with all of the relationships that hold for those objects, then knowing the truth value of an object is the same thing as knowing the true nature of that object in all respects, i.e. knowing the full meaning of that object. Unlike with mere binary true and false values, this implies we really do understand the truth in the deeper sense that we have been seeking.

This means that there are an indefinite number of truth values that exist, rather than merely two. Each object is like its own unique flavor of truth. Each tells us something specific about the universe through the relationships it holds to other things. If we were to somehow learn all of the relationships amongst all objects in the entire universe, then in essence we would have gained a complete understanding of all of existence. In other words, we would know the full truth of reality. The accomplishment of this task (or as close as possible) is the ultimate goal of science and philosophy.

To make it easier to remember that true and false are merely properties of objects, and that objects themselves are actually the real truth values, here is a short phrase you can perhaps use as a memory aid: "True is an adjective, not a noun."

Anyway though, let's make a formal definition for this concept:

Definition 15. *Every object that exists in the universe, or in some other arbitrary but consistent logical system, is its own distinct truth value. Such a truth value includes everything there is to know about the object and about all relationships that it has to the rest of the universe.*

On the other hand, objects that don't exist, whether through being self contradictory or through simply empirically not existing, can only have the truth value of false, i.e. the empty set.

We will refer to this concept of objects being truth values in and of themselves as **objectified truth**. *In contrast, we will refer to the opposite concept of truth values being separate entities from the objects to which they refer as* **nominal truth**.

For example, in nominal truth, we might say that it is true that a prism refracts light and treat this truth as existing separate from the prism itself. In contrast, in objectified truth, the fact that the prism refracts light would be part of the prism itself. In objectified truth, objects and truth are one and the same; every object is a truth value and conversely every truth value is an object.

Whether a system of truth is statement-based or existence-based, and whether it is nominal or objectified, are partially independent factors. At the very beginning of the book, when we were talking about the traditional views on what truth is, we were essentially exploring a nominal statement-based truth system. After taking a closer look at our preconceptions, we then transitioned to a nominal existence-based truth system and used it to solve the liar's paradox. To further expand the power of our system though, we will be transitioning again, this time to an objectified existence-based truth system.

While we're at it, let's clarify our notion of how we use and understand the empty set a little bit by making a related definition for it:

Definition 16. *In existence-based truth: the concepts of falsehood, the empty set, emptiness, nothingness, and meaningless all refer to the same exact thing and are indistinguishable. We will refer to this concept as* **void falsehood**. *There are many different words and symbols that can represent these. The following informal equation can be used as a memory aid:*

$$false = \emptyset = \{\ \} = empty\ set = emptiness = nothing = meaningless$$

In objectified truth, objects are their own truth values and hence the true/false label we attach to such objects feels a bit superfluous. Also, there seems to be something fundamentally different about expressing truth as a fully understood object versus merely labeling that object as true or false. This is no accident. The reason that true and false feel somewhat superfluous in objectified truth is because we are actually conflating two different things.

Although any set that exists is indeed true, the act of mentally labeling it as such is not an expression of its truth value. Only the set itself is the truth value. In other words, what we are really expressing when we label something as true or false is not the object's truth value, but rather it is our *knowledge* of whether or not it is true (i.e. of whether or not it exists validly).

Do you remember way back when I pointed out that statement-based truth commits an equivocation fallacy, wherein it conflates the symbolic and the semantic sets of a statement? Do you remember how I said that I don't blame people for not realizing there was an equivocation fallacy in the traditional statement-based definition of truth? That sure was an unlikely claim, wasn't it? Surely such an unlikely thing wouldn't happen again, would it?

Guess what's about to happen again...

It actually turns out that the symbolic-semantic equivocation fallacy is *not* the only logical fallacy in the traditional statement-based definition of truth. There is not just one but *two* fallacies in the way most people define truth (i.e. in statement-based truth). Surprise! "Common sense" has betrayed us yet again... I daresay it has a real talent for that.

Anyway, it's time for a new definition:

Definition 17. *The knowledge that an object is true is not the same thing as the truth value of that object. Knowledge values and truth values are not equivalent. To treat them as the same thing is to commit an equivocation fallacy. Whenever a system of logic treats the knowledge that something is true as the same thing as its truth value we refer to this as* **truth-knowledge equivocation**.

So what exactly are knowledge values though? Well, you've already encountered two of them. Whenever we encounter a specification for a set and mentally note whether or not it exists we are generating a knowledge value. If we see that the set does exist, then we label it as being known-true. Likewise, if we see that the set does not exist, then we label it as being known-false.

However, these are not the only possible knowledge values. There are four basic knowledge values in total[19]. Can you guess what the other two knowledge values are, besides known-true and known-false? I'll give you a hint: Think about the limitations of the human mind and of logic itself. If someone presents a claim to you, then what are the four possible knowledge values that you might respond to the claim with?

Here's the answer: The four knowledge values are known-true, known-false, unknown, and unfalsifiable. Let's give the terms official definitions:

known-true If we successfully determine that a set does exist (i.e. that it is non-empty), then that set is therefore known-true.

known-false If we successfully determine that a set does not exist (i.e. that it is empty), then that set is therefore known-false.

unknown If we do not currently know whether or not a set exists, and it is not known to be unfalsifiable, then that set is therefore unknown.

unfalsifiable (a.k.a. untestable or indeterminate) If we have proven that it is logically impossible to determine whether or not a set exists, then that set is therefore unfalsifiable.

Why I'm using sets here to define these values, instead of some other kind of objects, will gradually become clear later, as we progress through the book. Just humor me on that point for now. Anyway though, let's also define some symbols for each of these four knowledge values too, just for the heck of it. The more terms and symbols we have, the easier it is to reason about our thoughts. Even if we don't use them, someone else still might. Here's a table of two different possible symbol notations for these four knowledge values:

Knowledge Value	Notation 1	Notation 2
Known-True	T	KT
Known-False	F	KF
Unknown	?	UK
Unfalsifiable	¿	UF

What are some examples of each of these cases? The known-true and known-false cases are rather obvious, so I'll just give two brief examples. An example of a known-true claim would be the claim that humans need oxygen to live. An example of a known-false claim would be the claim that the sun is a giant light bulb in space powered by the

---

[19] If you take into account degrees of certainty, such as when dealing with probabilistic logic or fuzzy logic, then you can argue that there are more than four. However, here I will be considering only more primitive knowledge values, because it makes it easier to illustrate the point and makes it less complicated.

well-deserved deaths of people who don't use the oxford comma or who put adjacent punctuation inside nearby quotation marks[20].

Next let's consider the unknown case. Suppose that a mysterious taped-up box has just arrived in the mail. If someone then stated that the box contains a snow globe then this would be an example of an unknown truth value. We don't know whether the claim is true or false yet, but we can find out by opening up the box. For another example, consider the claim that alien life exists on some other planet in the universe. We don't know if it is true, but in principle it should be possible to find out with sufficient effort.

In contrast though, suppose I claimed that there is a tiny invisible and intangible man sitting on your shoulder eating a bucket of proportionally tiny popcorn. The fact that the tiny man is both invisible and intangible means that no attempt to see him or to detect him through any means[21] could ever succeed. This implies that the claim is provably unfalsifiable. It is logically impossible to ever justify any claim that the statement is true or false.

Unfalsifiable claims are very common in irrational belief systems and are frequently used as an unethical way of manipulating people. You have surely encountered such unfalsifiable claims many times before in your life, and will unfortunately surely continue to encounter many more in the future.

Charlatans and deceivers love unfalsifiable claims because once someone believes an unfalsifiable claim is true then the fact that the claim is unfalsifiable will make it difficult for the believer to ever stop believing in the claim. Unfalsifiable claims are often used when attempting to brainwash people. However, all unfalsifiable beliefs are always unjustified by definition, since it is logically impossible to ever justify them.

There are many different sources of unfalsifiable claims that one will often encounter in life. Some of the most common sources include religion, mysticism, pseudoscience, and grand theories of existence, just to name a few.

The reason why unfalsifiable claims are so effective at spreading nonsense is perhaps because most people haven't been adequately trained to defend themselves against such things. In fact, if anything, large portions of the current structure of society (e.g. religions, authoritarian governments, anti-intellectual sentiments, etc) are actually intentionally designed to *encourage* people to believe in things without sufficient justification, in order to thereby make those people much easier to manipulate, control, and exploit.

Too many people have simply not sufficiently disciplined and refined their minds enough to consistently be able to detect and defend against these kinds of unscrupulous mental attacks. Always remember though: if a claim is unfalsifiable then under no circumstances can it ever be justified to genuinely believe in it.

Since the dawn of history, humanity has been compelled to discover the underlying nature of existence and to understand why we are all here. Why does anything exist at

---

[20] I am joking of course... However, please use the oxford comma and stop putting punctuation marks inside places that they are not logically a part of. Putting punctuation inside the quotes does not "look better". Also, for the record, the brackets should never be omitted from conditional blocks when programming, the hanging part of toilet paper should always face outwards (away from the wall), and informal speech and emoticons are perfectly fine in professional publications. ☺

[21] For example, trying to detect his gravitational field wouldn't work either, because being intangible means that he does not have any detectable physical attributes.

all, etc? In the pursuit of this goal, people have constructed a vast number of different theories of existence posing different conceivable all-encompassing answers. However, regardless of how emotionally attached some people are to these ideas, all such grand theories of existence are essentially doomed to failure by definition, and I will explain why.

First lets start by listing a few of the most common grand theories of existence, so that we have some concrete examples to discuss:

Theism is the belief that there are one or more gods which have created the universe and/or rule over it. Most religions are examples of theism, each of which has its own wildly arbitrary and very obviously man-made doctrine attached to it. There is also a more generic form of theism that doesn't have any specific doctrine or narrative, and that also doesn't have any notion of clergy or of religious authority, which is called deism.

Simulated Reality is the idea that the existence we are now experiencing is nothing more than a mere simulation, such as in a computer. Another variant of this theory is the "Brain in a Vat" theory, which is nearly the same, except that instead of being computer programs we are brains in vats being controlled by biochemical signals. Amusingly, if you think about it carefully, simulated reality theories are actually very similar to theism, except that in simulated reality theories we are considered to be like computer programs instead of souls. In contrast though, computer programs are confirmed to exist, unlike souls[22], thus making this theory at least marginally more credible than theism, but still unfalsifiable and hence still worthless.

Solipsism is roughly the idea that the only thing that exists is one's own mind. In other words, it is the idea that one's mind is the entire universe itself. Solipsism is very closely related to philosophical idealism. There are also some other possible variations on the theme, such as a variant where the solipsist not only believes that the universe is all inside their mind, but also that they are in some sense the ruler of the universe.

Reincarnation is the belief that when we die our consciousness is reborn in the mind of a new living creature somewhere. For example, a human might die and then be reborn in the mind of a wolf. This sort of thing is a common belief in Buddhism and Hinduism, and in some other belief systems. It may or may not involve theistic elements. Unlike in the abrahamic religions, the "afterlife" for reincarnation is not an external higher realm of existence, but rather it is a return to reality in a new form. Sometimes there is also a concept of "leaving the cycle" though (e.g. "nirvana" in Buddhism).

String Theory is a mathematical theory created and explored by some physicists which seeks to find a unified theory of physics by posing the idea that the universe is

---

[22] Nonetheless, souls may still exist in some form. Consciousness is a very difficult question to resolve. In fact, this may be yet another question that is inherently untestable and impossible to answer.

composed of hyper-dimensional strings (i.e. strings that exist in more than 3 dimensions of space) interacting in various ways. It poses the idea that the universe has more dimensions than just the three that we see, and that higher-dimensional string-like objects act upon the universe. Despite its superficial scientific appearance, it is essentially untestable and hence is generally not real science.

Multiverse Theory is the idea that multiple universes exist and that perhaps these universes undergo a kind of natural selection process similar to biological natural selection and this supposedly creates "better universes" such as perhaps the one we are living in for example[23]. Of course, the problem with this theory is that it is impossible to test. By definition, the only things we can ever hope to test are things that exist within our own universe.

Dream World Theory is the idea that the world we are experiencing right now is all some kind of dream. For example, you might sometimes hear people say that they wonder where we go when we dream, and that maybe we are experiencing other people's lives while we dream, or something along those lines. There are many different entities that could conceivably be doing the dreaming. It could be us or it could be someone else. This idea is quite similar to simulated existence, except that it uses dreams instead of simulations.

In principle, we could go on forever listing out new theories of existence and variants of them. There are literally an infinite number of different theories of existence, and within each of those theories of existence there is generally yet another infinite set of conceivable variations on the general theme. For example, there are an infinite number of conceivable religions, none of which have any more credible supporting evidence than the others.

Theories of existence are very interesting to a lot of people. They can seem really compelling sometimes. Wars have even been fought over them. However, as I said before, all theories of existence that seek to give the final explanation of the ultimate nature of existence are doomed to failure. The reason is because all such theories are inherently unfalsifiable (i.e. untestable).

Notice that all of these theories share something strange in common: they all postulate the existence of something in a higher realm of existence that exists beyond our own concrete reality. The fact that this certain something exists beyond our own realm of existence means that it is effectively invisible and intangible in all respects[24], and therefore implies that it cannot be tested in any way.

Moreover, even if we were to grant all of the claims in any particular theory of existence as true then it would still not actually be proven true. For example, suppose

---

[23] Don't be misled by this idea into thinking that fine tuning of the universe is somehow necessary for life. It isn't. Evolution adapts to whatever universe it finds itself in, not the other way around. Life is adapted to the universe. The universe is not adapted to life. Be careful not to think backwards. Fine tuning arguments to "explain life" don't actually explain anything and are not logically valid.

[24] Just like the idea of the invisible and intangible tiny man sitting on our shoulder eating popcorn that we mentioned earlier as a thought experiment…

for argument sake that we somehow knew for sure that a god existed and created the universe.

Suppose we then walked up to this god and asked it this: "How do you know that you, and the entire universe that you have supposedly created, are not just part of a computer simulation? Suppose that the computer simulation also has you programmed to believe with absolute certainty that you are the creator of the universe. Isn't it impossible for you to deny the possibility that you are in fact a computer simulation and not actually a god at all?"

Notice that no matter how much we attempt to arbitrarily define "god" in this theistic belief system as being all-knowing and the ultimate cause of everything, the argument in favor of theism will still fall flat on its face regardless.

We can always pose the possibility that a different theory of existence is actually the real explanation for what we are observing, and no matter how hard we try we cannot prevent that doubt. That is because theism is fundamentally an unfalsifiable (i.e. untestable) claim. Only unfalsifiable claims display that kind of behavior.

It gets even worse then that. Given any theory of existence we can always wrap it inside yet another different theory of existence. We can continue this process an infinite number of times. The nonsense never ends. It is all utterly pointless, right from the start. Thus, all ultimate theories of existence are inherently unfalsifiable and are doomed to fail by definition.

All grand theories of existence that have ever existed are unjustified. Furthermore, all grand theories of existence that will ever exist in the future will likewise be unjustified. Indeed, for example, no belief in any religion has ever been justified, nor will any such belief ever become justified. It is literally logically impossible to ever justify such a thing. Every drop of blood ever shed in the name of any religion has been in vain. Such conflicts just result in needless suffering. It is all just one big pointless tragedy.

Indeed, since no worldview based on unfalsifiable theories of existence can ever succeed in actually justifying any of its beliefs or actions, devoting any time whatsoever to trying to justify or spread any such belief system is guaranteed to be a complete waste of time. Moreover, generally speaking, people who think that it is acceptable to believe things without evidence can often be convinced to do virtually anything if you apply a strong enough appeal to their emotions, regardless of how irrational the action is or how much suffering it will cause for others.

That is simply the natural consequence of not having any kind of sane standard of evidence. There is nothing really stopping such people from being convinced of all manner of horrible and violent things. Indeed, such things have happened repeatedly throughout history. Such is the true nature of blind faith. It is pure weakness. It functions primarily as just an endless fountain of human suffering. The potential for needless human suffering is enormous, both in history and going forward into the future. Only embracing evidence-based thinking has any chance of saving us from that.

I'm all for complete freedom of thought and expression though, so if someone wishes to believe in things without evidence then that is the their choice. I just hope that the people who do make that choice do not get our entire species killed off by the numerous unintended negative consequences of doing so, which is looking like a very strong possibility going forward, unfortunately. How many more people have to suffer and die all just for beliefs that are literally logically impossible to ever justify? It is truly an immense human tragedy.

Anyway though, there is one worldview amongst all others which is genuinely special. Only one worldview exists that does not make any unfalsifiable claims: scientific realism. In real science, we only make claims about the parts of the universe that can actually be tested. Anything beyond that is a waste of time and resources. Hypothetical ideas can still be explored of course, but only so long as we remind ourselves that such ideas are merely speculation until proven conclusively.

Sure, higher-dimensional or otherwise imperceptible things might exist, but that is irrelevant. Even if some higher realm of existence were discovered, it would do nothing to falsify the fact that the truths we have so far discovered would continue to hold within *this* realm of existence, which is the only realm of existence in which we ever made the claims. Even if the laws of nature suddenly changed, then all that would mean is that we would have to re-adjust our principles.

The point is this: Science is the best possible perspective a human being can ever hope to have on the truth of the universe. Don't be tempted by any flashy but ultimately hollow and superficial unfalsifiable theories. Life is too short for that and we have better things to do. There is essentially zero reason to concern oneself with what the ultimate nature of existence is. It has no bearing on anything that actually matters. Don't waste any of your precious little time in life on that kind of nonsense.

However, the fact that science is the only reliable way to justify beliefs has imbued it with a level of credibility that no other domain of human endeavor can match. Unfortunately, the credibility of science is sometimes parasitically exploited in order to give unscientific ideas the air of science without actually providing sufficient logical or evidentiary justification. Never treat an idea as being more credible just because it has been packaged with a bunch of superficial characteristics that are common in scientific culture (e.g. white lab coats, big words, statistics, authoritative-sounding organization names, meticulously detailed descriptions, etc).

Such superficial characteristics add literally zero inherent credibility to an idea. Science is not wizardry from some fantasy novel. Scientists don't just sit around waving magic wands that cause new discoveries to pop into existence out of thin air[25]. The only thing that ever makes an idea scientific is that it is testable and is backed by some combination of reasonably airtight logic and verifiable evidence. Everything else is irrelevant. Learn to ignore all of the other factors. It will make it more difficult for people to deceive and manipulate you.

The human mind's capacity for pattern matching and association, while very useful, also leaves it vulnerable to manipulation. Basing your trust on the superficial characteristics of a trustworthy source greatly increases the chances that later on someone else will exploit those same superficial characteristics in order to get you to trust them, without ever actually earning that trust in any way.

---

[25] Unfortunately, many people who are unfamiliar with how science is actually done pretty much view science as happening like this (i.e. arbitrarily, as if by magic). This may be part of why some people feel that they are justified in denying scientific truths despite not having any credible evidence to support that denial. They don't understand that, unlike certain other beliefs they've been brought up on, scientific claims are *not* arbitrary and were *not* just pulled out of some magic hat.

Much of the public's occasional mistrust of science may actually originate from the public's unwitting exposure to fake science. Greedy corporations, new age quacks, and lazy journalists besiege the public on a daily basis with a vast barrage of unscientific garbage dressed up in the verbiage of science. People thus sometimes come to distrust "science" when in reality the object of their distrust should actually be the charlatans who take advantage of insufficiently educated people's inability to discern the difference between real science and fake science.

Always be on guard for conflicts of interest. As such, the next time you see "news" about the "health benefits of product X" (or whatever else) then you should strongly consider the possibility that it may actually be propaganda from the X industry that you are seeing, and not real scientific news. Likewise, for example, because most psychiatric studies are funded by the multi-billion dollar pharmaceutical industry, you should automatically be at least moderately suspicious of any of their claims. Similar things can be said of other industries. I think you get the idea.

In contrast though, studies on things like gravity, electricity, algorithms, evolution, mathematics, and many other subjects in the hard sciences and STEM[26] fields much more often have little to no financial incentive to spread misinformation and thus are far more likely to be trustworthy by default.

When I say that the hard sciences and STEM fields tend to have less financial incentives to deceive the public and hence less corruption I am referring to their incentives for the act of making and publishing discoveries itself, not to the indirect industrial profit potential. For example, someone might discover a new property of semiconductors, and might even intend to profit from it in the manufacturing industry, but there is little to no profit to be made merely *from convincing people that this property of semiconductors is in fact true*. The vulnerability of a branch of science to corruption tends to rise in proportion to how much it stands to gain from *belief-based* profit if it were to deliberately deceive the public for leverage. This is a useful measure by which to estimate how cautious you should be in trusting any particular branch of science.

Notice the difference. For example, the psychiatric industry stands to gain enormous profits from convincing as many people as possible that they are clinically depressed, even if they are not, and even in cases where non-psychiatric treatments would be superior. They stand to profit *directly* from convincing as many people as possible that something is true, regardless of whether it actually is true. This is belief-based profit. It corresponds closely with potential corruption.

In contrast, the discovery of a new property of semiconductors only has *indirect* profitability potential. There is probably almost nothing to gain from merely convincing people that this newly discovered property of semiconductors is in fact true. Therefore, there is little to no reason to suspect a conflict of interest in the case of the discovery of a new property of semiconductors.

Basically, whenever a claim has a potentially large financial incentive associated with it, then you should treat it with a lot of extra suspicion. You should lean more towards "guilty until proven innocent" than towards "innocent until proven guilty" in such

---

[26]**STEM** stands for "Science, Technology, Engineering, and Math".

cases. More generally though, treat all claims as untrustworthy until some combination of reasonably airtight logic and verifiable evidence convinces you otherwise.

If there is not much transparency about how the relevant data was gathered and analyzed then you often shouldn't trust it. However, even if the data and methodology is transparent then that is still not necessarily sufficient reason to trust it. There are many factors that go into making a theory trustworthy. The only perfect authority on truth is airtight logic and verifiable evidence. Human beings can never qualify as perfect authorities. All humans sometimes make mistakes. As the old saying goes: To err is human.

It is wise to learn to recognize the hallmarks of fake science. Some fake science is subtle and hard to detect, but other times it has a certain special stink that makes it easy to detect for anybody with even a small amount of knowledge of the subject matter. For example, new age mysticism frequently uses scientific terms without much regard for the actual meaning of those terms.

A new age quack might try to convince you to buy a "regeneration crystal" that uses "ions" to improve your body's "quantum energy field", or to walk barefoot instead of wearing shoes because it "aids your feet in their natural process of releasing electro-magnetic radiation" and so on. There's basically no such thing as credible "new age" anything. The entire movement rests on a foundation of moral bankruptcy and exists primarily to swindle naive and uninformed people. In fact, all of mysticism in general is basically one giant con job.

Hype and sensationalism is often a red flag for fraud. The more implausible or "too good to be true" something sounds, the more evidence you should require before you believe it. As we like to say in science: Extraordinary claims require extraordinary evidence. When in doubt, always err on the side of skepticism. Real science hinges on introspection, re-examining past assumptions, and a detached willingness to criticize one's own beliefs. In contrast, fake science tends to not be self-critical, because that would increase the risk of being disproven, and fake science certainly has no interest in that. Real science embraces criticism. Fake science runs away from it.

By definition, if a claim is based on blind faith (i.e. blind assumptions) then it is not scientific, hence there can be no such thing as faith-based science. The very idea of faith-based science is self contradictory. Science does make mistakes of course, but unlike certain other worldviews it corrects those mistakes once they are identified.

Anyway, those are just some examples to illustrate what unfalsifiable claims are, why they matter from an intellectual standpoint, and how they tend to encourage counterproductive behavior. One could write an entire book just on debunking the vast plethora of unjustified beliefs people have. To fight against the anti-science propaganda machine in the hopes of a more prosperous and peaceful future for humanity is truly a noble goal and one that is surely worth pursuing. However, that is not the primary subject of this book. I'll leave this somewhat tangential discussion as it stands and return once more to the original subject matter we were previously discussing.

As I was saying, there is not merely one fallacy hiding in statement-based truth, but rather there are two. The two fallacies of statement-based truth are symbolic-semantic equivocation and truth-knowledge equivocation. We were talking about truth-knowledge equivocation when I described the four basic knowledge values and was giving some examples of them to make the concept more concrete and understandable.

This idea we have been talking about of there being four knowledge values might have reminded you of something we talked about earlier. Remember when we were talking about dialetheism and I mentioned in passing that there were also some people who even believe that there are four possible truth values: true, false, both true and false, and neither true nor false. I mentioned (on page 43) that these four hypothetical truth values were sometimes referred to as the Catuskoti in Buddhist and Indian logic, and also could conceivably appear in some form in multi-valued logic and paraconsistent logic.

Well, the Catuskoti as originally conceived as including "both true and false" and "neither true nor false" don't really make much logical sense. However, notice that our four knowledge values (known-true, known-false, unknown, and unfalsifiable) do have a certain striking resemblance to this notion of the Catuskoti. Indeed, it may be that the Catuskoti are actually an incorrect attempt to formulate the four knowledge values we have given here, one that got derailed by not fully understanding the nature of truth and by conflating certain concepts together.

As such, perhaps one could argue that "unknown" and "unfalsifiable" would be better choices for those other two truth values. This is just a random thought that I wanted to mention here though. It's just something for you to maybe consider, as a matter of interest. We'll actually have much more to say about systems with more than two nominal truth values later, once we get to the chapter on non-classical logic, but now is not the time for such things.

1.10 What's next?

So, to review: We have two different ways of thinking about what kinds of things truth is really a property of (i.e. statement-based truth versus existence-based truth), of which existence-based truth seems to be conceptually superior, but we also have two different perspectives on what a truth value itself is. One perspective is that of objectified truth, in which the existence or non-existence of sets and the relations that those sets have to everything else that exists is what constitutes the truth of something. The other perspective though is that of nominal truth, such as exemplified by the concept of knowledge values, which represent mental notes we have made as to whether or not a particular set has been confirmed to exist or what the state of our knowledge regarding it is more generally.

With objectified truth, every object is its own distinct truth value, which encompasses all the underlying relationships of each object. In this way, a truth value becomes much more than merely a binary true or false value, but rather it becomes a complete representation of everything there is to know about that object. Hence objectified truth captures that sense of deeper meaning and clarity of insight that the binary true or false values of nominal truth fail to capture.

When we seek the truth, what we really seek is this kind of deeper fundamental understanding of some part of the world. Objectified existence-based truth succeeds in capturing this deeper sense of meaning expressed in the concept of truth, whereas nominal statement-based truth fails to do so. Through careful thought experiments, we have cleared up some confusions about the nature of truth and by doing so we have even been able to defeat the liar's paradox and to explore a number of different interesting concepts related to truth.

Interesting as all of this may be, we have so far only discussed what truth is, rather than how to go about finding it. There's far more to the pursuit of truth than merely defining what truth is successfully. In order to take the next step and to learn how to actually systematically discover new truths we will have to explore a new subject: logic.

In this chapter, we started out with our existing notions of what truth is and gradually refined our ideas to find new insights. In the next chapter, we will do the same, except this time we will be exploring the concept of logic.

Through studying logic, we will learn to navigate the vast oceans of truth that lay all around us. Much like we have examined our intuitive notions of truth here with a critical eye, we will do the same for our intuitive notions of logic in the next chapter. In the process, we will discover that our preconceptions about what logic is have led us astray, just as our preconceptions of what truth is did at the beginning of this chapter.

In this first chapter, armed with a new concept of what truth is, we were able to confront and destroy certain anomalies that had long plagued the theory of truth, such as the liar's paradox. Likewise, in the next chapter and beyond, as we explore a fresh perspective on the nature of logic, we will be able to vanquish certain anomalies in the theory of logic as well.

It's time to unify the various different branches of logic into a single more coherent system. This will be our principle task for the remainder of most of this book: acquiring deep insight into the fundamental nature of logic and then utilizing that insight to help guide the creation of a new branch of logic called **unified logic**.

Chapter 2

What is logic?

2.1 What makes us believe something?

Human beings are compelled to seek the truth, but how is it that a person comes to believe that they have discovered a truth? In our daily lives we encounter many different beliefs about reality. Some beliefs are generated internally by our own mind. Some beliefs come from communication with other people. Some beliefs come from observation of the natural world or of artificial systems. How is it that people decide what they believe? Are some approaches to deciding what to believe better or worse than others?

First, to help properly frame our thoughts on these matters, let's take a moment to consider the nature of belief itself. If we are to genuinely understand the nature of beliefs, then it seems reasonable that we should first understand what belief itself really is.

Beliefs are things that exist inside our minds. Anything which exists inside our mind in a form that we can perceive and manipulate can roughly be referred to as a thought. Thoughts include things like ideas, concepts, beliefs, recollections of memories, and so on. Are ideas and concepts the same thing as beliefs? Well, let's consider a concrete example and see. Suppose I asked you to consider the idea of gravity being a force of repulsion instead of a force of attraction.

The mere fact that I have said this has caused the idea of gravity as a force of repulsion to be implanted in your mind, where it now resides in some form. However, would you say that you *believe* in the idea that gravity is a force of repulsion? You of course would not. Anyone can clearly see that gravity attracts, not repulses, otherwise you would be flung into outer space immediately. Hence, merely residing inside the mind as an idea or concept is not enough to qualify a thought as a belief.

Can beliefs exist outside of our minds? Can a rock be a thought? No, because thoughts are fundamentally defined as being part of a mind. We can think about rocks, but that does not make a rock a thought. It is necessary that a belief is a thought, but merely being a thought is not sufficient to make something into a belief.

Also, a thought can only exist in a mind, but what kinds of things qualify as minds?

Clearly human beings qualify as having minds, but what about other things? For example, it seems perfectly reasonable that if aliens exist somewhere in the universe who are as sentient as we are then they would also qualify.

What about artificial intelligence? Do computer programs have minds? Well, for our purposes here we will only concern ourselves with whether they have "beliefs" of any vaguely resemblant form, rather than even attempting to touch the very difficult and controversial topic of whether or not it will ever be possible to create a strong AI[1].

For example, suppose we have a robot on an assembly line that has a camera which it uses to detect whether products moving down a conveyor belt have a particular type of defect or not. When such a robot detects a product defect doesn't it seem reasonably fair to say that the robot *believes* that it has detected a product defect? It has data in memory which specifies this estimation. How is that substantively really much different from the same belief instead residing in an organic mind?

The robot isn't sentient, yet it seems to have beliefs in at least this sense. For example, it can believe that its core temperature has risen to over 100 degrees Celsius, which is the temperature at which water begins to boil. A belief is essentially a knowledge value, a truth label with some potential uncertainty attached to it. Thus, in other words, it does not appear to be necessary for something to be a higher-level independent self-aware mind in order for it to have beliefs in this sense. Merely having such estimations of reality (i.e. such beliefs) does not qualify something as being sentient. However, not being sentient does not eliminate the possibility of having these kinds of beliefs, at least not in the broadly encompassing way we are interpreting the concept here.

To say that one believes in some idea is simply to say that one thinks that whatever that idea refers to is true, which is the same as saying that one thinks that the objects and relationships described by the idea do exist in some physical or logical reality as specified by the given criteria. In other words, a belief is a knowledge value stored in some arbitrary mind, whether organic or artificial. Let's give an official definition:

Definition 18. *Whenever a mind thinks that an idea is not merely hypothetical, but actually exists in some physical or logical reality as specified by the criteria implied by the idea, then that idea is called a **belief**. Beliefs are fallible knowledge values (such as known-true, known-false, unknown, unfalsifiable, and degrees of certainty) that are stored inside a mind corresponding to the associated ideas.*

For the purpose of this definition, things that have minds include human beings but also include any other entities capable of at least some form of thought or decision making, such as many other organisms[2] and all artificial intelligences.

Broadly speaking, there are three different ways in which a mind can come to believe in a claim. These three possible bases upon which a particular belief might rest

---

[1] A "strong AI" is essentially an artificial intelligence that qualifies as a sentient being, capable of its own independent thought and genuine creativity. This is in contrast to a "weak AI", which is any lesser AI. Weak AI are abundant. Lots of different software already has weak AI embedded in it. For example, every time you interact with software that appears to be capable of any form of automated decision making or perceptual awareness whatsoever then you are interacting with (at least) a weak AI.

[2] A biological virus is an example of an organism that is incapable of having thoughts or making decisions. In contrast, any organism that has a brain will be capable of thinking and making decisions and hence will have beliefs about its environment.

are: assumption, intuition, and reason. Together these bases for belief form a rough continuum of degrees of certainty. Let's define an official term for that continuum:

Definition 19. *Every belief has a degree of certainty associated with it. This degree of certainty can be represented approximately with a number between zero and one (or alternatively with percentages). We will refer to this continuum of degrees of certainty as the* **spectrum of justification**.

Low certainty numbers represent beliefs that are poorly justified. High certainty numbers represent beliefs that are strongly justified. Assumption tends to yield poorly justified beliefs, intuition tends to yield moderately justified beliefs, and reason tends to yield highly justified beliefs.

It is important to remember however that in addition to these three primary bases for beliefs, there is a fourth related but somewhat independent mechanism as well. For while it is possible for the thought processes of a mind to be restricted to only the beliefs that the mind actually does hold, it is often extremely advantageous and convenient to be able to explore contingent lines of thought by acting as if one "believes" things that one does not necessarily actually believe. Let's define a term for this highly useful mechanism for exploratory thought:

Definition 20. *In order to explore an idea and to get a better sense of its potential credibility, it is often useful for a mind to treat the idea as if it were true, temporarily, in order to investigate what its consequences would be, regardless of whether or not the mind actually has sufficient reason to genuinely believe in it. Such a contingent exploratory assumption may be referred to variously as a* **premise**, *a* **postulate**, *a* **supposition**, *a* **hypothetical**, *or an* **axiom**.

Those are probably the most common synonyms you'll encounter. It would also be apt to refer to it as an **operating assumption**, *a* **contingent assumption**, *or a* **conditional assumption**. *You might also hear it referred to simply as just an assumption, but referring to it as such could occasionally run the risk of making it sound more like you actually believe in it, which may or may not be true.*

Which term you choose to use is generally arbitrary and people tend to be inconsistent about it, much like how natural laws in science are referred to variously as laws, theories, or principles (etc) with little difference in meaning, if any.

It's just one of those quirky historical and cultural things. The word choice is more often superficial than meaningful. Don't let the excess verbiage intimidate you. A lot of the complicated sounding wording distinctions are actually just smoke and mirrors.

One of the above terms tends to be used a bit more strictly however, and that is the term *axiom*. Generally the term axiom is reserved for premises which are so primitive and foundational that it seems impossible to reduce them to any simpler set of premises from which the axiom might be derived. However, even this rule is broken fairly often, especially in the study of formal systems, where it is not unusual to see someone declare a non-primitive statement to be an additional axiom that is assumed to be true.

The safest and most general of all these terms is perhaps *premise*. It has the least additional baggage in terms of connotation. When in doubt as to which term you should use to describe a tentative assumption, you can just call it a premise. It's a safe choice. It's also perhaps the best choice if you want to reduce the amount of redundant verbiage and cognitive load caused by switching between the different terms arbitrarily.

Anyway though, as I was saying, roughly speaking there are three bases upon which a belief may rest: assumption, intuition, and reason. It is worth taking a moment to consider what exactly it is that each of these respective things really is.

Definition 21. *Whenever we come to believe that an idea is true even though we do not have sufficient evidence to justify it, we call this process* **assumption**. *Pure assumption (also known as blind faith) is belief that has no supporting evidence whatsoever, and hence is always inherently unjustified.*

However, it is important to distinguish between assumptions that express genuine belief in a claim without any evidence (which are generally bad), and "assumptions" that are merely premises being used to hypothetically explore an idea, to simplify a thought exercise or calculation, or to precisely specify the foundation of a formal theory (which are generally good).

Definition 22. *Whenever we come to believe that an idea is true, or probably true, based upon some kind of heuristic method of thought, perhaps involving mental short-cuts, evolved adaptive instincts, or analogies to past experiences or to similar cases for example, we call this process* **intuition**.

Real life often throws a torrent of overwhelming information at us, and it is often too difficult or time-consuming to analyze all this information fully. We therefore are often forced to use intuition in order to process it all in a timely matter without becoming paralyzed by information overload. For computer programs, the closest analog of intuition is probably heuristic algorithms.

Definition 23. *Whenever we come to believe that an idea is true because we have acquired direct physical or logical evidence of it, we call this process* **reason**. *Reason is the foundation of all of science. Utilizing it properly requires discipline, a steady hand, practice, independent thought[3], curiosity, persistence, skepticism, and an awareness of (and trained resistance to) cognitive biases and logical fallacies.*

Reason is a skill. Simply having a mind does not make you reasonable, just as simply holding a paintbrush does not make you a painter. Common sense is to reason what stick figure drawing is to landscape painting.

One of the most important things to keep in mind for understanding the spectrum of justification correctly is that beliefs and premises are two very different things. A premise always exists in a contingent form, whereas a belief is something that has been internalized as if it is a fact.

When we assume something is true as a *premise*, it is as if we are attaching a disclaimer that says "if the premise is true" to every relevant statement that follows thereafter. In contrast, when we assume something is true as a *belief*, we are asserting that

---

[3] i.e. independent thought as in anti-authoritarianism essentially... a mind that thinks freely, based on justified principles and hard evidence instead of merely based on conformity or random whims

it is actually part of some physical or logical reality and not merely a convenient basis for reasoning.

Another pitfall to look out for is giving a premise less credit than it deserves. For example, sometimes you'll hear people say that the standard axioms of logic and mathematics are things that just have to be taken as true without evidence, as if the axioms are somehow baseless assumptions. In actual fact though, many of the most standard axioms of logic and mathematics are actually very highly correlated with the available evidence and are far from arbitrary.

For example, the axiom of geometry which says that between any two points a straight line can be drawn is mostly based on direct evidence involving dealing with points in real life. Yes, non-euclidean geometry and curved space-time complicate things, but for the most part the straight line axiom is evidence-based and is not merely some kind of blind assumption.

Thus it is not really fair to say that the fundamental axioms of logic and mathematics rest on blind assumptions. With regards to the most important and useful branches of logic and math (i.e. the ones that have real-life applications), the axioms are typically very strongly backed by direct evidence that you can observe in the real world.

It is therefore often inappropriate to treat the foundations as merely being arbitrary assumptions, even though this attitude is sometimes expressed. Blind faith has little to nothing to do with any of it. Math tends to either mirror reality or else explore arbitrary but highly structured logical spaces, neither activity of which qualifies in any way as being based on blind faith.

There is also nothing about saying that *if* some premises were true *then* certain consequences would follow (i.e. about the entire way that all of mathematics and logic are structured) that in any way really qualifies as resting on faith. The conditional nature of premises is widely and openly acknowledged and hence most premises in logic and math are generally not genuine beliefs except when also backed by physical evidence.

Anyway though, to answer the earlier question of whether or not some bases for believing things are better than others: the answer is of course yes. Of the three bases for belief (assumption, intuition, and reason) it is clear that reason is always the strongest basis for forming beliefs.

In fact, not only is reason a superior basis for belief, it is literally the essence of justification itself. Saying that a belief is justified and saying it has sufficient reason are synonymous. By definition, assumption and intuition will always be inferior in terms of justification strength. To say otherwise would contradict the very definitions of the terms and would thus make no sense.

Belief assumptions are defined by their general lack of sufficient justification. Therefore, in a certain sense, there cannot really be any such thing as a justified blind faith assumption. If an assumption is authentically justified then it is no longer merely an assumption, but is in fact based on evidence.

Again though, I warn you not to confuse belief assumptions with premise assumptions. The word "assumption" is often equivocated and used ambiguously to refer to both belief assumptions and premise assumptions even though they are actually logically distinct. In fact, let's go ahead and give an official definition for this distinction to make it easier to not equivocate the terms in the future:

Definition 24. *Whenever an assumption represents an actual internalized belief that the claim it makes is in fact true, then we will call this a **belief assumption**. In contrast, whenever an assumption merely represents a contingent supposition that the claim is true for the purposes of exploring the consequences, whether or not we think it is actually true, then we will call this a **premise assumption**.*

As an interesting side note, just as people can have beliefs that range anywhere on the spectrum of justification, a little thought shows that computer programs can also have beliefs that range anywhere on the spectrum of justification as well. When a program treats a set of conditions as true, but does not test whether those conditions actually are true, then we may say that those conditions have been hard-coded into the program.

Hard-coded parts of a program correspond to assumptions. Just as people that make unjustified assumptions often suffer negative consequences from those assumptions, computer programs that make unjustified assumptions often suffer negative consequences from those assumptions too.

The hard-coded sections of a program are often the most brittle, bug-filled, and unstable parts of that program. Likewise, the portions of a person's mind that have the most unjustified assumptions are typically the most dangerous and counterproductive to that person's health, well-being, and future outcomes, and also to the health, well-being, and future outcomes of society as a whole.

Computer programs can also be made to perform fuzzy or vague statistical analysis or to apply techniques that don't always work exactly (i.e. heuristic algorithms), and this corresponds roughly to intuition. Some computer programming problems cannot be solved in a timely manner without loosening the constraints enough to allow efficient computation. This corresponds to the way people use intuition as a shortcut when they lack the time for thinking through a situation completely.

On the other hand, computer programs can instead be made to perform strict logical proofs where they systematically guarantee that no errors have been made and that the structure of the code guarantees correctness. Even if a program can't produce an actual proof of correctness, well-designed programs will often run a large battery of concrete test cases (e.g. unit tests, etc) that produce tangible evidence that things appear to be working properly, and as much as possible operate with a careful awareness of the state of their environment. Such well-designed parts of programs correspond to reason-based beliefs.

The real world is filled with people who want to convince you of a wide variety of wildly arbitrary claims. They'll often try to push you to accept claims without providing sufficient supporting evidence (or indeed often without providing any supporting evidence at all) and yet will often take great personal offense if you deny their claims. Even just stating the blatantly obvious fact that only beliefs that are based on evidence or logic can ever be considered justified can potentially be met with intense hostility.

It is one of the great tragedies of our time that basic intellectual honesty still has not yet been fully acknowledged as the absolute necessity that it is. The basic necessity that all beliefs must be backed by evidence and logic as much as possible should ideally be as widely acknowledged as the basic necessity that we must all eat food to survive. Luckily though, we seem to potentially be moving in that direction, so I am hopeful that humanity will nonetheless eventually achieve intellectual enlightenment and world peace, perhaps sometime within the next thousand years maybe. Fingers crossed.

However, even just getting humanity to finally be firmly intellectually honest is not quite enough. It's not just the nonsense that people are trying to push on us all the time that we have to guard against. We also have to guard against our own internal cognitive biases and logical fallacies, as well as the tendency of natural phenomena to be tangled up in such a way that we are vulnerable to making erroneous or incomplete observations about such phenomena, even when we reason carefully.

Often multiple factors are at play in natural and artificial systems, and when multiple factors overlap it can be hard to detect and properly distinguish them. We can't always control the system as much as we would ideally like to in order to conduct proper experimentation and tests. However, we have to start somewhere and work our way up from there.

How do we take a set of directly observed facts or logical premises and then derive as many correct conclusions as we can from them? How do we take a set of things that we know to be true and then use those things to produce additional new facts, insights, and perspectives on the system as a whole? How do we look at a set of rules for a system and determine how they interact, or indeed whether they fit together coherently at all?

The answer to these questions is that we need to construct a precise system of thought that allows us to take any set of facts or premises and to infer or deduce what the consequences of those are and to understand how the underlying structures of the system relate to each other. What we need, in other words, is *logic*.

2.2 How did the formal study of logic begin?

Logic has always existed throughout human history. For as long as humans have been capable of abstract thought, they have surely been using logic in some form or another. From tracking wild game for food, to inventing and constructing simple tools, to navigating the social complexities of society without being ostracized, to planning harvests upon which they critically depend for survival: humankind has always had a need for the ability to reason and to make inferences beyond what is merely self-evident.

It is therefore impossible to credibly claim that anyone in history ever invented logic. Logic is an inborn gift of all humankind. However, without training and discipline, most people's ability to consistently use logic correctly is typically quite poor. For someone who has not practiced logic deliberately and with care, perfect reasoning is a rarity at best, especially for more complicated cases such as for theorizing about how and why the natural world operates as it does.

You would think perhaps that the formal study and practice of logic would go back to the very beginning of human history, seeing as it is so essential for correct thinking and is so useful for practical applications. However, logic as a coherent and deliberate field of study only really emerged relatively recently, roughly two and a half millenia ago (i.e. roughly 2500 years ago), probably sometime between the years 500 BCE and 300 BCE.

Although glimmers and fragments of logic began appearing prior to this period, it was mostly between the years 500 BCE and 300 BCE that logic began to coalesce into a solid systematic discipline rather than merely a collection of fuzzy informal concepts and speculative philosophy. It appears that several different cultures may have independently invented systems of logic during roughly this time period (give or take a few hundred years) namely: Greece, India, and China. However, the most influential of these systems of logic would probably be that of Greece.

Ancient Greece had a number of factors working in its favor that potentially made it a far more ideal environment for the emergence of logic as a formal discipline than any of the other cultures of the ancient world. Why Greece though? What exactly were the factors that made it into an exceptionally fertile nesting ground for rational thought?

Well, perhaps the most significant supporting factor was that during the fifth century BCE (i.e. between 500 BCE and 400 BCE) the Greek city-state of Athens became the world's first significant democracy. In addition, the influence of religion within Greece was relatively weak compared to other regions, although maybe not very weak in absolute terms. Consequently, people were able to express themselves significantly more rationally than in many other preceding or concurrently existing societies, instead of being constantly under the relentless control of authoritarian politicians or oppressive clergy.

In an environment where governance was based on direct democracy, it suddenly became more necessary for people to actually have a convincing substantive argument in order to influence policy and society, rather than to merely have a position of arbitrary authority from which to dictate according to one's random whims or mood. Thus, whenever important decisions where to be made, people would have debates over the merits of the respective sides.

The fact that suggestions were now met with opposition made it so that decisions were much more likely to be based on real merit than ever before. Suddenly, finding the truth started to matter a lot more. To wield power in this environment, you often had to actually convince other people that the truth was on your side. However, while this raised the standard of evidence higher than it had been before, this does not mean that objectivity reigned supreme. Deceptive and manipulative rhetoric that had little or nothing to do with the merits was still abundantly common, just as it is even to this day.

In addition to political debates, it also became common for people to have debates on philosophical questions about what certain concepts and words actually mean, about what the nature of the world is, and about how best to live one's life, among other things. Mathematics (especially geometry) had also recently been making considerable progress and was rapidly becoming more systematic.

Debates over a variety of different topics became increasingly common and were treated like a kind of sport. Audiences would gather to see two opponents argue about some matter of philosophy or fact, and in order to be successful one had to learn to be convincing. Some arguments were won on the basis of sound reasoning, whereas other arguments were won on the basis of logical fallacies.

Because decisions were now made on a democratic basis, it became important to win arguments through correct reasoning rather than through fallacies. Whenever the public was convinced by a fallacious argument it effectively put society as a whole at risk. Manipulation of public opinion was a major threat to democracy and to freedom of thought in Greece, just as it continues to threaten the free world to this day.

During this time period, the systematic pursuit of truth and of the techniques necessary to consistently form valid arguments became more important than they had probably ever been in any other time or place in human history before. Greek culture matured along these lines (somewhat turbulently) for a while. Some schools taught argumentative techniques, but the practice of logic largely remained informal.

However, in the year 384 BCE, a man named Aristotle was born in northern Greece in the city of Stagira. Aristotle began studying at Plato's Academy in Athens around 366 BCE, to which our history and our prosperity probably owe much. Aristotle would ultimately prove to be one of the most profoundly important people to ever live. Were it not for the impetus he gave to the development of logic, science might not have come into existence when it did and thus neither would the corresponding luxuries and high quality of life we enjoy today.

Plato's Academy was a prominent philosophical and educational institution in the ancient world and it is likely here that the initial seed was planted that would one day grow into one of the greatest minds in human history. Aristotle died in 322 BCE at the age of 62, but wrote prolifically and made many major influential contributions to logic, science, and philosophy.

Aristotle studied and taught a wide variety of subjects in philosophy and science over his lifespan, but his most significant work, for which he is most remembered, is a collection of books and notes called *The Organon* among which is a book entitled *The Prior Analytics* in which Aristotle essentially created the first comprehensive system of formal logic in human history.

The title *The Organon* literally means "The Tool". Aristotle viewed logic as a tool or instrument to be used to consistently arrive at correct conclusions when investigating something. The collection was named *The Organon* by Aristotle's followers, not by Aristotle himself. The logical system Aristotle created in *The Prior Analytics* is today referred to variously as syllogistic logic, term logic, or Aristotelian logic. In this book I will use the term *syllogistic logic* to refer to Aristotle's logic in general, including both the original system and the later medieval variants as well.

Syllogistic logic is so named because it involves inferences made using what are called syllogisms, which are a type of logical argument that have a specific kind of structure. A syllogism is an argument in which two statements are combined in order to infer a third statement as a conclusion. Each statement is about a pair of "terms" and how they relate to each other. A "term" from a syllogistic perspective means essentially a set of objects that satisfy some property, i.e. a "class" or "concept" if you will. Terms were also sometimes referred to as "categories".

For instance, "men" would be an example of a term in syllogistic logic and would mean essentially the set of all male human beings. Another example of a syllogistic term would be "mortals", which would consist of the set of all living beings who one day die. Yet another example of a syllogistic term would be "Socrates", which would consist only of a set containing the ancient philosopher Socrates and nobody else. However, in syllogistic logic, terms were thought of more as concepts than as sets of objects and the notion of sets was not as mature as it is in modern set theory.

The most commonly used example of a syllogism is perhaps the following set of three statements:

1. All men are mortal.

2. Socrates is a man.

3. Therefore, Socrates is mortal.

Look at the final line of the argument, where the logical conclusion of the syllogism is. In a syllogism, the grammatical subject of the verb (i.e. the object "doing" the verb) on this line is called the minor term and the grammatical object of the verb (i.e. the object "receiving" the verb) on this line is known as the major term. For example, in this case "Socrates" is the minor term and "mortal" is the major term. The remaining term is called the middle term. In this case, "all men" is the middle term.

It's basically a transitive sequence of subset relations. In other words: the set containing only Socrates is a subset of the set of all men, the set of all men is a subset of the set of all mortals, and therefore the set containing only Socrates is a subset of the set of all mortals, because subset inclusion is a transitive relation. Thus we are able to conclude that Socrates is mortal. That being said, please keep in mind that the concept of a subset in the modern sense came much later and the original works of syllogistic logic do not use terms from set theory such as "subset".

All of the statements in the syllogism above involving Socrates are examples of what are called "universal affirmatives" in syllogistic logic. A universal affirmative is any syllogistic statement that has the form "All A are B", wherein "All A" can also be replaced by a single object rather than a set of multiple things and it is still considered to be a valid universal affirmative. Syllogistic logic also considers universal negatives, particular affirmatives, and particular negatives.

A universal negative is a statement that says that all of some set of objects do *not* have some property, i.e. it is a statement of the form "All A are not B". In contrast, a particular affirmative is a statement that says that some of a set of objects have some property, i.e. it is a statement of the form "Some A are B". Notice however, that this particular affirmative does *not* say that we know that "Some A are not B" is also true. In fact, both "Some A are B" and "All A are B" might be true, but we cannot decide without more information than just "Some A are B". As you probably have guessed, a particular negative is a statement of the form "Some A are not B".

Although set terminology did not exist during the era of syllogistic logic, understanding that syllogisms are really about transitive subset relations nonetheless does make it much easier to remember which of the terms are the minor term, the middle term, and the major term. There are two kinds of valid syllogisms: (1) those with two universal premises and (2) those with one particular premise and one universal premise.

For syllogisms that involve two universal premises: The minor term is always a subset of the middle term, and the middle term is always a subset of the major term. In other words, the minor-middle-major terminology expresses the relative sizes of the subsets. The minor term is always the smallest set, the middle term is always the middle-sized set, and the major term is always the largest set. Thus the connotation of the words exactly matches their role in the transitive subset relation upon which syllogistic logic is based. Syllogisms of this form have conclusions that are universals (i.e. that express subset relations).

For syllogisms that involve one particular premise and one universal premise: The particular premise essentially states that the two sets involved in it have a non-empty set intersection (i.e. that they share elements in common). In such cases, the set which participates in the particular premise but which does not participate in the universal premise is the minor term. The lesser set in the subset relation in the universal premise is the middle term. Therefore, the remaining set is the major term. Syllogisms of this form have conclusions that are particulars (i.e. that express intersection relations).

Aristotle studied the various different possible syllogisms, i.e. groups of statements composed of any combination of universal/particular and affirmative/negative statements, and then deduced which forms led to valid inferences and which did not. This process yielded a large number of inference rules which syllogistic logicians essentially had to memorize and was tremendously unwieldy and artificial compared to modern logic.

Syllogistic logic is essentially the ancient ancestor of set theory. Set theory is the systematic study of groups of objects and how they relate to each other. Set theory is much more streamlined and easy to use than Aristotle's original syllogistic logic. The reason why is because set theory factors out all of the redundancy and memorization involved in Aristotle's original classification of inferences into universal/particular and affirmative/negative statements.

Set theory also removed the limitation of only being able to reason from two premises to one conclusion that syllogistic logic suffered from. In contrast to syllogistic logic, set theory can easily work with any number of sets, with arbitrarily complicated relationships, and with no need to treat each possible combination of universal/particular and affirmative/negative statements as a tediously memorized special case.

People frequently talk about Aristotle's syllogistic logic as if it is the ancient ancestor of classical logic, but it is not really accurate to treat it as such. Syllogistic logic is a crude form of set theory. Syllogistic logic studies groups of objects, which is the same thing that set theory studies. In contrast, classical logic studies statements and how the truth values of those statements stand in relation to each other. In short: set theory studies objects, whereas classical logic studies statements. This distinction will prove to be important beyond just syllogistic logic, but more on that later.

The obvious question to ask at this point is of course: Where did classical logic come from then? Well, shortly after Aristotle's time, another school of thought about logic arose to maturity, one that had a different perspective on what it thought the main object of study in logic should be. This historical alternative line of thought is known today as *stoic logic* and its most prominent proponent was the ancient philosopher Chrysippus of Soli.

Stoicism was both a philosophy of life and a philosophy of knowledge. It was founded in Athens around 300 BCE by the philosopher Zeno, of Zeno's paradox fame. Stoicism as a philosophy of life emphasized living in accordance with nature, wisdom, and strength of will. The Stoics emphasized living with a sense of rational calm and perpetually insightful perspective on one's circumstances, so that one would neither fall into despair during times of hardship nor lose control to overindulgence during times of prosperity.

Ideally, regardless of what misfortune afflicted a Stoic, a Stoic would always strive to maintain a fresh and nuanced view of their circumstances, one that allowed them to remain calm and contented no matter what happened to them. This tendency to always be calm and level-headed caused the word "stoic" to gradually mutate to its modern meaning of roughly "remaining calm regardless of circumstances" or "being emotionally passive".

As a philosophy of knowledge, Stoicism's perspective on logic was descended from another school of philosophy called the Megarian school, from which stoic logic got the idea of propositions (i.e. statements) being the object of study in logic. The philosophers responsible for the earliest substantive discussions of propositional logic were perhaps Diodorus Cronus and Philo, both of whom were Megarians.

However, the person who largely formalized stoic logic into a more coherent framework was Chrysippus, so he is generally seen as the most significant figure involved in early propositional logic. Chrysippus was the third leader of Stoicism. Cleanthes was the second leader of Stoicism, and the immediate successor of Zeno, but Cleanthes is apparently generally seen as less historically important. Despite the fact that Chrysippus was actually the third leader of Stoicism, he is sometimes referred to as the second founder of Stoicism as a reflection of his importance.

Chrysippus lived from 279 BCE to 206 BCE and died at the age of 73, which means his life took place roughly 100 years after Aristotle (who was born 384 BCE and died 322 BCE at the age of 62). Unlike Aristotle's syllogistic logic, stoic logic was able to deal with conditional statements, which used to be known as "hypothetical syllogisms" but which are now most commonly known as "implications" or "conditionals". Also unlike Aristotle's syllogistic logic, stoic logic has the concept of logical connectives, such as logical conjunction (e.g. "*A* and *B*") and logical disjunction (e.g. "*A* or *B*").

Stoic logic also had the concept of proving things by reducing them to certain tautological forms. The Stoics had five such tautological forms, which they referred to as the "indemonstrable forms". The five Stoic "indemonstrable forms" were, along with Euclid's axioms of geometry, some of the earliest examples of the idea of the axiomatic method being used as a foundation for proofs.

By the way, Euclid lived roughly in the same time period as the Stoics. He lived from 323 BCE to 283 BCE. It appears that the intellectual climate at the time was ripe for the early development of the concept of axiomatic proofs. It is plausible that the axiomatic method was independently invented several times during this era, but it is hard to know because so many philosophical documents from this era were lost or destroyed.

The lack of large-scale printing technology (such as the printing press for example) meant that relatively few copies of many of the documents from that time period ever existed to begin with, and thus some documents were inevitably lost purely due to chance. However, many other documents were destroyed quite deliberately and maliciously.

The abrahamic religions, such as Christianity in particular as it spread into Greece and Rome, often systematically destroyed any documents that disagreed in even the slightest way with their faith or that promoted independent rational thought in any form. Nothing is more threatening to blind faith than insightful introspection, intellectual diversity, and careful rational thought. If only they had remained calm enough to realize the tremendous damage they were inflicting upon history.

By destroying so many priceless documents in this manner, Christianity may have single-handedly set human wisdom and knowledge back hundreds of years. If what little of Greek philosophy that was successfully preserved is any indication, some of the philosophical documents that were destroyed probably contained very interesting insights into how to live a good life and also many other important pieces of lost knowledge.

The loss of so many documents from this time period is an immense and often underappreciated human tragedy. Many of the greatest advances in human knowledge and technology could have potentially happened hundreds of years earlier than they did if it were not for so much intellectual progress being set back by religiously motivated censorship. The modern scientific era might have even come to fruition before the medieval era even began if so much knowledge had not been so callously destroyed.

Anyway though, let's continue our discussion of stoic logic and related matters. A conditional statement is any statement of the form "if A then B" or "A implies B", where A and B are themselves complete statements that could stand on their own. The statement A is typically called the "antecedent", but I prefer to call it the "condition". Likewise, the statement B is typically called the "consequent", but I prefer to call it the "consequence". I don't like unnecessary technical words when simpler and clearer words that are more readily understood by the general public already exist. I believe in merit over tradition.

One of the most significant ideas to come out of the Megarian and Stoic schools was the concept of the material conditional, from which modern classical logic's notion of the material conditional is ultimately derived. The nature of conditional expressions in logic was debated by the Megarian philosophers Diodorus Cronus and Philo. Philo was Diodorus's student.

Diodorus believed that a conditional is only true if the rule it expresses holds true at all times, such that any time the condition is true the consequence is true also. Thus, if there was any circumstance where the conditional statement could conceivably not hold, then Diodorus considered that conditional statement to be false. To Diodorus, conditionals had to be universally valid rather than circumstantial.

In contrast, Diodorus's student Philo believed that a conditional statement is false whenever the condition is true and the consequence is false, but that it is true in all other cases. Thus even if the condition and consequence were both true in this moment purely by chance, yet would not be true at other times, then Philo would still consider the conditional statement to be true. To Philo, conditionals were true purely based on the current truth values of the respective statements, regardless of whether the rule was universally valid or not.

Chrysippus commented on this debate between Diodorus and Philo, years after Diodorus and Philo were dead, and disagreed with both of them. Chrysippus's view was similar to Diodorus, except that Chrysippus allowed conditionals that refer to the future to be true, whereas Diodorus required conditionals to be true at all times. Of the three philosophers. Philo's view was the least restrictive and Diodorus's was the most restrictive. Chrysippus's view was more restrictive than Philo's view but less restrictive than Diodorus's view.

Philo's concept of conditional statements was what we now call "truth-functional", which is to say that the truth value of any conditional statement in Philo's system depends solely on the truth value of the respective statements, regardless of what relationships the statements have to each other. This concept of a conditional statement is what is known today as "material implication". Classical logic uses Philo's concept of material implication, but analogs of Diodorus's and Chrysippus's formulations may arguably still exist in certain non-classical logics.

I'll have much more to say about material implication later. It may seem like a perfectly innocent concept right now, but as you will see later, it is anything but. Although material implication was ultimately adopted as the standard definition of implication in modern logic, it was not without conflict. Logicians who lived much later, relatively recently in fact, would one day strike up a new debate on it.

However, even those more recent discussions did not actually succeed in closing the issue. It turns out that all the treatments of the concept of implication up to the present day have failed to actually adequately capture the core essence of real implication and the different possible perspectives on it. The new system of logic we will be building in this book, called unified logic, strives to do exactly that: to capture the true essence of the concept of implication and to bring the controversy to a close (among other things).

Anyway, having now discussed the ancient logics (syllogistic logic and stoic logic) to a sufficient extent for our purposes in this book, I think it is about time to move on to one of the more modern forms of logic. The most prominent and common type of formal logic used today is what is known as classical logic. It is pretty much considered the standard logic, and if you say "formal logic" the image that will come to most people's minds is that of classical logic. If you have some memory of being taught some kind of formal system of logic that used variables and expressions that resulted in values that could be either true or false (and nothing else), and you don't remember what it was called, it was almost certainly classical logic.

Does the word "boolean" or the phrase "truth tables" sound familiar to you? Anyway, that's along the lines of what we'll be covering next. A proper understanding of classical logic will be necessary in order to be able to understand the context and significance of the new system of logic that we will be developing in this book. I will be providing extra commentary related to the material as I go along that will sometimes be more than merely a review of the basic mechanics of classical logic. I therefore do not advise skipping any of it even if you are already familiar with classical logic. You are free to make your own choices though, of course.

2.3 What is classical logic?

As I alluded to in the previous section, the ancient ancestor of classical logic is arguably actually stoic logic and not syllogistic logic. It's fair to say that syllogistic logic was the first significant system of formal logic to be created, in that it provided the first useful rigorous account of a subject that had previously been informal and vague.

It's also fair to say that syllogistic logic did indeed influence classical logic nonetheless. Indeed, syllogistic logic indirectly influenced practically all of logic and science to follow thereafter. Some people even consider Aristotle's creation of syllogistic logic to roughly mark the birth of science. Other people consider science to have come into existence earlier or later than that, depending on what aspects they consider to be most essential to the core of science.

When exactly science was actually "created" is kind of fuzzy and unclear. Science was formed from an accumulation of many factors from many different eras spread across a long span of time. Some principles of modern science only truly became incorporated into the system and culture of science relatively recently.

Perhaps a good rough estimate of the formative period of science is the time span from Aristotle's creation of syllogistic logic to Francis Bacon's work on inductive reasoning in his influential book the *Novum Organum Scientiarum*. This time span ranges approximately from the 4th century BCE to the 17th century CE (i.e. from 400 BCE to 1700 CE, which is slightly more than two millenia). To put that in perspective, the 17th century is also when calculus was invented by Isaac Newton and Gottfried Leibniz. Science grew a lot during this period and began maturing much faster thereafter. I am certainly not a historian however, so if you want a more detailed and nuanced account of the history of logic and science please refer to actual historical sources.

Anyway though, modern classical logic as we know it today took shape mainly through the work of George Boole, Augustus De Morgan, and Gottlob Frege during the 19th century (i.e. approximately between 1800 CE to 1900 CE). However, many modern logicians and mathematicians probably wouldn't immediately recognize the older notations as classical logic, unless they are already sufficiently familiar with the history.

Early classical logic experimented with several different notations before it settled on the modern notation familiar to us today. George Boole actually originally used a notation that visually looked identical to plain old numeric algebra. For example, Boole would write $xy = 1$ to indicate that both statement x and statement y were true. Boole's system used 0 to represent false and 1 to represent true. Thus $xy = 1$ would only be true if both $x = 1$ and $y = 1$ because if either were zero then the multiplication would also yield zero.

The modern term "boolean" is derived from George Boole's name. Modern day computer programmers refer to variables that store true or false values as "booleans" and sometimes classrooms introduce the study of standard classical logic by referring to it as "boolean algebra". However, while classical logic may utilize boolean algebra, the two concepts are not the same. The criteria that determine whether any given system is considered "classical" or "boolean" are technically different.

However, the standard true-false logic that most people are taught in school is indeed both classical and boolean. Boolean algebra focuses more on studying certain algebraic properties of operators, independent of interpretation, whereas classical logic focuses more on the theory of classical truth values and the associated proof construction techniques. Despite these differences though, classical logic and boolean algebra very frequently overlap. The term "boolean algebra" often refers specifically to *classical* boolean algebra, but not always.

In contrast to Boole's notation though, Gottlob Frege used a notation for logic that would likely be even more difficult to recognize for those only familiar with the modern notation. Frege invented a rather peculiar and quirky notation for logic which he called his "concept script". The notation used branching connected lines and roughly resembled a left-justified tree diagram or electrical circuit diagram with symbols for logical variables attached at various points.

Frege's notation was too unwieldy and eccentric and is now extinct in modern usage. That's natural selection for you though. Awkward and wasteful systems tend to die out and get replaced by more graceful and efficient alternatives over time. However, notation aside, Frege still made some major contributions to the axiomatic treatment of classical logic. Frege is also well-known for his attempt at deriving all of the laws of arithmetic from purely logical principles, in his book series *Grundgesetze der Arithmetik*.

This resulted in a famous incident wherein Bertrand Russell (another influential logician and mathematician) discovered that Frege's axioms (the central assumptions of his theory) would lead to a paradox. The paradox Russell cited is known today as Russell's paradox and is significant in its effect on the development of modern mathematical theory.

As for Augustus De Morgan, he is perhaps most widely known for deriving several especially important laws of classical logic known as De Morgan's laws. Along with George Boole, De Morgan was highly influential in helping to break the stagnation surrounding the historically dominant syllogistic logic and in making logic more flexible. De Morgan was the first president of the London Mathematical Society and one of its founders. De Morgan also coined the term "mathematical induction" and gave the first rigorous account of it.

The modern notation for classical logic originated during the early 20th century, via gradual improvements to the clarity of the notation and with occasional introduction of new symbols. The use of truth tables to define logical relations in a convenient form was not even common until the early 20th century. The modern approach to logic is in fact surprisingly young, considering how it seems so simple and obvious now.

Considering how surprisingly young modern systematized logic actually is, it is perhaps astounding how quickly modern technology such as programmable computers were invented. Such staggering progress, seemingly in the blink of an eye, is a testament to humanity's relentless productivity in this era of rapid technological advancement and sky-high quality of life. We are all extraordinarily lucky to live in such a prosperous time period. Rapid growth comes with its own share of problems of course, but overall we live very easy lives compared to most past humans.

Anyway, that covers a basic overview of the people who were probably the most influential in the development of classical logic. Now that we have that out of the way, it's about time that we discussed the actual mechanics of classical logic itself. Central to classical logic are a set of three fundamental laws that originated in Aristotle's time and which arguably form the innermost core or backbone of classical logic. These three fundamental laws are known as the laws of thought.

The three laws of thought are as follows:

1. **The Law of Identity:** Every statement is equivalent to itself.

2. **The Law of Non-Contradiction:** Every statement cannot be both true and false at the same time.

3. **The Law of the Excluded Middle:** Every statement must be either true or false.

The laws of thought are so named because it is widely believed that without these principles to guide us coherent logical thought would be impossible in some sense. There are a minority of logicians who believe that one or more of these laws are not necessary, but I'll have more to say about that later. Most logicians generally treat these laws as fact. Physical reality seems to obey these laws too, as long as all the concepts you use in how you think and communicate are given completely unambiguous and precise definitions.

Logicians who deny one or more of the laws of thought are considered to be non-classical logicians. However, not all non-classical logicians deny any of the laws of thought. In contrast, all classical logicians do accept the laws of thought because any logic that denies the laws of thought by definition wouldn't be classical. In other words, the laws of thought are an essential characteristic of classical logic.

Classical logic has its own set of operators, relationships, and rules. These operators, relationships, and rules can be leveraged to systematically understand a wide variety of logical scenarios, many of which would be much too hard to keep track of in your mind without having such a system at your disposal. Human memory and focus are often far too unreliable and imprecise to handle in-depth logic unassisted.

A system for manipulating concepts via symbols into a more desirable form is, broadly speaking, called an algebra. The algebra of logic is closely analogous to the more traditional algebra of numbers with which you may perhaps be more familiar. Much like in an algebra of numbers, an algebra of logic uses variables to represent values and connects them with a variety of different operators to indicate the relationship between them.

In classical logic in particular, variables represent simple true or false values corresponding to specific statements. For example, I could say that N represents the statement "It is nighttime." and that L represents the statement "I can see lightning strikes in the distance." and so on, binding statements to variables in any arbitrary way. If I happen to be inside a building standing by a window and observe that it is night and that I cannot see any lightning in the distance then we would say that N is true and L is false.

Next, let's discuss what the basic classical logical operators and relations are, so that we can form logical expressions to represent a more diverse set of different possible scenarios. The following are the most common operators in classical logic:

Negation: Negation is also known as the logical complement or the "not" operator. In symbols, negation is represented by "$\neg$" which resembles a numeric negative sign but has a little vertical segment at the end to make it visually distinct from numeric negation. Negation is a unary operator. For example, if we have a statement named A

then the negation of A would be written as $\neg A$. The expression $\neg A$ is most typically read as "not A". The negation of a true statement is false, whereas the negation of a false statement is true.

If A represented "It is dusk." then $\neg A$ would represent "It is *not* dusk.", meaning it may be any time other than dusk. It is generally unwise to refer to the negation of a statement as the "opposite" of the statement, as doing so can easily create confusion. For example, if I said "the opposite of A" then some people might interpret that as meaning "It is dawn.", which is certainly not the correct interpretation of $\neg A$, even though dawn is indeed the "opposite" of dusk.

Disjunction: Disjunction is also known as or the "or" operator. In symbols, disjunction is represented by "$\vee$". Disjunction is typically a binary operator, although variants that accept any number of parameters exist. For example, if we have a statement named A and a statement named B then the disjunction of A and B would be written as $A \vee B$. The expression $A \vee B$ is most typically read as "A or B". The disjunction of two statements is true whenever at least one of the statements is true, otherwise it is false.

Conjunction: Conjunction is also known as the "and" operator. In symbols, conjunction is represented by "$\wedge$". Conjunction is typically a binary operator, although variants that accept any number of parameters exist. For example, if we have a statement named A and a statement named B then the conjunction of A and B would be written as $A \wedge B$. The expression $A \wedge B$ is most typically read as "A and B". The conjunction of two statements is true whenever both statements are simultaneously true, otherwise it is false.

Material Implication: A material implication is sometimes referred to simply as a "conditional" or as an "implication". However, the terms "conditional" and "implication" are ambiguous due to the fact that there are at least several other forms of implication besides material implication, such as formal implication and non-classical versions of implication for example. Material implication is also sometimes referred to as the "if-then" operator.

In symbols, material implication is represented most often by "$\rightarrow$". It is also sometimes represented by "$\Rightarrow$" or "$\supset$", but I don't recommend using these alternatives, because "$\Rightarrow$" sometimes represents formal implication and "$\supset$" usually represents a proper superset relation from set theory, thus creating ambiguity and confusion.

Material implication is a binary operator. For example, if we have a statement named A and a statement named B then a material implication between A as the condition and B as the consequence would be written as $A \rightarrow B$. The expression $A \rightarrow B$ is most typically read as "if A then B" or "A implies B". A material implication between two statements is false whenever the condition is true and the consequence is false, otherwise it is true.

If A represented "There is a dark cloud on the horizon." and B represented "It is going to rain." then $A \rightarrow B$ would represent the statement "If there is a dark cloud on the horizon then it is going to rain.". Notice it could also have been written "Dark clouds on the horizon imply it is going to rain." and it would still have the same meaning. If indeed there is a dark cloud on the horizon and it rains then $A \rightarrow B$ is true. On the other hand, if there is a dark cloud on the horizon but it does not rain then $A \rightarrow B$ must have been false.

The weird case is when the condition is false. Any time the condition in a material implication is false then the material implication is considered true, regardless of whether the consequence is true or false. For instance, in our example, if there was no dark cloud on the horizon, then regardless of whether it rains or not the material implication would still be considered true. This behavior is the source of much confusion and conflict. I will have a lot of interesting things to say about it later.

Finally, for every implication (whether material or not) there is another implication related to it called its converse. The converse of an implication is the same implication except with the condition and consequence swapped. Thus, for example, the converse of $A \rightarrow B$ would be $B \rightarrow A$. Incidentally, $B \rightarrow A$ can also be written as $A \leftarrow B$, thereby using the arrow symbol to indicate which direction the implication goes. Notice that an implication and its converse are not conceptually the same and can easily have different truth values.

Equivalence: Equivalence is also known as the biconditional or the "if and only if" operator. In symbols, equivalence is represented by "$\leftrightarrow$". Equivalence is a binary operator. For example, if we have a statement named A and a statement named B then an equivalence between A and B would be written as $A \leftrightarrow B$. The expression $A \leftrightarrow B$ is most typically read as "A is equivalent to B" or as "A if and only if B".

$A \leftrightarrow B$ can alternatively be written as A *iff* B if you prefer. The "*iff*" part is a common shorthand for the phrase "if and only if". Personally I prefer to use the "$\leftrightarrow$" symbol, because it is more visually consistent with the various other logical operators. However, "*iff*" is common enough that you need to know how to read it.

An equivalence between two statements is true whenever either both statements are true or both statements are false, otherwise it is false. Two statements A and B are logically equivalent if both $A \rightarrow B$ and its converse $A \leftarrow B$ hold true. Notice that if you imagine merging "$\rightarrow$" and "$\leftarrow$" into a single symbol then it becomes "$\leftrightarrow$", which is the symbol for equivalence. This is where the symbol for equivalence comes from.

In order to prove that two statements are logically equivalent, you must prove that both $A \rightarrow B$ and $A \leftarrow B$ are true before you can claim that $A \leftrightarrow B$ is true. An implication in one direction and its converse in the other direction often have very different characteristics and often require very different approaches to prove.

It is not safe to assume that the converse of an implication is true just because the original implication is true, although untrained members of the general public often make this error. This kind of error in thinking is usually referred to as "affirming the consequent" but I prefer the simpler term "converse fallacy" because the meaning seems clearer to me.

Here's a real-world example: Supposing that "Most rapists are male." is true it would not follow that the converse "Most males are rapists." is true. Nonetheless, some people subconsciously act as if this kind of chain of reasoning is valid to some degree[4]. Such people are unwittingly committing a converse fallacy. Many forms of discrimination against particular groups of people might actually originate from converse fallacies[5]. Thus, poor reasoning skills provably contribute to human suffering. This is just one among many ways that happens though.

Exclusive Disjunction: Exclusive disjunction is also known as nonequivalence or the "exclusive or" operator. I think it would also be apt to refer to it as the logical "difference" operator, although this term for it does not appear to be in common use. Exclusive disjunction is often abbreviated as "xor". The abbreviation "xor" is pronounced "ex-or", like the letter "x" followed by the word "or".

In symbols, xor is perhaps most often represented by "$\oplus$". Xor is typically a binary operator. For example, if we have a statement named A and a statement named B then the xor of A and B could be written as $A \oplus B$. The expression $A \oplus B$ is most typically read as either "A xor B" or "A exclusive or B". The xor of two statements is true whenever the truth values of the two statements differ, otherwise it is false. In other words, xor is the negation of equivalence.

Xor is most often interpreted as representing a special "or" expression wherein only one of the two possible options can be true for it to be true. This is in contrast to normal disjunction, which is true when either one or both input statements are true. For this reason, normal disjunction is sometimes referred to as "inclusive disjunction" when comparing it to xor. Xor is also sometimes interpreted as representing an assertion that two statements are not logically equivalent. Both interpretations are correct.

(**WARNING**: Incoming verbose tangent... Skip ahead past the rest of this xor stuff, by going to page 90 to the paragraph that starts with "Anyway", if you don't care about the notation much. All you really need to know is that I also introduce "$\not\leftrightarrow$" and "$\require{enclose}\enclose{}{\vee}$" below as additional synonyms for the xor operator and that I prefer them over "$\oplus$".)

An obvious question to ask here is: Why was the "$\oplus$" symbol chosen to represent xor? It seems like a strange choice. Well, one possible explanation for this convention might be that the symbol could have been derived from the way that xor is used in computer engineering when creating electronic circuits in order to implement addition. A single value of either 0 or 1 stored on a computer is called a bit. If you imagine that false is represented by the number 0 and true is represented by the number 1, then the xor of two bits on a computer corresponds to the result of adding those bits for that digit (ignoring the carry).

Computer engineers connect together tiny devices called logic gates on electronic circuits in order to construct the basic features of a computer processor. The logic gates of which computers are built roughly correspond to the logical operators we have been describing here, and hence computers probably would not have been invented without logic having laid the foundation beforehand.

---

[4]Here's another (less morbid) converse fallacy example: "If I'm a good person then people will be nice to me. People are being nice to me. Therefore, I must be a good person." (This is an invalid conclusion. You might still be a good person, but this argument lends no support to it.)

[5]Here's a generalized form of the argument to make this clearer: "People with bad quality X often/always have coincidental quality Y, therefore people who have coincedental quality Y often/always have bad quality X." (See how the structure closely resembles common prejudices?)

Anyway, by combining a xor-gate with an and-gate in a certain way a computer engineer can create something called a half-adder, which can then be repeated and chained end to end in order to add arbitrary sequences of binary digits. That's more or less what computers do to add binary numbers together.

Thus, because xor can be thought of as "addition modulo 2" they perhaps decided to represent it as "⊕" to indicate that it is sort of like addition but different. As a mnemonic device that I came up with, you can think of the circle around the "+" symbol as representing the fact that the addition "circles around", much like how any operation modulo some value will cycle through the same range of values.

Another possible explanation for the origin of the "⊕" symbol comes from the fact that the "+" symbol was originally used to represent disjunction before the "∨" symbol was invented to distinguish the numeric addition operator "+" from the logical disjunction operator "∨". Therefore, the circle around the "+" symbol may have simply been intended to indicate that "⊕" is a variant of disjunction, specifically xor.

Versions of xor that accept more than two parameters exist, but there are two conflicting ways of interpreting what a xor that accepts more than two parameters should mean. This can lead to some confusion if which one you are using is left unspecified.

One version nests each successive pair of inputs with a binary xor and then chains them together, with the end result that the final output will be true whenever there are an odd number of true inputs and otherwise it will be false. This interpretation is more akin to the computer engineering way of interpreting xor. It's like you're adding a whole bunch of 0s and 1s together for a single bit and then returning the result modulo 2.

Alternatively, the other way of interpreting a xor of more than two inputs is to focus on the idea of the mutual exclusivity of the inputs. In other words, if exactly one of the inputs is true then the output is true, otherwise it is false. Thus, this interpretation correctly implements the notion of mutual exclusivity that corresponds to how sometimes we want to specify that only one of a certain set of options may be chosen.

For example, you may only be allowed to pick one of three prizes if you win some contest. This is arguably the more correct interpretation of the concept of mutual exclusion of options, but the other interpretation of xor is also useful and meaningful and is common in computer engineering, hence these two interpretations of generalized xor to this day remain in conflict with each other.

Besides "⊕" there are several other symbols that are sometimes used to represent xor. The most common of these alternatives are perhaps "∨" and "↮". The idea behind the "∨" symbol is probably to be analogous to the difference in notation between "<" and "≤" and similar symbols. The problem is that putting the line under "∨" in this manner implies the exact opposite of the intended connotation.

An operator that has a line under it is supposed to be *more* inclusive than the corresponding operator without a line. Notice however that xor is actually *less* inclusive than a normal disjunction, thus implying that "∨" would be a misnomer for xor. The line under an operator typically means "or equal", thus it broadens the scope of the operator to which it is attached.

In contrast, the alternative symbol "↮" is actually a really smart choice. Since xor is logically equivalent to the negation of equivalence, it makes a lot of sense to represent xor as a slashed-through equivalence symbol instead of having a new symbol for it.

Almost universally in logic and mathematics, when an operator is slashed-through in this manner it indicates the logical negation of that operator. Thus the meaning of it is very clear.

One could make an argument that the "↮" symbol is a superior choice to "⊕" because it makes the relation between exclusive-or and non-equivalence more readily apparent. However, "⊕" seems to perhaps be more commonly used. Thus, there's a strong argument for both symbols. Either of "⊕" or "↮" are fairly good symbol choices. However, I do not recommend using "⊻", because it seems more likely to confuse people.

Although these symbols are certainly adequate, they do however obscure the relationship xor has to regular disjunction somewhat. Therefore, I will introduce a new symbol for xor which captures that connection more clearly. The new symbol is "⩛". The xor of two statements A and B with this symbol will be written as $A \mathbin{⩛} B$. This symbol more clearly hints at the direct connection between or and xor. The "×" is an obvious mnemonic device for "**e**xclusive".

Thus, placing the "×" over "∨" as in "⩛" represents the concept of xor in perhaps the most obvious way possible, since "ex-or" is a more commonly used than "nonequivalence". Another reason that it fits well is because, if you look at where the "×" touches the "∨", it looks like two diverging paths, like a fork in a road, which is like an analogy for the concept of xor. You can only choose one path on a fork in a road, just as you can only choose one input to be true in a xor in order for it to be true. All that being said, the symbol "↮" is still a great choice for xor though, if you don't like my idea here.

Finally, so you're aware, xor is not nearly as frequently used as the other basic logical operators above are. Generally, the other logical operators tend to be easier to work with for manipulating logical expressions and are more widely understood. You won't see xor used very often in most documents. Nonetheless though, it is worth knowing.

Anyway, that covers the most common logical operators of classical logic. These operators are often viewed as being the most fundamental and natural set of operators on which to base classical logic. They are easy to understand and easy to build more complicated logical expressions with. Technically though, they are slightly redundant, and alternative sets of operators capable of representing the same expressions do exist. However, these are the operators in standard use and they are generally less awkward than the alternatives.

On a side note, two other logical operators you may encounter are joint denial and alternative denial, which are more commonly known as "nor" and "nand" respectively. In symbols, the nor operator is represented by "↓" and the nand operator is represented by "↑". As you might imagine, nor is the negation of disjunction and nand is the negation of conjunction. Thus it might also be apt to call nor "negated disjunction" and nand "negated conjunction", although neither term appears to be in use.

The reason why "↓" represents nor (I'm guessing) is probably because the head of the arrow looks like "∨", which is the symbol for logical or, but the vertical line disrupts it, thus implying a variation on logical or, specifically logical nor. The reasoning behind "↑" is probably analogously the same. Any time you see the symbols for nor (↓) or nand (↑), and you want to figure out which one you are looking at, simply ask yourself "Which of ∨ or ∧ does the arrow head look like?". If the arrowhead looks like disjunction (∨) then it is nor. Otherwise, if it looks like conjunction (∧) then it is nand.

The nor and nand operators have the unusual property that any classical logical expression can be written in an equivalent form consisting only of nors or only of nands. Thus nor and nand are both universally expressive in classical logic. However, writing every expression only with nor or only with nand is hard to read. They are interesting operators but aren't that relevant for our purposes in this book, so I will only mention them in passing.

Now that we have some basic logical operators, we can build more complicated expressions just as promised. For example, suppose we wanted to express the logic behind the idea that if it is winter then we will stay inside, otherwise if it is not winter then we will go outside. Suppose that W represents "It is winter." and I represents "We will stay inside.".

The expression in classical logic that fully captures this scenario is $(W \to I) \land (\neg W \to \neg I)$. Parentheses are used to group variables and operators together so that which parts they are associated with is unambiguous. In essence, this expression describes a list of conditions that the logical system must meet, where each of those conditions is joined by conjunction. The $W \to I$ part expresses the constraint that if it is winter then we will stay inside. In contrast, the $\neg W \to \neg I$ part expresses the constraint that if it is any season other than winter then we will not stay inside.

By definition, in order to not stay inside we must go outside, hence $\neg I$ is equivalent to "We will go outside." even though it is not directly phrased that way. Finally, the conjunction ensures that both of these conditions are applied to the system, rather than only one or the other, thus making the entire expression as a whole equivalent to the given logical scenario.

For another example, suppose we have a passenger plane flying from Japan to New Zealand and its fuel tank is punctured at some point during flight. In this scenario, suppose we know that either the plane will run out of fuel and crash in the ocean or the plane will not run out of fuel and will reach New Zealand safely. Clearly the plane cannot both crash in the ocean and yet still reach New Zealand safely.

Suppose F represents "The plane will run out of fuel.", C represents "The plane will crash in the ocean.", and Z represents "The plane will reach New Zealand safely.". The expression in classical logic that corresponds to this scenario is $(F \land C \land \neg Z) \lor (\neg F \land \neg C \land Z)$. This expression is only true when one of the two cases it describes are true. Thus, for example, a scenario where the plane doesn't run out of fuel and yet crashes anyway would not be consistent with these constraints, because $\neg F \land C \land \neg Z$ is not one of the possible outcomes described by the system.

Let's do one more. Suppose you are hiking in a forest with a friend, trying to reach some specific scenic destination, and come to a fork in the path ahead of you. Your companion turns to you and says "I remember that one or both of these paths will ultimately lead us to our destination". However, you believe that in fact neither of these paths lead to the destination, and that you instead need to walk off the beaten path to reach your destination. How do we represent your belief as a logical expression?

Well, suppose that A represents "Path 1 leads to the destination." and B represents "Path 2 leads to the destination.". In this case, your response could be represented as

$\neg(A \vee B)$, which incidentally is also equivalent to $\neg A \wedge \neg B$. This equivalence happens to be one of De Morgan's Laws, which describe how negation distributes over other logical operators.

In natural language, $\neg(A \vee B)$ could be read directly as "It is not the case that one or both of path 1 and path 2 lead to the destination.". The equivalent form $\neg A \wedge \neg B$, where the negation has been distributed via one of De Morgan's Laws, could be read as "Both path 1 and path 2 do not lead to the destination." which sounds more natural.

As we know, the negation of disjunction is equivalent to the nor operator. Hence $\neg(A \vee B)$ is also equivalent to $A \downarrow B$. Thus it could also be read as "Neither path 1 nor path 2 lead to the destination.". Clearly all three natural language forms are equivalent, just as the corresponding symbolic forms are. You can use whichever you prefer. However, sometimes when trying to simplify complex logical expressions you will find that one form or the other makes it easier to reduce the surrounding expression to a simpler form.

That's enough examples of expressions in classical logic for now. In most modern explanations of the basic logical operators, each explanation of each operator is accompanied by something called a truth table. A truth table is a table that lists every possible combination of inputs to a logical expression, along with the resulting truth value of the expression as a whole corresponding to each of those parameter sets.

Although I'm pretty sure you understand the operators by now anyway, truth tables are a highly effective mechanism for discussing the properties of logical operations, so I'm going to cover the truth table forms of the logical operators too. Truth tables are a particularly useful tool for understanding the behavior of more complex logical expressions. Furthermore, knowing what truth tables are and how to use them will be useful later on when we're comparing and contrasting the different branches of logic beside classical logic.

It is common to represent true and false with single symbols in order to save time and energy. Here are the three most commonly used notations for truth values in classical logic:

truth notation 1: "T" represents true. "F" represents false. This is the notation that I prefer and that I will use in this book. In my experience, it appears to be the most common notation here in the United States and perhaps also in the rest of the English-speaking world too. It is an English-specific memory aid though, and thus lacks some international generality.

truth notation 2: "⊤" represents true. "⊥" represents false. The idea behind "⊥" is obviously that it is an upside down true symbol, and hence represents false. The idea is probably to have only one essential symbol for truth, so as to make the connection between the true and false symbols more immediately apparent. In my experience, this notation is seldom seen outside academia and other specialized sources. The "⊥" symbol also looks practically identical to the symbol for perpendicular from geometry, thus creating unnecessary ambiguity.

truth notation 3: "1" represents true. "0" to represents false. In other words, it is simply binary notation, much like you'd see on a computer. However, binary notation for truth values like this actually predates computers. George Boole's original work in creating boolean algebra actually used 0s and 1s instead of other symbols for true and false. This is a good generalized notation choice, but using "T" and "F" for working with logic is more common.

If you're also working with degrees of certainty between 0 and 1 instead of only with absolute truth or falsehood however, then this binary notation is definitely the best choice. Using 0 for absolute falsehood and 1 for absolute truth, and everything in-between for degrees of certainty, is an elegant and ingenious notation that makes the relationship between absolute and fuzzy/probabilistic truth values easy to understand.

In order for a truth table to accurately represent the behavior of a logical expression, it is essential that you remember to list every possible combination of the input variables[6]. Any given input variable can be either true or false, which means there are two possibilities for every variable. Hence every time you add a new variable to a logical expression the number of possible combinations of all inputs will double.

For example, any logical expression that has only one input will result in a truth table with exactly two rows of values. An example of precisely this case can be seen in the truth table for negation, which is located at Table 2.1a on page 99. Likewise, any logical expression that has two inputs will result in a truth table with four rows of values. All of the tables on page 99, except the table for negation, are examples of truth tables with four rows of values.

If we had a table with three inputs, then it'd have eight rows of values. Likewise, if we had a table with four inputs, then it'd have sixteen rows of values, and so on. In other words, given that there are n input variables to a logical expression, we can infer that there will be 2^n rows of values in any complete truth table for that logical expression.

What that means in practice is that the size of a truth table grows exponentially in proportion to the number of input variables. This means that computing the complete truth tables for logical expressions with more than just a few logical variables is a lot of work, and indeed can become intractable even for a computer for expressions with large numbers of input variables. Nonetheless, truth tables are extremely useful for studying the behavior of logical expressions. It can be easy to spot important properties of a logical expression when you are looking at its truth table.

For example, regardless of how complex a logical expression may be, if a truth table only ever has true values in its output column then you can infer that the expression is a tautology: a statement that is always true. Likewise, if a truth table only ever has false values in its output column then you can infer that the expression is a contradiction: a statement that is always false.

These are useful facts to know. If something is a tautology or a contradiction, then changing the inputs will never change the truth value of the expression. In other words, the expression is true or false by form alone, regardless of its content. Thus, if a logical

---

[6]My favorite way to guarantee that every possible combination of inputs is listed in a truth table is to treat each input like a binary digit of a binary number and then imagine adding 1 to that number every time a new row is listed. Think of false as 0 and true as 1. This method eliminates all the guess work in filling out truth tables and is easy to remember.

expression is a tautology, then it indicates that the expression is a universally valid law of logic. Since the expression's behavior is always the same regardless of content, then you can safely apply it to any other logical expression that has a compatible form, and thereby manipulate that expression's form while still preserving that expression's truth value.

Remember our discussion of the laws of thought on page 84? Well, like all laws of classical logic, the laws of thought are tautologies too. Look at the corresponding truth tables on page 100, specifically at Tables 2.2a, 2.2b, and 2.2c. Their truth tables contain only true values for the output column.

I have also placed a few additional laws of classical logic besides the laws of thought on page 100 so that you can see some other examples of valid laws in classical logic. Notice that all of the tables have output columns that consist only of true values, thus proving that each of the corresponding expressions is a tautology and hence a law of logic. Please be aware that there are tons of valid laws (i.e. tautologies) in classical logic. I am only showing a tiny fraction of them here.

You might ask at this point: How do we actually use a given law of classical logic to manipulate a logical expression or to make valid inferences? Well, one helpful idea is to look for the presence of a material implication or equivalence in the law. Anytime you see one of these operators in a law (i.e. in an always true expression) it indicates a direct inference that you are allowed to make at any time.

Whenever a law of logic has an implication in it then that means that every time you see an expression matching the form of the condition then you can either (1) replace it with the form of the consequence in the expression, or (2) infer that the consequence is true if you already somehow know that the condition is true.

For example, if X and Y are any arbitrary expressions and if $X \rightarrow Y$ is a valid law of logic, then it follows that any time you see X in an expression you can replace it with Y. Alternatively, instead of substituting, if you know that X is true then you can infer that Y must be true also. Basically, you can think of "$\rightarrow$" as pointing in the direction of an allowed transformation or inference, but only when it occurs in a law of logic (i.e. a tautology). Outside of laws of logic though it is nothing more than an operator, which could be true or false. Without more information we don't know.

Anytime you make a series of correct inferences from some set of statements that you already know to be true to a new set of statements that you didn't know to be true, then you have constructed what is known in logic and mathematics as a *proof*. If the laws of logic used to construct a proof are indeed correct, then the proof guarantees that your conclusion must be correct. In this way, logic and mathematics are capable of attaining a level of certainty that other disciplines seldom can.

Because of the fact that equivalence is the same thing as the conjunction of two implications, each the converse of the other, it follows that the same substitutions and inferences that are enabled by implications are also enabled by equivalences. The difference is that when you are working with equivalences you can apply them in either direction instead of only in one direction.

For example, take the law of conjunctive identity, which can be found in Table 2.2e on page 100 and has the form $(A \wedge T) \leftrightarrow A$. Suppose we are given the expression

$A \to (B \wedge T \wedge C \wedge D)$, where T has the usual meaning of true and the other letters are arbitrary logical variables.

The law of conjunctive identity tells us that any time we see an expression that matches the form of $A \wedge T$ we can substitute it with A (and vice versa) without changing the truth value of the expression. Notice that $B \wedge T$ has the same form as $A \wedge T$, when A is substituted with B. Hence we can use conjunctive identity to infer that $A \to (B \wedge T \wedge C \wedge D)$ is equivalent to $A \to (B \wedge C \wedge D)$, thereby simplifying it. Alternatively, if we knew in advance that $B \wedge T$ were true, then we could infer that B is true also. In the absence of such information though, i.e. if we don't know whether B is true or false, then only substitution of form may be used.

Incidentally, these two different ways of applying a law of logic, (1) substitution of form and (2) inference from what is known or assumed, correspond to two different ways of constructing proofs. If a proof is constructed by repeated substitutions of forms, then that proof may be said to use "transformative deduction". In contrast, if a proof is constructed by repeated inference from what is known or assumed, then that proof may be said to use "evaluative deduction".

Actually, these are not standard terms. These terms ("transformative deduction" and "evaluative deduction") are ones that I thought of myself while writing this book. The distinction between these two approaches is important, and therefore I thought it merited giving each one a precise name. Both approaches take a set of axioms[7] (assumed rules and givens) and then use them to reach conclusions about the system under study. However, the approach that transformative deduction and evaluative deduction each take to this process is somewhat different.

Reasoning with transformative deduction is independent of the truth values of the statements you are working with. Transformative deduction only deals with the *form* of expressions and knows nothing about what their actual values are. Because it deals only with the form, it doesn't need to know anything about the values.

One of the most common applications of transformative deduction is in the algebra of numbers, or, in other words: in the various substitutions you use when manipulating a numeric equation. For example, transforming $2x = 4$ to $x = 2$ is a transformative deduction. It works by applying an equivalence from the laws of the algebra of numbers which states that an expression of the form $ax = b$ may be replaced by an expression of the form $x = b/a$, while still retaining the same numeric value. In symbols, the rule would be something like $(ax = b) \leftrightarrow (x = b/a)$.

The algebra of numbers works by preserving the *numeric* value of expressions, whereas the algebra of classical logic works by preserving the *truth* value of expressions. Do you see the deeper connection between different algebras more clearly now? Do you see how it's all just substitutions based on either one-way implications or two-way equivalences? In essence, all algebras are simply rule sets for transforming expressions from one form to another while preserving some property of those expressions. Any such system of rules qualifies as an algebra, in some sense.

---

[7] An axiom is typically a low-level law of logic, i.e. it is a law of logic that is built up from little to no other assumptions. However, whether something is low-level enough to be called an "axiom" doesn't really matter much. Laws of logic function the same regardless of whether they are low-level or high-level. High-level assumptions are still sometimes referred to as "axioms".

You can even create your own systems of substitution rules with whatever arbitrary properties you desire. Often such rules are useful in and of themselves, but sometimes you will want to derive additional higher-level rules in order to understand the system better or to make working with the expressions more efficient.

For example, you could make an algebra where ♈ → ♋♊ and ♊ → ♌♈ are the only substitution rules you are initially aware of, and then you could study the system's behavior and derive additional laws and so on. You can even just mess around to see what kinds of crazy things happen. For reference, so that you know how to verbalize them, the names of these symbols are aries (♈), gemini (♊), cancer (♋), and leo (♌). They are astrological birth signs. I chose them purely because they look aesthetically interesting. Astrology itself has zero credibility of course.

Speaking of crazy things, with this example, if you have ♈ and then repeatedly apply the substitution rules then you end up with an endless sequence of "♋♌♋♌♋♌...". Likewise, if you have ♊ then you end up with "♌♋♌♋♌♋...". In other words, this rule set naturally forms infinite alternating patterns. On the other hand, if you started with just ♋ or ♌ then nothing would change, due to the fact that neither ♋ nor ♌ have any associated substitution rules.

Merely substituting things according to some set of rules like this may seem relatively dry and uninteresting at first glance, but you might be surprised at what kinds of amazing things you can do with such a simple mechanism. For example, look up "L-systems". L-systems are basically just another term for any arbitrary substitution rule set for expressions, like I've been describing here, plus a few extra constraints.

It happens that if you interpret the stream of output symbols from an L-system as being drawing instructions for a computer graphics program, then certain L-systems (i.e. substitution rule sets) generate shockingly realistic drawings of real-life plants. In fact, L-systems were created by a biologist named Aristid Lindenmayer in order to model the growth patterns of plants.

It turns out that the system Lindenmayer devised is essentially identical to the basic techniques of transformative deduction, when you cut away all the extraneous bits. In other words, it appears plausible that plants use a method similar to transformative deduction in order to grow the complex structures of which they are built[8], which in turn gives the world we live in so much of its natural beauty. Hence this same beauty that exists in nature also must exist in logic. Cool right?[9]

In contrast to transformative reasoning, reasoning with evaluative deduction is dependent on the truth values of the premises you are working with. Evaluative deduction uses both the form and the *value* of the logical system. You start with some set of premises that you either know are true or assume are true, and then you make whatever valid inferences you can from them. If no contradictions arise, then all of the logical expressions you inferred along the way must be true.

---

[8] The idea is that plants seem to use repeated substitutions of one biological component for another according to a simple rule set, gradually generating a larger scale structure as a consequence. This process is essentially the same thing as transformative deduction if you think about it. Thus large scale beauty spontaneously emerges from trivially simple rules, just as evolution predicts.

[9] Thus, naturally, L-systems would be a very direct and easy way to illustrate "the beauty of logic and math" to the uninitiated. Dry pedantry in contrast probably convinces almost nobody who isn't already a logician or mathematician.

On the other hand, if contradictions do arise, then one or more of your premises must be false. Evaluative deduction often explores multiple branches of possibilities in order to determine what the system does under different sets of assumptions. Evaluative deduction feels like a tree growing up from an initial seed. All the branches and leaves that it reaches are analogous to proven facts. Anything it doesn't reach is either unknown or false.

Because of the fact that evaluative deduction is able to use *both* the form and the value of a logical system in order to reach its conclusions, it is sometimes able to reach those conclusions with much greater ease and simplicity. The reason is because it has strictly more resources and options available to it than transformative deduction has. Both have access to form, but only evaluative deduction has access to value.

However, evaluative deduction does have the additional overhead of either having to know the truth value of some of the components of the logical system in advance or having to otherwise make suppositions of their values in the course of exploring the different possible inference outcomes. Thus there are times where it is better to use transformative deduction. The two systems have different pros and cons.

Generally speaking, transformative deduction flourishes in an environment where lots of substitutions and manipulations of the form of expressions need to be made, such as in the algebra of equations (e.g. "Solve $x^2 + 5x + 6 = 0$ for x.") or in the algebra of classical logic (e.g. "What is an equivalent form for $\neg(A \wedge B \rightarrow C)$?").

In contrast, evaluative deduction flourishes in an environment where you need to construct proofs efficiently. Proofs that use evaluative deduction sometimes require fewer steps in order to reach a conclusion. The reason for this is because evaluative deduction can leverage its ability to know the values of a system, or to assume them for argument's sake, in order to reduce the search space and to have more techniques at its disposal for attacking the problem.

At this point one might feel compelled to ask: What about the cases where a law of logic doesn't have an implication or an equivalence in it, such as in the laws of thought for example? How do we use those laws? Well, basically such a law indicates that any expression that matches its form is simply true. In other words, instead of substituting in a new expression or making an inference from the truth of one expression to the truth of another, we can simply immediately infer it to be true.

In practice, the laws of thought are most often used to detect that an argument can't possibly be true, and that therefore one of the assumptions or inferences made somewhere along the way must be wrong. For example, if you begin by assuming $\neg A$ is true and then derive a contradiction that proves A is true, then this is called a "proof of A by contradiction". Alternatively, if you instead begin by assuming A is true and then derive a contradiction that proves $\neg A$ is true, then this is called a "*dis*proof of A by contradiction". Notice the subtle distinction.

Suppose you assume that some statement A is true, but as a consequence of that it follows that $B \wedge \neg B$ is true, where B is some arbitrary statement that may or may not equal A. We know from the law of non-contradiction that $B \wedge \neg B$ cannot possibly be true because it is a contradiction. Therefore our initial assumption that A is true must be wrong. If a truth value is not true then it must be false, because according to the law of the excluded middle every truth value is either true or false. Therefore, A must be false, or equivalently ¬A must be true. This is an example of a disproof of A by contradiction, or equivalently of a proof of ¬A by contradiction.

Having now covered the basic mechanisms of inferences and proofs, it is now probably becoming clearer to you what logic actually is, if it wasn't already. Beyond the basic intuition and informal reasoning abilities that all human beings have to some extent, formal logic allows a level of consistency and depth in reasoning that would be difficult to achieve otherwise. Perhaps now that we have more experience, we can revisit the basis of logic once more to see if we can extract additional insights. Let us ask ourselves once more: What exactly is logic?

A	$\neg A$
F	T
T	F

(a) Negation (not)

A	B	$A \vee B$
F	F	F
F	T	T
T	F	T
T	T	T

(b) Disjunction (or)

A	B	$A \wedge B$
F	F	F
F	T	F
T	F	F
T	T	T

(c) Conjunction (and)

A	B	$A \rightarrow B$
F	F	T
F	T	T
T	F	F
T	T	T

(d) Material Implication (if-then)

A	B	$A \leftrightarrow B$
F	F	T
F	T	F
T	F	F
T	T	T

(e) Equivalence (iff)

A	B	$A \veebar B$
F	F	F
F	T	T
T	F	T
T	T	F

(f) Exclusive Disjunction (xor)

A	B	$A \downarrow B$
F	F	T
F	T	F
T	F	F
T	T	F

(g) Joint Denial (nor)

A	B	$A \uparrow B$
F	F	T
F	T	T
T	F	T
T	T	F

(h) Alternative Denial (nand)

Table 2.1: Truth Tables for the Basic Classical Logic Operators

A	$A \leftrightarrow A$
F	T
T	T

(a) The Law of Identity

A	$\neg(A \wedge \neg A)$
F	T
T	T

(b) The Law of Non-Contradiction

A	$A \vee \neg A$
F	T
T	T

(c) The Law of the Excluded Middle

A	$(A \vee F) \leftrightarrow A$
F	T
T	T

(d) Disjunctive Identity

A	$(A \wedge T) \leftrightarrow A$
F	T
T	T

(e) Conjunctive Identity

A	B	$(A \vee B) \leftrightarrow (B \vee A)$
F	F	T
F	T	T
T	F	T
T	T	T

(f) Commutativity of Disjunction

A	B	$\neg(A \vee B) \leftrightarrow \neg A \wedge \neg B$
F	F	T
F	T	T
T	F	T
T	T	T

(g) Disjunction Negation

A	B	$\neg(A \wedge B) \leftrightarrow \neg A \vee \neg B$
F	F	T
F	T	T
T	F	T
T	T	T

(h) Conjunction Negation

Table 2.2: Truth Tables for Some Laws of Classical Logic

Well, we started this chapter with this sort of vague notion that in order to consistently reason correctly and to discover as many truths as possible we would need a system of thought that would allow us to take any set of facts or premises and infer or deduce what the consequences of them are and how the underlying structures of the system relate to each other. Such a system of thought is known as logic.

In our pursuit of logic, we have gradually deepened our understanding of its nature. So what exactly is the real nature of logic then? We already have this vague notion that logic is a system for deriving new facts from old facts. That is certainly true. Logic does indeed do that. However, true as that may be, is that the whole truth?

Just because something fits some criteria, does not necessarily mean that the criteria completely captures the essence of that thing. A leaf may be green, but that does not imply that greenness is the essence of what it means to be a leaf. In order to truly capture the essence of a thing, you must capture its qualities fairly completely.

Indeed, our concept of logic is currently incomplete. Our criteria fit, but merely fitting is not sufficient. What are we missing? Well, I'm going to explain what we're missing, but in order to do so I'm going to go off the beaten track for a little bit to describe a seemingly unrelated real-world scenario. Although the connection will be unclear at first, it will actually serve well to illustrate the point. Here we go:

Imagine you're standing in a cavern deep underground. The cavern consists only of one central chamber. There are no corridors or passageways connected to the room. It's just one big cavern surrounded by rock walls on all sides. There's an airshaft overhead that can also be used for supplies. You have everything you need to survive for any amount of time.

You're an underground mining engineer. It is the mission of you and your team to explore the surrounding rock for valuable minerals by drilling and digging tunnels in whatever arbitrary way seems best to you. Normally, when digging through so much rock there would be a risk of cave-ins and other disasters. To simplify the thought experiment though, assume that as long as you follow the rules then it is impossible to have a cave-in or any other disaster. Also, assume that you have an automatic conveyor system that moves all the debris up and out of the mine, such that debris never accumulates as you dig.

Naturally, since the only room you have access to at the beginning is the central cavern chamber, any tunnel you start digging at the beginning will have to be connected to the central cavern chamber. Occasionally, you might tunnel into another underground cavern like the first one, but if you do then as long as you follow the rules and transform the cavern into a safe environment then you can continue digging at any accessible location.

Suppose the mining project continues for many months and that eventually you and the team have created quite an extensive system of underground tunnels and rooms. However, one day an apprentice mining engineer confronts you, saying that excavating so many tunnels poses an unaccounted for danger. The apprentice is worried that, given how complicated the tunnels are becoming and given how many of the tunnels there already are, there is a danger that parts of the tunnel system will eventually become disconnected from each other.

The apprentice is concerned that if any two parts (or more) of the tunnel system become disconnected from each other then it will become possible for people to get stuck in one of the disconnected parts of the tunnel system and hence never be able to return to the central cavern chamber, where the only elevator capable of returning to the surface is now located.

Here's the question though: Assuming that all the constraints on this mining system we've specified are correct and that the mining team has consistently followed the proper procedures that make cave-ins and similar events impossible, do the apprentice's concerns have any merit?

Well, under the circumstances, because no cave-ins or similar events are possible, we can deduce immediately that old sections of the tunnel system could never become disconnected. What about new sections? Can the tunnel system become disconnected from itself through the process of digging? Take a moment to consider how the tunnel digging works.

Isn't it true that in order to dig a new tunnel the new tunnel must be connected to an already existing one? A little bit of thought shows that clearly this must be true. What about the older tunnel to which the new tunnel is connected? Well, the older tunnel also must have been initially dug out via yet another already existing tunnel, and hence must be connected to it.

If you think about it, regardless of where you are located in the tunnel system, if you start mentally tracing through the tunnels starting at your location and mark off every tunnel that is connected to the set of tunnels you already know are connected, then you will eventually mentally mark all of the tunnels in the entire tunnel system as connected. For reference, this process of marking all regions connected to a starting point is known as a "flood fill". Look it up if you want exact details.

Anyway, as you can see, given any location in the tunnel system, there is guaranteed to be a path back to the central cavern chamber where the elevator to the surface is located. Therefore the apprentice mining engineer's concerns have no merit, at least not under the assumptions we have made. Why is it that the tunnels stay connected though? Fundamentally, the reason is because every act of digging a new tunnel guarantees that the new tunnel will be connected to an old tunnel. In other words, tunnel digging is an operation that conserves the connectedness of any cave system upon which it operates.

That basically wraps up the underground mining scenario. Now, you may be wondering what exactly the point of that thought exercise was. What does it have to do with logic? Sure, we may have used a bit of reasoning to deduce that the tunnel system would always remain connected to itself, but how is that supposed to give us any new genuine insight into what the nature of logic is?

Let me give you another relevant example, besides just the underground mining scenario. Are you familiar with physics? Have you heard of the law of conservation of mass and energy? Well, the law of conservation of mass and energy basically says that whenever the laws of physics act upon the universe that matter and energy can never be created or destroyed, but instead matter and energy can only ever change forms.

In other words, nothing can ever just pop into existence. Instead, matter and energy can only change from one form into another. Anytime something appears to be created from nothing, what actually happened is that you simply failed to account for all the multitude of different forms of matter and energy in the system and where they went. As far as I am aware, no exception has ever been found to this rule, provided of course that the matter and energy are being correctly accounted for with sufficient competency by the person performing the measurements.

Physics conserves the mass and energy of the universe it acts upon. It also conserves the mass and energy of any isolated subset of that same universe. Does this sound familiar to you? Do you remember what the take away was from the underground mining scenario? It was this: Tunnel digging is an operation that conserves the connectedness of any cave system which it acts upon. Are you starting to see any similarities here?

Here's some more examples: Suppose a person jumps into a pool and thus becomes wet. If the person stays in that pool then they will remain wet for at least as long as they stay in the pool. In other words, pools conserve the wetness of whatever is in them.

Suppose a delusional person tries to use their mind to telekinetically move an object across a table. No matter how hard one might try, the human mind does not have the ability to perform telekinesis. In other words, the nature of the human mind conserves the property of not being able to perform telekinesis, regardless of the amount of effort put into it.

Likewise, suppose you are playing some video game, such as an RPG for example. Suppose this RPG has an experience point system where every time you defeat an enemy you always gain a positive amount of experience points. Because defeating an enemy can only ever yield a positive amount of experience points, rather than a negative amount, it is therefore logically impossible for defeating enemies to cause your experience point total to decrease. In other words, the experience point system of this hypothetical RPG game conserves the strictly increasing property of the player's experience point total.

What is it that all these examples have in common? Do you see the pattern now? Here's two more especially significant examples of systems that behave similarly to the above examples: Classical logic conserves the truth value of the logical expressions that it operates on. The algebra of numbers conserves the equality of the numeric equations that it operates on. Sounds familiar right?

The fact that all of these examples revolve around conserving certain properties of their respective systems is not some mere coincidence. These commonalities between these examples provide a powerful insight into the real nature of logic. With a fresh perspective and an alert mind, a little bit of thought will soon show us what the connection is. Once we see what the connection is, we will be able to formulate a more complete notion of what logic is than merely a system for solving things.

All of these systems share the characteristic of conserving some kind of value or property when they act upon objects in their domain. We should define a new formal definition for this. The power of words to enable new avenues of thought should not be underestimated. Even the most subtle of new distinctions can have power.

A concept without a word to match it is slippery and hard to grasp. Yet, with a word to name it, it is as though the gateway of exploration is opened. The river of the mind expands from a mere trickle to a torrent. It is like marking a trail: it lets others see a new path where once they were blind to it.

Anyway, here's our new definition:

Definition 25. *When a system, or a subset of a system, has the characteristic that whenever it acts upon the objects in its domain it consistently conserves some particular property or set of properties of the objects upon which it acts, then such a system may be called a* **conservation system**.

Conservation systems can be physical (e.g. part of the real world), virtual (e.g. part of a video game), or abstract (e.g. part of a logical system). Conservation systems occur naturally as part of the fabric of the universe, and thus they can't just be some mere quirk of our minds. In other words, all of reality effectively has logic embedded in it. Logic in this sense would exist even if our minds did not.

However, what is the underlying mechanism or principle that enables conservation systems to behave as they do and to have such useful properties? To understand this fully, we need to also understand the connection between the individual actions of which the conservation system is comprised on the one hand and the larger scale conservation effect on the other. To this end, let us make another new term:

Definition 26. *If an action always conserves a particular property of an object every time it is applied to that object, then it follows that no matter how many times the action is applied to that object the property will still be conserved. This is also true of any set of actions applied to any set of objects where any particular set of properties is conserved. In other words, the idea applies to both individuals and groups.*

For example, if tunnel digging never breaks connectedness, then no matter how many times you dig tunnels in a tunnel system the system will always remain connected. Similarly, if a system of logic conserves truth with every inference, then no matter how many inferences you make from known truths the system will always remain true. Regardless of whether a system is physical, virtual, or abstract, this idea of conservation of properties will always be valid.

This notion, that if a property is conserved during every individual action upon an object then the property will continue to remain conserved no matter how many such actions occur, will be called the **principle of property conservation**. *Alternatively, for brevity, if you prefer to use an acronym then you may refer to it as the* **PPC**.

So, we have the principle of property conservation now, but how is it supposed to give us insight into what the exact nature of logic is? Remember how I said that we'd use these examples and the concepts derived from them in order to form a more complete picture of what logic actually is beyond merely a system for making inferences? Well, now that we have these new terms and ideas we can do that.

Logic is any truth conservation system that is able to mirror a corresponding tangible system, thereby allowing one to explore consequences within that system. Interestingly, notice that physical reality is *also* a truth conservation system just like logic is. The only difference is that logic *mirrors* some other arbitrary truth conservation system, whereas physical reality actually *is* a tangible manifestation of a truth conservation system.

Whether its the laws of physics or the laws of logic, it's the same basic structure. Both the laws of physics and the laws of logic are simply conservation system rule sets. It is thus probably fair to say that there is not much substantive difference between the kinds of laws you find in reality and the kinds of laws you find in some arbitrary made-up logical system. The core essence of logic is truth conservation, and truth conservation is also the core essence of reality itself.

Thus it would be fair to say that reality and logic are inseparably related. They both operate according to the same foundational principle: property conservation. Thus, the evidence is actually quite strong that logic really is in some sense a part of the fabric of the universe. It cannot merely be something we made up. It has to be real.

Why? Well, any universe that exists must have some set of laws that govern it, but the act of applying such laws to all the objects that exist within that universe effectively creates a conservation system. Anything that is a conservation system can be mirrored by a system of logic, because that's exactly what logic itself is: an arbitrary one-size-fits-all conservation system.

We should define a new term for this idea that logic is embedded in reality itself, so as to enable efficient debate and discussion. Here's our definition:

Definition 27. *The belief that logic is embedded in the fabric of the universe, such that even if our minds did not exist then logic would still exist within the substance of the universe, will be called **logical naturalism**. The name logical naturalism is derived from the idea that logic occurs naturally within the universe, rather than merely being an artificial construct of our minds.*

We shouldn't be one-sided with our definitions though. A concept with a name is much easier to defend than one without. We should define the opposing term as well, so that it has a fighting chance at fair consideration. Here's the opposing term:

Definition 28. *The belief that logic is merely a figment of our own minds and only exists within our own minds, such that if our minds did not exist then logic would no longer exist, will be called **logical artificialism**. The name logical artificialism is derived from the idea that logic is merely an artificial construct of our minds, regardless of how well it correlates to or predicts the natural world.*

There are people out there who are very emotionally invested in the idea that it should be possible for something to exist outside the reach of logic. Often this is because of a misguided notion that being within the reach of logic and science somehow cheapens things. It doesn't. Being logical doesn't in any way decrease one's ability to appreciate things on an emotional level. That is a popular misconception, and a malicious stereotype against rational thinkers. If anything, logic and science deepen one's appreciation for all aspects of life, including emotions, not reduce it.

Mysticism and religion often rely on saying "logic doesn't apply here" in order to attempt to dodge questions and to not address any substantive criticisms they are faced with. This is probably one of the most common reasons why some people like the idea of logical artificialism. It lets them believe completely arbitrary things without any evidence, because any time their beliefs are confronted they just shrug and say "Oh, well I guess logic just doesn't apply to this case then." instead of actually addressing the merits.

Denying logic encourages a hand-wavy mentality that practically guarantees that no progress will be made in a debate. Without logic, the entire concept of an argument essentially becomes meaningless. If arbitrary special cases can be declared to be immune to logic, then it effectively gives everyone permission to believe anything, regardless of the merits. It makes coherent thought impossible. Denying the applicability of logic is never a valid move when debating anything, despite how popular a tactic it nonetheless is. Not having enough info or clarity to decide yet is fine though.

You can believe whichever you want. However, the idea of logical naturalism probably has much more intellectual merit than logical artificialism does. The fact that logic is so closely tied to the notion of a conservation system, and the fact that all conceivable universes must be conservation systems in order to behave coherently[10], is strong evidence in favor of the idea that logic is part of the fabric of the universe. Hence there is good reason to believe that logical naturalism is true and logical artificialism is false.

Anyway, let's put logical naturalism aside now, and return to our discussion of conservation systems and the principle of property conservation again. As I was saying, we have this idea that if an action conserves a property on an object then that property continues to be conserved no matter how many times we apply that action. Such an effect is the basis of a commonly used technique in mathematics: the principle of mathematical induction, or the PMI for short.

The way the principle of mathematical induction works is that you start with a given object for which you already know some property holds true. You then prove that if the property holds true for the given object then it will also hold true for the next object in some sequence. This effectively sets off a chain reaction, like a bunch of dominoes falling down, thereby causing the property that held true for the given object to be proven to also be true for all the other subsequent objects in the sequence.

The principle of mathematical induction is actually a limited form of the much broader principle we have defined here: the principle of property conservation. In contrast to the principle of mathematical induction, the principle of property conservation covers not only this type of argument, but also every other proof in existence. You literally cannot even do logic or math without implicitly assuming the principle of property conservation.

All proofs rely upon conserving truth through each successive step, and hence all proofs are instances of the principle of property conservation being applied. In contrast, not all proofs are examples of the principle of mathematical induction being applied. In other words, the principle of mathematical induction is a subset of the principle of property conservation.

Remember when I said you can't even do logic or math without implicitly assuming the principle of property conservation? Don't you think it's interesting that anywhere you see logic you always see the PPC riding along with it? Well, when two things always occur together it is a sign that those two things *may* in fact be equivalent. In other words, the principle of property conservation may not merely be a part of logic, but rather it may actually *be* logic. Logic and the PPC may be synonyms.

Are logic and the PPC equivalent to each other? How can we find out? Well, to find out we need to determine whether logic has any other components to its operation

---

[10]For example, in order for you to say that any objects exist in a universe, that universe must conserve the property of existence for those objects. In order to do so, the universe must be a conservation system, hence logic must apply. In other words, even if you try to strip away various other properties in an attempt to make logic not apply, it will still apply regardless. Logic always wins.

besides the PPC. If logic does have components other than the PPC, then logic cannot be equivalent to the PPC. It's like that analogy I made earlier: A leaf may be green, but that does not imply that greenness is the essence of what it means to be a leaf.

So can we think of anything that logic has that the PPC does not have? Well, yes we can, sort of: An example would be the fact that logic includes an arsenal of techniques by which to approach problems, whereas the PPC is more of an underlying mechanism that enables logic to operate. Thus, it would appear that logic and the PPC are actually not equivalent. However, this was a useful thought exercise nonetheless.

Because every act of logic is in essence an application of the PPC, we can say that logic and the PPC are in some sense *mechanically* equivalent to each other. However, logic is a field of study whereas the PPC is a single principle that enables the operation of that field of study. The two terms clearly have distinct meaning to us as human beings. The semantics are different.

In essence, the principle of property conservation may be thought of as the raw substance from which logic is forged. Just as a sword is forged from iron, so too is logic forged from the principle of property conservation. The principle of property conservation is the elemental substance, the primordial ooze, from which logic arises and evolves.

All this talk of logic and inferring consequences through the PPC reminds me of something though. Do you remember back on page 71 in Definition 20 where I mentioned that there were a bunch of synonymous terms for the notion of a premise? As you may recall, premises are tentative assumptions upon which we build additional inferences to reach conclusions.

Well, it occurs to me that I should also mention the numerous different terms used to refer to the conclusions we are led to by such a sequence of inferences. That's the other side of the coin, after all. On the one hand we have our initial assumptions, and on the other hand we have the inferences that grow out of them. We might as well cover it while we are still speaking of logic in such broad and fundamental terms. Here's the associated terms:

Definition 29. *Any set of logical inferences that collectively implies that a specific conclusion must be true is called a **proof**. A proof may also be referred to as a **theorem**, a **corollary**, or a **lemma**. Each word has slightly different conventions and connotations for use.*

The term theorem is typically used for "normal" proofs of non-trivial importance or size. The term corollary is typically used for proofs that are based heavily on an already proven theorem. Corollaries are often short and cover additional peripheral consequences of the theorem upon which they depend. The term lemma is typically used for relatively small proofs that are used as intermediate proofs for building up larger proofs in a more manageable way.

However, all of these terms are used inconsistently and the distinctions between them are fundamentally subjective in nature. These naming conventions cannot really be trusted. Moreover, the value added by distinguishing them in this manner is low, or possibly even negative. The core essence of all of these terms is that they are simply arbitrary proofs. There is nothing fundamentally different about these various terms, thus the distinctions between them mostly just cause unnecessary confusion.

I personally recommend that you disregard the terms corollary and lemma, and only use the terms proof and theorem. Doing so eliminates the inherent subjectivity and unnecessary redundancy involved. It reduces the confusion caused by labeling proofs with so many different arbitrary words.

Many people reading documents about mathematics become confused by these terms and end up thinking there must be some kind of special meaning to them. There isn't. These terms are superficial synonyms. You are better off using only the most objective and most universal of the terms: proof and theorem. Use theorem to mean any proof of special interest, regardless of its size or role. Sometimes a theorem will end up being called a principle or a law if the theorem seems important enough to merit the title though.

Anyway, we've kind of been going off on a partial tangent here with all this discussion of transformative deduction, evaluative deduction, the nature of logic, the principle of property conservation, and all the various synonymous terms for a proof. These concepts in fact apply to logic in general, rather than only to classical logic in particular.

As you may recall, we began this section of this chapter discussing specifically classical logic. I branched out from that to discuss the broader nature of logic as a whole (whether classical or not) because it is important to understand. These insights into the nature of logic will prove to be useful. One should have a firm grounding in what logic actually is before one attempts a proper comparison of the different approaches to it.

True insight into the nature of logic seems surprisingly sparse sometimes, even among logicians. Too many texts on logic are far too dry and bereft of the original motivations, insights, inspirations, and examples from which the discoveries were born. A more principled approach, one that emphasizes a fundamental understanding of the real mechanisms by which reason operates, is far more productive and fruitful than one that is merely comprised of rote memorization or thoughtless adherence to traditions.

More on that later though. I would like to now refocus our attention, from general contemplation of the nature of all logic, back to specific contemplation of the nature of classical logic in particular. That is, after all, the primary subject of this section of this chapter, and one that we will have to give more thought to in order to properly understand its full significance and relation to other systems of logic.

So, to refresh our memory, in classical logic we have a bunch of different logical operators that we combine together to form more complex logical expressions corresponding to different logical scenarios. We may then evaluate any such expression to see if it is true or false. Here is a brief table to remind you of what the operators we covered were:

Operation	Example Symbols	Example Readings
negation	$\neg A$	not A
disjunction	$A \vee B$	A or B
conjunction	$A \wedge B$	A and B
material implication	$A \rightarrow B$	if A then B
equivalence	$A \leftrightarrow B$	A if and only if B
exclusive disjunction	$A \veebar B$	A xor B
joint denial	$A \downarrow B$	A nor B
alternative denial	$A \uparrow B$	A nand B

Classical logic is "truth-functional", which means that the truth value of an expression is entirely determined by the truth value of each participating statement, independent of what relation any of the statements may have to any of the others. For example, if *A* represents "A massive rainstorm is approaching." and *B* represents "The river will flood.", then any expression in classical logic involving both *A* and *B* would be evaluated purely based on the separate truth values of each respective statement. Any special relationships that might exist between *A* and *B* would not be considered.

Up to this point everything we have explored in classical logic has seemed all well and good. We can form useful complex expressions from the various operators. We can evaluate those expressions and get simple true or false values back that are easy to understand. We can use classical logic to construct proofs. All of the laws (i.e. tautologies) that we saw for classical logic seem to be valid and reasonable. Everything seems to be going just fine.

Unfortunately, appearances can be deceiving. All is not well in classical logic land. In fact, as we will see, classical logic sometimes generates complete blithering nonsense. There are at least several flaws in the design of classical logic. For one, classical logic uses a statement-based truth system, and as you may recall, we already showed in Chapter 1 (*What is truth?*) that statement-based truth is not an adequate representation of the concept of truth. However, putting that particular flaw aside for the moment, I want to instead focus our attention on a particularly glaring source of problems in classical logic: material implication.

What's so wrong with material implication? Isn't it just an arbitrary truth table? How can it be wrong, as such? Well, it is true that material implication can be evaluated by simply looking up the truth value corresponding to the inputs in the truth table. However, one should not assume that our representation is conceptually correct. There is a difference between the *consistency* of a representation of a concept and the *correctness* of a representation of a concept. Does the representation actually do justice to the original concept it purports to represent or not?

As you may recall, when we first discussed the basic operators of classical logic, I mentioned in passing during the section on material implication on page 87 that any time the condition in a material implication is false then the material implication is considered true, regardless of whether the consequence is true or false. I also mentioned that this behavior is the source of much confusion and conflict and that I would have interesting things to say about it later on. It's time we talked about that a bit more.

Take a moment to think about the implications of this behavior. For material implication, whenever the condition is false the implication is considered to be true, regardless of what the consequence represents or whether it is true or false. What are the implications of this behavior? Isn't it strange that an if-then statement with a false condition would always be true? Does that make sense logically?

Let's consider some concrete examples to make the point much clearer. Suppose, for example, that statement A is "The Earth has 57 suns and all of them are visible in the sky right now." and statement B is "It is dark outside.". Suppose we evaluate $A \to B$ under the rules of classical logic. Because A is false it follows immediately that $A \to B$ must be true, regardless of the value of B. In other words, according to classical logic "If the Earth has 57 suns and all of them are visible in the sky right now, then it is dark outside." is a true statement. Do you see a problem here?

If there really were 57 suns visible in the sky where you were standing, do you think this would imply that it would be dark outside? Of course not. This conclusion is clearly meritless. Let's think back to what I was saying earlier about the difference between consistency and correctness. Does material implication correctly represent the concept of implication? Does it fully capture the essence of what implication should mean, or does it instead fall short?

This bizarre behavior of material implication where false conditions imply anything has a name. It is known as the paradox of entailment. The paradox of entailment is just one of a broader family of similarly nonsensical seeming laws of classical logic known collectively as the paradoxes of material implication. The paradox of entailment can be written in several different ways, each of which mean essentially the same thing.

It can be written as (1) $F \to B$, (2) $A \wedge \neg A \to B$, or (3) $\neg A \to (A \to B)$, where A and B are any arbitrary statement truth values. If you examine the paradox of entailment's truth table, you will see that it is a tautology (i.e. is always true) and hence is a law of classical logic. The paradox of entailment is thus an unavoidable consequence of the way the classical operators are defined. Here's a brief formal definition:

Definition 30. *In classical logic, due to the nature of the definition of material implication, a false condition implies any consequence. This law of classical logic is known as the* **paradox of entailment**. *Another alternative name (not currently in use), might be to call it the* **paradox of false conditions**.

Why does it have to be this way though? Well, the only way to make material implication work at all in classical logic is to define it this way. The fact that classical logic only allows two truth values (true or false) forces us to. How would you create a conditional relationship in classical logic if material implication wasn't defined this way? You couldn't. Here's why:

If you tried to define material implication any other way in classical logic then you'd either (1) cause implication to become the same thing as equivalence, (2) cause implication to be determined entirely by the truth value of the consequence, or (3) cause implication to become the same thing as conjunction. None of these outcomes is acceptable. Try changing the truth value of $A \to B$ for the cases where A is held false and B is allowed to vary and you'll see what I mean. Thus, classical logic must accept the paradox of entailment as a law, despite the seemingly nonsensical consequences of doing so.

The paradox of entailment is extremely closely related to another problematic property of classical logic known as the principle of explosion. The principle of explosion says that if you accept two contradictory premises as true then anything follows, thereby causing an proliferation of nonsensical conclusions whereby any and every statement in existence can seemingly be "proven" to be true. This resulting *explosion* of nonsense is where the principle of explosion gets its name.

The paradox of entailment and the principle of explosion are in fact nearly the same thing, arguably, but the principle of explosion is usually introduced using an example that relies on disjunctive syllogism and proof construction, whereas the paradox of entailment is usually introduced using an example that appeals directly to the definition of material implication. Ultimately, they are slightly different manifestations of the same essential conceptual design flaw in classical logic.

The principle of explosion is often summed up as "from falsehood anything follows". The antiquated Latin phrase "ex falso sequitur quodlibet" means the same thing and is sometimes used instead, but is unnecessarily pedantic and obscure and will only serve to waste the audience's time. When there are equivalent words available in a living language, these kinds of antiquated Latin phrases serve little to no actual purpose, except perhaps to make the speaker seem more profound or intelligent than they actually are.

Anyway though, earlier I mentioned that the paradox of entailment is just one of a larger family of problematic laws of classical logic known as the paradoxes of material implication. Let's take some time to thoroughly explore the paradoxes of material implication, so that we can more fully understand the consequences of defining implication in the way that material implication does.

Any expression in classical logic that is a tautology (i.e. that is always true) which focuses on material implication and generates seemingly nonsensical results can be considered a paradox of material implication. As such, there are technically an infinite number of paradoxes of material implication that can be constructed. However, a small subset of them exemplify the essential characteristics of the paradoxes of material implication. Those will be our focus.

As we have already discussed, in material implication a false condition implies any consequence and this is known as the paradox of entailment. However, the paradox of entailment also has a sibling paradox which relates to the consequence instead of to the condition. In particular: a true consequence is implied by any condition. Unfortunately, this sibling paradox does not seem to have a name, despite how important it is. If the paradox of entailment has a name, then it stands to reason that its closely related sibling should also have a name. Thus, we should define a new term:

Definition 31. *In classical logic, due to the nature of the definition of material implication, a true consequence is implied by any condition. This law of classical logic will henceforth be referred to as the **paradox of consequence**, so that it may be more easily discussed in relation to the paradox of entailment and the other paradoxes of material implication. Alternatively, you may refer to it as the **paradox of true consequences**,* *if you prefer the additional descriptive clarity this affords but aren't bothered by the increased verbiage*[11].

Because they capture the most critical aspects of what makes material implication behave so strangely, the two most important paradoxes of material implication are the paradox of entailment and the paradox of consequence. Both can be written in several different essentially equivalent forms. For easy reference, let's list those forms. In the expressions that follow, A and B represent any arbitrary statement truth values and F and T represent false and true respectively, as usual.

Here are the different ways of expressing the paradox of entailment:

1. $F \to B$

2. $(A \land \neg A) \to B$

3. $\neg A \to (A \to B)$

Likewise, here are the different ways of expressing the paradox of consequence:

1. $A \to T$

2. $A \to (B \lor \neg B)$

3. $B \to (A \to B)$

If you check the truth tables for all of these expressions you will see that they are all tautologies, and therefore are indeed laws of classical logic. These variants of each paradox may look like they have different meanings and may read differently, but they all really express the same concept if you think about it carefully. Let me explain this in more depth just to be sure that you understand why.

For example, $(A \land \neg A)$ is always false as we know from the law of non-contradiction, and therefore $(A \land \neg A) \to B$ is effectively just one specific case of $F \to B$. Likewise, $(B \lor \neg B)$ is always true as we know from the law of the excluded middle, and therefore $A \to (B \lor \neg B)$ is effectively just one specific case of $A \to T$.

The expression $\neg A \to (A \to B)$ for the paradox of entailment literally means "If A is false then A implies B", regardless of what A and B represent, which is the same as saying that a false condition implies any consequence, which is the same as saying $F \to B$.

Likewise, the expression $B \to (A \to B)$ for the paradox of consequence literally means "If B is true then A implies B", regardless of what A and B represent, which is the same as saying that a true consequence is implied by any condition, which is the same as saying $A \to T$.

---

[11] For this book, I will use "paradox of consequence" instead of "paradox of true consequences", because "paradox of consequence" sounds more structurally similar to "paradox of entailment", which itself is a much more widely known term than the corresponding alternative term "paradox of false conditions". However, the longer terms here are probably better in principle, from a descriptive standpoint.

Sometimes when people discuss the paradoxes of material implication they will write out some of these variants of the paradox of entailment and the paradox of consequence as if they are different paradoxes when in reality they are redundant forms of the same paradox. This is an understandable discrepancy considering the forms look so different at first glance.

The list of paradoxes in the Wikipedia article on the paradoxes of material implication (as of when this book was first written) is an example of such. A few of the items in the list are redundant. Let's take a look at the list. I've changed the variable names from P, Q, and R to A, B, and C to make them match the conventions I use in this book. I also made some other minor tweaks for consistency and clarity. Here's the list:

1. $(A \land \neg A) \to B$
2. $B \to (A \to B)$
3. $\neg A \to (A \to B)$
4. $A \to (B \lor \neg B)$
5. $(A \to B) \lor (C \to A)$
6. $\neg(A \to B) \to (A \land \neg B)$

As you can see, item 1 and item 3 are just different forms of the paradox of entailment. Likewise, item 2 and item 4 are just different forms of the paradox of consequence. In contrast, item 5 and item 6 are genuine examples of new paradoxes of material implication. However, just as the paradox of consequence didn't have a name, neither do item 5 and item 6. We should fix that to make them easier to discuss and investigate.

Let's give an official definition for the paradox expressed in item 5:

Definition 32. *In classical logic, due to the nature of the definition of material implication, any time you have two implications of the form $A \to B$ and $C \to A$ then one of them must be true. In symbols this is expressed as $(A \to B) \lor (C \to A)$. This law of classical logic will henceforth be referred to as the **paradox of participatory dichotomy**.*

The reason why the paradox of participatory dichotomy is a tautology is as follows: A must either be true or false. Therefore, either $(A \to B)$ is true because of the paradox of entailment or else $(C \to A)$ is true because of the paradox of consequence. Either way, one of them must be true. Therefore $(A \to B) \lor (C \to A)$ must be true.

In other words, regardless of whether there are any meaningful logical connections between what any participating statements refer to, the nature of material implication nonetheless creates a dichotomy where every statement will always either imply or be implied by all other statements. Participation in this dichotomy is unavoidable in classical logic, no matter how nonsensical the resulting implications may be. That's why I chose to name this the paradox of participatory dichotomy.

Likewise, let's give an official definition for the paradox expressed in item 6:

Definition 33. *In classical logic, due to the nature of the definition of material implication, any time you deny that an implication of the form $A \to B$ is true then it follows that A must be true and B must be false. In symbols this can be expressed as $\neg(A \to B) \to (A \wedge \neg B)$. This law of classical logic will henceforth be referred to as the **paradox of manifest denial**.*

I gave it this name because any time you deny a material implication, the logical negation of that implication essentially becomes tangibly manifest as true. It is as if merely denying a hypothetical if-then statement causes the participating statements to suddenly acquire concrete truth values. The condition magically becomes true and the consequence magically becomes false, regardless of whether this makes any sense or not.

Let me explain why the paradox of manifest denial exists in classical logic. First, notice that $A \to B$ has the same truth table as $\neg A \vee B$. Therefore, we can substitute $\neg A \vee B$ for $A \to B$ in $\neg(A \to B)$ which yields $\neg(\neg A \vee B)$, thereby proving that $\neg(A \to B)$ is equivalent to $\neg(\neg A \vee B)$.

Next, to evaluate $\neg(\neg A \vee B)$, we first apply the law of disjunction negation (one of De Morgan's Laws). This yields $(\neg\neg A) \wedge \neg B$, which we can then apply the law of double negation to. This yields $A \wedge \neg B$. Therefore, we can conclude that $\neg(A \to B) \to (A \wedge \neg B)$ is true, which is exactly the same as the form of the paradox of manifest denial.

Anyway, that basically covers the paradoxes of material implication that were listed in Wikipedia. However, item 5 in the list (i.e. the paradox of participatory dichotomy) is actually just one specific form of a more general form of paradox of material implication. There are in fact two other specific forms of this more general form of paradox of material implication. I cannot find any of these other forms documented anywhere on Wikipedia. In fact, I cannot find them documented anywhere else on the internet either.

As such, we should give these missing forms new definitions and properly document them. While all three of these paradoxes are close relatives, their structure is nonetheless different in a subtle but meaningful way, therefore I think they still merit being discussed individually. First though, let's discuss the more general form of these paradoxes:

Definition 34. *Due to the fact that $A \to B$ is equivalent to $\neg A \vee B$ in classical logic, it follows that the expression $(A \to B) \vee (C \to D)$ is equivalent to $\neg A \vee B \vee \neg C \vee D$. Therefore, because of the law of the excluded middle, if at least two of $\neg A$, B, $\neg C$, and D are logical negations of each other then $(A \to B) \vee (C \to D)$ will be true. This law of classical logic will henceforth be referred to as the **paradox of dichotomy***

If all of A, B, C, and D are distinct variables, where none of them is equal to any of the others, then $(A \to B) \vee (C \to D)$ will never be a tautology (i.e. will not have a truth table that is always true). However, when you substitute in values that cause the number of distinct variables to drop to three or less, then $(A \to B) \vee (C \to D)$ can become a tautology. In total there are three such specific cases worth mentioning, ignoring some cases that are either too trivial or too redundant to care about. Each of the paradox of dichotomy's special cases will be given distinct names.

A dichotomy is a division of a whole into two subsets that together cover all possibilities. Any expression of the form $X \vee \neg X$ is therefore one example of a dichotomy.

The paradox of dichotomy appears whenever $(A \rightarrow B) \vee (C \rightarrow D)$ contains at least one such dichotomy, i.e. whenever at least two of $\neg A$, B, $\neg C$, and D are logical negations of each other. That is why I named it the paradox of dichotomy.

Anyway, now that we've covered the general case of the paradox of dichotomy and the special case of the paradox of participatory dichotomy, we should move on to covering the other two special cases of the paradox of dichotomy. Here are our new definitions for them:

Definition 35. *When the paradox of dichotomy has the form $(A \rightarrow B) \vee (\neg A \rightarrow C)$ then it will henceforth be referred to as the **paradox of conditional dichotomy**. The reason why the paradox of conditional dichotomy is a tautology is as follows: One of A or $\neg A$ must be false. Therefore, one of $(A \rightarrow B)$ or $(\neg A \rightarrow C)$ must be true because of the paradox of entailment. Therefore $(A \rightarrow B) \vee (\neg A \rightarrow C)$ must be true.*

I gave the paradox of conditional dichotomy this name because the mechanism that explains why it is a tautology hinges on the behavior of the condition in material implication. Hence the adjective "conditional" is an appropriate choice. Obviously, the "dichotomy" part of the name comes from the fact that it is a specific case of the paradox of dichotomy.

Definition 36. *When the paradox of dichotomy has the form $(A \rightarrow B) \vee (C \rightarrow \neg B)$ then it will henceforth be referred to as the **paradox of consequential dichotomy**. The reason why the paradox of consequential dichotomy is a tautology is as follows: One of B or $\neg B$ must be true. Therefore one of $(A \rightarrow B)$ or $(C \rightarrow \neg B)$ must be true because of the paradox of consequence. Therefore $(A \rightarrow B) \vee (C \rightarrow \neg B)$ must be true.*

I gave the paradox of consequential dichotomy this name because the mechanism that explains why it is a tautology hinges on the behavior of the consequence in material implication. Hence the adjective "consequential" is an appropriate choice. Obviously, the "dichotomy" part of the name comes from the fact that it is a specific case of the paradox of dichotomy.

By the way, if you try to systematically discover all of the possible variants of the paradox of dichotomy by searching for all the possible pairings of $\neg A$, B, $\neg C$, and D that are logical negations of each other then you might discover a combination that looks like $(A \rightarrow B) \vee (B \rightarrow C)$. You might then be tempted to say that I have missed one. However, closer inspection shows that $(A \rightarrow B) \vee (B \rightarrow C)$ is actually the same form as the paradox of participatory dichotomy.

The variable names and the ordering of the expression are creating an illusory distinction. I'll prove it to you: First, swap the two sides of $(A \rightarrow B) \vee (B \rightarrow C)$. This results in $(B \rightarrow C) \vee (A \rightarrow B)$. Then, replace B with A, C with B, and A with C in parallel. This results in $(A \rightarrow B) \vee (C \rightarrow A)$, which is the same thing as the paradox of participatory dichotomy, thus proving my point.

$(A \rightarrow B) \vee (B \rightarrow C)$ is redundant. Nonetheless, if you still want a distinct name for it, then I suggest calling it the **paradox of transitive dichotomy**, due to the fact that its form is reminiscent of a transitive relation. I suppose that one can argue that giving both forms names has some benefits. It might make it a little easier to recognize them on sight because you don't have to perform the transformations between the two forms.

Anyway, now that we've identified the redundancies in the Wikipedia list of paradoxes of material implication and also discovered and named some paradoxes that were missing from the list, let's re-list them in an improved format. With our new definitions and our refactoring of the paradoxes we can produce a cleaner, more comprehensive, and more manageable list. Here's our new list:

1. **Paradox of Entailment:**
    a) **Form 1:** $F \rightarrow B$
    b) **Form 2:** $(A \wedge \neg A) \rightarrow B$
    c) **Form 3:** $\neg A \rightarrow (A \rightarrow B)$

2. **Paradox of Consequence:**
    a) **Form 1:** $A \rightarrow T$
    b) **Form 2:** $A \rightarrow (B \vee \neg B)$
    c) **Form 3:** $B \rightarrow (A \rightarrow B)$

3. **Paradox of Dichotomy:** $(A \rightarrow B) \vee (C \rightarrow D)$, where at least two of $\neg A$, B, $\neg C$, and D are logical negations of each other
    a) **Paradox of Participatory Dichotomy:** $(A \rightarrow B) \vee (C \rightarrow A)$
    b) **Paradox of Conditional Dichotomy:** $(A \rightarrow B) \vee (\neg A \rightarrow C)$
    c) **Paradox of Consequential Dichotomy:** $(A \rightarrow B) \vee (C \rightarrow \neg B)$

4. **Paradox of Manifest Denial:** $\neg(A \rightarrow B) \rightarrow (A \wedge \neg B)$

As you can see, there are at least four different kinds of paradox of material implication, but they all hinge upon the same essential truth-functional behavior of material implication. The paradox of dichotomy is a direct consequence of the paradox of entailment and the paradox of consequence. The paradox of manifest denial is a direct consequence of the truth table for material implication being equivalent to that of the disjunction $\neg A \vee B$. These paradoxes all ultimately originate from the context-independent nature of material implication. Material implication does not actually check whether a real relationship exists between the participating statements or not.

We now have a clearly defined named set of definitions for the paradoxes of material implication that will make discussing them much easier, but we shouldn't stop there. When all we have is abstract definitions there is a high risk that we will not actually fully grasp the meaning or the weight of the subject matter we are considering. Proper learning and understanding requires concrete examples. Therefore, let's explore some.

Let's begin with the most well-known of the paradoxes of material implication: the paradox of entailment. As you may recall, the paradox of entailment has the form $F \rightarrow B$ and means that a false condition will imply any consequence. Thus, to construct an example of this paradox, we'll need a false statement followed by any other arbitrary statement, which are then put into the form of an implication. Here are some examples:

1. If rain water is made of salt, then drinking rain water will make you less dehydrated.

2. If Isaac Newton was not human, then he must have been a gopher.

3. A human having naturally blue hair implies that they also have wings and can fly.

4. If you can bend 5 meter thick steel beams by hand, then your mother was a hamster and your father smelt of elderberries.

5. If an object is both all-natural and all-artificial, then it must be made of sapphires.

In classical logic all of these statements are considered logically valid, due to the paradox of entailment. Pretty crazy right? I should mention however that, although these statements are considered logically valid, they are *not* considered logically sound. An argument is considered to be logically valid only if the conclusion would be true under the assumption that the premises were true. In contrast, an argument is considered to be logically sound only if, in addition to also being logically valid, the premises actually *are* true.

However, one of the points I will be making in this book is that not only are these kinds of statements not logically sound, they are not even logically valid either. The conclusions in fact do *not* follow from the premises, even if we were to grant them. The idea that false premises imply anything has become almost universally accepted and is taken for granted in logic and sometimes even in popular culture, but, however popular and standardized this belief may be, we will nonetheless see that it is not really justified.

Classical logic defines a logically valid argument roughly as any argument that "cannot have both true premises and a false conclusion". In essence, in classical logic, the definition of validity and the definition of material implication are one and the same. Contrary to how popular this definition is however, it is actually conceptually inadequate as a model of real implication. Thus, if you were to try to counter this part of my argument in this book by merely appealing to the classical definition of validity then you would be guilty of circular reasoning.

You would in effect be arguing that the classical definition of validity is correct for no other reason than because it says so. I know that classical logic is considered the standard and is very entrenched. However, I will be describing a very compelling alternative approach as we gradually progress through the book. Wait until you fully understand my argument before you pass judgment on it. I am well aware of the standard rationalizations for why material implication is defined the way it is.

Anyway though, let's get back to exploring more concrete examples of the paradoxes of material implication. We just finished giving five concrete examples of the paradox of entailment. Every one of the paradoxes of material implication feels a bit different, so we should make sure to find concrete examples for all of them. It's hard to go wrong by being thorough, whereas it is easy to go wrong by rushing.

Let's proceed to the next paradox: the paradox of consequence. As you may recall, the paradox of consequence has the form $A \rightarrow T$ and means that a true consequence is implied by any condition. Thus, to construct an example of this paradox we'll need any arbitrary statement followed by a true statement, which are then put into the form of an implication. Here are some examples:

1. If igneous rocks are formed by cooling magma or lava, then touching poison ivy causes rashes.

2. If the moon influences the behavior of werewolves, then the moon influences the behavior of the tides.

3. If steel is an alloy of iron and carbon, then photosynthetic cells require light to produce energy.

4. The fact that apple pies are filled with watermelon slices implies that computers require electricity to operate.

5. If you have the ability to fall asleep at will, then you have the ability to avoid reading an academic research paper.

As you may have noticed, some of these don't sound quite as crazy as the examples we listed for the paradox of entailment. The reason for that is because, unlike examples of the paradox of entailment, examples of the paradox of consequence do not require that at least one participating statement is false. In other words, examples of the paradox of consequence can have both a true condition and a true consequence.

Notice, however, that even when the statements have both a true condition and a true consequence they can still be nonsensical. Consider item 1 and item 3 above. Notice that the condition and consequence for both of them are true, yet the idea that each of these conditions implies the corresponding consequence is ridiculous. What does the process by which igneous rocks are formed have to do with the effects of poison ivy? What does the composition of steel have to do with how photosynthetic cells generate energy?

As you may recall, all of the examples of the paradox of entailment we listed earlier were considered to be valid but not sound under classical logic. Interestingly however, this is not the case for the examples of the paradox of consequence. Some examples of the paradox of consequence can be both valid and sound under classical logic. In particular, any example of the paradox of consequence that has a true condition will be considered both valid and sound. In contrast, any example of the paradox of entailment will always be considered valid but not sound.

The fact that examples of the paradox of consequence can be both valid and sound is a bad sign. While there are of course many examples of statements of the form $A \to T$ that are perfectly reasonable, there are even more examples of statements of the form $A \to T$ that are not reasonable. Only a limited number of pairs of statements will be genuinely related to each other. The vast majority will not be related.

A definition of implication in which the condition has no relation to the consequence is a definition which fails to capture the essence of the very thing which it purports to define. Suppose we have two arbitrary statements A and B. What would it take to convince you that B is a consequence of A? If I were to tell you that A and B had no relation to each other would that *increase* or *decrease* your confidence that *A* causes *B*? Do you think it's possible for *A* to cause *B* if *A* and *B* have no relationship to each other? Isn't causality a relationship? If causality is a relationship, then isn't it true that only related things can imply each other?

How can we trust a definition of implication that indicates that pairs of statements that have no substantive relation to each other can nonetheless be not only valid but also sound? The paradox of consequence tells us that every true statement is implied by every other statement in existence. In effect, it says that everything is the cause of everything. Pick literally any two random true statements and classical logic says that each one implies the other. If everything implies everything else, then what does it even mean for one thing to be a condition and another thing to be a consequence in such a system? Doesn't this effectively render the entire concept of causality meaningless?

Anyway, let's proceed to the next paradox: the paradox of dichotomy. As you may recall, the paradox of dichotomy has the form $(A \rightarrow B) \lor (C \rightarrow D)$, where at least two of $\neg A$, B, $\neg C$, and D are logical negations of each other. Rather than bore you by constructing a full list for every single one of the three variants of the paradox of dichotomy, let's investigate two examples of each variant, for a total of six examples for the paradox of dichotomy as a whole. We'll list them in the same order as before: participatory, conditional, and consequential. Now let's see what they look like:

1. Either an object being made of plastic implies that it is made of gold or else an object being made of gold implies that it is made of silver.

2. Either being an electrical engineer implies that you are a medical doctor or being a blacksmith implies that you are an electrical engineer.

3. Either being immortal makes you able to play the piano or being mortal makes you able to use telekinesis.

4. Either cats are primates or anything that is not a cat is a reptile.

5. Either having red hair makes you a wizard or being able to shoot lightning from you fingertips makes you not a wizard.

6. Either leaves turning brown and falling to the ground implies that it is spring or flowers blooming implies that it is not spring.

Just as with the previous sets of examples, all of these examples here are considered to be valid reasoning in classical logic. However, just as before, none of these examples actually seem reasonable. Underneath the more complicated structure of these examples, you can see that the fundamental source of these problems is once again the context-independent nature of material implication. The paradox of entailment and the paradox of consequence allow far too many things to be valid.

Anyway, let's proceed to the final paradox in our list: the paradox of manifest denial. As you may recall, the paradox of manifest denial has the form $\neg(A \rightarrow B) \rightarrow (A \land \neg B)$. Thus, to construct an example of this paradox we simply need to take any implication, negate it, and then see what it implies. Let's see some examples:

1. If it isn't true that your grandmother having the ability to spit fireballs would imply that she could make French toast quickly, then it must be the case that she in fact *can* spit fireballs but *cannot* make French toast quickly.

2. Suppose that someone claims that if dark clouds appear on the horizon then it is going to rain. Suppose that you deny their claim. This denial can only be true if dark clouds in fact *are* on the horizon but it will not rain. It says nothing about any causal link.

3. Imagine that someone claimed that if a statue is made of clay then it can conduct electricity, but you denied the claim. Then the statue must be made of clay and cannot conduct electricity. This ignores all other other possibilities (e.g. a non-conductive statue made of plastic).

4. If it is false that a giant meteor heading for the Earth implies that humanity will go extinct, then a giant meteor must be heading for the Earth but humanity will not go extinct.

5. If it is not the case that working hard implies that you will be successful, then you must be working hard and yet will not be successful.

The nonsense never ends does it? As you can see, the paradox of manifest denial generates some astoundingly absurd results. By now the full weight of just how poorly material implication represents the real concept of implication should be starting to sink in hopefully. By exploring some concrete examples we have acquired a better sense of how the paradoxes of material implication actually effect reasoning with classical logic.

These paradoxes we've covered above are strongly representative of the paradoxes of material implication, but they are not exhaustive. Other paradoxes of material implication exist as well. Attempting to cover all of the possible variants though would probably be a waste of time. However, I want to finish this section by briefly defining just two more paradoxes of material implication. These two new paradoxes have some interesting consequences. After that though, we're going to move on to a new subject.

Anyway, here are the two new definitions:

Definition 37. *In classical logic, the following expression is a tautology and is therefore always considered true:*

$$[(A \to B) \land (C \to D)] \to [(A \to D) \lor (C \to B)]$$

*This law of classical logic will henceforth be referred to as the **paradox of consequence swapping**. The paradox of consequence swapping gets its name from the fact that it implies that in classical logic if multiple implications are true at the same time then they can have their consequences swapped and be put into a disjunction and the resulting expression will still be considered true.*

Definition 38. *In classical logic, the following expression is a tautology and is therefore always considered true:*

$$[(A \land B) \to C] \to [(A \to C) \lor (B \to C)]$$

*This law of classical logic will henceforth be referred to as the **paradox of disregarded conjunction**. The paradox of disregarded conjunction gets its name from the fact that it implies that in classical logic if multiple conditions must be true at the same time in order for a particular consequence to be true, then the requirement that both of the conditions must be true at the same time can be disregarded, and instead either one of the conditions can be considered independently sufficient to conclude that the consequence is true.*

These paradoxes clearly have insane consequences. The paradox of consequence swapping lets you arbitrarily swap consequences from one implication to another regardless of what relation the statements may have to each other. The paradox of disregarded conjunction lets you ignore the entire concept that some consequences require more than one condition in order to be true. Just like with the other paradoxes of material implication, the outcome appears to be catastrophic.

You might be wondering though, that if all of these paradoxes of material implication have such catastrophic outcomes, then why is it that modern logic and mathematics are able to function so well and are able to produce so many objectively correct observations when so much of the literature is nominally founded on classical logic. Well, to answer that, let me tell you a dirty little secret: Classical logicians don't actually believe many of the rules of their own system of logic.

Oh sure, if you ask a classical logician if, for example, they believe that a false condition implies any consequence then they will say yes. In fact, ask them whether any given paradox of material implication is considered valid in classical logic and, if they are sufficiently competent, they will say yes. However, if you instead look at what they do when they construct proofs or when they reason about real situations then you will notice that they almost never apply any of the rules mandated by the paradoxes of material implication.

You see, the human mind's innate awareness that the paradoxes of material implication are wrong is so strong that even when someone nominally convinces themselves that the paradoxes are valid then they will still tend to almost always avoid applying them. That is why the existence of the paradoxes of material implication has not created a catastrophe. That is why most existing results of logic and math are still valid. Classical logicians simply ignore the problematic rules and apply only the safe ones.

Classical logicians wanted a system of logic in which the truth values of expressions were determined solely by the true or false values of each respective participating statement. As we have previously discussed, this is known as a "truth-functional" system of logic. It also happens to be a statement-based truth system, as opposed to an existence-based truth system.

When you try to construct a system of logic under such constraints it will inevitably gravitate towards being at least somewhat like classical logic. Material implication may have many absurd consequences, but it is also one of the only truth-functional definitions of implication that even vaguely approximates the concept of implication.

Basically, classical logicians saw material implication as the best overall compromise for the properties they wanted in a system of logic. They wanted the convenience of a truth-functional logic and were willing to pay a high price for it. Plus, nobody really seemed to know of a better way of doing things. Other approaches inevitably introduced other undesirable side effects, and so classical logic settled in as the dominant form of logic.

Classical logicians then retroactively rationalized their own design decisions so that the deeply unsettling nature of the paradoxes of material implication would seem less threatening. They blamed the strangeness of material implication on informal natural language, claiming that in fact material implication was actually the correct representation of implication and that in contrast the informal natural language concept of implication must just be the result of the sloppy thinking of the uneducated masses.

Ironically however, by choosing a definition of implication that ignores the idea that two things must be related for one to imply the other, classical logicians actually constructed something which is merely a feeble shadow of what it was intended to represent. The "uneducated masses", who classical logicians viewed as not understanding the true nature of implication, actually had a more correct idea of what the true nature of implication was, even if they could not articulate the mechanisms by which it could be implemented. That, in fact, is one of the goals of this book: to try to finally articulate those mechanisms, and thereby create a definition of implication which is in harmony with both natural language and formal language, instead of being in conflict.

While it is true that one must be wary of people's intuitions, it is also true that one should not be too dismissive of them. When one is too dismissive of instinct, one loses out on the benefit of the much higher computational power of the subconscious mind. Subconscious instincts can guard against threats that your conscious mind would be far too slow and clumsy to ever even perceive. Grant instinct due respect, but not too much.

When we are too quick to dismiss the masses, we are sometimes blinded to hidden wisdom. Such are the consequences of being too rigid-minded to appreciate the power of natural decentralized systems (such as "the uneducated masses") to spontaneously converge upon efficient solutions and deep insights. Natural systems can sometimes silently evolve and adapt with a speed and precision that no conscious mind can easily match. It is wise to respect the power of nature.

In fact, nature will even have a role to play in correcting the definition of implication. Do you remember our discussion of conservation systems and the principle of property conservation? Do you remember how I said that because the universe is a conservation system it therefore essentially has logic embedded within it? Well, that means that logic and nature are interrelated. When we redefine implication in a form that is more in harmony with the nature of truth, then everything will start to fall into place. The paradoxes of material implication will evaporate.

The next time we discuss material implication in detail, we will be dismantling it, and in its place we will construct a new definition of implication, one which, among other things, is immune to all of the paradoxes of material implication. We will come to understand exactly what went wrong with the definition of material implication and what we can do to fix it. We will start building a bridge between the world of natural language and the world of formal logic that will yield a logic that is simultaneously both more natural and more rigorous than classical logic.

Our time with classical logic is coming to an end, for now. Before we can solve the problem of implication though, we will have to broaden our perspective and learn about some other branches of logic and mathematics too. Whereas in this section we studied relationships between statements, in the next section we will study relationships between objects. It's time for us to learn about set theory.

2.4 What is set theory?

Let's begin with some history. Earlier in this book, in the section entitled "How did the formal study of logic begin?", starting on page 75, we discussed the origins of the study of formal logic. We saw that while informal logic has essentially always existed, formal logic itself did not really exist as a rigorous discipline of study until a man named Aristotle wrote a book, now known as *The Prior Analytics*, which described perhaps the first ever systematic framework for making consistently valid logical inferences. For reference, our first significant discussion of Aristotle began on page 77.

As you may recall, I mentioned (on page 79) that Aristotle's syllogistic logic is essentially the ancient ancestor of modern set theory. However, syllogistic logic is much more limited than modern set theory in the cases it can handle. In particular, syllogistic logic is only capable of addressing a specific form of logical argument known as a syllogism, wherein two set relationships that we already know to be true are used to prove that a third set relationship must also be true. The two set relationships used to reach the conclusion in a syllogism consist of either (1) two subset relations or (2) one intersection relation and one subset relation. These two cases behave differently.

For over a millenia, Aristotle's syllogistic logic was seen as being essentially complete and incapable of being extended or enhanced in any meaningful way. However, as we now know, this sentiment was wrong. Syllogistic logic is in fact quite crude and is only really capable of handling a single type of logical argument among countlessly many other forms. Narrow-minded adherence to syllogistic logic, to the abandonment of all other alternative lines of thoughts, was the norm for a very long time. The medieval era is especially notorious for intellectual stagnation in these respects.

The person who finally disrupted this long period of intellectual stagnation in logic was a man named George Boole. Boole was born in 1815 in the city of Lincoln in eastern England. We have actually already mentioned Boole once before, in section 2.3 which is entitled "What is classical logic?". Our discussion of Boole began on page 83, where we briefly discussed his notation, his influence, and his system of reasoning.

Boole was highly influential in the development of both classical logic and set theory and is one of the most important logicians to ever live, being comparable in importance even to Aristotle himself perhaps. In 1847 Boole published a groundbreaking book called *The Mathematical Analysis of Logic* in which he created the first ever symbolic and algebraic system for logic. Prior to this publication, the concept of using variables and algebra to evaluate logic didn't even exist. Previous work in logic was far more tedious, often relying on lengthy descriptions written in natural language that discussed the subject matter via numerous unwieldy memorized rules for syllogistic logic.

Boole appreciated the value of the system that Aristotle had created, but sought to refactor and extend it so that it could be both more streamlined and more expressive. Boole examined syllogistic logic and saw the hidden underlying substructures that others had previously failed to recognize. He created a new system of reasoning called boolean algebra which combined concepts from numerical algebra with concepts from logic. Boole converted logic into a form that could be manipulated in equations using algebra, and thereby succeeded in factoring out much of the redundancy from syllogistic logic, thus creating a much more general and easy to use system.

After some years passed, as Boole gradually came to better understand the system he had created, he eventually came to view the original book *The Mathematical Analysis of Logic* as an imperfect explanation of his new theory. Thus, in 1854, he published a follow-up book entitled *The Laws of Thought* in which he further refined and extended his theory. This second book, *The Laws of Thought*, is considered the most definitive expression of the original theory of boolean algebra.

George Boole's book *The Laws of Thought* is available for free at Project Gutenberg (gutenberg.org), as are many other old books that are now in the public domain. Consider the following passage, taken from page 28 of the Project Gutenberg version of Boole's *The Laws of Thought*, and notice how the theory described in it shows signs of becoming a generalized theory of sets. I have made some minor edits to improve the clarity and phrasing of the text. Here's the passage of interest:

> ...Let us then agree to represent the class of individuals to which a particular name or description is applicable, by a single letter, such as x. If the name is "men", for instance, let x represent "all men", or the class "men". By a class is usually meant a collection of individuals, to each of which a particular name or description may be applied; but in this work the meaning of the term will be extended so as to include the case in which but a single individual exists, answering to the required name or description, as well as the cases denoted by the terms "nothing" and "universe", which as "classes" should be understood to comprise respectively "no beings" and "all beings".
>
> Again, if an adjective, such as "good", is employed as a term of description, let us represent by a letter, such as y, all things to which the description "good" is applicable, i.e. "all good things", or the class "good things". Let it further be agreed, that by the combination xy shall be represented that class of things to which the names or descriptions represented by x and y are simultaneously applicable. Thus, if x alone stands for "white things", and y for "sheep", let xy stand for "white sheep"; and in like manner, if z stands for "horned things", and x and y retain their previous interpretations, let zxy represent "horned white sheep", i.e. that collection of things to which the name "sheep", and the descriptions "white" and "horned" are together applicable. Let us now consider the laws to which the symbols x, y, z, etc, used in the above sense, are subject.
>
> First, it is evident, that according to the above combinations, the order in which two symbols are written is unimportant. The expressions xy and yx equally represent that class of things to which the names or descriptions x

and *y* are together applicable. Hence we have:

$$xy = yx$$

In the case of *x* representing white things, and *y* representing sheep, either of the sides of this equation will represent the class of "white sheep". There may be a difference as to the order in which the conception is formed, but there is none as to the individual things which are comprehended under it. In like manner, if *x* represents "estuaries", and *y* represents "rivers", then the expressions xy and yx will indifferently represent "rivers that are estuaries", or "estuaries that are rivers"…

While it is true that Boole also investigated statements (a.k.a. propositions), the original impetus for Boole's work was to refine and extend Aristotle's syllogistic logic, which dealt not with statements but with collections of objects. In this endeavor he succeeded more than anyone had previously thought possible. Indeed, the publication of *The Mathematical Analysis of Logic* and *The Laws of Thought* marks the main transition point from the stagnant era of syllogistic logic (which lasted for more than 1000 years) to the fresh new era of set theory. Therefore, if there is any one person that should be credited with founding set theory then it is probably George Boole.

The key phrase here however is unfortunately "should be credited", because in fact George Boole is not currently credited with founding set theory despite the fact that he seems to deserve the credit. Instead, when people ask who it was that founded set theory they are met with an ironically overconfident assertion that set theory was the single-handed creation of a man named Georg Cantor.

Georg Cantor was born in 1845 (just 2 years prior to Boole publishing *The Mathematical Analysis of Logic*) in the city of Saint Petersburg in Russia, but moved to Germany later in life. Although Cantor is credited with single-handedly creating set theory, in reality he mainly only created a relatively small esoteric subset of set theory known as the theory of transfinite numbers. He also studied number theory and trigonometry (etc), but is primarily known for his influential work on the properties of infinite sets.

Cantor created some new perspectives on tricky concepts such as infinity. He is most well-known for two closely related articles that he wrote concerning the relative size of infinite sets. Both of the articles were written in German and investigated the cardinality of infinite sets. The cardinality of a set is a measure of the number of elements it contains. Oh, but before we continue be aware that "natural numbers" means positive whole numbers (e.g. $1, 2, 3, 4, \ldots$) and "real numbers" means numbers on the continuum (like on a smooth number line, e.g. 0.345, 0.763, $97/100$, 1, π, etc).

The first article was published in 1874 and was entitled *Über eine Eigenschaft des Inbegriffes aller reellen algebraischen Zahlen*, which translates in English to *On a Property of the Collection of All Real Algebraic Numbers*. Mathematicians believe that in this paper Cantor successfully proved that the number of real numbers that exist is greater than the number of natural numbers that exist, even though both sets are infinitely large. In other words, the argument implies that some infinities are larger than other infinities.

The second article was published much later in 1891 and was entitled *Über eine elementare Frage der Mannigfaltigkeitslehre*, which translates in English to *On an El-*

ementary Question of the Theory of Manifolds. This was the article that introduced Cantor's famous diagonal argument technique, which he used to construct another apparent proof that the number of real numbers that exist is greater than the number of natural numbers that exist.

Each of these articles represents a different approach to reaching the same conclusion. Of the two approaches, the one in the second article (the one that uses the famous diagonal argument technique) is the most common way to introduce this concept to people. Most people have an easier time understanding it. However, Cantor's theories were the subject of much controversy and were opposed by at least several other notable mathematicians, and also by some people outside of the mathematics community.

Leopold Kronecker, Henri Poincaré, Hermann Weyl, and Luitzen Brouwer were among the most prominent mathematicians to oppose Cantor's theory. Outside of mathematics, the famous philosopher Ludwig Wittgenstein also opposed Cantor's theory. In addition, many logicians and mathematicians who lived before Cantor's time held views that likely would have stood in opposition to Cantor's views. For example, both Carl Friedrich Gauss and Aristotle would likely have not viewed Cantor's use of infinity as acceptable, because both believed that infinity could have at most a potential existence and thus that infinity could only ever be approached but never actually reached.

Cantor's theory of infinities is one of the beloved darlings of the mathematics community. It's one of those things that puts sparkles in the eyes of many mathematicians. The idea that some infinite sets could contain a larger number of elements than other infinite sets is so counterintuitive that it's shocking. It's the kind of thing that makes a great story, and since the dawn of history people have always loved a great story. It's like the mathematics equivalent of a campfire legend.

However, just because something makes a great story does not make it actually useful, nor does it even make it meaningful. You see, one of the problems with transfinite numbers is that it is physically impossible to ever actually construct one. Not only that, but it is also impossible to ever find a practical real-world application for one. In other words, transfinite numbers cannot ever generate any tangible external value for our fellow human beings. They are, in effect, worthless.

In stark contrast, George Boole's foundational studies into the basic properties of sets and how they interact with each other are profoundly useful and indeed one can scarcely even open one's eyes without seeing examples of the tangible value those theories have generated for humanity. Every time you turn on a computer, every time you perform a search on the internet, and every time you read almost any modern text on logic or mathematics, you owe at least some debt of gratitude to George Boole.

Computers might not even exist if it were not for the algebraic truth value system that Boole created. Whenever you search for something on the internet, the search engine is essentially identifying sets of websites that correspond to particular parts of your query, which it then ties together using set operations in order to discover what set of pages most closely matches the meaning of your search query as a whole.

For example, if you search for "president elected 1860" a search engine might convert that into a calculation of what webpages are common to all of the following four sets: the set of things related to the nation in which the user issuing the search query resides, the set of things related to presidents, the set of things related to elections, and the set of things related to 1860, and thereby it would return a set of likely results.

So if Cantor's theory is so esoteric and impractical, whereas Boole's theory is so groundbreaking and useful, then why is it that Cantor is the one credited with founding set theory? Well, there are a number of likely contributing factors. One of the most influential contributing factors though was probably intellectual elitism. Over time, much of the mathematics community has unfortunately become tainted by a sort of toxic fondness for "non-triviality" and "purity", i.e. by a fondness for complexity for complexity's sake regardless of any consideration of tangible merit.

Despite the fact that Boole's work was what allowed the study of logic to finally break free from syllogistic logic, and the more than 1000 years of intellectual stagnation that accompanied it, later logicians and mathematicians likely viewed Boole's work as too simple and therefore too "trivial" to deserve the credit for founding set theory. It was only after Cantor's publication of his theory of transfinite numbers that mathematicians finally came to view set theory as "non-trivial" enough to merit crediting anybody with founding it. In other words, Cantor's work tickled mathematicians' egos more than Boole's work did. Thus, Cantor was unjustly credited with founding set theory.

Another point to consider is that set theory is a branch of logic and, if you think about it, logic is really about studying the underlying laws and principles by which valid inferences can be made. Yet, if you look at Cantor's theories, you will notice that they are not even about the laws and principles of valid inference. They are about properties of infinite sets. Therefore, Cantor cannot possibly be the founder of set theory.

Transfinite set theory may be useless in real-world applications, but many mathematicians still believe that it expresses a fundamental insight into the nature of infinity. It is hard to find a mathematician who has not encountered Cantor's diagonal argument. It is one of the most beloved arguments in math. It is actually far more questionable than popularly believed though. It relies primarily on philosophically naive and sloppy use of the concept of infinity[12]. Indeed, there are actually many logicians and mathematicians who consider the idea of multiple levels of infinity to be nonsense (or at least unproductive and pointless), such as many finitists and constructivists for example.

Anyway though, before we move on from discussing the history of set theory to discussing set theory itself, there's one more person I'd like to mention who contributed to set theory as we know it today, and that person is Giuseppe Peano. Peano is probably most well-known for creating a system of axioms known as the Peano axioms, from which part of the theory of arithmetic can be derived entirely through formal logical inferences, without reference to any preexisting notion of what a number is. However, Peano also made significant contributions to set theory and to logic in general.

As you may have noticed from the earlier excerpt from Boole's *The Laws of Thought*, Boole's notation was essentially the same notation as that of numeric algebra, except that he reinterpreted it so that it could be used for logic and set theory instead. You might

---

[12] Briefly, for example: Cantor's diagonal argument is only possible by assuming the existence of "complete infinities". However, by mathematical induction, no amount of adding new elements to a set can ever make the set reach a complete infinity. At every moment the set remains finite, no matter how long the process is performed. Allowing "complete infinities" can thus be argued to be self-contradictory, a violation of mathematical induction, and hence nonsensical. Completeness and infiniteness are actually inherently incompatible properties, contrary to popular belief. Groupthink currently prevents progress though, unfortunately.

therefore be wondering at what point did the modern set notation, which in contrast to Boole's notation is quite visually distinct from numeric algebra, come into use. The answer is that modern set notation was mostly the invention of Giuseppe Peano.

Peano created the symbols for union (∪), intersection (∩), membership (∈), and existence (∃). Various other mathematicians also made contributions to the modern notation of set theory, but Peano was the one who gave it its initial form. Thus, Peano's initial notation choices are ultimately why modern set theory notation looks the way it does. Peano's notation established the look and feel to which later contributions to set theory notation would conform. In other words, Peano planted the initial seed from which modern set theory notation gradually arose.

Therefore, having now sufficiently discussed the history of set theory, we can now finally get a proper sense for its real origins. In summary: Set theory is a highly generalized and streamlined descendant of Aristotle's syllogistic logic. The person who most deserves credit for enabling the transition from syllogistic logic to set theory is George Boole. The modern notation for set theory was initiated by Giuseppe Peano. However, Georg Cantor is frequently falsely credited with single-handedly creating set theory, even though what he actually created was a small esoteric subset of set theory that has essentially zero practical value and which is based on very questionable foundations.

With the history of set theory having now been adequately covered, it is now time for us to proceed to an investigation of the actual mechanics of modern day set theory itself. We now ask ourselves: What are sets? What operations can be performed on them? What do those operations mean? What laws of inference are valid in set theory? How does set theory relate to other theories?

As you may remember, we actually already did discuss sets some in chapter 1 but only at a basic level in order to facilitate the discussion. We did not truly discuss the nature of a set in depth, nor did we discuss how different sets can relate to each other and what operations can be defined for them. For reference, the first definition of a set I gave was on page 28 and the first description of set notation I gave was on page 34.

Here was the first definition of a set that I gave:

> A **set** is an arbitrary collection of objects. The order in which the objects are listed is irrelevant and duplicates don't matter. If a set includes at least one object (i.e. if it is non-empty), then we will say that the set exists and is therefore true. Whereas, if a set does not include any objects (i.e. if it is empty), then we will say that the set does not exist and is therefore false. There is more to the definition of a set than just this, but this partial definition will suffice for now.

This definition was adequate in chapter 1 when our focus was on comparing and contrasting statement-based truth with existence-based truth, but it was not really a complete definition of what a set is. Also, normally a definition of a set would not focus so much on whether the set is empty or non-empty. I emphasized whether or not the set was empty in our first definition of a set because it helped focus attention on what was relevant to the discussion of statement-based truth versus existence-based truth. Now though, it's time to give a more complete definition of what a set is:

Definition 39. *A set is an arbitrary collection of objects, such that the identity of the objects included in it is the only thing that matters in determining the identity of the set. This implies two things: (1) the order in which objects are listed inside a set is irrelevant and (2) the number of copies of any object that exist inside a set is also irrelevant. In effect, sets are unordered arbitrary collections of uniquely identifiable objects.*

There are at least six different ways to specify a set. Each method has different advantages and disadvantages, and different situations in which it is more or less suitable. We've already used a few of these methods before. As with most math notation, there are sometimes slight variations in the notation within each method, between different locales, between different communities, and between different authors. I will use the notation I am most familiar with. Here are six different ways to specify a set:

1. **Natural Language:** Specifying a set using natural language means describing it in words using a naturally occurring spoken or written language. Some natural languages are more commonly used in scientific endeavors and terminology than others. For example, English, German, French, Greek, and Latin have all historically been especially highly influential. Some examples of sets described in natural language include things like:

    a) "the set of all people who are engineers"

    b) "all integers less than 60"

    c) "days when it rained in London"[13]

    d) "sales numbers for every print-on-demand book ever published"

2. **Predefined Reference:** When we specify a set using a predefined reference, it means simply that we refer to it by an already understood formal name or symbol. It is generally considered good practice to specify what a particular word, phrase, or symbol means in the context of your document in order to avoid any possible naming conflicts. However, some uses of particular terms and symbols are so common that authors don't bother specifying what they mean because it will likely be understood by the audience anyway. As such, it is seldom wise to reassign a common term to a new meaning. Do so only with strong justification. Some examples of predefined references include:

    a) $\mathbb{Z}$, meaning the set of all integers (i.e. all whole numbers, whether positive, negative, or zero)[14]

    b) $\mathbb{R}$, meaning the set of all reals (i.e. all numbers on a continuum, such as on a number line)

    c) $\varnothing$, meaning the empty set (i.e. the set containing nothing)

    d) $\mathbb{Q}$, meaning the set of all rational numbers (i.e. all integer fractions, i.e. all numbers that can be written in the form a/b where a and b are integers)[15]

---

[13] Or, equivalently, as a cheeky Brit would say: "all days".

[14] The symbol $\mathbb{Z}$ is derived from the German word for numbers: "Zahlen".

[15] The symbol $\mathbb{Q}$ is derived from the word "quotient", which means "the result of dividing one thing by another".

3. **List Notation:** This method is arguably the most direct way to specify a set. When we specify a set using list notation it means that we specify it by writing down a complete list of what objects are included in it. The advantage of this method is that it is completely explicit and leaves nothing to the imagination. The disadvantage is that listing all the objects included in a set can be impractical if the set is too large. In fact, for infinite sets it is more than merely impractical; it is impossible. To indicate that a list is a set, rather than an ordered list, we surround it with curly brackets, which look like this: { }. Some examples include:

    a) $\{2, 3, 5, 7\}$

    b) {Marcus Aurelius, Epictetus, Zeno, Seneca}

    c) {poison ivy, hemlock, nightshade}

    d) {algebra, geometry, statistics, calculus}

4. **List Notation With Omission:** This method is almost the same as the basic list notation above, except that we allow the list to contain an omission of some finite or infinite number of elements, if it is still clear from the context what the missing objects should be. The omitted objects in the set are represented by an ellipsis (...). Some examples include:

    a) $\{0, 2, 4, 6, \ldots\}$

    b) $\{5, 10, 15, \ldots, 90, 95, 100\}$

    c) $\{1, 2, 4, 8, 16, 32, 64, \ldots\}$

    d) $\{(x+1), (x+1)^2, (x+1)^3, (x+1)^4, \ldots\}$

5. **Set Builder Notation:** Set builder notation specifies a set by stating what logical properties objects must satisfy in order to be included in that set. Set builder notation has the form $\{x : P(x)\}$ or $\{f(x) : P(x)\}$ where x is any arbitrary dummy variable, $P(x)$ is any condition applied to x to test whether it is included in the set or not, and $f(x)$ is any arbitrary function applied to whatever objects pass the test given by $P(x)$, in order to transform those objects in some useful way. Multiple variables and conditions are allowed. The ":" is read as "such that". $P(x)$ is often expressed using logical operators. Some examples include:

    a) $\{n : (n = 2k) \land (k \in \mathbb{Z}) \land (k \geq 0)\} = \{0, 2, 4, 6, \ldots\}$

    b) $\{b : \texttt{IsBook}(b) \land \texttt{IsRare}(b)\}$

    c) $\{(x, y) : (x \in \mathbb{R}) \land (y \in \mathbb{R}) \land (xy = \pi)\}$

    d) $\{s^2 : (s \in \mathbb{Z}) \land (s \geq 0)\} = \{0, 1, 4, 9, 16, 25, 36, \ldots\}$

6. **Set Operation Expressions:** Building every set from scratch like in some of the previous methods can quickly become tedious. For that reason, it is often helpful to be able to specify a set according to how it logically relates to other sets that you already know about. To accomplish this, we can use set operations. The set operations will be explained later, but here's a few examples anyway:

a) $\mathbb{R} \cap \neg \mathbb{Q}$
b) $(A \cap B) \cup (C \cap D)$
c) $\neg(X \cap Y \cap Z) \cap \neg\{1, 2, 4, 8, 16, \ldots\}$
d) $\mathbb{C} \cap \{x : (x = ab) \wedge ((ab)^2 = -a^2) \wedge (a \in \mathbb{Z})\}$

There are some other more unconventional ways to specify a set, ranging from high tech methods, such as by dynamically generating and displaying a set using some kind of data visualization program on a computer, to low tech methods, such as having an actual physical set of objects and gesturing towards it. However, the six methods listed above are generally far more practical for most use cases in logic and math.

Set theory is technically a branch of logic, even though the name "set theory" might lead you to believe that it is something outside of logic. In fact, a strong alternative term for "set theory" would be "set logic", which arguably more clearly illuminates the relationship between it and the other branches of logic. However, I will use the standard term "set theory" in this book though, for the sake of communication, but I think perhaps "set logic" could potentially be a superior term, maybe.

Anyway, now that I've given you an overview of the different possible ways to specify a set, let's learn about the basic set operators and how they can be applied. There are really two different types of set expressions: (1) statement expressions and (2) value expressions. A statement expression is an assertion that a particular relationship is true. In contrast, a value expression returns a value corresponding to whatever matches the constraints it specifies, but does not make any assertions about anything.

In set theory, statement expressions are interpreted either (1) as simple assertions with no return value or (2) as expressions that return a truth value that depends on whether or not the assertion is true. Which interpretation is used depends on the context in which the statement expression occurs. In a context where the statement expression stands on its own it is generally interpreted according to (1), whereas in a context where the statement expression is located somewhere where a truth value is expected it is generally interpreted according to (2).

For each of the operators of set theory, I will indicate whether they form statement expressions or value expressions, and what they mean. To begin, let's start with an operator that will allow us to talk about what is or is not included within a particular set. I have already used this operation in some of the examples, so you might already have inferred what it means, but here's our official definition:

Definition 40. *The **membership** operator, also known as the **element-of** operator, allows us to assert or test whether or not an object is in a set. An object that is included in a set is called an element of that set. The membership operator is a statement expression. It represents either (1) an assertion that an element is in a set or (2) a truth value corresponding to whether or not an element is in a set, depending on the context in which it occurs. The symbol for the membership operator is "$\in$" and it is a binary operator that takes an object on the left side and a set on the right side.*

For example, suppose we have the expression "$x \in \mathbb{Z}$". This expression may be read as "x is a member of $\mathbb{Z}$", "x is an element of $\mathbb{Z}$", or "x is in $\mathbb{Z}$". It either (1) asserts that x is an integer or (2) returns true if x is an integer else false. The operator can also be written backwards, in which case the operands are also flipped. For example, "$x \in \mathbb{Z}$" could alternatively be written as "$\mathbb{Z} \ni x$". However, this is rarely seen in practice and "$x \in \mathbb{Z}$" is almost always the format that is used.

Let's do some examples:

1. Suppose that "Mortals" is the set of all mortals, that "Immortals" is the set of all immortals, and that "Socrates" is the famous Greek philosopher of the same name. Then "Socrates $\in$ Mortals" is true and "Socrates $\in$ Immortals" is false.

2. Suppose "Primes" is the set of all prime numbers. Suppose I then asserted that "$x \in$ Primes". Then this would be an assertion that x is a prime number.

3. Consider the set $\{n : (n = 2k) \wedge (k \in \mathbb{Z}) \wedge (k \geq 0)\}$ from an earlier example. This set includes all objects for which all three of the criteria it lists are simultaneously true. The expression "$k \in \mathbb{Z}$" is true whenever k is an integer. Combine this with the criteria that $n = 2k$, which specifies that n is twice whatever k is, and the criteria that $k \geq 0$, which specifies that k is positive or zero, and the net effect is that $\{n : (n = 2k) \wedge (k \in \mathbb{Z}) \wedge (k \geq 0)\}$ specifies the set of all even numbers.

4. Consider the set $\{(x, y) : (x \in \mathbb{Z}) \wedge (y \in \mathbb{Z})\}$. This is the set of all pairs of objects (x, y) such that both x and y are integers. In other words, this set is the set of all integer coordinates on a two dimensional plane. For example, it includes the coordinates $(0, 0)$, $(4, 7)$, and $(-12, -1)$, but does not include the coordinates $(5, \sqrt{2})$, $(2\pi, 0.6925)$, and $(1/2, 7/13)$.

Anyway, now that we've got a way to talk about whether or not an object belongs to a set, let's move on to the other operators. Next let's discuss some of the most common value expression operators for set theory. There are not very many basic set operators, but that's okay because, just like with classical logic, even this small number of operations can be surprisingly expressive. Here are the new operator definitions:

Definition 41. *The **complement** operator, henceforth also known as the **set negation** or **set not** operator, allows us to construct a new set that includes all of the objects that another set does not include. It is effectively the set theory equivalent of the logical negation operator. It is a value expression, and hence it simply returns a value without asserting anything about that value. The complement of an arbitrary set A is often written as A^C or as A', but I prefer to use the notation $\neg A$.*

The reason why I prefer $\neg A$ is because (1) it illustrates the close relationship between the set complement operator and the logical negation operator better and (2) the logical negation symbol "$\neg$" is more distinct. The C and prime (ʹ) symbols have many other uses in mathematics besides the set complement. The prime symbol in particular is an extremely overused symbol.

The prime symbol often means roughly "related or slightly modified variant of". Thus, A' could mean a related variant of some other set named A, without intending the set complement. As such, use of the prime symbol for the set complement is an unacceptably poor choice. The letter C is also a poor choice because you could easily have a set named C, which you might later want to take the complement of, leading to the confusing expression C^C. Another reason it is a bad choice is because C could potentially be any arbitrary letter variable standing for any arbitrary value, in addition to representing the set complement, thereby creating naming conflicts.

Anyway, suppose for example that we have an expression $\neg A$. This expression may be read as "not A", "A complement", or "the complement of A". It returns a set which contains all of the objects that are not contained in A, taken from some implicitly or explicitly understood universal set that represents "all things under consideration".

Definition 42. *The **intersection** operator, henceforth also known as the **set conjunction** or **set and** operator, allows us to construct a new set that includes all objects that satisfy the membership constraints of both of two other sets simultaneously. It is effectively the set theory equivalent of logical conjunction. It is a value expression, and hence it simply returns a value without asserting anything about that value. The intersection of two sets A and B is written as $A \cap B$ and is read as "A intersect B" or as "A and B". An object x will be in the set $A \cap B$ only if both $x \in A$ and $x \in B$ are true.*

Definition 43. *The **union** operator, henceforth also known as the **set disjunction** or **set or** operator, allows us to construct a new set that includes all objects that are in either or both of two other sets. It is effectively the set theory equivalent of logical disjunction. It is a value expression, and hence it simply returns a value without asserting anything about that value. The union of two sets A and B is written as $A \cup B$ and is read as "A union B" or as "A or B". An object x will be in the set $A \cup B$ whenever either $x \in A$ or $x \in B$ (or both) are true.*

There are two special sets with unique properties that are critical for a full understanding of set theory. It seems wise to first discuss them before we explore more examples of constructing sets using the set operators. Within any particular fixed system of sets, these two special sets represent the smallest and the largest sets in existence within that system[16]. In effect, within any given fixed domain, set theory is bounded by a least set and a greatest set. This is in stark contrast to number systems that allow infinite numbers, wherein least or greatest bounds may not even exist. We have already discussed one of these two special sets at great length. However, we haven't discussed the other special set much. Anyway, here are the definitions for them:

Definition 44. *The **universal set**, also sometimes referred to more briefly as simply the **universe**, is the set of all things that exist within a given system under consideration. In set theory the universal set essentially represents the concept of "everything". The universal set is also the closest analog to the value true from classical logic, in that the universal set behaves the most like the value true would in classical logic, relative to expressions of set theory and classical logic that are structurally similar. The universal set is usually represented by some symbol that is a variant of the letter U. The most*

---

[16] This statement might be alarming to a mathematician who is already familiar with the standard axiomatic treatment of set theory, known as ZFC, wherein the axioms effectively forbid the construction of a greatest set. However, later in this book we will modify set theory in such a way that this problem disappears. The distinction won't matter until later, so just humor me for now.

common representation is ⋃, but I prefer the symbol $\mathcal{U}$ because it makes it more visually distinct from the union operator (∪) and hence easier to keep track of with your eyes.

It is permissible to constrain the universal set to include only the set of objects you are interested in, thereby excluding objects from outside that set, so that there are less things you have to think about, and so that set negation is constrained to a relatively limited domain so as to make it more manageable. For example, if $\mathcal{U}$ was completely unconstrained and included all conceivable objects, then a set like $\neg \mathbb{Z}$ would actually include more than just numbers. It would even include things like cats, dogs, buildings, and planets, because all such things are indeed not integers.

Thus, it is common to implicitly or explicitly constrain the universal set to a more limited set than simply "the set of everything". Alternatively, instead of limiting the universal set, you can use set conjunction in order to constrain each individual set negation. For example, $\mathbb{R} \cap \neg \mathbb{Z}$ would constrain the set negation $\neg \mathbb{Z}$ so that the result is limited to numbers on the continuum that are not integers, rather than to all conceivable objects that are not integers. People often fail to precisely specify what universal set they are using though, so you will often have to infer it from context.

Definition 45. *The **empty set** is a set which includes no objects. Because there is only one way for a set to be empty, the empty set is therefore unique. In set theory the empty set essentially represents the concept of "nothing". The empty set is also the closest analog to the value false from classical logic, in that the empty set behaves the most like the value false would in classical logic, relative to expressions of set theory and classical logic that are structurally similar.*

The empty set is usually represented by either ∅, ∅, or { }. I prefer to use the symbols ∅ or { }, because ∅ looks too much like the number zero. The empty set remains the same regardless of what the universal set is. However, the set negation of the empty set changes depending on what the universal set is. The set negation of the empty set is always equal to whatever set is currently considered to be the universal set.

We first discussed the empty set in chapter 1 on page 36. However, in contrast, this is more or less our first time discussing the universal set, although we did briefly discuss the concept of a "universe of discourse" in chapter 1, which is a related concept. Whereas most sets we concern ourselves with in set theory are completely arbitrary and context-dependent, the empty set and the universal set are unique and fundamental. It is difficult to fully discuss the laws and properties of set theory without having the concept of the empty set and the universal set at one's disposal.

Anyway, now that we have some more operators and concepts to work with, let's do some more examples:

1. Suppose that X is $\{a,b,c,d\}$. Suppose also that Y is $\{c,d\}$. Then $X \cap Y$ is $\{c,d\}$, $X \cup Y$ is $\{a,b,c,d\}$, and $X \cap \neg Y$ is $\{a,b\}$.

2. Suppose that A is the set of all scientists, that B is the set of all physicists, and that C is the set of all chemists. Then $A \cap (B \cup C)$ is the set of all scientists who are either physicists or chemists (or both), and $(B \cap C) \cup \neg A$ is the set of all people who either qualify as both physicists and chemists or who are not scientists at all.

3. Suppose that M is the set of all moons, that S is the set of all man-made satellites, that E is the set of all objects orbiting Earth, and that J is the set of all objects orbiting Jupiter. Then $M \cap E \cap J$ would be the set of all moons orbiting both Earth and Jupiter, but the set would be empty because no such moons exist. In contrast, $S \cap E$ would be the set of all satellites orbiting Earth and $M \cap \neg J$ would be the set of all moons that don't orbit Jupiter, and both would be non-empty.

4. Suppose that P is the set of prime numbers, that E is the set of even numbers, that O is the set of odd numbers, and that S is the set of square numbers[17]. Then $P \cap E$ would be the set of even primes (i.e. {2}), $P \cap S$ would be the set of square primes (i.e. { }), and $O \cap S$ would be the set of odd square numbers (which would be non-empty and fairly diverse).

5. Suppose that the universal set $\mathcal{U}$ is the set of all phone calls, that T is the set of all telemarketing phone calls, and that A is the set of all annoying phone calls. Then $T \cap \neg A$ is the set of all telemarketing phone calls that are not annoying. As most people know from experience, this set is the empty set. In contrast, $\neg T$ would be the set of all phone calls that are not telemarketing phone calls. Without our explicit definition of $\mathcal{U}$ though this might not be clear and we'd have to guess.

These operators cover a lot of cases, as you can see from the examples given above, but there are still at least several more operators we will need to discuss in order to have a more complete vocabulary for expressing basic set relationships. The next few operators we will be discussing are critical for enabling one to talk about the structural relationships between sets and are especially useful. Here are the new definitions:

Definition 46. *The **subset** operator, also occasionally called the **inclusion** operator, allows us to claim that all of the objects that are included in one set are also included in another set. It is a statement expression, and therefore makes an assertion about how sets relate to each other. It represents either (1) an assertion that one set is a subset of another or (2) a truth value corresponding to whether or not the assertion would be true, depending on the context in which it occurs.*

If one set A is a subset of another set B, then we write $A \subseteq B$, which is read as "A is a subset of B" or as "A is included in B". If any member of A is not in B then $A \subseteq B$ must therefore be false. The empty set $\emptyset$ is a subset of every set, including itself. The universal set $\mathcal{U}$ is not a subset of any set, except itself.

*If A and B are equal then $A \subseteq B$ is still true. The operator requires only that all objects included in A are included in B, which does not exclude the possibility that the two sets are also equal. However, there is a related operator called the **proper subset** operator which requires that the set A is also not equal to B. The proper subset operator is represented by the symbol "$\subset$". Thus, to say that A is a proper subset of B, you would write $A \subset B$, which is read as "A is a proper subset of B". Notice that the symbols used for subset ($\subseteq$) and proper subset ($\subset$) are closely analogous to the symbols used for less than or equal ($\leq$) and less than ($<$) respectively.*

---

[17] A square number is any integer that is equal to some other integer squared. Some examples include 25, 16, and 100, which are equivalent to 5^2, 4^2, and 10^2 respectively.

Definition 47. The **superset** operator allows us to claim that one set includes all of the objects in another set. In other words, it is effectively the same thing as the subset operator, except that it is stated in terms of the larger set instead of in terms of the smaller set. Any subset relation can be expressed as a superset relation, and vice versa. The choice of which one to use is a matter of taste and generally depends on whether you want to draw more attention to the subset or to the superset.

The superset operator is a statement expression. Unsurprisingly, if one set A is a superset of another set B, then we write $A \supseteq B$, which is read as "A is a superset of B" or as "A includes B". If A does not include every member of B then $A \supseteq B$ must therefore be false. The empty set $\emptyset$ is not a superset of any set, except itself. The universal set $\mathcal{U}$ is a superset of any set, including itself.

If A and B are equal then $A \supseteq B$ is still true. Just like with the subset operator, there is a related operator called the **proper superset** operator which requires that the set A is also not equal to B. The proper superset operator is represented by the symbol "$\supset$". Thus, to say that A is a proper superset of B, you would write $A \supset B$, which is read as "A is a proper superset of B". Notice that the symbols used for superset ($\supseteq$) and proper superset ($\supset$) are closely analogous to the symbols used for greater than or equal ($\geq$) and greater than ($>$) respectively.

Definition 48. The **set equivalence** operator, also known as simply the **equality** or **equivalence** operator, allows us to claim that two sets include exactly the same objects, i.e. that two sets are in fact the same set. It is a statement expression, and therefore makes an assertion about how sets relate to each other. It represents either (1) an assertion that two sets include the same objects or (2) a truth value corresponding to whether or not the assertion would be true, depending on the context in which it occurs.

If one set A is equivalent to another set B, then we write $A = B$, which is read as "A equals B", "A is equal to B", or "A is equivalent to B". If either of A or B includes an object which the other does not include then the two sets cannot be equal. Every set is equal to itself. If both $A \subseteq B$ and $B \subseteq A$ are true, then $A = B$ must also be true. Likewise, if $A = B$ is true, then both $A \subseteq B$ and $B \subseteq A$ must be true.

As usual, we should also explore some concrete examples to help verify our understanding and to clarify concepts. When one has not taken the time to explore concrete examples, it is easy to trick oneself into thinking that one understands something when in fact one does not. As such, let's use these new operators in some examples:

1. Suppose that X is $\{a, b, c, d\}$. Suppose also that Y is $\{c, d\}$. Then $Y \subseteq X$, $Y \subset X$, $X \supseteq Y$, $X \supset Y$, and $X \neq Y$ are all true.

2. Let S be the set of only Socrates, let M be the set of all men, and let I be the set of all immortals. We know that $M \subseteq \neg I$ (i.e. that all men are mortal) and that $S \subseteq M$ (i.e. that Socrates is a man). Therefore, we may also infer $S \subseteq \neg I$.

3. Let W be the set of all wealthy people, let P be the set of all poor people, and let E be the set of all ethical people. Then, if one is not blinded by prejudice, it is easy to see that $W \cap E$, $W \cap \neg E$, $P \cap E$, and $P \cap \neg E$ are all large non-empty sets.

From the properties of the sets, it is also easy to see that $W \cap P = \emptyset$, $E \cap \neg E = \emptyset$, $(W \cap E) \subseteq W$, $(W \cap \neg E) \subseteq W$, $(P \cap E) \subseteq P$, and $(P \cap \neg E) \subseteq P$.

4. Let A be the set of all places where the average daily chance of rain is higher than 20%, and let B be the set of all places where the average daily chance of a significant amount of droplets of water falling from clouds in the sky is higher than 20%. Then, because $A \subseteq B$ and $A \supseteq B$, we can conclude that $A = B$.

5. In traditional mathematical notation, $\mathbb{N}$ represents the set of all natural numbers (i.e. positive whole numbers), $\mathbb{Z}$ represents the set of all integers (i.e. whole numbers), $\mathbb{Q}$ represents the set of all rational numbers (i.e. integer fractions), $\mathbb{R}$ represents the set of all real numbers (i.e. numbers on a continuum, such as a number line), and $\mathbb{C}$ represents the set of all complex numbers (i.e. numbers that are the sum of a real number and an imaginary number). Therefore, the following composite statement is true: $\mathbb{N} \subset \mathbb{Z} \subset \mathbb{Q} \subset \mathbb{R} \subset \mathbb{C}$. (Technically this is notation abuse, since we did not actually define how to chain subset expressions together like this. However, it is much more readable and concise in this format.)

We have now discussed definitions and examples for all of the following set concepts: membership ($\in$), complement ($\neg$), intersection ($\cap$), union ($\cup$), universal set ($\mathcal{U}$), empty set ($\emptyset$), subset ($\subseteq$), proper subset ($\subset$), superset ($\supseteq$), proper superset ($\supset$), and equivalence ($=$). These are the most basic concepts of set theory. However, there are also some other less common set operators that exist. We might as well strive for completion and define these less common ones too:

Definition 49. *The **set difference** operator, henceforth also known as the **set removal** operator, allows us to specify a set that includes all objects which are members of one set but which are not members of another set. It is sort of the set theory equivalent of subtraction, but with some important differences. It is a value expression, and hence it simply returns a value without asserting anything about that value.*

The set difference of two arbitrary sets is commonly written either as $A - B$ or as $A \setminus B$. However, I prefer the notation $A - B$ because (1) it more clearly indicates the analogy between set difference and subtraction, (2) because the division-like symbol "$\setminus$" has no apparent conceptual relationship to the concept of set difference, and (3) because the meaning of $A - B$ is more easily intuitively grasped by those unfamiliar with it and is more memorable. The expression $A - B$ can be read as "A minus B", "A excluding B", "A except B", or "A without B".

The set difference $A - B$ is equivalent to the expression $A \cap \neg B$, therefore making it redundant. In order to not have to memorize the special properties of $A - B$, it is often useful to convert any instances of $A - B$ into instances of $A \cap \neg B$. Converting $A - B$ to $A \cap \neg B$ often makes it easier to work with, because it can thereafter be treated as simply a set conjunction rather than as a separate unique operator. Unlike numeric subtraction, which allows negative numbers, the set difference can never return a result with fewer members than $\emptyset$. Also, the set difference $A - B$ ignores the part of B that is not in A. However, the set difference nonetheless feels very similar to subtraction.

Definition 50. The **symmetric difference** operator, henceforth also known as the **set exclusive disjunction**, **set exclusive-or**, or **set xor** operator, allows us to specify a set that includes all objects that are members of either of two sets but are not members of both sets. It is effectively the set theory equivalent of the classical exclusive-or operator. It is a value expression, and hence it simply returns a value without asserting anything about that value. The symmetric difference of two sets A and B is historically written inconsistently as $A \triangle B$, $A \ominus B$, or $A \oplus B$. It can be read as "A xor B", "A exclusive-or B", or "the symmetric difference of A and B".

The symbols for this operator are not well standardized, in part because the operator is rarely used, and in part because these symbols are all very arbitrary and are not very fitting or memorable. Therefore, I will introduce a new symbol that is a more natural and intuitive fit for representing the concept of the set exclusive-or: "⊌". Under this new notation, the set exclusive-or of two sets A and B will be written as $A ⊌ B$.

Notice that this symbol is the same symbol as the one used for union (∪), except that it has an "×" placed inside it to indicate that it is the exclusive variant of a set or. This symbol is much more memorable and its relationship to the other operators is much more obvious than the other historically used symbols. The historical term "symmetric difference" also has a less clear meaning than the other alternative terms, such as "set exclusive-or" and "set xor", and thus should probably not be used.

The set exclusive-or $A ⊌ B$ is equivalent to the expression $(A \cup B) \cap \neg(A \cap B)$. It is also equivalent to $(A \cap \neg B) \cup (B \cap \neg A)$. Just like with the set difference operator, it is often more useful to reduce the set exclusive-or operator to an equivalent expression that uses only intersection (∩), union (∪), and negation (¬) so that you do not have to treat set exclusive-or as a unique new operator whose rules must be memorized.

Definition 51. The **set nor** operator, henceforth also known as the **set negated disjunction** operator, allows us to specify a set that includes all objects that are members of neither of two sets. It is effectively the set theory equivalent of the logical nor operator. It is a value expression, and hence it simply returns a value without asserting anything about that value. The set nor of two sets A and B is written as $A ↓ B$, and can be read as "A nor B" or as "the negated disjunction of A and B".

This set operator actually does not appear to exist anywhere in the literature, even though it is a rather obvious extension of the logical nor operator into an analogous operator for set theory. As such, this may be the first time it has been formally defined and given a symbol. As you may recall, the logical nor operator is represented by the symbol "↓". The reason why "↓" represents nor is probably because the head of the arrow looks like "∨", which is the symbol for logical or, but the vertical line disrupts it, thus implying a variation on logical or, specifically logical nor. I have applied similar reasoning, by analogy, in choosing the symbol "⍗" to represent set nor. The "arrow head" of "⍗" looks like the union (a.k.a. set or) operator. Choosing the symbol in this way makes it more memorable and intuitive.

The set nor $A ⍗ B$ is equivalent to the expression $\neg(A \cup B)$. As is typical of these kinds of redundant operators, it is often more useful and convenient to reduce any instance of this operator to its constituent basic set operators. Instances of $A ⍗ B$ should generally be converted into instances of ¬(A ∪ B) in order to eliminate any need to learn unique new rules for manipulating expressions involving set nor.

Definition 52. *The **set nand** operator, henceforth also known as the **set negated conjunction** operator, allows us to specify a set that includes all objects that are not members of both of two sets. It is effectively the set theory equivalent of the logical nand operator. It is a value expression, and hence it simply returns a value without asserting anything about that value. The set nand of two sets A and B is written as A ⇑ B, and can be read as "A nand B" or as "the negated conjunction of A and B".*

Like the set nor operator, this set operator does not appear to exist anywhere in the literature, even though it is a rather obvious extension of the logical nand operator into an analogous operator for set theory. As such, this may be the first time it has been formally defined and given a symbol. As you may recall, the logical nand operator is represented by the symbol "↑". The reason why "↑" represents nand is probably because the head of the arrow looks like "∧", which is the symbol for logical and, but the vertical line disrupts it, thus implying a variation on logical and, specifically logical nand. I have applied similar reasoning, by analogy, in choosing the symbol "⇑" to represent set nand. The "arrow head" of "⇑" looks like the intersection (a.k.a. set and) operator. Choosing the symbol in this way makes it more memorable and intuitive.

The set nand A ⇑ B is equivalent to the expression ¬(A ∩ B). As is typical of these kinds of redundant operators, it is often more useful and convenient to reduce any instance of this operator to its constituent basic set operators. Instances of A ⇑ B should generally be converted into instances of ¬(A ∩ B) in order to eliminate any need to learn unique new rules for manipulating expressions involving set nand.

Definition 53. *The **cardinality**, henceforth also known as the **count**, of a set is the number of objects that are included in that set. It is a value expression, and hence it simply returns a value without asserting anything about that value. Unlike the other set operators we have discussed so far, this one returns a number instead of a set. The meaning of the term "count" is more clear and intuitive than "cardinality", even though "cardinality" is the more common term. Finite sets can obviously be counted, but even for infinite sets "infinite count" would be clearer than "infinite cardinality".*

Obscure words shouldn't be used when simpler ones already exist that are more widely understood. Avoiding obscurity and embracing clarity is one of the defining characteristics of good communication. Notation should be based on expressiveness and clarity, not merely on blind adherence to tradition and social momentum. Thus, I will henceforth usually refer to the "cardinality" of a set as the "count" instead.

The count of a set A is most often written as |A|. There are also a few other miscellaneous notations that are occasionally used. Although very common, the notation |A| is not a very good notation. The vertical bars in |A| are frequently used to represent the absolute value of a number, i.e. the magnitude of that number, ignoring the positive or negative sign. Also, using a parenthesis-like delimiter syntax with vertical bars, instead of using a normal unary operator or function notation, is unnecessary here. The other miscellaneous notations for the count aren't that good either. Therefore, I will define a new notation convention of simply writing Count(A) *to mean the count of A. There is no reason to obfuscate such a simple concept.*

For finite sets, the count of a set is just the number of objects included in that set. However, for infinite sets, the count is harder to define and requires care. All infinite sets do include an infinite number of objects, and hence the count is indeed infinite. However, if you remember, earlier I mentioned that many mathematicians believe that some infinities are greater in size than other infinities, based on Cantor's diagonal argument or some other similar argument. Many finitists and constructivists disagree though.

Definition 54. *The **cartesian product** of two sets is the set of all possible binary pairings of the members of two sets. In other words, it takes each individual member of the first set, pairs it to every member of the second set, and returns a set of objects that represent all of those possible pairings of the members of the two sets. It is a value expression, and hence it simply returns a value without asserting anything about that value. The cartesian product of two sets A and B is written as $A \times B$, and can be read as "the cartesian product of A and B".*

In a cartesian product, each pair of arbitrary objects a and b, where a is a member of A and b is a member of B, are bound together as a pair. Such a pair is typically represented as (a, b). The pairs created in this way are ordered pairs, thus the pair (a, b) is not the same as the pair (b, a), unless $a = b$. If set A and B are finite, and set A has m members and set B has n members (i.e. if $\text{Count}(A) = m$ and $\text{Count}(B) = n$), then $A \times B$ will have mn members (i.e. $\text{Count}(A \times B) = \text{Count}(A) \cdot \text{Count}(B) = mn$).

There is also a variant of cartesian product that accepts an arbitrary number of input sets, which is often written as $A_1 \times A_2 \times A_3 \times \ldots \times A_n$, and which is intended to return ordered lists corresponding to all of the possible combinations of the members of the participating sets, yielding a set containing lists of the form $(a_1, a_2, a_3, \ldots, a_n)$.

However, contrary to popular belief, the way the cartesian product is defined by pairing actually makes this generalized operation ambiguous. If you performed $A \times B \times C$ as $(A \times B) \times C$ in the same way as in the original binary cartesian product operator then you would actually end up with ordered pairs of the form $((a, b), c)$. Notice that this is actually a binary ordered pair whose first member is itself another binary ordered pair. Similarly, $A \times (B \times C)$ would result in pairs of the form $(a, (b, c))$. In contrast though, you could instead interpret $A \times B \times C$ as yielding ordered lists of the form (a, b, c), which is probably more desirable in most cases.

In order to make it possible to distinguish between these two interpretations I will split the definition of the cartesian product into two new operator definitions. I will also give each of these new operators names and symbols that are more descriptive, more memorable, and more fitting than that of the cartesian product. Clarity is important.

*The first variant will be called the **binary pairing** operator, and will only yield binary pairs. It will not generate flat lists. The binary pairing of two sets A and B will be written as $A \mathbin{\unicode{x00F8}} B$, and can be read as "A binary paired with B". The symbol "$\unicode{x00F8}$" was chosen because it looks like a pair of parenthesis bound together, which is reminiscent of the parentheses we use to enclose binary pairs as we bind them together. When chained, as in $A_1 \mathbin{\unicode{x00F8}} A_2 \mathbin{\unicode{x00F8}} A_3 \mathbin{\unicode{x00F8}} \ldots \mathbin{\unicode{x00F8}} A_n$, each binary pairing operator will be evaluated individually from left to right, successively generating new sets of binary pairs.*

*The second variant will be called the **concatenation** operator, and will join each pair of members of the two input sets together via flat list appendment instead of nested* binary pairing. The concatenation of two sets A and B will be written as A ∽ B, and can be read as "A concatenated with B", "A appended by B", or "B appended to A". The symbol "∽" was chosen because it is visually reminiscent of two things being connected seamlessly together. When chained, as in $A_1 \backsim A_2 \backsim A_3 \backsim \ldots \backsim A_n$, each concatenation operator will be evaluated individually from left to right, successively generating new sets of arbitrarily long ordered lists.

With the original sloppy definition of cartesian product behind us, we can now properly specify our intent by correctly selecting either binary pairing or concatenation. Allowing a definition to remain ambiguous greatly increases the chances of committing an equivocation fallacy, confuses readers, and obscures opportunities to explore new concepts. It is important to always strive to have precise and unambiguous definitions.

Definition 55. *The **power set** of a set is the set whose members are every possible subset of that set. It is a value expression, and hence it simply returns a value without asserting anything about that value. The power set of a set A is usually written in function notation using some variant of the letter P. Thus, for example, one might write P(A) or 𝒫(A) to represent the power set of A. However, the letter P is a common variable name in mathematics. The letter P could stand for practically anything, and thus it is a poor choice of notation. A good notation should be descriptive, memorable, and unambiguous. Therefore, I will instead write the power set of A as* PowerSet(A).

The power set operator is used rarely enough that having a longer name for it will generally be acceptable. Clarity of communication is usually more important than extreme conciseness, in most cases, in good communication. The balance generally only tips in favor of extreme conciseness over clarity when the operation is so common that the speed at which you can express it becomes the limiting factor in its usefulness. I've never been a big fan of the culture of obscurity (e.g. too many cryptic single-letter names, etc) that unfortunately pervades much of the mathematics literature.

Anyway though, if a set A contains n members then the power set of A will contain 2^n members. In other words, $\text{Count}(\text{PowerSet}(A)) = 2^{\text{Count}(A)}$. *The reason why is because, for any given subset in the power set, each member of A is either included or not, which means there are two options for every member of A: (1) include that member in the subset or (2) don't include that member in the subset. Thus, determining the number of subsets in the power set is the same as determining how many different ways there are to include or not include all the members of the set in a subset.*

Having now covered the most basic operators and concepts of set theory, plus a bit extra, our vocabulary for set theory is now more diverse and useful. However, there are still a few other pedantic technicalities in the literature that are often considered important for having a proper in-depth understanding of set theory that we have not discussed here. For example, I have not discussed the ZFC axioms.

However, such technicalities are not necessary for most of the things we will be discussing in this book, nor is the implied ideology embodied by the standard view even logically necessary. There are other viable alternatives. We do not want to make premature assumptions in this book. We want to keep our perspective fresh, unbiased, and unencumbered by fallacies of authority, tradition, or popularity. I will only mention esoteric technicalities such as ZFC (etc) when such things become relevant.

In any case, as usual, before we move on to new material we should first make sure that our understanding of what we have just introduced is accurate by going through at least a few concrete examples. Theory is all well and good, but concrete examples are necessary to make sure that ideas truly sink in. There's a limit to how well abstract language can hone in on a precise concept. Concrete examples fill that gap quite effectively. Anyway, here's our next batch of examples:

1. Let C be the set of all people who went to college for any length of time. Let D be the set of all people who dropped out of college. Let T be the set of all people who went to college only for training purposes and who were not actually pursuing degrees. Let G be the set of all people who graduated college. Then $(C - D) - T$ is the set of all people who went to college pursuing degrees but did not drop out. From our knowledge of the nature of G, we can therefore infer that $(C - D) - T = G$. We could also convert the set difference operators in this expression to intersections and negations. Thus, we can infer that $(C - D) - T = G$ is equivalent to $(C \cap \neg D) \cap \neg T = G$. Intuitively, this makes sense.

2. Let M be the set of all clothing designed to fit men. Let W be the set of all clothing designed to fit women. Therefore, the set $M \cap W$ is the set of all clothing designed to fit both men and women, i.e. the set of all unisex clothing. Let $U = M \cap W$ to represent this set. Suppose that a dance is being held, and suppose that many of the people attending are hoping to attract a member of the opposite gender. Under such circumstances, it is probably wise to wear clothing designed to suit one's gender as well as possible, and thus unwise to wear unisex clothing. Therefore, people attending the event should probably wear primarily clothing from the set $M \uplus W$, which represents clothing that is either designed to fit men or designed to fit women, but not designed to fit both.

3. Suppose that a married couple has just sat down at a table at a fancy restaurant. The husband's name is Albert and the wife's name is Elsa. Suppose that Albert and Elsa want to choose their meals in such a way that if either of them does not finish their meal, it will still be something that the other person would like, thereby reducing the chances of wasting food. Suppose that Albert hates all seafood and that Elsa is allergic to anything containing peanuts. Let F be the set of all food available at the restaurant. Let S be the set of all seafood. Let P be the set of all food containing peanuts. In order for Albert and Elsa to satisfy their criteria for meals, they need to select food which is in neither of the two sets S and P, but which is still available on the menu. Therefore, Albert and Elsa need to select food from the set $F \cap (S \downarrow P)$, which represents all food available at the restaurant that is neither seafood nor contains peanuts.

4. Suppose that there is an employee who works at a used book warehouse whose job it is to determine whether or not a given used book is of acceptable quality for resale. The rules of the company that owns the warehouse state that a used book should only be considered to be of acceptable quality if it does not have both a broken spine and lots of highlights. In other words, used books with only a broken

spine or lots of highlights are considered acceptable, but used books with both broken spines and lots of highlights are not. Let B be the set of all books received by the used book warehouse. Let S be the set of all books with broken spines. Let H be the set of all books with lots of highlights. Then $B \cap (S \uparrow H)$ represents the set of all books that will be considered acceptable by the used book warehouse.

5. Suppose that P is the set of all known chemical elements in the periodic table of elements as of the time this book was written. Then $\text{Count}(P) = 118$.

6. Suppose that we have two square 10x10 arrays of computer nodes, each of which has a little microprocessor that controls a corresponding light bulb on the 10x10 array. Let's call them array A and array B. Each computer node in the array will generate a digital signal unique to that particular computer node once its activation conditions are met. Both arrays have outlets on the back, one for every computer node, where cables can be connected in order to transmit the signal generated by each node to some other arbitrary destination. The lights on each of the computer nodes turn on when receiving the correct input.

Array A is powered by a big cable running across the floor, and all of the computer nodes on it are active and are generating corresponding unique digital signals. Array B however is different. None of the nodes in array B are active and none of them will become active unless the correct unique digital signal is connected to it from array A. The two arrays control an electronic lock on a door that you want to pass through, but the door will only open if all of the nodes in array B are powered up at the same time. In other words you need to match each node in array A to the corresponding node in array B and then connect them together until all the lights activate on array B. You'll need to test all of the possible connections between the two arrays. What is the set of all such possible connections?

Let S_A represent the set of all computer nodes in array A. Let S_B represent the set of all computer nodes in array B. We need a way to identify each of the nodes in the arrays, so let's exploit the two dimensional grid structure of the arrays and specify each node according to a pair of integer coordinates. Thus, we will treat S_A and S_B as sets of integer coordinates, i.e. as containing objects of the form (x, y) where x and y are integers between 1 and 10 inclusive. The set of all possible cable connections between array A and array B will therefore be $S_A \emptyset S_B$, which yields a large set of binary pairs of the form $((a_x, a_y), (b_x, b_y))$ where a_x, a_y, b_x, and b_y are any arbitrary integers between 1 and 10 inclusive. Notice that this is a more natural representation of this scenario than (a_x, a_y, b_x, b_y) would be, thus justifying our choice of binary pairing over concatenation.

Each array has $10 \cdot 10 = 100$ nodes. Each of the 100 nodes on each of the arrays could be connected to any of the other 100 nodes on the other array. Therefore, the total number of objects in $S_A \emptyset S_B$ will be $100 \cdot 100 = 10{,}000$. In other words, $\text{Count}(S_A \emptyset S_B) = 10{,}000$. Thus, we can see that there are many different combinations that will need to be tried, but it is at least still humanly feasible (although extremely tedious and time-consuming) to complete the task.

7. Let's return to a new scenario with the married couple, Albert and Elsa, eating at the fancy restaurant again sometime after the visit from example #3. This time the fancy restaurant is holding a special event. During the event the restaurant will be serving a three course dinner separated into an appetizer, a main course, and dessert. Albert and Elsa are wondering what all the different possible dinners are and how many of them there are. To determine this, Albert and Elsa find some copies of the menus. Albert and Elsa notice that the appetizer menu has 6 items, the main course menu has 20 items, and the dessert menu has 12 items. What is the set of all possible dinners? How many possible dinners are there?

   Let A be the set of all appetizers. Let M be the set of all main courses. Let D be the set of all desserts. Then $A \multimap M \multimap D$ is the set of all the possible dinners. By multiplying the counts of each of the menu sets, we can also determine how many different dinners are possible. Thus, there are $\text{Count}(A) \cdot \text{Count}(M) \cdot \text{Count}(D) = 6 \cdot 20 \cdot 12 = 1{,}440$. Every object in the set $A \multimap M \multimap D$ has the form (a, m, d), where a, m, and d are items from the respective menus.

8. Suppose that a real estate development company has begun construction of a new residential zone. The developers are seeking to find buyers for at least some of the properties in advance, so that they have more opperating money during the project and so that fewer of the houses will be vacant. The developers want to avoid paying any unnecessary maintenance and land ownership fees. To help entice new home buyers, they offer prospective buyers a list of options so that each buyer can customize the house to their own tastes. In this case, all of the options available are binary and independent, meaning that every option is either selected or not (with no middle ground) and none of the options constrain any of the other options. What is the set of all possible custom house configurations a prospective buyer can pick from and how many are there?

   Well, if you think about it, this is really a question about going through every possible option and either selecting it or not selecting it. So, if we let H be the set of all possible custom options, then $\text{PowerSet}(H)$ is the set of all possible subsets of H, which is the same thing as all possible ways of selecting options from the set H. For instance, H might include things like "balcony", "freezer/refrigerator", "skylight", "better AC system", and so on. Therefore, $\text{PowerSet}(H)$ will include things like {balcony, better AC system} and $\emptyset$ for example.

   The set {balcony, better AC system} means the house will have a balcony and a better than normal AC system. In contrast, $\emptyset$ means none of the extra options will be included, which a buyer might select if they want to save some money or if they don't like the options given by the developers and want a clean slate on which to build their own personal customizations. How many different possible custom house configurations exist under this system? Well, every option is either selected or not, so we have two possibilities for every option. If there are n options then there must therefore be 2^n possible configurations of those options.

9. Suppose that $L = \{a, b, c, d, e\}$, $A = \{a, b\}$, and $B = \{b, c\}$. Then all of the following are true:

a) $A - B = \{a\}$

b) $A ⊍ B = \{a, c\}$

c) $L \cap (A ↓ B) = \{d, e\}$

d) $L \cap (A ↑ B) = \{a, c, d, e\}$

e) $\text{Count}(L) = 5$

f) $\text{Count}(A) = 2$

g) $\text{Count}(B) = 2$

h) $\text{Count}(A \cup B) = 3$

i) $A ⦻ B = \{(a, b), (a, c), (b, b), (b, c)\}$

j) $A \multimap B = \{(a, b), (a, c), (b, b), (b, c)\}$

k) $(A ⦻ B) ⦻ B = \{((a, b), b), ((a, c), b), ((b, b), b), ((b, c), b),$
$((a, b), c), ((a, c), c), ((b, b), c), ((b, c), c)\}$

l) $(A \multimap B) \multimap B = \{(a, b, b), (a, c, b), (b, b, b), (b, c, b),$
$(a, b, c), (a, c, c), (b, b, c), (b, c, c)\}$

m) $\text{PowerSet}(A) = \{\emptyset, \{a\}, \{b\}, \{a, b\}\}$

n) $\text{PowerSet}(B) = \{\emptyset, \{b\}, \{c\}, \{b, c\}\}$

o) $\text{PowerSet}(A \cup B) = \{\emptyset, \{a\}, \{b\}, \{c\}, \{a, b\}, \{a, c\}, \{b, c\}, \{a, b, c\}\}$

Anyway, we now have definitions and symbols for all of the standard set operations, as well as some new definitions for some more obscure set operations. We also have definitions for a few special sets. We know at least several different notations for specifying sets and how to read them. We've been discussing these various operations and the underlying concepts for a while now, and therefore it would be wise to briefly refresh our memories as to what we have covered. As such, here is a summary table:

Concepts	Symbols	Readings
empty set	$\emptyset$	nothing
universal set	$\mathcal{U}$	everything
membership	$x \in S$	x is an element of S
subset	$A \subseteq B$	A is a subset of B
proper subset	$A \subset B$	A is a proper subset of B
superset	$A \supseteq B$	A is a superset of B
proper superset	$A \supset B$	A is a proper superset of B
equivalence	$A = B$	A is equivalent to B
complement	$\neg A$	not A
union	$A \cup B$	A union B
intersection	$A \cap B$	A intersect B
set difference	$A - B$	A without B
set exclusive-or	$A ⊍ B$	A xor B
set nor	$A ↓ B$	A nor B

145

set nand	$A \uparrow B$	A nand B
cartesian product	$A \times B$	cartesian product of A and B
binary pairing	$A \lozenge B$	binary pairing of A and B
concatenation	$A \multimap B$	concatenation of A and B
power set	$\text{PowerSet}(A)$	power set of A
count	$\text{Count}(A)$	count of A

As you may recall, the cartesian product is actually an ambiguous operation. It could plausibly be interpreted as either binary pairing or concatenation, but it is always one or the other. I included it in the table just for reference. It should be avoided in favor of either of the unambiguous variants (binary pairing or concatenation).

Besides the empty set and the universal set, I have also discussed a few other common sets in passing, such as $\mathbb{N}$ (the set of natural numbers), $\mathbb{Z}$ (the set of integers), $\mathbb{Q}$ (the set of rational numbers), $\mathbb{R}$ (the set of real numbers), and $\mathbb{C}$ (the set of complex numbers). I will have more to say about those sets near the end of this section, but before that I want to address some other things.

With the above table for reference, you should now have a fairly strong sense for what we have covered. It probably has not escaped your attention that there are certain parallels between the operations and concepts of set theory and those of classical logic. Think back to our discussion of classical logic. Compare that to our discussion of set theory. The two theories clearly have some similar operations and structure, but differ in what their object of study is. Classical logic studies truth-functional statements, whereas set theory studies collections of objects.

Notice, for example, how extraordinarily similar classical logic's disjunction operation is to set theory's union operation. Likewise, notice how extraordinarily similar classical logic's conjunction operation is to set theory's intersection operation. The parallels are quite striking. They feel almost like they are expressing the same underlying concept. This similarity is no mere coincidence. It is indicative of a far deeper connection between the two systems, one that, in the proper light, will prove to be profoundly significant. More on that later though. For now, I want to focus on exploring some examples of the parallels between the systems so that you can get a proper sense for just how similar in structure classical logic and set theory really are.

Do you remember when we were discussing the concept of a tautology in classical logic? A tautology is any expression in logic which is always true, by virtue of form alone, regardless of what the values of any individual participants in the expression are. Because every tautology is something that is always true, no matter what the context might be, we can therefore conclude that every tautology expresses a law of logic.

Discovering such laws of logic, by identifying tautologies, allows us to build up a vocabulary of universally valid logical rules. We can then use this vocabulary of universally valid logical rules to consistently reason about complex logical scenarios. These are essentially the fundamental operating principles of logic. We discover and study valid rules of inference and then carefully apply them in such a way as to conserve truth while simultaneously expanding the scope of our knowledge about a system.

In the section on classical logic, we discussed some specific examples of laws of classical logic and showed that all of them are indeed tautologies and do yield true for every possible combination of input values. The truth tables for those examples can be found on page 100, if you'd like to refresh your memory. For example, we saw that $(A \lor B) \leftrightarrow (B \lor A)$ is a tautology and hence a law of logic. This law tells us that the order of the operands in a disjunction doesn't matter and that we can freely swap the two operands on either side of a disjunction with each other without effecting the truth value of the expression in which they participate.

As you may recall, when material implications occur in laws of classical logic, it indicates that any time we see an expression matching the form of the condition then we may transform it into the form of the consequence while still conserving the truth of the expression for all of the cases where it was true before the transformation. Note however that it does not necessarily conserve the truth value of the false cases. That's why a condition must be true before we can infer consequences from it.

When a material implication and its converse are both true, then the two expressions on either side are equivalent. As such, whereas with a one-way material implication we may only perform the substitution in one direction, with an equivalence we may perform it in both directions. The law of classical logic $(A \lor B) \leftrightarrow (B \lor A)$ is such an example. Any expression of the form $(A \lor B)$ may be safely replaced with $(B \lor A)$ and vice versa, where A and B are any arbitrary well-formed expressions of classical logic.

It is clear that there are also some equivalences in set theory, just as there are in classical logic. That's why I gave an overview of the concept of implication and equivalence from classical logic again here: I wanted to refresh your memory of it. The reason is because we're going to be comparing the laws of classical logic to the laws of set theory, because this is one of the most effective ways to clearly illustrate the close parallels between classical logic and set theory.

When we first discussed laws of logic, in the section on classical logic, we only explored a small handful of laws in order to merely illustrate what those laws look like, how they behave, and how to recognize them. However, this time we will be exploring a much larger number of laws from both classical logic and set theory. The reason for this is so we can get a firmer grasp of the extent to which the two systems are similar. By covering numerous examples of the laws of both classical logic and set theory, in a format that is easy to compare and contrast, we can get a good bird's eye look at how classical logic and set theory relate to each other.

It seems to me that the best format for this comparison is to simply list laws of classical logic and laws of set theory side by side in a table, pairing each law with its closest structural analog from the other system. There's probably not much reason to make it any more complicated than that. So let's get to it. Every entry in the following table is a law (i.e. is universally valid regardless of context) in each respective system:

Classical Logic	Set Theory
$A \leftrightarrow A$	$A = A$
$\neg(A \land \neg A) \leftrightarrow T$	$\neg(A \cap \neg A) = \mathcal{U}$
$(A \lor \neg A) \leftrightarrow T$	$(A \cup \neg A) = \mathcal{U}$
$\neg(\neg A) \leftrightarrow A$	$\neg(\neg A) = A$
$(A \lor T) \leftrightarrow T$	$(A \cup \mathcal{U}) = \mathcal{U}$

$(A \wedge F) \leftrightarrow F$ $(A \cap \emptyset) = \emptyset$
$(A \vee F) \leftrightarrow A$ $(A \cup \emptyset) = A$
$(A \wedge T) \leftrightarrow A$ $(A \cap \mathcal{U}) = A$
$(A \vee A) \leftrightarrow A$ $(A \cup A) = A$
$(A \wedge A) \leftrightarrow A$ $(A \cap A) = A$
$(A \vee B) \leftrightarrow (B \vee A)$ $(A \cup B) = (B \cup A)$
$(A \wedge B) \leftrightarrow (B \wedge A)$ $(A \cap B) = (B \cap A)$
$((A \vee B) \vee C) \leftrightarrow (A \vee (B \vee C))$ $((A \cup B) \cup C) = (A \cup (B \cup C))$
$((A \wedge B) \wedge C) \leftrightarrow (A \wedge (B \wedge C))$ $((A \cap B) \cap C) = (A \cap (B \cap C))$
$(A \vee (A \wedge B)) \leftrightarrow A$ $(A \cup (A \cap B)) = A$
$(A \wedge (A \vee B)) \leftrightarrow A$ $(A \cap (A \cup B)) = A$
$\neg(A \vee B) \leftrightarrow (\neg A \wedge \neg B)$ $\neg(A \cup B) = (\neg A \cap \neg B)$
$\neg(A \wedge B) \leftrightarrow (\neg A \vee \neg B)$ $\neg(A \cap B) = (\neg A \cup \neg B)$
$(A \vee (B \wedge C)) \leftrightarrow ((A \vee B) \wedge (A \vee C))$ $(A \cup (B \cap C)) = ((A \cup B) \cap (A \cup C))$
$(A \wedge (B \vee C)) \leftrightarrow ((A \wedge B) \vee (A \wedge C))$ $(A \cap (B \cup C)) = ((A \cap B) \cup (A \cap C))$

Remarkably similar aren't they? As you can see, the two columns are for all practical purposes structurally identical. Each law on either side of the above table can be transformed into the analogous form on the other side by simply replacing each operator or object in it with the most similar operator or object from the other theory.

To be clear, not all laws of either system have equivalents in the other system. Some laws of classical logic don't exist in set theory, and vice versa. This table shows only a subset of all laws applicable to these systems. You should be careful when using analogies to reason about the two systems and should always test for validity. Nonetheless, the similarities between classical logic and set theory are uncanny. Clearly this is not merely coincidental, but rather is indicative of a strong relationship.

Notice, for example, that set theory even has its own equivalents of the laws of thought. If you look at the first three rows of the table on the classical logic side, you can see that the expressions there have the same fundamental form as the laws of thought from classical logic. In case you've forgotten, the laws of thought are (1) the law of identity, (2) the law of non-contradiction, and (3) the law of the excluded middle, and they are positioned in the table above in that same order.

You might notice that I added on "$\leftrightarrow T$" to the law of non-contradiction and the law of the excluded middle (rows 2 and 3). The reason I did that was to make it easier to show that those laws are indeed structurally similar to the corresponding laws on the set theory side. Usually when people write the law of non-contradiction and the law of the excluded middle they don't write the "$\leftrightarrow T$", because it is considered redundant.

Any form that is a law is a tautology. Tautologies are always true, and therefore a tautology can always be substituted with the literal value true in classical logic. That's why the "$\leftrightarrow T$" is considered redundant. Note however that you should not confuse substitutions that *apply* a law with those that *are* a law.

For example, if $X \rightarrow Y$ were a law then it would indeed be valid to transform X to Y, or to transform $X \rightarrow Y$ to T, but in contrast it would generally *not* be valid to transform X to T or to transform Y to T on this same basis. Don't get confused. A form matching the form of one of the conditions of a law is not conceptually the same as a form matching the form of the entire law itself.

Anyway, as you can see, set theory has its own versions of the laws of thought, which have the forms $A = A$, $\neg(A \cap \neg A) = \mathcal{U}$, and $(A \cup \neg A) = \mathcal{U}$. Kind of an interesting parallel isn't it? Well, it turns out the parallels go even deeper than that. All of the basic operators of classical logic can be emulated in set theory, if you have enough insight to see the relationship that will enable you to do it.

In particular, look for example at the correlation between $(A \vee T) \leftrightarrow T$ and $(A \cup \mathcal{U}) = \mathcal{U}$ and between $(A \wedge F) \leftrightarrow F$ and $(A \cap \emptyset) = \emptyset$. Notice how $\mathcal{U}$ seems to fill the same role in set theory as T does in classical logic. Likewise, notice how $\emptyset$ seems to fill the same role in set theory as F does in classical logic.

In fact, this pattern continues. You can mirror the behavior of all of the basic classical logic operators in set theory by using $\mathcal{U}$ and $\emptyset$ to stand as true and false respectively. Let's put aside the issue of what role all the other countless arbitrary sets between $\emptyset$ and $\mathcal{U}$ have to play in this picture, because that is something that we will be addressing later on in this book. For now, focus on just $\mathcal{U}$ and $\emptyset$ and think about how they would behave when used as substitutes for true and false in any arbitrary classical logic expression.

Consider for example the union operator, which corresponds to the disjunction (logical or) operator in classical logic. We have two inputs, each of which can be two possible things ($\mathcal{U}$ or $\emptyset$), and therefore have four possible outcomes. What are the output values for each of the possible inputs? Let's build a table and see, and put it side by side with the classical logic version for comparison:

A	B	$A \vee B$	$A \cup B$
F $\emptyset$	F $\emptyset$	F	$\emptyset$
F $\emptyset$	T $\mathcal{U}$	T	$\mathcal{U}$
T $\mathcal{U}$	F $\emptyset$	T	$\mathcal{U}$
T $\mathcal{U}$	T $\mathcal{U}$	T	$\mathcal{U}$

Apply this idea to all the other basic classical logic operators and you will get similar results. Negation and complementation, disjunction and union, conjunction and intersection, and equivalence all can be mirrored in this way by using $\mathcal{U}$ and $\emptyset$ as stand-ins for true and false. However, set theory can also do more than just mirror classical logic in this respect. It can also deal with arbitrary sets in-between the two extremes of $\emptyset$ and $\mathcal{U}$. This means that set theory is effectively a superset of classical logic, with respect to these particular operators.

You will notice however that one of the basic operators of classical logic has been left out of this comparison: material implication. The reason for that is because implication is special and deserves its own in-depth treatment, which will come later in the book. Understanding that a connection between implication and set theory exists, and fleshing it out properly, is part of the key to revealing the hidden secrets that enable the creation of unified logic.

Before we discuss how to create unified logic though, there is some poorly chosen terminology used in set theory (and also in mathematics in general) which I would like to improve upon. At this point we have covered all of the essentials of set theory, excluding the technical details of a more axiomatic approach to set theory (e.g. ZFC) because such an approach seldom actually adds any significant value to real-world applications, is dreadfully boring, and is out of the scope of this book mostly. As such, we can now afford to take some time to clarify our terminology in order to reduce the cognitive load involved in understanding what we're saying.

In my experience, people often greatly undervalue improvements in clarity in terminology and are too quick to treat them as "trivial", "unimportant", or "redundant". In reality, improvements to clarity are some of the most impactful changes you can ever make to a system. When you reduce the cognitive load of a system, by clarifying terms or refactoring, it often makes the system much easier to spot patterns in, much easier to understand, much faster to work with, and much more likely to not discourage newcomers. Newcomers are the lifeblood of any discipline. Never forget that.

Every improvement to the clarity and expressiveness of a system makes every other aspect of that system easier to deal with, less costly, and more extensible. Even the smallest improvements in clarity often yield huge gains in productivity over time, especially if the improvements to clarity are applied to fundamental or commonly used components within the system.

It is with this perspective in mind that I suggest that we stop wallowing in the status quo of our current terminology and our current way of doing things and instead proactively take steps to improve the system. There is no point in waiting for anyone else's permission to make such changes. One should instead push for the changes one wants to see and then let natural selection take its course. When design flaws in a system are not fixed, it has a perpetual cost that countless generations of human beings will endlessly pay, needlessly, and for no other reason than because we were too petrified to simply dare to challenge the status quo.

You can become an authority whenever you choose to be one. People have authority only because other people give it to them. Authority is ultimately just a social convention. You can take it away, or even give it to yourself, at any moment, so long as you have a substantive logical or evidentiary basis on which to do so rightfully. If you want true intellectual freedom, you need to realize that there isn't really any such thing as a real authority figure. You don't need a permission slip to think freely. You can make your own authority and grant your own permissions, within reason. Permission-seeking behavior is often intellectually counterproductive.

If you have any interest in being in control of your own beliefs and in being resistant to manipulation by others, then this is an important perspective to keep in mind. Being strong-willed (when merited) is good for your mental health. It means your mind has a good immune system. Yet, there are some who would chastise you for being willful and for questioning their assertions. They may try to spin it as you "not being nice" or "not being sociable", when in reality it is *them* "not being nice" by trying to destroy your intellectual independence in order to make it easier to bend you to *their* will.

Notice the hypocrisy. A person who is giving you a hard time about being strong-willed is, by that very same act, being strong-willed in trying to control you. This shows that, in reality, it is actually more often about the other person seeking power and control over others, than it is about any kind of personal flaw on your part. "You are being stubborn." is often actually code for "You are not obeying me.", especially if your perspective is rooted in reason and evidence rather than mere selfishness.

Anyway, now that I've properly emphasized the importance of not continually mindlessly adhering to traditions, to "authority" figures, and to the status quo, it is time that I discussed what specific change in terminology I have in mind for set theory here. In particular, I suggest that we change the names of nearly all of the most common standard sets of mathematics, owing to the fact that almost all of them have absolutely horrible names which are in almost every case deeply non-descriptive and opaque. The symbols used to represent them are also generally poor choices as well.

When I say "the most common standard sets of mathematics" here, I am referring specifically to the following: the set of all natural numbers ($\mathbb{N}$), the set of all integers ($\mathbb{Z}$), the set of all rational numbers ($\mathbb{Q}$), the set of all real numbers ($\mathbb{R}$), the set of all complex numbers ($\mathbb{C}$), and the set of all imaginary numbers (no standard symbol).

If you've had a lot of exposure to formal mathematics, you might be so familiar with these that you can no longer see why most of them are terrible choices of terminology. Experts often become blinded by familiarity and lose the ability to perceive things from a layperson's perspective. Don't worry though, I will explain it in terms that even an expert can understand. I have no worry that the laypersons' will understand, as they are not as blinded by familiarity or cultural dogma, nor do they have a vested interest in avoiding the work necessary to make the changes.

So anyway, let's get to work. We'll start with the first and perhaps most conceptually simple set: the set of all natural numbers. Imagine you've never encountered this term before. Naturally, whenever one comes across a new term one immediately wonders what it is that that term might mean. Without prior knowledge of the term, the best one can do is get a best estimate by interpreting that term descriptively according to one's best understanding of the constituent words that make it up.

So, we've got this new term called "natural numbers", and now we ask our selves: What does it mean? Well, what does it mean to be "natural" and what does it mean to be a "number" and moreover what does it mean to be both? Let's work through it and try to estimate what it is probably intended to mean.

When we say that something is "natural" we typically mean either (1) that it is something that occurs in nature without human intervention or (2) that it is something that is somehow harmonious, graceful, or otherwise well suited for a given situation or environment. Numbers are of course pretty much anything involving quantities of some kind, such as things like the number of items in a box or the length of a specific road for example.

So then, interpreting the term "natural numbers" in the most sane and descriptive sense possible, what does it mean? Well, let's think. One interpretation of "natural" is whatever occurs in nature. What kinds of numbers occur in nature? Clearly any count of the number of distinct individuals of some type qualifies as occurring in nature. For example, you can have 5 electrons, 44 apples, or 2 grizzly bears, and clearly all of these things occur in nature. Surely these are natural numbers then.

Another kind of number that occurs in nature though is any kind of continuous measure of length. For example, we can measure an animal's height, we can measure the temperature of a river, and we can measure the total mass of a piece of gold. Clearly all of these things are natural as well, so surely these too must be natural numbers. It would be irrational to think otherwise given what the word "natural" means.

What about other objects, like vectors? When an object is moving in some specific direction at a certain speed this can readily be represented as a vector. Lots of naturally occurring objects have vectors corresponding to their movement, and therefore clearly vectors are natural as well. However, while vectors may occur in nature and may *contain* numbers, that doesn't make a vector itself a number. Therefore, since vectors aren't really numbers per se, it seems that vectors can't really be "natural numbers".

Alternatively, what about the interpretation where "natural" means "harmonious, graceful, or otherwise well suited for a given situation or environment"? This is a more subjective qualifier perhaps, but it would be reasonable to interpret it as roughly corresponding to elegance, efficiency, and expressiveness in mathematics. For example, the numbers we use to count distinct objects and the numbers we use to measure continuous lengths both qualify. So do many other types of numbers. Surely then, logically speaking, all of them must be "natural numbers" in this sense.

So, in summary, interpreting "natural numbers" descriptively, one would estimate that a "natural number" is any number which occurs in nature or is otherwise "harmonious" in some sense, such as by being elegant, efficient, and expressive. Thus, one would think that pretty much any number could be a "natural number" then.

The problem is of course that "natural number" does not mean that. It is a far more restrictive term than it seemingly indicates. According to the standard definition, a "natural number" is a quantity that represents how many distinct objects there are in some set. In other words, a "natural number" is simply a count of something (a positive "whole number"). You can't always represent a measure of a continuous space with a "natural number", because you can't count a continuous space in the conventional sense, at least not like you can count the number of apples in a basket.

It is clear then that "natural number" is in fact a poor name for the set of objects we are referring to. From what I understand, the reason for this term choice was because mathematicians considered the numbers that we use to count to be the "most natural of all numbers" due to the fact that such numbers were probably the first numbers that humans discovered and hence perhaps came the most "naturally" to humankind. Unfortunately, while this might make some sense from a historical perspective, "natural numbers" is still an extraordinarily ambiguous and opaque term. There is no reason why anyone would expect it to mean the numbers we use to count.

A good term should be descriptive, memorable, and unambiguous. The term "natural numbers" has none of these qualities. Only by sheer rote memorization can one become accustomed to using it. It is perhaps one of the worst terms you could possibly pick for what it represents. Even a random made-up word would be better, because at least then it wouldn't be ambiguous. Luckily though, we don't need a random made-up word. There is already a blatantly obvious superior term for these types of numbers. Just call them *counting numbers*.

Think about it. If you wanted to explain what a "natural number" was to someone who didn't already know, how would you explain it? You would tell them that "natural numbers" are counting numbers, i.e. positive whole numbers, which we use to represent how many of some discrete amount of objects we have. But why ever go through the trouble of explaining that? Why not cut out the middle man and just refer to them as *counting numbers* in the first place?

Unlike "natural numbers" the term "counting numbers" is descriptive and almost unambiguous. If I went up to some random person on the street who has no formal training in logic or math whatsoever and at some point mentioned "counting numbers" they would almost certainly immediately understand what I was referring to.

In contrast, if I said "natural numbers" then they would only understand that I was referring to numbers of some kind, but they wouldn't really have any idea what kind exactly. Clearly then, "counting numbers" is a much superior term. Whereas the term "natural numbers" does not satisfy any of the criteria of good terminology, the term "counting numbers" fits quite well.

The next set that we should discuss is the set of all integers. Integers are like counting numbers, except that negative numbers and zero are included. I actually don't have much of a problem with this term, although it could still perhaps be improved slightly anyway. The word "integers" is admittedly opaque, but it is at least extremely unambiguous. The word "integer" is used to refer to essentially nothing else except integers. It has zero ambiguity for all practical purposes, which compensates fairly well for the fact that the term is totally opaque to anyone who hasn't encountered it before.

When I said that "almost all" of these standard set terms were bad, the set of integers was the exception I was referring to when I said "almost". However, there is nonetheless an alternative term that could possibly be superior (depending on your tastes), and that term is "whole numbers". If you were talking to a layperson and mentioned "whole numbers" at some point, then they would probably understand what you meant. In contrast, only someone with mathematical training would understand "integers".

In my own experiences, "whole numbers" is typically interpreted to mean *any* whole numbers, whether positive, negative, or zero. Admittedly though, there is a moderate danger of someone interpreting "whole numbers" as meaning only positive numbers. Thus, ultimately, using the term "whole numbers" makes it more immediately understandable to most laypersons, but at the cost of making it potentially somewhat more ambiguous. For that reason, I offer this particular terminology suggestion on a "take it or leave it" basis. Either of "integers" or "whole numbers" is a good term for the concept. The two terms have different pros and cons. Pick your poison.

One possible compromise is that you could use "integers" when talking to anybody you expect to already be familiar with the concept, while you could instead use "whole numbers" when talking to anybody who might not know what an "integer" is. On the other hand, it may be easier to just quickly explain the meaning of "integers" in such cases, so that you can take advantage of the total lack of ambiguity of the term. It's a toss-up really. I'm on the fence on this one.

Next up is the set of rational numbers. A rational number is any integer fraction, i.e. any numbers that can be written in the form a/b where a and b are integers. People who have not studied formal mathematics much usually just refer to such numbers simply as "fractions". Anyway though, what is it that makes "rational numbers" a bad term?

Let's apply a thought process similar to the thought process we used to understand why "natural numbers" was also a bad term. Imagine you've never encountered the term before. What does a "rational number" sound like it means? From a descriptive standpoint, it sounds like it means something which is both "rational" and a "number". We already know what a number is. That at least is crystal clear.

However, what does it mean for something to be "rational"? Well, to say that something is "rational" is to say that it is *reasonable*, i.e. that it rests on a sound logical and evidentiary basis. So, what then is a "rational number"? Can you think of any numbers in logic or mathematics which do *not* rest on a sound logical or evidentiary basis yet are still considered valid objects?

In formal logic and mathematics, the answer is essentially that there are no such objects. If an object does not have a sound logical or evidentiary basis then it cannot be considered a valid object. Thus, all objects that are studied by logic and mathematics are required to be "rational" in this sense. Anything that isn't defined in a "rational" way is not considered to be a valid object of study. Therefore, under this interpretation, saying "rational numbers" is redundant and ambiguous and has more or less the same meaning as just saying "numbers". It has very little descriptive value.

Why was it then that such a vague term was chosen to describe this set of numbers? The reasoning for calling them "rational numbers" probably goes something like this: Fractions express ratios of a numerator to a denominator. We want a name for the set of all such objects. The word "rational" has the word "ratio" inside it. Ratios are closely related to fractions. Therefore, let's call them "rational numbers", even though "rational" is overwhelmingly used as a synonym for "logical" and the fact that "rational" contains "ratio" is coincidental.

Clearly this is a terrible choice of words. If a student asked why the term was chosen this way, perhaps a reference to the fact that the word "rational" contains "ratio" would be given, as if it somehow justifies the choice well (it doesn't). Add in some social momentum and lazy thinking, and that's basically why these kinds of poor terms are still with us. Alternatively, we could just exert the small amount of effort needed to fix it by simply giving it a better name.

Oh wait, but that would require that you give yourself permission to think freely, which most people apparently don't do because they're always waiting for some magical "authority" figure to someday tell them that they've finally "put in their time" and so now they're finally allowed to think *slightly* freely. Apparently, instead of improving, let's all just continue doing what we're currently doing regardless of the merits.

Let's let it all fester so that every generation of human beings from now until eternity will continue paying the price for it, summing up to an enormous human cost over time, all because we're so fixated on the current system being "the way things are done" that we all just in effect curl ourselves up into balls, sucking on our thumbs like infants, horrified of exercising even the slightest effort necessary to make even the most simple and obvious improvements. Hooray for social inertia. Can you detect my sarcasm?

Anyway though, where was I? Ah yes, I was just about to give an alternative for the term "rational numbers". Are you ready for this? Here it is: Just call them "fractional numbers", or simply "fractions" for short. Gasp! Imagine that, using the most simple and obvious descriptive terms to describe things. Now, I know what some of the professional mathematicians in the audience are thinking, deep in their heart of hearts:

"But Jesse?! How will we ever manage to massage our egos with such simple and easily understood terms? We need terms that are fancy, unnatural, and difficult for people to understand. Otherwise, those people might get wise to the fact that what we are doing is simpler than it appears, and then we couldn't churn our terminology into a proper bubbly froth with which to lather our brains, in a kind of bubble bath for the ego that let's us feel superior to the uninitiated and inflates our sense of self-worth artificially. And isn't that what it's all really about, ultimately, Jesse? Don't you want a frothy egotistical bubble bath for your brain too?"

OK... Perhaps I'm exaggerating here a bit, but I think you get my point. If it sounds like I'm bitter about people often not being reasonable enough to make even the most obvious of improvements to systems, then you are correct in your perception. I am. I've seen a lot of situations where counterproductive decisions have been made based purely on mindless social inertia instead of objective merit.

That's why I'm making a point of discussing it now, in the hopes that doing so will prevent that same kind of thoughtless outcome from happening here. I'm not just saying it for the sake of this one specific term though. That would be ridiculous. I am saying it as a general principle and a thought exercise, applicable to many different situations in life, so that people will think about these kinds of things more deeply and carefully.

Anyway, that covers the "rational numbers". Let's move on to the next set: the "real numbers". By now there's a good chance that you're starting to see the pattern in all of these bad terms (if you weren't already) and may already be accurately guessing what I'm going to say next. Even if you probably already know what I'm going to say next though, it is still best that we discuss it carefully and in-depth anyway.

Clarity is very important to me, and I've always found the idea that covering simple concepts carefully and in-depth could somehow be "insulting" or "above someone" or "too dumbed down" to be extremely toxic and counterproductive. The fundamentals and basics of any system have by far the most leverage of any of its components, and even the slightest improvement in clarity or understanding of them can easily yield massive gains in expressiveness, usefulness, or skill.

Sometimes some people have a backwards attitude about this though. Thorough coverage of the basics and a high degree of concern over clarity are sometimes treated dismissively, as if focusing on making things easier or clearer were somehow an indicator of an inferior mind. It isn't. The reality is quite the contrary. Indeed, it is actually conceptual clarity and mastery of the basics that is the most important and most valuable thing to focus on. In contrast, being comfortable with obscurity and focusing on esoteric highly derived "advanced" concepts actually tends to minimize value and maximize waste. Yet, ironically, this far less valuable endeavor often receives far more praise and prestige, much to the detriment of progress.

As such, let's now return to the topic of what is wrong with the term "real numbers". The meaning of the "numbers" part is already clear, so there's not really much to say about that part. It only gets interesting when we combine it with the constraints specified by the other part, the "real" part. If you think about it, the word "real" is actually quite a vague term. What does it mean for something to be "real" exactly?

One thing that seems to be pretty unambiguously real is the physical world in which we live of course. So that's one example. What's another? Well, what about virtual objects, such as objects in a computer simulation or video game? Are those real objects? Certainly the physical data stored on the device simulating the objects is real. Is the surface-level perspective the user sees real as well? Can the simulation itself be considered real? I would lean towards yes, but it is not entirely clear.

What would it mean for a number to *not* be "real"? The opposite of real things is imaginary things. Doesn't that then imply that all numbers that are not "real numbers" must therefore be "imaginary numbers"? Thinking about the term descriptively, that is what it would imply. Having (hypothetically) not encountered the term "real numbers" before, we might think to ourselves: "Oh, so that's why they must have defined this other term 'imaginary numbers' over here. These 'imaginary numbers' must be the set of all numbers that are *not* real numbers. Makes sense."

Furthermore, in the world of logic and mathematics, isn't every logically consistent object you conceive of automatically considered "real" within the world of logic and mathematics, even if it isn't "real" in the physical world? Wouldn't that therefore imply that *all* valid objects we study in logic and mathematics are ultimately "real"? It would then follow that saying "real numbers" is effectively the same as saying "all numbers", but with extra emphasis on the fact that one wants to be sure to exclude any logically inconsistent concepts of numbers perhaps.

As you can see, interpreted descriptively, the term "real numbers" is borderline meaningless. How is a "real number" different from any other number? Who is to say which numbers are "real" and which aren't? It's yet another example of an abysmally poor choice of words. I can scarcely even imagine picking a worse name. I feel like you might have to try quite hard to find one.

So what then, in fact, is a "real number" defined to be? Well, the easiest way to think about it is that a "real number" is essentially any number which can represent any position on a continuous space, i.e. on a space which has no breaks or missing parts in it. It is always possible to find another "real number" between any two given "real numbers". This is in contrast to integers for example, where not all pairs of integers have more integers between them. If you want to measure the length or volume of something, you would typically use a "real number" to do so. A "real number" is essentially any arbitrary point on a complete number line.

This clearly is not at all the same thing as what the actual wording of the phrase "real numbers" implies. There's nothing about referring to a number as being "real" that would lead a rational human being to conclude that it probably means a number on a continuum. It's a complete non-sequitur.

Moreover, if we look at the terms descriptively, the logical negation of "real numbers" seems like it should be "imaginary numbers", but if we look at the actual standard definitions used in mathematics then we can see that these sets are not actually logical negations of each other. For example, the set of numbers that are not imaginary numbers includes many hypercomplex numbers, such as quaternions and octonions (among others), but the hypercomplex numbers are certainly not a subset of the real numbers, thus proving that the logical negation of the set of imaginary numbers is not equal to the set of real numbers.

Now that we've established that the term "real numbers" is a bad choice, we need to find an alternative that is more descriptive and apt. Here's an idea: "Real numbers" represent positions on a continuum, so why not just call them *continuum numbers*? That is, after all, what they represent. Why not simply derive the name from that?

Remember how renaming "natural numbers" to "counting numbers" made the term instantly understandable to the uninitiated? Well, much the same applies to our new term "continuum numbers". Lots of people understand what a continuum is, even outside of technical fields. For example, you might hear an art teacher talking about a "continuum of color" or you might hear someone say that someone else "falls somewhere on a continuum". The point is, if I say "continuum number" to someone unfamiliar with mathematics then it is much more likely that they will understand what I am talking about than if I say something descriptively opaque like "real number".

So that covers the "real numbers". The next poorly named set I want to discuss is the "complex numbers", and also their close relatives the "hypercomplex numbers" and the "imaginary numbers". These numbers have some amazingly useful properties and broad applications, but you might not know it from the way they are often poorly taught, nor would you expect a number labeled as "complex" to be so simple to use. In fact, not only are "complex numbers" simple to use, but they are very useful for simplifying certain otherwise much more complex problems. What a hilariously backwards name.

Anyway, let's take a moment to consider the meaning of the term from a descriptive standpoint. As before, we already know what a number is. It is the "complex" part that we have to decipher. If you had never head of the term "complex number" before, and you encountered it, what would you naturally interpret it to mean? Well, what does it mean to be "complex"? Let's ignore the less common interpretations of "complex", such as "a group of associated buildings" or "a type of mental condition", since these interpretations clearly aren't the intended meaning in a mathematical context.

If something is "complex" then it generally means that it is made up of many different interrelated parts, usually with the connotation that it is hard to understand or operate without considerable practice or skill, or that it is otherwise overwhelming in some way. What kinds of numbers fit this description? Who decides which numbers are "complex" and which are not? It sounds highly subjective doesn't it?

Few people would probably consider small and simple numbers like 1, 7, or 100 to be complex. What about numbers like π and e? The number π is the most commonly used circle constant[18], and represents the ratio of a circle's circumference to its diameter. The number e is the base of the natural logarithm, and is thus sometimes confusingly referred to as "the natural number e"[19], even though it isn't a natural number (i.e. isn't a counting number). Are π and e "complex" numbers? Are they simple or are they complex? It seems hard to say. Any opinion either way would feel inherently subjective.

---

[18] Although, interestingly, despite its popularity, π is actually *not* the best choice of a circle constant. Twice π, i.e. τ is. We will briefly discuss this idea more, starting on page 730, in the "Extras" chapter appended to the end of the book.

[19] This is yet another reason why "natural numbers" is a poor choice of terminology for the counting numbers. Incidentally, "the natural number" is also a terrible term for e. This just goes to show how widespread poor terminology is in math, to the point where poor terminology occurs in abundance even in the most basic and fundamental concepts. Individually these poor design choices may seem insignificant, but collectively they contribute greatly to making math seem much harder than it actually is.

Are continuum numbers more complex than integers? At first glance it may seem so, but it is not really clear. The fact that you can always find another continuum number between any two continuum numbers actually has tremendous simplifying properties for many applications. On the other hand, there are other aspects of continuum numbers, especially with respect to foundational details about them, which are perhaps more complex and difficult to deal with than the corresponding theories of integers. Also, from the natural universe's perspective, aren't they both simple things and perhaps equally fundamental? The integers and the continuum numbers seem to simply be different, i.e. they seem incomparable as far as which is more "complex" than the other.

I could go on, but I think you get my point. It seems impossible to rigorously categorize numbers on a scale of complexity. This renders the adjective "complex" effectively worthless as a way of differentiating types of numbers from each other. So what then does "complex number" actually mean in mathematics? Well, a complex number is defined as the sum of a continuum number and an imaginary number.

We already know what a continuum number is. What is an imaginary number then? An imaginary number is any continuum number multiplied by $\sqrt{-1}$, where $\sqrt{-1}$ is usually written as i. Either (or both) of the continuum part and the imaginary part can be zero, in which case that part is typically not explicitly written. The set of all continuum numbers and the set of all imaginary numbers are both subsets of the set of all complex numbers. For example, all of the following are valid complex numbers:

1. $1 + 5i$

2. $\sqrt{3/4} + (1/2)i$

3. 8

4. $3i$

5. $e + \pi i$

In much the same way that "complex number" is not a good descriptive name for what it represents, neither is "imaginary number". From what I hear, the term "imaginary number" was originally coined as a derogatory term for the concept, before the concept was fully understood by mathematicians. The idea of an imaginary number was originally controversial when it was introduced[20]. Considering that imaginary numbers are valid mathematical objects, and are logically consistent and well-defined, how can we really justifiably deride them as being merely "imaginary"? Aren't they just as "real" as any other arbitrary logically coherent concept?

Taking all of this into account, it is clear that "complex numbers" is a poor choice of terminology. Before I introduce my alternative though, we will need to discuss what complex numbers are and what they are useful for in a more meaningful sense. We have already discussed what they *technically* are (i.e. sums of continuum numbers and

---

[20]Incidentally, zero and negative numbers were also initially quite controversial when they were introduced. That may be hard to believe today, but it is true. It just goes to show how much the intellectual climate of humanity has changed over time.

imaginary numbers), but that doesn't give much (if any) insight into what they are in a more meaningful sense.

Our current education system often introduces complex numbers in an utterly dry and uninteresting way that seems to pay almost zero attention to why complex numbers are actually valuable. It is taught in a way that seldom gives people any good sense for what the *fundamental nature* of complex numbers really is. Only once you understand the true nature of a complex number will you be able to appreciate why my alternative terminology is so much more apt.

The easiest way to illustrate the true nature of complex numbers is probably something like this: Notice that every complex number has two parts, namely the "real" part and the "imaginary" part, and has the form $a + bi$. This can be thought of as a vector, such that corresponding to every complex number of the form $a + bi$ there is a vector (a, b) that can be drawn on a 2-dimensional coordinate grid. Well, any time you multiply one complex number C_1 by another complex number C_2 then it returns a new complex number that has the net result of taking vector C_1 and rotating it around the origin by an amount equal to the angle that C_2 makes with the x-axis and then shortening or lengthening C_1 by a factor equal to the magnitude of C_2.

What does this mean? It means that complex numbers allow you to re-orient other complex numbers (or vectors) without ever performing any trigonometry. In effect, you can rotate a vector by an angle without even knowing what that angle is. It is also an extremely computationally efficient operation compared to using trigonometric functions. It only requires distributing the two complex number sums over each other, such as in $(A_x + A_y i) \cdot (B_x + B_y i) = (A_x B_x - A_y B_y) + (A_x B_y + A_y B_x)i$.

One interesting difference between how complex numbers represent rotation and how angles represent rotations is that, if you ignore the length scaling effect of complex number multiplication, complex numbers represent an orientation *uniquely* whereas angles represent an orientation *redundantly*. For example, the angles 45°, 405°, 765°, and −315° all represent the same orientation, but of all the complex numbers that are of length 1 only $\sqrt{1/2} + \sqrt{1/2}i$ represents this same 45° orientation.

Why am I ignoring the length scaling effect of the complex number multiplication when considering redundancy? After all, surely all multiples of $\sqrt{1/2} + \sqrt{1/2}i$ count as redundancies as well right? The reason is because for complex number multiplication the length scaling effect is logically independent of the rotation effect, whereas for angle-based rotations the various redundant representations of the same orientation are logically dependent on each other and effect the same conceptual dimension.

Although correct, that probably still sounds somewhat hand-wavy, so let me explain it in a more convincing way, one that makes the special distinction between angles and complex numbers crystal clear: Angles may represent *orientations* redundantly, but they most certainly do *not* represent *rotations* redundantly. I know what you're thinking: "But hey! Aren't orientations and rotations the same thing?". Nope. Sorry, but they aren't. That is a popular misconception, much like how some people think that points and vectors are conceptually identical when they in fact aren't[21].

---

[21] For example, you can add vectors together, but you can't add *points* together. The reason is because vector addition only has meaning relative to a frame of reference given by a coordinate system with a well-defined origin. In contrast, the meaning of

Let me explain: Consider for example the angles 45° and 405°, which, interpreted as orientations, represent exactly the same orientation. What do I mean when I say that angles represent *orientations* redundantly, but do not represent *rotations* redundantly? Well, suppose that instead of just thinking about fixed orientations of 45° and 405°, we introduced an element of *time*.

In other words, let's have rotations of 45° and 405° happen as *changes over time* and see what happens. Suppose, for example, that we have two wheels mounted side by side on axles, separated from each other so that they don't bump each other. Suppose we then set these two wheels in motion, rotating at a uninterrupted constant rate.

If orientation and rotation were identical, then we would expect that if we rotated these two wheels at 45° per second and 405° per second respectively then they would both behave in exactly the same way and have exactly the same orientation at all moments in time. Is this what would happen? Is rotating at a rate of 45° per second conceptually the same thing as rotating at a rate of 405° per second?

The answer is of course no. A wheel that rotates at 405° per second will be rotating $405/45 = 9$ times faster than a wheel that rotates at 45° per second. These are clearly not the same thing. In fact, if you think about it, *all* rotations which have different numeric values (but still the same units) will *always* represent conceptually different rotations. Angles are only redundant when they aren't moving in time. The moment you make an angle change *over time* then all angles suddenly become conceptually unique. This proves that orientation and rotation cannot possibly be the same thing.

Now, let's try to apply the same thought experiment to complex numbers and see what happens. What we want to do is find two complex numbers such that when applied *over time* they result in different rotational speeds. Consider, for example, the complex number $\sqrt{1/2} + \sqrt{1/2}i$ which represents the same orientation as the angle 45° does. Can you find a complex number which represents this same orientation yet which would rotate at a different speed in time?

The answer is no. There is no such complex number. It is impossible to store different rates of rotation over time inside a complex number. Of the two, only angles allow you to do that. The only redundancy that a complex number has is with respect to length scaling, which, as I said before, is logically independent of the rotational effect, whereas different angles representing the same orientation are inter-dependent and exist in the same conceptual dimension.

Both angles and complex numbers can be used to represent orientations. Likewise, both angles and complex numbers can be used to represent rotations. Nonetheless, their behavior within these contexts is different. Often in mathematics multiple different types of objects can be used to represent the same thing. However, just because an object can represent something doesn't make it the most conceptually appropriate choice of representation. Different objects have different affinities for certain environments. Some representations of a thing are more conceptually correct than others.

Here's an analogy: It's kind of like mathematical objects have "homes" to which they are native and in which they are the most comfortable. Nonetheless, mathematical objects tend to be quite flexible and can be used in many different contexts, including

---

adding points would change wildly depending on where you place your frame of reference, coordinate system, and origin.

outside their natural environment. While an object might be most comfortable in a certain environment, it is often useful to use it outside of that context to take advantage of certain properties that the object may have. In these respects, angles are naturally suited for rotations, whereas complex numbers are naturally suited for orientations. However, it is still often useful to use either one in the environment of the other.

Here's another way of looking at it. To truly understand what a particular type of object is, you need to understand what it is that that type of object captures the essence of. So, you can think about it like this: "Natural numbers" capture the essence of counting. "Rational numbers" capture the essence of fractions. "Real numbers" capture the essence of measurements on a continuum. So, what then do "complex numbers" capture? "Complex numbers" capture the essence of *orientation*.

Anyway, having now made this connection to what a complex number really is at a conceptual level, we are now finally ready to give the concept a new and vastly more appropriate name. After all, you can't expect to give something a good descriptive name if you don't even have a strong grasp on what it conceptually is. Alas, none of this lengthy elaboration would probably even be necessary if the schools did a better job of conveying what complex numbers are good for and why they are awesome.

It also doesn't help that the phrase "complex number" practically screams "I am boring, have lots of moving parts, am hard to understand, and only apply rarely.", when in reality complex numbers are one of the most interesting types of numbers, have few moving parts, are easy to understand, and are applicable to many different situations.

If you've been paying attention to the pattern in how I've been creating these new improved names for sets, then there's a good chance you already have a good guess for what new name I'm going to give to the set. Are you ready? Here it is: "Complex numbers" embody the concept of orientation, so why not call them *orientational numbers*? It wouldn't be appropriate to call them rotational numbers, because that's what angles represent and complex numbers can't represent all of the possible rotations when you put them in the context of change over time.

It's that same old classic principle of good terminology in action. Instead of choosing opaque, ambiguous, non-descriptive terms (like "complex numbers") for things, it is far wiser to give clear, unambiguous, and descriptive terms (like "orientational numbers") whenever possible.

Before we move on to other things, we should also discuss two other things related to orientational numbers: "imaginary numbers" and "hypercomplex numbers". These terms should also be assigned new names. Otherwise, our new terminology system would be incomplete and hence not as useful and not as likely to be adopted. Coming up with appropriate new terms for them is not that hard though.

As you may recall, an "imaginary number" is any number which can be represented by a continuum number multiplied by $\sqrt{-1}$. $\sqrt{-1}$ is more commonly written as i because it makes the notation cleaner and faster to write. Anyway, the core idea in an imaginary number is this notion of taking a square root of a negative number.

When we take the square root of a number (such as in $\sqrt{x}$) it means we are trying to find a number which when multiplied by itself yields the number under the square root sign (such as x in $\sqrt{x}$). The problem with trying to do this with negative numbers is that any two negative numbers multiplied by each other always yield a positive number, thus making it seemingly impossible to find the square root. That is why i was invented. It allows you to act as if the square root of a negative number is well-defined and then manipulate the result in a logically coherent way.

So, what then should be the new name for "imaginary numbers"? One idea I initially had was to perhaps simply call them "negative square roots", but the problem with this idea is that the meaning is ambiguous. For example, $\sqrt{4}$ has two different possible answers, namely 2 and -2. Doesn't the phrase "negative square roots" sound like it could actually refer to something like -2 in $\sqrt{4}$, rather than to an imaginary number? That's why "negative square roots" is ultimately probably not a good term for imaginary numbers. It's also a bit too verbose and pedestrian.

After considering a bunch of different options, I've decided that "orientational roots" is the best alternative I've been able to come up with so far. I still wonder if there is some deeper insight into the nature of imaginary numbers and the related concepts in mathematics (such as hypercomplex numbers) which would provide inspiration for a better name. Ultimately though, "orientational roots" was the best that I came up with when trying to balance the relevant factors of good terminology. That's what I'll be using, since it is at least better than the term "imaginary numbers".

The other thing related to orientational numbers that we should discuss is the "hypercomplex numbers". The hypercomplex numbers are essentially a family of higher-dimensional generalizations of the concept of a complex number. The complex numbers we have been discussing so far have all been 2-dimensional, but the hypercomplex numbers can have more than just two dimensions. For example, there is a 4-dimensional generalization of complex numbers called the quaternions. Interestingly, the quaternions are quite useful for performing rotations in 3-dimensional space and are used frequently in game engines and simulations.

Another example of a hypercomplex number set is the octonions, which is an 8-dimensional generalization of the concept of complex numbers. In contrast to the quaternions, which are widely used in many practical applications, I'm not sure if there are any practical applications for octonions. As the number of dimensions of a hypercomplex number set increases, the tedium involved in trying to use them also increases.

Anyway, for hypercomplex numbers, it seems fitting to me that you could refer to different hypercomplex number sets according to how many dimensions they have, so long as doing so is unambiguous. For example, if someone just said "orientational numbers" then you could assume by default that they are talking about 2-dimensional complex numbers. However, you could prefix the phrase "orientational numbers" by "nD" where "n" is the number of dimensions in order to differentiate them.

Under this system then, "2D orientational numbers" would mean the same thing as the complex numbers, "4D orientational numbers" would mean the same thing as the quaternions, and "8D orientational numbers" would mean the same thing as the octonions. I think parameterizing it this way, so that you don't need to create a whole new unique name every time you explore a new orientational number system would be useful. In effect, "orientational numbers" would be broadened to include all of the hypercomplex numbers, but would default to meaning the complex numbers if no dimensionality specifier is prefixed to the front.

However, I am no expert on hypercomplex analysis. In fact, the only two complex or hypercomplex sets I have ever actually used myself in real life are the complex numbers and the quaternions. It appears that there may in fact be some additional variants of hypercomplex numbers that do not fit cleanly into the classification system I have specified here. In such cases, it might be wiser to stick to the original name for those sets. Alternatively, somebody who is more familiar with hypercomplex analysis than I am could perhaps kindly come up with a better naming and classification system and publish it somewhere for the benefit of anyone with an interest in such things.

Anyway, that covers what I wanted to say about complex numbers, imaginary numbers, and hypercomplex numbers. We have now introduced new alternative terms for the most common number sets of mathematics: the natural numbers, the integers, the rational numbers, the real numbers, the complex numbers, the hypercomplex numbers, and the imaginary numbers. The term "integers" is a fair choice, but all the other standard set names were clearly poorly chosen.

Did you notice a common pattern in the bad terminology choices? The common pattern among all of these poorly chosen terms is a combination of ad-hoc subjectivity, non-descriptive wording, and premature narrow-minded assumptions. Each of the standard terms is instilled with the premature biases of the time period from which it came, as well as with just plain old carelessness and a lack of concern for clarity.

Take "natural numbers" for example. People call them "natural" because counting numbers were the first to be discovered and seemed like the easiest to understand. But how does that justify labeling them as being more "natural" than other numbers? Similarly, notice how "imaginary numbers" were labeled as somehow "not real" and were mocked as a ridiculous idea before a full rigorous analysis had even taken place, rendering the judgment premature. With the exception of the integers, not even one of these sets was given a truly good name. All of these standard names were primarily the product of biased thinking and premature judgments.

Logic and mathematics are supposed to be the pinnacle of unbiased thinking and due diligence. Yet here we are, even after all this time, still labeling our most basic concepts with names that are ironically derived almost entirely from biased thinking and premature judgments. Isn't it about time to remove these eyesores from our terminology? Wouldn't you prefer terms that are based on careful rational thought instead of merely on historical momentum and biased thinking? Rigor is the bedrock foundation of logic and mathematics. Our terminology should mirror and reflect that lineage. We should not be slaves to the inertia of the status quo, nor should our history cement our future. We're better than that, aren't we?

The momentum of convention is not as powerful as one might believe. Simply having the intellectual integrity to realize that you don't need anyone else's permission to enact changes is probably most of the battle. A small change enacted now, regardless of whether someone else gives you permission or not, can easily ripple outward in time and create change faster than you might expect. It's like how a physical lever allows you to lift an object with much less force than you would otherwise require. In other words: Time is like a lever by which even the greatest things can be moved. Opportunity is the fulcrum and patience is the beam.

Speaking of time, I think it's about time to create a summary table of the new terminology, so that we can get a good overview of what the new term suggestions are and how they compare to the old terms. Long-form exposition is good for soaking in the details, but it's also important to periodically take a step back and get a better vantage point on one's surroundings. You need to know where you are to know where you are going. As such, here's a comparison table of the old terms versus the new ones:

Old Term	Old Symbol	New Term	New Symbol
natural	$\mathbb{N}$	counting	`Coun`
integer	$\mathbb{Z}$	integer / whole	`Int / Wh`
rational	$\mathbb{Q}$	fractional	`Frac`
real	$\mathbb{R}$	continuum	`Cont`
complex	$\mathbb{C}$	orientational	`Ori`

As you can see, I put two different suggested options in the row for integers. Both the old term "integers" and the alternative term "whole numbers" are good choices of terminology, both with their own pros and cons. Because the term "integers" is slightly less ambiguous, and because I want to give credit where credit is due, I will mostly be using the standard term integers instead of the alternative term "whole numbers" for the rest of the book. I still wanted to put the suggestion for "whole numbers" out there though, for anyone who might prefer it. I'm personally on the fence on it, and thus I am defaulting to the more widely standardized term as a tiebreaker.

You probably also noticed that I added in columns for what shorthand symbols one could use to represent each of the sets. I know I didn't mention these symbols for the new terms during the discussion of why the old terms are so bad and what the alternatives are, but I figured now would be a good time to introduce them anyway, since we're building a table and tables are an ideal format for displaying such things.

Notice that the new symbols I have given for the new terms are not one-letter variable names. I did this for a very good reason, even though it goes against current traditions and aesthetics in mathematics. In particular: The main reason why only a tiny number of objects in logic and mathematics have dedicated symbols is because when you restrict your names so that they can only include single-letter variables then it vastly reduces your ability to create an extensive shared vocabulary of symbols.

The current tradition of typically only allowing single-letter variables in math is quite frankly crippling. With such a small selection of names to choose from, only a tiny number of the most fundamental or most prominent concepts will ever have any hope of being given standardized names (e.g. π, e, $\mathbb{N}$, $\mathbb{Z}$, etc). This is an unacceptable and unsustainable state of affairs. The number of logical and mathematical concepts that exist to explore is growing faster and faster every day, and the current system of math terminology is buckling under the pressure and has little capacity left.

Mathematicians now have to resort to paying close attention to contextual clues, and nearly any discussion must be preceded by an extensive and time-consuming process of redefining terms in order to ensure that the discussion is not ambiguous. While computer programmers are building up massive libraries of reusable functions that provide an extensive and expressive vocabulary that ensures that wasted time is minimized, mathematicians are stuck with only a tiny number of reliable pre-defined symbols and must waste huge amounts of time constantly re-inventing the wheel. It is ridiculous.

Logicians and mathematicians should put an end to this tradition of only allowing one-letter variable names. It greatly hinders progress. With such a small selection of symbols, the meanings of the names become increasingly ambiguous and conflict-prone, and thus precise and efficient discussions become more and more difficult as a consequence. Readers must carefully memorize what it is that conceptually corresponds to every symbol, instead of simply being given meaningful self-descriptive terms.

To be blunt, mathematics is no longer the most rigorous human discipline. Computer science now deserves that title. The mathematics community has often shown a stunning inability to adapt to the modern age. Mathematics has become rife with ambiguity, vague intentions, unquestioned traditions, and unnecessarily laborious context sensitivity. The terminology sometimes borders on being indecipherable, and for an outsider it often feels like hardly any attention is paid to clarity or to making things easier to understand or keep track of.

Lifting the restriction on variable names so that not all variables use single-letter names would be an important step in the right direction. It would enable more descriptive names and make it much easier to keep track of the meaning of expressions. It also would greatly increase how many terms could be standardized and made part of the basic vocabulary. That way, people wouldn't have to spend so much time redefining all the basic buildings blocks of a theory every time they want to say something about anything.

That's why I used names that have more than just one letter in them for the "new symbol" column in the terminology table. Allowing people to use multi-letter variable names is an essential step in enabling a more expressive and powerful notation. Even with such abbreviated names as I have given here in the "new symbol" column, it is vastly more difficult to forget what these names mean compared to the old symbols.

Having multi-letter names also makes it possible to define a much larger number of standard sets. In the old system, with symbols like $\mathbb{N}$, $\mathbb{Z}$, $\mathbb{Q}$, $\mathbb{R}$, and $\mathbb{C}$ you dare not add on new names for any new sets, for fear of accidentally creating a name collision with some other set which you are unaware of. For example, you might be tempted to define $\mathbb{P}$ as the set of all positive numbers, but what if that set already means the set of all prime numbers? One-letter variable names can mean practically anything. That is why they are so often terrible for reading comprehension.

The danger of naming conflicts in such an environment is extremely high. Moreover, even when a naming conflict doesn't occur, the chances of finding an unused letter that matches well to a term and is memorable is very low. This causes the system to remain perpetually petrified, with only a tiny number of standardized symbols ever being successfully adopted. With multi-letter naming in contrast, we can easily create a vast arsenal of useful standardized names that can save us lots of time and effort.

In fact, let's go ahead and define some additional common sets as an example. If you've spent a long time exposed to the culture of formal mathematics then it is easy to lose perspective and mislead yourself into thinking that $\mathbb{N}$, $\mathbb{Z}$, $\mathbb{Q}$, $\mathbb{R}$, and $\mathbb{C}$ are the only sets worthy of universal standardization, but really this is not true at all. There are in fact many other common sets missing from that list. For example, here's a table of some other common sets and some suggestions for their corresponding symbolic names:

Number Set	Symbol
all numbers	Num
prime numbers	Pri
positive numbers	Pos
negative numbers	Neg
square numbers	Squ
even numbers	Even
odd numbers	Odd
repeating numbers	Rep

One of the great things about building up a vocabulary of sets like this is that now we can combine these sets with other sets using set operators. This allows us to express many arbitrary sets with very little hassle. For example, if I want to talk about the set of all positive even numbers then I can just write Pos ∩ Even, instead of having to go to the trouble of redefining what positive numbers and even numbers are.

Likewise, if I wrote ¬Odd∩Pri then you would know that I am talking about the set of prime numbers that are not odd, i.e. the set containing only the number 2. Similarly, Cont ∩ Rep is the set of all continuum numbers that have repeating digit sequences, Num∩¬Ori is the set of all objects that are not orientational numbers yet are still numbers of some kind, and Frac ∩ ¬Int is the set of all fractions that are not also integers. Any set expression is valid. See how much more readable and flexible this system is? Multi-letter variables really open the flood gates on clarity and expressiveness.

Oh, and while were at it, I think it would be useful to also provide simple plural forms of the set terms so that one does not always have to say "<adjective> numbers" when one is reading the sets or talking about them. An example will make what I'm talking about here clear. Consider, for example, the difference between the phrase "real numbers" and the corresponding simple plural form "reals". The phrase "real numbers" can feel like a mouth-full compared to "reals", especially if you are saying it a lot or talking about lots of different types of numbers. As such, it would be nice to have analogous simple plural forms for each of the new set terms, for the sake of brevity.

Another example is "integer numbers" versus simply "integers". A few of the other common sets, such as the set of natural numbers ℕ and the set of rational numbers ℚ, also have simple plural forms that are sometimes used. For instance, "naturals" is occasionally used as a synonym for "natural numbers", and likewise for "rationals" and "rational numbers".

On the other hand though, some common sets, such as the set of complex numbers ℂ have no simple plural form. Nobody ever says "complexes" as a synonym for "complex numbers". That's OK though, because our new terms will all be given consistent simple plural forms. This seems to me like the kind of information that would be most naturally conveyed by a table. So, let's get to it. Let's define some simple plural forms corresponding to each of the full names of each of these common sets:

Full Name of Set	Simple Plural Form
counting numbers	counts
integer numbers	integers
whole numbers[22]	wholes
fractional numbers	fractions
continuum numbers	contins
orientational numbers	oris

See? Wasn't that easy? Notice that all these simple plural forms are either already intuitive or else are made-up and will eventually become intuitive with time and exposure. Does this mean that I've just created two new words ("contins" and "oris") out of thin air without consulting the dictionary for permission? Yes. Yes it does, and I would do it again with no apologies. I trust that you remember my disdain for permission-seeking behavior in intellectual endeavors, and that I need not repeat it here.

Now that we have these new simple plural forms it will be easier to form natural and comfortable sounding sentences when talking about these sets. Don't underestimate the human factor in these kinds of things. Logic and math may be heavily technical endeavors, but that does not mean that surface-level nuances of communication should be ignored. A streamlined and fully fleshed out system of communication is always valuable when one's mind becomes stretched thin, as is likely to happen in any technical endeavor. The more we can reduce cognitive load in the foundations, the more cognitive energy we will have available for higher-level tasks that are of interest or value to us.

Anyway, with these new simple plural forms at our disposal we can now say things like "Multiply these three oris together.", "The integers are a subset of the contins.", "Not all contins contain repeating sequences of digits.", and so on. The point is that these abbreviated forms are less of a mouthful and make our vocabulary more complete and more flexible. Any of these simple plural forms can be replaced with the full name if you prefer.

We sure have been talking about this terminology stuff for a while now, haven't we? Well, it's now time to switch to a new topic. If you recall, we originally started discussing all of these terminology improvements on page 149, after we had just finished discussing all of the operators of set theory and exploring them thoroughly.

We were talking about how $\emptyset$ and $\mathcal{U}$ could emulate F and T, and about how close the parallels between the laws of classical logic and the laws of set theory were. I drew your attention towards the fact that material implication was missing in the comparison between the laws of classical logic and the laws of set theory, and that finding the relationship between implication and set theory would have special significance for the successful creation of unified logic. I said that I would cover it later on. We're getting closer to that goal, but to get there we will first have to take another detour.

Unified logic is a new branch of logic that we will be constructing which seeks to satisfy the primary motivating criteria of all of the most prominent philosophically distinct branches of formal logic simultaneously, without suffering significant negative

---
[22]Whole numbers are the same thing as integers. I included this row for the sake of completeness.

trade-offs for doing so, while also breaking new ground in other ways. That's the plan. However, in order to construct unified logic, we must first understand what all the other branches of logic are. You can't unify what you don't understand.

Before we do that though, it would seem wise to first establish a base framework by which to analyze the properties of other systems of logic in a neutral way, one which makes only the bare minimum of assumptions. To accomplish this goal, we will create an ultra-minimalistic[23] logical framework that we will call *transformative logic*.

Once we have constructed transformative logic, we will move on to discussing the non-classical logics. The non-classical logics do not obey the same rules as classical logic or set theory. As we explore the various non-classical logics, we will identify the core motivations of each one and thereby come to better understand the nature of the existing problems with classical logic and how systems of logic could be made to behave differently. We will eventually show how the core principles of each of these branches of non-classical logic can all be integrated (at least partially) into unified logic.

Transformative logic will allow us to make statements about what inferences are allowed in a given situation without actually imposing any assumptions on how truth is supposed to work. This is important because some systems of logic may have different concepts of truth than others. Truth doesn't necessarily work the same way in one system of logic as it does in another. It's easy to overlook that if (like most people) you have only ever been exposed to classical logic.

Be that as it may, for many of the things we will be discussing it is not actually necessary to have transformative logic. However, transformative logic is a useful tool to have available when the situation calls for it, and it is better to prepare in advance for such inevitable moments, rather than defining things on the spur of the moment in a more awkward way that may not give us adequate time for full coverage or proper perspective. Anyway, let's get started.

---

[23] Here I mean minimalistic in terms of making very few assumptions, not minimalistic in terms of vocabulary or capabilities.

Chapter *3*

What is transformative logic?

3.1 The surprising strength of mindless forms

Do you remember when we talked about something called "transformative deduction" earlier in the book? Our discussion of it started on page 95. We talked about two different ways of applying a law of logic to an expression.

The first way to apply a law of logic to an expression was called "transformative deduction" and dealt only with substitutions of form. In contrast, the second way to apply a law of logic to an expression was called "evaluative deduction" and involved inferences from what was known or assumed true. Transformative deduction is limited to substitutions of form, whereas evaluative deduction is able to utilize both form and value for making inferences. In this way, evaluative deduction seems to have a more flexible perspective, but at the cost of requiring assumptions about truth values to be made or to be known in advance.

Well, as you might guess from the names, the idea of transformative logic is inspired by transformative deduction. The term "transformative logic" is something that I created for this book due to the fact that there was not any existing term that perfectly captured the exact meaning of the concept that I wanted to express. However, the core idea of transformative logic is not unique for the most part. Transformative logic is very similar to a few already existing concepts.

Specifically, it is similar to: (1) formalism and (2) syntactic consequence. Transformative logic is largely based on the same kind of ideas. However, in transformative logic I have trimmed away some of the associated extraneous assumptions, notational problems, and subtle ambiguities (for my purposes at least). I have clarified certain concepts, making them more precise, and I have even added on some new capabilities that seem to be absent from the traditional frameworks. Before we get to that though, we need to understand what formalism and syntactic consequence are.

When I say "formalism" here I am referring to a specific philosophical position regarding the foundations of math, rather than to the much broader concept of "formalizing" any arbitrary concept to make it more rigorous. Formalism is a mathematical philosophy which claims that all of mathematics is in fact just exploration of arbitrary sets of rules for manipulating symbols, i.e. that math is essentially nothing more than a "game of manipulating symbols" which has no inherent meaning of its own, but rather only acquires external meaning if we choose to interpret it in some special way.

As a philosophy, formalism arose in part as a reaction to the tendency of traditional mathematics (at the time) to focus predominately on a narrow set of conventional concepts to the exclusion of all other logically conceivable concepts, and also in part as a reaction to increasing pressure to put mathematics on a more firm and more rigorous foundation by making the reasoning process as mechanical, as fool-proof, and as closed to subjective interpretation as possible.

Formalism played an important role in the history of logic and mathematics. It helped to establish a more rigorous foundation for much of mathematics and opened people's eyes to new logical possibilities. By drawing attention to the pure form of logical and mathematical expressions, and to how different forms could be made to relate to each other, it enabled people to explore new arbitrary rules for logical systems.

These explorations led to a better understanding of the true scope of logic and mathematics, and a deeper appreciation for how complex systems can be expressed purely in terms of uninterpreted forms, and the various relationships that can be imposed upon them. By forcing oneself to express everything explicitly in terms of relations between forms, one can avoid many pitfalls associated with interpreting things implicitly or subjectively. This improved the clarity and specificity of many concepts and thereby enabled a multitude of new discoveries.

Formalism is a philosophical view, not an actual logical framework for performing calculations. There isn't a logical algebra corresponding directly to formalism, and formalism has no official operators of its own. For example, classical logic is a branch of logic that *can* be treated in a way that adheres to the philosophy of formalism, but treating it as such is not strictly necessary.

The goals of transformative logic are basically the same as formalism. However, we want transformative logic to be an actual full-fledged logical framework capable of performing calculations in a mechanical way. A mere philosophical viewpoint such as formalism is not enough to satisfy these goals. We therefore have to either find a system that already does satisfy these goals or else create one of our own.

This brings us to the concept of syntactic consequence. As far as I can tell, syntactic consequence is probably the closest concept to transformative logic in the current literature. Syntactic consequence is closely related to another concept called "semantic consequence", which is in some sense its counterpart. Syntactic consequence is essential to a subdiscipline of mathematics known as "proof theory", whereas semantic consequence is essential to a subdiscipline of mathematics known as "model theory".

What is syntactic consequence? Syntactic consequence is based on the idea of deriving a conclusion from some set of initial premises purely by applying some particular set of symbolic rules of inference until the conclusion is reached. In other words, one claim B is a syntactic consequence of another claim A if claim B can be derived from claim A by constructing a proof which consists entirely of reasoning steps taken only from some specific chosen set of rules of inference, written in terms of raw symbols and syntax and without regard to how those symbols are to be interpreted. If such a conclusion can be successfully reached, then we say that *B* is a syntactic consequence of *A* within the formal system of rules of inference under consideration.

Just because a claim is derivable from another claim under one set of rules of inference does not mean that it will be derivable under other sets of rules. Changing what rules you allow changes what can be derived within the system. Syntactic consequence does not account for any implicit values that the subject matter may have. It is only aware of information that is expressed in the syntactic rules.

Semantic consequence is different. Semantic consequence uses the implicit values of the objects under consideration to determine whether an implication is universally valid. To determine if a claim *B* is a semantic consequence of a claim *A*, ask yourself this: "For all conceivable cases where the claim *A* could be true, would the claim *B* also be true?". If the answer is yes, then claim *B* is a semantic consequence of claim *A*.

For example, let's test one of the laws of classical logic to see if semantic consequence applies to it. Consider the law of classical logic which says that $(A \vee B) \leftrightarrow (B \vee A)$, i.e. that the order of operands in a logical disjunction does not matter and therefore $(A \vee B)$ can be substituted with $(B \vee A)$ and vice versa at any time.

Let's test the left-to-right direction of this equivalence, i.e. $(A \vee B) \rightarrow (B \vee A)$. Is $(B \vee A)$ a semantic consequence of $(A \vee B)$? To test this we ask ourselves "Would $(B \vee A)$ be true in all conceivable cases where $(A \vee B)$ is true?". If we construct a truth table for each of these two expressions then we can see that this is indeed the case. Therefore, we can conclude that $(B \vee A)$ is a semantic consequence of $(A \vee B)$.

As you can see, these concepts of syntactic consequence and semantic consequence have parallels to the concepts of transformative deduction and evaluative deduction that we covered earlier in the book starting on page 95. In fact, I wonder if they may even be in some sense equivalent, but I lack sufficient familiarity with proof theory and model theory to determine that. Perhaps an interested reader can investigate. It seems like a subtle question. On the other hand, "deduction" is a process whereas "consequence" is a relationship, so perhaps that is the nature of the distinction between them.

Syntactic consequence and semantic consequence both have some standardized symbols associated with them. To say that *B* is a *syntactic* consequence of *A*, we may write $A \vdash B$. In contrast, to say that *B* is a *semantic* consequence of *A*, we may write $A \vDash B$. These sideways T-shaped symbols are referred to as turnstiles. The similarity between them makes them pretty easy to remember.

These operators seem pretty reasonable don't they? You might be wondering then why I feel inclined to not reuse syntactic consequence. Why am I creating a new framework (called transformative logic) when the syntactic consequence operator seems maybe pretty close to what I need? Well, there are at least a few reasons. One reason is that I have too little knowledge of the field of proof theory, so I feel uncomfortable reusing the syntactic consequence operator when there is a danger that there are subtle aspects of how it is used or interpreted that I do not understand. Another reason is that current conventional use of the turnstiles permits an abuse of notation that I do not like, specifically when $\vdash$ or $\vDash$ is used without a left operand to introduce an assumption. I also want to avoid any other subtle notational abuses and weirdness that the current conventions may have.

Perhaps the most important reason though is that syntactic consequence seems to define itself in terms of proofs of truth-valued statements, which thereby automatically introduces an implicit notion of truth of at least some kind. In contrast, the concept of transformative logic that I have in mind is so generalized that it does not even have a built-in notion of truth of any kind.

Transformative logic can emulate truth, but it can also deal with subject matter where the notion of truth and proofs is (in a sense) meaningless or irrelevant. It would therefore seem irresponsible to attempt to use syntactic consequence to represent this concept, owing to the fact that syntactic consequence seems inherently bound to the concept of truth throughout the literature. A custom system is a safer choice.

Do you remember our discussion of L-systems, which started on page 96? As you may recall, L-systems were designed to model the growth patterns of plants. L-systems specify a set of rules for repeated substitutions of form within an expression. Once the expression has undergone some number of substitution iterations (i.e. transformations), the symbols in the expression are then reinterpreted as instructions for how to draw the plant. If the rules are selected well, this can result in intricate plant-like drawings that can be surprisingly realistic.

While L-systems may have originally been intended to model plants, L-systems are in fact a much more general type of system. L-systems use a process which is nearly identical to transformative deduction except for that L-systems often add on a few additional constraints such as requiring every possible substitution to be done in parallel during each iteration. In contrast, transformative deduction is more general in that all substitutions of form are optional and can happen in any order and at any time. L-systems are in effect a special case of transformative deduction.

One odd thing you may have noticed though, if you were paying close attention, is that L-systems have nothing to do with truth or proofs (not directly anyway). They are purely structural. You aren't really trying to "prove" anything with an L-system. An L-system is just a repeated process of *transformations* taken on the basis of some set of consistent rules. In fact, L-systems are an example of exactly the kind of system that syntactic consequence is probably not appropriate for. The only truly fitting representation for an L-system has to be a system that has no preconceived notion of truth in it. As you will see, transformative logic is such a system, and indeed the focus on *transformation* rather than on truth is where transformative logic gets its name.

In addition to frameworks like L-systems, where the concepts of both truth and proof are irrelevant, there are also frameworks which deal with proofs yet are not always capable of assigning a truth value to an expression. One example of such a framework happens to be one of the most famous alternatives to classical logic: intuitionistic logic.

Intuitionistic logic is similar to classical logic, except that it denies the law of the excluded middle ($A \lor \neg A$) and the law of double negation elimination ($\neg\neg A \rightarrow A$) and determines the validity of expressions via constructive proofs instead of via truth values. What is a constructive proof? A constructive proof is a proof that demonstrates an explicit method by which an object that is claimed to exist can actually be constructed. In contrast, non-constructive proofs provide no such method and prove nothing more than that an object exists.

As you may recall, in classical logic every expression resolves to a definite truth value (either true or false). Even if you don't know which truth value a specific statement in classical logic has, you know that regardless of what that truth value actually ends up being, the overall expression will always end up being either true or false. In contrast, in intuitionistic logic it not always possible to even determine if a specific statement can ever be proven or disproven.

For all we know, there might be some statements which can neither be proven nor disproven. Since intuitionistic logic deals with constructive proofs instead of truth values, it inherits the same potential indeterminacy that proofs have. Thus, there are times when intuitionistic logic effectively cannot assign a value to certain symbols in an expression. We must therefore have a means by which to talk about expressions in intuitionistic logic which allows for the possibility of treating the expression purely as an uninterpreted form. It would not be possible to deal with such situations in a coherent way in a system that requires all values to be fully determined.

Truth tables don't work for intuitionistic logic. For example, if you tried to use the truth tables for disjunction ($\vee$) and negation ($\neg$) from classical logic in intuitionistic logic, then it would result in a self-contradiction. The truth tables of disjunction and negation both inherently imply that the law of the excluded middle and the law of double negation are valid[1], yet intuitionistic logic requires them to be treated as *invalid*. How then do we even define intuitionistic logic in a usable way? The answer is that we define all of the rules of inference that we want to allow in intuitionistic logic purely in terms of forms, instead of using truth values and truth tables.

We need to be able to deal with forms as forms. Only a system that does not depend upon knowing the value of every form can operate in an environment where the implicit values of those forms are unknown or non-existent. Intuitionistic logic is the most well-known member of a broader family of logics collectively known as *constructive logic*. All constructive logics share the trait of requiring constructive proofs before they will treat any claim as credible. Constructive logic is sort of like the formal logic version of the informal everyday principle of "I'll believe it when I see it." in that it requires tangible examples that have been fully constructed and which leave nothing to the imagination. It does not accept intangible proofs.

Anyway, L-systems and intuitionistic logic are just two examples of non-classical logics. There are quite a few other types of non-classical logics. Later on, we will explore some of the most important non-classical logics, each of which has different core motivating principles. Then, as we construct unified logic, I will show how unified logic will satisfy the core motivations of all of these branches of non-classical logic simultaneously. However, before we do that, we'll want to define transformative logic first so that we'll have a sufficiently flexible framework by which to talk about the properties of any arbitrary branch of logic.

My purpose in discussing L-systems and intuitionistic logic here was to open your eyes to the necessity of having a system of logic available that doesn't require conventional truth values. When you've only ever been taught to think in terms of classical logic and binary truth values it can be hard to even conceive of how non-classical logics are supposed to work. It just doesn't click. It's like a color-blind person trying to under-

---

[1] e.g. If $\neg$ makes T into F and F into T, then surely $\neg\neg A \leftrightarrow A$ must be true, right?

stand a description of some colorful object. You must first unlearn your preconceptions so that you can stop seeing things as you presume they ought to be, and start seeing things as they actually are.

One sidenote I want to mention is that earlier in the book we actually briefly used something that was bordering on being transformative logic, but I didn't tell you at the time because it would have distracted from the point. In particular, I am referring to our earlier discussion (starting on page 96) of how you can create arbitrary "algebras" based on the notion of transformative deduction. As you may remember, the example I gave defined four symbols and two rules of inference. The four symbols were all astrological birth signs, namely aries (♈), gemini (♊), cancer (♋), and leo (♌). The two rules of inference were ♈ → ♋♊ and ♊ → ♌♈.

Notice these are arbitrary forms with no specified truth values or interpretation. We observed that this particular set of symbols and rules of inferences had a tendency to generate infinite alternating sequences, citing the examples where ♈ would become "♋♌♋♌♋♌…" and ♊ would become "♌♋♌♋♌♋…". This is an example of a higher-level property derived from a set of rules of inference. A step-by-step demonstration that this behavior will always occur is essentially analogous to a "proof" within that system, even though it has no defined notion of truth.

Technically we were abusing notation when we specified the rules of inference as ♈ → ♋♊ and ♊ → ♌♈. We had already defined "→" as representing the truth-functional material implication operator from classical logic, but that is certainly not how we were actually using it in ♈ → ♋♊ and ♊ → ♌♈. None of ♈, ♊, ♋, or ♌ had truth values, so it couldn't have been material implication. I used it like this anyway because it helped convey the underlying nature of transformative deduction more clearly. I did it knowing full well that I would correct it later on (i.e. now).

Clarity of communication requires focus. The more I remove information noise from a discussion, the easier it is to teach someone the real information. Thus, when I was focused on introducing transformative deduction, I ignored the detail that we were technically abusing notation because it mostly amounted to information noise. For example, suppose you wanted to teach a child what the words for various objects were. Would you (1) point out specific objects to the child and name them, or (2) sit the child in front of a talking head on a television screen and expect them to derive the meaning of the words that way? Which sounds more effective?

For instance, would it be easier to learn what a basketball is by (1) having someone simply hold one up and say "basketball" out loud or (2) by listening to a basketball connoisseur ramble on about a half dozen different aspects of basketball, interspersed with random nostalgic stories about major events from the past or famous players, in an effort to convey a full overview of the subject all at once, in a way that is accurate to the nuance and spirit of it? Can you imagine trying to deduce the meaning of the word "basketball" from such a rainbow vomit of basketball related verbiage?

Unfortunately, academia often suffers from a culture that too frequently prefers to take the rainbow vomit approach to explanation instead of the more empathetic clarity-oriented approach. Public disdain for science and intellectual endeavors, and all of the associated threats it poses to humanity's survival and prosperity, likely cannot be eradicated until more academics learn to understand the critical value of clarity in communication. Many other factors would also need resolved too though of course.

Anyway, we got off on a bit of a tangent there. As you may recall, just prior to our diversion, we were talking about L-systems and intuitionistic logic and how they are examples of systems that demonstrate the necessity of having a system of logic available that doesn't require conventional truth values. We saw that it would be impossible to coherently define all of the different systems of logic that one might conceive of without having some way of talking about such systems in a way that does not contain any implicit assumptions about how truth is supposed to work.

Having provided sufficient motivation for why we would want an ultra-minimalistic framework of logic, it is now finally time to define transformative logic and to describe how it operates. Don't worry though, transformative logic might sound like it would be esoteric, but it is not. As you will see, it is quite easy to understand. In fact, in some ways it is even simpler than classical logic. The most basic use cases for transformative logic require only a single elemental operator and a convention for parsing expressions and for defining groups of rules of inference. That's all there is to it.

Before we define the primary elemental operator upon which all of transformative logic is based we should first discuss what exactly we mean by a "form" in a more rigorous sense than we have previously used the term. We have used the term "form" as more or less a synonym for "any arbitrary symbolic expression".

For example, we have spoken of forms like $(A \vee B)$ and $\neg(A \wedge B)$, and of how a form can be transformed into a new form if a matching rule of inference is considered to be valid in the system we are studying. For instance, under classical logic $(A \vee B) \leftrightarrow (B \vee A)$ and $\neg(A \wedge B) \leftrightarrow (\neg A \vee \neg B)$ are both considered valid rules of inference. Thus, any time we see the form $(A \vee B)$ we may transform it into $(B \vee A)$ and vice versa, and likewise for the forms $\neg(A \wedge B)$ and $(\neg A \vee \neg B)$.

However, is this all there is to what a form is? Is our definition too narrow? Is it too broad? Are we being precise enough? Well, "any arbitrary symbolic expression" is fairly close to being a full definition but it isn't quite there. It is unclear what qualifies as a "symbolic expression" and what does not. It needs to be made more precise. We should define what exactly we mean by "symbol" and "form". Here are the definitions we will use:

Definition 56. *A **symbol** is any distinct object that serves to communicate a single unit of information. Some symbols have an associated meaning or interpretation, but others do not and are purely mechanical. When operating upon a symbol in transformative logic we will always treat the symbol as if it has no associated meaning or interpretation. This is done in order to prevent implicit assumptions from creeping into the system, which would damage the system's generality and expressiveness if it was allowed.*

A symbol can be conveyed over any arbitrary medium of communication and it will still be treated as a legitimate symbol regardless of the nature of the medium. Whether a symbol is communicated via text, binary data, sound, stone carvings, sign language, or an assortment of miscellaneous household knick-knacks makes no real difference as long as the symbol still acts as a single unit of information.

Definition 57. *A **form** is any arbitrary structured collection of objects, such that each object acts as a symbol, and the collection as a whole forms a coherent unit in some* *sense. What those objects actually physically are is irrelevant so long as they can be coherently interpreted as acting as a collection of symbols, tied together in a well-defined and structured way. A form may or may not have an implicit meaning, but in either case we will treat it as an uninterpreted object.*

Although sometimes forms can be transformed into other forms, no form should ever be treated as merely "standing for" something else. Every form is a value in and of itself, and its value consists exactly of the literal structured collection of symbols of which it is composed. Thus, for example, $2 + 3$ is a form and its value is itself. It is not the same form as 5. If the rules permit it, then it might be possible to transform $2 + 3$ into 5, but even then that does not make 5 the same form as $2 + 3$.

The expressions $2 + 3$ and 5 are fundamentally different forms, just as 4 and 8 are fundamentally different numbers. A form is itself and nothing more. This is contrary to the way forms are traditionally treated in logic and mathematics, wherein all forms are typically evaluated eagerly and thus not treated as values in and of themselves. However, as odd as it may seem to not give forms any implicit interpretation, it is actually profoundly more powerful and flexible than the conventional view.

Forms have the same relationship to symbols that words have to letters or that phrases have to words. Just as a word is made up of one or more letters, so too is a form made up of one or more symbols. Notice that this implies that a single symbol standing on its own can be referred to both as a symbol and as a form. Likewise, just as chaining words together to create a phrase can convey a higher-level piece of information than any of the individual symbols in isolation, the same is true of chaining symbols together to make forms.

Notice that I did not say that a form has to be text. Previously in this book, every time we talked about forms we were referring to strings of symbols written in text. This is not necessary. Text just happens to often be the most convenient and least labor intensive way of creating a form and communicating it to other people. Characters written in ink on a page (or displayed on a device) are not the only kind of symbols that exist. We can sometimes lose sight of that in the endless fog of words and text that constitute modern everyday life and work.

For example, suppose you go out for a walk in a rainstorm and come across a stone cross standing on an altar in the decrepit ruins of some ancient long-forgotten church. Regardless of how imperfect the environment is, the altar and the cross you see before you are still just as much symbols as any other symbols you might see.

Just as every symbol you are reading in this book can be either treated as a meaningless physical object or as a meaningful symbol, so too can countless other different objects we encounter in life serve as symbols, and hence as forms. Furthermore, just as you might shuffle symbols of text around to convey some meaning or carry out some logical argument, you could do the same with other kinds of objects.

The same notions of creating rules of inference that can be applied to transform one form into another, in addition to being applicable to textual symbols, can be applied to any arbitrary structured collection of objects. For example, are you familiar with the board game known as chess? Well, the entire chess board (including the positions of all the pieces on the board) can be thought of as one giant logical form. The rules for how you are allowed to move the pieces on the board, and under what circumstances, constitute the rules of inference for chess when it is conceived of as a logical form[2].

Notice that a chess board is a perfectly valid form, yet it does not have the same structure as a linear sequence of symbols that you'd find in text. You don't have to think linearly like that. It is often convenient, but you don't have to. It is sometimes more useful to broaden your perspective. This is important to know so that you don't end up having an overly narrow and rigid notion of what kinds of things are admissible as objects of study in logic.

Grids of symbols, such as chess or tic-tac-toe or matrices, are just as much valid forms as any more conventional strings of textual symbols are. It doesn't have to be a linear string of symbols or a grid though. Literally any kind of structure is admissible as a form, as long as it is structured enough that it can be operated on in a consistent way. Another way to think of it is that a form is like a complete generalization of the concept of a quoted piece of text or a literal value, where instead of being restricted to only textual characters it can apply to anything at all.

Interestingly, although you are not required to use text to represent a system, all arbitrary systems can nonetheless be fully described using nothing but text if you put forth sufficient effort to do so. This may seem counterintuitive at first, but I assure you that it really is true. Why? Well, the reason is because text can be used to describe any conceivable algorithm or set of relationships among objects. In fact, programming languages are essentially based on this idea. Programming languages wouldn't be capable of universal computation if it wasn't true.

Even if the conceptual space you want to talk about is hyper-dimensional[3] it can still be expressed in the linear 1-dimensional format of text. This is a pretty interesting fact if you actually stop and think about it. It says that any conceivable space can be crammed into a sufficiently structured 1-dimensional space, which is kind of surprising.

However, keep in mind that I never said that the resulting text would be easy to read. It might be, but it might not. Alternative formats like grids are sometimes nice, but text is often easier to work with. Ultimately, the linear format of text is not actually a limiting factor. This is a fact that we should all be very grateful for. Defining arbitrary rule sets and computer programs would be much harder if it wasn't true. Narrative texts (e.g. novels) are also dependent on this property.

By the way, our definition here for a form bears a certain resemblance to a definition I gave much earlier in the book, namely the definition of a symbolic set, which can be found on page 28. Our definition of a symbolic set has the same kind of indifference to the medium through which it is communicated as our definition of a form has. However, the emphasis of these terms is different.

The term "symbolic set" emphasizes the medium by which a statement is conveyed, regardless of what internal symbolic structure that statement has as a form. In contrast, the term "form" emphasizes the relative structure of the symbols within the form, and we use it to talk about precise transformations between forms rather than merely about

---

[2]On a personal sidenote, I honestly find chess to be an incredibly boring and tedious game. It is far too reliant on brute force memorization and anticipation. I gave it as an example here just because it illustrates the point well and is widely known. These days video games offer vastly more enjoyable gameplay and vastly more variety than any of the ancient traditional board games.

[3]i.e. even if it has more than the three dimensions of space that we live in: width (x), height (y), and depth (z)

the existence and interpretation of statements. I admit that this distinction is maybe paper thin, but it nonetheless feels potentially useful from a communication standpoint perhaps, despite the partial redundancy.

The way that math has generally been traditionally taught has a strong tendency to ingrain certain implicit assumptions in the way that people read symbolic expressions. These habits have often become so ingrained in people's minds that people often aren't even consciously aware that they are making so many implicit assumptions and that things could be different. However, as long as we remain unaware of our implicit assumptions it will be impossible to maximize the expressiveness of our language.

Transformative logic won't work properly if we make implicit assumptions. It strictly requires that all forms be treated as if they are uninterpreted. Traditional training will often get in the way of this. Old habits die hard. Thus, we should explore this concept of uninterpreted forms a bit more just to make sure that the full scope of its implications is understood.

For example, in numeric algebra we typically see that each expression contains a mix of three different types of participants: values, variables, and operators. Each of these three types of participants has a corresponding set of symbols that is used to represent members of that type. For instance, values are represented by numbers (e.g. 7, 0, -12, 60, etc), variables are represented by letters (e.g. x, A, B, δ, etc), and operators are represented by other special symbols (e.g. $+$, $-$, $\div$, $\times$, etc). However, this distinction does not hold in transformative logic.

Almost all symbols are treated equally in transformative logic: almost none has any special status over the others and almost all behave in exactly the same way. To do otherwise would be to introduce an implicit interpretation, which is not allowed. There is generally *no inherent distinction* between values, variables, and operators in transformative logic. The only possible exceptions are perhaps a few of the special built-in operations of transformative logic itself, and even those are somewhat debatable sometimes maybe.

Transformative logic sees no fundamental difference between x and 7, nor does it see any difference between $+$ and 2, nor between $/$ and y. They're all treated as just arbitrary forms. In numeric algebra, we are used to thinking of x as a variable, as something that does not yet have a concrete value until something else is substituted into it. This is not how we think about things in transformative logic. In transformative logic, x could be just as much a concrete fully-constructed value as the number 7 is.

Likewise, for all we know, the symbol 7 might be functioning in the same way as a variable, thus only to later on be replaced by some other form, such as $+$ or 2 perhaps. In the same sense that the rules of numeric algebra might permit us to substitute 2 into x in $x^2 - 1$ (yielding $2^2 - 1$), the rules of some other system might permit us to substitute $+$ into 7 in 174 (yielding $1 + 4$).

What transformations are allowed depends on what rules are currently in effect. I know these concepts might sound weird, but once you get used to transformative logic it will really open your eyes to new possibilities. Not only that, but it may even start to feel *more* natural than the traditional implicit way of doing things. You will come to see that forcing everything into three different implicit types of participants (values, variables, and operators) often actually accomplishes nothing more than weakening the scope of what you are able to say and making the system more brittle.

This is not to say that anything goes. Indeed, if a traditional rule set such as that of arithmetic or the algebra of equations is currently in effect, then nonsensical transformations will still not be allowed. To do otherwise would be madness. The purpose here is not to make absurdities valid, but rather it is to make you realize that you are carrying around an implicit rule set in your head that you are often not even aware of.

By being consciously aware of those implicit rules and making them explicit it suddenly becomes possible to modify them, thereby enabling you to discover entire new realms of logical possibilities. That's the goal here. We want to widen our perspective as far as possible. To do that we'll need to let go of any unnecessary preconceptions we might have. You can't free yourself from your preconceptions if you've already assumed those preconceptions as givens. You have to make as few assumptions as possible.

Whether or not any particular form is acting as a value, a variable, or an operator (or maybe even something else entirely) will depend on what the rules in each specific case are. As you will see, we will be able to define essentially any arbitrary rule set that we want. You may become quite confused if you forget that transformative logic thinks of everything in terms of raw uninterpreted forms. There will only be a few exceptions. For transformative logic, you should assume by default that every form I write should be read as being uninterpreted unless something I say implies otherwise.

Resist your urge to automatically collapse or substitute expressions. For example, if I write $2+3$ then make sure that you *don't* automatically think of it as being the same thing as 5. Similarly, x is not necessarily a variable. It could be filling the same role as a value. It might even be an operator for all we know. Heck, it could even be *both* a value and an operator, or even none of the above. The only information we will know about x will be what the rules explicitly tell us. Don't mess this up. This is essential. You won't understand transformative logic if you don't grasp this concept.

3.2 Transformative implication basics

Anyway, let's get started. It's time for a more concrete example. However, rather than just doing a typical dry example, let's try something fresh and unconventional. A colorful example will probably be more entertaining. Just as interior design and structural engineering can coexist in perfect harmony in architecture, so too can personality and rigor coexist in perfect harmony in logic and math. Adding some flavor doesn't hurt. As such, here's the first major example we'll be exploring in transformative logic:

Suppose we're playing a survival horror video game of some kind (in the spirit of say *Resident Evil* for example) and we come to a dead end at the end of a dark hallway, wherein we find ourselves standing in front of an eerie old alcove with a life-size statue of a warrior carved in the style of ancient Greek or Roman sculpture work. After a moment of observation, we notice that its right hand is open, such that it seems designed to allow an object to be placed in it. After some exploration of the other rooms, we find three objects which seem to be suitable for placement in the statue's grasp: a book, a sword, and a shield.

We recall that earlier in a different room we noticed a Renaissance era painting depicting a figure identical in character to that of the mysterious statue of the warrior. Within the painting, the warrior stands upon an outcropping of rock, among the chaos of some ancient battlefield, with right hand outstretched and holding up a particular object towards the sky. Strangely though, rather than holding a weapon in hand, which one would think would be more suitable to a battlefield, the warrior is holding up a book.

As it turns out, this hypothetical survival horror video game is designed so that placing the sword or the shield in the statue's hand will trigger a trap. Only if we place the book in the statue's grasp will we be safe, and in doing so a secret door next to the alcove with the statue will open up, allowing us to progress to a new area in the game. Now what, you may ask, was the point of me going to the effort of describing all of that? What does a statue puzzle in a survival horror video game have to do with transformative logic?

Well, this statue that we've been talking about in our hypothetical survival horror video game is conceptually an example of a form that permits a set of well-defined transformations to be applied to it. This is exactly the kind of system that transformative logic is designed to deal with. The statue is inert before we place an object in its grasp.

However, once we place an object in its grasp, the statue system transforms from its initial form (a statue with an empty hand) to a new form (a statue with one of the three objects in its grasp). Which of the three potential forms it transforms into will determine whether or not the trap is activated. Conceptually, this is not so different from how $x+3$ might transform into $2+3$ and then from there transform to 5. It is a simple substitution and resolution system. This should be pretty straightforward to model.

I could have chosen to do a more mundane example here (e.g. something from standard mathematics), but I chose this unconventional statue puzzle idea instead in order to raise your awareness of how broadly applicable these kinds of ideas about transformations of forms really are. These kinds of systems are all over the place. You interact with them every day. Focusing too much on traditional mathematical examples would likely blind you to the true scope of the idea.

It generally takes at least a few diversely selected concrete examples of a principle before people will really start to understand how broadly applicable the principle really is. Theory is all well and good, but a few concrete examples selected specifically to help ourselves get a better sense for the true scope of an idea can go a long way. It is easy to say to ourselves "yeah yeah, I understand, let's move on" when we hear an abstract theory, and yet not actually fully grasp the idea on a deeper level. Diverse examples go a long way towards fighting this effect.

There are many different possible ways to model this statue system. We could verbally describe it in natural language, like we are right now. We could simulate it on a computer using a game engine by producing some suitable art assets and code. Heck, if we were really hardcore we could even build a real-life version of it using actual mechanical engineering (e.g. using an elaborate system of gears, hydraulics, weighing scales, etc... whatever it takes). However, for our purposes it will be more efficient to represent this system with textual symbols and rules of inference instead.

Remember our old example, in the section on transformative deduction on page 96, where we posed a hypothetical "algebra" (a system of symbols and rules for manipulating them) wherein the only symbols were aries (♈), gemini (♊), cancer (♋), and leo (♌), and the only allowed transformations were ♈ → ♋♊ and ♊ → ♌♈? Well, let's do something similar to represent the statue scenario from our hypothetical survival horror video game. As you will see, it is not difficult at all.

First, we will need some symbols to represent each individual concept involved in the scenario. I have found some suitable symbols from some obscure packages that are available for the typesetting system I'm using to write this book[4]. Let's go through the symbols one at a time:

1. To represent the statue, we will use a symbol for the scales of justice (⚖), because the statue judges the player according to what the player places in its hand.

2. To represent the statue's open hand, in which we can place an object, we will use a symbol for a gear (⚙), because placing an object in the statue's hand activates a mechanism, and a gear is a good generic symbol for the concept of a mechanism.

3. To represent the book, we will use a symbol for a book (📖), for obvious reasons.

4. To represent the sword, we will use a symbol for two crossed swords (⚔), for obvious reasons. We'll just ignore the fact that there are two swords instead of one. Symbols are supposed to be abstract anyway, so it's ok. It's good enough.

5. To represent the shield, we will use a symbol for a shield (⛉), for obvious reasons.

6. To represent the trap being activated, we will use a symbol for a skull (☠), because a skull is a common symbol for danger or death.

7. To represent the correct object being placed in the statue's hand, resulting in the player's continued safety and the secret door opening next to the statue, we will use a symbol that looks like an angel (☦), because angels represent safety[5].

Next, we will need to define some transformation rules that capture what all the possible transformations between the various forms are in this system. We'll first frame things in terms of our old notation and then gradually refine the system from there. There is a certain degree of arbitrariness in how we go about doing this, and multiple different approaches to representing it will work. The key is to make sure that whatever method we choose to represent the system captures the essential characteristics relevant to our study of the system. Certain details can often be safely ignored, but others are necessary. It depends on what you care about. As such, here is what I decided on for our transformation rules:

1. ⚙ → 📖

---

[4] The name of the typesetting system I'm using to write this book is LaTeX. If you've had any experience writing professional scientific documentation then you probably already guessed that though. Indeed, for typesetting, LaTeX is practically the *lingua franca* (the shared language) of the scientific publishing world, especially in math.

[5] This is not what the package I grabbed this symbol from intended this symbol to represent. The package says it (☦) is supposed to represent something "commercial". I have no idea what it is actually supposed to be an image of and I don't really care.

2. ⊕ → ✕

3. ⊕ → ⊘

4. 🗿⋓ → ☦

5. 🗿✕ → ☠

6. 🗿⊘ → ☠

Finally, there is one more thing we'll need to specify in order for the scenario to be fully represented in our notation: the initial form. Just like in classical logic or in numeric algebra, it's not enough to just have a list of all of the valid transformation rules. We also need to have a form that we start with and which we subsequently apply transformations or inferences to.

One possible analogy would be that it's kind of like the way a fireplace works. An empty fireplace is indeed designed for handling fires, but it's not going to do anything unless you put something combustible in it to light on fire. An initial form is like that. For instance, in numeric algebra we might have $x^2 + y^2 = 1$. Suppose we were then asked to solve for x. To do this we could transform $x^2 + y^2 = 1$ in steps, like so:

$$(x^2 + y^2 = 1) \to (x^2 = 1 - y^2) \to (x = \pm\sqrt{1 - y^2})$$

Using our fireplace analogy, the laws of numeric algebra are like the fireplace and the initial form $x^2 + y^2 = 1$ is like the combustible material. Similarly, we will need an initial form to represent the starting configuration of the statue puzzle from our hypothetical survival horror video game. Can you guess what the initial form will be? If you look carefully at the structure of the set of rules I gave, it shouldn't be too hard to guess what the initial form should be. It should be this:

$$🗿⊕$$

Do you see why this is what it has to be, given what the rules are? The only way we can reach the outcome states (either ☦ or ☠) is by going through one of the bottom three transformation rules in the list. The only way of getting to the forms of one of those conditions is if we start with 🗿⊕ and then transform ⊕ according to one of the top three transformation rules. Each of these top 3 rules in the list represents a different choice of which object to place in the statue's hand.

The reason why I put the empty hand symbol (⊕) to the right side of the statue symbol (🗿) is because, if you recall, I said in the original scenario that it is the statue's *right* hand that is open. However, even if I had written the ⊕ on the left side in all the transformation rules it wouldn't have really mattered, because which of the statue's hands was open was essentially information noise. It's an example of the kind of detail that can be ignored. Keep in mind though that one could easily conceive of a modified scenario wherein which hand is chosen actually does matter. Like I said earlier, it depends on what you care about.

Notice also that I have multiple rules for ⊕. Specifically we have all three of ⊕ → ⋓, ⊕ → ✕, and ⊕ → ⊘ being applicable. Did that bother you? If it did, then it's probably because you weren't accounting for the fact that transformation rules are *optional* and can be taken *at any time*. However, while transformation rules are optional and can be taken at any time, that doesn't mean you can perform multiple applicable rules for a specific form simultaneously.

Imagine if you tried to apply both ⊕ → ⚔ and ⊕ → 🛡 simultaneously. What would the resulting symbol be? It can only be one symbol. Is it the sword (⚔)? The shield (🛡)? The answer is that it's nonsense. It's a non-existent self-contradictory symbolic object. You can't have one symbol that is equal to two different individual symbols simultaneously (not to be confused with a set of two symbols, which in contrast is perfectly valid). Two things that are different can never be equal[6].

Anyway, it is clear that the player's goal in this system is to somehow transform the starting form ♟⊕ into the desired form ♀. This is not difficult to do. The transformation path we must take to accomplish this looks like this:

$$♟⊕ \rightarrow ♟◫ \rightarrow ♀$$

Remember that transformation rules can be applied to any matching form regardless of where that form occurs, even if it is contained inside another form. That is why the rule ⊕ → ◫ is our justification for transforming ♟⊕ into ♟◫, even though the ⊕ → ◫ rule says nothing about ♟ symbols.

Notice the parallels to our earlier numeric algebra example. In the numeric algebra example, we started with $x^2 + y^2 = 1$ and transformed it to $x = \pm\sqrt{1 - y^2}$. The reason we knew our solution was guaranteed to be correct was because we performed *only* steps justified by the rules of numeric algebra. Similarly, in the statue puzzle example, we started with ♟⊕ and transformed it to ♀. Likewise, the reason we know our solution for the statue puzzle is guaranteed to be correct is because we performed *only* steps justified by the rules. These processes are fundamentally the same. I want the wildly diverse and powerful nature of this kind of thinking to really sink into your mind so that you genuinely understand it, instead of merely regurgitating mechanical rules that were taught to you in school by rote memorization and academic necessity.

This system was very easy to solve. The pathway we had to take to get from our starting point (♟⊕) to our goal (♀) was readily apparent. However, this will not always be the case. Many systems will be much harder to solve. Indeed, for some problems, it will be unclear whether a pathway to certain final forms even exists at all. The great unsolved problems of mathematics are examples of such cases. Some problems are like hellish mazes, where it often happens that you are moving along, thinking you are making great progress, only to find that the path you've been following for the last 100 steps was actually a dead end and led you nowhere. Time to start over!

On the other hand, sometimes we aren't even interested in solving things. Sometimes our interests are instead exploratory and undirected in nature. Sometimes we just want to let the rules act upon the system to see what intricate higher-level beauty might

---

[6]Note however, that we are talking about strict logical equality here. Human rights equality is a different concept. I don't want someone reading this thinking that "two things that are different can never be equal" justifies discriminatory beliefs like racism or sexism (etc). Equal rights are about equal opportunity and overall net value, whereas logical equality is a much narrower concept. Don't equivocate these concepts.

spontaneously emerge from the interactions of the lower-level components. L-systems are an example of such an endeavor, as are many other objects of study such as fractals and cellular automata. Video games are another example, an absolutely massive and complex example, but an example nonetheless.

Our statue notation is pretty straightforward and easy to understand. However, if you were paying close attention you may have noticed that we're still abusing our notation in these examples. The symbol "$\to$" has been defined to represent material implication, which is an operation that deals exclusively with truth values (T and F), yet we are using the operator as if the statue symbols have unique values all of their own.

It was useful to gloss over this detail temporarily so that we could remain focused on the basic idea of transforming forms arbitrarily, without having to define a new vocabulary in the heat of the moment while we were still discussing other things. However, I think it is now time to correct this abuse of notation. To do so, we will define the main operator upon which all of transformative logic is essentially based:

Definition 58. *To be capable of expressing arbitrary frameworks of logic, in which assumptions from classical logic may not hold, we need an operator that expresses transformation rules (a.k.a "rules of inference" or "laws") in such a way that those rules deal only with forms and make no implicit assumptions about the nature of truth. To fulfill this role, we now define a new operator named* **transformative implication** *(or "trans implication" for short) and represent it with the symbol "$\twoheadrightarrow$".*

For example, to say that the form A can be transformed to the form B, we may write $A \twoheadrightarrow B$. This arrow operator representing transformative implication can also be written in the other direction if so desired, in which case $A \twoheadrightarrow B$ could be written equivalently as $B \twoheadleftarrow A$. Unlike material implication, transformative implication has no truth value. Whereas $A \to B$ will always evaluate to either true or false, $A \twoheadrightarrow B$ will always evaluate to itself (i.e. its own form is its own value).

Notice that while this operator is an arrow and closely resembles the material implication symbol ($\to$) it is not the same symbol. Care should be taken to ensure that the two operators, material implication ($\to$) and transformative implication ($\twoheadrightarrow$), are visually distinct when written. From this point on, we will no longer use the material implication symbol to represent anything other than the truth-functional classical logic operation it was originally intended to, except if explicitly stated otherwise.

Although the transformative implication operator is already sufficient on its own to cover the use cases for transformative logic, we will also define a symbol for two-way transformative implication for convenience, drawing inspiration from the design of equivalence symbols from other branches of logic. Here is our definition for it:

Definition 59. *When a transformation rule $A \twoheadrightarrow B$ and its converse $A \twoheadleftarrow B$ are both valid, we want a way of concisely writing this as a single rule instead of as two rules. To do this, we define a new operator named* **transformative equivalence** *(or "trans equivalence" for short) and represent it with the symbol "$\leftrightarrow\!\!\!\!\!\leftrightarrow$". For example, to say that the form A can be transformed to the form B, and vice versa, we write $A \leftrightarrow\!\!\!\!\!\leftrightarrow B$. Unlike material equivalence, transformative equivalence has no truth value. Whereas $A \leftrightarrow B$*

will always evaluate to either true or false, A ↔ B will always evaluate to itself (i.e. its own form is its own value).

With these new operators now at our disposal, we can see that the way we wrote the rules for the survival horror statue puzzle on page 181 was technically wrong. Just to make sure everything is crystal clear, let's re-write that list of rules in the more correct form so that we can be certain that we understand it. Here's the new re-written rules:

1. ⊕ ⇸ 🍶
2. ⊕ ⇸ ✗
3. ⊕ ⇸ ⊘
4. ⚔⊕ ⇸ ☥
5. ⚔✗ ⇸ ☠
6. ⚔⊘ ⇸ ☠

Likewise, previously we wrote that the path from the initial form (⚔⊕) to the desired form (☥) looked like ⚔⊕ → ⚔🍶 → ☥. However, with our new operators, we can see that actually it should have been written as ⚔⊕ ⇸ ⚔🍶 ⇸ ☥. Similarly, the rule set for the weird alternating astrological sign system that we discussed in the section on transformative deduction on page 96 was written as ♈ → ♋♊ and ♊ → ♎♈, but now we see that it actually should have been written as ♈ ⇸ ♋♊ and ♊ ⇸ ♎♈.

Notice that the empty hand symbol (⊕) in the statue puzzle rule set is acting sort of like a variable. It allows itself to be transformed into three different other symbols (✗, ⊘, or 🍶). However, at the same time, the empty hand symbol is not actually merely an abstraction. We're used to thinking of all variables as being abstractions rather than part of the concrete reality of a system. However, the empty hand is just as much a part of the concrete reality of the statue puzzle system as any of the other forms are.

Thus, the empty hand symbol here is in a certain sense *both* a value and a variable. These kinds of systems are in fact quite common and are perfectly valid and logically consistent. The traditional distinction between values, variables, and operators (etc) is essentially superficial at the highest level of generality. Nonetheless, we need to understand how to work with the concept of variables in transformative logic since it is a very common pattern of behavior for many systems.

We often think of variables as being able to be substituted with "anything", but this is in fact often not true. It is necessary to carefully define the scope of each variable (i.e. its domain) so that the things we substitute in for that variable don't end up being nonsense. For example, in classical logic, suppose we had the expression $A \vee B$. Imagine what could happen if we allowed ourselves to substitute anything that we wanted to into A and B. We could end up with something like $2\pi \vee \mathbb{Z}$, which would make no sense.

We need a way to precisely specify the domain of any given variable. Interestingly, we already have a way of doing so. All we actually need to do is to use transformative implication. We will define two different ways of doing this. One way is the longhand way, which involves exhaustively writing out every possible substitution for a specific named variable. The other way is the shorthand way, which compacts multiple different possible substitutions into a single transformation rule. Let's take the expression $A \vee B$ from classical logic as an example. How would we define the domains of the variables A and B? Recall that all variables in classical logic can be either true (T) or false (F) and nothing else. As such, here's how we would define the domains the longhand way:

1. $A \twoheadrightarrow F$
2. $A \twoheadrightarrow T$
3. $B \twoheadrightarrow F$
4. $B \twoheadrightarrow T$

Look familiar? Stylistically speaking, it is very similar to the transformation rules we defined for the survival horror statue puzzle system and the alternating astrological signs system. The only difference is that it has different symbols and different definitions. Stylistically and structurally it relies on the same mechanisms that the other systems did. Does that strike you as strange? Why does this work as a way of defining domains for variables? Why are we able to reuse our transformative implication operator to also define domains?

Well, that's because all transformation rules in logic are *optional* and can be taken *at any time*, remember? Those are exactly the same criteria that variable substitutions would require, so that's why transformation rules also serve as domain definition rules. This also has unifying value. Instead of having two different syntaxes for implications and domains we have one, which keeps things more clean and more general.

However, as you may have noticed, this longhand notation for defining domains is going to become very tedious very quickly. Imagine if instead of working with boolean variables, we were working with variables that could be any of a large number of different values. In order to write the domain of such variables in longhand form, we would have to write out a long list of transformation rules corresponding to each and every possible value of the variable. That would be incredibly tedious and time-consuming. Therefore, we need a means by which to specify numerous different possible substitutions as a single rule. To do so, we will use the concept of sets. For example, let's re-write the rules for specifying that A and B are classical truth values by using set notation to make it more concise:

1. $A \twoheadrightarrow \{F, T\}$
2. $B \twoheadrightarrow \{F, T\}$

These rules state that A and B may each be substituted with either of F or T. However, while this is certainly an improvement on the previous way of specifying domains (i.e. with one rule per possible value), sometimes even this relatively convenient method is still far too tedious. Manually listing out all elements in a set is only feasible for relatively small sets. There is still the issue of what to do when we have an infinite number of values in a set, which would make listing them all impossible.

We could maybe solve this problem by relying on one of the other methods of specifying a set, such as predefined reference (e.g. Even), listing with omission (e.g. $\{\ldots, -4, 2, 0, 2, 4, \ldots\}$), and set builder notation (e.g. $\{x : (x = 2k) \wedge (k \in \text{Int})\}$), etc, but let's be careful and take this one step at a time. We need to be careful that these methods will actually make sense and not be too ambiguous before we permit their use. For reference, we introduced the above methods of specifying sets on page 129.

With the transformation rules for A and B now specified, $A \vee B$ has the effect of representing every possible binary disjunction in classical logic and corresponds to the following list of substituted forms: $F \vee F$, $F \vee T$, $T \vee F$, and $T \vee T$. Sets that are written into expressions in transformative logic have the effect of being transformable into any member of that set in that position in the expression. Thus, when we have a bunch of different variables in an expression, then that expression can represent every conceivable variation on that expression allowed by the constraints of the domain.

In this way, we can specify variations on forms that include not just conventional variables as most people think of them, but also any other arbitrary symbol within a form. For example, consider the two laws of classical logic that state that disjunction and conjunction have operands that are order independent (commutative). This could be expressed in symbols in transformative logic as $A \vee B \leftrightarrow B \vee A$ and $A \wedge B \leftrightarrow B \wedge A$. Since this order independence is a trait that both of these operators ($\vee$ and $\wedge$) share in common, it turns out that in transformative logic we can abstract over not just A and B, but also over $\wedge$ and $\vee$. For example, the following rule set expresses the order independence (commutativity) of both disjunction and conjunction simultaneously:

1. $A \twoheadrightarrow \{F, T\}$

2. $B \twoheadrightarrow \{F, T\}$

3. $\square \twoheadrightarrow \{\vee, \wedge\}$

4. $A \square B \leftrightarrow B \square A$

As you can see, we are using the $\square$ symbol as a generic operator variable that can stand for either disjunction ($\vee$) or conjunction ($\wedge$). In this way, we can simultaneously state all operators to which a given algebraic property applies within a system by simply adding those operators to the list. Cool trick right? It also serves as yet another exercise to remind you of the level of generality of the system we are dealing with. The disjunction symbol ($\vee$) and the conjunction symbol ($\wedge$) are just arbitrary symbols and therefore are as equally amenable to abstraction as any more traditional variable or value is.

However, for the example above, keep in mind that A and B are not the same thing as the set of things they can transform into. A and B are just arbitrary symbols, not sets. If X appears in a condition of a transformation rule then *only* forms matching X can apply that transformation. Thus, as written, this rule set permits transformation of $A \vee B$ into $B \vee A$, but not transformation of $T \vee F$ into $F \vee T$ directly. There are multiple ways to define these rules and they are subtly different. This distinction will become important later. There actually is a way to define this rule set so that transforming $T \vee F$ into $F \vee T$ directly would also be permitted, but it requires a few mechanisms that we haven't discussed yet[7]. More on that later.

Anyway, now that we've introduced this set notation into transformative logic, we can rewrite the rules for the statue puzzle system in a more concise form. For reference, the previous version of the transformation rules for the statue puzzle system can be found on page 185. Here is how we would rewrite it if we wanted to take full advantage of set notation to reduce the number of rules we have to write:

1. ⊕ → {◫, ✕, ∅}

2. ⚔◫ → ✝

3. ⚔{✕, ∅} → ☠

This is basically as tightly as you could possibly express the statue scenario. It really distills the essence of the situation and makes it easy to see what is going on. However, eliminating variables in this way will not always make things better. Sometimes named variables will be easier to deal with, and sometimes anonymous inline sets will be easier to deal with. It depends on the situation and the scale of the abstraction (e.g. on how many values a variable abstracts over, etc).

You can actually use these kinds of transformation rule sets to create little inferential puzzle games. All you need to do is create an interesting rule set, indicate what initial form the player has to start from, and then tell them what the target form is and see if the player can successfully transform the initial form into the target form. Our toy example here (the statue puzzle) is very easy to solve, but it's not hard to see how a system with much more complex rules could pose a daunting challenge for a player.

In the same sense that people make crossword puzzle books and other kinds of paper-based games, you could collect together the most interesting inferential puzzles like this that you can find or create into a book. One of the great things about transformative logic (as you will see) is that it is so flexible that you'll probably never run out of new properties to explore.

In the main branches of mathematics it is perhaps easier to run out of new things to discover because the territory is so well explored, but this scarcity may be much less true in transformative logic, because of how much more arbitrary it is. The potential for emergent complexity from simple rules is quite large. The arbitrariness of transformative logic makes it easy to create custom rule sets that have unconventional properties.

You could leave your custom puzzle game in a purely symbolic form, or you could maybe even put it into a video game somehow. You could even generate initial forms or rule sets procedurally in order to create an infinite number of puzzle variations within one game. Or, you could go the L-systems route and just use rule sets to create pretty pictures. There's a lot of potentially fun stuff you can do here. Logic doesn't have to always be cut and dry.

---

[7] specifically transience delimiters and selective transformative implication

Logic is actually very flexible and diverse. It supports both technical and creative endeavors. Properly understood, logic is simply the exploration and study of all conceivable consistent systems. Thus, for example, all videos games, music, works of art, novels, and films (etc) can in a certain sense be considered to be applied logic in disguise, at least insofar as they manage to stay internally logically consistent.

It's important to realize that the concept of a variable applies not just to individual symbols, but also to arbitrary forms of any size. We can just as easily conceptualize a complex form (e.g. $A + B \cdot C$) as a single variable as we can a single symbol standing on its own (e.g. A). Both the condition and the consequence of a transformation rule can consist of one or more symbols. In other words, any possible combination of forms is allowed on either side of the transformative implication operator (→), regardless of how many individual symbols each form is composed of. Anyway, let's explore another arbitrary rule set as practice:

1. ᚱ → ᚱᛗ

2. ᚱ → ᛗᚱ

3. ᛗ → ᚱᚱ

4. {ᚱᚱ, ᚱᚱ, ᛗᛗ} → X

Initial Form: ᚱᚱᛗ

Target Form: XXX

For entertainment value, I decided to use some ancient runes as the symbols here. These runes were taken from the Anglo-Frisian runic alphabet, which is also known as the *futhorc*. The futhorc alphabet was used to write Old English before the introduction of the Latin alphabet. The Latin alphabet gradually evolved until it became our modern day alphabet. Who knows, in an alternate history we might still be using something derived from a runic alphabet if Latin hadn't won out. Oh, and in case you're wondering, I have no idea what these runes actually mean.

As you probably noticed, I added an initial form and a target form into the list this time and labeled them as such. I put those in there so that you could have something to play with to help you explore the behavior of the system. The initial form and target form are not really part of the rule set itself necessarily.

If you want to generate new puzzles of your own design to challenge other people with, a helpful technique if you're having trouble coming up with ideas is to start with a random initial form and then repeatedly apply various rules to it until it happens by chance to become an interesting form in the course of those modifications. At that point, simply record the initial form and the target form and give it to someone else to solve since you now know that that particular form is indeed reachable in the system. You should also record all the steps you took to get there so that you'll have an answer key. Speaking of which, here's one way to solve the puzzle I posed for the rune system:

ᚱᚱᛗ → ᚱᛗᚱᛗ → ᚱᛗᚱᛗᛗ → ᚱᛗᚱᚱ → ᚱᚱᚱᚱᚱ → XXX

As you can see, this puzzle is harder than the statue puzzle but is still relatively straightforward. Keep in mind that you don't generally have to restrict yourself to 1-dimensional strings of symbols like we do in these text-based examples. You can expand into higher numbers of dimensions with your puzzles. In fact, the Rubik's Cube can be thought of as a 3-dimensional example of such a puzzle. Likewise, those sliding number board puzzles where you only have one empty space open for swapping number tiles around in is a 2-dimensional example of such a puzzle. Discovering what the underlying rules and invariants are in a puzzle may help you solve it.

Another thing to keep in mind with these kinds of puzzles, or with any system of rules of inference really, is that each new derivation you make can be added to the rule set as a shortcut. In this way, you can build up a larger vocabulary of rules, which makes it easier to reason about the system in any future encounters you have with it. For instance, because we have discovered that ᚷ ᚷ ᛗ can be transformed into X X X using only the given rules, we can now add it to the list of rules as a new shortcut, thereby giving us the option to apply it any time we encounter a matching form.

In other words, we have proven that ᚷ ᚷ ᛗ → X X X is valid in this system. We often refer to such derived rules as "theorems", and much of the goal of logic and mathematics is simply deriving as many useful theorems as possible. Theorems effectively provide a way to build up your vocabulary and intuition for a system, even for systems that initially seem non-intuitive and opaque.

Theorems in mathematics are analogous to subroutines or functions in programming. Every new shortcut you discover in a logical system allows you to compartmentalize part of the lower-level details of the system into higher-level rules. The more of these you build up, the more powerful your vocabulary becomes, and the easier it becomes to bend the system to your will. If you've always struggled to get a grip on complex systems, then there's a good chance that it's because you still haven't fully internalized the fundamental nature and power of modular thinking.

You need to cut down on the number of things that you have to keep in mind at once. The human mind is fundamentally easily fatigued and short on mental bandwidth. Getting overwhelmed easily is definitely the norm. It's how you respond to it that actually makes the difference. Don't expect yourself to be able to juggle a million pieces of complex information at once. That's not the trick to being smart. The trick to being smart is to take complex things and make them simpler. Instead of imposing unrealistic standards on yourself, accept the fundamentally limited nature of the human mind and then work within those limitations in a strategic and mindful way.

Don't stretch yourself thin like a child struggling to reach an object on a shelf. Instead, stop acting like a fool and just find a stool or a stepladder. If no stool or stepladders are available, then *make one*. You'll get a lot further in life when you stop expecting other people to have already provided you with all the tools you'd ever need. Don't turn problem solving into an ego game. Intelligence is more a game of patience than a game of juggling. People don't care about intellectual pissing contests. They only care about what value you add to society. Never be afraid to "dumb things down". It's the smartest thing you could ever do.

It is a common misconception that "smart people" are just inherently better at dealing with lots of complexity and chaos in a problem than the rest of us are. In reality, most of being "smart" actually just involves systematically and carefully deriving higher-level reusable principles in order to make seemingly complex problems into progressively simpler and simpler problems until the system as a whole has been dumbed down so much that our primitive minds are finally able understand it.

In addition to simplifying problems by deriving higher-level rules, it is also often useful for us as human beings to create physical real-world metaphors as analogies for abstract concepts and ideas. Most of the human brain evolved during times when physical constraints, such as hunting or staying alert for predators, were the majority of what we had to think about. Intense abstract symbolic reasoning probably only became a major part of human culture relatively recently on the evolutionary timescale.

As such, our minds seem to have a much easier time remembering and understanding things when those things are posed as analogies to some kind of physical reality. This is perhaps due to the fact that our brains probably have much more evolutionary hardware in them optimized for dealing with physical constraints than for dealing with abstract or symbolic reasoning.

For example, I used to not quite get the intuition behind how to consistently and correctly use subtraction when creating a new formula to model some special phenomenon or behavior for which there wasn't already an exact formula available (e.g. for game mechanic design). Subtraction was just a cut-and-dry mechanical operation to me, until an insightful physical analogy occurred to me. It occurred to me that $b - a$ represented the displacement it would take to get **from** a **to** b.

The metaphor here is that it's like we are moving from point a to point b, just like we would in physical reality, and measuring the displacement as we move in a straight line from point a to point b. Ever since I made this connection I have tended to read expressions of the form $b - a$ as "the displacement from a to b" instead of merely as "b minus a". Notice how the second way of reading it ("b minus a") gives little to no insight into the underlying nature of the operation.

I'm always looking for new analogies like this. They are extremely empowering. I have found that discovering them often makes it massively easier to just spontaneously create new math formulas to suit any conceivable situation or idea I want to implement. You can chain these insights together until it gets to the point where you can read math formulas as easily and intuitively as other people read English.

For example, if I see $(d - c)/(b - a)$, I don't think "$(d - c)$ divided by $(b - a)$" but rather I instead think "the displacement from c to d in terms of the displacement from a to b". In other words, I see that this expression actually means "what would the displacement from c to d be if it was written in terms of units that have the same magnitude and direction as $b - a$?".

In fact, in general, subtraction and division have nearly identical semantics. Both $b - a$ and b/a can be read as "what it takes to get from a to b", it's just that in the case of subtraction the difference is additive, whereas in the case of division the difference is multiplicative. In effect, division is "multiplicative difference" in exactly the same sense that subtraction is "additive difference". Subtraction lets you find out how to travel from point a to point b via addition, whereas division lets you find out how to travel from point a to point b via multiplication. Think of them as expressing different forms of *movement*.

Logical systems are like physical spaces, where different places in that space are connected by different operations. Each of these operations permits a different form of movement or transformation in that space. Just as trains and elevators permit different kinds of movement within a city, so too do subtraction and division permit different kinds of movement within a mathematical system. In fact, *all* logical operations can be thought of in these terms (i.e. as movements or transformations). Navigating a mathematical space is not fundamentally much different from navigating a city. All math expressions are just maps that tell you how to get from one place to another. See? It is often a lot easier to figure out what a mathematical expression actually means when you think about it in terms of physical analogies.

Once this really clicks, you begin to realize that, just as any arbitrary sequence of words in natural language can be strung together to form an arbitrary narrative, the same is true of expressions in mathematics. The real difference between math and natural language is that mathematical expressions are so precise that they can be put into computer programs and thereby become manifest in physical reality as automated simulations. Unlike natural language, mathematics enables you to create living worlds that you can reach out and touch and experience directly via a computer. Math is alive. In contrast, natural language is limited to a person's imagination.

Systems described in natural language (e.g. English) aren't precise enough to become manifest in physical reality. Thus, in a sense, mathematics is a language that *allows you to play god*. Mathematics is the language of magic. Nobody ever said that playing god would be easy though. That's why it takes so much effort and study to do things like create video games and simulations. In fact, you really shouldn't be surprised that it's so difficult.

Completely mastering logic, mathematics, engineering, and computer science (to the point of knowing everything that could ever be known about each of those respective disciplines) would effectively allow you to become the closest possible thing to a real-life wizard that physical reality allows. It would mean that you simultaneously understand the structure of all conceivable logically consistent universes, know how to simulate them on a computer, and can sometimes even (when possible) engineer devices to make them manifest in physical reality as more than just a simulation.

Mathematics and programming are the closest thing to magic that humanity has ever invented. In principle, all you need to know is the perfect combination of magic words (i.e. mathematical expressions and computer code) and anything you could ever dream of could be made a part of reality via a computer program and some hardware. The only limits are computational feasibility, available resources, and your imagination. The overwhelming expressive power and untapped potential here is so huge that it defies description.

However, as natural as some logical and mathematical systems may be to work with, not all support direct intuitive analogies to the physical reality in which we live. In fact, some are quite odd and are only really expressible in terms of themselves. However, this does not mean all is lost. Even the most arbitrary and unnatural system can be made relatively intuitive and comprehensible with sufficient effort. The key is to derive as many useful theorems (i.e. shortcut rules) as you can so that you can begin to grasp the higher-level structure of the space. In order to accomplish this, we will need a strong vocabulary for talking about transformation paths and what properties they have. There are several different independent factors we will want to consider. Let's begin.

3.3 Transience delimiters

Let's define some vocabulary for how to talk about "how far along" (in some sense) a form is on its available transformation paths. These new terms will make it easier to describe the current state of any given form relative to what kinds of transformation options are still available to it. As you will see, besides just being useful from a descriptive standpoint, having these new terms defined will also enable us to further broaden the expressive scope of transformative logic so that it supports a few new fundamental operators. The more words and symbols we have to label different places in a conceptual space, the easier it generally is to navigate that space.

What then is this new terminology I have in mind? Well, it has occurred to me that there is a certain different character in the way that forms behave relative to whether or not they have any transformations available to them. There is also the case where the only transformations available to a form are two-way. It seems like we should have some terms for distinguishing these cases, terms that indicate what kind of transformations (if any) are available to a given form. That brings us to our new definitions:

Definition 60. *If a form does not match the condition form of any of the transformation rules in the rule set then we say that the form is **trans-closed**. A trans-closed form has no way of ever changing under the current rule set. A trans-closed form has essentially reached a dead-end on a one-way transformation path, or never even had a transformation path in the first place. Transformative implications from a form to itself don't count. A form with only transformative implications to itself is still trans-closed.*

Definition 61. *If the only transformation paths that are applicable to a form are two-way transformative equivalences then we say that the form is **trans-equal**. Whether or not the transformation rules are written as one rule, as in $A \leftrightarrow B$, or as two rules, as in $A \to B$ and $B \to A$, is irrelevant. Both are transformative equivalences. A trans-equal form can only ever change to other transformatively equivalent forms.*

If a form matches a condition form for a one-way transformation rule, and if it would have no way of transforming back to itself along some other path if it were to apply more transformation rules, then it cannot be trans-equal. Transformative equivalences from a form to itself don't count and at least one such non-trivial equivalence must exist for a form to be trans-equal. "Non-trivial" here means that at least one of the transformations must differ by at least one symbol.

Networks of forms that all connect back mutually to each other through some arbitrary system of paths can still be trans-equal even if they don't look like it at first glance. Be aware that if a network of mutually connected forms has a pathway of escape, wherein upon entering that transformation path the form cannot return to the network of mutually connected forms from which it came, then none of those mutually connected forms can be trans-equal.

In order for a form to be trans-equal, there cannot be any irreversible transformation paths from itself to any other form. In other words, if you check every possible transformation path from a form, and none of them are irreversible, and at least one of them is non-trivial (i.e. differs by at least one symbol), then that form is trans-equal. Note that a form with only trivial (i.e. self-referential) transformations available to it is trans-closed, not trans-equal.

Definition 62. *If a form matches the condition form of at least one one-way-only transformation rule then we say that the form is **trans-open**. A trans-open form is a form that is still open to some kind of potential irreversible change. This is in contrast to trans-equal forms, which only have two-way transformation paths available to them, and to trans-closed forms, which have no transformation paths available to them.*

Trans-open forms often act kind of like "unresolved" entities, but this is not always the case. It could just be that the open transformation paths are deliberately not being taken because we are already where we want to be with the form. Any form which has at least one irreversible transformation path available to it is trans-open. In contrast, any form which has no irreversible transformation paths available to it can never be considered trans-open.

Notice that all of these types of forms have a close relationship with what kinds of changes are permitted for a form and whether or not those changes are reversible. In particular, notice that trans-closed forms permit no changes, trans-equal forms permit only reversible changes, and trans-open forms permit at least one irreversible change.

The easiest way to keep these three types straight in your head is to always think in terms of exploring all possible transformation paths originating from the form whose type you are trying to classify. No change possible implies trans-closed, only reversible change possible implies trans-equal, and at least one irreversible change possible implies trans-open. This is the rule we use in practice. The full-length definitions for each type given above were mostly just exploratory and were intended to illustrate the concepts to get you thinking.

A form might satisfy the criteria of one of these types in one rule set and yet not satisfy the criteria in another rule set. These type classifications are always relative to a rule set. Notice also that these three types are mutually exclusive. A form which is a member of one of these types cannot possibly be a member of either of the other two. In addition to defining a name for each of these types, we should also define a name for the axis of classification as a whole:

Definition 63. *Every form within a given rule set can be classified into one of three mutually exclusive types having to do with what kinds of transformation options are available to that form. These three types are named trans-closed, trans-equal, and trans-open. We refer to the type to which a form belongs as its **transience type**. Thus, if someone asked you what the transience type of a form was then you could answer either trans-closed, trans-equal, or trans-open. These three types collectively exhaust all possibilities, thus all forms can be classified into one of these three types.*

Oh, and it just occurred to me that it would also be useful to have some nested subset variants of these terms. Having the mutually exclusive terms has its advantages, but so does having terms that can be nested inside each other as subsets. If you think about it, trans-closed and trans-equal forms both have a certain similar behavior. Both trans-closed and trans-equal forms feel like they have resolved to some kind of stable entity. It's just that in the case of trans-equal forms the form can take one of multiple forms, whereas in the case of trans-closed forms the form only has one representative form.

It seems like there should be a term for any form that meets either of these qualities, since it seems likely that sometimes you might not care whether a form is trans-closed or trans-equal as long as it is at least one of them. Likewise, there should also be a catch-all term for any type of transience whatsoever. Thus, we have two more definitions:

Definition 64. *If a form is trans-closed or trans-equal, then we may say that it is **trans-fixed**. All trans-closed or trans-equal forms are also considered to be trans-fixed. Trans-fixed forms often feel "resolved" in some sense.*

Definition 65. *If a form is trans-closed, trans-equal, or trans-open, i.e. if it has any transience type whatsoever, then we may say that it is **trans-free**. All forms can be considered trans-free. By itself, this term is not very useful and sounds a bit weird. However, it will become useful later on once a corresponding related operator is defined.*

I know that you might be *transfixed* by these transience type definitions for the moment, but let's move on. Notice that the trans-closed, trans-fixed, and trans-free types form a nested chain of subsets. All trans-closed forms are trans-fixed forms and all trans-fixed forms are trans-free forms, hence constituting a subset chain. With these new terms now defined, one could say that we now have five transience types instead of just three.

However, this is somewhat a matter of taste and depends on whether or not you want to count non-exclusive types like trans-fixed and trans-free as being genuinely different types, or if you would instead rather treat them as redundant types. Let's refer to the mutually exclusive trio of transience types (trans-closed, trans-equal, and trans-open) as the **mutex transience types**, and to the subset-related trio of transience types (trans-closed, trans-fixed, and trans-free) as the **subset transience types**.

So far in transformative logic, with only a few special exceptions, we have always treated forms literally, i.e. as uninterpreted collections of symbols. Anytime we see symbols in the condition or consequence of a transformative implication, we don't impose any implicit interpretation upon them, and instead only allow exactly the transformations that the rules have specified to be performed upon them.

If a form does not match the exact symbols of the condition of a transformative implication, then the transformation cannot be applied to it. Stated more precisely, given a transformation rule of the form $X \to Y$ for some arbitrary forms X and Y, we currently have no way of saying that any form *derivable* from X can be transformed to Y, but rather we can only say that the literal form X can be transformed into the literal form Y. Notice the distinction.

Even with only the ability to work with literal symbols the system is still quite powerful. However, it would be useful if we also had some way to match all forms *derivable* from X, rather than being limited to matching only the literal symbol X itself. In order to extend transformative logic so that it gains this ability, we will need to create some new operators corresponding to each of the transience types. Why transience types are a good basis for this extension will become clear with time. As it turns out, there's not just one way of interpreting what "forms derivable from X" should mean. Transience types will help us establish that distinction.

To define these new operators, we are going to use something called a *delimiter*. A delimiter is a special mark or boundary used to logically separate an object from its surroundings in some way. We've used a few different types of delimiters already, without explicitly referring to them as such. For example, parentheses and quotation marks are examples of delimiters. It's difficult to unambiguously define expressions without having at least a few delimiters. Different delimiters do different things.

How do delimiters work? Well, delimiters essentially mark the beginning and ending of some part of an expression so that it can be considered as a whole of some type. For instance, when you place a piece of text inside quotation marks, it blocks that text from being interpreted as *meaning something* and instead forces it to be interpreted as *pure text*. Delimiters allow you to disambiguate the contextual interpretation of an object and force it into a specific interpretation. This sort of mechanism seems highly suitable for distinguishing between a form itself (on the one hand) and all things derivable from that form (on the other hand), so it's what we're going to use. Anyway, here are our new definitions:

Definition 66. *Given any form $\mathcal{F}$ and any transience type $\mathcal{T}$, chosen arbitrarily and possibly unrelated, there is a corresponding set of forms $\mathcal{D}$ containing all forms reachable from form $\mathcal{F}$ that have the transience type $\mathcal{T}$. This set $\mathcal{D}$ is called the* **derivation set** *of form $\mathcal{F}$ relative to transience type $\mathcal{T}$.*

Corresponding to every possible transience type $\mathcal{T}$, we will define a pair of delimiters that indicate that only forms of transience type $\mathcal{T}$ should be allowed into the derivation set $\mathcal{D}$. Thus, if an arbitrary form $\mathcal{F}$ is enclosed by the pair of delimiters corresponding to transience type $\mathcal{T}$ then it will be interpreted as meaning the set of all forms derivable from $\mathcal{F}$ that have transience type $\mathcal{T}$. Since we have defined five distinct possible transience types, we will need to define five new pairs of delimiter symbols. We will refer to such pairs of delimiters as **transience delimiters**. *Here are the symbols that we will use to represent them:*

$$
\begin{aligned}
\textit{trans-free:} &\quad \mathopen{)\!\!} \mathcal{F} \mathclose{\!\!(} \\
\textit{trans-fixed:} &\quad \triangleleft \mathcal{F} \triangleright \\
\textit{trans-closed:} &\quad \cdot| \mathcal{F} |\cdot \\
\textit{trans-equal:} &\quad \blacktriangleleft \mathcal{F} \blacktriangleright \\
\textit{trans-open:} &\quad \mathopen{(\!\!} \mathcal{F} \mathclose{\!\!)}
\end{aligned}
$$

Thus, for example, if we wanted to express the trans-free derivation set of form X then we would surround X with trans-free transience delimiters, resulting in $\mathopen{)\!\!} X \mathclose{\!\!(}$. This expression $\mathopen{)\!\!} X \mathclose{\!\!(}$ effectively means "the set of all forms derivable from X" and thus if we wrote $\mathopen{)\!\!} X \mathclose{\!\!(} \to Y$ then it would mean that *all* forms derivable from X (including X itself) are allowed to transform into Y. This is in stark contrast to the original example where we had $X \to Y$, which only allowed exact matches to the literal symbol X to transform into Y.

Notice the way the shapes and shading are used to make it easier to remember what each of the delimiters means. Each mutex transience type's delimiter is distinguished from its corresponding subset transience type's delimiter by being a shaded-in version of the same symbol. Thus, you can remember the shaded-in versions as meaning the mutually exclusive version of that level of transience, whereas the unshaded versions are subset-related. As for the trans-closed delimiter, notice that it has a different structure and visually looks like it could belong to either shading group. This was done to make it clear that it belongs to both trios of delimiters, rather than just to one or the other.

Let's do an example to test our understanding. Suppose we have a system that we are trying to model, wherein we want to be able to easily describe transformations involving capitalized letters of the English alphabet, but without having to laboriously write out the set every time we use it. This is a perfect use case for transience delimiters. It is quite common to want to talk about an entire category of symbols at once like this.

Perhaps the letters themselves are what we are interested in (i.e. letters as "values"), or perhaps the letters merely stand for something else (i.e. letters as "variables"). Either way though, the underlying mechanism for reuse will be the same: transience delimiters. Anyway, suppose we want α to represent the set of capitalized English letters. Here's what this would look like as a transformation rule:

$$\alpha \rightarrow \{A, B, C, \ldots, Z\}$$

Suppose we wrapped α in some transience delimiters. How would the result differ depending on which type of transience delimiters we used? Here is the answer:

1. $⟅ \alpha ⟆$ would have the same effect as writing $\{\alpha, A, B, C, \ldots, Z\}$

2. $\triangleleft \alpha \triangleright$ would have the same effect as writing $\{A, B, C, \ldots, Z\}$

3. $\cdot | \alpha | \cdot$ would have the same effect as writing $\{A, B, C, \ldots, Z\}$

4. $\blacktriangleleft \alpha \blacktriangleright$ would have the same effect as writing the empty set $\{\ \}$

5. $⦅ \alpha ⦆$ would have the same effect as writing $\{\alpha\}$

Do you understand why these are the results? If not, then there's something wrong with your understanding of transience delimiters. You may want to review the definitions again, or perhaps practice with some examples or thought experiments, otherwise you're likely to get confused later on. By the way, don't worry that the trans-fixed and trans-closed sets are the same here. This is pure coincidence. Sometimes transience delimiters will return the same set, but this will only happen on a case-by-case basis. In general, different transience delimiters do not return the same sets.

Transience delimiters are not just some esoteric extension to transformative logic. It may actually sometimes be impossible to express certain things without them. Without them, we would not be able to use variables as a shorthand to stand for multiple different forms in a transformation rule. Any symbol we picked to fill such a role would end up being interpreted as a raw uninterpreted form (a literal), and therefore would not actually be able to fill the role.

Transience delimiters allow us to overcome this limitation. For example, suppose we had some huge set of values that we wanted to allow to be substituted into a bunch of different places. Without transience delimiters, we would have to write out the entire set every time we used it in an expression, otherwise any variable we chose to stand for the set would be interpreted as a raw symbol. Consider the following rule set:

1. $A \twoheadrightarrow \{1, 2, \ldots, 99, 100\}$

2. $B \twoheadrightarrow \{-100, -99, \ldots, -2, -1\}$

3. $f(\cdot\mid A\mid\cdot,\ \cdot\mid B\mid\cdot) \twoheadrightarrow 0$

This example is contrived, but it still illustrates the point well. The actual meaning of the symbols used in this rule set is technically undefined[8], but one way you could interpret it is as a statement about the behavior of a function named f. Let's assume that f is indeed a two parameter function. Under this interpretation, the third rule basically says that whenever the function f receives a first input that's an integer between 1 and 100 and a second input that's an integer between -100 and -1 then it always returns 0. The expression $\cdot\mid A\mid\cdot$ means the same thing as if we had written out $\{1, 2, \ldots, 99, 100\}$ longhand, and similarly $\cdot\mid B\mid\cdot$ stands for $\{-100, -99, \ldots, -2, -1\}$. However, what would this rule set look like if we didn't have transience delimiters? It would look like this:

1. $f(\{1, 2, \ldots, 99, 100\},\ \{-100, -99, \ldots, -2, -1\}) \twoheadrightarrow 0$

Not exactly the pinnacle of readability or brevity, is it? It's pretty clear that if we didn't have some mechanism by which to let symbols stand for multiple different forms in an expression then things would get out of hand very quickly. The notation wouldn't scale. Transience delimiters allow us to modularize our thoughts. It's similar to why programmers separate programs out into a bunch of different smaller functions that they can then combine together in various ways. The human mind gets overwhelmed very quickly, even by relatively small amounts of complexity. Modularity is crucial for making complex systems understandable. Incidentally, here's another way we could have written the rule set (using transience delimiters again):

1. $A \twoheadrightarrow \{1, 2, \ldots, 99, 100\}$

2. $B \twoheadrightarrow \{-100, -99, \ldots, -2, -1\}$

3. $\cdot\mid f(A,\ B)\mid\cdot \twoheadrightarrow 0$

Notice that this time we have written the trans-closed delimiters around the entire expression $f(A,\ B)$ instead of only around each of A and B individually. For this rule set, as written, this has the same net effect as $f(\cdot\mid A\mid\cdot,\ \cdot\mid B\mid\cdot)$ did. Why? Well, the reason is because $\cdot\mid f(A,\ B)\mid\cdot$ means the set of all trans-closed forms derivable from the form

---

[8]Or, put differently, the rule set itself *is* the meaning. It is only the structure of the rules, and how the participating forms stand in relation to each other, that is truly substantive. The rest is essentially just arbitrary symbols. How we choose to interpret those symbols is up to us.

$f(A, B)$, but the only possible transformations available to the form $f(A, B)$ are those enabled by A and B.

Thus, in this case, there is no net difference between $f(\cdot|A|\cdot, \cdot|B|\cdot)$ and $\cdot|f(A, B)|\cdot$. However, notice that if there were other transformation rules available, such as for f or for the parentheses, then $f(\cdot|A|\cdot, \cdot|B|\cdot)$ and $\cdot|f(A, B)|\cdot$ might not be equivalent. Get the idea? Which parts of a form you wrap in transience delimiters, and what type of transience delimiters you use, allows you to very precisely specify your intent. Transience delimiters generalize what it means for something to "stand for" something else.

3.4 Selective transformative implication (part 1)

Now that we have the ability to specify both literal forms and derivation sets, and considering that transformation rules are totally arbitrary and can include whatever we want them to, you may be tempted to think that we have pretty much a complete system for transformative logic now. Well, not so fast. It turns out that, in keeping with my long-standing tradition of temporarily oversimplifying things in order to make it easier to learn, I have not yet disclosed a major missing component of this system to you. You might have noticed it if you were paying close attention to what I *didn't* say. One of our earlier rule sets was actually ambiguous, although others were not and were fine.

You see, transformative implication actually comes in variations. The version we've been using so far is just the most straightforward one. There's a good chance you didn't even notice that one of our rule sets was actually ambiguous and ill-defined. The reason for that is because the human mind is good at filling in implicit details using contextual info without even consciously realizing it. However, good rigor requires that we make everything explicit so that we don't accidentally end up stepping on our own toes or miscommunicating our intent.

For a mechanism as simple as transformative implication, it sure looks like we've defined everything we could ever possibly need though, doesn't it? Looks can be deceiving. Speaking of looking, let's look at one of our most basic rule set examples from earlier and see if we can detect a problem with it. In particular, let's look at the bare-bones rule set that we used to specify that A and B range over the domain $\{F, T\}$. Think about how we might try to use these rules in the context of classical logic. Here's the rule set:

1. $A \to \{F, T\}$

2. $B \to \{F, T\}$

For the purpose we intend it for, this rule set is ambiguous. Can you see why? Does it surprise you that such a simple rule set could be ambiguous? I'll give you a hint: It's hard to see why this rule set could be ambiguous until you look at an expression like $(A \land A) \lor (B \land B)$ and think about all the different ways you could substitute values into it. Still don't see it? Let me make it crystal clear: There is absolutely nothing in the way we have specified transformative logic so far that would stop you from transforming the expression $(A \land A) \lor (B \land B)$ into something like $(T \land F) \lor (T \land F)$.

Do you see it now? At no point did I ever say that all instances of a given symbol in an expression have to be transformed into the same target value in parallel. I *did* say that one individual instance of a symbol can't be substituted with multiple different symbols at the same time, because it's physically impossible to write one symbol instance that is simultaneously strictly equal to two different symbols. However, I never said that different instances of the same symbol in different positions couldn't receive different substitutions.

As written, there is nothing to stop you from turning the A on the left side of $(A \wedge A)$ into T while you simultaneously turn the A on the right side of $(A \wedge A)$ into F, and similarly for B. These symbols in $(A \wedge A) \vee (B \wedge B)$ may *look* the same, but the mere fact that they have been placed in different locations in the expression could in itself potentially change their identity.

Position is part of an object's identity. Objects that have different positions in space should not automatically be assumed to have the same identity. Nobody ever said that every symbol in every arbitrary logic one could ever conceive of has to be substituted for the same value in parallel. It is *not* safe to assume that an arbitrary symbol couldn't in fact legitimately behave differently depending on its position in an expression. I did this intentionally.

However, it is clear that allowing different instances of A and B to take on different truth values when they occur in different positions in expressions in classical logic is unacceptable. It doesn't match our intent. Any given variable in classical logic can only ever be exclusively true or exclusively false, never both at the same time.

If we have any hope of transformative logic being able to emulate any arbitrary logic, then surely it must also at least be capable of emulating classical logic. However, it would be impossible to emulate classical logic unless we define a mechanism by which to distinguish these different ways of treating symbols. That's why there are in fact, as you will see, multiple different variations of transformative implication.

Those variations are what will allow you to distinguish between the case where a symbol needs to be substituted in parallel (thus taking on the same value in all places in which it occurs in an expression) and the case where a symbol may be substituted with any of its values anywhere it occurs (thus making it act more like a grab bag of possibilities than like a fixed value), and all the cases in-between. "The cases in-between?! What does that even mean?" you ask? Yes, the cases in-between… They exist too. Don't worry, it may sound weird now but it'll seem obvious and intuitive by the time we're done with it.

It feels like we have quite a conundrum on our hands, doesn't it? How are we going to address this dilemma? How do we create a system that allows both strict substitutions and loose substitutions at the same time? If you think about it, basically what we really need is a method of binding (or not binding) the identity of variables to particular values. At what level of scope should these bindings occur? Should the binding be per expression, or per symbol? Consider, for example, the earlier expression $(A \wedge A) \vee (B \wedge B)$. When we substitute in the value of A or B should it only be required to be the same value *within this expression*, or should it be required to be the same value *globally* within the entire logical system?

Let's think about it. Think back to classical logic. What do variables like A and B

represent in classical logic? They represent the truth values of specific statements. Now ask yourself: Is the truth value of a statement dependent on what expression it occurs in, or is it instead independent? If you think about it, it is independent. If I say that the ocean is a body of water and represent that statement by A, then there is no reason to think that the truth value of A should be different in $A \land B \land C$ than in $\neg A \lor B$. Thus, as you can see, A is independent of the expression in which it occurs. Therefore, it seems that we should bind our symbols globally in order for them to behave properly for the kinds of constraints we want to specify.

So, how should we go about actually doing this binding? Should we add on another delimiter to indicate which way it should be interpreted? That might work. However, merely marking a form with another delimiter will not necessarily be sufficient to enforce a global binding on that form. Any such delimiter would be attached to a specific expression, which would make it local to that expression. We want the binding to apply to the form itself, rather than to just specific individual instances of it.

Our hypothetical delimiter would have to indicate which interpretation (strict substitution or loose substitution) applies to a given variable. Imagine if we had a delimiter that meant "strict substitution" and used it in a bunch of different expressions. That might work to make sure the substitution *within one expression* is done properly, but it wouldn't guarantee that the variable would behave the same across *all* expressions. For that reason, implementing the "strict vs loose" binding distinction via a new delimiter seems like it would probably not be the right choice.

We'd like our choice of how we design this system to correlate as closely as possible to what the actual underlying concepts in play are. The more natural and conceptually correct our system is the better off we are. Our first idea of distinguishing substitution behavior by delimiting each variable to indicate which type it is came to mind pretty quickly. However, as we saw, it ultimately doesn't seem to be the right choice here. That's where the variations on transformative implication come in. Creating variations of transformative implication seems to probably be the more conceptually correct way to approach the problem.

Let's think back to our old definition of the domain specifications for A and B in the context of classical logic. With the rules as currently written (i.e. $A \twoheadrightarrow \{F, T\}$ and $B \twoheadrightarrow \{F, T\}$) they are ambiguous. But notice this: If we instead had exact assignments of which truth values A and B would end up being, such as only $A \twoheadrightarrow T$ and $B \twoheadrightarrow F$ (and not also $A \twoheadrightarrow F$ and $B \twoheadrightarrow T$), then there would be no ambiguity and any application of these rules would guarantee that A and B would have consistent substitution values wherever they occur in any expression. It is only when we leave the final bindings unspecified that it becomes ambiguous. However, we nonetheless still need to be able to say that A can be either F or T, so we cannot actually go so far as to omit $A \twoheadrightarrow \{F, T\}$. How might we be able to leverage this effect?

Let's see if we can figure out the nature of this relationship a bit more clearly to see if that helps us some. How are $A \twoheadrightarrow F$ and $A \twoheadrightarrow T$ related to $A \twoheadrightarrow \{F, T\}$, if you really stop and think about it? Well, they are *selections* of objects taken from the set $\{F, T\}$. So here's an interesting idea: What if, instead of just writing $A \twoheadrightarrow \{F, T\}$, we created a new variation on transformative implication (i.e. a new variation on $\twoheadrightarrow$) that specifies how many items we are allowed to select from the set $\{F, T\}$, and then used it to generate new transformation rules that embody those selections?

Wouldn't that basically solve our problem? If we have a variant of transformative implication that specifies that we can only select *one* of the members of the set, then isn't that the same thing as the "strict substitution" concept? Likewise, if we have a variant of transformative implication that specifies that we can select *any* of the members of the set, then isn't that the same thing as the "loose substitution" concept? So, suppose we have a variable that trans implies any of a set of n forms, i.e. something of the form $X \to \{x_1, x_2, x_3, \ldots, x_n\}$ where X is our variable and x_1 through x_n are any n arbitrary forms.

What we've been calling "strict substitution" is the same thing as allowing selection of only 1 member of the set at a time. Likewise, what we've been calling "loose substitution" is the same thing as allowing selection of any of the n members of the set at any time. So here's an interesting question for you: What happens if you select k members from the set, where $1 < k < n$? Well, then you get something *in-between* the two cases of strict substitution and loose substitution. See? I told you it wouldn't actually end up being that weird.

It's basically just combinatorics. After you have made your selections, you generate new rules that bind to the subsets you have chosen. You can even use the laws of combinatorics to calculate exactly how many different selections are possible for any given selection of k members from a set of size n. In any case, we now have the general idea of what we need to do. Now it's time to get into the specifics.

In particular, we now have three main questions to confront: First, how are we going to represent the different variations of transformative implication? We'll need some symbols for it. Second, how exactly is this process of "generating new transformation rules to embody selections" supposed to work? Third, what should our terminology for talking about this stuff be? The gears of effective communication turn slowly without names by which to refer to things. Words are the grease in the cogs of intellectual progress.

Let's start with the question of how we are going to symbolize the different variations of transformative implication. This is probably the most straightforward of these three questions. Thinking back to what we said earlier, we need to be able to specify exactly how many members of the set are being selected. However, we also need to be able to omit the selection number when we don't care about it, so that we can still support our existing use cases for it. We sometimes won't want to restrict our selection in any way, not even to all n members of an individual rule's consequence set[9].

This means that if a set has n members then there will be $n+1$ possible corresponding variations on transformative implication, one for every possible number of members one might choose to select, plus one more to account for the unrestricted variant. It would be ridiculous to try to design unique arrow symbols for every possible number of selections, seeing as this would entail creating an infinite number of unique symbols. Thus, the obvious approach here is to somehow use numbers to indicate which variation

---

[9] It turns out that a transformation rule that selects from all n members of an individual rule's consequence set won't actually be equivalent to an unrestricted transformation rule, even though it may seem like they'd be equivalent at first glance. You'll understand this point of nuance better once we get into the implementation details.

we are using, and that's exactly what we are going to do. Here's our new definition:

Definition 67. *Transformative logic is designed to make as few assumptions about the way a system of logic is supposed to work as possible, and therefore must be capable of dealing with multiple different possible interpretations of what substitution could mean. Sometimes the substitution value of a symbol must be the same everywhere, whereas other times the substitution value of a symbol could vary by position. More generally however, substitution behavior ranges over a spectrum that varies based on a combinatorial selection of the members of the domain set. If a set contains n members then there are $n+1$ different ways substitution over that set could behave.*

The first of these variations of transformative implication, the one we have already studied, is the unrestricted variant, which we will henceforth refer to as **normal transformative implication**. *A transformation rule that uses normal transformative implication can be referred to as a normal transformation rule. Normal transformative implication is represented by the symbol $\rightarrow$, just as it was previously. It imposes no restrictions on how the transformations it describes may be applied to an expression.*

Thus, any time an expression contains a form that matches the condition form of any normal transformation rule, then any instances of that form in that expression may be replaced with any combination of forms in the consequence set of the normal transformation rule. Normal transformation rules may be mixed freely and will never exclude each other, unlike the selective variants of transformative implication, as you will see.

To each of the other n different substitution variations, we must assign a new symbol. If k is the number of members of the domain of size n that are allowed to be selected in the substitution process for a particular variation of transformative implication, then $\rightarrow_k$ is the symbol we use to represent that variation. All such variants may be referred to as **selective transformative implication**. *More specifically however, we may refer to each individual variation as a* **k-form transformative implication**, *where k is the specific number corresponding to that variation. Note that $1 \leq k \leq n$ will always be true, because it is impossible to select more members than exist in the domain.*

The unrestricted case and the cases where $k = 1$ or $k = n$ will probably be the most common. If $k = n$ then we typically prefer to refer to that variation as **n-form transformative implication** *(where n is* not *replaced with the specific number of elements to be selected) rather than specify its numeric k value explicitly. The reason for this is simply convenience. Doing so prevents us from having to count the number of elements in the consequence set every time we want a selective transformation rule to include all of them, which would be tedious. In such cases, we also write the symbol as $\rightarrow_n$ (literally) rather than specify the exact numeric value of n.*

Just as with the normal transformative implication operator, the selective transformative implication operator can be written facing the other direction, in which case it looks like $_k\leftarrow$. In effect, the k subscript is always placed at the arrow head, on the side with the set being selected from.

For example, if we had a rule in transformative logic of the form $X \rightarrow_1 \{a, b, c\}$ then it would mean that X could bind to only one of a, b, or c at a time. Thus, in an expression such as $X + X + X$ the only possible substitutions would be $a+a+a$, $b+b+b$, and $c+c+c$. In contrast, if it was instead $X \twoheadrightarrow_2 \{a,b,c\}$ then interpretations such as $a+b+a$, $d+c+c$, and $c+c+c$ would all be possible, but interpretations that involve any more than 2 of $\{a,b,c\}$, such as $a+b+c$, $c+a+b$, or $b+c+a$, would be disallowed. If, on the other hand, it was normal transformative implication or n-form transformative implication, as in $X \twoheadrightarrow \{a,b,c\}$ or $X \twoheadrightarrow_n \{a,b,c\}$, then any conceivable substitution drawn from the set $\{a,b,c\}$ would be allowed in $X+X+X$. However, bear in mind that the behavior of $\twoheadrightarrow_n$ might differ from $\twoheadrightarrow$ if there were multiple transformation rules for X.

It's important to understand that n-form transformative implication (written as $\twoheadrightarrow_n$) and normal transformative implication (written as $\twoheadrightarrow$) are not the same, despite the fact that they may seem like they'd be the same at first glance. If we had $X \twoheadrightarrow_n \{a,b,c\}$ and $X \twoheadrightarrow_n \{d,e,f\}$, then only substitutions taken from *one* of the sets $\{a,b,c\}$ and $\{d,e,f\}$ could be applied to X in $X+X+X$. If, on the other hand, it was normal transformative implication, as would be the case if we had $X \twoheadrightarrow \{a,b,c\}$ and $X \twoheadrightarrow \{d,e,f\}$, then any conceivable substitution drawn from any combination of either set would be allowed in $X+X+X$. Thus, having $X \twoheadrightarrow \{a,b,c\}$ and $X \twoheadrightarrow \{d,e,f\}$ has the same net effect as if you had $X \twoheadrightarrow \{a,b,c,d,e,f\}$. In contrast though, having $X \twoheadrightarrow_n \{a,b,c\}$ and $X \twoheadrightarrow_n \{d,e,f\}$ is not the same as having $X \twoheadrightarrow_n \{a,b,c,d,e,f\}$. Why this distinction exists will make more sense later on, once we cover the details of selective transformative implication in greater depth.

In the definition for selective transformative implication, we created a way of referring to each specific variation of selective transformative implication by its number by prefixing it with "k-form". A form can have any number of different transformation rules applicable to it, and each one of those different rules can potentially have a different k number. However, it is common for many forms to have only a single transformation rule with one specific k number. We should have some terminology for this, since it will often serve as a convenient shorthand by which to refer to certain types of forms. As such, here's a new definition:

Definition 68. *Every form could potentially have any arbitrary number of rules, and each of these rules could have a different k number with respect to selective transformative implication. However, for the case where a form has only one transformation rule, with only one specific k number, we may refer to such a form as being a **k-form**, where k has been replaced with the specific number. For example, if $k = 1$ then we may refer to that form as a 1-form. Likewise, if $k = n$ then we may refer to that form as an n-form. More broadly, any form to which at least one selective transformation rule is applicable may be referred to as a **selective form**, without having to specify a specific k value.*

*Generally, forms with multiple applicable transformation rules do not have simple k-form names, due to the fact that having multiple rules would typically make it impossible to classify such a form using a single k value. However, there are two exceptions: (1) forms with only normal transformation rules and (2) forms with only 1-form transformation rules. Regardless of how many rules a form with only normal transformation rules has, it can always be referred to as a **normal form**. Similarly, regardless of how many rules a form with only 1-form transformation rules has, it can always be referred* to as a **1-form**.

The reason for this is because 1-forms and normal forms behave the same regardless of how many transformation rules they are split up into[10]. The rules behave like a simple union of the consequence sets in such cases. However, these are special cases, and this property does not hold for other cases. For any given form with multiple applicable rules, if not all of the applicable rules are normal transformation rules, or if not all of the applicable rules are 1-form transformation rules, then no single k-form name will apply. In such cases, splitting the rules up will potentially change the behavior.

Alternative terminology must be used in such cases. As such, if multiple transformation rules for a form exist, and all such transformation rules share the same k value, then that form may be referred to as a **split k-form**, where k is the specific number. Generally, split k-forms do not behave the same as k-forms that have only one rule, despite sharing the same k value. It often won't act like a simple union.

Like all forms with multiple applicable selective transformation rules, applying any one selective transformation rule to a split k-form will have the net effect of forcing all your substitutions in a given expression to come from one of the selective transformation rules. This behavior is simply a consequence of how the selective transformation process works, which we'll be learning about soon. Just trust me for now. Also: By convention, forms with only n-form transformation rules applicable to them may still be referred to as **split n-forms**, even though the underlying k values may differ.

The most complicated case is when the k values differ. If a form has multiple transformation rules available to it, and at least one of them has a different k value than another, then we may refer to that form as a **hybrid selective form**. A hybrid selective form cannot be classified by a single number. It has multiple associated k values, so we would need to use multiple numbers to classify it. Typically, we will not even bother specifying these numbers in any way, and will instead simply use the broad term "hybrid selective form" to refer to such forms.

However, if you wish, you may specify the numbers by listing them in ascending order separated by dashes, with no redundancies for rules that share the same k value. It will look something like **hybrid k_1-k_2-...-k_{m-1}-k_m form**. Thus, for example, if a form has three different selective transformation rules available to it, with k values of 3, 5, and 7, then such a form could be referred to as a hybrid 3-5-7 form. If one or more (but not all) of the rules is a normal transformation rule, then an asterisk (i.e. *) should be appended to the dashed number list.

Thus, if we added a normal transformation rule to the "hybrid 3-5-7 form" example then it would become a "hybrid 3-5-7-* form" instead. Similarly, if at least one n-form rule exists for a hybrid selective form, then an n should be added to the dashed list, placed after all the numbers but before any asterisk, as in "hybrid 3-5-7-n-* form". Just as with the numbers, no more than one of n or * should ever occur in any such name. Alternatively, you could use the actual k value instead of n, but then you'd have to count the set.

---

[10]By the way, this is an example of a broad mathematical phenomenon known as *degeneracy*, which is when a subset of a system exhibits simpler behavior than the rest of the set, due to special properties the subset possesses. The phrase "degenerate case" is often used to refer to such cases.

Technical Note: Normalization binding and selective substitution (two processes that you will learn about later) do not count towards determining a form's classification as a k-form, split k-form, or hybrid selective form. The normal transformation rules generated by these processes do not have the same condition form as the original form, and therefore are irrelevant to the original form's classification.

Don't worry if that sounded like too much info. Just keep reading anyway. The examples will make things much clearer. For example, if we had a rule of the form $X \rightarrow_1 \{a, b, c\}$ then we could say that X is a selective form, and more specifically that it is a 1-form. If a form is a 1-form then that would imply that any expression with that form in it could have that form bind to only 1 member of the domain set at a time.

Similarly, if we had $Y \rightarrow_4 \{2, 3, 5, 7, 11, 13, 17\}$ then we could say that Y is a 4-form. Like all 4-forms, Y could bind to up to 4 members of the domain set at a time in any expression in which Y occurs. In other words, Y could be substituted with up to 4 different values in the same expression, and it still would not be considered inconsistent. This is quite unlike most uses of variables in mathematics, which are typically 1-forms and which therefore must be substituted with the same value across an entire expression.

If the only transformation rule for X in a rule set was $X \rightarrow_1 \{a, b, c\}$ then referring to X as a 1-form would be valid. However, if we had both $X \rightarrow_1 \{a, b, c\}$ and $X \rightarrow_2 \{d, e, f\}$ in the same rule set then we could no longer refer to X as a 1-form. In such a case, X would be a hybrid selective form and thus could no longer be classified by a single k value. You could call it a hybrid 1-2 form though.

If we had the expression $X + X + X$ under this rule set, then the X symbols in the expression could be resolved into either exactly 1 of $\{a, b, c\}$ or some selection of at most two of $\{d, e, f\}$. Thus, on the one hand, we could transform $X + X + X$ into $a + a + a$, $b + b + b$, or $c + c + c$. Or, on the other hand, we could transform $X + X + X$ into forms like $d + e + e$, $f + d + f$, $d + d + d$, or $e + f + f$ (etc). However, we could *not* transform $X + X + X$ into something that mixes elements from both $\{a, b, c\}$ and $\{d, e, f\}$, such as $a + a + d$, $c + e + c$, or $f + e + b$ (etc).

What about *n*-forms and normal forms? Well, if we had $X \rightarrow_n \{a, b, c\}$ and $X \rightarrow_n \{d, e, f\}$ then we could say that X is a split *n*-form. Under this rule set, substitutions for X in $X + X + X$ could only come from one of the two sets. Thus, $X + X + X$ could be transformed into forms like $a + b + c$, $b + a + c$, $d + e + f$, or $f + e + d$, but could not be transformed into forms like $a + d + f$ or $d + b + a$. In contrast, if we had $X \rightarrow \{a, b, c\}$ and $X \rightarrow \{d, e, f\}$ then we could say that X is a normal form. In that case, any conceivable substitution would be permitted, including even substitutions that mix elements from both sets, such as $a + d + f$ or $d + b + a$. Notice that non-split *n*-forms (i.e. forms with only one rule: a single *n*-form transformation rule) are the only case where *n*-forms behave the same as normal forms.

I also want to point out that the same kind of rule-exclusive behavior that applies to hybrid selective forms and split *n*-forms also applies to split *k*-forms of all kinds, even though split *k*-form rules all share the same *k* value. I mentioned earlier that if we had both $X \rightarrow_1 \{a, b, c\}$ and $X \rightarrow_2 \{d, e, f\}$ then $X + X + X$ could only apply one rule or the other, not both at the same time. The same applies to split *k*-forms, and indeed to all forms with selective transformation rules. Thus, for example, if we instead had

$X \twoheadrightarrow_2 \{a, b, c\}$ and $X \twoheadrightarrow_2 \{d, e, f\}$ then the X symbols in $X + X + X$ would have to take their substitutions from *one* of either $\{a, b, c\}$ or $\{d, e, f\}$, never from both. Hence, $X + X + X$ could transform into things like $a + b + b$, $c + c + c$, $d + f + d$, or $f + f + e$, but never into things like $a + e + e$ or $d + c + d$.

What would happen if we *didn't* make the substitution behavior for selective transformations rule-exclusive like this? The answer: logical incoherence. You see, in order to make reasoning in this system practical we want a system where adding duplicate copies of the same rule to a rule set would not change the behavior of that rule set. However, if substitution during selective transformation wasn't rule-exclusive then redundant copies of the same rules could easily change the behavior of a rule set. For example, consider the following redundant rule set:

1. $X \twoheadrightarrow_2 \{a, b, c\}$

2. $X \twoheadrightarrow_2 \{a, b, c\}$

What would happen if we allowed substitutions of X to take values from *both* rules in the same expression? Well, in effect, it would allow us to ignore the selective constraints of the k values, rendering such constraints immediately pointless. Not only that, but because sets ignore duplicates, every rule set can implicitly be thought of as having an infinite number of copies of its already existing rules, even if those copies aren't written out explicitly, but this surely should not change the behavior of the rule set since it is still the same rule set.

Yet, if we *didn't* make the substitution process rule-exclusive, then it *would* change the behavior. The result would be nonsense. Thus, only one selective transformation rule can be applied to a given form in an expression at once, on pain of otherwise degenerating to logical incoherence. That's why I made this design choice. Rule-exclusivity is necessary for coherent selective transformation[11].

Anyway, it would be wise to have an official term for the number subscript that we write near the transformative implication symbol ($\twoheadrightarrow$) for each variant of selective transformative implication. We've been calling it the "k value" up until know. As we know, it represents the number of members of the domain set to which a form could bind at any moment of time. We should create a more specific name for it, one that is less liable to have naming conflicts than "k value" is. What would a good name for it be? It'd be best to have a name that captures the essential nature of it in some sense. To fill that role, here is our next definition:

Definition 69. *Every variant of selective transformative implication has a number attached to it that indicates which variant it is and how it behaves. This number represents the maximum number of members of the domain to which it can bind at any single moment of time. In a k-form transformative implication, any combinatorial selection of up to k members of the domain set of size n is allowed. We may refer to this number k*

---

[11] However, be that as it may, the way that rule-exclusivity is actually enforced is via normalization binding and selective substitution (two processes that you will learn about later). Rule-exclusivity is merely a natural consequence of these processes. It is not actually a separate policy.

as the ***individuality number*** (*or alternatively as the **selection number***) *of that variant of selective transformative implication. The reason why it's called the "individuality" number is because it determines how much like an "individual" a form subject to the transformation rule will behave during substitution.*

Individuality numbers are sort of like rankings. Just as competitors in contests who are given lower ranking numbers are considered to be better competitors than those given higher ranking numbers (e.g. as in "I placed #1 in the tournament!"), *k*-forms that are given lower individuality numbers are considered *more* like individuals than those with higher individuality numbers. You can think of the individuality number as sort of a measure of how many "individuals" a form stands for, hence the name and the associated connotation.

3.5 Tracing notation

On a sidenote, let me explain why I chose to write the individuality number as a subscript under the arrowhead, instead of at any other location near the arrow. As for why I put the number near the head of the arrow, that's because doing so makes it clearer that the operator is selecting that many members from the set on that side. However, the more important point I want to make is why I chose to write the number as a subscript instead of writing it above the arrow.

You see, the problem with writing the number above the arrow would be that sometimes people write numbers above implication arrows in order to indicate which of a numbered list of inference rules that particular transformation step was justified by. This can make following the line of reasoning much easier and is very useful. For example, let's take another look at the rule set for the rune puzzle system. The rune puzzle rule set originally appeared on page 189, but I'm going to re-write the list here to make it easier to refer to:

1. ☆ → ☆ ⋈
2. ☆ → ⋈ ☆
3. ⋈ → ☆ ☆
4. {☆☆, ☆☆, ⋈⋈} → X

Initial Form: ☆☆⋈

Target Form: XXX

As you may remember, I gave the solution for how to get from the initial form to the target form under this rule set. After that, we talked about how you can easily create custom inferential puzzles by simply coming up with a list of arbitrary symbol transformation rules designed to have interesting properties, and then creating pairs of initial forms and target forms to embody specific puzzles under that rule set. Furthermore, I mentioned how solving this puzzle allowed us to derive a new shortcut rule of the form ⚹⚹☒ → ✕✕✕ and that we call such derived rules theorems. Anyway, let's try re-writing our solution with numbers above each arrow to indicate which transformation rule is being applied at each step in the process:

$$⚹⚹☒ \xrightarrow{2} ⚹☒⚹☒ \xrightarrow{1} ⚹☒⚹☒☒ \xrightarrow{4} ⚹☒⚹✕ \xrightarrow{3} ⚹⚹⚹⚹✕ \xrightarrow{4} ✕✕✕$$

See? This makes it a lot easier to trace what the reasoning was behind this proof. Furthermore, as you derive new rules for a system you can add them to the list of rules and give them a new number, so that you can also refer to your own derived theorems easily by number. This can make systematically solving things much easier. Instead of trying to keep all the properties you figure out for a system up in your head, which is quite difficult, you can instead just continually add on new numbered rules of inference to your list. Every new derived rule you discover will increase your chances of solving new problems you encounter for the system.

If I had chosen to notate the individuality number (a.k.a. the k value or selection number) of selective transformative implication by writing the number above the transformative implication arrow, then it would have created a conflict with this useful numbered step-tracing notation, thereby making it impossible to use it safely. We wouldn't want that.

In addition to numbering the steps though, there's also another thing you can do to make the reasoning easier to follow. You can mark which parts of each form match the form of the condition for the next transformation step. One idea would be to use parentheses. Using parentheses for this could work maybe, but some logical systems use parentheses for grouping, so it probably wouldn't ultimately be a good choice. Instead, we will use underlines, which are relatively seldom used in math notation and therefore relatively safe:

$$⚹\underline{⚹}☒ \xrightarrow{2} ⚹☒\underline{⚹}☒ \xrightarrow{1} ⚹☒⚹\underline{☒☒} \xrightarrow{4} ⚹\underline{☒}⚹✕ \xrightarrow{3} \underline{⚹⚹⚹⚹✕} \xrightarrow{4} ✕✕✕$$

See? It's even easier to follow now. However, we can take it even one step further if you're a real stickler for clarity and for making the steps as easy to trace as possible. How? Answer: In addition to using underlines, we can also use overlines to mark which parts of each new form came from the previous transformation. In a sense, overlines are sort of the opposite of underlines. Underlines mark the condition of each step, whereas overlines mark the consequence of each step. Thus, by combining the two, we can maximize the traceability of the steps. Here's what it would look like with both underlines and overlines, in addition to the step-tracing numbers:

$$⚹\underline{⚹}☒ \xrightarrow{2} ⚹\overline{☒\underline{⚹}}☒ \xrightarrow{1} ⚹☒⚹\overline{\underline{☒☒}} \xrightarrow{4} ⚹\underline{☒}⚹\overline{✕} \xrightarrow{3} \underline{\overline{⚹⚹⚹⚹}✕} \xrightarrow{4} \overline{✕✕✕}$$

This might look a little weird at first, but once you get used to it then it makes it a lot easier to see exactly what's happening without having to pause to refer to the list of transformation rules or to carefully study the forms. It's a pretty useful set of notations. In fact, I think we should give each of these different tracing notations specific names.

You know what I like to say about the power of words. Having a few new words to refer to what these are could really help the ideas stick and aid in communication.

First off though, we'll need to give a name to the base object which we are embellishing with these different notations. Specifically, I am referring to the chain of forms and arrows which collectively embody a description of each and every step in the transformation process. Let us formally refer to such chains of transformations from now on as **transformation paths**.

Thus, any expression of the form $X_1 \to X_2 \to \ldots \to X_n$ (where X_1 through X_n stand for any arbitrary forms) is an example of a transformation path. Furthermore, let us refer to the embellishments we place on a transformation path to aid in tracing its behavior as **tracing marks**, and to each individual variation of tracing mark notation as a **tracing type**. Having defined those base terms, we are now ready to define the different tracing types:

1. **untraced:** Any transformation path which bears no tracing marks whatsoever may be referred to as an *untraced* transformation path.

2. **step-traced:** If a transformation path has numbers written above each transformative implication symbol ($\to$) to indicate which numbered transformation rule justified that step, then it may be referred to as a *step-traced* transformation path.

3. **pre-traced:** If a transformation path has underlines written below each form to indicate which parts of those forms match the conditions of each successive transformative implication, then it may be referred to as a *pre-traced* transformation path.

4. **post-traced:** If a transformation path has overlines written above each form to indicate which parts of those forms match the consequences of each preceding transformative implication, then it may be referred to as a *post-traced* transformation path.

5. **half-traced:** If a transformation path is both pre-traced and post-traced, but not step-traced, then it may be referred to as a *half-traced* transformation path.

6. **full-traced:** If a transformation path is pre-traced, post-traced, and step-traced simultaneously, then it may be referred to as a *full-traced* transformation path.

Notice that I added in some possible variants that we didn't directly discuss for the rune puzzle system. Specifically, so far we've written the rune puzzle solution using four different tracing variants: (1) untraced, (2) step-traced, (3) both step-traced and pre-traced, and (4) full-traced. Notice, for example, that it is possible to use just post-tracing, even though I didn't cover that case for the rune example.

One interesting variant I want to draw attention to though is the half-tracing variant. The interesting thing about half-tracing is that when you combine pre-tracing and post-tracing then it has the effect of implicitly expressing the transformation rules that govern each step without actually having to refer to them by number. This could be useful if you don't have a numbered list or if you don't want to bother with numbering but still want a high degree of clarity.

My personal favorite tracing methods here are untraced, step-traced, half-traced, and full-traced. I feel like these four methods have the best overall advantages in terms of clarity versus effort put into them. Untraced has a certain minimalistic elegance because it doesn't create as much visual noise as the traced variants do. However, half-tracing and full-tracing are great for making it as easy as possible to keep track of what you are doing and for anyone reading your work to check whether it is actually justified by the rules or not. Step-tracing is good for when you want to track your justification but don't want the labor involved in underlining and overlining. Each of these tracing methods have different pros and cons. Pick your poison.

If we don't specify what type of tracing a transformation path uses, for example if we just refer to it as "a transformation path", then it could be using any tracing type or none at all. If it isn't specified then you should assume the general case. Also, there are some notations that this system will conflict with. For example, overlines are used in geometry above pairs of point names to indicate a line segment between them. Thus, for example, $\overline{AB}$ in geometry would mean a line segment between point A and point B. If you encounter conflicts, try using a tracing type that doesn't conflict with the other notations you are using. If all else fails, you can always revert back to using untraced notation.

As great as these tracing notations are, remember that often when communicating your results it is more effective to explain them in plain old prose, i.e. in a natural language like the one I'm writing this sentence in right now for example. This is especially true when you are writing for a wide general audience. This book, for example, is targeted towards a wide general audience. That's why you'll never see me write out giant cryptic lists of purely symbolic arguments in this book.

Purely symbolic arguments are typically only seen at the higher levels of logic and math, such as in academic research, and also in calculation-heavy endeavors such as in homework assignments, in verification of other people's work, and in ad-hoc back-of-the-napkin explorations. Augmenting your work with prose is often far better for clarity and comprehension than using only symbols though.

Writing for a general audience favors regular prose pretty heavily, and for good reason. Don't underestimate the value of natural language. On the other hand, symbol-heavy writing tends to become mandatory once the situation becomes too complex for natural language to unambiguously or efficiently express it. In practice, you'll have to dip into using symbols at least some of the time, regardless of how much you are trying to write for a wide general audience. Symbols are more expressive for some things, whereas natural language is more expressive for other things. You need to try and strike the right balance between the two.

3.6 Selective transformative implication (part 2)

Anyway though, we kind of got off on a tangent there for a while. It was a productive and useful tangent, but a tangent nonetheless. If you recall, before we went off on a tangent (starting on page 208) that led to an in-depth analysis of why I chose to write the individuality number as a subscript under the trans implication arrowhead and to defining a bunch of useful tracing marks, we were discussing selective transformative implication, selective forms, and some other related terminology.

However, before *that*, when we had just had our core insight about how to best go about distinguishing the different ways one could make substitutions (on page 202) I mentioned that when these new transformative implication variants selected some members of the domain set to bind to that it would "generate new transformation rules that embody those selections". This is an important implementation detail. Earlier we gave examples of how selective transformative implication would effect the interpretation of specific expressions. For example, we said that $X \twoheadrightarrow_1 \{a, b, c\}$ would cause the only possible interpretations of $X + X + X$ to be $a + a + a$, $b + b + b$, and $c + c + c$. In contrast, different individuality numbers could result in substitution binding behaviors that differ from this.

Well, it's now time that we discuss exactly how this binding behavior actually works in-depth. As you may recall, I said that selective transformative implication would "generate new transformation rules that embody those selections". However, I wasn't being totally precise when I said this. It isn't necessarily true that it will need to generate new transformation rules, although it often is. A more correct statement would be that if you can't find a compatible transformation rule that could embody your selection, then you must generate a new one.

What do I mean by "finding a compatible transformation rule that could embody your selection" here? Well, basically, in order for a normal transformation rule to be considered compatible with a selective transformation rule it needs to meet two criteria: First, the set it trans implies must be a subset of the set that the selective transformation rule trans implies. Second, the set must have no more than k members, where k is the individuality number of the selective transformation rule. Let's give this kind of compatibility a formal definition to make it more precise and to make it easier to refer to:

Definition 70. *Suppose we have two arbitrary forms A and B. Suppose that each of these forms trans implies a corresponding set (each with any arbitrary number of members), represented by S_A and S_B respectively. Furthermore, suppose that A is a selective form with individuality number k, that B is a normal form (i.e. a form with no selective rules), and that $A \twoheadrightarrow_k S_A$ and $B \twoheadrightarrow S_B$ are held to be valid. Under these circumstances, if the set S_B is a subset of the set S_A and the set S_B has at most k members, then we say that B has **selective compatibility** with A.*

For example, suppose that $A \twoheadrightarrow_1 \{F, T\}$ and $B \twoheadrightarrow F$. As you can see, $\{F\}$ is a subset of $\{F, T\}$ and the number of members in $\{F\}$ is less than or equal to the individuality number 1. Therefore, we can conclude that B has selective compatibility with A. Similarly, if $A \twoheadrightarrow_2 \{1, 2, 3, 4\}$ and $B \twoheadrightarrow \{2, 4\}$, then B would be selectively compatible with A. In contrast though, neither of the pairs $A \twoheadrightarrow_2 \{a, b, c, d\}$ and $B \twoheadrightarrow \{e, f\}$, nor $A \twoheadrightarrow_1 \{\pi, e, i\}$ and $B \twoheadrightarrow \{\pi, i\}$, have selective compatibility. In the former case they are incompatible because $\{e, f\}$ is not a subset of $\{a, b, c, d\}$, whereas in the later case they are incompatible because $\{\pi, i\}$ does not have at most k members (because in this case $k = 1$).

Now we know what compatibility for substitution binding really means. However, how exactly does the binding process itself work? How do we go about actually performing selective transformation on a rigorous basis? Well, there are three phases to the process of performing selective transformation. The first phase is called *normalization binding*, the second phase is called *selective substitution*, and the third phase is called *normal substitution*. The process is very simple and straightforward as long as you keep the distinctions between this three phases clear in your mind. Otherwise, you might get confused or make careless mistakes. Let's create some formal definitions for these phases:

Definition 71. *The first phase of applying a selective transformation rule is to find (or, if necessary, create) a normal transformation rule which is selectively compatible with the selective transformation rule we intend to apply. Suppose that $A \twoheadrightarrow_k S_A$ is our selective transformation rule, where S_A is any arbitrary set. Then, according to the definition of selective compatibility, we must either find a transformation rule $B \to S_B$ where $S_B \subseteq S_A$ and $\mathrm{Count}(S_B) \leq k$, or we must create one. We call this process of either finding or creating a normal form to act as our substitution binding for a selective transformation rule* **normalization binding**. *You know you're done with this phase when you have identified a normal form that is selectively compatible with the selective transformation rule.*

Each selectively compatible normal form that we find or create is, in effect, like a distinct identity within the domain set. For example, for a 1-form X such that $X \twoheadrightarrow_1 \{a, b\}$ we would consider each of a and b to be distinct individuals or identities upon the domain. Interestingly however, it is also true that for a 2-form Y such that $Y \twoheadrightarrow_2 \{a, b, c, d, e\}$ we would consider all of $\{d, b\}$, $\{a, b\}$, and $\{e, c\}$ (etc) to be distinct individuals or identities upon the domain as well. Notice the nuance of this. What it means to be a distinct "individual" or "identity" for the domain set changes depending on the individuality number of the transformation rule. This should also make the reason why I chose the term "individuality number" as the term for this a little bit clearer. Anyway, time for the next phase:

Definition 72. *The second phase of applying a selective transformation rule is that we must replace every instance of the selective form, in the expression in which that form occurs, with a specific normal form corresponding to the identity for which we wish to evaluate it (i.e. with the identity we chose during the normalization binding phase).*

Suppose that A was the selective form and B was the normal form, and that we wanted to evaluate an expression that contains instances of the symbol A. Then, every single instance of A in that expression must be replaced with B, with no exceptions and regardless of the individuality number of A. We call this process **selective substitution**. *You know you're done with this phase when the expression you are evaluating no longer contains any instances of the selective form and now contains only the corresponding normalized form embodying our selection.*

For example, suppose that $X \twoheadrightarrow_1 \{a, b, c\}$ is our selective transformation rule. In that case, X would be our selective form. Suppose we then wanted to evaluate the expression $X + X + X$. Specifically, suppose we wanted to evaluate the case where we select b from the set and we do not yet have a binding for this selection. Suppose we then create a binding for it of the form $\beta \twoheadrightarrow b$ and then apply selective substitution. This would result in $X + X + X$ being transformed to $\beta + \beta + \beta$. In contrast, transforming $X + X + X$ partially, such as into $X + \beta + \beta$ for example, would not be allowed. The selective substitution process requires all instances of the selective form to be replaced with the normal form that was chosen during normalization binding, otherwise correct behavior would not be guaranteed. Next is the final phase:

Definition 73. *The third phase of applying a selective transformation rule is to simply apply whatever normal transformation rules we want to. After normalization binding and selective substitution have finished, we are left with a form that contains no selective forms. All of the applicable rules that were previously written in terms of selective transformation rules have been converted to a form that uses only normal transformation rules.*

This renders the system foolproof. We may now perform any transformations the normal transformation rules specify without having to worry that we might misinterpret the intended selective substitution behavior. We call this process **normal substitution**. *You know you're done with this phase when you have finished applying whatever arbitrary normal transformation rules are necessary to get to the final form you want.*

Note that normal substitution can take place any time after you have completed selective substitution. You do not immediately have to apply the applicable normal transformation rules. You are free to wait as long as you wish, and may continue manipulating the expression in other ways. Just because a transformation rule exists doesn't mean you have to apply it. Transformations in transformative logic are always at-will. You don't have to transform expressions until you want to, even if that means never doing so. In this way, abstractions may be maintained for as long as you need.

In effect, this three step process of rigorously resolving selective transformation rules is just a way of converting selective transformation rules into normal transformation rules so that they are easier to reason about and less error prone. Normal transformation rules are basically a special case of the process of selective transformation, wherein you simply immediately skip to the third phase (normal substitution) because the first two phases (normalization binding and selective substitution) are already done.

Remember the example from earlier where we had $X \twoheadrightarrow_1 \{a, b, c\}$? Let's finish it up. As you may remember, we started with $X \twoheadrightarrow_1 \{a, b, c\}$. We then applied normalization binding which resulted in the creation of a new transformation rule of the form $\beta \twoheadrightarrow b$. After that, we then applied selective substitution to the expression $X + X + X$, making sure to replace *all* instances of X with the normalized variable β, which resulted in $\beta + \beta + \beta$. Finally, we will now apply normal substitution, resulting in $b + b + b$. Here is what our work would look like in symbols:

1. $X \twoheadrightarrow_1 \{a, b, c\}$

2. $\beta \to b$

3. $X + X + X \to \beta + \beta + \beta \to b + b + b$

This is pretty laborious isn't it? It seems like an excessive amount of work for such a simple task. Indeed it is. Moreover, I have used a whole lot of words here to describe something that is really quite simple in practice. I had a reason for going into so much detail about this though. I wanted to make sure you fully understood the underlying rigorous basis for how selective transformation works and why it is a legitimate thing to do. In practice however, it is often more convenient to not bother writing out the bindings and intermediate forms for selective transformations and to instead simply be careful to substitute the correct way.

Understanding the underlying rigorous basis for this process helps you understand precisely what you can and cannot do. This internalized understanding makes it easier to be sure that you are being safe, even when you aren't actually bothering to write out the bindings and intermediate forms. The rigorous version of selective transformation is also more foolproof and is a good fallback method for when you get confused when using selective transformation rules. Thus it was still important to cover it.

One important thing to note about this process is that there is nothing to stop you from performing it multiple times for the same rule set, thereby creating and evaluating multiple different normalization bindings. A selective transformation rule does *not* say that the condition form can only transform to one specific selection of k forms, but rather it merely says that the condition form can only stand for k of those forms at any given moment of time.

Thus, there is nothing to stop you from evaluating all three possible transformations of $X + X + X$ under the rule $X \twoheadrightarrow_1 \{a, b, c\}$ by creating three different normalization bindings $\alpha \to a$, $\beta \to b$, and $\gamma \to c$ that would in turn result in $a+a+a$, $b+b+b$, and $c+c+c$ being derived. In fact, after deriving these, you could even add in $X+X+X \to a+a+a$, $X + X + X \to b+b+b$, and $X + X + X \to c+c+c$ to the rule set as shortcut rules if you wanted to and it would be correct.

As a sidenote, I want to point out that if we wanted to represent the transformation path for the $X + X + X$ example with step-tracing then we would need to extend our step-tracing notation slightly. Why? Well, look at the selective substitution step in the transformation path (i.e. at $X + X + X \to \beta + \beta + \beta$). How could we represent that step using only the usual numbered rule references? We couldn't. The selective substitution step has a different nature than most transformation steps.

We need a way to indicate that something a bit different is going on at that step. In order to do so, we will refer to both the selective transformation rule and the normal transformation rule. Instead of just writing one number, we will write $S_{m,n}$, where m is the list number of the selective transformation rule and n is the list number of the normal transformation rule embodying our selection. The S in $S_{m,n}$ stands for "selective substitution". Thus, if we wanted to write a step-traced version of the $X + X + X$ transformation path (where $X \twoheadrightarrow_1 \{a, b, c\}$ is rule #1 and $\beta \to b$ is rule #2) then we would write it like so:

$$X + X + X \xrightarrow{S_{1,2}} \beta + \beta + \beta \xrightarrow{2} b + b + b$$

By far the most common types of transformative implications you will encounter are 1-form transformative implication (e.g. $A \twoheadrightarrow_1 B$) and normal transformative implication (e.g. $A \twoheadrightarrow B$). Luckily, these two types are also the easiest to reason about without having to do the full-scale rigorous process for selective transformative implication. As for the cases in-between, you are more likely to get confused by them, especially when the individuality number is relatively high, and hence it might be safer to use the rigorous process for them if you feel uncomfortable.

In general though, the selective transformation process is quite easy. Don't let these verbose descriptions of the underlying rigorous mechanics trick you into thinking it's more complicated than it really is. Oh, but you do need to make sure that the new names you give to forms generated by normalization binding don't conflict in any way with any existing names. That's one little gotcha that might bite you from time to time if you're careless about how you write your rules.

The normalized rules generated by the selective transformation process should never interfere with or change the behavior of any other rules. If they do, then you probably inadvertently introduced a name conflict somewhere. Making sure that the normalized rules generated by the selective transformation process never use symbols from the same alphabet as the rest of your rules is one way to prevent this problem from ever happening.

Anyway, it's finally time to return to the original example we used to demonstrate that transformation rules can be ambiguous in transformative logic if we don't have a method by which to distinguish the different possible ways of performing substitution. The example I am referring to was discussed on page 199. We were examining the very simple domain set specifications for the pair of boolean variables A and B when we came to the startling discovery that they were ambiguous as written, despite being so simple. Here are the corrected versions of those rules:

1. $A \twoheadrightarrow_1 \{F, T\}$

2. $B \twoheadrightarrow_1 \{F, T\}$

As you may recall, earlier we considered what might happen if we had applied the original rules defining the domains of A and B to the expression $(A \wedge A) \vee (B \wedge B)$, and we concluded that there was nothing stopping us from substituting into the variables in ways that are not allowed in classical logic. In particular, we could have transformed it into something like $(T \wedge F) \vee (T \wedge F)$, thereby giving inconsistent values to A and B. However, now that we have corrected the original rules so that they use selective transformative implication, there is no longer any danger of this. The expression $(A \wedge A) \vee (B \wedge B)$ will now always be evaluated as intended.

So, that covers the simple boolean example. What about the other rule sets we explored? Were any of those ambiguous too? To refresh your memory, the other rule sets we explored were (1) the alternating astrological sign system, (2) the survival horror statue puzzle system, and (3) the rune puzzle system. Well, luckily, none of them were ambiguous. They all function exactly as intended. However, if you added on more transformation rules or more constraints, or if you wanted to be able to handle a wider variety of initial forms in a conceptually correct way, then the use of selective transformative implication could easily become necessary.

Oh, on a side-note, in addition to normalization binding and selective substitution, the opposite processes are also allowed and will be referred to as *denormalization binding* and *denormalizing substitution* respectively. The processes are similar but have slightly different rules. Let's make some formal definitions for them:

Definition 74. *In addition to being able to convert selective forms into normal forms, the opposite process is also possible. We call this process **denormalization of forms**. It's purpose is to allow us to intentionally restrict ourselves to a more narrow set of possibilities than what a normal form allows. For example, we can force ourselves to look at only the transformations that a 1-form would allow even if that form is actually a normal form. The reason why we are allowed to do this is because normal forms allow us to choose any conceivable substitution behavior and all transformation rules are always optional, therefore there is nothing to stop us from willingly restricting which selections we want to explore for a form.*

Definition 75. *In order to be able to denormalize a form, a property similar to selective compatibility must be true. However, it is slightly different. We call this required property **denormalizing compatibility**. Suppose that A and B are arbitrary forms, that S_A and S_B are arbitrary sets, and that $A \to S_A$ and $B \to_k S_B$. Then, B has denormalizing compatibility with A if S_B is a subset of S_A.*

The value of k is irrelevant, unlike in selective compatibility. Notice also that the order has changed. Denormalizing compatibility is a property of a normal form considered as a candidate for being replaced by a selective form, whereas selective compatibility is a property of a selective form considered as a candidate for being replaced by a normal form.

Definition 76. *Before performing denormalizing substitution of a form A, you must first either find or (if necessary) create at least one form B which has denormalizing compatibility with A. This process of either finding or creating at least one form B that has denormalizing compatibility with A is called **denormalization binding**. It is in some sense the opposite of normalization binding.*

Definition 77. *Suppose we have some arbitrary form X, which contains some number of instances of normal form A and that we want to apply denormalization to these instances of A in X. Suppose furthermore that we have n forms $B_1, B_2, \ldots, B_n$ that each have denormalizing compatibility with A. Then, we may optionally substitute any of the instances of A in X with any of the forms $B_1, B_2, \ldots, B_n$ according to our whims.*

*Unlike selective substitution, not all instances of A have to be replaced with anything and furthermore different instances of A in X can be replaced with different selective forms from $B_1, B_2, \ldots, B_n$. We call this process **denormalizing substitution**. It is in some sense the opposite of selective substitution. All denormalizing substitutions are optional, whereas all selective substitutions are mandatory.*

In step-tracing notation, denormalizing substitution should be represented with $D_{m,n}$ instead of $S_{m,n}$. Anyway though, we won't really be paying any more attention to denormalization of forms in this book. I'm including it mostly just for the sake of completion and to broaden your perspective a bit. Putting all this talk of denormalization aside, do you remember when we briefly discussed an example of how to state the order independence (commutativity) of both classical disjunction and classical conjunction simultaneously using transformative logic?

The discussion occurred earlier in the transformative logic section, on page 187. Do you remember how I said that the rule set that I gave wouldn't (as written) actually allow you to directly transform $T \vee F$ to $F \vee T$ (etc)? Do you remember how I mentioned that there was a way to do it, but that it involved operators we had not yet defined? Well, we now finally have the operators we need (transience delimiters and selective transformative implication), so we're going to define the complete version of that rule set now. Here it is:

1. $A \rightarrow_1 \{F, T\}$

2. $B \rightarrow_1 \{F, T\}$

3. $\square \rightarrow_1 \{\vee, \wedge\}$

4. $⟅ A \square B \leftrightarrow B \square A ⟆$

Notice that the normal transformation rules for A, B, and $\square$ from before have been replaced with selective transformation rules and that the entirety of the last rule has been surrounded with trans-free delimiters. The last rule here is the most likely to give you pause. It might be a bit confusing. Let me explain.

Basically, the $⟅ A \square B \leftrightarrow B \square A ⟆$ means that we are treating $A \square B \leftrightarrow B \square A$ as a form in and of itself, and then saying that we want to add *all* rules derivable from that form to our list of rules. Thus, the rule $⟅ A \square B \leftrightarrow B \square A ⟆$ has the net effect of enabling the user of the rule set to perform both transformations like $A \square B \leftrightarrow B \square A$ and transformations like $T \vee F \leftrightarrow F \vee T$, and so on, instead of being restricted to only forms involving the literal symbols A and B (i.e. instead of being restricted like it was originally, as written on page 187).

3.7 Transformative languages and interpretation injectors

Anyway, for practice, let's take one of our earlier scenarios and modify it in a way that causes selective transformation to become necessary. And, just to keep it interesting, let's modify it in a way that requires the use of one of the in-between variants of selective transformation (i.e. not normal and not 1-form). That way, you'll get some idea of what an in-between variant feels like in a real situation. Which system should we modify though? Well, I came up with a plausible modification of the survival horror statue puzzle system that fits reasonably well with the idea of in-between selective transformation, so that's the one we're going with.

We first described the original survival horror statue puzzle system on page 179. As you may recall, the way the player in our hypothetical survival horror video game is able to solve the puzzle is by noticing that in one of the other rooms there is a Renaissance era painting depicting a figure identical in character to that of the mysterious statue of the warrior and that in that painting the character is holding up a book instead of a sword or a shield.

Let's mess with the scenario now. Instead of having one statue, let's have six. Instead of the statue being located in one solitary alcove, the statues will now be lined up in a row along one side of a dark hallway. All six statues will each have one open hand for you to put an object in, much like with the one statue from the original scenario. Imagine that every one of these statues is very intricately sculpted to have a life-like resemblance to a unique person that each one is intended to represent. As such, all of the statues are thus easily distinguished by their features.

Additionally, instead of having only one of each item (book, sword, or shield) available to you, you now have an effectively unlimited number (at least 6) of each type of item potentially available to you. However, there is a catch. The items are stored in one of three special computer-controlled cabinets, each of which supplies only one of the three types of item. The moment the computer detects that you have opened any two of the three cabinets at least once each then it will permanently lock the remaining cabinet.

The Renaissance era painting has changed now as well. This time the painting depicts two distinct sides fighting in a bloody small-scale skirmish. Each warrior is wearing one of two different uniforms, corresponding to which nation that particular warrior has pledged his or her allegiance to. There's no particular pattern to what weapons they are using or to what they are holding. Some have swords, some have bows, some have spears, some have maces, some have shields, and some have had their weapons knocked entirely out of their hands. One thing you do notice though is that all of the warriors wearing one type of uniform have an emblem of a sword engraved on their armor, whereas all the other warriors have an emblem of a shield.

So, here's the puzzle the player is presented with in this scenario: You have three types of objects you can place in each statue's hand (book, sword, or shield). Which statues should receive which items? At the end of the hallway is a lever. If you pull the lever when you have the wrong combination of objects in the statues' hands, then acid will spray out of the pipes in the ceiling above the dark hallway, leading to your inevitable demise. On the other hand, if you get the correct combination, then a storage chest nearby will unlock and inside it you will find a secret weapon. What should you do?

The answer is not too hard to figure out. You should use the distinctive features of the various warrior statues in the hallway to match them up to what kind of uniform they were wearing in the painting, and then from that match them up to which kind of emblem they were wearing. Once you know which type of emblem each warrior was wearing, you simply place the corresponding object in their hands. What does the rule set look like for this system though? Well, here's one way of implementing it:

1. $\oplus \rightarrow_2 \{📖, ✗, \varnothing\}$

2. ⟨🛡⟩ → {🛡,✗,∅}

3. ⟨✗⟩ → {🛡,∅}

4. ⟨∅⟩ → {🛡,✗}

5. ⚖∅ ⚖∅ ⚖✗ ⚖∅ ⚖✗ ⚖✗ → ✝

6. ·|⚖∅ ⚖○ ⚖○ ⚖○ ⚖○ ⚖○|· → ☠

7. ·|⚖∅ ⚖∅ ⚖○ ⚖○ ⚖○ ⚖○|· → ☠

8. ·|⚖∅ ⚖∅ ⚖✗ ⚖○ ⚖○ ⚖○|· → ☠

9. ·|⚖∅ ⚖∅ ⚖✗ ⚖∅ ⚖○ ⚖○|· → ☠

10. ·|⚖∅ ⚖∅ ⚖✗ ⚖∅ ⚖✗ ⚖○|· → ☠

11. ·|⚖∅ ⚖∅ ⚖✗ ⚖∅ ⚖✗ ⚖✗|· → ☠

Initial Form: ⚖⊕ ⚖⊕ ⚖⊕ ⚖⊕ ⚖⊕ ⚖⊕

Target Form: ✝

I've arbitrarily designed the rule set so that ⚖∅ ⚖∅ ⚖✗ ⚖∅ ⚖✗ ⚖✗ is the correct solution for the placement of the items in the statues' hands. The solution could have been any other combination of swords and shields. It's kind of like a binary password code, except instead of 0s and 1s it uses swords and shields. Also, arguably, we could have removed all of the ⚖ symbols in order to make these forms less tedious to write, which would have resulted in ∅∅✗∅✗✗ being the solution form. However, I decided to leave the statues in anyway (as inert as they may admittedly be) so that this version of the puzzle resembles the original version more closely.

Notice that there are two different major ways the player could screw up the puzzle. The first way is by picking the wrong two cabinets to open, thereby causing the remaining one to lock and making it impossible to solve the rest of the puzzle. The second way is by placing at least one wrong item into any of the statues' hands. Admittedly the cabinet scenario is kind of contrived, but it nonetheless illustrates the concept of selective transformation fairly well.

The thing that probably sticks out the most about this solution though is that we have to specify a ton of extra rules in order to correctly express the failure state rule. Do you see why we have to specify so many rules? We need to express the fact that every other form of the statue scenario besides the one that leads to success will lead to failure. The problem with this is that the only method we have of expressing forms and transformation rules is by expressing them directly. We have nothing in our vocabulary for saying that in order for a form to match a condition of a transformation rule it has to *not* match a certain form. Basically, we currently can only create direct transformation rules, never indirect transformation rules, and this causes us to have to work around the problem in a laborious way.

Let's fix this problem so that these kinds of rule sets are not so tedious to specify. How should we go about fixing it though? We could just add in one new special delimiter that encloses a form and matches to anything that does *not* have that form. It would work like parentheses or quotation marks, except it would mean "any form that doesn't match this form" instead. That would be the obvious thing to do. It's the first idea that comes to mind.

However, we can actually do a lot better than that. Let's not rush into this decision, and let's instead take a moment to really think about what the underlying cause of this weakness in our notation might be. Solutions that are conceptually correct and that actually correlate to the underlying nature of a system are often vastly more powerful than ad-hoc spur-of-the-moment solutions.

Let's think. What is the real source of our system's poor expressive capabilities? Why is it so difficult to express the concept of form negation in our current vocabulary? We've dealt with logical negations before in other frameworks of logic and it never gave us any trouble then. It was super easy to define negation for both classical logic and set theory. What changed?

What is it about transformative logic that is so different from previous frameworks of logic that it causes the concept of negation to not be directly expressible in it? In the natural sciences, when you are faced with a mysterious new behavior in a system, it is often useful to ask yourself: What changed in the time between when the system did not exhibit that behavior and when it first started to? We will use this same principle as inspiration for our thoughts.

Well, let's just try to force a negation operator into transformative logic in the most naive way possible and see what happens. We already know that it won't work, but let's momentarily suspend our disbelief and just give it a try anyway. Maybe it'll help us gain some more perspective on the nature of the problem. We're going to basically just recklessly cram the negation symbol from classical logic and set theory (¬) into transformative logic and see what happens. Here's what the failure state rule for the multi-statue puzzle system might look like if we did this:

$$\neg(🗿\varnothing \; 🗿\varnothing \; 🗿✗ \; 🗿\varnothing \; 🗿✗ \; 🗿✗) \to ☠$$

The idea here is that this rule is supposed to match to any form that is *not* of the form 🗿∅ 🗿∅ 🗿✗ 🗿∅ 🗿✗ 🗿✗. Naively speaking, this looks like it should work. There's just one little problem: Transformative logic is so general that any symbol could represent anything at all. You see that ¬ symbol there? That could be your grandmother for all you know. Perhaps it's a pictogram of grandma's bad back, all bent over at a right angle like that. Maybe she stopped by the ol' zombie mansion to dust off some of her grandson's creepy statues in between watching her soap operas and blowing off the heads of the undead with her trusty 12-gauge. That ¬ symbol is just an uninterpreted form and could mean anything... including a militarized grandmother. I wonder if she can still spit fireballs. That would be ideal, what with the zombies and all.

And sure, we could just reserve ¬ as a special character and say that it can't be used to mean anything other than negation in transformative logic, but there are problems with that approach too. For example, if we reserved ¬ to mean negation in transformative logic then we would no longer be able to use it for talking about classical logic's or set theory's versions of ¬ via transformative logic. Why not? The answer is because then we wouldn't be able to talk about ¬ as an uninterpreted symbol anymore. Talking about it as an uninterpreted symbol is exactly what we would have to be able to do if we wanted to specify the laws of classical logic or set theory in terms of transformative logic.

That's not the end of our problems though, because if we start reserving symbols for every special operator we want to reuse in transformative logic then all the corresponding symbols would become unavailable for use in describing other systems of logic via transformative logic. In effect, we would now have to define a bunch of different redundant versions of the symbols in order to talk about them properly. Also, the more special operators we pull into transformative logic, the less like a generalized ultra-minimalistic logic it would be, and the more like a specific narrow type of logic it would become. We don't want that. We want transformative logic to be elemental in nature and to make as few arbitrary assumptions as possible.

This thought experiment has revealed what it is that is so different about transformative logic, as compared to classical logic and set theory, that makes it so much more difficult to define a negation operator for. It works purely in terms of uninterpreted forms. That's what's different. That's why it's tricky to define a negation operator for it. If we wanted to do a negation in transformative logic then we would need to interpret the negation symbol as a meaningful operator.

Yet, transformative logic requires that almost all symbols be treated as pure uninterpreted forms. Thus, we have a dilemma. We can't implement negation directly without violating the sanctity of transformative logic, but at the same time we can't increase our expressive power without having a means by which to do negation. We want to negate pure forms, in effect matching for any other form which does *not* have that form. We need to somehow straddle the lines between these two worlds.

If you really think about it, the initial form ⛩⊕ ⛩⊕ ⛩⊕ ⛩⊕ ⛩⊕ ⛩⊕ acts kind of like a name for a set. It represents the set of all possible placements of objects in the statues' hands. The solution form ⛩⊘ ⛩⊘ ⛩✗ ⛩⊘ ⛩✗ ⛩✗ is just one of the many members of that set. So, when we say that we want to "negate" the solution form what we are really saying is that we want to specify the same set as the initial form except *without* including the one member that represents the solution form.

If you remember from our discussion of set theory, this operation is called a set difference and if A and B are two sets then $A - B$ represents this set difference and yields a set that includes all members of A that are not in B. For reference, our discussion of set theory began on page 123, and the definition of the set difference operator in particular was on page 137. It is actually this set difference operator that we want to use to shorten the rule specification for the multi-statue puzzle system, not negation. The set difference operator is similar to pure negation, but is not quite the same thing. Negation uses an implicit universal set, and is actually a special case of set difference. Thus, in our set theory notation for this book, $\mathcal{U} - A$ and $\neg A$ actually mean exactly the same thing.

All this talk of set operators sparks an interesting thought though. Negation and set difference are both operators of set theory, but why implement only them? Why not also implement all of the other set theory operators as well? If we had all of the oper-

222

ators of set theory available to us, then it would enable us to express any conceivable set relationship for a form. That would be far more powerful than only adding on negation or set difference. However, like I said, we don't want to pollute the namespace of transformative logic with a bunch of extraneous operators. Is there some way to work around this problem in order to get the best of both worlds?

Luckily, as it happens, yes there is. However, before we can fully understand the nature of that solution, we need to also broaden our understanding of its scope even more. We started out just trying to extend our notation with the concept of negation, yet it then suddenly occurred to us that it would be better to go ahead and somehow find a way to safely permit *all* of the operations of set theory to be expressed. Having the full expressive capabilities of set theory at our fingertips for specifying constraints on forms would be immensely powerful.

Why stop at set theory though? Why not provide a means to express and interpret *any* conceivable logical system at will? What we are really missing is a means by which to specify *which* of any already existing sets of transformation rules should apply to any given form at any given moment of time. *That* is the lightning strike of insight that cuts to the very core of what the real problem with the expressiveness of transformative logic so far actually is. *That* is the key.

Not being able to express negation or set difference in transformative logic is just a surface-level symptom of a much deeper problem. The real problem is that in our current vocabulary for transformative logic we have no way to specify what rule set a form should be interpreted under. Currently, the only rules we can apply are the ones that come from the same set of transformation rules that we are already working inside. What we need is a way to force a form to be interpreted according to a foreign external rule set instead of always being stuck with only the rule set we are currently working with. Essentially, we need a way to switch between different languages in a consistent way.

Now that we know what we need, how should we go about implementing it? Well, let's start the same way we usually start. Let's define new terms until the problem becomes precise enough that it is easy to solve. Ambiguity is poison to scientific progress. As long as a problem remains "fuzzy" it will often seem unsolvable. Anyway, if you think about it, before we define a way of switching between different interpretations of forms, it would be wise to first specify what exactly it is we're switching between:

Definition 78. *Every set of transformation rules in transformative logic represents a different way of interpreting forms. In effect, each set of transformation rules is like its own unique language and system of reasoning. Through the relationships and structures that a set of transformation rules embodies, it implements the required behavior of the objects it is intended to model. We refer to each such set of transformation rules as a* **transformative language**. *Every transformative language has its own namespace and foreign transformative languages cannot intrude upon that space unless explicitly invited.*

For example, the alternating astrological sign system, the single statue puzzle system, the rune puzzle system, and the multiple statue puzzle system are all examples of transformative languages. Classical logic, set theory, and numeric algebra are *also* examples of transformative languages, although admittedly we haven't defined them in terms of transformative implications. A little bit of thought will show that it should be possible to describe the properties of all of these systems in great detail via transformative logic. In fact, it may even be that they can be *completely* defined in terms of just transformation rules, but since we don't have time to investigate that I will stop short of actually making that claim.

This concept of a transformative language gives us a useful term for talking about different possible systems of transformation rules. However, its utility as just a label is limited. It's time we defined how this mechanism of forcing a form to be interpreted according to a foreign transformative language is supposed to work. How should this process of "inviting" a foreign interpretation actually work? This is the subject of our next definition:

Definition 79. *Normally in a transformative language the only rules that can be applied within that system are its own rules. However, this limitation can be stifling and can result in an excessive amount of labor being necessary to express certain concepts. Foreign transformative languages often have useful transformation rules for dealing with such cases, but we do not want to have to add those same rules to our own list because doing so would often pollute the conceptual space of whatever transformative language we are currently working in. Therefore, we need a way to use a foreign transformative language in a controlled way, without actually bringing it into the current namespace. To accomplish this task, we will create a new type of delimiter called an* **interpretation injector**.

Our notation for interpretation injectors will be similar to function notation. Any time we wish for an arbitrary form X to be interpreted according to the rules of a foreign transformative language we will write it as **LanguageName**(*X*), *where LanguageName is an abbreviated name for that foreign transformative language. In this book, we will always write the language name part in bold font for interpretation injectors in order to make it clear that something special is going on. The language name could be anything, and therefore there are actually an unlimited number of different interpretation injectors, one for every transformative language that you have defined.*

You'll need to specify a name to use for each transformative language if you plan on using it for interpretation injection. Multiple different interpretation injectors can coexist inside the same expression, and they can even be nested inside of each other. The innermost nested interpretation injectors are evaluated first, just like with many other delimiters such as parentheses or transience delimiters. **LanguageName**(*X*) *means the same thing as the set of all forms derivable from X under the rule set of LanguageName.*

We now have what we need to make the rule set for the multi-statue puzzle system much more concise. All we need to do is give the transformative language for set theory a name to be used for interpretation injection and we should be ready to go. Let's go with the obvious choice and name our interpretation injector for set theory **Set**(…). Here's one way the multi-statue puzzle system's rule set could be re-written to leverage the power of this mechanism:

1. ⊕ →₂ {⛉, ✗, ∅}

2. I → 🗿⊕ 🗿⊕ 🗿⊕ 🗿⊕ 🗿⊕ 🗿⊕

3. S → 🗿∅ 🗿∅ 🗿✗ 🗿∅ 🗿✗ 🗿✗

4. ·|S|· → ✞

5. **Set**(·|I|· − ·|S|·) → ☠

As you can see, in addition to using an interpretation injector, we are also using the variables I and S. This was done in order to prevent us from having to repeatedly write out those long forms. The reason why I chose the names I and S here is because I is the first letter of "initial form" and S is the first letter of "solution form". By writing **Set**(·|I|· − ·|S|·) → ☠, we are able to directly specify that any form derived from the initial form I except for the solution form S will lead to the failure state. In contrast though, ·|S|· → ✞ specifies that anything which matches the solution form 🗿∅ 🗿∅ 🗿✗ 🗿∅ 🗿✗ 🗿✗ may be transformed into the target form ✞ (i.e. the success state).

As I said before though, we can specify a lot more than just set differences with this system. For example, if we had two variables A and B, which each pointed to arbitrary sets of forms, and we then wrote **Set**(·|A|· ∩ ·|B|·) in the condition for a transformation rule, then that rule would match only to trans-closed forms that can be derived from *both* A and B. Similarly, **Set**(·|A|· ∪ ·|B|·) would match only to trans-closed forms that can be derived from *either* (or both) of A or B.

For practice, let's actually implement one of these other arbitrary set expressions in a rule set, just to demonstrate that it does indeed work. For example, suppose there was a mistake made in the design and engineering of the multi-statue puzzle mechanism that caused the storage box to also unlock if the player placed a shield in the hands of every statue, in addition to when given the originally intended solution. We could implement the rules for this system by using a combination of both set difference and set union. Here is one such implementation:

1. ⊕ →₂ {⛉, ✗, ∅}

2. I → 🗿⊕ 🗿⊕ 🗿⊕ 🗿⊕ 🗿⊕ 🗿⊕

3. S_1 → 🗿∅ 🗿∅ 🗿✗ 🗿∅ 🗿✗ 🗿✗

4. S_2 → 🗿∅ 🗿∅ 🗿∅ 🗿∅ 🗿∅ 🗿∅

5. S → **Set**(·|S_1|· ∪ ·|S_2|·)

6. ·|S|· → ✞

7. **Set**(·|I|· − ·|S|·) → ☠

This is a lot less work than what we had to do before we had interpretation injection available to us. This system will really go a long way towards helping to keep us sane when we specify more complicated or tedious rule sets. Let's do one more variation on this. Let's show a case that uses set intersection instead of set union. Suppose that we modify the multi-statue system so that now the only thing that matters for success is that there are a shield at each endpoint and two swords in the center. It isn't actually difficult to express this without using set operations, but let's show what it could look like using set intersection anyway, just for demonstration purposes:

1. ⊕ →₂ {🛡,✗,∅}
2. I → 🗿⊕ 🗿⊕ 🗿⊕ 🗿⊕ 🗿⊕ 🗿⊕
3. S_1 → 🗿∅ 🗿⊕ 🗿⊕ 🗿⊕ 🗿⊕ 🗿∅
4. S_2 → 🗿⊕ 🗿⊕ 🗿✗ 🗿✗ 🗿⊕ 🗿⊕
5. S → **Set**($\cdot | S_1 |\cdot \cap \cdot | S_2 |\cdot$)
6. $\cdot | S |\cdot$ → ♱
7. **Set**($\cdot | I |\cdot - \cdot | S |\cdot$) → ☠

Notice how the set intersection operation combines the constraints specified on S_1 and S_2 so that it matches to any form that satisfies the criteria of both simultaneously. Admittedly though, it would be simpler in this case to just write the solution form as

$$S → 🗿∅\ 🗿⊕\ 🗿✗\ 🗿✗\ 🗿⊕\ 🗿∅$$

. However, some expressions would be much harder to write without set intersection. You can use set intersection to say "give me the set of all forms that are derivable from both initial form A and initial form B" and in fact that is exactly what we are really doing when we say **Set**($\cdot | A |\cdot \cap \cdot | B |\cdot$). There is no requirement that A and B be similarly structured. The scenario could potentially be much more complicated and nuanced than this example was. Keep this in mind as yet another tool in your toolbox.

Yet, as powerful as this may be, set theory is just one of countless possible transformative languages that you could inject into an expression. In fact, transformative logic allows us to safely construct heterogeneous structures built up out of multiple different languages and re-interpreted in any way that one could imagine. One of the interesting things about transformative logic is that it only specifies how transformations on forms that it *recognizes* are supposed to work. When a transformative language is confronted with foreign symbols for which it has no matching rules, it is simply not allowed to touch those symbols in any way. It just ignores anything that it can't understand. It doesn't panic.

This makes it easy for different transformative languages to interoperate with each other relatively seamlessly. Traditional mathematics doesn't do this. Traditional mathematics would just say "ERROR" the moment it got any foreign input, and then it would give up and sit there helplessly, doing nothing. In contrast, transformative logic can ignore the forms that it doesn't recognize while still being able to continue operating on the forms that it does recognize. In these respects, whereas traditional math is more brittle and breaks the moment it encounters unknown input, transformative logic is more flexible and can tolerate any amount of unknown input.

For example, we could take the output of the rune system from earlier and then reinterpret it in arithmetic expressions. Let's name the interpretation injector for the rune system **Runes**. Likewise, let's name the interpretation injector for basic arithmetic **Arith**. Let's also define a new intermediate rule set called **RToN** whose sole purpose is to translate between the rune language and the arithmetic language. We first listed the rules for the rune system on page 189. Let's list the rules for the rune system again here for easy reference, followed by the rules for the translator:

Runes

1. ᚱ → ᚷᛗ
2. ᚷ → ᛗᚱ
3. ᛗ → ᚱᚷ
4. {ᚱᚱ, ᚷᚷ, ᛗᛗ} → ᛉ

RToN

1. ᚱ → 1
2. ᚷ → 2
3. ᛗ → 3
4. ᛉ → 4

We will assume that the transformative language for arithmetic has already been defined, and will rely on our pre-existing understanding of how arithmetic works to do the transformations for that system. It would likely be fairly tedious and challenging to define all of the transformation rules for arithmetic in transformative logic. Such a task is probably non-trivial and well beyond the scope of this book. We don't have time for it. In fact, I haven't even checked to make sure that it can be done, but I suspect it probably can. To add a little more credibility to the idea though, notice that you could easily brute force the creation of a bunch of correct arithmetic results, such as $3+4 \to 7$ and so on. That could help you get the ball rolling. There's probably a more elegant solution though.

Now that we've defined what rule sets we'll be using, let's come up with an example of an initial form to test it with. For example, consider the expression

Arith(RToN(Runes(ᚱᚷᛗ)) + RToN(Runes(ᚱᚷᚷᚱ)))

. This expression would result in the set of all numbers that can be derived by first performing any arbitrary chain of transformations on the forms ᚱᚷᛗ and ᚱᚷᚷᚱ allowed by the rune system rules, then translating them into numbers, and then summing those numbers together. Obviously, the rules we gave for how to translate from runes to numbers are just one possible way of doing it. There are lots of other mappings that you could do.

Now that you have this expression, you could ask yourself some interesting questions about it. For example, you could ask yourself what the set of all possible numbers derivable from this process is. What kind of structure do they have? I imagine it's probably a difficult problem. There's bound to be patterns in a system like this, but whether or not those patterns will be easy to reason about is another matter entirely. Emergent effects are often hard to predict. Our ability as humans to perceive what kinds of higher-level complexity will emerge from the interaction of all of the lower-level rules of a system is limited. On the other hand, that means that lots of unexpected and interesting things are possible.

In addition to being able to re-interpret expressions in a cohesive way like this, you could also simply allow them to coexist in their own little bubble worlds right alongside each other, but tied together into one higher-level expression. For example, we could chain together completely unrelated content in the algebra of equations, classical logic, and set theory into a single expression. For example, consider the following expression:

$$(x^5 - 7x^2 + 3x = 0) \ \square \ ((\neg A \lor B) \leftrightarrow (A \to B)) \ \square \ (\alpha \cap \neg(\beta \cup \gamma \cup \neg \delta))$$

We could then feed this expression into any of the transformative languages corresponding to the algebra of equations, classical logic, or set theory. Because each transformative language only understands symbols from its own domain, it could potentially be set up so that they can't mess with each other's stuff. Alternatively, you could enclose each separate part in an interpretation injector if you wanted to be safer by confining each expression to only be interpreted according to the corresponding language.

For instance, suppose that the expression above was $A \ \square \ B \ \square \ C$ for brevity. Let's say that **Eq**(…) is the injector for the algebra of equations, that **Cla**(…) is the injector for classical logic, and that **Set**(…) is the injector for set theory. We could then re-write $A \ \square \ B \ \square \ C$ as **Eq**$(A) \ \square \ $**Cla**$(B) \ \square \ $**Set**$(C)$ if we wanted to specify the interpretation of each part. Perhaps some other transformative language could come along later to interpret what the $\square$ symbols mean, but we could also just allow them to sit around as meaningless symbols. Basically, you can work with symbols however you want to in transformative logic. It's a matter of choice.

3.8 Well-formed formulas and rule set adherence

Why you'd want to do this kind of stuff is anybody's guess. It's basically arbitrary and context-dependent. Use it however you want. However, another thing you can do with rule sets is use them to define what a well-formed formula should look like for any given transformative language. The idea is that you start with an initial form and then define a set of rules in such a way that no matter how you apply those rules to the initial form the resulting expression will always be one that is considered valid in that language.

In mathematics we may say that such a definition of well-formed formulas is *inductive*, keeping in mind that the word "inductive" has a different meaning in mathematics than it does in the other sciences[12]. We could alternatively say that this is an application of the more general principle of property conservation (defined on page 104). Essentially, adherence to the properties specified by the transformation rules is what is being conserved during each transformation step. We could say that the system is "conservative" instead of saying it is "inductive" if we wanted to avoid the naming conflict with what the term "inductive" means in the other sciences.

Let's do a partial example of a well-formed formula rule set to illustrate the point. Previously, we have always just started with an arbitrary initial form that we knew in advance would work in the rule sets we designed. We technically could have used any initial form at all, even including ones that don't have any symbols that are recognized by the transformative language. Transformative languages are inert when confronted with something they don't have in their vocabulary.

Thus, sometimes we don't care much whether an initial form has a certain structure or not. However, sometimes we may want to specify that certain initial forms are not allowed. The way we do this is by defining some transformation rules that specify the requirements for a well-formed formula, and then allow only forms that have been derived from those rules to be used as our initial form when working with the system. Here's an example:

1. $w \rightarrow \{A, B, C, \ldots, Z\}$

2. $w \rightarrow (w \odot w)$

3. $w \rightarrow (w \circledast w)$

In this rule set, the symbol w (lowercase fancy W) represents any arbitrary well-formed formula of our hypothetical language. As you can see, the rules specify that every w can either be converted into a capitalized letter in the range of A to Z (perhaps representing a variable) or else converted into a parentheses delimited binary operation of one of two types. Remember that parentheses are just as much arbitrary symbols as any other symbols are and could potentially mean anything. Strictly speaking, you need to be careful to always think in terms of raw uninterpreted forms.

Notice that the rules are effectively recursive. They are written in terms of themselves. Just because a symbol occurs in the condition of a transformation rule doesn't mean that it can't also occur in the consequence. Nothing in the rules for transformative logic says that you can't transform a symbol into multiple copies of itself. It's totally a legitimate thing to do if it fits what you are trying to model.

Notice also that the fact that we allow normal transformative implication in transformative logic, instead of only restricting ourselves to 1-form transformative implication as in traditional logic, makes these rules much easier to write. It is yet more evidence

---

[12]When the other sciences say that a property is "inductive" they mean that it is a statistically uncertain generalization built up from observations. In contrast, when mathematicians say that a property is "inductive" they mean that it can been proven beyond any doubt using an airtight technique called "proof by induction". In terms of the vocabulary of the other sciences, mathematics deals *only* with deductive reasoning and never with inductive reasoning, except for perhaps in the study of statistics and such.

that the conventional strictly "parallel" way of substituting for variables (i.e. 1-form transformative implication) is in fact *not* the only legitimate way of performing substitutions.

Anyway, what kinds of forms do you think this rule set will generate? Well, let's think. Anything derived from the form of an already well-formed formula via a rule set that conserves well-formedness in every rule will inevitably end up also being a well-formed formula. The principle of property conservation guarantees that this will be true. How do we get started though?

Well, we need to pick a pre-existing form that we already know is a well-formed formula. As it happens, the symbol w is *defined* as being a well-formed formula and therefore we can start with it as our initial form. In effect, anything that can be derived from the symbol w by following any arbitrary transformation path allowed by the rule set will end up being a well-formed formula. Here are some examples of some valid transformation paths originating at w:

1. $w \twoheadrightarrow (w \odot w) \twoheadrightarrow (A \odot B)$

2. $w \twoheadrightarrow (w \circledast w) \twoheadrightarrow ((w \odot w) \circledast w) \twoheadrightarrow ((Z \odot Y) \circledast X)$

3. $w \twoheadrightarrow N$

4. $w \twoheadrightarrow (w \odot w) \twoheadrightarrow ((w \circledast w) \odot (w \circledast w)) \twoheadrightarrow ((T \circledast U) \odot (V \circledast W))$

These transformation paths prove that $(A \odot B)$, $((Z \odot Y) \circledast X)$, N, and $((T \circledast U) \odot (V \circledast W))$ are all well-formed formulas of this transformative language. We don't actually know what the $\odot$ and $\circledast$ symbols are supposed to mean, but that doesn't really matter. In fact, for all we know, the sole purpose of these expressions might be aesthetic and they may have no meaning at all. If we wanted to give them more meaning, we could append additional transformation rules to the rule list to define more about how the well-formed formulas are actually interpreted after they are created. For example, here's one such extension of the rule set:

1. $\alpha \twoheadrightarrow \{A, B, C, \ldots, Z\}$

2. $w \twoheadrightarrow \alpha$

3. $w \twoheadrightarrow (w \odot w)$

4. $w \twoheadrightarrow (w \circledast w)$

5. $\{\alpha_1, \alpha_2, \ldots, \alpha_n\} \twoheadrightarrow_1 \cdot |\alpha|\cdot$

6. $\cdot|(\alpha_1 \odot \alpha_2) \twoheadrightarrow \textbf{Arith}((\alpha_1 \cdot \alpha_2) + \alpha_1)|\cdot$

7. $\cdot|(\alpha_1 \circledast \alpha_2) \twoheadrightarrow \textbf{Arith}(\alpha_2 \cdot (\alpha_2 + \alpha_1))|\cdot$

Notice that I have added in α above our w rules. Why did I do this? Well, if you think about it, what alphabet we are using, and what our definition of a well-formed formula is, are really two conceptually different things. Our definition of a well-formed formula just coincidentally happens to accept lone members of our alphabet as being well-formed formulas. It could have been different. It is also useful to be able to reuse our alphabet in other expressions in our rule set.

Notice also that I have used a convenient little trick to say that all subscripted α symbols stand for 1-forms (i.e. forms that must be replaced with the same value everywhere in an expression in parallel). This is very useful when you could potentially have an unlimited number of 1-form variables being used in your expressions. By the way, the symbol α is the Greek letter "alpha", which also happens to be the first part of the word "alphabet", thus making it a fitting and memorable name for the set.

Anyway, the idea here is that the user is supposed to specify numeric values for some of the variables in $\{A, B, C, \ldots, Z\}$ and then use the other rules in the system to evaluate what the result of the $\odot$ and $\circledast$ operators would end up being. As you can see, we have used an interpretation injector to specify the arithmetic interpretations of these operators. For example, here's what a few evaluations with some concrete values might look like:

1. $A \twoheadrightarrow 2$

2. $B \twoheadrightarrow 3$

3. $(A \odot B) \twoheadrightarrow (A \cdot B) + A \twoheadrightarrow (2 \cdot 3) + 2 \twoheadrightarrow 6 + 2 \twoheadrightarrow 8$

4. $(A \circledast B) \twoheadrightarrow B \cdot (B + A) \twoheadrightarrow 3 \cdot (3 + 2) \twoheadrightarrow 3 \cdot 5 \twoheadrightarrow 15$

Are you wondering where the well-formed formula steps are in these transformation paths? They're there too, in some sense, but for the sake of brevity we seldom actually write them. Basically, we tend to think of the well-formed formula rules as having already generated all of the corresponding concrete rules for the expressions in advance. However, for completeness, I'll explain what the full derivation chain for one of these results would look like in order to make sure that you understand it. Consider the expression $(A \odot B)$ for example. Here's the missing first half of its path:

$$w \twoheadrightarrow (w \odot w) \twoheadrightarrow (\alpha \odot \alpha) \twoheadrightarrow (\alpha_1 \odot \alpha_2) \twoheadrightarrow (A \odot B)$$

One step that might be confusing you here is the $(\alpha \odot \alpha) \twoheadrightarrow (\alpha_1 \odot \alpha_2)$ step. Where is this step coming from? Where is it justified in the rules? Well, basically it is an instance of denormalizing substitution and therefore it works a little bit differently than the other transformation rules. It doesn't show up explicitly in the rule set. The reason why it is justified is because the sets α_1 and α_2 are both subsets of α, and therefore α may be replaced by either of them via the denormalizing substitution process. Still confused? Perhaps you forgot that the definition of a subset includes the possibility that the two sets are equal (see page 135).

Technically, it isn't necessary to use denormalizing substitution here. We could have gone straight from $(\alpha \odot \alpha)$ to $(A \odot B)$ instead. The reason why I included the denormalizing substitution was to make the intent more clear. It is a debatable decision. The user always has a choice of constraining a normal form to a more narrow subset of the allowed transformations. This allows you to explicitly distinguish the identities in a form more clearly.

Let's put this extended version of the rule set aside though, and refocus our attention on the original well-formed formula rule set. We've discussed some examples of forms that *are* well-formed formulas under this rule set, but we should also take a moment to consider what kinds of forms are *not* well-formed under it. Can you think of any? Remember, the only forms that count as well-formed formulas for this system are those that can be derived from w. Here's the original rule set again for reference:

1. $w \rightarrow \{A, B, C, \ldots, Z\}$
2. $w \rightarrow (w \odot w)$
3. $w \rightarrow (w \circledast w)$

It's clear that there are a vast number of conceivable forms that cannot be derived from w. In fact, the overwhelming majority of conceivable forms don't satisfy the criteria of w. For example, any form that contains even a single symbol that is not taken from the set $\{w, A, B, C, \ldots, Z, \odot, \circledast, (,)\}$ is instantly disqualified from being a well-formed formula under this rule set. Also, even though parentheses are permitted here, not all conventional uses of them are actually allowed by this rule set. We may say that any form that is not well-formed is ill-formed. Consider the following list of forms (relative to this rule set) and make sure you understand *why* they are ill-formed:

1. (N)
2. $(A \odot B \odot C)$
3. $((A \wedge B) \rightarrow C)$
4. $((X \circledast Y))$
5. $U \circledast V$
6. $(A \circledast (B \odot C) \circledast D)$
7. ᛋᚴᛏᛏᛦᛯᛘᛋᛐᛦᛪ

What would happen if we tried to allow expressions like $(A \odot B \odot C)$ in our arithmetical extension of the well-formed formula system? How would you interpret the expression? The order in which you carried out the operations could change the result. If a transformation isn't explicitly specified in the rule set then we can't do it. If we are to have any hope of being able to represent all conceivable logical systems then we must always make our intentions explicit.

If we allowed implicit assumptions to creep in then it could create ambiguities that would be impossible to resolve. Implicit assumptions place a ceiling on the maximum expressive capabilities of any language. After a certain point, the only way to increase expressiveness is to remove implicit assumptions and to make everything more explicit. This is an inherent property of all communication, not just of logic and mathematics.

Another thing I want to draw your attention to is the fact that redundant parentheses around expressions are not allowed in this rule set as written. For example, as you can see from the above list of examples of ill-formed expressions, the forms (N) and $((X \circledast Y))$ are not well-formed formulas. However, unlike the other examples of ill-formed formulas I have given in this list, there's no deeper reason why we couldn't make them into well-formed formulas by just adding on a few new rules. The other expressions in the list of ill-formed formulas in contrast are either ambiguous or from a different transformative language entirely. What would it look like if we wanted to extend the well-formed formula rule set so that it was able to handle redundant parentheses? It would look like this:

1. $w \twoheadrightarrow \{A, B, C, \ldots, Z\}$

2. $w \twoheadrightarrow (w \odot w)$

3. $w \twoheadrightarrow (w \circledast w)$

4. $w \twoheadrightarrow (w)$

5. $((w)) \twoheadrightarrow (w)$

Do you see why we didn't add in a rule of the form $(w) \twoheadrightarrow w$ and instead only allowed $((w)) \twoheadrightarrow (w)$? If we allowed $(w) \twoheadrightarrow w$ as a transformation then it would make it possible to transform an expression like $((A \odot B) \circledast C)$ into $(A \odot B \circledast C)$, thereby causing the resulting expression to become ambiguous in terms of the order of operations implied by the parentheses. Instead, we are only allowing $((w)) \twoheadrightarrow (w)$ so that there is no way for the transformation rules to remove that last essential pair of parentheses. The $((w)) \twoheadrightarrow (w)$ still enables us to remove truly redundant parentheses though.

Actually, there is still one more case of redundant parentheses we should ideally allow the rule set to remove. It might not be safe to remove the last pair of parentheses from a general well-formed formula whose actual form we don't know in advance, but there is actually one special case where we can safely remove the last pair of parentheses. Can you see what it is? I'll give you a hint. It has something to do with our alphabet of allowed variable names. Here's what the rule set looks like when we extend it to handle this additional special case of parentheses removal:

1. $\alpha \twoheadrightarrow \{A, B, C, \ldots, Z\}$

2. $w \twoheadrightarrow \alpha$

3. $w \twoheadrightarrow (w \odot w)$

4. $w \twoheadrightarrow (w \circledast w)$

5. $w \twoheadrightarrow (w)$

6. $((w)) \twoheadrightarrow (w)$

7. $(\alpha) \twoheadrightarrow \alpha$

Can you see why it's ok to remove the last pair of parentheses from the alphabetic form but *not* from the more general case of any arbitrary well-formed formula? In other words, can you see why $(\alpha) \twoheadrightarrow \alpha$ is a safe rule whereas $(w) \twoheadrightarrow w$ would not be? Also, did you notice that we don't have a rule of the form $\alpha \twoheadrightarrow (\alpha)$? The reason is not because it would be illegitimate. In fact, it would be perfectly acceptable to add $\alpha \twoheadrightarrow (\alpha)$ into the rule set. It would do no harm. The reason we didn't is because it would be redundant. Adding parentheses in this way to a member of α is already covered by the $w \twoheadrightarrow (w)$ rule by virtue of the fact that w could be any well-formed formula, which includes lone alphabetic symbols from the α set.

Admittedly, these extra parentheses rules aren't actually that useful in this system. In fact, in this case they basically accomplish nothing except making our rules needlessly more complex. Why even bother with redundant parentheses when we can just not allow them to begin with? However, that being said, I still had a good reason for showing you this. I wanted to show you what kind of stuff you'll need to do if you want to add in support for traditional parentheses-like constructs into a transformative language that you are designing. This is a pretty common use case, so it pays to know how to actually do it correctly.

3.9 Quotation delimiters

Speaking of parentheses and similar constructs, I have a confession to make. We have so far actually been abusing our notation for transformative logic slightly. Remember how I said that parentheses are just arbitrary symbols and could potentially mean anything? Well, it turns out that we've actually been abusing our notation for one of the other parentheses-like pairs of symbols that we've been using. Can you guess what pair of symbols I'm talking about? If you've been paying close attention then you may have noticed it. There's been one pair of symbols that we've been breaking our own rules for. That pair of symbols is the curly brackets { }, which we've frequently used to notate sets in transformation rules.

Those curly brackets are just as much arbitrary symbols as anything else is. Thus, if you think about it very strictly, we have actually been abusing our notation by giving these symbols special meaning instead of just treating them like raw uninterpreted symbols like we do with everything else. Thus, you might be wondering what's the correct way to write those curly bracket delimited sets without abusing our notation. The answer is that you technically need to wrap them inside an interpretation injector for set theory. Thus, for example, if we wanted to define the domains of two boolean variables for classical logic, named A and B, then we technically should perhaps write:

1. $A \twoheadrightarrow_1 \mathbf{Set}(\{F, T\})$

2. $B \twoheadrightarrow_1 \mathbf{Set}(\{F, T\})$

However, even though this is arguably technically more correct, we will generally never write sets this way and will continue to abuse our notation. The reason for this is because this notation for sets is so common and so useful that it is worth allowing it as a special case even though it technically breaks the rules. Basically, we will treat any expression of the form {...} as implicitly being shorthand for **Set**({...}). Anything other than the special operators of transformative logic (e.g. transformative implication, set notation, transience delimiters, interpretation injectors, etc) will be treated as just raw uninterpreted symbols.

Even though we will continue using this set shorthand, it was still worth discussing what the technically correct way of writing it is. Previously, the curly bracket notation for sets may have seemed like "magic" and thus it may have confused you about the fact that I had said that everything in transformative logic should be treated as raw uninterpreted forms. Discussing this inconsistency, even though we are going to continue using it, helps to further clarify the underlying nature of transformative logic. It helps ensure that you have a conceptually correct understanding of what is going on.

If you genuinely need to use those curly bracket symbols as uninterpreted symbols, then you have two options. The first option is to state that you are not following the implicit convention of treating { } as indicating sets and that you are instead just using the curly brackets as raw symbols. The second option is to use a quotation delimiter. When a form is wrapped in quotation delimiters, it indicates that everything inside those quotation delimiters should be treated as uninterpreted symbols instead of as meaning something. To accomplish this, we will need to define some symbols to represent quotation within transformative logic.

The first option that comes to mind is to use the standard quotation marks from English (and many other languages) to represent this new delimiter. However, these quotations marks have the unfortunate characteristic that on many typesetting systems and in many fonts they have no visual directionality (e.g. ' and "), thus making it sometimes impossible to distinguish an opening quote from a closing quote. Good delimiters have directionality and clearly distinguish between beginning and ending. Thus, instead of English quotes we will be using Japanese quotes (a.k.a. *kagikakko* or "hook brackets"). Unlike English quotes, Japanese quotes are always directional. This makes them a generally superior choice for quotation symbols. Here's the new definition:

Definition 80. *Sometimes it is useful to be able to force special symbols in forms to be treated as being uninterpreted. To accomplish this, we need to use* **quotation delimiters**. *Suppose we have some arbitrary form $\mathcal{F}$. Then, the quotation of $\mathcal{F}$ will be represented by:*

$$\ulcorner \mathcal{F} \lrcorner$$

Actually, true Japanese quotation symbols were not readily available in the typesetting system I'm using to write this book (LaTeX), so I just picked something very similar instead. Real Japanese quotes have an appearance that is somewhere in-between what you see here and the floor and ceiling symbols of mathematics. In other words, if I had the typographically correct symbols available then it would actually look like something in-between $\ulcorner \mathcal{F} \lrcorner$ and $\lfloor \mathcal{F} \rfloor$. From these two available pairs of symbols in my typesetting system, I picked the more stubby pair to represent quotation delimiters because they have less chance of being confused with the floor and ceiling functions.

In addition to Japanese quotation marks, the Japanese language has a bunch of other useful visually distinct delimiters. It would be nice to have these symbols available, but unfortunately the typesetting system I'm using just doesn't seem to support Japanese very well, as far as I can determine. There's supposedly a way to set it up, but it seems like it might conflict with some of the packages I'm using already, and I don't want to risk breaking my setup, so it's probably not worth the hassle.

Personally, I think that the LaTeX typesetting system should import all of the missing Japanese delimiter symbols into the standard library and make them available for general use. The more visually distinct delimiter symbols we have available the better. They can be used for all sorts of helpful things, like making nested parentheses easier to read (by switching between different symbols), defining new operations, or whatever else you feel like doing with them.

Anyway, now that we have quotation delimiters defined, we can escape from our implicit interpretation of set notation whenever we want to. For example, if we wrote $X \to \ulcorner \{a, b\} \urcorner$ then it would mean that the literal symbol X could be transformed into the literal form $\{a, b\}$. In contrast, if we wrote this transformation rule without the quotation delimiters, as in $X \to \{a, b\}$, then it would mean that X has the option of transforming into either of a or b.

Notice the difference. Notice also that the outermost quotation symbols disappear after being evaluated. The quotation delimiter symbols only serve to specify how we should interpret the form. To talk about transformations involving quotation symbols themselves, we would need to have multiple levels of quotation. For example, $\ulcorner \ulcorner X \urcorner \urcorner$ would mean the same thing as $\ulcorner X \urcorner$ treated as an uninterpreted form, instead of as a quotation delimiter being applied to X.

Transformative logic is very strict about treating the overwhelming majority of things as uninterpreted forms. Only a small number of special symbols have an implicit interpretive meaning. Consequentially, the meaning of most forms will not change at all when surrounded by a single level of quotation delimiters. For example, there is no difference whatsoever between $A + B \to C + D$ and $\ulcorner A + B \urcorner \to \ulcorner C + D \urcorner$ in transformative logic, except for that the later takes more effort to write.

However, quotation delimiters are useful for making statements about the properties of the special operators of transformative logic from within transformative logic itself. This allows you to talk about the "meta-logic" of transformative logic itself within its own terms. Thus, for example, $\ulcorner A \to B \urcorner \leftrightarrow \ulcorner B \leftarrow A \urcorner$ would allow you to say that any transformation rule of the form $A \to B$ could be re-written as $B \leftarrow A$, and vice versa. Keep this trick in mind in case you ever need to do this. It can come in handy sometimes.

Anyway, with that out of the way, let's return to our discussion of interpretation injectors and well-formed formulas. Let me take a moment to point out that, like all forms, well-formed formula specifiers can be used with set theory interpretation injectors in order to stipulate that a form must be derivable from that well-formed formula in addition to any other criteria you wish to specify. Thus, for example, if we wanted to say that a transformation rule condition matches to any form that is both a well-formed formula under w and also derivable from x then we could write **Set**($\cdot | w | \cdot \cap \cdot | x | \cdot$).

Thus, **Set**($\cdot | w | \cdot \cap \cdot | x | \cdot$) $\to y$ would mean that only forms that could be derived from the criteria of both w and x would be allowed to use the rule to transform into y. The reason I'm telling you this is just to keep you aware of the flexibility of the system so that you don't start thinking too rigidly. Set interpretation injectors make it easy to specify multiple different *types* of well-formed expressions or other criteria and then logically combine them using any arbitrary set theory expression.

Oh, on a random side-note, in case you haven't consciously realized it yet, notice how transformative logic fully supports many-to-many transformation rules, in addition to one-to-one, many-to-one, and one-to-many. Thus, for example, $\{a, b\} \to \{c, d\}$ would mean exactly the same thing as writing out all four of $a \to c$, $a \to d$, $b \to c$, and $b \to d$.

When creating rule sets you need to be careful that your transformation rules are doing what you think they are. It is fairly easy to make subtle mistakes. Using transience delimiters incorrectly is one of the easiest mistakes to make. We are used to thinking of certain symbols (e.g. x and y) as automatically being "variables", but in transformative logic this assumption does not hold, so we need to be careful to not introduce implicit unwritten assumptions. For example, consider the following rule set:

1. $x \to 1$
2. $y \to 2$
3. $z \to 3$
4. $x + y \to z$

A naive user of this rule set, one who is accustomed to automatically treating letters as implicitly being numeric variables, is in danger of thinking that this rule set allows more transformations than it actually does. Under this rule set, if you were given the initial form $1 + 2$ then you might instinctively be tempted to think that you are allowed to transform it into 3, but doing so would be wrong. This rule set allows no such thing. It's true that if you were given $x + y$ then you could transform it into $1 + 2$ if you chose to do so. However, nothing in the rule set says that you can transform $1 + 2$ into 3.

The last rule only says that the literal form $x + y$ can be transformed into the literal form z. In fact, once you transform anything in $x + y$ into anything other than what it already is then the pathway to z will immediately close. The two transformation paths $x + y \to 1 + 2$ and $x + y \to z \to 3$ are two fundamentally separate paths. Taking one excludes the other. You need to be wary of your traditional math training sometimes. It can mislead you in transformative logic. If you wanted to also allow transformations like $1 + 2 \to 3$ for this rule set though, then here's one way of implementing it:

1. $x \to 1$
2. $y \to 2$
3. $z \to 3$

4. $(\!(x+y)\!) \to (\!(z)\!)$

However, notice that this implementation would also allow partial transformations such as $x + 2 \to z$ and $1 + x \to 3$, in addition to the more conventional transformations like $x + y \to z$ and $1 + 2 \to 3$ that you may have been thinking of when you wrote it. This may or may not be what you want. It just goes to show that you need to be careful about specifying your intent precisely. If, instead of writing $(\!(x+y)\!) \to (\!(z)\!)$, we wrote $\cdot |x + y| \cdot \to \cdot |z| \cdot$, then the 4th rule would have the effect of only permitting $1 + 2 \to 3$. Transformations such as $x + y \to z$ wouldn't be allowed. See? Small changes can have a big impact on what a rule set actually means.

The danger of miscommunicating your intent doesn't end there though. It is also easy to underestimate just how far reaching a seemingly well-contained rule can be. Consider, for example, the original rule set where we had $x + y \to z$. It's clear that this rule will allow you to transform $x + y$ into z anywhere you find it in a form. This *looks* relatively innocent at first glance, but this rule is actually capable of some pretty unexpected behavior. For example, as written, this rule set actually also permits you to transform $xx + yy$ into xzy. Do you see why?

This rule set never specified any kind of implicit order of operations. Someone designing a rule set like this may be thinking subconsciously that a transformation like $xx + yy \to xzy$ shouldn't be allowed, but the system has no way of knowing that, nor does anyone else reading the rule set. Nothing in the rule set says it follows conventional grouping rules.

If you wanted to change the system so that these kinds of unexpected transformations weren't allowed, then the easiest solution would probably be to require parentheses around all operations so that the interpretation and scope is always unambiguous. Thus, if we changed $x + y \to z$ to $(x + y) \to z$ then it would prevent unintended transformations like $xx + yy \to xzy$ from ever being allowed.

Not using at least some kind of delimiters (e.g. parentheses) in your rule sets is one of the easiest ways to introduce subtle logical errors and unexpected behavior into a system in transformative logic. Ignore this advice at your peril. There are many simple systems for which delimiters may not be necessary, such as the statue puzzle systems from earlier in the book for example.

However, more complicated systems increasingly tend to require some form of delimiting symbols in order to behave predictably. When in doubt, use parentheses. If you're experimenting with a rule set and its behavior starts becoming unpredictable then there is a good chance that it is because you are not delimiting the concepts you are trying to model properly. You need to clearly express what each coherent "unit" or "whole" is within your system, and delimiters are generally how you do that.

Transformative logic is also capable of handling empty sets. If a transformation rule has an empty set in its condition or consequence then it has the same net effect as if that transformation rule did not even exist. Thus, any transformation rule whose condition or consequence is an empty set of forms will not actually enable any transformations to occur. Such rules are completely inert and harmless.

In the case of the condition being an empty set, no form could ever match the condition for the transformation since no conditions exist for anything to match to. In the case of the consequence being an empty set, even if a form matched the condition of the transformation, it would be like saying that it is adding on zero new forms to the list of possible forms that the condition form can transform into and thus there'd still be no new transformations enabled by an empty consequence set. That's why transformation rules with empty condition sets or empty consequence sets have no effect.

Here's another way of thinking about it to help you understand why the rules behave this way. Think about transformation rules in terms of how they would expand into individual transformation rules if we didn't allow set notation. For instance, if we had $\{a, b\} \to \{c, d\}$ then it could be expanded to the four rules $a \to c$, $a \to d$, $b \to c$, and $b \to d$. Notice that if the condition set has m elements and the consequence set has n elements then they will expand to $m \cdot n$ individual transformation rules.

However, what would happen if either of the sets were empty? How many individual transformation rules would it expand to then? That's right. It would expand to *zero* individual transformation rules. It's like a cartesian product. If either participant has zero elements then the product will also have zero elements. The product's elements in this case are the transformation rules themselves, and thus a transformation rule having an empty set as a condition or consequence is logically equivalent to that transformation rule not existing at all and hence has no effect.

It is important to note that a transformation rule whose consequence is an empty set does *not* have the effect of making it possible to delete the condition form. Empty sets are not the same thing as empty forms. In transformative logic, an empty set represents an absence of any forms whereas an empty form represents an absence of any symbols in one specific form.

For this reason, in transformative logic we need to explicitly distinguish between two different types of "emptiness" in forms. Leaving either the condition or consequence blank is ambiguous and thus is not allowed. We already have a way of representing empty sets of forms by using the { } symbols without anything inside them[13]. We have not yet officially defined a way to represent the empty form, but we should. That brings us to our next definition:

Definition 81. *To represent a form that contains no symbols, we write empty quotation delimiters, as in* ⌜ ⌟*, and we call it the* **empty form**. *Thus, for example, if you wrote* $X \to$ ⌜ ⌟ *then it would have the effect of making it possible to delete the form X wherever you find it, whenever you choose to do so. When the empty form is placed inside non-empty forms, it can be removed or added at will, regardless of where in the non-empty form it is placed.*

Thus, for example, the form X⌜ ⌟⌜ ⌟Y⌜ ⌟ is equivalent to the form XY, and so on. There would be no difference between what X⌜ ⌟⌜ ⌟Y⌜ ⌟ and XY would match to if placed in the condition of a transformation rule. A difference would only show up if you wrapped these forms in another layer of quotation delimiters, such as in ⌜X⌜ ⌟⌜ ⌟Y⌜ ⌟⌟ and ⌜XY⌟. The empty form is completely harmless when placed inside non-empty forms. It has no substance of its own.

---

[13] As you may recall, the symbol ∅ is another way of writing the empty set in set theory. However, for convenience, in transformative logic, we will treat the symbol ∅ as just a pure symbol. This will make it possible to talk about the empty set without using quotation delimiters and we'll be using it that way later in the book.

In stark contrast however, think about what would happen if you wrote an empty *set* into a non-empty form. For example, think about the difference between $X\ulcorner\lrcorner Y$ and $X\{\ \}Y$. The expression $X\ulcorner\lrcorner Y$ would mean the same thing as XY, but in contrast $X\{\ \}Y$ would mean the same thing as $\{\ \}$. In other words, if the empty set appears anywhere in a form then it will effectively cause the rest of the entire form to evaporate and become equivalent to the empty set (i.e. to nothingness). This is the correct behavior.

We want to be able to use set expressions in such a way that $\{a, b\}c$ means either of ac or bc, but by extension if we want to use sets this way then placing an empty set anywhere in a form will have to have the same effect as erasing that entire form from existence. That's the only logically consistent way to do it. It may seem weird at first glance, but it is actually the most natural and useful way of treating the sets. If you have a set expression in part of a form that ends up not matching to anything, and hence generating an empty set there, then erasing the entire form is in fact probably the best way of reacting to it. It makes combining different expressions together to express arbitrary concepts in a logically coherent way much easier.

As you may recall, early on in our discussion of transformative logic I mentioned that the transformative implication operator was sort of similar, in some ways, to the syntactic consequence operator (represented by a turnstile: $\vdash$) which is already in use in the mathematical literature. However, I also mentioned that the syntactic consequence notation is frequently abused in traditional mathematics literature in that people often use blank conditions with syntactic consequence to indicate that the consequence is an "axiom", when what they really mean is that it is an initial form as opposed to a transformation rule.

Thus, in the traditional notation, you will often see people write something like $\vdash X$ to mean that X is assumed as a given (as an "axiom"). However, what this really means in the general case is that X is an initial form which is assumed to already have a certain property of interest (it could be truth, or it could be something else entirely) and that we will accept only forms that can be derived from these specified initial forms ("axioms") and the transformation rules ("rules of inference") as "true" under our system. It's basically an application of the principle of property conservation to a set of initial forms ("axioms") that we know (or assume) already possess the property we want to conserve. Sometimes that property is "truth" but other times it is not.

Well, our discussion of the empty set and the empty form, as they relate to transformative logic, and our discussion of why it's never safe to leave conditions or consequences blank (because its ambiguous), should help make it clearer to you now what one of the reasons why I was so firm on avoiding the traditional syntactic consequence notation was. I didn't want to deal with any of its associated baggage.

Besides just the horrible abuse of notation involved in leaving the condition blank for "axioms" though, the syntactic consequence as used in the traditional literature also has some other implicit assumptions (e.g. sequent notation and implicit assumptions about the nature of truth) that I did not want to be burdened with. This is in addition to the fact that I have little to no experience with proof theory and model theory, and thus wanted to avoid any possibility of accidentally using syntactic consequence incorrectly. So really, there were a lot of reasons why I avoided using syntactic consequence and chose to define my own system instead.

Anyway though, why write initial forms in this weird $\vdash X$ format when you can just use the initial forms as starting points directly instead? It makes a lot more sense that way. Both initial forms and transformation rules can be considered to be axiomatic assumptions, and thus the term "axiom" as used in this context of syntactic consequence is a misnomer. Both "axioms" and "rules of inference" are axioms. It doesn't create a real distinction and thus it's confusing. Hopefully our discussion of this has cleared it up some though. Making our system of reasoning unambiguous and intuitive, with minimal unexpected side-effects and quirks, is important if we want it to be maximally useful.

User experience matters. Good mathematics is not just about proving things, it's about proving things well. Notation should minimize friction and cognitive load. Look at the difference between mathematical productivity before symbolic variables were invented and afterwards to get a better sense of the colossal magnitude of how much of a difference a good user experience can make in enabling people to see patterns better. Clarity matters.

A clarification of a foundational piece of notation is often far more valuable than a high-level advanced proof. The deeper into the foundations of a system an improvement is made, the more impactful the benefit will tend to be. Thus, ironically, when mathematicians dismiss "small" issues like fixing ambiguities in their most basic notation because they think it's "unimportant" compared to advanced research they are thinking exactly backwards.

The foundations of a theory are the most important part. Advanced research and obscure results have value, but you seldom use them on a daily basis. A strong foundation is the key to acquiring an expressive vocabulary. The more basic an operation is, and the more contexts it fits in, the more powerful it will tend to be for combining different components together to express higher-level arbitrary concepts. In this respect, transformative logic has really started to show its real strength. It started out as a much more primitive system that was only capable of expressing the most basic transformations, but as we've defined new operators for it its expressive capabilities have expanded considerably.

3.10 Miscellaneous fundamentals

At this point, transformative logic is really starting to feel a lot more complete than it did when we first started discussing it. However, be that as it may, there are actually still quite a few more features that would seem wise to include and discuss. As it stands, most of the power of transformative logic has already been integrated into the system at this point. However, we want to strive for completion. Some aspects of the remaining basic features we have yet to discuss are quite interesting.

As you may recall, we were recently discussing the concept of the empty form on page 239. Well, as it happens, there is also another concept corresponding to the empty form which is sort of its opposite counterpart. Just as we have the empty set and the universal set in set theory, we also have the empty form and the universal form in transformative logic. It does pretty much what you'd imagine it does. Whereas the empty form represents a form containing no symbols, the universal form represents a form containing any conceivable combination of symbols. Here is our new definition for it:

Definition 82. *To represent a form that could be any conceivable form, we write ⌑, and we call it the **universal form**. The reason why this particular symbol (⌑) was chosen to represent a universal form is because it has a nice generic boxy shape and it also looks kind of like a little explosion of some kind, like a magic portal or a singularity from which any conceivable thing could emerge.*

In addition to the universal form, it occurs to me that there is one more way of defining what the "opposite counterpart" of the empty form could be. The universal form matches to any conceivable form, which even includes the empty form, and thus the empty form acts like a subset of the universal form. However, just as it was useful to have both mutually exclusive and subset-related variants defined for the transience delimiters, the same might be true of the empty form and its opposite counterpart. We already have the subset-related variant (the universal form), so we should also define the mutually exclusive variant. Here's the corresponding new definition:

Definition 83. *To represent a form that could be any conceivable non-empty form, we write ⊟, and we call it the **non-empty form**. The symbol here (⊟) was chosen to resemble a chest of drawers. Chests of drawers are usually not kept empty, and thus the symbol serves as a reminder that it represents a non-empty form.*

These two forms (the universal form and the non-empty form) are rather extreme in what kinds of forms they match with. Consequentially, you will probably not be using them that often. Be that as it may, it's still a good policy to define them anyway, just for the sake of completeness if nothing else. They are potentially useful for defining very all-encompassing rule sets. For example, consider the following two rule sets:

Rule Set #1

1. ⌑ → ⌑

Rule Set #2

1. ⌐⌙ → ⊟
2. ⊟ → ⌐⌙

These two rule sets have the same net effect. Both of them allow you to transform anything into anything. In effect, they both define a transformative language without any constraints, where anything is allowed and nothing has any kind of consistent meaning. These two rule sets aren't the only way of defining this transformative language though. Since this transformative language has special properties, let's define a term for it. I don't have any uses for it in mind, but we might as well.

Definition 84. *If a transformative language allows all conceivable transformations, such that it in effect doesn't have any real rules or constraints, then we will refer to that transformative language as being an instance of the **chaos language**. There are many different ways to create a rule set that is equivalent to the chaos language, but the following rule set is the easiest way:*

$$\unicode{x2A00} \rightarrow \unicode{x2A00}$$

The chaos language illustrates an important point about transformative languages: We actually don't want a transformative language to be capable of anything. Transformative languages have virtue only insofar as they have at least some constraints. Too much freedom is effectively the same thing as chaos, and renders the entire system worthless. Thus, in a certain sense, pursuing "maximum freedom" in a transformative language is inherently a fool's errand. Infinite freedom might sound good on paper, but in practice it actually drains the world of all meaning and color. Meaning only exists in the presence of constraints. This is true not just of transformative logic, but of life in general.

This is why, for example, programmers who pursue infinite expressiveness often end up just wasting time and creating nothing of value. It's like searching for the pot of gold at the end of a rainbow. It's compelling, but also inherently fruitless. A blank page of source code sitting open on a compiler has infinite freedom, but it also has no value. The moment you write any code onto the page then by definition you will be eliminating at least some possibilities, and hence reducing your freedom.

It's not even possible to create value without also creating constraints. Thus, if you want to be productive, you should think more in terms of maximizing how interesting or useful a system is, rather than merely maximizing how much freedom it has. All we can do as humans is move from systems with worse constraints to systems with better constraints. This is true both in technical endeavors and in social structures. The idea of eliminating all constraints is nonsense, owing to the fact that it is incompatible with the pursuit of value.

A high degree of freedom is necessary in order for society to have a high quality of life. However, there is a breaking point where increasing freedom leads to decreasing quality of life. Total freedom would actually create a hellish nightmare. In a world of total freedom, there would be nothing to stop people from murdering each other, nor would there be anything to stop corporations from destroying the Earth for short-term profit. Complete freedom is in fact a very bad idea.

Putting that discussion aside, one thing that the universal form and the non-empty form are especially useful for though is creating form meta-variables. If you want a variable to stand for "any one specific arbitrary form" then the universal form and the non-empty form provide a convenient way to do this. The trick is to combine them with selective transformative implication. For example, consider the following rule set:

1. $A \rightarrow_1 \unicode{x2A00}$

2. $B \rightarrow_1 \boxplus$

With this rule set defined, *A* and *B* can now be used to express arbitrary forms where we don't care what *A* and *B* ultimately actually are, but we do care that they have consistent values across an expression. Thus, for example, ⟅ *BAB* ⟆ under this rule set would mean "any form surrounded on both sides by the same non-empty form". For instance, forms like *xyx*, *xx*, and □(sin(*x*) + cos(*x*))□ would match. In other words, the universal form and the non-empty form are useful for defining variables that range over the set of all forms instead of some more specific set.

In addition to it being useful to be able to talk about the set of all conceivable forms though, it would also be useful if we had a convenient way to refer to all of the initial forms of the current rule set at once, without having to list all of them manually. Having such a mechanism would enable us to say "all forms derivable from any of the initial forms of the current rule set" (and similar expressions) much more easily. As such, here is a definition designed to fill this role:

Definition 85. *As a convenience, it is often useful to have a form that enables us to refer to all of the initial forms of the current rule set all at once. To accomplish this, we will define a special symbol which always implicitly points to all of the initial forms of the current rule set via transformative implication. We will call this form the* **root form** *and represent it with the symbol* ⊬. *The symbol* ⊬ *was chosen because it resembles the roots of a tree growing underground, which helps remind the user that it represents the root form.*

More precisely though, the way the root form works is like this: Suppose that our rule set has some arbitrary set of initial forms that are assumed to already be valid, in addition to whatever transformation rules it has. Let $I_1, I_2, \ldots, I_{n-1}, I_n$ represent all n of these initial forms[14]. *Then, an additional unwritten transformation rule of the form* ⊬ → $\{I_1, I_2, \ldots, I_{n-1}, I_n\}$ *may be treated as existing in the rule set.*

This implicit rule thereby enables us to use ⊬ *to refer to all of the initial forms of the current rule set as a group much more conveniently. Besides the one special implicit transformation rule implied by the use of* ⊬ *though,* ⊬ *should otherwise be treated just like any other arbitrary form. Thus, the form* ⊬ *can still be used in other rules and in other forms in the rule set if the user chooses to do so and accepts the consequences, just as with any other arbitrary form. Indeed,* ⊬ *is just a convenience, and the same effect as* ⊬ *can be accomplished manually by the user if they simply assign a similar kind of rule to whatever other arbitrary form they choose.*

One of the most common use cases for the root form is to wrap it in transience delimiters so that you can easily extract the set of all forms derivable from the initial forms of a rule set that match a specific transience type. Thus, for example, if you wanted to say "all forms derivable from any of the initial forms of the current rule set" in the most open-ended sense possible, then you could write ⟅ ⊬ ⟆ to accomplish it.

The form ⊬ would also be included in the resulting set of forms though, which may or may not be ok with you. You could alternatively use ◁ ⊬ ▷, ⊣ ⊬ ⊢, or ◂ ⊬ ▸ to avoid the inclusion of ⊬ in the set[15]. Or, if that isn't sufficiently precise, then you could use set

---

[14]The list could also be infinite too, and that would be fine.

[15]I'm assuming here that you didn't add in additional rules that cause ⊬ to no longer be trans-open.

operations to specifically exclude the symbol ✦ (and any other forms you don't want) in the result, e.g. such as by writing **Set**(◐ ✦ ◑ − {✦}) or whatever else you want.

In addition to it being useful to be able to talk about the set of all forms derivable from the initial forms and transformation rules of the current rule set though, it would also be useful to be able to test whether or not a form or set of forms is in this set (i.e. in ◐ ✦ ◑) and to use this as a filtering mechanism. To accomplish this, we will define the following new operator:

Definition 86. *It is sometimes useful to be able to require that a form is derivable from the initial forms and transformation rules of the current rule set (i.e. is valid). As such, whenever we wish to filter a set of forms by testing for validity, then we will wrap that set of forms inside a **validation filter**. We will represent this validation filter in symbols as **val**(X), where X is the set of forms we want to filter.*

*Another set of forms will be returned. Only forms in the validation filter which are indeed derivable from the initial forms and transformation rules of the current rule set (i.e. are valid) will be allowed to pass through. If none of the forms are valid though, then the empty set will be returned. If the X in **val**(X) is just a single form standing on its own (i.e. not written as a set using {…}), then it should still be treated the same as {X} and processed accordingly.*

Thus, for example, if we were working in the transformative language of true equations (according to the rules of standard mathematics), then **val**(2 + 3 = 23) would return { } (i.e. the empty set), since 2 + 3 = 23 is not valid in standard mathematics. Similarly, if $w \to (w \odot w)$ was our only transformation rule and w was our only initial form, then

$$\mathbf{val}(\ \{(w \odot (w \odot w)),\ w,\ \lambda x.(x + 1)\}\ )$$

would return $\{(w \odot (w \odot w)), w\}$, since only $(w \odot (w \odot w))$ and w in this set are derivable from the given initial forms and transformation rules. The other expression in the set, $\lambda x.(x + 1)$, is actually from a different transformative language, specifically lambda calculus[16]. Often, an efficient way to validate a form is to follow the applicable transformation rules backwards[17] to test whether or not you can reach the root form. If you can, then the form must be valid. Also, notice that ◐ ✦ ◑ and **val**(⌑) refer to the same set (i.e. the set of all valid forms under the current rule set), but do so via different computational processes. Do you see why they both refer to the same set? Also, can you see which one would be more computationally efficient?

Anyway though, that covers the empty form, the universal form, the non-empty form, the root form, and the validation filter. We also have transience delimiters, selective substitution, and interpretation injectors. We've got quite a good vocabulary for talking about forms now. It's a powerful system. However, one thing you may have

---

[16] Well, technically this expression could have any conceivable interpretation, because transformative logic works in terms of uninterpreted forms. However, lambda calculus is what most people familiar with the material would recognize it as. Lambda calculus is an interesting system, by the way. It's worth a look if you're curious about it.

[17] This is done by applying a process called *transformative integration* to the form. I'll have more to say about that later though, when we get to that part.

noticed is that our transformation rules allow transformations to be applied regardless of where in another form the condition form is located.

Usually this is a useful way for the system to behave, but sometimes we may want to constrain it more. For example, sometimes we may want a rule to match only forms that stand on their own, isolated from any other forms and not contained inside any other forms. To accomplish this, we will need to delimit form boundaries. This brings us to our next definition:

Definition 87. *Sometimes we want a transformation rule to be restricted to only being applied to isolated forms, such that it cannot be applied unless the form in the condition represents the entirety of the form that the transformation is being applied to. Given an arbitrary form $\mathcal{F}$, we can indicate that it should be considered in isolation from other forms by surrounding it with* **form boundary delimiters**, *which look like ⦃ ⦄. Thus, the form $\mathcal{F}$ considered in isolation from other forms would be written as ⦃$\mathcal{F}$⦄.*

The symbol "⦃" means the left-side boundary of a form, and likewise the symbol "⦄" means the right-side boundary of a form. These delimiters can be used separately if desired (unlike most delimiters), but pairing both of "⦃" and "⦄" together and wrapping them around a form will probably be the most common use case. In the case of ⦃$\mathcal{F}$ it means that the form $\mathcal{F}$ must occur at the beginning of the form in which it occurs. Similarly, in the case of $\mathcal{F}$⦄ it means that the form $\mathcal{F}$ must occur at the end of the form in which it occurs

Keep in mind that the form boundary delimiters are not themselves visible symbols, but rather they merely indicate the boundaries of a form. Consider, for example, the difference between $X \to Y$ and ⦃X⦄ $\to Y$ in terms of what they allow. In the case of $X \to Y$, if we were given $X + X + X$ as an initial form then we would be able to transform as many of the X symbols in the expression as we like into Y symbols. However, in the case of ⦃X⦄ $\to Y$, if we were given $X + X + X$ as an initial form then no operations upon $X + X + X$ would be permitted. The rule ⦃X⦄ $\to Y$ only says that X can be transformed into Y when it stands in isolation. Thus, the rule permits transforming an isolated X into Y, but says nothing about $X + X + X$.

We've now got a pretty good number of different ways of specifying constraints on forms. By combining things like set theory, transience delimiters, quotation delimiters, and form boundary delimiters we can accomplish quite a lot. However, if that's not enough for you, keep in mind that you can also import even more new capabilities by using interpretation injectors. For example, have you heard of *regular expressions* from computer programming? Regular expressions, or *regex* for short, are a way of concisely describing very precise and flexible criteria for text to match. Let's create a new interpretation injector for it. Naturally, we'll call it **Regex**. Here's an example of it in use:

1. $P \to$ **Regex**(`[a-zA-Z0-9_]+`)

2. ⫶$P + P$⫶ $\to$ `SumOfIdentifiers`

The expression **Regex**(`[a-zA-Z0-9_]+`) means any form consisting of at least one (`[...]+`) of the following: lowercase letters (`a-z`), uppercase letters (`A-Z`), numbers

(0-9), and underscores (_). No whitespace is allowed. This is (roughly) the standard criteria for what is allowed in variable names in many programming languages (ignoring Unicode and non-English languages). Thus, the variable *P* in the above rule set is bound to "any programming language variable name". As such, the expression ·| *P* + *P* |· effectively means "any two arbitrary programming language variable names being summed together in an expression". Therefore, the rule ·| *P* + *P* |· → SumOfIdentifiers has the net effect of allowing you to transform any such pair of variable names being summed together into the literal form SumOfIdentifiers. Let's do one more regex example. Consider the following rule set:

1. /* **Regex**([a-zA-Z0-9_\s]*) */ → ⌐⌙

2. ⌐⌙ → /* **Regex**([a-zA-Z0-9_\s]*) */

When added on to any other arbitrary rule set, these rules have the net effect of allowing the user to insert C-style comments anywhere they want into any arbitrary form. By "comments" I mean sequences of text that have no effect on the behavior of a form, thus allowing the user to insert arbitrary commentary. By "C-style" I mean that this particular style of creating comments by delimiting text with /* ... */ originates from a family of programming languages known as the "C family" which are all descended from a very popular and historically important programming language named "C". Unlike the previous time we used regex, this time the regex expression matches to zero or more of the indicated characters ([...]*) and also allows arbitrary whitespace (\s).

Why does this work? Why does this pair of rules allow you to add comments on to pretty much any other arbitrary rule set? Well, the reason why it works is because the rules allow you to remove the comments at any time, thereby ensuring that they can't actually block any of the preexisting transformation paths of the rule set that you are modifying. That's the idea anyway. There might be some problems with it in certain contexts, depending on what other rules are in the rule set. It's just for illustration purposes. Anyway, that's enough of regex now. I think you get the idea. Just remember that these kinds of things are available to you if you ever feel inclined to make use of them.

Through our most recent extensions, we now have empty forms, universal forms, non-empty forms, form boundary delimiters, and regex available to us. However, another thing I want to make available to us is a few extensions to transience delimiters. Transience delimiters are fairly powerful and work well most of the time, but they also sometimes overreach and grab more forms than we really want them to. For that reason, it would be useful to have at least a few more ways of constraining them. Here is one such way:

Definition 88. *The regular transience delimiter types sometimes return forms that we don't want included in the derivation set. Thus, we may want a method by which to limit the scope of a transience delimiter so that it cannot reach forms that are more than a certain distance away from the base form (the form wrapped in the transience delimiters). To do this we may use a* **transformation delta**.

A transformation delta is an equality or inequality written as a subscript of a transience delimiter pair that indicates how far away (in terms of transformation steps) from the base form the derivation set is allowed to reach. This distance is represented by the Greek letter delta (δ). The subscript will look something like $\delta = k$ or $a \leq \delta \leq b$ (or any other arbitrary inequality), where k, a, and b are arbitrary non-negative integers.

For example, if we wanted to retrieve the set of all forms reachable from form X within only a single transformation step then we could write ⟅ X ⟆$_{\delta=1}$. Notice that this set would *not* return X itself in the resulting set. It would only return forms that can be reached from X in exactly one transformation step. In order to also include X in the resulting set then you would need to write ⟅ X ⟆$_{0 \leq \delta \leq 1}$ or ⟅ X ⟆$_{\delta \leq 1}$. The reason why ⟅ X ⟆$_{\delta \leq 1}$ is also allowed is because transformation deltas are always non-negative integers, so the left side of the inequality in $0 \leq \delta \leq 1$ can be considered redundant in this case.

For the purposes of transformation deltas, backtracking is not allowed when counting the number of steps away a form is from the base form. Not applying any transformations to a form and transforming a form trivially back to itself are also not allowed. In other words, you need to maintain a set of "forms that have already been visited" and use it to check whether each new form you visit was already visited and if so ignore it. It ends up working kind of like a breadth-first search[18], except that instead of searching for some specific element we are searching for all forms that exist within some arbitrary transformation depth range.

Furthermore, when evaluating transformation deltas, we ignore any transformation path shortcuts that haven't been explicitly specified. Otherwise, the concept of a transformation delta wouldn't actually make any sense. Given any arbitrary transformation path $f_1 \twoheadrightarrow f_2 \twoheadrightarrow \ldots \twoheadrightarrow f_n$, a new transformation rule that just transforms f_1 directly to f_n (as in $f_1 \twoheadrightarrow \twoheadrightarrow f_n$) can always be created as a shortcut, i.e. as a "theorem".

If this was allowed for transformation deltas though, then in effect all forms would always be just one transformation step away, because any arbitrary transformation path could always be shortened to one step. Thus, whenever transformation deltas are used, you need to treat the current rule set as being set in stone and as not allowing any shortcuts to be created. This imposes some inherent limitations on the concept of a transformation delta, causing it to be somewhat arbitrary and reducing its usefulness.

Keep in mind that any arbitrary inequalities are allowed and thus ⟅ X ⟆$_{3 \leq \delta \leq 5}$ would mean the set of all forms derivable from X that are between 3 and 5 transformation steps away. Basically, transformation deltas are a rough approximation to what "distance between forms" might mean in transformative logic. Sometimes it might be useful for expressing certain things, but be very careful with it. The fact that you can't take shortcut rules unless they are explicitly specified can lead to some subtle conceptual problems. However, luckily, the next extension to transience delimiters we will be dis-

---

[18] A breadth-first search is a search algorithm from computer science. Look it up if you want to know more precise details, but you can kind of think of it like a ripple on a pool of water or an ink blot spreading across a sheet of paper. The search starts at some specific location and emanates outwards from that point, spreading to all of the locations adjacent to already-visited locations each iteration. It is also very closely related to the concept of a "flood fill".

cussing doesn't have this weakness. Here are the new definitions that we will need for it:

Definition 89. *In mathematics, when we want to indicate that something is limitless in quantity we often use the infinity symbol ∞. However, sometimes it might also be useful to instead indicate that something is finite but that we don't necessarily know what exact quantity it is. To indicate this, we will use the symbol ⊣ and we will call it **finity**. The reason why the sideways* F *symbol was chosen here is because it serves as a memory aid, because* F *is the first letter of the word "finite" and drawing it sideways makes it more reminiscent of the ∞ symbol.*

Definition 90. *As we know, corresponding to every form there is a set of forms that can be reached by following the transformations allowed on that form, which we call the derivation set of that form. This derivation set will have some number of distinct forms in it, whether finitely or infinitely many. We call this count of the number of elements in the derivation set of a form the **terminacy count** of that form. There are two types of terminacy counts: finite and infinite. We refer to such types as **terminacy types**. Thus, every given form has either **finite terminacy** or **infinite terminacy**, depending on whether the form can reach a finite or infinite number of distinct forms by applying any applicable transformation rules.*

If we merely want to return the terminacy count of a form, then we simply reuse the standard count operator from set theory (a.k.a. the "cardinality" operator) on a trans-free delimited instance of the form. Thus, for example, if X was the form whose terminacy count we wanted to determine, then computing Count(⟨ X ⟩) *would give us its terminacy count. As a shorthand notation, we may instead write* Ter(X) *and this is equivalent to writing* Count(⟨ X ⟩).

*It may sometimes be useful to specify that we only want to return forms that have a certain terminacy count or terminacy type from the derivation set of a transience delimited form. To accomplish this, we may use a **terminacy subscript** attached to the transience delimiters surrounding the base form. In such subscripts, we represent the terminacy count with the letter t and we specify constraints upon it using arbitrary equalities or inequalities. Terminacy counts are positive integers.*

The terminacy count t of a form is always at least 1, because every form is always capable of reaching itself. However, if we only care about what terminacy type the derived forms have, and not about what the specific count is, then we can just write ⊣ or ∞ in the subscript corresponding to which terminacy type we want to allow. Terminacy subscripted transience delimiters are evaluated as the conjunction of the transience delimiter type constraint and the terminacy count constraint, as applied to every form in the derivation set of the delimited form.

Thus, for example, if we wrote ⟨ X ⟩$_{t \leq 10}$ then it would mean the set of all forms derivable from X that themselves are only able to reach at most 10 distinct forms. In other words, for every form X_k derivable from X, if Ter(X_k) $\leq$ 10 then X_k will be included in the set returned by ⟨ X ⟩$_{t \leq 10}$, otherwise it won't.

On the other hand, if we wrote ⟨ X ⟩$_⊣$ then it would mean the set of all forms derivable from X that themselves can only reach a finite number of distinct forms, i.e. all forms X_k derivable from X such that $\text{Ter}(X_k) = \neg$. If we instead wrote $⟅ X ⟆_\infty$ then it would only include forms derivable from X that are capable of reaching an infinite number of distinct forms, i.e. all forms X_k derivable from X such that $\text{Ter}(X_k) = \infty$.

Infinite terminacy is typically caused by the presence of recursive transformation rules in a rule set. Thus, in a sense, you can think of finite terminacy as meaning "non-recursive" and infinite terminacy as meaning "recursive". Notice that the terminacy count constraints are based on the derivation set of *each* form reachable from X, not only based on the derivation set of X itself. It's important to remember the difference between terminacy counts and terminacy subscripts. Terminacy counts are just descriptive values for a form (integer counts), whereas terminacy subscripts are search constraints.

By the way, you could define other variants of terminacy counts based on the other transience delimiters (e.g. $\text{Count}(·| X |·)$ instead of $\text{Count}(⟅ X ⟆)$), but I didn't really see much point to doing it here. It seemed like overkill perhaps. I didn't want to overburden our notation and terminology here. If I did define those variants of terminacy counts though, then I would refer to them as **<transience type> terminacy counts** and would define corresponding **<transience type> terminacy subscripts** for them, where "<transience type>" is replaced by the specific transience type. Thus, given a form X, the terminacy count variants would be:

1. trans-free terminacy count: $\text{Count}(⟅ X ⟆)$

2. trans-fixed terminacy count: $\text{Count}(◁ X ▷)$

3. trans-closed terminacy count: $\text{Count}(·| X |·)$

4. trans-equal terminacy count: $\text{Count}(◀ X ▶)$

5. trans-open terminacy count: $\text{Count}(◐ X ◑)$

However, this is just an idea I felt like mentioning. There are some notational issues with creating the corresponding terminacy subscripts for these variants that I don't feel like solving, partly because we won't be using them anyway. Thus, for the remainder of this book, the terminacy count of X will always refer to $\text{Count}(⟅ X ⟆)$, as before, for the sake of keeping things simple.

Anyway, unlike transformation deltas, terminacy counts are uneffected by whether or not shortcut transformations are allowed. This arguably makes terminacy counts more conceptually stable and universal than transformation deltas. However, both concepts have their uses. The terminacy count of a form gives you a measure of how many transformation options a form truly has available to it. Anyway, here's another extension to transience delimiters that I'd like to throw into the mix, just in case it ever becomes useful for anyone:

Definition 91. *Associated with every form is a set of forms that can be reached from that form within only a single transformation step. We will call this set of forms the **directed adjacency set** of the base form. Furthermore, we will call the number of distinct forms in the directed adjacency set the **directed adjacency count**. Given a form X, then we* may write $\text{DirAdjSet}(X)$ to represent its directed adjacency set and $\text{DirAdjCount}(X)$ to represent its directed adjacency count.

A form X is directionally adjacent to another form Y only if X has a transformation path that would allow it to transform into Y in one move. Directional adjacency is one-way, not two-way. Thus, if Y did not have a transformation path to X then it could not be considered directionally adjacent to X, even if X was directionally adjacent to Y via $X \twoheadrightarrow Y$.

If both of two forms can be transformed into each other in one step though, then we may say that those forms are **bidirectionally adjacent**. As such, we may write $\text{BiAdjSet}(X)$ to represent a form's bidirectional adjacency set and $\text{BiAdjCount}(X)$ to represent its bidirectional adjacency count. Two forms X and Y are only bidirectionally adjacent to each other if both can transform into the other, i.e. if both $X \twoheadrightarrow Y$ and $Y \twoheadrightarrow X$ exist as rules.

If, on the other hand (speaking more loosely), one or the other (or both) of two forms can be transformed into the other in one step then we may say that those forms are simply **adjacent**. Thus, all directionally adjacent and bidirectionally adjacent forms are also adjacent in this more general sense. As such, given a form X, then we may write $\text{AdjSet}(X)$ to represent its adjacency set, and $\text{AdjCount}(X)$ to represent its adjacency count, as one might expect.

In order to be able to constrain a derivation set to only forms that have a specified range of adjacency counts, we may write an arbitrary equality or inequality expressing this constraint in the subscript of a transience delimiter, and we will call it an **adjacency subscript**. We represent the adjacency count in such subscripts by a different name depending on which type of adjacency we want to constrain. Directional adjacency is represented by dj, bidirectional adjacency is represented by bj, and general adjacency is represented by j. All such adjacency counts (dj, bj, and j) are non-negative integers.

Thus, for example, if we wrote $\triangleleft X \triangleright_{4 \leq dj}$ then it would mean the set of all forms derivable from X that are each directionally adjacent to at least 4 other distinct forms. Likewise, $\triangleleft X \triangleright_{dj \leq 1}$ would mean the set of all forms derivable from X that are each directionally adjacent to at most 1 other distinct form. Notice that just like with transformation deltas, adjacency counts only really make sense if shortcut rules that have not been explicitly written out are ignored. Thus, adjacency counts have some of the same artificial quality and conceptual weakness that transformations deltas have. Nonetheless, adjacency counts still have their potential uses as long as you freeze the rule set.

Anyway, there's a good chance this talk of adjacency counts may have reminded you of another object of study from mathematics, especially if you are familiar with computer science or discrete mathematics. What other "object of study" am I referring to here? I am referring to graph theory, the study of connected networks of nodes. The system also has a relationship to finite state machines as well, since those are a specific type of graph. Transformative logic has much the same structure as graph theory, although transformative logic has quite a different focus and purpose. Each form is like a node and each transformation rule is like a directed arrow.

A consequence of this close relationship between transformative logic and graph theory is that many of the ideas from each field could potentially be applied to the other. For example, you could apply some of the shortest-path algorithms from graph theory to rule sets in transformative logic if you wanted to, if you ignored the implicit shortcut rules, and if it made sense in the context. The concept of adjacency count shows up all the time in graph theory of course, and we have just now introduced it into transformative logic as well.

However, that being said, adjacency is typically far less important in transformative logic than in graph theory. Adjacency is part of the nuts and bolts essentials of working in graph theory, whereas adjacency is much more of a fringe concern in transformative logic. In transformative logic we tend to care far more about whether or not a transformation path exists between two given forms than about what forms another form is immediately adjacent to.

One notable difference between graph theory and transformative logic is that in transformative logic forms can be embedded inside of each other and the same transformation rules applicable to each individual embedded form will still be applicable, despite each individual embedded form now being part of a new whole. It's kind of like a specialized union of all outgoing arrows.

In contrast, graph theory doesn't really have an equivalent concept of "node embedding" as far as I'm aware, or at least doesn't emphasize it as much if it does. Thus, in transformative logic, the transformation "arrows" for a specific form are not generally centralized in one location, attached to a single node, but rather are spread out all over the place, anywhere that the matching form occurs inside any other form. In contrast, nodes in graph theory are usually treated as if they are set in stone and the arrows applicable to one node are applicable only to it and to no other context.

For this reason, the concept of an "identity" in transformative logic with respect to the subset of a rule set applicable to a specific form is a bit different. In graph theory, all of the transformations available to an "identity" (i.e. all outgoing arrows from a node) are attached to that one instance of that identity (i.e. attached to that one node), whereas in transformative logic all of the transformation rules applicable to one specific "identity" (i.e. to one specific form with associated transformation rules) are attached to all forms that contain at least one instance of that "identity" inside themselves. Thus, in transformative logic, in order to treat an "identity" as one complete and distinct entity, rather than as a bunch of disorganized specific cases, you must gather all of the related forms together into one set. Here's a new definition for this set:

Definition 92. *In transformative logic, the set of rules that are applicable to some arbitrary form X are actually also applicable to any form Y that happens to contain at least one X. Therefore, the form X itself is not necessarily a complete representation of the "identity" of that form with respect to the transformations applicable to it. If we want to gather together all forms to which the transformation rules of X apply, then we actually have to gather together all valid forms that contain at least one X. We call such a set of forms the **identity set** of X, and we write **id**(X) to represent it.*

*The identity set is always relative to whatever set of transformation rules and initial forms are currently in effect. If a form cannot be derived from the given initial forms by applying the given transformation rules, then it cannot be in the identity set. Hence, when evaluating **id**(X), if no forms can be found which contain X, then **id**(X) will be* *empty. The identity set of a form is in some sense the "meaning" or "significance" of that form under the current transformative language. The identity set* $id(X)$ *captures all objects and relationships in which the form X participates, and hence captures every possible aspect of the relevance and impact of X within the system being studied. Valid (i.e. "meaningful") forms have non-empty identity sets, whereas invalid (i.e. "meaningless") forms have empty identity sets.*

In systems where we concern ourself not merely with just forms as structures, but with forms that express *truth*, then the identity set of a form essentially becomes the same thing as the set of all true relationships in which that form participates. For example, in the algebra of equations, the identity set of the number 2 (i.e. $id(2)$) is the same thing as the set of all true relationships involving the number 2. Thus we have something like:

$$id(2) = \{2+2=4, \ 6/3=2, \ 2^3=8, \ 2+1=3, \ ...\}$$

Thus, $id(2)$ contains everything you could ever possibly want to know about the number 2. Every conceivable form that expresses a truth about the number 2 will be in this set. However, if we changed what transformative language we were working under then the result could be different. For example, suppose that instead of working under the transformative language of the algebra of equations, we were instead working under the transformative language of arithmetic. In that case, the result of $id(2)$ would be quite different. It would look more like this:

$$id(2) = \{2+6, \ 2\cdot 2, \ 4/2, \ 2^{1/2}, \ 2+2, \ 2^{32}, \ 2/16, \ ...\}$$

Notice how this time there aren't any equals signs in this set of forms. None of these forms are making any assertions about truth. These are all simply numeric value expressions. Here, instead of meaning "all true relationships involving the number 2", the set $id(2)$ means "all conceivable arithmetic expressions involving the number 2".

As such, what transformative language we are currently working under clearly has a large impact on the behavior of the identity set of a form. After all, transformative languages are all totally arbitrary anyway, so it should come as no surprise then that the same symbol could mean completely different things under different transformative languages. Even such a common symbol as 2 could conceivably be reused to mean something else that has absolutely nothing to do with the conventional concept of the number 2.

The obvious danger here is that we might misinterpret what the identity set of some form is intended to refer to. We need to know what context (i.e. what transformative language) it belongs to in order to understand it unambiguously. Luckily, the answer to this problem is easy. We will simply use interpretation injectors to specify our intent. If we intend for the identity set of some arbitrary form X to be relative to the current transformative language in which we are working then we will simply write $id(X)$. If, on the other hand, we wish to specify some other interpretation for $id(X)$ then we will write **LanguageName**($id(X)$), where **LanguageName** is the name of the intended transformative language.

Thus, for example, the two different ways of interpreting the identity set of 2 that we previously discussed could be distinguished explicitly by writing **Eq**(**id**(2)) and **Arith**(**id**(2)) respectively. In case you've forgotten, **Eq** and **Arith** are the interpretation injectors for the transformative languages of the algebra of equations and of arithmetic, respectively. Otherwise, if we just wrote **id**(2) then it would be interpreted as the identity set of 2 under whatever transformative language is currently in effect.

By definition, in transformative logic you always have to be working under at least some transformative language, otherwise everything you did wouldn't have any context. Thus, there is relatively little danger of not knowing what transformative language an identity set should ultimately be interpreted under, as long as the author is not omitting necessary information.

In addition to supporting simple forms, the identity set notation also supports arbitrary transformative logic expressions. Any expression that can occur in the condition or consequence of a transformation rule can also occur inside the parenthetical delimiters of the identity set. In this way, you can use identity set notation to evaluate multiple different identity sets at once and treat them as a group.

The expression inside the identity set parentheses is evaluated according to the normal rules of transformative logic, yielding a corresponding set of forms. Each such form is then separately evaluated as an identity set, and each of the resulting identity sets are then combined together as a union of sets. Thus, for example, under the algebra of equations, we could evaluate the identity sets of both 2 and 3 simultaneously as follows:

$$\mathbf{id}(\{2, 3\}) = \mathbf{id}(2) \cup \mathbf{id}(3)$$
$$= \{2 + 2 = 4,\ 4/2 = 2,\ \ldots,\ 3 + 3 = 6,\ 9/3 = 3,\ \ldots\}$$

You're not restricted to just the basic set notation though. You can use any notation you'd normally use in transformative logic. For example, if you had a well-formed formula rule set whose initial form was w then you could write **id**(⊢w⊢) to mean the union of all identity sets of all trans-closed forms derivable from w. For another example, if you wrote **id**(⌑) then it would mean the union of all identity sets of all conceivable forms.

However, since identity sets are always relative to a specific transformative language, **id**(⌑) would end up only including the identity sets of all *valid* forms under that given transformative language. The reason is because all of the forms generated by ⌑ that aren't part of the vocabulary of the given transformative language would have empty identity sets and hence not appear in **id**(⌑). Thus, in effect, **id**(⌑) means the same thing as the set of all valid forms under the rule set, i.e. all forms derivable from at least one of the initial forms via the given transformation rules.

The set returned from **id**(⌑) does *not* include any forms that cannot be reached from the assumptions of the rule set. For example, even if some arbitrary form X had some transformation rules that could hypothetically be applied to it under the rule set, if X would never appear inside any form derivable from the given initial forms under the given rule set then X would not be included in **id**(⌑) in any way. Indeed, **id**(X) would be an empty set in such a case.

id(⌑) may mean the same thing as the set of all valid forms under a rule set, but it is certainly not the only way of expressing that. Indeed, as you may recall, we have already discussed other mechanisms that are capable of expressing that. Specifically, ◖✢◗ and **val**(⌑) also return the same set as **id**(⌑)[19]. However, ◖✢◗ is the most computationally direct and efficient, since all the forms derived from ✢ constructively build up from ✢ according to a well-defined process, whereas **val**(⌑) and **id**(⌑) are more computationally intractable and require you to examine every conceivable form in existence regardless of whether each form is part of the vocabulary of the current rule set or not. Thus, ◖✢◗ is computationally preferable, even though both **val**(⌑) and **id**(⌑) theoretically refer to the same set. In summary: **id**(⌑) is the most convoluted way of expressing this set, **val**(⌑) is somewhat less so, and ◖✢◗ is the most direct.

Regardless though, the concept of an identity set will become extremely important later in the book. It is actually one of the most powerful concepts we have introduced in this section, but you won't get to see why yet until later. Besides this brief introduction, we won't be discussing identity sets again for quite a while. I will say though that identity sets are essential to the creation of some of the more advanced features of unified logic, as you will see once we get to that part.

Oh, and there's one more really basic thing that I forgot to mention before this point, because we haven't needed it yet. In particular, I want to make sure that you're aware that you can work with multiple initial forms at once *as a single expression* via set notation if you choose to do so and if the circumstances permit it. Doing so is often highly convenient and useful. It allows you to very concisely express multiple transformation operations in a single expression.

For example, if you were working in the transformation language of arithmetic, and you wanted to compute both $1+1$ and $2+1$, then it would not be necessary to transform them each separately. There's a shared factor here, so we can therefore instead do both of them at once via a single expression. Specifically, we could just write $\{1,2\} + 1$ instead of $1 + 1$ and $2 + 1$ separately. We could then transform $\{1,2\} + 1$ into $\{2,3\}$ by applying the transformation rules of arithmetic to each form in the set wherever applicable, at our discretion.

The reason this is permitted is because transformative logic treats set notation as special, as being shorthand for multiple different possible forms existing at a given location in another form. Thus, expressions like $\{1,2\}+1$ just end up naturally behaving like we are working with multiple forms at once, any one of which we could apply a transformation rule to at any time.

Note that even though this way of working with forms involves transforming multiple forms at once, it is ultimately still just a normal application of transformative implication. It's just shorthand for multiple different forms being separately transformed in arbitrary ways via transformative implications. It's just that we can treat these conceptually separate forms as a group, for brevity. It is often convenient, but doesn't always scale well. Keep it in mind as yet another tool in your toolbox. Oh, and let's define some corresponding formal terms, just for ease of communication:

---

[19] i.e. the set of all forms derivable from the initial forms and transformation rules of the current rule set, which is conceptually the same thing as the set of all valid forms under the rule set

Definition 93. *When we apply transformation rules to only one form at a time, then we may say that we are performing* **individual transformation**. *For instance, transforming 1 + 1 to 2 would be an example of individual transformation.*

Definition 94. *When we apply transformation rules to multiple forms at once, expressed as a single expression, then we may say that we are performing* **collective transformation**. *For instance, transforming $\{1, 2\} + 1$ to $\{2, 3\}$ would be an example of collective transformation.*

Note that collective transformation may be performed in any arbitrary way. Each transformation of each form is optional, as always in transformative logic. Thus, for example, $\{1, 2\} + 1$ could also just be transformed into $\{1 + 1, 3\}$, instead of into $\{2, 3\}$. In other words, we can apply transformations as completely or as partially as we decide to, as long as we do so correctly. Each form in the set is still conceptually distinct and can be reasoned about separately.

3.11 Transformative derivation and integration

Anyway, up until this point we've always followed transformation rules in the forward direction, from the condition form at the tail of the transformation arrow to the consequence form at the head of the transformation arrow. In other words, given an arbitrary transformation rule of the form $A \to B$, we have always moved from A to B. We have referred to the set of all forms reachable from A in this manner as the derivation set of A, relative to some transience type. For example, the broadest possible derivation set based on A would be the trans-free delimited A, written as ⟨ A ⟩. We already have a term defined for the derivation set, but let's also define a term for the more general process as well:

Definition 95. *The process of applying transformation rules to forms in the forward direction, as in transforming A to B using the transformation rule $A \to B$ (where A and B are any arbitrary forms), will be called* **transformative derivation**. *When we say that a form has undergone transformative derivation we are simply saying that it has undergone some arbitrary sequence of transformations in the forward direction. It does not imply that all possible transformations paths are being explored. Transformative derivation can be used to just transform one specific form into another, but it can also be used more broadly to explore any arbitrary subset of the set of all possible forms derivable from a form. Any such subset of all forms derivable from a form is what we have been calling a derivation set.*

Transformative derivation is what we've been doing throughout our explorations of transformative logic up to this point. However, interestingly, the opposite process is also possible. It is possible to follow the transformation rules backwards, but doing so changes what the result means. Both transformative derivation and its opposite process are quite useful. Let's now define a term for the opposite process:

Definition 96. *Normally we apply transformation rules in the forward direction and we call this process transformative derivation. However, the opposite process is also pos-*

sible. We call this opposite process ***transformative integration***. It's behavior is almost the same as transformative derivation, except for that everywhere where transformative derivation follows transformation arrows forwards transformative integration instead follows them backwards[20]. In effect, relative to any given rule set, transformative integration acts kind of like the rule set's mirror image and enables the transformations of the rule set to be undone. The direction of the transformation arrows must still always be consistent with the original rule set though, to avoid confusion.

Whereas the result of transformative derivation provides you with what forms are derivable from a given form, transformative integration provides you with what forms would be able to *reach* the given form if transformative derivation were applied to those forms. Furthermore, just as we have a set of transience delimiters defined to specify different derivation sets under transformative derivation, we also have an opposite set of delimiters that apply the same process but in reverse for transformative integration. Here are the relevant definitions:

Definition 97. *Any set of forms derived by applying transformative integration to a form may be referred to as an* **integration set** *of that form.*

Definition 98. *Corresponding to each of the transience delimiters used for transformative derivation, there is also an analogous delimiter used for transformative integration, which works in the opposite direction and which we will therefore call a* **reverse transience delimiter**. *All of the behaviors of these reverse transience delimiters are the same as for the normal transience delimiters except for that the transformation arrows are followed backwards. We represent the reverse transience delimiters using symbols that are a mirror image of the normal transience delimiters. Let $\mathcal{F}$ represent any arbitrary form. Then the reverse transience delimiters would be written like so:*

$$
\begin{aligned}
&\textit{reverse trans-free:} && \mathcal{D}\,\mathcal{F}\,\mathcal{C} \\
&\textit{reverse trans-fixed:} && \triangleright\,\mathcal{F}\,\triangleleft \\
&\textit{reverse trans-closed:} && \vdash\mathcal{F}\,\dashv \\
&\textit{reverse trans-equal:} && \blacktriangleright\,\mathcal{F}\,\blacktriangleleft \\
&\textit{reverse trans-open:} &&)\,\mathcal{F}\,(
\end{aligned}
$$

This notation makes it very easy to remember how to write each of the reverse transience delimiters for transformative integration, relative to each of the corresponding transience delimiters for transformative derivation. You just mirror them. Admittedly though, reversing the direction of these delimiters like this could potentially create problems for unambiguously parsing expressions. However, we will use be using this notation for transformative integration in this book regardless, despite this danger.

The main reason for this choice is because we won't actually be using transformative integration anywhere except right here, so even if it ends up creating major problems it still won't end up mattering that much given the limited scope of this book. I'm only introducing the concept to you here so that you are aware of its existence, to broaden your perspective on what transformative logic is capable of.

---

[20] Selective transformation requires special treatment to reverse correctly though.

The other reason is because, even though this notation might potentially introduce expression parsing ambiguities, the context and visual appearance of the delimiters will often make the intent clear anyway. In addition, the memory-aiding qualities of simply reversing the transience delimiter symbols is quite appealing and elegant. I'm not sure if this is the best notation choice, but it's what we're going with regardless since it's so intuitive.

Anyway, let's do an example. Consider, for example, the concept of well-formed formulas from earlier in the book. Rule sets that define what it means to be a well-formed formula are highly suitable for illustrating the practical utility of transformative integration. It's one of the most likely use cases for it. The first well-formed formula rule set we discussed was on page 229. Let's list it again here for ease of reference:

1. $w \to \{A, B, C, \ldots, Z\}$

2. $w \to (w \odot w)$

3. $w \to (w \circledast w)$

What would transformative integration look like under this rule set? Well, let's think about that. Consider, for example, how we might apply transformative integration to a form like $((T \circledast U) \odot (V \circledast W))$ under this rule set. All the same rules are available to us, it's just that now we have to follow them in reverse. Transformation paths for transformative integration have the same basic structure as transformation paths for transformative derivation, except that we write the transformation arrows backwards as we move across the page, i.e. typically with the arrow head on the left and the tail on the right as we write from left to right. Thus, applying transformative integration to $((T \circledast U) \odot (V \circledast W))$ would look like this:

$$((T \circledast U) \odot (V \circledast W)) \leftarrow ((w \circledast w) \odot (w \circledast w)) \leftarrow (w \odot w) \leftarrow w$$

When thinking in terms of transformative integration, we read this path from left to right. As such, this transformation path says that $((T \circledast U) \odot (V \circledast W))$ can be derived from $((w \circledast w) \odot (w \circledast w))$, which in turn can be derived from $(w \odot w)$, which in turn can be derived from w. In essence, for a well-formed formula rule set like this one, if a given form X can be transformed backwards until it becomes just w then it means that X is indeed a well-formed formula under w. It says that X is reachable from w. Thus, transformative integration can be used as a well-formed formula detector.

Transformative integration essentially tells you what the potential parent forms of any given form might have been. In this case, for $((T \circledast U) \odot (V \circledast W))$, it unambiguously resolves to just one final parent form. However, in other cases there could be many different potential parent forms. The real parent form from which a form may have originally come is thus somewhat uncertain, much like how integration in calculus[21] often has an element of uncertainty of a constant additive factor.

This analogy is in fact part of why "derivation" and "integration" are the terms we use for these processes. Regardless though, notice that transformative integration

---

[21] the calculus of change, a.k.a. Newton's and Leibniz's calculus, often referred to ambiguously as just "calculus"

itself is typically just as straightforward of a process as transformative derivation is. To determine what you are allowed to do at each step during transformative integration, you simply read the rule set backwards.

How do the reverse transience delimiters work? Well, for example, in the case of performing transformative integration on $((T \circledast U) \odot (V \circledast W))$, the trans-closed delimited form $\vdash((T \circledast U) \odot (V \circledast W))\dashv$ would be equivalent to a set containing only w. In contrast, the trans-free delimited form $\triangleright((T \circledast U) \odot (V \circledast W))\triangleleft$ would be equivalent to a set containing all forms that can reach $((T \circledast U) \odot (V \circledast W))$, which would be the same set as all possible forms you could get along the way when transforming $((T \circledast U) \odot (V \circledast W))$ back into w, including $((T \circledast U) \odot (V \circledast W))$ itself. $\triangleright((T \circledast U) \odot (V \circledast W))\triangleleft$ would also be equivalent to a set containing just w, since there aren't any trans-equal forms involved here.

Note also that transformation deltas, terminacy counts, adjacency counts, and all the other concepts that are applicable to transformative derivation may also be applied to transformative integration. However, the direction of the behavior is reversed, as always. For instance, if we had $\triangleright X \triangleleft_{\delta \leq 3}$ then it would mean the set of all forms that can reach X within no more than 3 transformation steps. Similarly, if we had $\triangleright X \triangleleft_{\infty}$ then it would mean the set of all forms that can reach X that themselves can reach an infinite number of parent forms. In other words, $\triangleright X \triangleleft_{\infty}$ would mean the set of all forms that can reach X that are themselves recursive in the backwards direction. Get the idea? Everything is the same in terms of mechanics. It's like a mirror world.

For another example, consider the rune rule set, whose rules we first listed on page 189. If you recall, I posed the challenge of trying to transform the initial form ᚠᚠᛘ into the target form ᚷᚷᚷ using the transformation rules of the rune system. Here's a question for you though: Is ᚠᚠᛘ the only form in this rule set that is capable of transforming into ᚷᚷᚷ (besides the other forms on the ᚠᚠᛘ → ᚷᚷᚷ path)? Are there more? What exactly are they? Do they differ significantly from ᚠᚠᛘ?

If you attempt to think of this problem in terms of transformative derivation (i.e. in the forward direction) then it might seem daunting at first glance. You'd be tempted to try to explore all of the possible forms you can derive from other initial forms by brute force to see if you can find any other transformation paths leading to ᚷᚷᚷ. That would likely take a lot of effort. However, such an ordeal is completely unnecessary. Transformative integration makes solving the problem very easy. Can you see why? Here's the rule set for the rune system again, for ease of reference:

1. ᚠ → ᚠᛘ

2. ᚠ → ᛘᚠ

3. ᛘ → ᚠᚠ

4. {ᚠᚠ, ᚠᚠ, ᛘᛘ} → ᚷ

Instead of trying to guess what initial forms might lead to ᚷᚷᚷ, it is far easier to just take the target form ᚷᚷᚷ and perform transformative integration on it. By definition, any form that you can derive from ᚷᚷᚷ through the process of transformative integration is guaranteed to be a form that is able to reach ᚷᚷᚷ. Thus, for example, the following transformative integration path gives us another initial form that can reach ᚷᚷᚷ:

ᚷᚷᚷ ↞ ᛟᛟᚷᚷ ↞ ᛟᛟᛝᛝᚷ ↞ ᛟᛟᛝᛝᛟᛟ ↞ ᛟᛞᛝᛟᛟ

This transformation path proves that ᛟᛞᛝᛟᛟ is able to reach ᚷᚷᚷ. In fact, this method of discovering new initial forms that can reach ᚷᚷᚷ is totally foolproof and mechanical. When you read the rune rule set backwards, and apply those backwards transformations to the form ᚷᚷᚷ, then the result at every step of the transformation path is always guaranteed to be a form that can reach ᚷᚷᚷ. This is vastly easier than the brute force approach of guessing the correct initial forms and then testing whether or not they can reach ᚷᚷᚷ.

Transformative integration is very useful for performing these kinds of calculations. Keep it in mind as another tool in your toolbox. For example, when trying to test whether a transformation path exists between any two given forms under a transformative language, it is sometimes useful to try performing both transformative derivation and transformative integration. Basically, you apply transformative derivation to the initial form, and also apply transformative integration to the target form, and then use the information you glean from these two separate processes to get a better sense for how the two sides of the transformation path will behave.

This might give you insight into how you should go about somehow connecting the two forms together. The more you understand the behavior of both sides, the more likely you are to see how to create a bridge between the two forms. Plus, you can even build up a vocabulary of shortcut rules on both sides, i.e. intermediate theorems moving in opposite directions, and this may help considerably in making the problem more tractable.

As easy as transformative integration usually is, there is unfortunately one potential complication that must be taken into account. Specifically, care must be taken when applying transformative integration to selective transformation rules. A selective transformation rule can only be reversed if doing so would create a form which is still capable of returning to what it was before the integration.

Many different transformative integrations are often possible, but you have to test that the reversal is valid. A path back to the original form that you started the transformative integration from must always exist. Every time a selective transformation rule is integrated you must test that this path still exists. As long as the path still exists, then it will always be a valid integration, regardless of how arbitrary it is, but otherwise it will not.

For example, suppose that $X \twoheadrightarrow_1 \{a, b, c\}$ was our only rule. Suppose we were then given the form $a + b + b$ and wanted to apply transformative integration to it. It would be valid to integrate $a + b + b$ to $X + b + b$. We know this because it clearly would be possible to transform $X + b + b$ back into the original form $a + b + b$ by applying $X \twoheadrightarrow_1 \{a, b, c\}$ to it.

However, while $a + b + b \twoheadleftarrow X + b + b$ may be valid, integrating again to create $X + X + X$ would not be. The path $a + b + b \twoheadleftarrow X + b + b \twoheadleftarrow X + X + X$ isn't a valid integration for this rule set. It would be impossible to transform $X + X + X$ into

$a + b + b$ under this rule set. The reason why is because X in this context is a 1-form (as a consequence of $X \rightarrow_1 \{a, b, c\}$) and thus must be given the same substitution value (only one of a, b, or c) everywhere in any expression in which it occurs.

There are only four possible integrations of $a + b + b$ under this rule set: $X + b + b$, $a + X + X$, $a + X + b$, and $a + b + X$. Do you see why? The reason is because (1) they can all potentially be transformed into $a + b + b$ and (2) they collectively exhaust all of the safe ways of doing so. The fact that $X + b + b$, $a + X + X$, $a + X + b$, and $a + b + X$ could also be transformed into other things besides just $a + b + b$, such as transforming $a + X + X$ to $a + a + a$ or $a + X + b$ to $a + a + b$ is irrelevant.

Transformative integration only cares whether a form is reachable, not whether it is the only reachable form. Thus, integration often yields a form that has more options than necessary to reach the target form still open to it. Indeed, as I previously mentioned, this is roughly analogous to the additional unknown constant that is frequently added in when performing integration in calculus.

To apply integration to selective transformation rules properly, you have to stay aware of the history of the form being integrated. Transformative integration is supposed to yield only forms that are capable of reaching the original form being integrated, so any chain of inferences that doesn't satisfy this criteria couldn't possibly be valid. Many arbitrary integrations are possible.

Notice that integration of $a+b+b$ does *not* require that all instances of a form in the expression be reverse substituted at once, otherwise $a + X + b$ and $a + b + X$ wouldn't have been valid integrations but they nonetheless are. Normal transformation rules, in contrast to selective transformation rules, can always be reversed mindlessly. It's only once per reverse application of a selective transformation rule that you have to check that the transformation back to the form being integrated is still valid. Otherwise, in all other steps, you may continue as usual, without needing to concern yourself with such things.

3.12 Rules about rules (meta implication)

Anyway, now that we understand both transformative derivation and transformative integration, we have a lot more tools available to us for working with forms. However, sometimes we don't want to talk about transformations directly, but rather we want to talk about transformations on the meta level, i.e. sometimes we want to talk about "rules about rules" instead of just about direct transformations. The quotation delimiter provides us with some of this ability to create "rules about rules", such as in the case of $\ulcorner A \rightarrow B \urcorner \leftrightarrow \ulcorner B \rightarrow A \urcorner$ for example, which we briefly mentioned earlier in the book on page 236. However, using quotations delimiters in this way is inconvenient in the general case. Thus, to work around this problem, we will be defining a new specialized variant of transformative implication designed to make it easier to specify "rules about rules" and similar constructs.

If you really think about it, "rules about rules" are actually just transformation rules whose condition and consequence forms are themselves rule sets. We have usually represented such rule sets as numbered lists and have treated these lists as being somehow special, but really these lists themselves can also be considered as just being arbitrary forms. A rule set is just as much an arbitrary form as any other form is. There are also multiple different ways of representing such sets. Vertical lists of rules separated by newlines is one way, but so is traditional set notation using { } delimiters and commas.

Take $\ulcorner A \to B \urcorner \leftrightarrow \ulcorner B \to A \urcorner$ for example. It says that given the rule $A \to B$ we can transform it into the rule $B \to A$, and vice versa. This works perfectly fine for individual rules, but what if we wanted to talk about the collective effect of multiple rules in a rule set? Quotation delimiters would be inconvenient for that. It's still doable, since rule sets are just as much arbitrary forms as anything else is, but it takes extra effort. For example, consider what we would have to write to express the transitivity of transformation rules, i.e. to express that from $A \to B$ and $B \to C$ you can infer $A \to C$. How would we do this using standard transformation rules and quotation delimiters? We would have to write something like this:

$$\ulcorner \{A \to B, \ B \to C\} \urcorner \to \ulcorner \{A \to B, \ B \to C, \ A \to C\} \urcorner$$

A, B, and C here represent arbitrary forms. Notice that this approach requires us to write the rules $A \to B$ and $B \to C$ twice, once in the condition and once in the consequence. This is redundant. As the size of our rule sets increase, this system will not scale well. It will become quite tedious to work with. The set $\{A \to B, \ B \to C\}$ could also conceivably contain more than just these two rules. Arguably, we should be required to write out any other rules that are in the rule set as well, besides just $A \to B$ and $B \to C$, even if they aren't relevant to the logical inference being made here, i.e. in this case even if they aren't relevant to transitivity. This problem is what our new variant of transformative implication is intended to solve. Here's the definition for it:

Definition 99. *In order to make it easier to create "rules about rules" and similar constructs, we need some way of eliminating the redundancy involved in creating transformation rules about entire rule sets. Otherwise, the system would be too tedious to work with in many cases. To fill this role, a more specialized variant of transformative implication will be used, which we will call* **meta implication**. *The way meta implication works is by stipulating that if a set of forms contains the forms in the condition set of the meta implication then it must also contain the forms in the consequence set of the meta implication.*

We will define two different notations for writing meta implications. One notation will be called **vertical meta implication**. It will separate each individual form in a set by a newline, and it will separate the condition set and consequence set by a horizontal line. The other notation will be called **inline meta implication** and it will instead use traditional set notation combined with a new arrow symbol that looks like $\prec\!\!\!\cdot$.

The arrow symbol for meta implication can be written facing either direction, such that $S_1 \prec\!\!\!\cdot\, S_2$ and $S_2 \,\cdot\!\!\!\succ S_1$ are exactly equivalent, given any two arbitrary sets S_1 and S_2. Also, if either set contains only one element, then the surrounding set brackets for that set may optionally be omitted for brevity. The horizontal line in the vertical notation and the $\prec\!\!\!\cdot$ symbol in the inline notation will both be referred to equally as meta implication. Thus, for example, a rule expressing the transitivity of transformation rules using meta implication would look something like this in each respective notation:

$$\frac{\begin{array}{c}A \to B\\ B \to C\end{array}}{A \to C} \qquad \{A \to B,\ B \to C\} \lessdot \{A \to C\}$$

If you wanted to, you could alternatively rewrite $\{A \to B,\ B \to C\} \lessdot \{A \to C\}$ here as $\{A \to B,\ B \to C\} \lessdot A \to C$. Similarly, if you had $\{A \wedge B\} \lessdot \{A, B\}$ then you could instead write $A \wedge B \lessdot \{A, B\}$. Likewise, if you wanted to express that any classical disjunction $A \vee B$ where A is true will always return true, then you could just omit all the set brackets on both sides and simply write $A \lessdot A \vee B$[22].

Keep in mind though that these are all still arbitrary uninterpreted forms. The involved expressions indicate classical truth values only when you interpret them as such. However, as a reminder to you to always think in terms of sets when dealing with meta implication though, I will henceforth always include the set brackets on both sides in meta implications in this book, even where I could omit them.

Anyway though, if you have prior experience with logic and rules of inference then you might recognize the vertical notation. This vertical notation is in fact probably the most common notation for writing rules of inference in traditional logic. This is not merely a notational coincidence however. It's potentially far more important than that. You see, traditional logic seems to be essentially mostly blind to the broader concept of transformative implication, and seems to treat meta implication (as written here in the vertical notation) as being the only possible kind of rule of inference.

This is despite the fact that meta implication represents just a subset of what kinds of rules of inference are possible under transformative implication. This is perhaps a very consequential fact. Most traditional logicians seem unaware that they may actually be working within just a small subset of what is actually logically possible. It makes you wonder how much content might be missing from the literature, doesn't it? Considering that people seem to have been working in such a limited subset of what's possible (at least as far as I'm aware), then there may be a lot of unexplored territory.

In fact, long ago, logicians set out to attempt to reframe all of mathematics in terms of pure logic but ultimately failed to do so. This movement was known as *logicism*. Gottlob Frege, Richard Dedekind, Bertrand Russell, and Alfred North Whitehead were historically some of the main proponents of this theory. Within the framework of only meta implication it may not be possible to fulfill the goals of logicism. However, I see no apparent fundamental impediments to fulfilling the goals of logicism using transformative logic more broadly.

Transformative logic deals with arbitrary forms and whatever transformations are applicable between them. This allows you to theorize about both systems of truth and systems of any other kind of arbitrary object (e.g. numbers and arithmetic) equally well. Thus, the possibility of transformative logic fulfilling the goals of logicism seems quite plausible. I'm not familiar enough with logicism in-depth to say this with any certainty, but nonetheless my instinct is that it may be true.

---

[22] Alternatively, if you wanted to more strictly account for the fact that A and $A \vee B$ here are actually value expressions and *not* statement expressions, then you could write the rule more unambiguously as a statement expression as $A = T \lessdot A \vee B = T$. The literature on classical logic tends to be sloppy on this point however.

Anyway, meta implication is essentially just conditional set appendment. Using it on rule sets is perhaps one of the most common use cases for it, but it can also be applied to any arbitrary set of forms, even if those forms have nothing whatsoever to do with rules or truth.

For example, suppose we have three people named Alex, Bob, and Catherine and that we represent them in symbols as A, B, and C respectively. Suppose furthermore that we know that whenever Alex and Bob are together then Catherine is also with them. Maybe Alex and Bob act hilarious when they are in the same room together, and Catherine likes witnessing this spectacle, so she always tags along. We could represent this as a meta implication using either vertical notation or inline notation, like so:

$$\frac{\begin{array}{c} A \\ B \end{array}}{C} \qquad \{A, B\} \prec \{C\}$$

On at least one occasion I have seen it said that rules of inference written in the vertical style can only have one consequence. However, this claim is untrue. Conceptually speaking, there is clearly nothing stopping you from including any arbitrary number of forms in either the condition or consequence of a meta implication, regardless of whether it is written in vertical notation or inline notation.

The meaning is still perfectly coherent. It simply means that any set which contains all the forms in the condition set may also append all the forms in the consequence set. In both cases (vertical notation and inline notation) we would write it exactly as you would expect, i.e. by simply adding more forms to the respective parts. For example, if D represents someone named Douglas, who (like Catherine) also tags along with Alex and Bob whenever they are together, then we could represent this using symbols like so:

$$\frac{\begin{array}{c} A \\ B \end{array}}{\begin{array}{c} C \\ D \end{array}} \qquad \{A, B\} \prec \{C, D\}$$

With transformative implication, the meaning of chaining multiple transformative implications together is clear. It simply creates a multi-step transformation path. However, what about chains of meta implications? How should we interpret those? Luckily, the answer is again quite simple. Chains of meta implications are simply read cumulatively, such that each new meta implication in the chain is read as if its condition set contained all of the forms contained in all of the previous sets in the meta implication chain. Consider, for example, the following meta implication chain:

$$\{A, B\} \prec \{C, D, E\} \prec \{F, G\}$$

If we only had $\{A, B\} \prec \{C, D, E\}$ here then it would be clear that it would mean that any set that contains the members of $\{A, B\}$ could also append the members of $\{C, D, E\}$. However, for the chain of multiple meta implications above, how should we interpret the expression once we get to the $\{C, D, E\} \prec \{F, G\}$ part? The answer is that we interpret that part the same way as if it had been written as $\{A, B, C, D, E\} \prec \{F, G\}$.

That's what I meant when I said that you read chains of meta implications "cumulatively". You accumulate all of the forms in all previous sets in the chain as you move from left to right when reading a meta implication chain. The same applies to vertical notation as well, except that it uses newlines and horizontal lines instead of set notation and meta implication arrows ($\prec$). Note that $\{A, B\} \prec \{C, D, E\} \prec \{F, G\}$ can be rewritten as the following equivalent set of single meta implication rules:

1. $\{A, B\} \prec \{C, D, E\}$

2. $\{A, B, C, D, E\} \prec \{F, G\}$

Notice also that set notation isn't interpreted the same way for meta implication as it is for transformative implication. When set notation appears in a transformative implication rule, and it is not surrounded by quotation delimiters, then it effectively creates multiple rules corresponding to each possible combination of forms in the condition set and the consequence set. Thus, as you may recall, $\{a, b\} \to \{c, d\}$ would be the same thing as writing out all of $a \to c$, $a \to d$, $b \to c$, and $b \to d$.

Meta implication doesn't work like this. Meta implication is a single statement about any set that contains the forms listed in the condition set. It doesn't expand to multiple rules in the same way that transformative implication does. Thus, writing $\{a, b\} \prec \{c, d\}$ would *not* be the same as writing all of $\{a\} \prec \{c\}$, $\{a\} \prec \{d\}$, $\{b\} \prec \{c\}$, and $\{b\} \prec \{d\}$.

However, the condition set of a meta implication *does* match to any conceivable set that contains all of the elements in the condition set, of which there could easily be many, so in that sense you could perhaps say that it stands for multiple rules. Notice though that this is a different sense of what "standing for multiple rules" means than for transformative implication. Anyway though, keep in mind that meta implication is just a specialized form of transformative implication. Thus, the following two rules mean roughly the same thing:

1. $\{A_1, A_2, \ldots, A_m\} \prec \{B_1, B_2, \ldots, B_n\}$

2. $\ulcorner\{A_1, A_2, \ldots, A_m\}\urcorner \to \ulcorner\{A_1, A_2, \ldots, A_m, B_1, B_2, \ldots, B_n\}\urcorner$

I said that these two rules are only *roughly* the same thing because of the fact that the one that uses transformative implication (#2) is technically treating the set delimiters and commas purely as uninterpreted symbols. We need to actually think of those set delimiters and commas in terms of set theory in order for the two rules here to truly be equivalent. The meta implication rule here is making a statement about all kinds of sets, whether represented with { } and commas or not. In contrast, the transformative implication rule here is technically making a statement specifically about sets represented with { } and commas.

For example, suppose we had a rule set consisting only of $\ulcorner\{A, B\}\urcorner \to \ulcorner\{A, B, C\}\urcorner$. Under this rule set, the expression $\cdot|\ulcorner\{A, B\}\urcorner|\cdot$ would resolve to the uninterpreted form $\{A, B, C\}$. In contrast though, the expression **Set**($\cdot|\ulcorner\{A, B\}\urcorner|\cdot$) would resolve to something that behaves like the actual set $\{A, B, C\}$, rather than merely to a meaningless uninterpreted form that just happens to use set delimiters and commas in this manner. Basically, we want all of the properties that apply to real sets to also apply here. This includes properties like order independence (i.e. ignoring the order of elements) and uniqueness (i.e. ignoring duplicates), etc.

Meta implication automatically treats its inputs as real sets, but transformative implication does not. By wrapping the resulting forms that the rule set generates in set interpretation injectors, or by duplicating all the rules of set theory inside the current transformative language, we can ensure that a form written in set notation is indeed actually interpreted as a set. It may not be a big deal though, in some contexts, so perhaps it's ok to not worry about this technicality too much. We may be splitting hairs here somewhat. I'm not sure.

Anyway, on a more miscellaneous note, before I forget, let me explain to you what the reasoning for choosing the $\prec\!\!\!\prec$ symbol to represent meta implication was. Like many of my symbol choices, it is essentially a pictogram designed as a memory-aid. When we have some set of transformation rules and initial forms, and we perform derivations upon that set to discover higher-level properties and other kinds of logical consequences, the set of transformation rules and initial forms acts kind of like a tree, where each new derivation from that tree is analogous to it growing a new branch.

Thus, I designed the symbol $\prec\!\!\!\prec$ to represent meta implication because it resembles a branching tree of implication arrows. Meta implication is all about describing the conditions under which such a tree would be able to sprout a new branch. The set of transformation rules and initial forms that we start with is like a seed from which a fully grown tree of consequences spontaneously emerges.

However, as powerful and convenient as meta implication is, it is fundamentally more limited than transformative implication. I have asserted this claim a few times before during our discussion of meta implication, but I haven't actually demonstrated it to you yet. Let's take a moment to actually do that. It isn't that hard to demonstrate really. Since we already know that meta implication can be expressed in terms of transformative implication, all we need to do now to demonstrate that meta implication is weaker than transformative implication (as opposed to equally strong) is to find a single example of something that transformative implication can do that meta implication can't. This is not difficult.

Consider, for example, what would happen if I asked you to create a rule that doesn't deal with sets. Meta implication deals *only* with sets, and therefore such a rule would be impossible to implement using meta implication. For instance, consider even an utterly trivial example like $2 + 2 \to 4$. How would you implement this using meta implication? Meta implication says that given that a set contains all the members of the condition set then it must also contain all the members of the consequence set. However, $2 + 2 \to 4$ has nothing to do with such things. It simply says that $2 + 2$ can be transformed into 4. Sure, you could write $\{2 + 2\} \prec\!\!\!\prec \{4\}$ and it would mean that given that a set contains $2 + 2$ then 4 could also be added to that set, but this is not the same thing as actually transforming $2 + 2$ into 4.

Another weakness of meta implication is that it seems to not be able to do self-referential structural recursion. It deals only with forms in their entirety. This is in contrast to transformative implication, which permits transformations to be applied to any matching form embedded within another form. For example, think back to the original well-formed formula rule set, which we defined on page 229. How would you implement this rule set using meta implication instead of transformative implication? You couldn't. See if you can figure out why. For ease of reference, here's the original well-formed formula rule set again:

1. $w \twoheadrightarrow \{A, B, C, \ldots, Z\}$

2. $w \twoheadrightarrow (w \odot w)$

3. $w \twoheadrightarrow (w \circledast w)$

Consider what would happen if you tried to implement some of this behavior using meta implication. For example, suppose that you were given $\{w\} \prec \{(w \odot w)\}$ as your only rule and $\{w\}$ as your initial set of forms. If this were transformative implication and you were given $w \twoheadrightarrow (w \odot w)$ and w, then you could easily generate an infinite number of derived expressions recursively. However, all you could do with $\{w\} \prec \{(w \odot w)\}$ and $\{w\}$ would be to derive $\{w, (w \odot w)\}$. After that point, the derivation chain would stop. There would be no more inferences permitted by the meta implication rule.

Thus, as you can see, meta implication seems to be a more limited form of transformative implication. That being said, meta implication can still be applied numerous times to a rule set to derive countless different consequences. This can feel sort of recursive, but it isn't really recursive in the same sense as transformative implication is. It's not truly structurally self-referential. It's just repeated application of a rule some arbitrary number of times, perhaps even an unlimited number of times. For example, here's what the transitivity rule of transformative implication might look like when written out a bit more formally and completely than we previously wrote it:

1. $A \twoheadrightarrow_1 \varkappa$

2. $B \twoheadrightarrow_1 \varkappa$

3. $C \twoheadrightarrow_1 \varkappa$

4. $\cdot | \{A \twoheadrightarrow B, \; B \twoheadrightarrow C\} \prec \{A \twoheadrightarrow C\} | \cdot$

Notice that instead of just informally saying "where A, B, and C stand for arbitrary forms" we are actually writing it all out completely in symbols this time. Notice how A, B, and C are all 1-forms that can each stand for any one particular arbitrary form at any given moment of time. Thus, the rule $\cdot | \{A \twoheadrightarrow B, \; B \twoheadrightarrow C\} \prec \{A \twoheadrightarrow C\} | \cdot$ here actually stands for an infinite number of different specific instances of transitivity being applied to specific combinations of forms. It actually stands for multiple rules. Previously, we had just written $\{A \twoheadrightarrow B, \; B \twoheadrightarrow C\} \prec \{A \twoheadrightarrow C\}$, but technically this rule (without the selective transformative implication and transience delimiters) would only be applicable to the literal symbols A, B, and C and nothing else, save for the fact that we informally stipulated that A, B, and C stood for arbitrary forms.

Given some arbitrary rule set, we can apply the transitivity property of transformative implication as many times as the system permits. This may result in a finite or infinite number of such applications. The transitivity rule of transformative implication is what allows us to create shortcuts through long transformation paths and thus to derive "theorems" that make it easier to reason about a system. This may feel a bit recursive, but it's not really. The chain of inferences can be applied as few or as many times as you wish. In general, transitivity can be summed up into a broader rule which we will call the **shortcut rule**. Supposing that f_k stands for any arbitrary form, where k is some positive integer, then the shortcut rule would look something like this when written using vertical notation:

$$\frac{f_1 \rightarrow f_2 \rightarrow \ldots \rightarrow f_{n-1} \rightarrow f_n}{f_1 \rightarrow f_n}$$

The choice of whether to use inline notation or vertical notation for a meta implication is arbitrary. Both notations have pros and cons. Inline notation fits better with other inline notations, such as transformative implication rules for example. However, vertical notation is often visually cleaner due to the fact that it uses newlines to separate elements of sets. With inline notation, there might be some confusion as to whether the commas in the set of forms should be treated as part of each element or as part of the set. Vertical notation doesn't have this problem, but also doesn't always mix as well with inline notations.

For example, how would you surround an entire vertical meta implication with transience delimiters? Transience delimiters are an inline notation. Also, the horizontal line in vertical notation takes proportionally more effort to write as the lengths of the forms in the condition and consequence sets increase, whereas the inline meta implication arrow always takes a fixed amount of effort to write regardless of the length of the forms in the condition and consequence sets. Each notation has pros and cons. Pick your poison. In traditional logic literature though, the vertical notation is probably the style you will see most often. It's one of the most common ways of representing rules of inference in the literature, although there are also some other notation variants that have somewhat different behavior, such as sequent notation and natural deduction notation, etc.

Oh, and by the way, as is typical for an implication operator, in addition to the one-directional version of meta implication, there is also a two directional version of the operator. It works exactly as you would expect. It has the exact same logical meaning as simply writing both possible directions of the implication separately and claiming them to be simultaneously valid. Here's our official definition for it:

Definition 100. *Given any two arbitrary sets S_1 and S_2, if you wish to assert that both $S_1 \prec S_2$ and $S_1 \succ S_2$ are simultaneously valid, then you may optionally instead choose to write $S_1 \bowtie S_2$. We will refer to this operation ($\bowtie$) as* **meta equivalence***. In the event that you become confused as to how this operator should be interpreted, always remember that it is simply the same as simultaneously asserting both $S_1 \prec S_2$ and $S_1 \succ S_2$ separately and that should clear up any confusion. The vertical notation version of meta equivalence will be represented by simply writing two horizontal lines instead* *of one. For example, here are two different representations of the same valid meta equivalence:*

$$\frac{A \twoheadrightarrow B}{B \leftarrowtail A} \qquad \{A \twoheadrightarrow B\} \bowtie \{B \leftarrowtail A\}$$

Anyway, putting all that aside, one thing you might be wondering about meta implication is: If transformative logic can be applied both forwards and backwards (i.e. transformative derivation versus transformative integration) does the same apply to meta implication? The answer is in part yes, but unlike reversing transformative implication reversing meta implication is not as symmetric. With a transformative implication of the form $A \twoheadrightarrow B$ we could just take a form matching B and apply the transformation in reverse to get back A.

However, in contrast, when applying a meta implication of the form $\{A\} \prec \{B\}$ in reverse you can't just take any set that contains B and claim that it must also contain A. That would be wrong. Instead, you have to say that *if* a set contains *both* A and B then B could be removed because it could be re-derived from A via $\{A\} \prec \{B\}$ later on if needed. In other words, reverse meta implication essentially removes derivative forms. Thus, if you applied reverse meta implication to a rule set to the furthest extent logically possible then you would essentially get back a minimal set of initial forms and transformation rules from which the rest of the consequences could be re-derived as needed. It would eventually reduce the rule set to just its basic axioms. Let's give a formal definition for this process:

Definition 101. *Just as transformative implication can be applied in reverse via the process of transformative integration, so too can meta implication be applied in reverse via a process that we will call* **meta integration**. *However, unlike reverse transformative implication, reverse meta implication is not entirely symmetric with respect to its forward counterpart. Given two sets S_1 and S_2, and a meta implication of the form $S_1 \prec S_2$, reverse meta implication can only be applied to sets that contain all of the members of the union of both S_1 and S_2. It has the net effect of removing all of the elements of S_2 from the set that previously contained all of the elements of both S_1 and S_2, due to the fact that those elements from S_2 could be re-derived from S_1 using $S_1 \prec S_2$ later on if needed.*

When performing meta integration, the same symbol ($\prec$) is still used, but instead of using the meta implication rule to conditionally add new forms to the existing set of forms, you perform the reverse process. Just as with transformative integration, you may adopt the notational convention of writing the arrow symbol backwards (e.g. $\{B\} \succ \{A\}$) to informally indicate your intent to perform integration a bit more clearly.

However, please be aware that every meta implication rule you write down simultaneously implies that *both* applying the rule *and* undoing it are logically valid. The same is true of transformative implication also. You can't say that a rule can be performed in one direction without also implying that it can also be reasoned about in reverse (i.e. undone). However, this is *not* to say that both directions of implications are always valid, which is in fact of course false. There's a big difference between applying a *converse* of an implication and *undoing* an implication. Don't get confused on that point. The fact that it is always possible to think in reverse does not imply that all implications are valid in both directions.

Oh, and since we have defined meta integration as a new term now, it makes sense for us to similarly refer to the normal version of meta implication (the one that works in the forward direction) as being **meta derivation**. That way, we will have symmetric terms for both the forward and reverse processes of transformative implication and meta implication. Thus, we have both transformative derivation and transformative integration on the one hand, and meta derivation and meta integration on the other hand.

Both of these integration processes are useful. Transformative integration tells us what forms another form might have been transformed from, whereas meta integration tells us how to simplify sets of forms by removing any forms that could be re-derived from other forms in that set. Thus, applying both transformative integration and meta integration to a rule set will allow you to distill that rule set down to its fundamental essence by removing any unnecessary transformation rules and initial forms.

However, be aware that doing so can often make a rule set much harder to use. Minimalistic rule sets may be elegant, but they are also often very tedious to work with. Remember that much of the whole point of logic and mathematics is to derive higher-level easier-to-understand or easier-to-use rules, and that by reducing rule sets to their most axiomatic form you are essentially undoing much of that work. Distilling a rule set to its bare essence can give you deep insights into its nature, but it can also make the resulting rule set impractical to actually use. It's best to strike a careful balance between minimalism and expressiveness, one that is powerful yet does not impose too much cognitive load on the user.

Anyway, I want to take a moment to explore two of the most commonly used rules in logic and how they relate to transformative implication and meta implication. The two rules I am referring to here are known as *modus ponens* and *modus tollens*. These two rules are ubiquitous in logic, and it is hard to even do logic without encountering them at some point. The phrases "modus ponens" and "modus tollens" originate from Latin. Translated into English, modus ponens means "the affirming way" and modus tollens means "the denying way". In classical logic (i.e. in the logic of simple true or false values), they work like this:

1. **modus ponens:** Given that we know that A is true and $A \rightarrow B$ is true, we can conclude that B is true. In classical logic this can be written in symbols as $(A \wedge (A \rightarrow B)) \rightarrow B$. Modus ponens is one of the most fundamental laws of logic and is the basis for all conditional reasoning in classical logic.

2. **modus tollens:** Given that we know that B is false and $A \rightarrow B$ is true, we can conclude that A is false. In classical logic this can be written in symbols as $(\neg B \wedge (A \rightarrow B)) \rightarrow \neg A$. Modus tollens is the reason why contraposition (transforming $A \rightarrow B$ into $\neg B \rightarrow \neg A$ and vice versa) is always valid in classical logic.

The truth tables of these expressions are tautologies (i.e. always true) in classical logic, which is why they qualify as valid expressions of these laws within the context of classical logic. Another way of writing these rules in the traditional notation, if you wanted to avoid truth tables and instead make a statement about the formal inference itself, would be to write "$A, A \to B \vdash B$" and "$\neg B, A \to B \vdash \neg A$".

However, as I have previously stated, we are avoiding syntactic consequence and sequent notation in this book due to the conceptual mud they might introduce into our discussions compared to using transformative logic. Using my own custom system helps avoid any possible miscommunication of my intent or any other conflicts. Anyway though, modus ponens and modus tollens don't mean the same thing in transformative logic as they do in classical logic. They're similar, but different.

The most obvious difference is the fact that transformative logic deals with arbitrary forms, whereas classical logic deals only with binary truth values. That's not the only difference though. In transformative logic we also need to distinguish between two different kinds of modus ponens, one of which is specific to transformative implication and the other of which is specific to meta implication.

For transformative implication, modus ponens is essentially just the same thing as applying a transformation rule to a form, causing it to become something else. Thus, if we have $A \twoheadrightarrow B$ as a transformation rule and A as an initial form, then we can transform A into B if we choose to do so and this is essentially the transformative implication equivalent of modus ponens.

However, if we used meta implication here instead of transformative implication then the meaning would be slightly different. Suppose we had $\{A, A \twoheadrightarrow B\} \prec \{B\}$ and were given $\{A, A \twoheadrightarrow B\}$ as our initial set. Then, we could apply meta implication to transform our set from $\{A, A \twoheadrightarrow B\}$ into $\{A, A \twoheadrightarrow B, B\}$. This is the meta implication equivalent of applying modus ponens. It essentially says that given the transformation rule $A \twoheadrightarrow B$ and the initial form A then we can add B to our list of initial forms. Thus, in addition to A, B could now also be used as a starting point for transformation paths. Here is what modus ponens for meta implication would look like when written in vertical notation (where A and B are interpreted as standing for arbitrary forms):

$$\frac{A \qquad A \twoheadrightarrow B}{B}$$

In contrast to the above, in modus ponens for transformative implication when we apply $A \twoheadrightarrow B$ to A we are actually erasing A and transforming it into B. Meta implication is talking about a set of forms, typically a rule set, whereas transformative implication is talking about one specific form. After applying meta implication you would still be able to use both A and B, whereas after applying transformative implication the particular instance of A that you transformed into B would no longer be available.

However, both transformative implication and meta implication can still be used to explore the full tree of all possible inferences that they entail. When I say that transformative implication "erases" A when $A \twoheadrightarrow B$ is applied to it, I mean only in the immediate sense (i.e. for a given instance of A in an expression), not in the abstract exploratory sense (i.e. not limiting the user's ability to explore all possible things that *A could* have been transformed into).

For clarity, we should define terms to distinguish these two different types of modus ponens in transformative logic. As such, when we are directly applying a transformation rule to a form, as in using $A \to B$ to transform A into B in an expression in which A occurs, then we refer to this as **transformative modus ponens**. In other words, transformative modus ponens is simply an instance of applying a transformative implication, i.e. just a transformation.

Transformative implication is an operator, whereas transformative modus ponens is an application of that operator (subtle difference). In contrast, when we are indirectly applying a transformation rule to a form, to add that form to the rule set as a new initial form that we are now allowed to use, as in using $\{A, A \to B\} \prec \{B\}$ to add B into our rule set if both A and $A \to B$ are already in our rule set, then we refer to this as **meta modus ponens**.

Notice that transformative modus ponens actually changes a form (i.e. is a "destructive edit"), whereas meta modus ponens merely adds another form onto an existing set of forms. The two operations are quite similar in spirit and are very closely related, but they are not technically the same, despite the fact that they do sort of feel like they are the same in some sense. These kinds of precise distinctions *do* often end up mattering in logic and mathematics, so it is generally best to play it safe by erring on the side of caution by carefully distinguishing between each case.

In the long term, it is very important for logicians and mathematicians to build up an increasingly strong and expressive vocabulary, one that somehow manages to simultaneously be both precise and streamlined. Such a vocabulary inevitably ends up being a perpetual work-in-progress, but nonetheless every new improvement makes the system at least a little bit more manageable and powerful. It is important not to let stagnation and mindless adherence to tradition set in. Logic and mathematics won't reach their full potential if they don't continue to evolve and improve. Clarity and precision are critical in this regard.

Anyway, that covers modus ponens. It's now time to discuss modus tollens. Modus tollens actually isn't valid for transformative implication, but it does have a rough equivalent in meta implication. To understand why modus tollens isn't valid for transformative implication, let's think about how modus tollens works in classical logic and see how it wouldn't be safe in the context of transformative implication. In classical logic, given that $A \to B$ is true, modus tollens allows us to use the fact that B is false to conclude that A must also be false, because otherwise if A were true then $A \to B$ could be applied to conclude that B is true, which would be self-contradictory. How would this work for transformative implication?

The interpretation of what modus tollens would mean for transformative implication hinges upon how the negations in the modus tollens are interpreted. The only such interpretation that would *directly* correspond to the structure of classical modus tollens would be to translate the negations in modus tollens into transformative logic as meaning "any form that is not this form".

Translating modus tollens into transformative logic in this manner would result in the following statement: "Given that $A \to B$ is a permitted transformation, then any form that is not *B* can be transformed into any form that is not *A*."... This is invalid. Do you see why? The rule would have very nearly the same effect as the chaos language ($\square \to \square$). The result would be far too aggressive and would render our rule set nearly useless. The only difference from the chaos language would be that *B* wouldn't be allowed to transform into *A*. It isn't rational to conclude from $A \to B$ that $\mathbf{Set}(\neg\{B\}) \to \mathbf{Set}(\neg\{A\})$ also holds.

However, modus tollens does have a reasonable translation into terms of meta implication. The key insight is to think of the negation in modus tollens not as meaning truth negation or set negation, but rather to think of it as meaning "not derivable", i.e. as being a statement about whether or not a transformation path exists between two forms. In order to make this work however, we will have to introduce a third participant into the modus tollens argument and also define a new operator corresponding to this concept of a form being "not derivable" from another form. As such, here is our new operator definition:

Definition 102. *To represent that a transformation path does not exist between two forms, we will use the $\not\to$ symbol. We will refer to this concept as **negated transformative implication**, or equivalently as **non-derivability**. Negated transformative implication doesn't actually allow you to do anything, it merely makes an assertion that no transformation path exists between two forms. Thus, if we wanted to assert that no transformation path exists between A and B then we would write $A \not\to B$.*

Care must be taken that the non-derivability relation is not used in a context where it generates self-contradictions. For example, if $A \to B$ exists in a rule set then $A \not\to B$ cannot, and vice versa. Always remember to account for the shortcut rule when making this determination. The existence of a lengthy path from A to B through many different forms still prohibits you from asserting $A \not\to B$ just as much as a direct path from A to B would.

Notice that non-derivability is a binary relation, not a unary relation. If someone says that something is "not derivable" then that statement only makes sense if there is some kind of associated context. Derivations can't happen out of thin air. There always has to be a basis from which a derivation originated. Sometimes this basis can be determined implicitly based on the context, but in general it should be specified explicitly. The question you have to ask yourself is: Non-derivable *from what*?

Now that we have this concept of non-derivability, we have everything we need to articulate the closest analog of modus tollens for transformative logic. Can you guess what the rule will look like? I already said that it will be a meta implication. I also said that the key idea is to interpret the negation in modus tollens as meaning "not derivable" and that we will need to introduce a third participant into the argument. Let *A*, *B*, and *X* stand for arbitrary forms. Here is what the rule will look like:

$$\frac{A \to B \quad X \not\to B}{X \not\to A}$$

Translated into plain English, this rule means that if a transformation path exists from A to B and a transformation path does not exist from X to B then a transformation path also cannot exist from X to A. This makes perfect sense if you think about it. The reason why it is valid is because if a transformation path from X to A *did* exist then we could transform X to A and then use $A \to B$ to reach B, but that would contradict our assumption that $X \not\to B$. Notice that this argument is perfectly analogous to classical modus tollens, except that it deals with derivability of forms instead of with classical truth values. From now on, we will refer to this rule as **meta modus tollens**.

Notice how this differs from modus ponens though: For modus ponens (insofar as it applies to transformative logic), both transformative modus ponens and meta modus ponens were valid, but for modus tollens (insofar as it applies to transformative logic) only meta modus tollens is valid. There is no such thing as transformative modus tollens. It would be invalid. Consequentially, contraposition has no direct application to transformative implication in the conventional sense.

Attempting to transform $A \to B$ into $\neg B \to \neg A$ as an "equivalent form" wouldn't make sense[23]. They are not equivalent. This is quite unlike in classical logic, where $A \to B \leftrightarrow \neg B \to \neg A$ (contraposition) is perfectly valid. At best, when you want to apply something like contraposition to a transformative implication, you can use meta modus tollens. However, transformative implication can still freely talk about other systems of logic that *do* support contraposition though, of course. Don't confuse rules about transformative logic itself with rules about other systems of reasoning that transformative logic just happens to be talking about.

Anyway, the idea of introducing a third participant (X) into the rule in order to make it possible to construct an analog of modus tollens for transformative logic was clearly effective. That in itself is already interesting enough on its own, but that's not even all there is to it. You see, meta modus tollens has a rather interesting and unexpected cousin, a related rule which happens to also be equally valid. This related rule becomes apparent when you consider what modus ponens would look like if it too involved a third variable just like meta modus tollens does. Naturally, let's call the third variable X again, just as we did for meta modus tollens. This is what the resulting rule would look like:

$$\frac{A \to B \qquad X \to A}{X \to B}$$

Look familiar? No? Try replacing X with A, A with B, and B with C, all in parallel, and then swap the first two rules, and look again. This results in the following equivalent rule:

$$\frac{A \to B \qquad B \to C}{A \to C}$$

---

[23] The symbol $\neg$ could mean literally anything in transformative logic. Transformative logic does not even have a negation operator, at least not a predefined one. Thus, conventional contraposition is not applicable in transformative logic. Meta modus tollens is the closest permissible analog.

Does it look familiar now? It's simply our good ol' friend the law of transitivity. In other words, in effect, transformative logic has the rather unexpected property that the opposite of its closest analog to modus tollens is not conventional modus ponens, but rather it is the law of transitivity. The law of transitivity can thus be viewed as a kind of generalized modus ponens, in a certain sense[24].

Weird, right? In effect, this demonstrates that when you think in terms of derivability and non-derivability from some other form X then it is actually the law of transitivity (not conventional modus ponens) and meta modus tollens (not conventional modus tollens) that govern conditional reasoning. I find this interesting. It makes me wonder what other relationships like this may be hiding in plain sight that we aren't aware of.

Various other rules of inference exist besides just modus ponens and modus tollens, some of which apply to transformative logic in general and some of which apply only to specific transformative languages. Most rules of inference in the literature will probably translate directly into meta implication rules with relatively little effort. Discussing all of them is outside the scope of this book. You can look them up if you want to. Just be sure to think about them first and to verify that they actually make sense in the context of whatever rule set you are currently dealing with. For example, $\{A \rightarrow B, B \rightarrow A\} \lessdot \{A \leftrightarrow B\}$ is universally valid in transformative logic, but in contrast $\{\neg\neg A\} \lessdot \{A\}$ (a.k.a. double negation elimination) is something specific to classical logic and other similar logics. It isn't valid in all logics.

In fact, for example, $\{\neg\neg A\} \lessdot \{A\}$ isn't a valid rule of inference in intuitionistic logic. Notice that the $\neg$ operator I am using here in $\{\neg\neg A\} \lessdot \{A\}$ is not being interpreted as the set complement operator from set theory, but rather it is merely being treated as an uninterpreted form. If I had wanted to instead make a statement about the properties of $\neg$ in set theory, and A referred to some arbitrary set defined by trans-closed transformation rules, then I could have wrote $\ulcorner \mathbf{Set}(\neg\neg \cdot | A |\cdot) \urcorner \rightarrow \ulcorner \mathbf{Set}(\cdot | A |\cdot) \urcorner$. In that case it would have been universally valid. Notice that removing the Japanese quotes ($\ulcorner \ldots \urcorner$) here would not be safe, due to the fact that the transformative implication operator behaves like a cartesian product of all possible transformation rules between its condition set and consequence set.

In fact, let me take this opportunity to clarify the behavior of transformation rules a bit more, just in case you haven't caught on to some of the nuances yet. Let's do an example designed to make the subtleties of how condition and consequence sets behave more apparent. It is important to remember that transformative implication behaves like a cartesian product of its condition set and consequence set, otherwise you are liable to inadvertently write down rules with unintentional effects. Consider the following two rule sets:

1. $X \twoheadrightarrow_1 \{a, b\}$

2. $\cdot | X |\cdot \rightarrow \cdot | X |\cdot$

1. $X \twoheadrightarrow_1 \{a, b\}$

2. $\cdot | X |\cdot \rightarrow X |\cdot$

---

[24] The fact that the law of transitivity can be viewed as a kind of generalized modus ponens can be seen more easily when it is written in the form $\{A \rightarrow B, X \rightarrow A\} \lessdot \{X \rightarrow B\}$ (or equivalently the corresponding vertical notation). However, even when written in its more typical form $\{A \rightarrow B, B \rightarrow C\} \lessdot \{A \rightarrow C\}$ it is still exactly the same law.

What's the difference in behavior between these two rule sets? As similar as they may appear, they don't do the same thing. The left rule set's $\cdot | X |\cdot \rightarrow \cdot | X |\cdot$ rule is equivalent to writing all of $a \rightarrow a$, $a \rightarrow b$, $b \rightarrow a$, and $b \rightarrow b$. In contrast, the right rule set's $\cdot | X \rightarrow X |\cdot$ rule is equivalent to writing only $a \rightarrow a$ and $b \rightarrow b$. In this case, writing $\cdot | X \rightarrow X |\cdot$ is pointless. Rules that merely transform a form back into itself essentially have no effect and might as well have not been written. However, there are many other cases where surrounding an entire transformation rule in transience delimiters in this manner is useful. In fact, we have used transience delimiters in this way at least several times.

The reason why this distinction exists is because sometimes we might want to apply selective substitution to one part of an expression separately from another part of an expression. Granted, the case where the selective substitution is applied in parallel across an entire transformation rule is perhaps more common, but that doesn't mean that the other case doesn't sometimes happen too. Transformative logic is intended to allow the user to express totally arbitrary systems of logic, so we can't really favor any one interpretation over another. As such, consider the following two rule sets:

1. $X \twoheadrightarrow_! \{a, b\}$
2. $\cdot | X \backsim X |\cdot \rightarrow \cdot | X |\cdot$

1. $X \twoheadrightarrow_! \{a, b\}$
2. $\cdot | X \backsim X \rightarrow X |\cdot$

This time, the left rule set's $\cdot | X \backsim X |\cdot \rightarrow \cdot | X |\cdot$ rule is equivalent to writing all of $a \backsim a \rightarrow a$, $a \backsim a \rightarrow b$, $b \backsim b \rightarrow a$, and $b \backsim b \rightarrow b$. In contrast, the right rule set's $\cdot | X \backsim X \rightarrow X |\cdot$ rule is equivalent to writing only $a \backsim a \rightarrow a$ and $b \backsim b \rightarrow b$. We have no idea what $\backsim$ is supposed to mean, so either of these rule sets is perfectly plausible.

For example, the a and b symbols might represent two different identity elements under some operation represented by $\backsim$. In the case of the left rule set, $\cdot | X \backsim X |\cdot \rightarrow \cdot | X |\cdot$ could be interpreted as meaning that a and b act as identity elements with respect to each other in any combination, whereas in the case of the right rule set, $\cdot | X \backsim X \rightarrow X |\cdot$ could be interpreted as meaning that a and b act as identity elements only with respect to themselves and not each other. Alternatively, the concept of "identities" may be totally irrelevant here. Who knows? It's all just arbitrary symbol manipulation rules ultimately. What it actually "means" depends on the context and what the system is intended model. As long as the logical structure that the rules express corresponds properly to what the rules are intended to model, then everything is fine.

These same principles can be applied to meta implication as well, as we already saw in the formal rule set for transitivity of transformative implication on page 267. Basically, transience delimiters treat their contents literally, kind of as if the form they delimit is surrounded by quotation delimiters, and then return the resulting set of all forms that can be derived from that form. Thus, wrapping an entire meta implication rule in transience delimiters is just as valid for meta implication rules as it is for transformative implication rules.

Hopefully that clarifies things a bit, if it wasn't clear already. Anyway though, I think it's time for us to move on from meta implication now. At this point, we've covered meta implication well enough for our purposes in this book. There's just a few more new things I want to introduce into transformative logic before we'll be done with it. We're getting pretty close to discussing the non-classical logics now.

Admittedly, many of these features we've added on to transformative logic in this section won't really be necessary for our discussion of the non-classical logics. However, many of these features have inherent value of their own and thus are still worth discussing and exploring. Sometimes the journey can be more valuable than the destination. You can miss a lot of serendipitous value if you're always in too much of a rush to reach your destination. Stop and smell the roses. Patience and a steady hand are often far more productive than rushing, often yielding a much greater value per unit of time, resources, and effort invested.

3.13 Inclusion directives

The next feature we'll be adding to transformative logic will allow us to use transformative languages in a way that we previously couldn't. Up until this point, when dealing with transformative languages (i.e. named rule sets), in order to define how a form should be interpreted we've always either used whatever rules were defined in the current rule set or else used interpretation injectors. This approach is useful and fully general. However, sometimes it might be convenient to be able to reuse rules from other transformative languages without having to write the corresponding interpretation injectors every time we do so. This brings us to our next definition:

Definition 103. *Sometimes in a transformative language it is desirable to be able to reuse another transformative language's rules without having to actually explicitly write interpretation injectors every time we do so. In essence, what we need is a way of copying all of the rules from another transformative language into the current transformative language. To fill this role, we will use what we will call an **inclusion directive**, and we will represent it by the symbol ⊕ . The ⊕ symbol is followed by the name of the transformative language whose rules you want to copy. Thus, to copy all of the rules from a transformative language named LanguageName into the current rule set, this is what you would write:*

⊕ LanguageName

Those of you with a background in computer science might already recognize this concept. It operates based on the same basic mechanism as package inclusion (i.e. copy-pasting or linking files together). Its effect in the context of transformative logic is also fairly similar to the concept of implementation inheritance from object-oriented programming. In fact, it even shares the same conceptual weaknesses as implementation inheritance, which is to say that it is a very heavy-handed and brittle way of reusing prior work, one that creates extremely strong dependencies between the two systems. Whereas with interpretation injectors all of the effects are tightly contained and there is no danger of interference with other rule sets, the opposite is true of inclusion directives.

Inclusion directives practically invite rule conflicts, due to the fact that any of the rules being copied into the current rule set by the inclusion directive could potentially interact with other rules in the current rule set in unintended ways. As such, in general, you should prefer interpretation injectors over inclusion directives. However, sometimes it may be convenient to actually copy all of the rules from one rule set into another, and that's why I've included it. It is sometimes a convenient shorthand. However, heavy use of inclusion directives will tend to backfire in transformative logic, just as heavy use of implementation inheritance will tend to backfire in object-oriented programming. Use it with care or not at all. Interpretation injectors are far more flexible and foolproof.

That being said, let's do a concrete example. Do you remember the rune puzzle system? Suppose that we wanted to keep all of the same rules from the original rune puzzle rule set, but also wanted to add on some new rules while still leaving the old rule set and its associated interpretation injector name intact. We first listed the rules for the rune puzzle rule set on page 189. It's interpretation injector name is **Runes**. Let's make a new variant of the rune rule set using an inclusion directive and call it **RunesExt**. The "Ext" is short for "Extension". Here is the old rule set (for ease of reference), followed by the new extended rule set:

Runes

1. ᛉ → ᛉᛟ
2. ᛋ → ᛟᛉ
3. ᛟ → ᛉᛋ
4. {ᛉᛉ, ᛋᛋ, ᛟᛟ} → X

RunesExt

1. ⇨ **Runes**
2. Xᛉ ↔ ᛉX
3. Xᛋ ↔ ᛋX
4. Xᛟ ↔ ᛟX

As you can see, I've added on a trio of new rules. The new rules here have the net effect of allowing you to relocate X symbols wherever you want. This will have a significant impact on the behavior of the rule set. In the original rune rule set, if we had something like ᛉXᛉ then we wouldn't have been able to apply ᛉᛉ → X to it because of the X in the middle blocking that move. Now however, we can shift that X aside, and then apply ᛉᛉ → X, as shown in the following transformation path:

$$ᛉXᛉ → ᛉᛉX → XX$$

Thus, as you can see, inclusion directives are a useful shorthand for experimenting with different variations on rule sets. One thing you might be wondering though, is how would step-tracing notation for transformation paths work now? One of the rules we're using in this transformation path isn't actually in the current rule set itself anymore but is instead being imported from elsewhere. How are we supposed to represent this in step-tracing notation?

The answer is pretty simple. We will use a system of numbering similar to how subsections in books are numbered. In other words, if *m* is the line number on which an inclusion directive occurs, and *n* is the line number of a specific rule from the included rule set, then the associated step-tracing number will be written as *m.n*, just like subsection numbering in a book. Thus, the step-traced version of the previous transformation path would look like this:

$$⚹ ✕ ⚹ \xrightarrow{2} ⚹ ⚹ ✕ \xrightarrow{1.4} ✕ ✕$$

This numbering system supports arbitrary nesting of inclusion directives. Thus, a number of the form *i.j.k* (where *i*, *j*, and *k* are arbitrary positive integers), implies that there are two levels of inclusion directive nesting. If *A* is the current rule set and it has an inclusion directive on line *i* referring to rule set *B*, and rule set *B* in turn has an inclusion directive on line *j* referring to rule set *C*, and the rule on line *k* of rule set *C* is the actual rule being used, then it will be referred to as rule *i.j.k*. In other words, you can just append however many numbers are necessary to unambiguously identify the specific rule you are referring to. This is a lot of words to explain such a simple concept really. It is quite simple in practice.

3.14 Intention marks

Anyway, moving on, the next three symbols we'll be covering are designed to make it easy for a user of transformative logic to clearly distinguish what kind of process each line in a rule set originated from. Previously, we have always kept our transformation rules and initial forms separate from any instances of our rules being applied to specific cases. We have always instead explained our intent using natural language and lengthy discussion. We have also always labeled initial forms and target forms using regular text, by writing out the full phrases "Initial Form:" or "Target Form:" etc. This system works, but it is sometimes inconvenient and laborious. We should have a more concise and convenient way of working with rule sets.

If you really stop and think about it, each time we write down some symbols on each line in transformative logic we are really doing one of three things: (1) introducing an assumption, (2) making an inference based on our assumptions, or (3) specifying a target form indicating what goal we are trying to reach. Thus, each line could potentially have a different intent than the others. Not labeling the intent of each line could potentially result in confusion as to what is really going on.

For example, suppose we were exploring some arbitrary transformation path under our rule set. Suppose we then wrote down that inferred transformation path right underneath the rule set we started with. Looking at the rule set now, after having appended this inferred transformation path, how would you distinguish the assumptions from the inferences? You couldn't.

For all we know, from the perspective of an outsider, the transformation path we've added to the list could've been an axiom, in which case it couldn't possibly be considered wrong even if it didn't obey any of the other rules. For instance, for the rule set of the rune puzzle system from page 189, if we added ᚠᚢᛗ → ᚷᚷᚷ to the next line right after the rule set, how would we know whether it was an axiom or just something that we inferred based on the existing axioms? We couldn't. Thus, we need some way of indicating our intent. This brings us to some new definitions:

Definition 104. *If we want to be able to concisely list assumptions, inferences, and target forms all together in the same list, without relying on verbose descriptions written in prose, then we will need to have some way of distinguishing the intent of each line. To accomplish this, we will prefix a special mark to each line corresponding to its type, which we will call* **intention marks**. *We will define three types of intention marks.*

Definition 105. *To indicate that a line is an assumption, such as an axiomatic transformation rule or an initial form, we will prefix it with* ☆, *which we will refer to as an* **assumption mark**.

There are three reasons why this symbol (the 5-pointed star) was chosen: (1) Because, if you superimpose the capital letter A over the star symbol, the primary lines of the letter A line up with the star symbol. A is the first letter of "assumption", which is what the mark indicates. (2) Because, stars symbols are often used to indicate that something is "special" and assumptions are certainly special. (3) Because, despite how complicated the ☆ symbol looks, it can actually be written very quickly using only a single stroke of the pen.

Definition 106. *To indicate that a line is an inference, based upon our existing assumptions up to that point, we will prefix it with* ∴, *which we will refer to as an* **inference mark**. *In other contexts in mathematics, this symbol* ∴ *is frequently used as shorthand for "therefore". The meaning of an inference mark is extremely similar to "therefore", and that's why the symbol was chosen here. However, the inference mark's connotation is maybe slightly different from conventional "therefore".*

Whereas a conventional "therefore" tends to be read as if it is implicitly justified by the most recent statements, an inference mark does not necessarily have as much of this connotation. Inference marks have a broader connotation. Inference marks can pull from any prior statements, and shouldn't be read as if implicitly justified by only the most recent statements. The inferences could be coming from anywhere, rather than from only the most recent statements. Otherwise though, inference marks and conventional "therefore" have essentially the exact same meaning. It's kind of a hair-splitting difference, but I thought it may be worth mentioning.

Definition 107. *To indicate that a line is a target form, meaning that we want to indicate to the reader that we are going to try to somehow reach that specific form, we will prefix it with* ⌖, *which we will refer to as a* **target mark**. *This symbol resembles the crosshair of a gun, which is used for targeting, hence making it a suitable and memorable symbol to represent the concept of a target mark. The target mark can be used with both individual forms and transformation paths.*

When prefixed to an individual form, the target mark indicates that we intend to reach that form, but we don't care how we get there as long as the inference is legitimate.

In contrast, when prefixed to a transformation path, the target mark indicates that we intend to transform the form at the beginning of the transformation path into the form at the end of the transformation path somehow.

If there are more than just two forms in the transformation path of the target marked line, then it also indicates that we intend to visit all of those intermediate forms in the path as well, in the same relative order as indicated. We don't care whether there are other intermediate forms in the path besides the indicated ones, as long as the relative order and the first and final forms are the same, as indicated by the line prefixed by the target mark.

Without use of target marks, it might be unclear to the reader what the aim of a sequence of inferences is supposed to be until the final moment. Target marks allow us to specify our goal in advance, thereby making the subsequent sequence of inferences more readable. It basically gives the reader an overview of the argument. For short arguments, indicating your intent in advance in this manner may not be worth the trouble, but for long arguments it is likely to be helpful. The target mark may also be useful for merely indicating what problem you want a reader to solve, such as in problem sets in a textbook for example. Just leave the subsequent proof blank in that case.

All of these intention marks are optional and aren't considered to be part of the forms they prefix. If you don't want to use intention marks then you can always make your intent clear using prose instead, although it may be more laborious. Furthermore, technically, as long as you don't make any mistakes, any new transformation rules or forms that you successfully derive from an existing rule set can be added to that rule set as if they were always part of it. Thus, the distinction between assumptions and inferences can often be ignored, as long as you never make any mistakes.

However, there is a catch. The catch is that if you *do* make a mistake in your derivations, and haven't distinguished your assumptions and your inferences from each other explicitly by using the corresponding intention marks, then it will become impossible for anybody who later reads your work to figure out which parts were errors and which parts were axioms. Thus, if you want to avoid confusion, it is generally better to use intention marks than to omit them, even though they are nominally optional.

Anyway, let's do an example. As always, it's not enough to merely talk about a concept in the abstract. It is far too easy for us to delude ourselves into thinking nonsense when we only think in the abstract. We need accountability. We must always test our assumptions. Only through exploration of concrete examples can we ever be truly certain of the validity of any concept, no matter how simple it may appear on the surface. As such, let's see what the original rune puzzle from page 189, where we were tasked with finding a transformation path from ᚢᚨᛗ to ᚷᚷᚷ, including the solution to it, would look like when completely rewritten to use intention marks instead of prose:

1. ☆ ᚨ → ᚨᛗ

2. ☆ ᚨ → ᛗᚨ

3. ☆ ᛗ → ᚨᚨ

4. ☆ {ᚨᚨ, ᚨᚨ, ᛗᛗ} → ᚷ

5. ☆ ᚢᚣᛙ

6. ⋄ ᚢᚣᛙ → ⵝⵝⵝ

7. ∴ ᚢᚣᛙ → ᚢᛙᚢᚢ → ᚢᛙᚢᛙᛙ → ᚢᛙᚢⵝ → ᚢᚢᚢⵝ → ⵝⵝⵝ

As you can see, this is much more concise than describing our intentions in English. Admittedly, it does require the reader to understand what the intention marks mean, thus making it less immediately understandable to the uninitiated reader. However, it is not like the meaning of these marks is hard to explain. The reader will have to learn the bulk of the basic symbols and operators of transformative logic either way, so they might as well learn to use the intention marks as well. It doesn't add that much burden to the user, and it really improves readability, so overall it is a pretty clear win[25].

Notice that ᚢᚣᛙ is listed explicitly among our assumptions (on line 5), even though it's not a transformation rule. Merely indicating our intent via the target path ᚢᚣᛙ → ⵝⵝⵝ would not technically be enough. Initial forms are just as much assumptions as transformation rules are. Marking an initial form as an assumption is essentially a way of asserting that the initial form is a form that already adheres to the rules of the rule set. Otherwise, our application of the principle of property conservation to the initial form wouldn't necessarily be valid. You can't conserve a property that you never started with. Thus, we have to stipulate that at least one form already has that property, the property implicitly embodied in the rules.

If we didn't include line 5 then we could still write line 7, but in that case line 7 would only have a hypothetical nature instead of a tangible nature, i.e. it would instead only mean that *if* we had ᚢᚣᛙ among our initial assumptions *then* we could transform it into ⵝⵝⵝ. In contrast, when line 5 (☆ ᚢᚣᛙ) is included then it implies that we can indeed actually reach ⵝⵝⵝ, by virtue of the fact that the transformation path on line 7 is also valid. In fact, our discussion here has made me realize that I should define some new terminology to make it easier to talk about these two different types of inference lines. Here are the new terms:

Definition 108. *If an inference line begins with an initial form that* can *be derived from at least one of the assumptions of the rule set, then we call that inference line a* **tangible inference**. *If we wish to keep track of which inferences are tangible then we may subscript the letter t to the ∴ symbol on that line, as in $\therefore_t$, but this is completely optional. All tangible inference lines are tangible regardless of whether they are marked as such.*

Definition 109. *If an inference line begins with an initial form that may not necessarily be derivable from at least one of the assumptions of the rule set, then we call that inference line a* **hypothetical inference**. *If we wish to keep track of which inferences are hypothetical then we may subscript the letter h to the ∴ on that line, as in $\therefore_h$, but this is completely optional. All hypothetical inference lines are hypothetical regardless of whether they are marked as such.*

---

[25] If you want, you can also add a vertical line dividing the intention marks from the rest of the forms, like a line number gutter in an IDE (a text editor for programmers, one usually also capable of compilation). This may help you keep the intention marks mentally separate from the forms. It's not necessary though. It's just another idea.

By the way, if you do use the subscript notation for your inference marks, be careful to do it correctly so that you don't mislabel any lines. When in doubt, just don't label them. In this sense, not labeling them is more foolproof, but on the other hand it may make the tangible versus hypothetical origins of the inference lines a bit harder to trace. You'll have to balance the pros and cons against each other. Such is life. It's not that important though, so don't fret much. There are more important factors to consider.

In particular, it is important to realize that tangible inferences and hypothetical inferences have different requirements for what initial forms they are allowed to start with. All tangible inferences must begin with a form which is derivable from an existing assumption[26]. In stark contrast though, all hypothetical inferences may begin with any arbitrary form, since it's all just hypothetical anyway. Make no mistake though: Hypothetical inferences are completely valid (although only in a conditional sense).

Hypothetical inferences are essentially just applications of the already existing rules, i.e. rules that simply describe possible ways that the rule set could respond to whatever arbitrary input forms you could give it. Hypothetical inferences are thus fully in compliance with the rules, even though they might not even be connectable to the assumed initial forms. Their validity is genuine, but conditional.

To make an rough analogy: Tangible inferences are like trees rooted in the ground, whereas hypothetical inferences are like islands floating in the air. Hypothetical inferences are internally consistent (like isolated little fantasy worlds), but are not necessarily reachable from the assumptions you currently have in place at the ground level, hence the floating islands versus trees analogy. Tangible inferences deal with reality, whereas hypothetical inferences deal with fiction. Both are valid though.

Anyway, although we have to stipulate that at least one form already has the property implicitly embodied in the rules if we want our target form to be reachable in more than merely a hypothetical sense, keep in mind that the choice of whether any given initial form satisfies the properties of a rule set you are designing is completely up to you. It's arbitrary. You can choose whatever you want.

Even if an initial form doesn't share any symbols in common with the rest of the rule set, you are still nonetheless permitted to forcibly declare that it has the property implicitly embodied in the rules regardless. However, it is always your responsibility when creating rule sets to correctly model whatever it is you intend to, which may constrain you. Assuming forms that share nothing in common with any of the other assumed forms in a rule set is allowed, but is generally pointless since you won't actually be able to apply any of the other rules to them.

By the way, in the intention mark example for the rune system, we also could have just written ✕✕✕ for the target intent instead of ᛉᛉᛘ → ✕✕✕ if we had wanted to. In that case then, it would change the meaning of our intent to be that we are seeking to transform *any* of the given initial forms into ✕✕✕ somehow. As such, if we had more initial forms than just ᛉᛉᛘ, one of them might have a viable path to ✕✕✕, and hence would also count as another valid solution to the target form.

Notice that just writing ✕✕✕ for the target intent instead of ᛉᛉᛘ → ✕✕✕ would be less redundant, given that we already listed ᛉᛉᛘ in our assumptions and given that

---

[26]Assumptions are of course also derivable from themselves, in case you've forgotten.

it is also our only assumption that is not a transformation rule. However, omitting the source form from the target transformation path like this wouldn't always mean the same thing in other cases. You should take this difference into account when deciding how to specify target intents.

By chance, this specific example for the rune puzzle system happens to use only one target mark and only one inference mark, but this is not necessary in the general case. A system may use however many instances of each intention mark as it wants. There is no limit on the number of each type of intention mark. Nor is there a limit on the number of initial forms, despite the fact that we have previously only ever focused on a single initial form for any given rule set. As such, for diversity, let's do an example that uses multiple initial forms, multiple inference marks, and multiple target marks now:

1. ☆ $x \to \{0, 2, 4\}$

2. ☆ $y \to \{1, 3, 5\}$

3. ☆ $x \to yy$

4. ☆ $y \to xx$

5. ☆ xyx

6. ☆ yxy

7. ✧ 102201

8. ✧ 013310

9. ∴ $x \to yy \to xxxx$

10. ∴ $y \to xx \to yyyy$

11. ∴ $xyx \to xyyyyx \to 102201$

12. ∴ $yxy \to yxxxxy \to 013310$

Notice how we are creating intermediate rules ($x \to xxxx$ and $y \to yyyy$) to make the system a bit easier to think about. In this case the system is rather easy to solve, so I just chose to solve it using direct transformative derivation. However, for more complicated systems, it may sometimes be useful to work backwards from the target form using transformative integration instead. In that case, you would simply write the transformative integration path prefixed by an inference mark, just like any other transformation path.

Transformative derivation and transformative integration are treated the same way, it's just that the transformation paths they embody are merely created in different ways. Thus, the following two paths, the first created by transformative derivation and the second created by transformative integration, would ultimately mean exactly the same thing and would be completely interchangeable, even though the thought process involved in their creation was different:

$$\therefore\ xyx \rightarrow xxxx \rightarrow xyyyyx \rightarrow 102201$$

$$\therefore\ 102201 \leftarrow xyyyyx \leftarrow xxxx \leftarrow xyx$$

Oh, and don't forget the bridging technique either. For solving truly difficult transformation paths, it can be quite a helpful technique. While transformative derivation or transformative integration alone are often sufficient to solve a problem, sometimes it is useful to attack the problem from both sides. Try building up a vocabulary of inferred higher-level rules based on both sides of the transformation path that you are trying to connect. Apply both transformative derivation to the initial form, and transformative integration to the target form, and use what you learn from this process to get a better sense for how the system behaves on both ends. You might also end up discovering some useful intermediate rules along the way that will help you connect the two sides together.

However, constructing a list of inferred forms for a rule set can become overwhelming once the list becomes too large. For short lists, it's often fairly easy to keep track of what the justifications for the inference steps are without having to actually explicitly write out the justifications for each step. For instance, as you can see, I didn't bother including step tracing numbers for the previous few example rule sets. I could have written numbers above each of the transformative implication arrows though, to make it easier to keep track of why each step is justified. Is our existing step tracing notation for transformation paths (i.e. numbers over $\rightarrow$ arrows) already sufficient for all our conceivable step tracing needs though?

The answer is no. The existing step tracing notation may be sufficient for tracing transformative implication paths, but meta implications work a bit differently sometimes. Sometimes we apply meta implications by directly adding new forms to the rule set list on a new line. However, by doing so we are no longer able to use the step tracing notation that we used for transformative implication paths to trace this action. There's no arrows to write the numbers over in such cases. For this reason, we need a vertical version of step tracing, one that lets us trace our justifications by referring to specific lines of the rule set in advance of an inferred form, instead of only being restricted to the horizontal number-over-arrow approach used for transformative implication. As such, here's a new definition designed to fill this role:

Definition 110. *Inference marks allow us to indicate to a reader of a list of rules and inferences that specific lines are inferences rather than assumptions or axioms. However, sometimes we also would like to indicate what exactly are the other lines that justify an inference line as valid. In this respect, what we require is a system that we will refer to as* **vertical step tracing**. *To indicate which other lines a specific inference line is justified by, we may prefix the inference mark symbol (the $\therefore$) of that line with a parentheses enclosed list of numbers and/or named abbreviations that indicate the justifications for that line.*

*Let's call each such parentheses delimited group of numbers and/or abbreviations a **justification note**. All justification notes are optional. They exist purely for clarity* *and readability. For example, writing "$(2,5,7) \therefore X$" would indicate that adding X to the rule set is justified by lines 2, 5, and 7. Keeping the parentheses on the left side of the inference mark helps keep the justification note visually separate from the actual forms and content of that line, which the justification note might otherwise be confused as being a part of. This design choice was made to prevent such potential confusion.*

Inferences can only be made based on already existing knowledge, hence justifications notes should ideally only refer to numbered lines whose numbers are less than the current line's number. However, such ordering is not strictly necessary since numbers can also serve merely as unique identifiers whose order is unimportant. Keeping lists of rules and inferences numbered and ordered at all times often makes keeping track of where inferences are coming from much easier than would otherwise be the case.

In most cases, this is what you should do. Numbers are very concise and eliminate the need to immediately think up new names for every new inference you make. However, some rules may be so common that you may want to give them specific names so that you do not have to keep listing them again in every new rule set. It is for this reason that you are permitted to add named abbreviations for specific rules to the justification note for an inference line.

For example, you could say that "MP" means "meta modus ponens"[27] since this is more a rule of transformative logic itself than it is something you'd want to bother writing over and over again for every rule set. You could also just use the abbreviations to give more descriptive names to rules, rather than merely using numbers, but for many rule sets creating lots of specific names for rules may become very tedious very quickly.

By the way, while you *can* include meta modus ponens ("MP") in the justification note for inferences if you want to, it is actually usually unnecessary and a waste of time. I just gave it as an example to illustrate named justifications. It is probably better to be more concise in this case, since it is such a common rule. You should usually just omit it. For example, consider the following two different ways of writing the same rule set, both of which use meta modus ponens but each of which notate it slightly differently:

1. ☆ $A \to B$
2. ☆ $B \to C$
3. ☆ A
4. $(1,3) \therefore B$
5. $(2,4) \therefore C$

1. ☆ $A \to B$
2. ☆ $B \to C$
3. ☆ A
4. $(MP, 1, 3) \therefore B$
5. $(MP, 2, 4) \therefore C$

As you can see, listing meta modus ponens ("MP") explicitly in the justification notes here doesn't really add any value, due to how ubiquitous and implicit the rule overwhelmingly is. Oh, and by the way, on an unrelated note, I think that inferences that use transformative implication are probably usually best written using the numbers-over-arrows approach to step tracing instead of vertical step tracing. We wouldn't be able to tell which steps in a chain of transformative implications were justified by which

---

[27] For reference, page 272 contains the first mention of meta modus ponens.

numbered lines very easily if we had all the numbers jumbled up and prefixed to the inference mark. Writing the numbers inline and over the arrows would be better in that case probably.

Basically, you should use vertical step tracing when applying meta implications to append new forms to the current rule set, but should still use numbers-over-arrows for transformative implication chains. Also, meta implication chains which use inline notation (e.g. $\{A\} \prec \{B\} \prec \{C\}$) can still use the numbers-over-arrows notation for tracing steps. However, notice the distinction between (1) applying a meta implication to the current rule set and (2) inferring a new meta implication rule and appending it to the current rule set. These are not the same thing. The former can't be expressed using numbers-over-arrows step tracing notation, whereas the later can. The former *applies* a meta implication rule, whereas the later *generates* a meta implication rule.

Anyway though, I think you get the idea behind intention marks now. If you want more practice though, perhaps try rewriting some of the various example scenarios we've discussed in the transformative logic section so that they use intention marks instead of prose. For example, try rewriting the survival horror statue rule set so that it uses intention marks instead of prose. It'll be easy. You can find the original transformative rule set on page 185, and the more concise version on page 188. Remember that transformation rules and initial forms can both be assumptions and that target marks support both individual forms and arbitrary transformation paths as possible goals. That should help you with the translation process. There's not much else left to say, so that concludes our discussion of intention marks.

3.15 The power of conceptual clarity and proper perspective

I think it's time to move on to the next subject. We are in fact just about finished with transformative logic. The next major subject we will be exploring is what the various branches of non-classical logic are and what their underlying core motivations are. However, before that, I have just a few more points about transformative logic that I want to discuss.

First, I want you to think back to when we encountered a major expressive roadblock in transformative logic when we noticed that it was extraordinarily labor intensive to try to express form negations without the vocabulary of set theory available to us. We were exploring how the statue puzzle system could be modified to make selective transformation necessary. This resulted in the creation of the multi-statue puzzle system. The description of the multi-statue puzzle started on page 218, but we didn't actually start discussing how tedious its rule set was to specify until page 220.

Our very first idea on how to fix this was to just add on one new operator that expressed the concept of form negation and to treat it as a special case. However, we decided not to do that. Instead, we took the time to think carefully about the underlying concepts in play so that we could get to the real core essence of what was wrong with the expressiveness of transformative logic as it stood at the time, which was that it was limited to only direct transformation rules defined within the current rule set, due to it being unable to defer to other transformative languages. Aren't you glad we took the time to fully understand this instead of just slapping on one new operator in an ad-hoc way?

If we had rushed, and not taken the time to consider the possibility that there was a deeper conceptual problem in play, then we might never have come up with interpretation injectors and the enormous flexibility and convenience that they provide us. I'm bringing this up because I want you to take a moment to appreciate the enormous power of conceptual correctness and working from first principles. Ad-hoc solutions might get you by in the heat of the moment, and may be better than nothing, but they are often the tip of the iceberg of hidden truths. Remember to take a moment to consider the possibility that something deeper is wrong whenever you notice a surface-level problem in a system. It's like the difference between treating the symptoms of a disease and curing the underlying condition.

I hope the point here sinks in and that it helps you to learn the value of patience, if you haven't already. When it comes to technical endeavors, such as logic, mathematics, computer programming, science, and any other kind of detective work in general, it really pays to be patient. Work methodically and strive to devise thought experiments and tests to evolve your understanding of the system. Don't be superstitious and don't rush. Work with a steady hand, going slow enough to make sure you understand the system well, as much as the time and resource constraints will allow, and always strive to understand the underlying cause and reason for the system's behavior as much as possible.

Steady work almost always ends up actually being faster than rushed work. If you wish to be maximally productive in any kind of technical, intellectual, creative, or artistic endeavor then you would do well to mind this principle. The more you try to rush things, the slower it will often actually end up going for you, and the less deep insights you will have. The less deep insights you have, the fewer leverage points you will see, and the more you will end up missing opportunities to get exponentially more benefit for the same amount of effort. Rote memorization is better than nothing, but a principled understanding is far more powerful and flexible. At any given moment, there might be a treasure chest of hidden insights buried beneath your feet. If you're in too much of a rush to see that the soil has been disturbed though, then you'll surely miss it.

Second, I want to revisit the notion of transformative deduction versus evaluative deduction to point something out to you. Our discussion of transformative deduction and evaluative deduction started on page 95. As you may recall, transformative deduction is reasoning that only deals with forms, whereas evaluative deduction also allows you to consider the values of those forms. Transformative logic is based on the concept of transformative deduction. You might therefore be tempted to think that it excludes the use of evaluative deduction. However, it is actually not so cut and dry.

You see, while it may be true that transformative logic can only deal with things in terms of transformations between uninterpreted forms, it is also true that transformative logic can emulate value systems by defining matching transformation rules. For example, consider classical logic. Transformative logic can of course model various different laws of classical logic in terms of transformations between pure uninterpreted forms. For instance, we could write $(A \vee B) \leftrightarrow (B \vee A)$ and that would be a valid way to express the order independence (commutativity) of classical disjunction in transformative logic. We can express any law of classical logic in terms of transformative logic in this way. However, we could *also* define the actual truth table valuations of classical logic in terms of transformative logic as well. Consider, for example, the effect of the following list of transformation rules:

1. $F \vee F \twoheadrightarrow F$
2. $F \vee T \twoheadrightarrow T$
3. $T \vee F \twoheadrightarrow T$
4. $T \vee T \twoheadrightarrow T$

This list of transformation rules effectively implements the entire truth table for classical disjunction in terms of transformative logic. Yet, by so doing, we have effectively intruded upon the territory of evaluative deduction, even though transformative logic was nominally supposed to only be about transformative deduction. Thus, the distinction between transformative deduction and evaluative deduction is actually rather fuzzy. Transformative logic can emulate arbitrary valuation systems if you define the proper transformation rules for it. Yet, even though it is emulating a value system, it is still technically thinking purely in terms of uninterpreted forms. Thus, there isn't actually much of a fundamental difference between transformative deduction and evaluative deduction, but rather it is more like they are just different perspectives on how to approach problems.

Interestingly, even though transformative logic can implement the complete truth tables for classical logic in terms of transformation rules, there's nothing stopping us from leaving this implementation incomplete or perhaps even changing it in fundamental ways. This would often be impossible to do in a truth-functional logic. For instance, consider our implementation of classical disjunction in terms of transformation rules, as seen above. There's nothing stopping us from deleting a few of the rules from it, if we wanted to. Consider, for example, what would happen to the behavior of the system if we did this instead:

1. $F \vee F \twoheadrightarrow F$
2. $T \vee T \twoheadrightarrow T$
3. $F \vee T \leftrightarrow T \vee F$

Notice that I have deleted the rules for $F \vee T \twoheadrightarrow T$ and $T \vee F \twoheadrightarrow T$, and inserted a rule for $F \vee T \leftrightarrow T \vee F$. Suppose that this is our entire rule set, i.e. that there are no other hidden rules for transformations involving the $\vee$ operator. How would this change the meaning of disjunction? Well, it would imply that we could no longer resolve expressions like $F \vee T$ and $T \vee F$ to specific classical truth values. However, we would still be able to push T and F values around in larger expressions (via $F \vee T \leftrightarrow T \vee F$) and at least would be able to remove some redundant copies of them via subsequent application of $F \vee F \twoheadrightarrow F$ or $T \vee T \twoheadrightarrow T$.

What would be the intuitive interpretation of this behavior? Well, perhaps we could think of expressions like $F \vee T$ and $T \vee F$ as now meaning "true or false, but we don't know which". This gives $\vee$ more the character of an uncertainty operator than truth table resolution. This is just a subjective idea for one way we might interpret this change. It might be inaccurate. There may be better interpretations. It may not even be a useful rule set at all. Who knows? I'm just putting it out there as a thought experiment basically.

Ultimately though, transformative logic is just about uninterpreted forms, i.e. is just about symbols that we push around in a certain way in order to model something. What it "means" depends on a combination of (1) the structural characteristics of the rules you choose and (2) what the forms are intended to represent relative to what you intend to model. The point is: I don't want you to think too rigidly. There's really quite a lot of unusual things you can do with transformative logic. I want to make sure you keep a broad perspective. Don't be a slave to preconceptions. Expand your intellectual canvas. Free it from the chains of tradition. Think from first principles, not from first precepts.

Anyway, incidentally, the fuzziness that exists in the distinction between transformative deduction and evaluative deduction may also exist in the distinction between syntactic consequence and semantic consequence. Syntactic consequence is clearly closely analogous to transformative implication. What of semantic consequence though? Well, if you remember, semantic consequence is this idea that if a claim B is true in all of the cases where A is true then B is a semantic consequence of A. This too could maybe be re-written in terms of pure transformation rules. If the truth tables for any operator can be written in terms of transformative logic, and semantic consequence works by simply evaluating those same truth tables, then it seems reasonable to think that it may be possible.

Speaking of possibilities, now that we have transformative logic in our arsenal, I think it's time that we discuss what some of the other possible logics are. So far we've mainly discussed a combination of classical logic, set theory, and transformative logic. We've seen how transformative logic can be used to really broaden our perspective on what kinds of logical frameworks can be created. Transformative logic has also helped us acquire a deeper and more fundamental understanding of what the real underlying mechanisms of logic and mathematics actually are.

By generalizing the concept of implication as far as possible, we have come to realize that there are actually a lot more possible transformation rules than merely what the standard set of algebra rules they teach you in school would lead you to believe. Logic truly is a diverse and bountiful landscape. There are many different viewpoints on how exactly truth should work. It is now time to experience a larger sample of that diversity. Thus, we now turn our discussion to the non-classical logics. The underlying motivating principles of each of these various different systems of non-classical logic will provide us with fresh new perspectives on what kinds of criteria one might want a system of logic to have. Knowledge of these possibilities will also later prove useful in the creation of unified logic.

Chapter 4

What is non-classical logic?

4.1 Why do we care?

Before we actually dive into the details of what the main types of non-classical logic are though, we should first refocus our minds on why we are even considering other types of logic in the first place. Let's start by taking stock of what resources we already have available to us. We have classical logic, set theory, and transformative logic. We did briefly mention a few other systems of logic from time to time, but we did not explore them enough for it to qualify as a real understanding of the nature of those systems. We also mentioned syllogistic logic and stoic logic, which are the ancient ancestors of set theory and classical logic respectively, but these are essentially redundant now since they have been completely eclipsed by their modern descendants in terms of versatility and scope.

Classical logic is pretty useful. It's also very simple and easy to understand. As long as you avoid actually using the paradoxes of classical logic to justify anything, it's also a fairly safe and reliable logic to use. One of its best properties is its ability to determine whether any given expression is a tautology (i.e. is universally valid) by simply exhaustively checking all possible cases in a truth table. Few other systems of logic are so amenable to such a mindless brute force approach.

In the general case though, in other logics, you will be required to construct actual proofs of claims and these proofs will often require significant cleverness and inventiveness to reach. In contrast, anybody with enough time to spare can do a brute force evaluation of all possible cases of a truth table for an arbitrary expression in classical logic. Of course, classical logic has proofs too, and these are generally preferred anyway, but it is still always nice to have a mindless mechanical fallback procedure for testing claims.

However, there are problems with classical logic. It is often way too permissive in what it allows you to infer. Think back, for example, to the section on the paradoxes of material implication. It started on page 110. The paradoxes of material implication are tautologies in classical logic, which means that they are universally true within it. Tautologies that contain implications express transformation rules that should be valid to apply at any time within that system, to any form matching the condition. However, performing any of the transformations that the paradoxes of material implication nominally permit will not generally result in sane reasoning. The truth-functional values of classical expressions may be conserved properly, but the underlying meaning of those expressions in terms of what they are supposed to represent may not be.

Classical logic is also narrow in terms of the scope of what it is capable of talking about. It can't talk about arbitrary forms. It has no understanding of where the truth values it manipulates come from, and thus is neither able to handle contextual information nor able to account for any special relationships that might exist between the participating statements in any given expression. It has little to no ability to deal with uncertainty or incomplete information.

Every input classical logic receives has to be definitively true or definitively false. Classical logic does not care whether a condition has any genuine relationship to what it supposedly implies. Classical logic only thinks in truth-functional terms. It usually isn't capable of talking about the structure of other logics. Attempts to talk about the structure of other systems of logic in classical logic will often immediately result in contradictions or nonsense, due to the fact that classical logic makes very narrow assumptions about the nature of truth and has no vocabulary capable of reaching outside those assumptions.

Once a single pair of contradictory claims has been mistakenly accepted as true in classical logic, then it becomes possible to fallaciously "prove" all conceivable statements as true simultaneously, even if they directly contradict each other, by using a property known as the principle of explosion. This means that classical logic is inherently fragile and error intolerant. A single error in a classical argument can easily spread like a plague, distributing nonsense far and wide throughout a theory. It has essentially zero error resistance.

Classical logic makes no distinction between statement expressions and value expressions, even though they are conceptually distinct. For example, if I just write $A \to B$ then you would have no idea whether I am asserting that B in fact does follow from A or if I am instead merely asking you to evaluate the truth-table for the expression, which could be true or false and which makes no actual claim about anything at all. Instead, we often have to rely on context and our knowledge of whether or not $A \to B$ is a tautology, but even then the intent is still ambiguous. If the expression is a tautology then it must also be a law of classical logic, and hence perhaps will be more likely to be an assertion, but even then we still don't know whether the intent actually is to assert it as a fact or to merely treat it as an expression to be evaluated.

There actually is a symbol that is sort of intended to provide this distinction, but it is used very inconsistently. That symbol is the double-lined version of the material implication symbol, which looks like $\Rightarrow$. This symbol can be read as "formal implication", as opposed to "material implication", with the idea being to assert that the consequence actually does follow from the condition. However, in practice, the symbols $\to$ and $\Rightarrow$ are usually treated as being completely synonymous representations of material implication.

Most authors seem to simply arbitrarily pick one of $\to$ or $\Rightarrow$ to represent material implication and then completely ignore the other. Sometimes, people use $\to$ when talking about the value expression version of material implication, and then use either syntactic consequence (⊢) or semantic consequence (⊨) for assertions. At least, that's my impression. Thus, currently, it seems best to maybe avoid use of ⇒ altogether, due to the fact that its usage is so inconsistent. Even when used as "formal implication", it can be unclear what *kind* of formal implication it is actually talking about (e.g. syntactic vs semantic consequence, etc).

As long as we're still talking about ambiguities, it occurs to me that there is another ambiguity in the terminology that I should also point out. Throughout this book, we have generally used the term "classical logic" to refer to one very specific logic. However, the term "classical logic" actually technically refers to an entire family of logics, consisting of any logic that satisfies a certain set of common properties. It's just that the version of classical logic that we've been talking throughout this book is overwhelmingly the most common one, to the point where it is often treated like it is practically the only real representative of the family. I am unfamiliar with any of the others, but what I've read seems to imply that they exist.

Alternatively, some people refer to what we've been calling "classical logic" in this book as "propositional logic", perhaps thinking that this term is more specific or perhaps just using it by habit. However, this term too is unfortunately ambiguous. As it turns out, propositional logic is *also* the name of an entire family of logics and not technically just one in particular. You see, the term "propositional" actually indicates only that a logic is about statements as atomic units and that it does not support quantifiers.

A logic that does not support quantifiers treats each statement as an indivisible unit and cannot speak precisely of how many of the internal set of objects the statement applies to. Propositional logic conceives of statements as the smallest atomic unit of study and thus cannot peak inside them. The classification "propositional" is just one point on a larger axis of classification having to do with how extensively an expression in a logic can be quantified. Oddly enough, there doesn't appear to be an official term for this axis, so let's make one:

Definition 111. *Every logic either does or does not support quantifiers. Moreover, there is a rough spectrum of how much support a logic has for quantifiers and what types of quantifiers it allows. We will refer to this spectrum as the **quantifier spectrum**, and to each individual place on the quantifier spectrum as a **quantifier order type**. In some cases, a quantifier order type may have a number associated with it, which we may call its order.*

It is these broader terms "quantifier spectrum" and "quantifier order type" which might be new here. In contrast, referring to these logics by what "order" they are classified under is nothing new and is quite traditional and common. For example, a logic which does not support quantifiers may be called a zeroth-order logic. Thus, "zeroth-order logic" is a synonym for propositional logic. There is also first-order logic, which is also known as predicate logic. Systems of logic with an order higher than 1 (i.e. higher than predicate logic) on the quantifier spectrum are generally referred to simply as "higher-order logics", but can also be referred to by a specific order number sometimes. I'll have more to say about what quantifiers are and what some of the various orders are later on, but for the time being let's refocus on what we we're previously discussing.

Anyway, as you may remember, we were talking about how both "classical logic" and "propositional logic" actually refer to entire families of logic rather than to one in particular. I mentioned that the term propositional logic was one alternative to the term classical logic, but that it isn't really any more specific than the term classical logic is. You see, there are countless other arbitrary logics that you could conceive of that may not use quantifiers. You could just make up some crazy new logic using transformative logic that doesn't use quantifiers and acts like it uses propositions, but which definitely isn't the specific logic we've been calling classical logic. Perhaps combining the two terms into "classical propositional logic" is clearer, but it still doesn't seem totally specific. Therefore, let's make a new term that doesn't have any of these ambiguities:

Definition 112. *To refer unambiguously to the most common form of classical propositional logic, we may call it* **standard classical logic**. *This refers specifically to only the simple truth-functional system of classical propositional logic that has overwhelmingly been the most highly standardized and most widely taught variant of classical logic as of the time this book was written. It is a specific system of logic with a specific notation and a specific behavior and set of conventions, rather than merely being any logic that meets a broader set of criteria. It cannot be construed as referring to an entire family of logics, unlike the terms "classical logic" or "propositional logic".*

Definition 113. *In contrast, to refer in general to any logic that meets the broad characteristics of being a classical logic (e.g. accepting the law of the excluded middle, etc), we may call it a* **classical-type logic**. *The purpose of this term is to provide a way to clarify when you are talking about the entire family of classical logics, rather than using the term "classical" in the narrower sense of standard classical logic. In order to not interfere with prior usages of terminology, we will leave the historical usage of the term "classical logic" undisturbed, such that it will continue to refer ambiguously to either standard classical logic or classical-type logic.*

Oh, and by the way, the term "boolean logic" also wouldn't be enough to totally pin down the meaning. There is in fact an entire family of boolean algebras, and they do not necessarily refer to standard classical logic. One of the few downsides of all the constant pushing for generalization that goes on in logic and mathematics is that sometimes things generalize so much that some of the earlier terminology starts to become ambiguous. I wanted to really pin down the term before we began our discussion of the non-classical logics. Better safe than sorry.

Speaking of non-classical logic, now seems like a good time to tell you that when I say that this section is about "non-classical logic" I mean that it is about systems of logic that are not standard classical logic. Notice that this is not quite the same thing as not being classical-type logic. Thus, for example, while some might consider standard predicate logic to qualify as a form of classical logic, in this section we will treat it as a non-classical logic because it is not standard classical logic, which does not allow quantifiers. If you don't like the term "non-classical logic" for this, you may prefer to use the term "alternative logic" instead. The term "alternative logic" de-emphasizes the classical part and instead puts more emphasis on exploring and comparing the various possible choices you have when picking a logic.

We will also be broadening the scope to include a few things that wouldn't normally be covered in a book like this, such as informal logic. If this sounds to you like its going to end up being way too much stuff to cover given the pace we've been moving at in this book, you are correct. It would probably take another entire book to actually cover all of these logics in-depth. Instead, this section will be focused on giving basic overviews of each non-classical logic, overviews designed to give you a sense for the core motivations and principles each logic is inspired by.

A few concrete examples will be given for some of the logics, when I feel it is within the scope of relevance for what we are trying to accomplish in this book, but not always. Admittedly, my knowledge of many of these branches of logic is actually quite limited, but ultimately it won't matter much because the main reason why we are exploring these non-classical logics is actually mostly just so that we will be able to show that many of the core underlying motivations of these non-classical logics will be satisfied by unified logic once we get around to creating it. The level of knowledge that would be necessary for me to cover all of these branches of logic in-depth and at an expert level would be prohibitively high, given my limited time and resources. Limited knowledge or not though, I will always strive to be rigorous and to not intentionally misrepresent any of these branches of logic, as best I can.

Studying these other branches of logic will give us new insights into different ways that a system of logic might work. It will give us more of an outline of what the parameters are. To make a rough analogy, it's kind of like a radio box or sound synthesizer: We want to know what the various little dials and switches actually are, so that we can understand what kinds of interesting things might happen if we adjust the settings in just the right way. Ideally, we'd want to tune the system until it became as close to optimal as possible, so that we could maximize the overall power of our system of reasoning. Each non-classical logic is kind of like one of these dials, in that each one tends to focus on one specific special change that it makes to the way our system of reasoning works.

What kind of changes can we make though? What are non-classical logics like? Well, we have actually already met several non-classical logics. I just didn't tell you that they were non-classical at the time. In particular, I am referring to syllogistic logic, set theory, and transformative logic. All three of these logics are non-classical. For example, none of them are truth-functional. All of them are multi-valued and aren't restricted to just true or false. None of them use truth tables, although admittedly transformative logic can emulate them. All of them can be used for some types of proofs, yet they do not even directly deal with the concept of truth, at least not in the same conventional sense that classical logic does. Many of the properties that apply to classical logic do not apply to any of these three logics.

Interestingly, transformative logic is not the only logic capable of emulating classical logic. Set theory can do it too. In fact, I basically implied this earlier in the book. Perhaps you remember. When we were talking about how there were a whole bunch of striking parallels between the laws of classical logic and the laws of set theory (on page 147), I pointed out an interesting correlation. Specifically, I showed (on page 149) that the behavior of all of the basic classical operators could be mimicked perfectly in set theory by using $\emptyset$ to act as F and $\mathcal{U}$ to act as T. The relationship between classical logic and set theory goes even deeper than this though, and people don't seem to realize just how important it actually is. There's a treasure chest of insight buried beneath the surface here, but more on that later.

As for syllogistic logic, it is merely a crude ancestor of set theory, so there is not really anything useful to say about it. We have already devoted an extensive amount of time and text to thoroughly covering the basics of all of these three non-classical logics (syllogistic logic, set theory, and transformative logic). Therefore, they will not reappear in this section of the book. However, you should keep in mind that they technically belong in here too.

Also, be aware that some people don't treat set theory as a branch of logic, but I think that this is a mistake. Consider, for example, the fact that set theory can perfectly mimic classical logic within itself. Isn't that pretty strong evidence that it is a branch of logic? There are also other deeper reasons why it should be considered a branch of logic. You'll see why later on, but for now we need to stay focused on other things.

One thing you may start wondering as we discuss the non-classical logics is why is it that so many of them are seldom actually used in practice. Why is classical logic, augmented with a few other things, so dominant? Classical logic, set theory, predicate logic, and informal logic are very widely used. Other logics are much less so. Why is that? Don't people want the beneficial special properties that the non-classical logics provide? The answer is, well, yes they do, but the problem is that all of these logics come with advantages and disadvantages.

The most popular branches of logic are the branches that tend to have the best overall balance of advantages and disadvantages and ease of use. Whenever you pick one of the lesser known branches, you will often have to give up some of your old advantages in order to gain the new ones. Some of the branches of non-classical logic are much harder to intuitively understand than classical logic. It's largely a matter of costs versus benefits. There are also a few branches of logic, such as probabilistic logic and fuzzy logic, which are specialized for specific use cases. Within those specific use cases, they may be common, yet they may not be nearly so common in other situations.

We will be exploring what the rationale is for each branch of logic, and during this process I will also try to provide some sense for what the problems with each of them might be. As for classical logic, its dominance is probably largely owed to its overwhelming ease of use and relatively intuitive nature, plus social momentum. Of all the different known logical frameworks, classical logic seems to maybe be the easiest to understand and use. However, not all of the non-classical logics are counterintuitive or difficult. Some (like set theory, transformative logic, and informal logic) are in fact very intuitive. Moreover, if a system of logic doesn't seem natural or sensible, then any argument constructed in it is unlikely to be genuinely convincing.

The fundamental reason why proofs are convincing in the first place is because they are based on chaining together applications of simple rules that are obviously true, such that the proof as a whole (being made up entirely of such simple rules) must thus undeniably be true. However, if those simple rules (the foundations of your logical system) are not themselves undeniably true, then any argument constructed in that system will likewise not be convincing, except perhaps hypothetically. Thankfully, as you will see, unified logic will end up being a fairly easy system to work with, while somehow still managing to satisfy the underlying motivations of many different branches of logic simultaneously.

Oh, there's one more little side tangent I want to go over before we actually start exploring each of the non-classical logics in turn. In particular, I want to clarify the meaning of two commonly used traditional terms that you will probably encounter fairly regularly in the literature when exploring formal definitions for various different systems of logic. The two terms I am referring to are "axiom" and "rule of inference". These two terms are often used to distinguish between two different ways of specifying rules for what the properties of a system of logic should be. However, unfortunately, despite how common these terms are, they are actually arguably misnomers.

Traditionally, when you want to define how a system of logic is going to work, you specify a set of at least one "axiom" along with a set of at least one "rule of inference". Together these two sets define how the system of logic is supposed to work. Why do you have to have at least one of each? Well, if you didn't specify at least one "axiom" then you wouldn't have a starting point for deriving new theorems. Likewise, if you didn't specify at least one "rule of inference" then the only theorems you could derive would be the "axioms" themselves, because you would have no way of making inferences from what you already know to generate new theorems. Does this process sound kind of familiar? Does it ring a bell? It should.

You see, what is traditionally called an "axiom" in the literature is analogous to what we've been calling an *initial form*. Likewise, what is traditionally called a "rule of inference" in the literature is analogous to what we've been calling a *transformation rule*. Do you see the relationship more clearly now? There is a little bit more nuance to this comparison than what I'm saying here, but this is roughly the correct idea. So, you might be wondering, what exactly did I mean when I said that the terms "axiom" and "rule of inference" were actually misnomers here?

Well, you see, the problem with these terms is that "axioms" and "rules of inference" are *both* axioms. An axiom is anything that we assume true as a given, typically something with an elemental or foundational character that is hard to imagine deriving from anything else, but not always. Both "axioms" and "rules of inference" in the context of these traditional rule definitions for a system of logic meet this criteria. That's why it's a misnomer. Rules of inference can always be thought of as axioms, and thus the wording of these terms does not actually create any real distinction.

Systems of logic tend to be defined in roughly one of two styles: either *axiom heavy* or *rule of inference heavy*. If a system of logic has been defined in an axiom heavy way then that indicates that it has numerous "axioms" (i.e. initial forms) and only a few "rules of inference" (i.e. transformation rules). In contrast, if the system has been defined in a rule of inference heavy way then the opposite is true, and it will have numerous "rules of inference" and relatively few "axioms". As you may recall, all of our rule set examples for the transformative logic section were rule of inference heavy. We always gave a big list of transformation rules and very few initial forms (usually just one) for each system.

Since we have only ever defined rule of inference heavy systems in this book, you might be wondering what the other approach looks like and why it works too. Basically, the idea of the axiom heavy approach is to have a lot of different starting points from which to form your derivation path, instead of having a lot of different transformation rules in your vocabulary. In effect, you have less flexibility while exploring each derivation path, but more flexibility when picking an initial form to start from. There's a reason why all the rule sets I gave were written in a rule of inference heavy style though. The axiom heavy approach has a tendency to be much harder to reason about.

Look up "Hilbert-style proofs" and read some of the examples. The most typical examples of Hilbert-style proofs consist of a list of multiple "axioms" (initial forms), along with one (or very few) rules of inference (typically just modus ponens), wherein everything looks like a bunch of material implication arrows ($\rightarrow$) strung together in a semantically nonsensical way. If you see a rule set that looks like that then you've probably come to the right place. See if you can follow the reasoning. Hilbert-style proofs are probably the most famous kind of axiom heavy proofs, and they are often wickedly difficult to read and make sense of.

As far as I can remember, I've never seen a Hilbert-style proof that made any kind of intuitive sense to me. They often seem borderline incomprehensible at first glance. Almost nothing about them matches natural human intuition or logical instinct. They don't seem to fit well into the physical analogy system that our minds have evolved to excel at. Sure, Hilbert-style proofs may indeed be formally well-defined, and that formal process itself may be relatively easily understood, but such proofs nonetheless tend to feel logically unnatural and contrived.

In contrast, rule of inference heavy systems seem to often more closely resemble the kind of thinking that we use in everyday life and as such tend to feel much more comfortable. When I see a Hilbert-style proof, the chain of reasoning often looks so convoluted that I find myself instinctively not trusting it. The method of proof may look even more doubtful than the claim it is purporting to prove. Even if you know that it is rigorously true, it may be hard to shake this feeling off. For these reasons, I generally advise avoiding Hilbert-style proofs when you can.

That being said, there's nothing wrong with exploring lots of different initial forms. In fact, part of the whole point of many systems of logic is to be able to deal with countless different initial forms in order to solve them, such as in the algebra of equations for example. Another reason to accept a lot of different initial forms is to provide a diverse set of seeds from which to generate a bunch of different structures, such as in L-systems or random map generators in games for example. It is also useful as a basis for creating a multitude of amazing special effects, such as fancy programmable shaders or particle systems or dynamic gameplay mechanics, all of which often require seed data of some kind. The sky is the limit really.

If that's the case, and accepting lots of different initial forms is so useful, then what is the real distinction here? Well, the real distinction is whether the system of logic has a lot of rules of inference rather than only a few. As long as a system of logic has lots of transformation rules, then no matter how many initial forms you also accept in that system it will still tend to feel natural to work with. That's the real reason why rule of inference heavy systems of logic tend to be easier to understand.

The reason is because they have so much more flexibility in how you can navigate the transformation path you are currently on. In contrast, axiom heavy systems of logic tend to front-load the burden of the proof more, which makes choosing the right "axioms" (initial forms) feel like it requires the use of magic, which is to say that it is hard to guess what form and structure will be required in advance. In other words, the reason why axiom heavy systems are stifling is not because they have a lot of initial forms, but rather it is because they have far too few transformation rules.

The terms "axiom heavy" and "rule of inference heavy" that we've been using so far are thus somewhat missing the point. As such, let's define some new terms that capture the sense and essence of what they refer to a bit better. Now that we've thought it through, we should be able to come up with some more meaningful names. If you think about it, this distinction is all about how much weight or burden we place on either the initial forms or the transformation rules of a system. Thus, I feel like maybe an analogy to the "weighting" of the system could be a good angle of attack for how to choose good new descriptive terms for these concepts. Here are the new definitions:

Definition 114. *If a rule set is "axiom heavy", meaning that it has lots of initial forms to choose from for deriving theorems but has very few transformation rules, then we may call it a **front-loaded rule set**. Front-loaded rule sets tend to place a lot of burden on picking the correct initial forms. The limited number of available transformation rules tends to give the user very few options for how to move within any given transformation path, and therefore tends to make the system more awkward to use and more difficult to understand.*

Definition 115. *If a rule set is "rule of inference heavy", meaning that it has a relatively large number of transformation rules, then we may call it a **distributed rule set**. How many initial forms a distributed rule set allows actually doesn't tend to effect the way it feels all that much. The key factor is that it has a lot of useful transformation rules, which thereby tends to make it more flexible. Although distributed rule sets may require more transformation rules compared to front-loaded rule sets, this tends to give the user lots of options for how to move within a transformation path, and therefore tends to result in rule sets that are more intuitive and easier to understand.*

The analogy here is drawn from how weight can be distributed on the chassis of a vehicle, or on some other heavy structure. When you are engineering a vehicle, or any other kind of structure that might be required to handle heavy loads, if you place way too much weight on one side of the object then you are liable to make it unstable or structurally unsound. An object that has too much weight on any one side is going to be more prone to tipping over in that direction.

That's why I chose to call "axiom heavy" rule sets "front-loaded rule sets". It's a good physical analogy for why they tend to be awkward. In contrast, the term "distributed rule sets" is analogous to when an object has had its weight distributed more evenly, so as to make it more stable and more easy to work with. Distributed rule sets spread out the user's ability to choose from a diverse set of possibilities so that the choice can happen at any time, instead of concentrating it all at the beginning, thus making the "distributed" analogy appropriate.

Anyway, before we got off on this side tangent about these two pieces of terminology, we were talking about exploring the rationale behind each branch of non-classical logic and about the trade-offs involved in making them work. Which branch should we start with though? Well, how about we start off where logic really first began: in informal logic. It may be true that syllogistic logic was the first *formal* logic, but it was certainly not the first form of logic of any kind.

Informal logic is in fact by far the oldest branch of logic. It has existed in at least some form probably since the earliest days of human civilization. Long before formal logic, humans still needed a means by which to reason about the world. Cities had to be planned, resources had to be carefully allocated, and skills had to be learned. In the most general sense, informal logic is a pretty all-encompassing term that has relevance to virtually all human endeavors. However, like most treatments, we will be focusing primarily on the subset of informal logic that studies argumentation and critical thinking in a society that is all too often rife with misinformation and poor reasoning. Let's begin.

4.2 Informal logic

Forgive me for breaking the fourth wall here for a moment, but it is necessary. You see, when I originally wrote this section on informal logic here, it was supposed to just end up being a handful of pages giving a very brief introduction to informal logic. After discussing the basic nature of informal logic, I planned to include a small handful of informal fallacies and cognitive biases to give a better sense for what informal logic feels like, in keeping with my tradition of including concrete examples for clarification purposes. Each example was intended to be no longer than roughly a paragraph. Things did not go as planned. I basically accidentally wrote a 2nd book, of quite some length, one about informal logic. It ended up being nearly 130 pages long. I didn't realize it was that long at the time, until it was too late to be worth stopping.

Consequently, the original version of this section became so utterly enormous that leaving it in place here would have been unwise. It would have completely broken your (the reader's) train of thought. You'd maybe have completely forgotten about transformative logic by the time you were done reading it, or anything else we were talking about for that matter. Thus, I have moved the entire original informal logic section out of this part of the book and into its own entire chapter located near the end of the book. I also added a bunch of new section headings to it that weren't originally in it to make it easier to navigate and digest.

If you would like to read that section, as it was originally written, then please proceed to page 591 and then return here once you're done reading it. Behold Chapter 6 (*A rant-filled introduction to informal logic*) in all its ranty glory ☺. Bear witness[1] to my profound inability to control feature creep, the truest hallmark of anyone having a background in game development. I look forward to your dazed and disoriented return, and I hope that the distracting effects that reading it will have upon the train tracks of your mind will not be so severe as to render your mind insolvent. If, on the other hand, you prefer to keep your train of thought intact, then continue reading from here, and wait to read the informal logic chapter (or should I say book?) until you come upon it in the natural course of proceeding forward from here.

---

[1] definition: a member of the Ursidae family that was present at the scene of a crime

No matter which path you choose though, please be aware that your brain is still going to turn to jelly by the end. This is an unavoidable tragedy, and one for which I offer no apologies, mostly because I'm a jerk. However, fret not, for I have a backup plan. Just pour some packets of flavored sugar into your ear canals before then[2]. That way, all that brain mush will taste like a sort of finely cultured and intellectually mature fruit smoothie.

Hipsters will love it. I'll make so much money, at least until it becomes popular. And so you see, there's a silver lining in everything. Its just like the old saying goes: When life gives you lemons, make lemonade, and, when small subsections accidentally turn into miniature books, turn your brain into a fruit smoothie. Such wisdom. Much enlightenment. Surely these must be the most insightful words I have ever uttered[3]. Putting hipsters aside though, seeing as "nobody likes them"[5], it is now time for us to move on to the next section.

4.3 Predicate logic

I suppose we have to return to doing responsible and meaningful work now, instead of just sitting around making more bad puns and abusing the footnote system. Alas... Whether or not we succeed in making this transition will be *predicated* on how well we manage our time. As such, it is now time to talk about predicate logic. Predicate logic is a rather straightforward and easy to understand logic, one that is very widely used in the logic and mathematics literature. Essentially, it is an extension of a system of logic to support quantification over variable domains. Typically, this extension is applied to classical logic, but it could probably also be applied to some of the other systems of logic as well. However, we will only be discussing *classical* predicate logic here. I've never really seen it used in any other context, as far as I can remember.

---

[2]Attention Idiots: Don't actually do this. You won't be able to hear everybody laughing at how stupid you are for doing this if it causes you to go deaf. We don't want that to happen. Social feedback seems important at this point, clearly.

[3]Mooooooo... "**utter**ed"... Get it? Haha... Such wit. Holy guacamole[4], I may have already outdone myself again. Fantastic. I have no shame. I gave all of it away to my family, so that they can feel it for me. Not really though. They love me. Everybody does. People come from all over the world to hear my bad puns. I'm basically the best. Yep. I said it. Feels good you know, admitting it. You may now commence the process of sulking over your own insecurities while you bask in the light of my greatness. Try to do it publicly though, on social media. It's more mentally healthy that way. Always make your self-worth dependent on others. That's the key to mental health. Complete emotional dependence on factors outside your control is the key. That's what most of the people who call themselves normal do. Hop on board the *Conformity Express*! It's the only socially acceptable form of codependency! Get it while it lasts! What could possibly go wrong?! #NoFilter #NoBrain #BecauseHipsterSmoothie.

[4]$500,000 per pound at Chipotle by the way, essential for avocado toast, totally worth being homeless for, a daily necessity for all us millennials, according to the opinions of our participation trophy sponsors, a.k.a. the older generation, a.k.a. "Baby Boomers", a.k.a. "the Me Generation" or "Generation Me" as their own parents were so fond of calling them, in reference to their alleged selfishness and lack of perspective / empathy / responsibility / etc... When are people going to realize that literally almost every generation unjustly discriminates against their children like this? #FootnoteInsideOfFootnote #GotAProblemWithIt #BiteMe.

[5]If we started liking hipsters then they'd become popular, and then they'd have to start hating themselves. So really, if you think about it, when we hate on hipsters, we're actually being as nice to them as we possibly can be, rationally speaking. (I'm just kidding by the way, in case you can't tell. I don't actually have a negative view of hipsters. Benign lifestyle choices are of no concern to me. Anything that's within the constraints of benevolence and objectivity is fine by me.)

Anyway, as I was saying, predicate logic adds on support for quantification to a system of logic. What is quantification though? Well, in the most general sense of the word, quantification refers to assigning numbers or mathematical models to a phenomenon. However, in the more narrow context of predicate logic, quantification refers only to coarse-grained specifications of what portion of a set of objects a certain property applies to. In particular, quantification typically stipulates whether a property either (1) applies to at least 1 (i.e. "some") of the members of a set or (2) applies to all members of a set. These two cases are known as existential quantification and universal quantification, respectively.

As you may recall, I briefly mentioned quantifiers in passing when we were discussing (on page 293) the fact that the term "classical logic" is technically a little bit ambiguous because it technically refers to a family of logics rather than to one specific one, of which the standard classical logic is overwhelmingly dominant. We also discussed the quantifier spectrum some, and I mentioned how predicate logic is also known as "first order logic" because it only allows quantification over the "first order" of entities in a system, i.e. it allows quantification over simple inert objects but not over general functions and predicates. Anyway, I just wanted to remind you of that. It's time to discuss how quantifiers actually work now.

However, before we discuss the quantifiers themselves, we'll need to understand what a *predicate* is. It would be a mistake to think that the "predicate" in "predicate logic" refers to quantifiers. Quantifiers are the primary distinguishing characteristic of predicate logic, and also what people tend to fixate on, but that's not where the name comes from. Simply put, a predicate is any function which takes one or more arguments as input and returns a truth value as output. It is essentially a test function, one that checks whether or not the given input objects satisfy the indicated property.

For example, in computer science, if we had a function named `IsRed(x)` that tested whether the input object x was colored red, and then returned true if yes and false if no, then the function `IsRed(x)` could be called a predicate. However, in contrast, predicates in mathematics are most often expressed with single letter names, even though this often damages conceptual clarity. The two main reasons why this is done is for brevity (math is often hand-written) and to avoid ambiguity (adjacent letters may indicate multiplication in math).

However, as I have stated before (in a discussion starting on page 164), I think multi-letter variable names would be very valuable and that math conventions should be changed in order to make multi-letter variable names no longer ambiguous with multiplication. Anyway though, as such, a test function like `IsRed(x)` would usually be written something like $R(x)$ in math notation.

Here's where quantifiers come in. We have our predicates now, and we know that we want to somehow be able to make statements that "quantify" what portion of a set of objects a property applies to. Predicates have us covered on the property part, but what about the quantification? What exactly are we supposed to quantify over? The answer: any arbitrary well-defined logical expression containing at least one predicate. Not all of the terms in the expression have to be predicates, although some people do seem to get mislead into thinking so maybe. The presence of at least one predicate with at least one input variable is all it takes to make a logical expression quantifiable though.

Every application of a quantifier is composed of three logically distinct parts: the quantifier, the domain expression, and the predicate expression. These three parts, considered as a whole, constitute a quantified expression. These are the terms we will be using to refer to each of these parts, to make the discussion easier to follow. However, be aware that the literature seems to have no official terms for these different parts of the structure, as far as I am aware, so these terms are something I had to define myself.

To that point, it often amazes me how many different pieces of basic terminology seem to be missing from the math literature. In computer science[6], the rigor of the compiler forces us to define almost everything, but mathematicians face no such constraints so they tend to be much more sloppy and much less complete in their approach and perspective. Traditional math notation would seldom compile. A lot of mathematicians think they represent the pinnacle of rigor, when in reality much of what they do is actually based on a plethora of hidden assumptions, untested claims, and intuition. Much of it is wildly underspecified, even the supposedly precise stuff.

Anyway though, let's walk through the structure of a quantified expression piece by piece now. First, we have the quantifier. There are two types of quantifiers in common use. The others are too obscure or too outside our concerns to be worth mentioning here. The first quantifier is existential quantification, which is represented by the symbol $\exists$ and is read as "there exists" or "for some". The second quantifier is universal quantification, which is represented by the symbol $\forall$ and is read as "for all". The quantifier is written first in every quantified expression.

Next, we have the domain expression. The domain expression is a logical expression that specifies what set of objects we will be testing against the criteria of the predicate. We have some property P and some set of objects D, and we want to claim that either some or all of the objects in D have the property P. The way we express D is using the domain expression. For example, if we wanted to say that the domain of objects we are testing is the set of all integers then we might write $x \in \text{Int}$ in the domain expression. The general form is $x \in D$, where D is any arbitrary domain (i.e. any arbitrary set).

Finally, we have the predicate expression. As you may recall, the predicate expression can be any arbitrary well-defined logical expression containing at least one predicate. This is typically written using the notation of classical logic, with perhaps some set theory and set builder notation mixed in. Combining all three parts together into a complete quantified expression, the general format will look something like this:

quantifier (domain expression) (predicate expression)

For example, suppose we had two criteria we wanted to simultaneously test against: (1) being rusty and (2) being a barrel. In computer science, we would likely express these condition tests as something like `IsRusty(x)` and `IsBarrel(x)`. However, in math we would probably use something more like $R(x)$ and $B(x)$, so that's what we'll do here. These are our predicates. Suppose furthermore that W represents the set of all objects in the world, and that this is our intended domain. Suppose we then wanted to claim that at least one object in the world is a rusty barrel. This is what the quantified

---

[6]I graduated college as a computer science major, but before that I almost finished a degree in math before I switched majors. I've also studied both subjects a lot in my free time though. Lifelong learning is a big part of my philosophy of life.

expression for this would look like:

$$\exists(x \in W)(R(x) \land B(x))$$

This expression says "For some object in the world, that object is rusty and is a barrel.". The parentheses convention varies according to author. The one I have chosen here is a bit non-standard. I have chosen to surround each logically distinct part with parentheses (except the quantifier), because doing so makes it easier to see the underlying 3-part structure of each quantified expression more clearly than otherwise. Anyway, given that there are plenty of metal barrels in the world, many of which are not rust resistant and will be exposed to water (hence oxidize), this statement is true. What would happen if we changed the existential quantifier ($\exists$) to a universal quantifier ($\forall$) though? The expression would look like this:

$$\forall(x \in W)(R(x) \land B(x))$$

In contrast to the existential expression, this expression says "For all objects in the world, that object is rusty and is a barrel.". Clearly this is false. If it were true then we'd be living in a "rusty barrel world", where every object in the world would be a rusty barrel. It would be like the big oil companies were put in charge of decorating the planet on Earth Day. It would be an absurdity. So clearly, the choice of which quantifier you use in a quantified expression is very important to get right.

It's usually not a very hard choice to make though. All you have to do is consider whether or not you intend the statement to apply to *some* or to *all* of the domain. See? Easy. The basics are quite simple to understand. There are also some more advanced topics, but I will be ignoring almost all of them. I don't want to linger too much on predicate logic. I'm just going to cover a few more topics for it and then we'll be moving on to the next branch of logic. We have better things to do, ultimately.

The next thing I want to mention is the fact that quantified expressions can be nested inside each other. Part of the reason why this is possible is because of the fact that the predicate expression requires only that it be a logical expression that contains at least one predicate. Any quantified expression contains at least one predicate, and thus is itself also eligible for inclusion in the predicate expression of another quantified expression.

This is the underlying structural basis for nesting of quantifiers. For example, suppose we wanted to claim that all people in the world are married. Let $M(a, b)$ represent that a and b are married. Since marriage is a symmetric relation, notice that whenever $M(a, b)$ is true then so is $M(b, a)$. Let P represent the set of all people. A quantified expression for this statement would look like this:

$$\forall(x \in P)(\ \exists(y \in P)(M(x, y))\ )$$

Notice that $\exists(y \in P)(M(x, y))$ is both the predicate expression of the outer $\forall$ expression as well as its own quantified expression, which is part of why nesting is justified, just as we discussed. There's also something else I want you to notice here though. Imagine, for a moment, just as a thought experiment, that $\exists(y \in P)(M(x, y))$ wasn't nested inside another quantified expression, but instead stood on its own, like so:

$$\exists(y \in P)(M(x, y))$$

Would this expression be able to exist outside of the outer ∀ expression that binds to x? Is this expression well-defined? The answer, it may surprise you, is yes. This expression is well-defined, even with the x hanging loose like this. You see, when the x here is left dangling like this it becomes what is known as a *free variable*, which is to say that it acts like a constant, like a specific predefined value, one that we may or may not know the value of depending on the circumstances. In contrast, when a variable is not a free variable, such as when it is connected to a quantifier in the surrounding expression, then it becomes what is known as a *bound variable*. Bound variables are just dummy variables used as labels to sweep out the portion of the set that the quantifier applies to.

The fact that x in $\exists (y \in P)(M(x, y))$ is treated as a free variable, i.e. as if it is some predetermined value for the purposes of evaluating $\exists (y \in P)(M(x, y))$, is an important factor in why the nested quantification works. Expressions in the innermost parentheses in math are evaluated first, but if $\exists (y \in P)(M(x, y))$ were not well-defined, due to x lacking a quantifier, then this would make the expression invalid, thereby making it unable to nest. You can't nest an error inside of an otherwise valid surrounding expression.

The fact that $\exists (y \in P)(M(x, y))$ *can* stand on its own, and still have a valid meaning, is precisely what makes it possible for the innermost expression $\exists (y \in P)(M(x, y))$ of the larger expression $\forall (x \in P)(\exists (y \in P)(M(x, y)))$ to be evaluated. In fact, if you think about it, having $\exists (y \in P)(M(x, y))$ treat x as some predefined value (a constant) is exactly what you'd need to make this system work.

The outer ∀ expression essentially keeps reassigning x to different values to evaluate, which the inner ∃ expression is unaware of but doesn't need to know about anyway, because from its perspective x is just a free variable, just a constant. Having explained this, you should now have a more complete grasp of why nesting of quantifiers is actually justified, and why the concepts of "bound variables" and "free variables" both need to exist in predicate logic.

Notice that $\forall (x \in P)(\exists (y \in P)(M(x, y)))$ is interpreted as "For all people, there exists at least one other person that they are married to.", i.e. that each person is married to at least one other person. This is a correct implementation of what we were trying to say. What would happen if we messed with the quantifiers some though? Consider the following three cases and ask yourself how their meaning might differ:

1. $\forall (x \in P)(\forall (y \in P)(M(x, y)))$

2. $\exists (x \in P)(\forall (y \in P)(M(x, y)))$

3. $\exists (x \in P)(\exists (y \in P)(M(x, y)))$

All of these three expressions have vastly different meanings than the original nested quantified expression we were exploring. The first case, $\forall (x \in P)(\forall (y \in P)(M(x, y)))$, means that all people are married to all other people, i.e. that given any two people selected at random from the whole world a marriage relationship will always exist between them. It would essentially be maximal polygamy. Clearly this is false.

The second case, $\exists (x \in P)(\forall (y \in P)(M(x, y)))$, means that there exists at least one person in the world who is married to everyone else. This is not as extreme as the previous case, but is still obviously false. It would be like everyone in the world fell in love with the same person, or were all forced to marry that person, or something like that. It would be the ultimate "alpha male" or "alpha female" scenario.

Finally, the third case, $\exists(x \in P)(\exists(y \in P)(M(x, y)))$, means that there exists at least one couple in the entire world that is married. This is clearly true. In fact, of all four of the variants of the marriage example that we have discussed, this last one is the only one that is true in the real world. The only other plausible sounding case was $\forall(x \in P)(\exists(y \in P)(M(x, y)))$, but even that case was clearly false, due to the fact that some people are single.

This exercise clearly demonstrates that great care must be taken in how you nest quantified expressions inside each other, otherwise you may end up making a claim that is very different from the one that you intended. It's not very difficult once you get used to it, but people who are new to predicate logic do sometimes get the order and choice of quantifier wrong. It's something that you need to be vigilant about.

Next, I'd like to discuss just a couple of laws that apply to predicate logic. As you may recall, in standard classical logic we discussed how certain classical forms are equivalent to certain other classical forms, such that either form may be transformed into the other without changing the corresponding classical truth value. In classical logic, laws are often proven via testing truth tables for tautology (i.e. for having all true outputs). A collection of truth tables for some of the classical tautologies we discussed back then can be found on page 100.

For instance, in classical logic we have De Morgan's laws, of which there are two, namely disjunction negation and conjunction negation. As you may recall, the law of disjunction negation has the form $\neg(A \lor B) \leftrightarrow \neg A \land \neg B$. Likewise, the law of conjunction negation has the form $\neg(A \land B) \leftrightarrow \neg A \lor \neg B$. By the way, in transformative logic we would instead write these as $\neg(A \lor B) \leftrightarrow \neg A \land \neg B$ and $\neg(A \land B) \leftrightarrow \neg A \lor \neg B$, if we wanted to think purely in terms of forms instead of in terms of binary truth values.

Well, as it happens, a similar pair of laws applies to predicate logic, namely existential quantifier negation and universal quantifier negation. Notice how in De Morgan's laws the act of applying the negation to each parenthetical expression causes the logical operators inside the parenthetical expression to flip and for each term to be negated. The same principle applies to existential quantifier negation and universal quantifier negation. Here are the quantifier laws:

1. **Existential Quantifier Negation**: $\neg\exists(x \in D)(P(x)) \leftrightarrow \forall(x \in D)(\neg P(x))$

2. **Universal Quantifier Negation**: $\neg\forall(x \in D)(P(x)) \leftrightarrow \exists(x \in D)(\neg P(x))$

Notice that both of these laws make perfect intuitive sense. In the case of existential quantifier negation, saying that there does not exist any x in D with property P is clearly the same as saying that for all x in D the property P does not hold. Likewise, in the case of universal quantifier negation, saying that not all x in D have property P is clearly the same as saying that there exists at least one x in D that doesn't have the property P.

This similarity between the behavior of De Morgan's laws and the quantifier laws is no mere coincidence. In fact, in a certain sense, De Morgan's laws and the quantifier laws are actually the *same* laws. The law of existential quantifier negation and the law of disjunction negation are based on the exact same underlying principle. You can think of existential quantifier negation as actually just being an application of the law of disjunction negation, or vice versa. The same applies to the law of universal quantifier negation and its relationship to the law of conjunction negation. That might seem a bit confusing, so let me explain.

Suppose that we know what all of the objects in the domain set D are, and that there are exactly n of them, such that we can list them. In that case, let $D = \{d_1, d_2, d_3, \ldots, d_{n-1}, d_n\}$. We know that $\exists (x \in D)(P(x))$ has the effect of claiming that *at least one* of the objects in D has the property P. Likewise, we know that $\forall (x \in D)(P(x))$ has the effect of claiming that *all* of the objects in D have the property P. Furthermore, remember that predicates are just functions that return classical truth values based on their input objects. In that light, consider the net effect of the following expressions:

1. $P(d_1) \vee P(d_2) \vee P(d_3) \vee \ldots \vee P(d_{n-1}) \vee P(d_n)$

2. $P(d_1) \wedge P(d_2) \wedge P(d_3) \wedge \ldots \wedge P(d_{n-1}) \wedge P(d_n)$

Isn't it true that expression #1 above returns true only when *at least one* of the objects in D satisfy P? And likewise, isn't it true that expression #2 above returns true only when *all* of the objects in D satisfy P? Indeed it is. And that, in fact, is the nature of the relationship between De Morgan's laws and quantifiers. In other words, the following two statements are true:

1. $\exists (x \in D)(P(x)) \leftrightarrow P(d_1) \vee P(d_2) \vee P(d_3) \vee \ldots \vee P(d_{n-1}) \vee P(d_n)$

2. $\forall (x \in D)(P(x)) \leftrightarrow P(d_1) \wedge P(d_2) \wedge P(d_3) \wedge \ldots \wedge P(d_{n-1}) \wedge P(d_n)$

That is to say, existential quantifications are equivalent to chains of disjunctions, and universal quantifications are equivalent to chains of conjunctions. What happens when we negate an entire chain of disjunctions or conjunctions though? That's right, De Morgan's laws. That's what happens. Thus, naturally, whenever a quantified expression is negated, the quantifier and the predicate expression in a quantified expression will flip in exactly the same manner as if you had applied one of De Morgan's laws to it.

In fact, if you wanted to implement the existential and universal quantifiers in a computer program, then the most straightforward way to do it would be to just iterate over the domain set while applying logical disjunctions or conjunctions to each predicate test of each member of the domain set. You would use a single boolean variable to accumulate the work-in-progress truth value of the quantified claim, while repeatedly assigning the boolean variable to be equal to the disjunction or conjunction of itself and the predicate truth value of the next element in the set, until the whole set has been evaluated in this way.

For nested quantified expressions, you would just do the same thing, except that each inner quantified expression would be an inner loop whose free variables have been assigned by the outer loop. The resulting truth value would tell you whether the overall quantified expression is in fact true or false. You could also do non-standard variants of quantifiers, such as "at least 5 objects have the indicated property" by simply using counters instead of boolean accumulation, or whatever other techniques may apply.

Anyway, that covers negation. There's just one more law for quantifiers that I want to discuss, and that's what to do if the domain set of a quantified expression is empty. Those of you who have already studied predicate logic will recognize this as a lead-up to the concept of vacuous truth. According to vacuous truth, any universally quantified expression whose domain is the empty set will be true. However, there is actually a wrinkle in that preconception, as it turns out.

You see, while it is true that most logicians and mathematicians accept vacuous truth, from a logical standpoint there is actually a strong argument to be made that perhaps one shouldn't. True, there is a set of relationships in classical predicate logic that allow you to argue that universally quantified expressions over empty domains should be true, but on the other hand there is also a strong argument that the result should be something else entirely (perhaps an undefined value, empty form, or empty set).

Before we discuss these two possible ways of handling empty domains however, we will first need to expand our vocabulary a bit. You see, up until now we have always been working with quantified expressions that restrict their bound variables to certain sets. However, there is another way in which quantified expressions are sometimes used which does not require the specification of any domain sets. This alternative way of creating quantified expressions requires only the quantifier, the variable bindings, and the predicate expression. No domain set is needed. The format looks like this:

quantifier (variable bindings) (predicate expression)

Notice that the second part now has a different nature than it did when it was a domain expression. Most treatments of predicate logic are sloppy and fail to point this out. The question now is this: How does this method of creating quantified expressions differ from the method that used domain sets? As it happens, the two methods are indeed quite similar in terms of interpretation and behavior, *but they are not the same*. Therein lies the catch. Many people overlook this. Let me explain. I am well aware that such an assertion potentially goes against the standard preconceptions. Anyway, before we continue the discussion, here's a couple of definitions that we'll need:

Definition 116. *Whenever a quantified expression contains a domain expression, then we may refer to that quantified expression as a **domain-bound quantified expression***.

Definition 117. *Whenever a quantified expression doesn't contain a domain expression, and instead contains only variable bindings where the domain expression would otherwise be, then we may refer to that quantified expression as a **domain-free quantified expression***.

If we wish to refer to all kinds of quantified expressions, rather than specifically to one type, then we will simply use the all-encompassing term "quantified expression". Having now clarified how we will be referring to these different approaches, let's continue the discussion. As such, the obvious question to ask now is: How do domain-free quantified expressions work? We already understand domain-bound quantified expressions, having already used them, but now we need to understand domain-free quantified expressions.

The key idea behind making a domain-free quantified expression work is this: Translate the domain criteria into terms of the predicate expression, so that it can now be expressed inside of the predicate expression instead. In addition to this translation, the universal set will now be treated as the implicit domain set, such that all domain-free quantified expressions effectively range over the entire universe of all objects in the entire logical system. For example, in computer science, this would be analogous to iterating over every single data field or object in the entire program every time you perform a search, instead of only iterating over one specific array of data.

It turns out that there is a way of translating the domain constraints into the predicate, so that the quantified expression *seemingly* has the same meaning as it did before. The translation is different depending on whether the domain criteria are part of an existential or universal quantification. In the case of existential quantification, the domain is rewritten as a conjunction with the predicate expression. In contrast, in the case of universal quantification, the domain is rewritten as a material implication with the domain as the condition and the predicate expression as the consequence. As such, the translation from domain-bound to domain-free looks like this:

1. $\exists(x \in D)(P(x)) \approx \exists(x)((x \in D) \land P(x))$

2. $\forall(x \in D)(P(x)) \approx \forall(x)((x \in D) \to P(x))$

The symbol $\approx$ here means "approximately equal" or "similar". The translation is imperfect. You will see why later. Notice that $x \in D$ is now being used as a predicate. Well, even in the domain expression it was always still sort of already a predicate, in a manner of speaking. Technically, in a domain-bound quantified expression, the domain expression is both a variable binder and a predicate, in typical sloppy mathematics fashion. However, when $x \in D$ was in the domain expression it was restricting the scope of the search space and now it isn't. That's important. It changes the meaning in a subtle way.

By the way, we could just have easily have written $x \in D$ as $D(x)$, if we wanted it to visually resemble a typical predicate more clearly. However, regardless of whether a predicate is written in infix notation (such as in $x \in D$) or in function notation (such as in $D(x)$) it is still the same predicate. For clarity though, here's what it would look like if we used this alternative notation:

1. $\exists(D(x))(P(x)) \approx \exists(x)(D(x) \land P(x))$

2. $\forall(D(x))(P(x)) \approx \forall(x)(D(x) \to P(x))$

Anyway, here's an interesting question for you: What happens when we expand these quantified expressions out to their equivalent forms as simple chains of disjunctions or conjunctions? Will the domain-bound and domain-free quantified expressions behave the same way? For instance, will $\exists(x \in D)(P(x))$ expand to the same thing as

$\exists(x)((x \in D) \wedge P(x))$? Likewise, will $\forall(x \in D)(P(x))$ expand to the same thing as $\forall(x)((x \in D) \to P(x))$? The answer is *no*.

The domain-free variants implicitly operate on the entire universal set, and thus will expand to chains of disjunctions or conjunctions on the entire universal set. In contrast though, the domain-bound variants will only expand to chains of disjunctions or conjunctions on the domain set. This will change how they react to the presence of an empty domain.

A domain-bound quantified expression with an empty domain will expand to *nothing*. In contrast, a domain-free quantified expression with an empty domain will expand to either a disjunction of $(x \in D) \wedge P(x)$ or a conjunction of $(x \in D) \to P(x)$, applied to every object in the entire universe. Add into this mix the paradoxical way that material implication handles false conditions and you have yourself a recipe for the possibility of different return values.

In the case of existential quantification, it is clear that we should interpret a quantified expression over an empty domain as false, seeing as an empty set cannot possibly contain any existent objects. The expansion may be different, but the interpretation either way is essentially the same. However, this is not true of universal quantification. A universally quantified expression will return true if it is domain-free (hence vacuous truth), but will behave quite differently if it is domain-bound. To see why, consider domain D. Suppose that $D = \{d_1, d_2, d_3 \ldots, d_{n-1}, d_n\}$, as before. Think back to what the expansions into disjunctions and conjunctions looked like. They looked like this:

1. $\exists(x \in D)(P(x)) \leftrightarrow P(d_1) \vee P(d_2) \vee P(d_3) \vee \ldots \vee P(d_{n-1}) \vee P(d_n)$

2. $\forall(x \in D)(P(x)) \leftrightarrow P(d_1) \wedge P(d_2) \wedge P(d_3) \wedge \ldots \wedge P(d_{n-1}) \wedge P(d_n)$

What if n is zero though, as it would be if D were the empty set? What would these expansions look like then? That's right, they'd be empty. There wouldn't be even a single term in the expression in that case. No predicates, no disjunctions or conjunctions, just emptiness of some kind. Thus, the correct return value for domain-bound quantified expressions over empty domains would seem to be an undefined value, empty form, or empty set of some kind. Notice that this behavior aligns better with existence-based truth, transformative logic, and set theory than it does with classical logic. More on that later in the book though.

In contrast though, notice that domain-free quantified expressions certainly do not expand to empty expressions like this. Instead, domain-free quantified expressions expand into absolutely massive chains of disjunctions or conjunctions that span the entire set of all objects in the entire logical universe under consideration. The paradoxical behavior of material implication ("from falsehood, anything follows"[7]) thereby causes domain-free universal quantifications to return true.

Thus, we have a conflict in definitions. One set of relationships implies that a universal quantification over an empty set should return some kind of undefined value, empty form, or empty set, whereas another set of relationships implies that a universal quantification over an empty set should return true. Consequently, the conventional

---

[7] a.k.a. the paradox of entailment, the principle of explosion, or the Latin phrase "ex falso quodlibet"

view that vacuous truth is necessarily the correct way of interpreting such cases cannot be trusted. Vacuous truth for quantifiers is a popular belief, but it is not justified. We have a dilemma. However, it is not an unsolvable dilemma. One of these interpretations is wrong, and the other is right. Once you've seen the alternative, and you realize that material implication is the wrong way of thinking about implication, you'll see that.

In fact, unified logic, the system of logic we are working our way up to, has the property of vacuous *falsehood*. This is the opposite property of vacuous truth. Now isn't that interesting? Did you notice that the domain-bound quantified expression returned emptiness, and that in an existence-based truth system empty sets are the same thing as falsehood? That's not just some random coincidence. Interesting insights await us. More about that later though. For now, let's refocus our attention on predicate logic.

There's one more point I want to drive home about this distinction between the domain-bound and domain-free approaches to quantification before we move on. In particular, I want you to consider this: What happens if the universal set itself is empty? How would a domain-free quantified expression respond to that? Remember, all domain-free quantified expressions implicitly use the universal set as their domain. What if it's empty though? What truth value would a domain-free quantified expression be able to return in that case? Would vacuous truth still work then, even in classical logic? Could it still legitimately return true? The answer is *no*.

In other words, the poorly defined nature of vacuous truth is *inescapable*. If vacuous truth were truly the correct approach, as most logicians and mathematicians currently seem to believe, then why does it fail catastrophically when the universal set itself is empty? A domain-free universally quantified expression operating in the context of an empty universe has no choice but to return an undefined value, empty form, or empty set. Why? Well, if the universe is empty, then the expansion of a domain-free universally quantified expression into its equivalent chain of conjunctions can't be anything other than empty. Isn't that interesting? Doesn't that support the idea that vacuous truth is wrong, and in fact returning an undefined value, empty form, or empty set is more rational?

By the way, the existential quantifier also has these same parallels in how it behaves, but the difference is that in the case of existential quantifiers the way the domain-bound and domain-free cases are interpreted is in harmony rather than in conflict. A domain-free existential quantified expression over an empty domain will expand to a disjunction involving all objects in the universe, all of which will evaluate as false since none of them are members of the domain of interest, resulting in a final return value of false. On the other hand though, a domain-bound existential quantified expression over an empty domain will expand to nothing, just like a domain-bound universally quantified expression over an empty domain would.

However, our thought experiments about existence-based truth from earlier in the book (see *void falsehood* on page 58) clearly indicate that empty sets themselves should be thought of as implying falsehood. Thus, in both the domain-bound and domain-free cases, existential quantification over an empty domain should be interpreted as false, whereas universal quantification faces a dilemma. However, notice that the existential and universal quantification over empty sets both gravitate towards vacuous falsehood in this way, rather than towards vacuous truth, thus perhaps suggesting the possibility of a more uniform framework of thought than what classical logic has to offer, attainable if we embrace existence-based truth and treat sets themselves as embodying truth rather than merely using crude binary truth values.

Anyway, we're almost done with predicate logic now. I want to cover just two more special points of interest. The first point is that predicate expressions don't have a truth value until the corresponding domain of objects has been given definite values. Consequently, the laws of predicate logic work a bit differently than in a propositional logic (where no quantifiers exist).

This gives predicate logic a more context-dependent character, and generally causes the set of out-of-context laws provable for it to be narrower than for a propositional logic, relatively speaking. In this sense, some predicate logic expressions can be thought of as not yet having truth values, which is part of what gives predicate logic a slightly non-classical character.

Secondly, I want you to notice that classical predicate logic is essentially a way of jury-rigging set theory onto classical logic, i.e. of augmenting standard classical logic with the ability to talk about sets somewhat. It's like duct-taping part of set theory onto classical logic. Prior to the introduction of quantifiers, classical logic could only talk about statements, and not about the objects participating in those statements. In this way, classical predicate logic is kind of a hybrid of standard classical logic and set theory.

Not being able to talk about sets of objects is a crippling limitation, one that classical logic had to overcome in order to become broadly applicable to the multitude of different kinds of systems one might want to logically model. Hence, in practice, quantifiers are a necessity. Why though? If standard classical logic is supposed to be the proper foundation for logic, as it is traditionally seen by many in logic and mathematics, then why does it almost immediately fail to be sufficiently expressive, such that it must be augmented with set theory (via quantifiers) lest it become nearly useless in many situations?

Do you remember how I said that $\mathcal{U}$ and $\varnothing$ could emulate true and false within set theory (see page 149), without any need to rely on concepts from classical logic itself? Well, to that point, have you considered the possibility that maybe the reason why standard classical logic gravitates so desperately towards needing to be augmented with set theory (via quantifiers) is because it may actually be a degenerate representation of something already expressible directly in set theory itself?

Combine this thought with our insights into the close parallels between the laws of classical logic and the laws of set theory related to $\mathcal{U}$ and $\varnothing$ (see the table on page 147) and the concept of existence-based truth (wherein truth is a property of *sets of objects* rather than of true-false statements) and a realization might be beginning to dawn on you.

I just figured I'd pose that thought experiment to you as a teaser. It's just something for you to think about. I'll have more to say about it later, during the main discussion of unified logic. All the connections will start coming together into a more cohesive whole once we get to that part. For now though, it is time to move on to a completely different subject. It is time to discuss the next branch of non-classical logic: multi-valued logic.

4.4 Multi-valued logic

As we know, standard classical logic is a binary (a.k.a. "bivalent") logic, meaning that it has only two truth values: true and false. However, what would happen if it weren't? What would happen if we added on more truth values to the system? The answer is that we would get what is known as a *multi-valued logic*. The term "many-valued logic" is also sometimes used instead, in place of "multi-valued logic", but I personally think that "many-valued logic" is perhaps a poorer choice of words. Whether or not there are "many" truth values in a system of logic (as opposed to "few") is a subjective and often irrelevant consideration. This subjective quality may not be the intended connotation of the phrase, but that's what it sounds like from a descriptive standpoint.

Thus, I personally prefer the term "multi-valued logic" instead of "many-valued logic". The term "n-valued logic" would also be a plausible contender to refer to this category of logics, but usually people fill in the n in "n-valued logic" with the actual number of truth values available to the specific system of multi-valued logic in question. On the other hand though, "multi-valued logic" and "n-valued logic" could both technically still refer to 2-valued logics, from the standpoint of descriptive connotation, which also is confusing in its own way I suppose. It's a bit of a subjective choice. Pros and cons. Pick your poison. As for me, I've chosen to use "multi-valued logic" in this book. Anyway though, putting aside these terminology issues, let's now shift our focus to the substance of the discussion.

Both truth-functional and non-truth-functional multi-valued logics exist. However, the literature on multi-valued logic seems to more often focus on the truth-functional multi-valued logics, to the point where anytime someone says "multi-valued logic" without further qualification then it is most likely that they are talking about truth-functional multi-valued logic. Some definitions and articles giving an overview of multi-valued logic may even completely omit the non-truth-functional case.

However, both cases definitely exist. In fact, transformative logic (as you have seen) and unified logic (as you will see) can both be considered as examples of non-truth-functional multi-valued logics, as can some other systems. Keep this in mind when reading the literature. Multi-valued logic isn't actually restricted to being truth-functional, although being truth-functional is more common in the literature and is more stereotypical.

Multi-valued logics are free to define their truth values and operations completely arbitrarily, even to the point of potentially having no coherent intuitive interpretation in terms of real truth. However, most well-known multi-valued logics strive to have some kind of coherent truth-like behavior. Like standard classical logic, all truth-functional logics have the unusual attribute of being able to have all of their operations expressed completely in terms of truth tables. Therefore, naturally, all truth-functional multi-valued logics (hence perhaps the majority of the literature labeled as "multi-valued logic") share this property as well. Most of our discussion in this section will focus on well-known examples of these truth-functional multi-valued logics, since that is where the bulk of the literature seems to currently reside.

As such, let's begin. Let's start with something that's easy to understand for someone who has a background in standard classical logic. By doing so, it should be easier to relate to the principles behind it, at least for people who have been exposed to the basics of standard classical logic. To this end, I can think of no more fitting example than Kleene's 3-valued logic, which is perhaps the most widely known and most classically intuitive of all multi-valued logics.

The basic principle of Kleene's 3-valued logic is simple: add on a third truth value that indicates "unknown" and then make the resulting truth tables adhere to classical principles as much as possible. We will represent this third truth value as U, and it will be read as "unknown". When we say that U represents an unknown truth value, what we mean is that it could be either true or false but we don't know which. Not all representations of uncertainty in multi-valued logics behave like this. Anyway though, here are the truth tables:

Kleene's 3-Valued Logic

A	$\neg A$
F	T
U	U
T	F

$A \vee B$		B		
		F	U	T
	F	F	U	T
A	U	U	U	T
	T	T	T	T

$A \wedge B$		B		
		F	U	T
	F	F	F	F
A	U	F	U	U
	T	F	U	T

$A \rightarrow B$		B		
		F	U	T
	F	T	T	T
A	U	U	U	T
	T	F	U	T

$A \leftrightarrow B$		B		
		F	U	T
	F	T	U	F
A	U	U	U	U
	T	F	U	T

Notice that I used a different table format than I usually do, except for in the negation truth table. Using this alternative table format for conjunction, disjunction, implication, and equivalence here makes them much easier to read. Doing so also makes it easier to see the underlying behavior and structure of these operators. Notice, however, that this special table format would only work for operators with exactly two inputs. In contrast, operations with exactly one input or with three or more inputs are typically best written in the more traditional table format (i.e. using parameterized columns, as in the $\neg A$ truth table above, or as in the classical truth tables on page 99).

Anyway, do you see what the reasoning was for filling out the truth tables like this? As you may recall, I said the idea behind Kleene's 3-valued logic was to extend standard classical logic with a third truth value for "unknown" but to otherwise behave as classically as possible. This is exactly what these truth tables do. Basically, these truth tables are exactly what you'd get if you asked yourself what you were still able to infer about an expression in classical logic if one or more of the inputs were unknown. Thus, for example, if we had $A \vee B$ and we knew A to be true but didn't know the value of B, then we could still nonetheless conclude that $A \vee B$ must be true, because any classical disjunction with at least one true input will always be true regardless of the other inputs.

Likewise, all the other values in the tables are defined according to similar reasoning.

From a classical perspective, these truth tables seem very reasonable. It's easy to think that this extension to standard classical logic would be good to have. Why then is it not used much? Why not use Kleene's 3-valued logic instead of standard classical logic? Kleene's 3-valued logic seems capable of more, so why not use it instead? Well, it turns out that there is a catch. Kleene's 3-valued logic is not as reasonable as it may seem at first glance. There are several problems with it.

One of those problems is its inability to handle tautological and contradictory forms correctly. Two of the most glaring examples of this are the law of the excluded middle and the law of non-contradiction. Take the law of the excluded middle for example, which has the form $A \vee \neg A$. Suppose that A is unknown, i.e. is U. What does the expression $A \vee \neg A$ resolve to in this case? Well, it resolves to U. Think about that though... Why would it resolve to unknown? Is the value of $A \vee \neg A$ really unknown? Not really, because $A \vee \neg A$ is logically guaranteed to be true even if we don't know the value of A. Similarly, $A \wedge \neg A$ would also evaluate to U if A were unknown, even though we know that it should always be considered false regardless.

Notice that this same bad behavior would happen irrespective of how you specified the tables. This is an inherent problem with *all* truth-functional logics, not just with Kleene's 3-valued logic. Truth-functional logics lack context awareness. This is the price you pay for being able to just plug in truth values for the inputs and mindlessly return the corresponding output values. No such logic can ever understand context. This is a high price for convenience, if you think about it.

The underlying reason why $A \rightarrow B$ can be true in standard classical logic even if A and B are totally unrelated, and also why $A \wedge \neg A$ can be treated as unknown even though it clearly must always be false, are ultimately one and the same. Both standard classical logic and Kleene's 3-valued logic are truth-functional, and therefore are inherently incapable of ever understanding context. Both systems actually have the same fundamental problem, it's just that Kleene's 3-valued logic makes it more readily apparent.

In contrast though, solving logical problems in real-life very often requires context sensitivity. This is a dark omen for the idea of truth-functional logic. It essentially proves that no truth-functional logic can ever be fully general, except in certain specific situations possessing suitable characteristics via augmentations (e.g. in conditionals in programming, which don't use material implication and which have context awareness by virtue of having alterable flow of control that allows passing between different contexts and freely modifying data and hence full computational generality).

This lack of context sensitivity is the core fundamental weakness of most multi-valued logics (which, as you may recall, tend to be truth-functional in the literature). It's why standard classical logic and other truth-functional logics often feel so disconnected and arbitrary, as if they don't even care whether or not any genuine relationships exist between the participants in an expression.

If a logic is expressed entirely in terms of truth tables, then how could it ever be aware of the structure of the relationships between entities? It couldn't. All it could ever know is how to map simple input truth values to simple output truth values. Notice that nothing in a truth table ever gives any indication of the nature of the participants, and thus it couldn't possibly ever take such natures into account.

Nevertheless, as futile as all truth-functional logics may ultimately be for solving this problem, let's explore some more examples anyway so that we may acquire a better sense for the diversity and arbitrariness of multi-valued logic. Kleene's 3-valued logic is not the only attempt at a 3-valued logic of unknowns. There are in fact numerous different ways of defining what an "unknown" value should be and how it should behave.

The two other most common approaches to augmenting standard classical logic with a third truth value designed to indicate an "unknown" of some kind are Bochvar's logic and Łukasiewicz's 3-valued logic. They both behave classically with respect to expressions that only deal with true and false values, but the way they treat unknowns is different than what you'd expect in classical reasoning. Of the two, let's look at Bochvar's logic first.

Luckily, Bochvar's logic is very easy to understand. In fact, it is arguably even easier to understand than Kleene's 3-valued logic is, even though Kleene's 3-valued logic is the most classically intuitive of the three. All you have to do for Bochvar's logic is to evaluate all of the truth tables the same way as you would for standard classical logic, except for that if even a single "unknown" value occurs anywhere in an expression then the whole expression is automatically treated as "unknown". Thus, the truth tables for Bochvar's logic look like this:

Bochvar's 3-Valued Logic

A	$\neg A$
F	T
U	U
T	F

$A \vee B$		B		
		F	U	T
	F	F	U	T
A	U	U	U	U
	T	T	U	T

$A \wedge B$		B		
		F	U	T
	F	F	U	F
A	U	U	U	U
	T	F	U	T

$A \to B$		B		
		F	U	T
	F	T	U	T
A	U	U	U	U
	T	F	U	T

$A \leftrightarrow B$		B		
		F	U	T
	F	T	U	F
A	U	U	U	U
	T	F	U	T

As you can see, this makes the truth tables for Bochvar's logic very easy to remember. All you have to remember to do is to always evaluate expressions having at least one U as input as always having U as output, but otherwise just apply the rules of standard classical logic as usual, so overall it adds almost nothing to your memorization burden compared to standard classical logic. One thing you might be wondering though, is why would someone want truth tables that behave like this? Well, from what I understand, the idea is to treat U as a kind of indeterminate value for which any conclusion is unsafe.

In other words, Bochvar's logic treats U like an error, and then propagates that error to everything that it touches. In computer science terms, it's kind of like throwing an exception, except that it can't be caught or handled in any way. Bochvar's logic is essentially an ultra-paranoid model of uncertainty. As such, perhaps "indeterminate" or "error" or "bad state" would be more apt than "unknown". However, I will continue using "unknown" here anyway, for the sake of comparison, as a thought experiment regarding how "unknown" may be interpreted in different ways.

Next is Łukasiewicz's 3-valued logic. Łukasiewicz's 3-valued logic is a bit more intuitively strange than either of Kleene's 3-valued logic or Bochvar's 3-valued logic. However, Łukasiewicz's truth tables for negation (¬), disjunction (∨), and conjunction (∧) are actually the same as for Kleene's 3-valued logic. Only the definitions of implication and equivalence differ. Here are the truth tables for Łukasiewicz's 3-valued logic, minus the truth tables it shares with Kleene's 3-valued logic:

Łukasiewicz's 3-Valued Logic

$A \to B$		B: F	B: U	B: T
A	F	T	T	T
A	U	U	T	T
A	T	F	U	T

$A \leftrightarrow B$		B: F	B: U	B: T
A	F	T	U	F
A	U	U	T	U
A	T	F	U	T

The only difference between Kleene's 3-valued logic and Łukasiewicz's 3-valued logic is that in Łukasiewicz's 3-valued logic $U \to U$ and $U \leftrightarrow U$ are both true, whereas in Kleene's 3-valued logic $U \to U$ and $U \leftrightarrow U$ are both unknown. The reason why this was done in Łukasiewicz's 3-valued logic is so that it will have truth table tautologies (i.e. expressions whose output value is always true). Kleene's 3-valued logic and Bochvar's 3-valued logic, in contrast, have no truth table tautologies whatsoever.

That being said, some rules of inference can still be applied to Kleene's 3-valued logic and Bochvar's 3-valued logic, it's just that truth tables aren't viable for determining laws anymore. For example, the order independence (a.k.a. commutativity) of disjunction, as given by $A \lor B \leftrightarrow B \lor A$, clearly still applies to both Kleene's 3-valued logic and Bochvar's 3-valued logic, even despite them not having any truth table tautologies.

Interestingly, from what I hear, Łukasiewicz's 3-valued logic is actually historically the first truth-functional multi-valued logic ever invented. Łukasiewicz invented it sometime around 1920. As such, truth-functional multi-valued logic is actually quite a young concept on the historical timescale, despite how simple of a concept it is. This may seem surprising, but then again, human beings do tend to take a long time to come up with new ideas sometimes.

The simplest principles are often the hardest to conceive of. Derivative ideas occur frequently, but truly original ideas sometimes take centuries or millenia to come to light, even though once the idea is known it may seem blatantly obvious. The human mind is much like fire kindling: passive and dull most of the time, yet explosively powerful and contagious the moment the right spark of insight strikes it.

Anyway, that covers most of what I want to say about the simplest multi-valued logics. If you're interested in seeing what a 4-valued logic looks like, with two different "unknown" truth values, representing "either true or false" and "both true and false" respectively, then lookup Belnap's logic. Various other less widely known multi-valued logics also exist. There are numerous different ways of defining how things might work.

In fact, remember that you could technically just fill out an input-output table with whatever arbitrary values you feel like, regardless of meaning, and it nominally could still be considered a multi-valued logic. If you made it so that your table is the only thing governing the output values (i.e. no context awareness), and made sure every possible input combination maps to an output, then your custom multi-valued logic would also be truth-functional. However, usually people prefer to design logics with more coherent meanings, because otherwise it wouldn't be much different from just having a bunch of arbitrary functions that don't mean anything in particular and can't really be used for reasoning.

Oh, by the way, one interesting thing I want to note is that some multi-valued logics have equivalent representations in terms of purely numeric values and numeric functions. For example, in Kleene's 3-valued logic, if you map F to 0, U to 1/2, and T to 1 (or any other similar mapping), then $A \vee B$ becomes equivalent to $\max(A, B)$, and $A \wedge B$ becomes equivalent to $\min(A, B)$, in terms of behavior[8].

Similar things happen to the other logical operators in Kleene's 3-valued logic. They can all be reframed into terms of equivalent numeric operations. However, whether these kinds of translations into numeric terms exist for a logic, and how they work, varies according to which logic you are talking about though, so be careful. You shouldn't necessarily assume that disjunction will always translate into the maximum function in every logic, for example.

Numeric representations of multi-valued logics are more than just an interesting coincidence though. A numeric approach sometimes enables a much more flexible system. Incidentally, all of the multi-valued logics we've been talking about so far have had a finite number of truth values. However, this is not necessary in the general case. Multi-valued logics are free to have infinitely many truth values if they wish. The two most widely known of such infinitely multi-valued logics are probably fuzzy logic and probabilistic logic. Both of these logics have infinitely many truth values, represented by numbers on the continuum.

Naturally, numeric truth values are a vastly easier way to give a logic infinitely many truth values than defining special symbols (e.g. F, U, T) would be. However, keep in mind that using numbers is not actually the only way to create an infinitely multi-valued logic. Set theory can also act as an infinitely multi-valued logic, even when it doesn't include numbers in the discussion. That being said, while set theory is objectively a multi-valued logic, it is often not conventionally viewed that way (yet). You'll understand the nuance of this point better by the end of the book. In the meantime, to lend credibility to the idea, recall our discussion of using $\emptyset$ to fill the role of F and $\mathcal{U}$ to fill the role of T, such as illustrated in the tables on page 147 and 149.

---

[8]Here, "min" means minimum and returns the lesser of the two input values. Likewise, "max" means maximum and returns the greater of the two input values.

Anyway, both fuzzy logic and probabilistic logic represent truth values as numbers between 0 and 1 (inclusive) on the continuum. Both logics also measure "uncertainty" in some sense. This causes fuzzy logic and probabilistic logic to sometimes be confused with each other. However, despite these superficial similarities, fuzzy logic and probabilistic logic are fundamentally different. They each measure a very different kind of "uncertainty". Fuzzy logic is for reasoning about vagueness (i.e. degrees of truth, e.g. a statement being somewhat true versus being very true, etc), whereas probabilistic logic is for reasoning about probabilities (i.e. about the chances of events occurring, etc).

Let's talk about fuzzy logic first. To understand fuzzy logic, it is best to first make an analogy to how set membership can be represented numerically, and then show how this notion can be extended to allow fuzzy or vague set membership. This involves using predicates that return numeric values instead of special truth symbols. For example, consider predicates (i.e. boolean truth functions) and set theory. Normally, we collect objects together into sets, and then we can say whether or not a given object is a member of that set.

However, we can also think of set membership purely in terms of predicates instead. For instance, we can say $S(x)$ to represent the claim that x is a member of set S, i.e. that it has the property S. We could then have $S(x)$ return the value F when it is false and T when it is true. However, we could instead use numbers for this. We could instead have $S(x)$ return 0 when it is false and 1 when it is true, and this would have the same net effect. Thus, instead of using an explicit list of members of the set S, we can just use the predicate $S(x)$ to independently test the membership of any given object x in S. Here's the interesting part though: What happens if we allow the return value of $S(x)$ to vary on the continuum between 0 and 1? The answer is fuzzy logic.

By allowing truth values that represent the degree of membership in a set, we have effectively made it possible to think in terms of vague truth. This is a useful trick, especially in computer programs, where control systems often have to evaluate conditions in a nuanced way in order to make optimal decisions[9]. If you don't like the way that standard classical logic tries to cram everything into only either strictly false or strictly true, but you still want a sane logical system that doesn't allow self-contradictions, then fuzzy logic is for you. Fuzzy logic is a rigorous conceptualization of vagueness. Unlike many other multi-valued logics, it is not very contrived. Fuzzy logic is broadly applicable to many real-world problems. This is unlike many other multi-valued logics, which in contrast were often designed arbitrarily, without truly sound principles.

The real world is full of things that may sometimes be difficult to effectively deal with in the black-and-white terms of strict true or false values. For example, suppose we want to say that an object is wet. How wet is wet? The atmosphere is full of water particles, so even a seemingly dry object may have a bit of water on it anyway. Does that make it wet? Is wetness merely the property of having water particles clinging to something? If so, where should we draw the line for what is wet and what is not? It wouldn't be very useful to say that something is wet if almost everything was considered

---

[9] In fact, if you're a programmer, there's a fair chance you have used fuzzy values like this without actually consciously realizing that "fuzzy logic" was the official term for what you were doing.

to be so by virtue of merely having a small number of water particles clinging to it.

To reason effectively with this concept of wetness, we'd benefit from thinking in vague terms. For instance, we could have wetness be a percentage of how much of an object's surface is covered in water, which would then be our fuzzy truth value. Thus, $W(x)$ might represent *how* wet an object is, as a number between 0 and 1, rather than merely a black-and-white evaluation of whether it is or is not wet.

We can then later use this fuzzy value to make black-and-white decisions if we need to. This fuzzy concept of wetness could also be applied to volumes, such as to measures of humidity. For example, a program controlling an air conditioning system in a data center might be programmed to activate a dehumidifying system once the amount of water in the air rises above 60%, i.e. whenever $W(x) \geq 0.6$ where x is the body of air[10].

Nonetheless, outside of fuzzy logic, you will sometimes see us use vague concepts in logic as if they had black-and-white interpretations. This is just for convenience. It isn't some kind of profound philosophical shortcoming when we do this, despite what some philosophers (who frequently lack much real science experience) seem to think. In logic and science, we often assume that the terms we are using implicitly have a precise underlying definition, but sometimes we just don't bother specifying it exactly, because it sometimes isn't important for making our point.

Thus, you might still see someone use "wet" as if it is a strictly true or strictly false concept, but you should know that when a logician or scientist says this they generally mean that there is some kind of implicit precise criteria that has just been left unspecified, e.g. that a wet object is any object that has at least 25% of its surface covered covered in water particles or whatever.

As rigorous and reliable as the basic principle underlying fuzzy logic is, there are a few aspects of fuzzy logic that are a bit more open to debate. In particular, it is perhaps not always clear what the most correct way of combining multiple different fuzzy predicate truth values into one logical expression should be. There are multiple different conceivable ways of doing it.

One of the most popular ways is to translate disjunction into the maximum function and conjunction into the minimum function, but this is not the only way. As such, let's compare and contrast a few of the possible systems. Let $P(x)$ and $Q(y)$ be arbitrary fuzzy predicates, i.e. functions returning numbers on the continuum between 0 and 1 (inclusive). Here are two of the most popular competing methods for defining the basic logical operators in fuzzy logic:

The Min-Max Method (a.k.a. "Zadeh Operators"):

1. $\neg P(x) \rightarrow 1 - P(x)$

2. $P(x) \wedge Q(y) \rightarrow \min(P(x), Q(y))$

3. $P(x) \vee Q(y) \rightarrow \max(P(x), Q(y))$

---

[10] I found a few websites that mentioned that a range of roughly 40% to 60% humidity is desirable for data centers.

The Product-Sum Method:

1. $\neg P(x) \to 1 - P(x)$

2. $P(x) \wedge Q(y) \to P(x) \cdot Q(y)$

3. $P(x) \vee Q(y) \to (P(x) + Q(y)) - (P(x) \cdot Q(y))$

In the product-sum method above, the justification for defining the disjunction as $P(x) \vee Q(y) \to (P(x) + Q(y)) - (P(x) \cdot Q(y))$ is based on an application of De Morgan's laws to the already existing definitions for the $\neg$ and $\wedge$ operators. Specifically, it exploits the law of conjunction negation combined with an additional negation. The reasoning is as follows: Whatever $P(x) \vee Q(y)$ should be, we'd like it to obey the law of conjunction negation, so that $P(x) \vee Q(y)$ should be equivalent to $\neg(\neg P(x) \wedge \neg Q(y))$. This expression $\neg(\neg P(x) \wedge \neg Q(y))$ uses only $\neg$ and $\wedge$, whose equivalents we have already defined. Thus, under this definition, $P(x) \vee Q(y)$ will be the same as $1 - ((1 - P(x)) \cdot (1 - Q(y)))$, which in turn can be rewritten as $(P(x) + Q(y)) - (P(x) \cdot Q(y))$. This justifies the definition.

This definition of disjunction will always yield numbers between 0 and 1 (inclusive), although it might not look like it at first glance. It is easier to see that this must be true by looking at the equivalent expression $1 - ((1 - P(x)) \cdot (1 - Q(y)))$ that we used during the derivation process. Since $P(x)$ and $Q(x)$ are both numbers between 0 and 1, $(1 - P(x))$ and $(1 - Q(y))$ must likewise always be numbers between 0 and 1. It's like reversing percentages (e.g. the complement of 30% is 70%).

Likewise, the final logical negation on the outside performs just yet another of these reversals. Thus, since $1 - ((1 - P(x)) \cdot (1 - Q(y)))$ and $(P(x) + Q(y)) - (P(x) \cdot Q(y))$ are equivalent, it follows that $(P(x) + Q(y)) - (P(x) \cdot Q(y))$ will always return numbers between 0 and 1 as well. The logical negation and conjunction operators will also always return numbers between 0 and 1, given that their inputs are also between 0 and 1. Thus, all of these "product-sum" operators do indeed stay within the 0 to 1 range, as required.

Anyway though, the obvious problem here is that we need to decide which system to use. Should we use the min-max method or the product-sum method, or perhaps even something else entirely? Which (if any) is the most rigorous and principled choice? Well, if you ask me, of the two, I think that the min-max method has more advantages and more conceptual correctness overall, even though there is still a weak argument that can be made in favor of the product-sum method in certain very limited respects.

There are multiple different ways that one might wish to account for fuzzy factors in reasoning, and hence there are potentially multiple valid techniques that one might wish to employ to do so. The min-max method was invented by Lotfi Zadeh, who also was the first to coin the term "fuzzy logic" apparently. I'm not quite sure about the origin of the product-sum method though.

The min-max method may seem oddly simple and arbitrary at first glance, but it actually behaves very well and has largely stood the test of time despite its early origin. I very much prefer it over the product-sum method. It usually behaves much better and is much easier to reason about and use. The min-max method is very useful for a diverse range of real-world applications.

Let's discuss the other option (the product-sum method) first though, so we can see why the ambiguity about which method should be used exists. Doing so will provide us with a better basis on which to later understand why the older min-max method is still much better overall, at least in most applications of fuzzy logic. Basically, the product-sum method has a few minor arguments in its favor, but it also has multiple severe weaknesses that weigh heavily against it. The min-max method is vastly more foolproof.

What may have motivated the creation of the product-sum method though, despite the fact that the min-max method already existed? Well, in particular, in logic, one might expect that a disjunction would become more and more likely to be true as you add more participants to it, and that a conjunction would become less and less likely to be true as you add more participants to it. The reason is because each participant in a disjunction can only provide *more* ways for the expression to be true, whereas each participant in a conjunction can only cause there to be *less* ways for the expression to be true.

However, while adding more participants to a logical expression under the min-max method does indeed cause it to be more likely to be highly true for disjunction and less likely to be highly true for conjunction, the min-max method does not vary continuously in value with every new participant added to a logical expression. Instead, the min-max method is only effected by the least and greatest participants, even though all participants are conceptually a part of the fuzzy truth being expressed by the entire collective expression.

One might find this fact troubling. Some may ask: Why should fuzzy operators behave in such an unresponsive manner to new inputs? One might suppose that even if a participant doesn't have the least or the greatest fuzzy truth value, it should still be able to influence the overall fuzzy truth of the encompassing expression at least somewhat. This is one of the primary objections against the min-max method that may cause someone to prefer the product-sum method instead. The return value of the product-sum method varies continuously with each added participant to a disjunction or a conjunction, instead of only changing if the least or greatest participants change.

Let's do a few examples to illustrate the point. For instance, suppose that we have some object x, and we want to test how well x fits the criteria expressed by some fuzzy predicates named P and Q. Suppose we calculate that $P(x) = 0.50$ and $Q(x) = 0.50$, and we want to evaluate the conjunction of these, i.e. we want to know that if $P(x)$ and $Q(x)$ are each 50% true separately then how true are they *jointly*?

In this case, the min-max method specifies that $P(x) \wedge Q(x)$ should be considered 50% true, whereas the product-sum method specifies that $P(x) \wedge Q(x)$ should be considered 25% true. Which of these (if any) is more correct? As you can see, the min-max method is totally uneffected by the presence of multiple copies of the least or greatest values, whereas the product-sum method varies continuously. Which of these behaviors is actually better though? It's seems hard to say. Only an in-depth analysis will illuminate the true nature of the difference.

Likewise, let's consider an example of disjunction. Suppose that we again have some object x and some fuzzy predicates named P and Q, such that $P(x) = 0.50$ and $Q(x) = 0.50$. This time though, we want to evaluate disjunction. What is the value of

$P(x) \vee Q(x)$ under each of the two methods? Well, under the min-max method $P(x) \vee Q(x)$ is yet again considered to be 50% true, just as it was for $P(x) \wedge Q(x)$. In contrast though, the product-sum method specifies that $P(x) \vee Q(x)$ should be considered 75% true. As you can see, under the product-sum method, the disjunction is trending in the opposite direction of the conjunction, even when all the participating fuzzy predicates have the same numeric value.

This behavior generalizes. In fact, under the product-sum method, if we had n fuzzy truth values of the form f_k, such that $0 < f_k < 1$ (intentionally excluding 0 and 1) for all positive integers k among n, then as n approaches infinity $f_1 \vee f_2 \vee \ldots \vee f_{n-1} \vee f_n$ will always converge arbitrarily close to 1, regardless of the specific values of the various f_k. Likewise, as n approaches infinity $f_1 \wedge f_2 \wedge \ldots \wedge f_{n-1} \wedge f_n$ will always converge arbitrarily close to 0, regardless of the specific values of the various f_k. In contrast though, the min-max method does not have this behavior.

For the min-max method, only the values of the various f_k will determine the outcome, whereas the product-sum method has an uncontrollable gravity that pulls it to converge towards 0 or 1 as the number of participants increases, regardless of the actual values of those participants generally. In this respect, the min-max method is more similar to standard classical logic than the product-sum method is, in that chains of disjunctions or conjunctions in both systems can only be determined by the values of the participants and do not exhibit this odd value-agnostic convergence behavior exhibited by the product-sum method.

When thinking about how disjunctions and conjunctions should behave, such as when implementing a fuzzy logic for them, it is useful to keep in mind what conjunction and disjunction are supposed to represent. Conjunction is supposed to represent the condition that *all* of the participants are true, whereas disjunction is supposed to represent the condition that *at least one* of the participants is true.

While $P(x)$ and $Q(x)$ may each be 50% true separately, if P and Q are independent factors then we would expect that it would be harder for both of them to be true, whereas it would be easier for at least one of them to be true. Both the min-max method and the product-sum method satisfy this expectation, but differ in what contexts they exhibit it in. The min-max method exhibits it whenever a new least or greatest participant is added, whereas the product-sum method exhibits it every time any participant whatsoever is added, even including when redundant copies of predicates that are already participating are duplicated.

However, even though the product-sum method does satisfy this expected criteria, just as the min-max method also does, it still has some other problems. Probably the biggest problem it has is that it lacks context awareness, just like all other truth-functional logics. The consequences of this fact are even worse for the product-sum method for fuzzy logic than for most other truth-functional logics though.

To see why this is a problem, consider this: What if the fuzzy predicates P and Q aren't independent criteria? What if the fuzzy truth of one is similar in behavior to the fuzzy truth of the other? For instance, suppose that P and Q were merely different names for the exact same criteria. Should $P(x) \wedge Q(x)$ really be less than each of $P(x)$ and $Q(x)$ under this condition?

The answer is no. If P and Q specify the exact same criteria, then we would expect

$P(x) \wedge Q(x) = P(x) = Q(x)$ to be true, but the product-sum method is not context-aware enough to catch this. This is yet another demonstration of the deep and fundamental weakness inherent in all truth-functional logics. No truth-functional logic can ever be context-aware, at least not without external augmentation that accounts for context somehow.

However, be that as it may, I will soon show you a method for partially compensating for this problem, one that will allow you to somewhat reduce the impact of this lack of context awareness when evaluating fuzzy logic expressions using the product-sum method. Regardless though, the product-sum method will still end up being much worse than the min-max method overall, even after these correcting factors have been added to it. More on that later though.

Notice that the min-max method is not as vulnerable to this lack of context awareness as the product-sum method is though. Both methods are truth-functional of course, and so can't possibly ever account for context. However, the damage caused by lack of context awareness when using the min-max method is much less than when using the product-sum method. The min-max method still correctly satisfies $P(x) \wedge Q(x) = P(x) = Q(x)$ whenever P and Q specify the exact same predicate criteria.

Imagine a huge chain of predicates that are all just different names for the exact same underlying criteria. Wouldn't it be absurd if the conjunction or disjunction of all of those predicates, each of which *express identical criteria*, converged to 0 or 1 regardless of the context? Yet, that is exactly what the product-sum method does. Thus, for example, under the product-sum method if $P(x) = 1/2$ then $P(x) \wedge P(x) = 1/4$ and $P(x) \wedge P(x) \wedge P(x) = 1/8$ even though $P(x) = P(x) \wedge P(x) = P(x) \wedge P(x) \wedge P(x) = 1/2$ should obviously be the expected behavior from a conceptual standpoint, since no new information is actually being added by the multiple copies of $P(x)$ in these conjunctions.

Despite these severe shortcomings, the product-sum method can still sometimes be made to behave tolerably well under certain very limited circumstances. When all of the participating fuzzy predicates are conceptually independent from each other, the product-sum method is at least somewhat plausible as a model of the fuzzy truth of logical expressions.

For example, when the number of input variables is low and the product-sum conjunction is close to 1 then it means that all of the participating conditions are probably highly true, and likewise when the product-sum disjunction is close to 1 then it means that at least one of the participating conditions is probably highly true. Even then it's not very reliable though. The product-sum method's behavior tends to degenerate to nonsense (especially convergence to 0 or 1 regardless of context) as more conjuncts or disjuncts are added to an expression. It only behaves sanely for very small numbers of inputs.

One important side-note: Whether or not the product-sum method's disjunction operator is order independent (commutative) and grouping independent (associative) is not immediately obvious. Without knowing this, we couldn't be sure that it is safe to chain product-sum disjunctions together arbitrarily. Without order independence (commutativity) and grouping independence (associativity), what order we write disjunctions in and how we group them might change the outcome. Classical disjunction certainly has both these properties (order independence and grouping independence), but we don't yet know whether the product-sum disjunction does. Luckily though, as it turns out, the product-sum disjunction is indeed both order independent (commutative) and grouping independent (associative). Let's prove it though, so that you can trust it.

First, let's prove order independence. To prove it, we need to demonstrate that $P(x) \lor Q(y)$ and $Q(y) \lor P(x)$ will always have the same value. This is not difficult. All we need to do is apply the product-sum disjunction definitions and then rearrange the two expressions until they become identical. For brevity and readability, let a represent $P(x)$ and let b represent $Q(y)$. Here's the reasoning:

$$a \lor b$$
$$= (a + b) - (a \cdot b)$$

$$b \lor a$$
$$= (b + a) - (b \cdot a)$$
$$= (a + b) - (a \cdot b)$$

$$\therefore a \lor b = b \lor a$$

Thus, as you can see, product-sum disjunction is indeed order independent (commutative). Next, let's prove that it is also grouping independent (associative). For brevity and readability, let a represent $P(x)$, let b represent $Q(y)$, and let c represent $R(z)$. Here's the reasoning:

$$(a \lor b) \lor c$$
$$= ((a \lor b) + c) - ((a \lor b) \cdot c)$$
$$= ([(a + b) - (a \cdot b)] + c) - ([(a + b) - (a \cdot b)] \cdot c)$$
$$= (a + b + c - a \cdot b) - ((a + b) \cdot c - (a \cdot b) \cdot c)$$
$$= a + b + c - a \cdot b - (a \cdot c + b \cdot c - a \cdot b \cdot c)$$
$$= a + b + c - a \cdot b - a \cdot c - b \cdot c + a \cdot b \cdot c$$

$$a \lor (b \lor c)$$
$$= (a + (b \lor c)) - (a \cdot (b \lor c))$$
$$= (a + [(b + c) - (b \cdot c)]) - (a \cdot [(b + c) - (b \cdot c)])$$
$$= (a + b + c - b \cdot c) - (a \cdot (b + c) - a \cdot (b \cdot c))$$
$$= a + b + c - b \cdot c - (a \cdot b + a \cdot c - a \cdot b \cdot c)$$
$$= a + b + c - b \cdot c - a \cdot b - a \cdot c + a \cdot b \cdot c$$
$$= a + b + c - a \cdot b - a \cdot c - b \cdot c + a \cdot b \cdot c$$

$$\therefore (a \lor b) \lor c = a \lor (b \lor c)$$

These facts are very useful to know. As a consequence of proving them, we now know that we can reorder and regroup homogeneous chains of product-sum disjunctions or conjunctions however we would like to, just as we would in classical logic. However, in contrast, please be aware that neither product-sum conjunction nor product-sum disjunction distribute over the other.

Thus, neither $(A \vee B) \wedge C = (A \wedge C) \vee (B \wedge C)$ nor $(A \wedge B) \vee C = (A \vee C) \wedge (B \vee C)$ are valid for the product-sum method's definitions of these operators. Consequently, you can't use the distributive law directly when working with fuzzy logic expressions that use the product-sum method, although of course any multiplications they decompose into will still distribute over addition. The min-max method in contrast *is* distributive, allowing both distribution of disjunction over conjunction and vice versa, in addition to being order independent (commutative) and grouping independent (associative).

One interesting thing to note is that the behavior of the product-sum method for defining fuzzy logic operators is very structurally similar to how calculating areas enclosed by unions and intersections of sets would work. Imagine, for example, a Venn diagram depicting two sets A and B side by side, with part of them intersecting. How would we calculate the area enclosed by the union of set A and set B given that we only know the area of each set A and set B separately and can compute $A \cap B$.

The answer is that we would sum the areas of A and B, and then subtract out the area of the intersection $A \cap B$. Subtracting out the area enclosed by $A \cap B$ is necessary because merely summing the areas of A and B would end up counting the area of the intersection twice. Thus, we have to subtract the area of $A \cap B$ once in order to renormalize our sum so that it correctly models the real behavior of the system.

Notice that this is exactly the same pattern that the product-sum method follows. The product-sum disjunction $a \vee b$ is defined as $(a+b)-(a \cdot b)$, which has the same underlying structural nature as the Venn diagram example. The $-(a \cdot b)$ term is essentially an error correction term. Without it, we'd actually be double counting part of the fuzzy truth value of $a \vee b$ (at least according to how the product-sum method views things). Hopefully this thought experiment has now made it clearer to you why $(a+b)-(a \cdot b)$, and not merely $a+b$, is the correct value for the disjunction in the product-sum method.

Notice also that any product-sum disjunction chain consisting entirely of fuzzy truth values f_k such that $0 \leq f_k < 1$ for all k will always have a fuzzy truth value at least slightly less than 1. This is as intended. No disjunction built up entirely of incomplete truths should ever yield a complete truth. In contrast, if we merely calculated the sum $a+b$ (perhaps with clamping to keep it in range, e.g. $max(a+b, 1)$), then disjunctions of incomplete truths would be able to yield complete truths, but that would imply more certainty than could actually ever exist in any disjunction of only incomplete truths. Only the presence of a predicate with absolute certainty (e.g. where $f_k = 1.0$) could ever justify a disjunction having a fuzzy truth value of exactly 1. Otherwise, the fuzzy disjunction should always have at least a small element of vagueness, or else it could not possibly be conceptually correct.

Also, notice that the analogy of the Venn diagram counting correction actually continues to be correct no matter how many sets you add into the Venn diagram. For example, consider the case where you have three intersecting sets in a Venn diagram, instead of only two. Look at what the product-sum disjunction definition does for the corresponding case involving three fuzzy truth values. We already calculated this in our proof that product-sum disjunction is grouping independent (associative). Specifically, we determined that $(a \vee b) \vee c = a \vee (b \vee c) = a + b + c - ab - ac - bc + abc$.

Notice that $a + b + c - ab - ac - bc + abc$ is exactly analogous to the calculation that you would have to do if you wanted to determine the area of the union of the three sets A, B, and C on a Venn diagram when you only knew the areas of each separate set. You would first sum the areas of each set together, but then you'd realize that you double counted the intersections $A \cap B$, $A \cap C$, and $B \cap C$, and also that you triple counted $A \cap B \cap C$.

You would then subtract out all of the areas of $A \cap B$, $A \cap C$, and $B \cap C$ once each, but this would have the side effect that you have now removed $A \cap B \cap C$ three times, reducing its total impact to zero. You would therefore now have to add in one copy of $A \cap B \cap C$ to compensate. This is exactly what the expression $a+b+c-ab-ac-bc+abc$ analogously does. The $-ab-ac-bc+abc$ part of the expression corrects for all of the double, triple, and zero counting errors that $a+b+c$ would otherwise introduce.

This patterns continues no matter how many new participants you add into a product-sum disjunction chain. By defining the disjunction $a \vee b$ as $(a+b)-(a \cdot b)$ we ensure that every successive application of disjunction will continue to correct for any error introduced by duplicate counting of the same parts of an expression's fuzzy truth. This automatic error correction extends to any number of dimensions, without you having to actually think about it. This is good news. The reasoning would get quite convoluted and tedious if you had to account for all these duplicate factors by hand.

On the other hand though, whether or not the product-sum method is even a sane model of fuzzy truth to begin with is very questionable. Indeed, I suspect that the product-sum method is actually fundamentally wrong, and that the min-max method is vastly superior. However, be that as it may, if the product-sum method were indeed an accurate model of fuzzy truth then this automatic correcting for duplicate counting would be a good thing. Even if the basic premises of the product-sum method were correct though, at least one thing that would still be obviously very wrong with the product-sum method is the way that it doesn't handle predicates with similar or identical criteria well.

Recall, for example, the fact that even when all the participants in a disjunction or conjunction defined by the product-sum method specify the exact same criteria, the value of the disjunction or conjunction can still nonetheless change. Thus, as a consequence, if $0 < P(x) < 1$ then $P(x)$, $P(x) \wedge P(x)$, $P(x) \wedge P(x) \wedge P(x)$, $P(x) \vee P(x)$, and $P(x) \vee P(x) \vee P(x)$ (etc) would all return different values even though the truth values of all of these expressions conceptually should not differ from $P(x)$ alone, since no new information or constraints are actually being added by the presence of multiple copies of $P(x)$. Yet, this cannot possibly be the conceptually correct behavior. As such, if the product-sum method is to be even remotely viable as an option, it must be modified to compensate for this problem.

In this respect, I have come up with a partial remedy: a modification to the product-sum method that will allow it to handle similar predicates more correctly. I was unable to find an adequate solution in the literature, and indeed all the few sources I saw didn't even seem aware that the problem existed at all, so I have had to formulate this modification on my own. However, before I can explain to you what the modification is, we will need to take a detour through an entirely different method of handling combinations of fuzzy predicates, one that happily has tremendous value in and of itself besides just being necessary for creating an improved version of the product-sum method.

This entirely different method of handling combinations of fuzzy predicates exists independently from both the min-max method and the product-sum method, and is useful in and of itself. It is neither analogous to disjunction nor analogous to conjunction. It is its own separate thing: an independent tool that you can use to your advantage regardless of whether you are using the min-max method or the product-sum method. It belongs as part of the vocabulary of fuzzy logic, a basic building block that enables you to express a different kind of concept.

What is it? Well, quite simply put, it is averaging, using either equal weight or arbitrary weight depending on our intent. As simple a concept as averaging may seem, it is a powerful piece of mathematical vocabulary. It is not merely a statistic. The true power of averaging only becomes apparent when you actually understand what it is on a deeper and more fundamental level, rather than merely thinking about it as a mechanical process.

You see, averaging enables the mathematical coalescence of parts into wholes. It allows us to express blending, much like a painter who mixes colors together in arbitrary proportions, thereby creating nuance and balance in their work. It also, for example, helps for expressing carefully chosen allocations of resources, e.g. such as when a businessperson decides they will put 80% of their resources into X and the other 20% into Y to account for priorities and risks etc. Averaging is all of these things and more, in a sense. It is a very flexible and expressive concept, when properly understood and generalized. Equally weighted averaging is just a special case of this much broader phenomenon of arbitrarily weighted averaging, i.e. of blending, of coalescence of separate parts into wholes.

One thing an average is *not* though, very much contrary to popular belief, is the "most typical" representative of a set. That is typically the domain of the median, not the average. The way most people use the phrase "the average person" is thus wrong in this sense. An "average person" would be a person who has a little bit of every trait of every person.

An "average person" would thus be part astrophysicist, part gymnast, part Hitler, part medical doctor, part comedian, part rocket scientist, and part masseur, as well as part countless other things. An average person would be an absurdity. There is no such thing as an average person. In contrast though, there are lots of people who are *close to the median* with respect to specific (totally-ordered) dimensions of who they are, and it is only in this sense that "the most typical person" of a set of people can really exist. Don't confuse "typical" with average. They are actually fundamentally different concepts.

Anyway, let's explore how averaging works and how it should be in interpreted in greater depth now. Naturally, let's start with the definitions. There are two types of averaging we will want to consider. One is the special case of equally weighted averaging, and the other is the general case of arbitrarily weighted averaging. All equally weighted averages can be considered to be arbitrarily weighted averages where all of the weights on each participant just happen to be the same. Like many special cases in mathematics, the equally weighted average is easier to compute than the more general case.

Let $v_1, v_2, \ldots, v_{n-1}, v_n$ be numbers on the continuum that represent the values that we want to average together in some way. Let $V = \{v_1, v_2, \ldots, v_{n-1}, v_n\}$, the set of all such values. Furthermore, let $w_1, w_2, \ldots, w_{n-1}, w_n$ be numbers on the continuum between 0 and 1 (inclusive), such that $w_1 + w_2 + \ldots + w_{n-1} + w_n = 1$. We call each such v_k a *value* and each corresponding w_k that value's *weight*. As such, here are the formulas for the equally weighted and arbitrarily weighted averages of the values of V:

Equally Weighted Average
$(v_1 + v_2 + \ldots + v_{n-1} + v_n)/n$

Arbitrarily Weighted Average
$(w_1 \cdot v_1) + (w_2 \cdot v_2) + \ldots + (w_{n-1} \cdot v_{n-1}) + (w_n \cdot v_n)$

Notice that the expression for the arbitrarily weighted average would become equal to $(v_1 + v_2 + \ldots + v_{n-1} + v_n)/n$ if $w_1 = w_2 = \ldots = w_{n-1} = w_n$ was true, because all the w_k could be factored out as $1/n$, thus showing that the equally weighted average is indeed a special case of the arbitrarily weighted average. Also, notice that the weights need to add up to 1. You need to make sure that you satisfy this criteria or else your weighted average may go out of bounds, which is usually not what you want.

Making sure that the weights always add up to 1, while simultaneously trying to give all of the values specific weights suitable for your needs, can sometimes make it a little bit tricky to choose a valid set of weights. An easy way around this problem though is to define your weights in "parts" of any magnitude on the continuum (in a special way) instead of as numbers in the range of 0 to 1.

There are two important requirements for doing this effectively though: (1) You must intuitively understand what a "part" means, and (2) once you have assigned how many "parts" each weight should get, you need to sum all of the weights together, and then divide each of them by that sum, so that you can translate them into weights in terms of numbers between 0 and 1 for performing the actual weighted average. Let $p_1, p_2, \ldots, p_{n-1}, p_n$ be the values of our desired weights in terms of parts. Let $P = \{p_1, p_2, \ldots, p_{n-1}, p_n\}$. Then, here is how to perform arbitrarily weighted averaging in terms of parts:

$\forall (p_k \in P)(\,(p_k \in \text{Cont}) \land (p_k \geq 0)\,)$
(i.e. "parts" are non-negative continuum numbers of any magnitude)

$\exists (p_k \in P)(p_k > 0)$
(i.e. at least one $p_k > 0$, to prevent division by zero)

$S = p_1 + p_2 + \ldots + p_{n-1} + p_n$

$w_1 = p_1/S$
$w_2 = p_2/S$
$\ldots$
$w_{n-1} = p_{n-1}/S$
$w_n = p_n/S$

Arbitrarily Weighted Average by Parts

$$(w_1 \cdot v_1) + (w_2 \cdot v_2) + \ldots + (w_{n-1} \cdot v_{n-1}) + (w_n \cdot v_n)$$
$$= (\tfrac{p_1}{S} \cdot v_1) + (\tfrac{p_2}{S} \cdot v_2) + \ldots + (\tfrac{p_{n-1}}{S} \cdot v_{n-1}) + (\tfrac{p_n}{S} \cdot v_n)$$
$$= (\,(p_1 \cdot v_1) + (p_2 \cdot v_2) + \ldots + (p_{n-1} \cdot v_{n-1}) + (p_n \cdot v_n)\,)/S$$

If we want to explicitly distinguish between these two ways of doing arbitrarily weighted averages, then we may refer to the original method that required all the weights to already sum to 1 as "arbitrarily weighted averaging by percentages", whereas we may refer to the method that uses weights that can be any non-negative continuum numbers as "arbitrarily weighted averaging by parts". If, in contrast, we merely say "arbitrarily weighted averaging" then we could conceivably be referring to either method. These phrases, descriptive as they may be, are kind of a mouthful though. As such, the shortened phrases "averaging by percentages" and "averaging by parts" may be used instead for brevity.

Although this distinction is useful for communication, the method of averaging by parts actually behaves identically to averaging by percentages whenever $p_1 + p_2 + \ldots + p_{n-1} + p_n = 1$. As such, the method of averaging by parts is actually a strict superset of the method of averaging by percentages. Thus, averaging by parts is generally preferred, because it is strictly more general and more expressive than averaging by percentages. Anything averaging by percentages can do, averaging by parts can do equally well. Any valid input for averaging by percentages is also valid for averaging by parts, and will return the same result. The reverse is not true though. It's easy to give averaging by percentages invalid inputs by accident.

Thinking in terms of parts instead of percentages tends to make assigning the weights far more foolproof and far less tedious. This is especially true if you expect to add in new weights and values in the future, because in that case using averaging by percentages will cause the weights to constantly fluctuate, making the system much more difficult to reason about and modify. In contrast though, weights defined in terms of parts will remain invariant regardless of what other new weights and values you add into the system. This makes averaging by parts much easier to work with in cases where the data set is large or open to change (or both). Maintaining a changing list of weights for averaging by percentages is a nightmare, whereas it is trivially easy for averaging by parts.

At first glance, percentages may seem easier to think about, and for small data sets undergoing weighted averaging this may indeed be true. However, thinking in terms of parts scales better. Percentages may feel very intuitive initially, and so you may be loath to abandon them, but parts are actually at least as intuitive (if not more) once you understand how to think about them. Allow me to explain. You see, the trick is to anchor your thoughts on one of the pre-existing weighted values in your data set, and then to think *relative to that* whenever you're adding in a new value and trying to decide what weight you want to give it.

Probably one of the easiest analogies I can make here is cooking recipes. In cooking recipes, you often see instructions that say you need to use "3 parts ingredient X for every 1 part ingredient Y". This is averaging by parts. This kind of common usage of the word "parts" is actually where I derived the term "averaging by parts" from. Notice that we are anchoring our thoughts about how much weight to give X in terms of Y.

This is what I was referring to when I said the trick is to anchor your thoughts in one of the pre-existing weighted values.

The beautiful thing about averaging by parts is that even if all the other data items change, both the value and the ratio of each unchanged pair will remain the same. This is what makes averaging by parts so easy to work with. Stated more formally, given any two fixed weights p_i and p_j defined in terms of parts, both the difference $p_j - p_i$ and the ratio p_j/p_i will remain constant regardless of any changes made to any of the other weights.

This allows you to localize your thoughts. Whereas with averaging by percentages you must understand the relative magnitudes of *all* weights in order to understand what you need to set a new weight to in order to achieve a desired effect, with averaging by parts you only have to understand how much you want each new data item to weigh relative to some other specific data item. You only need to think about what you care about.

Cooking is of course not the only example where thinking in terms of parts is useful. For example, if a game developer is trying to decide the chances of each of some set of random items of being placed at some location, it will generally be much easier to work in terms of parts than percentages, especially if there are a large number of items in the game or if changes to the weights happen frequently. The developer can simply say "a dagger should be 10 times more likely to be randomly selected than a sword" and encode this via averaging by parts.

The great thing about this is that the developer doesn't need to know anything about the percent chances of any of the other items anymore. This makes it vastly easier to create a very well-balanced item generation system. All the developer has to do to ensure a perfect mix is to look at the relative ratios of each weight compared to some other well-understood item[11], think about whether that ratio is desirable, and then adjust accordingly. These same principles also apply to item strengths, not just to item randomization. In fact, this technique can be applied to virtually any aspect of a game's underlying relationships. Averaging by parts is a very powerful concept. It is immensely useful for improving your ability to express arbitrary concepts.

Indeed, weighted averaging allows you to create hybrids of pretty much any set of behaviors. For instance, suppose you have two arbitrary functions f_1 and f_2. Want a function that behaves 50% like f_1 and 50% like f_2? Just use weighted averaging to blend them together accordingly (e.g. $0.50 \cdot f_1(x) + 0.50 \cdot f_2(x)$) and voila, you now have a function whose behavior is an equal hybrid of f_1 and f_2, with almost no effort. You can even vary the weights over time, while simultaneously continuing to blend the corresponding values.

In fact, the splines you see in vector art editing programs are essentially just specific applications of this broader concept. For example, Bézier curves are often actually generated by repeatedly applying weighted averaging to a set of simple linear interpolation functions (i.e. line segments) spanning a sequence of control points, where each successive line segment has been connected end to end. For more information, lookup

---

[11] i.e. p_j/p_i, where p_i and p_j have been chosen from whatever pairing of items is easiest for you to think about at any given moment of time

"De Casteljau's algorithm". This is not the only way of creating splines though. A vast number of different function pairs generate beautiful splines when you blend them together seamlessly over a continuum. Don't be afraid to experiment.

In some sense, the word "averaging" is actually kind of confusing. Lots of people have a really strongly ingrained habit of thinking about "average" as meaning "most typical". If this is you, then consider using the word "blending" in place of "averaging" instead. The real essence of averaging is actually blending, after all, so it makes more sense as an intuitive term for what it is you are really doing when you work with averages. Thus, you could say "blending by percentages" instead of "averaging by percentages", and "blending by parts" instead of "averaging by parts", and you may find this to be more intuitive and clear. I personally do, but I've chosen to use the term "averaging" a lot in this section anyway, because doing so could maybe help clarify that averaging does *not* really represent what many people think it does.

Averaging is all about blending, i.e. is all about *hybridization* based on combining separate parts. It is certainly *not* about finding the most typical member of a set. Rather, if you took an equally weighted average of a set of objects with n members, and named that average α, and then made n clones of α, then the total amount of "stuff" in those n clones of α would be the same as in the original set of objects.

This would not be true of a median though, which in contrast *does* represent the most typical member of a set generally. Thus, an equally weighted average is actually analogous to the *center of mass* of a set. The average is not "the most typical representative of a set", but rather it is more like a balancing point, like the point on the hilt or the blade of a sword where if you sat it on top of a narrow beam it wouldn't fall off to either side.

When you want to roll up some entities together according to some specific desired balance of factors, or want to reduce multiple concerns to one concern, then averaging (a.k.a. blending) is most often the technique you'll want to use to accomplish it. Averages may not really represent the "most typical" object of a set, but they can still act as representatives of that set in the sense that they are the most conceptually correct form of what that set would be if it were just one thing instead of many things.

Thus, for example, formulas in physics often rely on finding the center of mass of an object in order to describe its properties concisely and elegantly (e.g. moments of inertia for different shapes). This would often be prohibitively difficult to do otherwise (without a computer simulation), because if you didn't calculate the average of all of the mass to find the center of mass, then you would have to do all your calculations in terms of *all* of the atoms of which the object is composed. Averaging eliminates this problem by conceptually squeezing all of those atoms together to find what the closest representation of the object would be if it were just *one* object instead of many. One object is much easier to work with than many.

It is this sense in which averaging fuses multiple object together into one that is most necessary for understanding how to properly utilize averaging correctly in applications. In this respect, let us now turn our attention back to the subject that first got us started on this tangential discussion of averaging: fuzzy logic. As you may recall, we had been talking about the min-max method and the product-sum method for evaluating fuzzy logic expressions, and were evaluating and comparing their respective properties, when

I mentioned that another independent technique also existed, one whose meaning was analogous neither to disjunction or conjunction, and that this technique was averaging.

How exactly should averaging be used in the context of fuzzy logic? Well, it should be used precisely in the context where you want to fuse multiple different fuzzy truth values together so that you can think of them as one fuzzy truth value instead of many. After all, that kind of coalescence is the essence of what averaging is, so it makes sense that that's also how it would be used in fuzzy logic. One of the most effective applications of this principle is in weighting each of a set of concerns according to how much you care about them.

For example, suppose we have two fuzzy truth values a and b, but we care much more about the degree to which a is true than we do about the degree to which b is true, but both are still a factor in our decision making. For instance, suppose a is 80% of what we care about and b is 20% of what we care about. In this case, weighted averaging should be applied to a and b so that we now have a new fuzzy truth value c, such that $c = 0.80 \cdot a + 0.20 \cdot b$. We can now make our fuzzy decision based solely on the value of c, which will be much easier to do correctly than trying to think in terms of multiple different separate factors at once.

For example, suppose our scenario is that an artificial intelligence controlling a creature inside a video game is trying to decide when it should use a healing ability or not. Suppose that a represents the degree to which the creature is near death and b represents the degree to which the creature feels threatened by its surroundings. This could be a real-world example of where we might want to use $c = 0.80 \cdot a + 0.20 \cdot b$ for our decision.

This (i.e. using $c = 0.80 \cdot a + 0.20 \cdot b$ for this scenario) would have the net effect that the overwhelming majority of the creature's decision would be based on how close it is to death, but part of it would also be based merely on how dangerous its environment is. Thus, the artificial intelligence controlling the creature would end up using healing abilities sooner when injured in dangerous environments than in safer environments. Nice trick right? It's easy to think about too, which is why it's such a great technique.

However, you may alternatively want to use disjunction instead. It depends on the nature of the scenario. The meaning is subtlety but critically different. A fuzzy disjunction evaluates whether *at least one* participating condition is highly true, whereas a weighted averaging *fuses multiple conditions into one condition*. These will actually result in *very* different conceptual behaviors, so it is important to have a well-practiced understanding of the nuances of the differences between the two approaches. For instance, using a fuzzy disjunction in our creature scenario above may cause the creature to use healing abilities even when not significantly injured, based merely on whether it is near death *or* in a dangerous environment (or both). See the difference?

As yet another point of contrast, a fuzzy conjunction would require *both* conditions to be simultaneously highly true, i.e. would require the creature to be both near death and in a dangerous situation, which is again *not* the same thing conceptually as merging the conditions using weighted averaging. Moreover, fuzzy disjunctions, fuzzy conjunctions, and weighted averages of fuzzy truth values can all be combined together and nested in countless arbitrarily ways, all of which have subtly but critically different meaning from each other. Choosing exactly the right logical expressions, i.e. the ones that most closely match a conceptually correct and principled interpretation of what you are trying to do, is crucial for ensuring that an artificial intelligence that utilizes fuzzy logic will behave in a maximally pleasing way.

If you choose the wrong combination of fuzzy logic expressions, then you will notice that the artificial intelligence won't act the way you expect. The behavior will feel "mysterious" somehow, as if all you can hope to do is fiddle around randomly with the parameters until it vaguely seems to work. An artificial intelligence system that operates on the wrong principles may still superficially seem to "work", but you will often find that it seems oddly unresponsive to parameter tweaks and that the overall behavior breaks easily when changed.

These are classic hallmarks of a conceptually incorrect implementation. If you do things in a conceptually correct and principled way though, then it will seldom feel "mysterious". It will feel much more controllable when you're doing it correctly. It will tend to be relatively easy to make changes, and significantly tweaking the parameters will have clearly visible and predictable impacts on behavior.

Fuzzy negation, fuzzy disjunction, fuzzy conjunction, and weighted averaging of fuzzy truth values are all extremely important tools to be aware of when working with fuzzy logic. They are critical components of your vocabulary. They are the basic building blocks with which you will construct the vast majority of all fuzzy logic expressions. They are the fundamental elements — the air, earth, wind, and fire — of working with basic fuzzy reasoning.

By the way though, an analog of material implication exists for fuzzy logic (because $(A \rightarrow B) \leftrightarrow (\neg A \vee B)$ in classical logic), but we won't be talking about it here because it is not as useful for fuzzy logic as algorithmic conditionals[12] are. In any case, it is outside the scope of what I want to talk about. In fact, it is now time that we finally turn our attention back to what originally inspired this whole discussion of averaging: the product-sum method, its flaws, and how we might reduce the impact of those flaws.

As you may recall, starting on page 323, I mentioned that the product-sum method responds very badly to the presence of identical or similar predicates in product-sum disjunction and conjunctions. For instance, I eventually gave the example of $P(x) = 0.50$, and demonstrated how expressions like $P(x) \wedge P(x)$ and $P(x) \wedge P(x) \wedge P(x)$ and so on would keep changing value even though this makes no sense from a conceptual standpoint because no new information is being added besides what is already indicated by $P(x)$ alone. I also mentioned that there would be a way to compensate for this problem.

The conceptually correct way to compensate for this problem in the product-sum method is with averaging. Why? Well, let me explain. You see, when a set of predicates express identical criteria this amounts to saying that in fact these "multiple" entities are actually *one* entity in disguise. Therefore, if we want to compensate for this problem, then we need to have some way of recombining these multiple predicates back into one object instead of multiple. Think about what I just said though. What's a method we know of that lets you combine multiple things into one thing? We just talked about it. It's just averaging (a.k.a. blending).

---

[12] i.e. the if-then statements found in programming languages, which are a completely different concept from material implication and are also much better behaved

On the other hand, not all predicates will be the same. Sometimes we will need to treat them as distinct, and sometimes we will need to treat them as identical, and we'll need to handle all the cases in-between as well. Thus, we will actually want two layers of blending, one for how distinct or similar the set of predicates are overall, and one for treating the set of predicates as equally constituting one whole. We then want to combine that with the product-sum's operator definitions.

We will call this method the **similarity-adjusted product-sum method**, or the "sim-adj prod-sum method" for short. Like many things in this book, I was forced to invent this idea myself, since I couldn't find anything similar. I developed it out of pure curiosity while writing the book, to see if the product-sum method's severe flaws could be removed or reduced. I still very much prefer the min-max method regardless though, for reasons that will be explained later.

Anyway though, this will make a lot more sense once I show the actual formulas. In order to make this work though, we will need to introduce a third participant into every disjunction and conjunction. The purpose of this third participant will be to provide a user-specified estimate of how much similarity there is among the predicates in the expression. This is necessary so that we can know in what proportions we should blend the product-sum method's overall result. We will call this third participant the **similarity factor** and represent it by σ (the Greek letter sigma).

The similarity factor σ is a number on the continuum between 0 and 1 (inclusive). If $\sigma = 1$ then all of the predicates have identical criteria, whereas if $\sigma = 0$ then all of the predicates have independent criteria from each other. The cases in-between represent varying degrees of similarity. Thus, $\sigma = 0.80$ would mean that the participating predicate criteria are about 80% similar, and likewise $\sigma = 0.20$ would imply 20% similarity. The user must provide this estimate manually.

The formula cannot determine σ on its own, since no truth-functional logic can ever be context-aware. Thus, this system requires the user to provide three parameters to every disjunction and conjunction (i.e. two predicates plus a similarity factor), rather than only requiring the normal two parameters that we are used to. Deciding the precise value of σ is often a subjective decision, one with no firm rigorous basis. It's a matter of art, so to speak. It comes down to arbitrary tweaking until the desired behavior is produced. Here are the formulas:

The Similarity-Adjusted Product-Sum Method:

1. $\neg P(x) \rightarrow 1 - P(x)$

2. $P(x) \wedge_\sigma Q(y) \rightarrow (1 - \sigma)[P(x) \cdot Q(y)] + \sigma \left(\frac{P(x) + Q(y)}{2} \right)$

3. $P(x) \vee_\sigma Q(y) \rightarrow (1 - \sigma)[(P(x) + Q(y)) - (P(x) \cdot Q(y))] + \sigma \left(\frac{P(x) + Q(y)}{2} \right)$

The reason for the $\frac{P(x) + Q(x)}{2}$ part of these definitions is because it is an equally weighted average of the participating predicates, which is exactly what we need in order to conceptually blend all of the participating predicates together into one object. This blending together of the predicates into one object is what allows us to express how similar the predicates should be treated as. The reason why we are not using weighted averaging for the $\frac{P(x)+Q(x)}{2}$ part is because it would be prohibitively tedious to deal with. Even just being required to specify the similarity factor σ for every disjunction and conjunction has already made our work much more tedious. Adding additional weights to the mix would just be too unwieldy. In any case, the arbitrary choice of σ is already supposed to factor in how similar the predicates are to each other, so we don't really need or want any additional weights in that part anyway.

This method allows us to prevent some of the horrible behavior we witnessed earlier that happens whenever product-sum disjunctions and conjunctions are given identical or highly similar predicates. For example, consider the case where we had $P(x) = 0.50$ from earlier, and we noticed that $P(x) \land P(x) = 0.25$ even though this made no sense. Using a similarity-factor, we can now correct this. Consider instead what happens when we evaluate $P(x) \land_1 P(x)$. Notice that the $\land_1$ represents a similarity-adjusted product-sum conjunction with a similarity factor of 1. If $\sigma = 1$ this means all the participating predicates express exactly the same underlying criteria. Thus, it evaluates like so:

$$P(x) \land_1 P(x)$$
$$= (1 - \sigma)[P(x) \cdot P(x)] + \sigma \left(\frac{P(x)+P(x)}{2}\right)$$
$$= (1 - 1)[P(x) \cdot P(x)] + 1 \cdot \left(\frac{P(x)+P(x)}{2}\right)$$
$$= \frac{P(x)+P(x)}{2}$$
$$= \frac{0.50+0.50}{2}$$
$$= 0.50$$

The calculation for the average here may seem pointlessly laborious, and indeed it is in this case, but when the similarity factor σ has the property that $0 < \sigma < 1$ then the averaging will have a meaningful impact and will be important for conceptual correctness. For example, consider what would happen if we instead had $P(x) = 0.70$ and $Q(x) = 0.60$, and P and Q were highly but not perfectly correlated with each other, such that $\sigma = 0.80$ was a good estimate of the similarity of the underlying criteria. In this case, the averaging calculation would be meaningful. The calculation would go like this:

$$P(x) \land_{0.80} Q(x)$$
$$= (1 - \sigma)[P(x) \cdot Q(x)] + \sigma \left(\frac{P(x)+Q(x)}{2}\right)$$
$$= (1 - 0.80)[0.70 \cdot 0.60] + 0.80 \cdot \left(\frac{0.70+0.60}{2}\right)$$
$$= 0.20 \cdot 0.42 + 0.80 \cdot \left(\frac{1.3}{2}\right)$$
$$= 0.084 + 0.80 \cdot 0.65$$
$$= 0.084 + 0.52$$
$$= 0.604$$

Probably the trickiest part of using the similarity-adjusted product-sum method is figuring out what to set the similarity factor σ to. What exactly does "similarity" mean, in this context? Well, it means how close the predicate criteria are to being identical. It's not quite the same thing as dependence. For example, "hot" and "very hot" are highly similar, "warm" and "very hot" are somewhat similar, and "darkly colored" and "radioactive" are not similar at all.

However, just because two predicates operate in the same domain (e.g. warm vs hot) does not make them necessarily similar. For example, "cold" and "hot" arguably have little to no similarity as predicates for the purposes of determining σ, even though they are inversely related and hence have a very strong dependent relationship. Basically, similar predicates are predicates that tend to be true at the same time due to the nature of the criteria they express. In contrast, dissimilar predicates don't have positively correlated fuzzy truth values.

This technique (similarity adjustment) does indeed reduce the flaws of the product-sum method somewhat. However, even with these correcting adjustments, I still think that the product-sum method is a terrible choice for how to implement the basic operators of fuzzy logic. The min-max method[13] is far more principled and far easier to work with and to reason about, not to mention far less tedious as well. Exactly why this is true is what we are going to talk about next. In particular, there are at least six strong reasons to prefer the min-max method. The product-sum method is very badly behaved in comparison, and may in fact be outright conceptually wrong. Here are at least six reasons to prefer min-max over product-sum (in no particular order):

1. It is much easier to reason about the behavior of conditions expressed in terms of the min-max method than for the product-sum method. Constructing fuzzy logic expressions using the min-max method feels easy, predictable, and intuitive, whereas the product-sum method has multiple unexpected behaviors and reacts non-linearly to input. Designing finely-tuned and well-behaved expressions is vastly easier for min-max expressions than for product-sum expressions.

2. The min-max method is order independent (commutative), grouping independent (associative), and distributive all at the same time, just like the analogous operators in classical logic and set theory are. In contrast though, the product-sum method is not distributive. This makes a huge difference in how easy it is to manipulate expressions. Using the product-sum method, expressions that look like they should have identical meanings often don't, due to lack of distributivity. The min-max method in contrast always behaves exactly as you would expect in this respect.

3. The min-max method is much more computationally efficient than the product-sum method. Applying the minimum and maximum functions to pairs or lists of objects is extremely easy to do and involves very little labor. This is just as true of work done by hand as it is of work done by a computer. Anybody can look at a list of numbers and easily find the least or greatest element without having to think much. Similarly, the minimum and maximum functions are significantly

---

[13] See page 320 if you need to refresh your memory of how the min-max method for fuzzy logic works.

less computationally expensive to perform on a computer than the arithmetic operations of the product-sum method (especially multiplication and division). The less expensive an operation is for a computer, the more a computer will be able to perform vast numbers of that operation in a short span of time. Alternatively, the computer could use the extra computational headroom to reduce energy consumption or to generate less heat, which may help to save money and protect the environment, just like with any other efficiency gain.

4. The min-max method does not require similarity adjustment. Unlike the product-sum method, the min-max method always handles predicates that have underlying similarities in their criteria in a logically consistent way. Hidden similarities and relationships make no difference in what the min-max operators return, whereas the product-sum method is very much vulnerable to how similar the participants are. By not having to ever specify a similarity factor, the min-max method is also much less tedious to work with. The min-max method is foolproof, whereas the product-sum method is arbitrary and error prone, even when a similarity factor σ is used.

5. In the min-max method, the truth of a disjunction can only come from one specific condition, whereas the product-sum method allows it to come piecemeal from multiple weak partial truths (i.e. from multiple poorly supported and mostly false claims). This behavior of the product-sum method can be dangerous. A fuzzy disjunction will often be designed under the assumption that it should only be treated as highly true if at least one of the participating conditions is highly true. However, the product-sum method often returns highly true fuzzy truth values even for disjunctions where all of the conditions are more false than true. This behavior does not make much sense and is usually a huge liability.

6. The prod-sum method converges towards 1 as more participants are added to disjunctions, and converges towards 0 as more participants are added to conjunctions, at least so long as all participating fuzzy truth values f_k have the property that $0 < f_k < 1$. In contrast, the min-max method's result is determined only by the actual values of the participating conditions and not by how many of them there are. It makes no sense that a logical disjunction or conjunction would change its truth value based merely on the presence of more participants. The truth values of the participating conditions should always be the determining factor, not merely how many of them there are.

Clearly then, there are a lot of good reasons to prefer the min-max method. However, knowing which of two techniques is better is not of much use if we don't also know how to properly use and interpret that technique. We need to understand how to actually interpret min-max expressions on an intuitive level if we are to truly feel comfortable using them. Luckily, this is not difficult at all.

First, notice that the min and max functions can easily be generalized to work with more than two inputs and the meaning is clearly the same. As you would expect, generalized min and max functions simply return the least or greatest element among their list of *n* inputs. Thus, we could write *min(a, b, c)* to mean the least of three elements and so on, generalizing to $min(v_1, v_2, \ldots, v_{n-1}, v_n)$ for any arbitrary set of sufficiently ordered values.

As such, notice that nesting minimums inside minimums, or maximums inside maximums, has the same net effect as if it was just a generalized minimum or maximum with the same number of elements. Thus, for example, min(min(*a*, *b*), *c*) = min(*a*, *b*, *c*) and max(max(*a*, *b*), max(*c*, *d*)) = max(*a*, *b*, *c*, *d*) and so on. No matter how many input values a min or max function has, it will always simply return the least or greatest member. This is important for understanding the intuitions that will come into play.

What are those intuitions? Well, let's try an analogy. Have you ever heard of something called a high-water mark? A high-water mark is essentially the highest point a body of water reaches during high tide or flooding, in profile against any nearby cliff-faces or landmasses. A person passing by the area shortly after high tide or flooding may notice a dark line up to where the water reached, caused by the ground getting wet from contact with the water. Sometimes people who study the body of water or who are in charge of maintaining it for the public may even take note of this high-water mark and may permanently mark it on the cliff-side somehow in order to get a sense of how high the flooding can go for that particular body of water.

The behavior of the min-max method's operators is fairly closely analogous to this concept of high-water marks. The main difference is that whereas high-water marks measure the height of water, the min-max method measures the height of fuzzy truth. The other difference is that the min-max method also measures what would analogously be called the "low-water mark". More specifically, min-max disjunction is analogous to a high-water mark, whereas min-max conjunction is analogous to a low-water mark. Fuzzy logic expressions are thus sort of like a series of canals and gates designed to constrain, test, and measure the fuzzy truth flowing through them. It's not a perfect analogy, but it may help give you a general sense for how working with the min-max method's operators roughly feels.

In addition, to grasp the intuitions in play in the min-max method, it helps to also understand how fuzzy truth values are generally used. There are two main ways that fuzzy truth values are actually used in real-world applications: (1) to make binary yes-no decisions based on how high the fuzzy truth value is or (2) to proportionally adjust a system's behavior according to the magnitude of the fuzzy truth value. Let us refer to the first of these as **thresholding** and to the second of these as **proportional variation**

These two approaches are generally about equally easy to work with, but differ in how they behave and how they must be incorporated into a system. Thresholding is essentially just a way of converting fuzzy truth values into binary true-or-false values, typically so that you can make some kind of simple yes-or-no decision about whether or not you want to perform some action. Often the action has some kind of irreducible or chunky nature[14], but not always.

Proportional variation just means that we vary some parameter in the system in correspondence with the actual magnitude of the fuzzy truth value. The fuzzy truth value could be used in any arbitrary way, but something simple like $a \cdot f(x)$ (for some

---

[14]i.e. the action is "discrete", as we like to say in mathematics

fuzzy truth value *a* and some arbitrary function $f(x)$) is one of the most common cases. For example, *a* could be a fuzzy truth value representing how cold someone feels and $f(x)$ could be a function controlling how much they are shivering. Thus, as *a* increases they would shiver more, whereas as *a* decreases they would shiver less.

Anyway, back to the min-max method though. In our analogy of water flowing through a system of canals and gates, thresholding would be analogous to a dam. A dam only allows water to flow over it once the level of the water exceeds the level of the dam[15]. This is analogous to how a thresholding decision works in fuzzy logic. For example, if we have a fuzzy predicate $P(x)$, and we specify a threshold of ≥ 0.80 for it, above which we will take some action, then $P(x) \geq 0.80$ is our threshold condition and it behaves like a dam. The action will only happen when $P(x)$ is at least 80% true (i.e. by analogy, when the water has risen above the 80% level and thus is now flowing over the dam).

What about disjunctions and conjunctions though? Well, they could be seen as being analogous to sets of computer-controlled gates that have been linked together to a control module at the intersection of several input canals and one output canal. Each input canal has one gate and each of the canals are separate from each other so that they don't share water. At any given moment of time, only one of the computer-controlled gates connected to an input canal can be opened, and one always will be. The computer at each such station is connected to sensors in each of the input canals that tell it what the water levels are in each one.

In the case of a disjunction, the computer will open the gate only for the input canal that has the *highest* water level at any given moment of time. In contrast though, in the case of a conjunction, the computer will open the gate only for the input canal that has the *lowest* water level at any given moment of time. Multiple sets of these gateways can exist in a canal system, and the way they are connected together will correspond to different arbitrary logical expressions. All of these disjunction or conjunction gateways will eventually merge together into one final canal that will either be blocked by a dam (a threshold) or else be allowed to flow out of the system freely (proportional variation).

There's also a way to generalize the concept of a threshold so that instead of only outputting true or false (1 or 0) it could output anything else in the range of 0 to 1, or really anything at all if you want it to. My point here in mentioning this is just to remind you to not think too narrowly. Always think from first principles, in terms of what it is possible for you to do. The conventional uses of these concepts are typically the most useful and common, but there are all kinds of odd ways that you can modify the way things work. Having a knowledge base firmly grounded in first principles and a conceptually correct view of how things work will allow you to see where all of the *real* conceptual boundaries actually are.

Anyway though, let's explore the principles and intuitions behind the min-max method just a bit more before we move on. It's really important to have a firm grasp of the underlying nature of a system, otherwise you won't feel empowered to actually use it. The way the min-max method works is really quite easy to think about once it clicks.

---

[15] I'm assuming here that there are no vents in the dam. Some dams have spouts on the side that start venting water long before water ever reaches the top of the dam.

The waterway and canal system analogy is decently good, but there are also other ways to think about the min-max method that will help in developing an intuition for it.

For example, in some contexts, you can think of a threshold for a fuzzy logic expression as defining a margin of error in some sense. In real-world situations, you can't always expect to get a strong signal from something you are measuring, so it often helps to build in some error tolerance for systems that rely on those measurements for making decisions. As people, we sometimes want to be completely certain about our decisions, so we might feel an impulse to do nothing until we have complete evidence to make a judgment. However, delaying a decision is often actually far more dangerous or ineffective than simply taking an educated guess and acting on it. Sometimes something merely "seeming" to be true is more than enough reason to act.

Many situations that call for the use of fuzzy logic are not about "margins of error" though. For example, there is also the case of many things in the real world having "fuzzy boundaries" between distinguishing characteristics. For instance, you might see an object painted in a color that is somewhere between red and orange, and it may be more sensible to say that it is 'somewhat red" than to sloppily just label it as "red" or "orange". Proximity is another example. How close is close? How far is far? It is often helpful to clearly define these terms and then to calibrate the fuzzy thresholds and weights against your intuitions until the behavior feels right. The sky is the limit really. There are countless different situations in which being able to work with vagueness is advantageous.

Let's explore a few examples of nested min-max operators to make sure we have a good sense for their behavior. It's not enough to just understand how each operator behaves individually. We want to make sure we also understand how they behave when composed, because otherwise we might not understand how to interpret composite structures built up out of multiple operators. As such, suppose we have four fuzzy truth values named a, b, c, and d. Let's explore a few different ways of building fuzzy logic expressions with them.

Let's start with one of the easiest to understand examples. Consider the case where we have $(a \wedge b \wedge c \wedge d) \geq 0.70$. What does this mean? Well, it means that whenever all of a, b, c, and d are at least 70% true each then the overall expression will be evaluated as true, and otherwise false. What good is this? What can we use it for? Well, it fills pretty much the same role as a classical conjunction, it's just that it's more vague and tolerant now. Saying that something is "70% true" is like saying that it is "fairly true". Expressions we use in everyday language such as "somewhat" or "very" have pretty close analogs in fuzzy logic. What the precise correlations are is very arguable, subjective, and context-dependent, but here's one possible mapping:

natural language	fuzzy truth approx.
overwhelmingly false	0% to 5%
slightly true	5% to 20%
kind of true	20% to 40%
somewhat true	40% to 60%
fairly true	60% to 80%

very true	80% to 95%
overwhelmingly true	95% to 100%

Anyway, in contrast to our conjunction example, saying $(a \vee b \vee c \vee d) \geq 0.70$ would be like saying "at least one of a, b, c, or d is fairly true". What about the case where we mix operators though? Consider, for example, if we instead had something like $[(a \wedge b) \vee (c \wedge d)] \geq 0.90$. How would we interpret this one? Well, not too surprisingly, it would simply mean that at least one of $(a \wedge b)$ or $(c \wedge d)$ is very true (specifically ≥ 0.90).

What about if we flipped the operators and had $[(a \vee b) \wedge (c \vee d)] \geq 0.90$ instead? In that case, it would mean that at least one of a or b are very true and at least one of c or d are very true. Perhaps the user has two concerns, and some corresponding criteria that they want to be very true, but each of those two concerns can be fulfilled in two different ways, namely through a or b for the first concern and through c or d for the second concern. That would be a real-world scenario for why you might want a conjunction of disjunctions like this.

As for negations, it just reverses the meaning of the expressions, just as you would expect. Thus, if we had $\neg(a \wedge b) \geq 0.90$ then it would mean that we are testing whether the negation of $a \wedge b$ is very true, i.e. that we are testing whether at least one of a or b is very false. The expression $\neg(a \wedge b)$ is equivalent to $\neg a \vee \neg b$ under the min-max method. In other words, the min-max method obeys De Morgan's laws. This is in addition to also being order independent (commutative), grouping independent (associative), and distributive, which we already discussed earlier.

One subtle gotcha I want to point out here though is that $\neg(a \wedge b) \geq 0.90$ and $\neg(a \wedge b \geq 0.90)$ are *not* the same expression. You may have noticed that if you paid close attention to the way I worded the previous paragraph. I implied that "testing whether the negation of $a \wedge b$ is very true" is the same thing as "testing whether at least one of a or b is very false". In contrast though, if we had $\neg(a \wedge b \geq 0.90)$ then it would have been the same as testing whether a and b are not both highly true, i.e. as testing whether $a \wedge b < 0.90$.

Notice that the *inequality* is the only thing that flipped this time. Neither a nor b nor $\wedge$ flipped this time. The expression $\neg(a \wedge b) \geq 0.90$ is true only when at least one of a or b is very false, whereas the expression $\neg(a \wedge b \geq 0.90)$ is true whenever at least one of a or b is less than very true. Notice the very distinct difference. This kind of thing will bite you hard if you don't pay careful attention to it. Make sure when translating your thoughts to paper that your translation of what you think you are negating is actually the conceptually correct one.

When not applying thresholding to the result of a min-max fuzzy logic expression, the result simply tells you *how* true the expression is. Thus, for example, $a \wedge b \wedge c \wedge d$ would return a fuzzy truth value representing to what degree all of a, b, c, and d could be considered to be simultaneously true. Hence, 10% would mean they could be considered simultaneously at least slightly true. Likewise, 50% would mean somewhat true, 90% would mean very true, and so on.

You could then subsequently use the fuzzy truth value for proportional variation of some parameter in the system, if you felt inclined to do so. You could also use it for decision making processes that are more complicated than simple thresholding, or for merely observing the state of the system in terms of fuzzy truth. In any case though, the point is: Whether you use thresholding or not, the value of min-max fuzzy logic expressions still remains easy to think about regardless.

The behavior of the operators would not be even remotely as easy to think about if we were using the product-sum method. For example, suppose we were evaluating $a \wedge b \wedge c \wedge d \wedge e \wedge f$ under the product-sum method, and we knew that $a = b = c = d = e = f = 0.90$ was true. All of the participants in this conjunction are very true. How true is the conjunction though? The answer is 0.90^6 (i.e. 0.90 multiplied by itself 6 times, because there are 6 conjuncts involved) true, which comes out to be approximately 54% true. Thus, under the product-sum method, even though all of the participants are very true, the conjunction is still considered only somewhat true. This kind of non-linear behavior is confusing and often very difficult to control.

Of course, as we already discussed earlier, there's also the even worse case where the product-sum method converges infinitesimally close to 0 for conjunctions and 1 for disjunctions, based on essentially nothing more than the number of participants. This has the net effect of rendering the product-sum method virtually worthless for any expression containing a large number of disjunctions or conjunctions. Once the number of participants increases over a certain size, disjunctions will almost always return a value near 1, and conjunctions will almost always return a value near 0. Consequently, the number of participants for which the product-sum method still preserves logical distinctions is fairly small. This general pattern is true even when the product-sum method is augmented with similarity adjustment.

Basically, the behavior of the product-sum method is terrible. You will probably regret it if you try to use the product-sum method instead of the min-max method. Everything you try to do in fuzzy logic will be much harder if you do. It is a very difficult task to balance decision making thresholds for the product-sum method, whereas it is quite easy for the min-max method. The min-max method closely follows logical intuition, whereas the product-sum method has at least several unexpected gotchas. I know the min-max method might initially look a bit arbitrary at first glance, but as we have seen it is actually quite principled and well-founded.

The key to feeling comfortable working with the min-max method is simply to get a firm grasp on why the minimum and maximum functions were chosen, how they compose together to form higher-level expressions, and how to read and interpret those expressions in a coherent and intuitive way. Luckily, this is not hard. The min-max method is truly an elegant and powerful solution for how to think in terms of fuzzy logic. It is well worth learning and using. It probably doesn't currently get as much attention as it deserves. It is especially useful in computer programs, but is also very easy to perform by hand. It's pretty much an ideal system for working with rudimentary fuzzy logic.

However, there are a few specific application areas where you might conceivably want to use a specialized model for fuzzy logic, one that differs from the min-max method, but such contexts are in the minority probably. The obvious example is perhaps neural nets, which are computer simulations of neurons that attempt to simulate flexible learning. I don't know much at all about neural nets, as I've never really studied them, but it appears that there are many different models for how a neural network might work. I imagine that many of these models probably don't use the min-max approach, although some perhaps do. As for ordinary fuzzy rules though, like the kind you'd put into hand-written conditions in computer programs, the min-max method is probably the way to go, so I'd recommend you do that.

The min-max method seems to be the most human-readable model for fuzzy logic (as far as I'm aware). That's a big part of why it's my recommendation for most use cases for fuzzy logic. In contrast though, the product-sum method is so badly behaved that I suspect that it is actually outright conceptually wrong as a model of fuzzy logic. The initial argument in favor of the product-sum method looks superficially appealing at first glance, but as you analyze the product-sum method and discover its properties it becomes increasingly clear that it doesn't seem to fit.

For instance, the argument for why disjunction is defined as $P(x) \vee Q(y) \to (P(x) + Q(y)) - (P(x) \cdot Q(y))$ (see page 321), and the corresponding analogy for how it avoids double counting areas of sets (see page 326), while appealing in some ways, may just be smoke and mirrors. After all, what does the area of a set really have to do with fuzzy logic? Why would we assume that the product is necessarily a good model of conjunction? There are too many open questions for the product-sum method, and some of its properties seem unexpected and counterproductive. It just doesn't quite make sense. The min-max method seems to rest on far sounder principles.

The uncertainty surrounding which kind of system to use for fuzzy logic has, I think, slowed down its adoption rate and caused it to be used much less widely than it should be. When you make people face an ambiguous choice it often severely reduces how many people will be willing to make the choice at all. Thus, you end up with a situation where many people would benefit from a new technique, but don't bother because they are too bewildered by the options they have in front of them. Clear communication is essential if you ever want a new idea to become widely used. Even very small amounts of friction in the learning process will massively decrease how widely adopted a new idea will be.

It is not surprising then that ideas that are hidden away in cryptic academic journals seldom ever see wide use. If an idea is not communicated well, then it might as well not exist, because as far as the rest of humanity is concerned it already doesn't. That's just human nature, it seems. Many academic research articles are thus comprehensible and significant only to their own creators. Similarly, ambiguity in what method is actually the correct method for applying a concept will inevitably lead to not as many people using that concept.

In this respect, I hope that our discussion here has had sufficient clarity to remove this ambiguity in the choice of how to do fuzzy logic, if you have ever considered using it. Fuzzy logic is an empowering technique, one that should be used much more widely, and yet isn't. This is unfortunate. It really isn't even hard to understand. The only tricky part is realizing that the min-max method is the more conceptually correct way to do it. Fuzzy logic using the min-max method should be as widespread as standard classical logic, set theory, and predicate logic. It is an essential technique for dealing with certain common situations, and should be counted among the most basic and standard parts of logic and mathematics, yet isn't because of a quirk of history.

Fuzzy logic is a strict superset of standard classical logic in terms of capabilities, so in a sense if you're going to use a truth-functional logic then you might as well use fuzzy logic. When you want to only deal with the strictly false and strictly true cases then you can just restrict yourself to only using 0 or 1 as your truth values. Otherwise, whenever you want to handle something in-between strictly false and strictly true then you can just insert a fuzzy truth value into the intended expression and the system will still work seamlessly. The minimum and maximum functions are really not any more conceptually difficult to deal with than the regular disjunction and conjunction functions from standard classical logic, but are more powerful.

Anyway, we've covered a bunch of different ideas in this section on fuzzy logic, so lets take a moment to refresh our perspective before moving on to some different subject matter. As we have seen, there is a certain set of operations which form our basic vocabulary for fuzzy logic, i.e. our most essential tools for constructing a diverse array of arbitrary structures that utilize fuzzy reasoning. It's important to remember what these all are, so let's take a moment to list them here, collected into one place for easy reference and excluding the bad methods:

1. **the min-max method** ($\neg$ as complement, $\vee$ as max, and $\wedge$ as min)

2. **weighted averaging** (a.k.a. blending, i.e. mixing different criteria together in proportion to how much we care about them)

3. **algorithmic conditionals** (i.e. "if-then" as in programming... not as in material implication, which in contrast behaves badly)

4. **thresholding** (i.e. converting fuzzy truths into binary truths, typically for making discrete decisions)

5. **proportional variation** (i.e. varying behaviors of a system according to the magnitude of fuzzy truth values)

6. **fuzzy predication** (i.e. constructing maps from the raw information in the system to corresponding user-defined fuzzy truth values expressed as predicates, for whatever set of fuzzy criteria you might be interested in)

7. **application-specific methods** (e.g. neural nets, etc)

A few of these items in this list have not been given official names until now. For instance, previously in our discussion, we've always assumed that we already have pre-existing fuzzy predicates that just magically know what fuzzy truth values to return for any given input, but in reality you of course have to define this correspondence explicitly. We will refer to the act of defining such a correspondence as fuzzy predication[16].

For example, if you have a car moving down a street, you might need to create a fuzzy predicate that defines how fast "fast" is before you can decide whether the car is considered to be moving fast or not. This choice is somewhat arbitrary and context-dependent, so it's hard to give you a general process for it. A person moving down a

---

[16]This is a term I made up, based on an existing word with a similar meaning that fits the concept well.

sidewalk has a much different notion of "fast" than a car moving down a highway does. The exact nature of the gradient of the fuzzy boundaries is typically very arbitrary as well.

You'll need to make sure you calibrate your fuzzy predicates to the specific context of the environment in a reasonable way. Regardless of what technique you choose to use to do this, clamping your function output between 0 and 1 will guarantee that you don't accidentally mess things up by going out of bounds. Thus, when in doubt as to whether the range of your predicate function is between 0 and 1, always clamp. This process of fuzzy predication can take some time and care.

You'll want to make sure that the fuzzy truth values that your predicate function returns correlate strongly with your own estimates of what the fuzzy truth values of the inputs should be, and calibrate it accordingly. However, once properly defined, having these fuzzy predicates will significantly improve your ability to express "soft" criteria about the system in an easy way. This is useful for making computers behave in a somewhat less rigid manner, among many other things.

You may notice that several of the items in the summary list of fuzzy logic techniques above are related to each other. The reason for that is because fuzzy logic generally happens in multiple phases, each of which uses different techniques to move you closer to getting some real-world value out of fuzzy logic. There are three phases in total, some of which may be skipped sometimes. We will call the three phases (1) predication, (2) expression evaluation, and (3) utilization. Here's how the phases generally happen:

$$\text{predication} \rightarrow \text{expression evaluation} \rightarrow \text{utilization}$$

Note that the arrows I am using here (i.e. $\rightarrow$) are intended to just be illustrative arrows, not material implication and not transformative implication. The predication phase's only purpose is to reframe the relevant parts of the system into terms that fuzzy logic can understand and work with. The expression evaluation phase is where fuzzy truth values are tied together into arbitrary logical expressions, which are subsequently evaluated. Finally, the utilization phase is where the fuzzy truth values from the previous phases are put to use, usually in order to alter the state of the system. Here are the correspondences between the items in the fuzzy logic summary list and these three phases:

1. **predication phase:** fuzzy predication (obviously) and sometimes application-specific methods

2. **expression evaluation phase:** the min-max method, weighted averaging, and sometimes application-specific methods

3. **utilization phase:** algorithmic conditionals, thresholding, proportional variation, and sometimes application-specific methods

For some application-specific methods, such as neural nets, the phases may be blurred together and constantly changing due to ongoing learning processes. That's why I've included application-specific methods tentatively in all three phases here. The utilization phase is just where the fuzzy truth values are mapped into practical uses, like causing some action to happen, so it often isn't really a part of any learning processes per se. It's a bit debatable. If you ignore the application-specific methods though, then the differences between the phases are pretty clear cut generally.

Anyway, once you understand the basic workflow, fuzzy logic is really pretty easy to work with. However, when defining fuzzy predicates, it is important to always clearly understand that fuzzy logic is not the same as probabilistic logic or probability theory. Fuzzy logic does *not* deal with the uncertainty of events, and thus cannot be used to make any calculations about the chances of random events, but rather fuzzy logic only deals with vagueness and shades of truth.

It isn't generally safe to treat a probability value as if it is a fuzzy truth value. It might work sometimes, because the two concepts are indeed *somewhat* related. That being said, the best way to handle probabilities is of course to use probability theory. In this respect, it is important to realize that the laws of probability theory are *much* different from the laws of min-max fuzzy logic. If you want to learn more about probability theory and probabilistic logic though, then you'll need to look it up yourself.

I've decided to cut probability out of this section entirely, to save space, because the fuzzy logic discussion became much larger than originally planned, and also because my knowledge of probability and statistics is weak. If you do look up probability theory though, or are already familiar with it, then you will notice that it has a few structural similarities to the product-sum method that we discussed here for fuzzy logic. The difference is that in probability theory those ideas are *valid* in certain contexts, whereas in fuzzy logic those ideas seem to essentially be nonsense. Thus, in a sense, the product-sum method may actually be the result of people subconsciously (or consciously) confusing probability theory with fuzzy logic, leading to a misleading dead end: the product-sum method.

Fuzzy logic and probability theory are among the most rigorous, conceptually well-founded, and useful of all multi-valued logics. In contrast, all of the really widely known finitely multi-valued logics such as Kleene's 3-valued logic, Bochvar's 3-valued logic, and Łukasiewicz's 3-valued logic don't seem to have anywhere near as much practical utility. In fact, frankly, in my opinion, it just doesn't even seem worth bothering to use any of these 3-valued logics usually. The models of uncertainty used by the 3-valued logics are far too crude and limited, and don't even always behave in a conceptually correct way. In contrast, fuzzy logic (under the min-max method) and probability theory will serve you much better under most circumstances.

The concept of multi-valued logic allows you to expand your domain of consideration beyond merely the strictly true or strictly false values of standard classical logic. However, as we have seen, many multi-valued logics suffer from poor context awareness and counterintuitive behaviors. Among the least vulnerable to these negative effects are fuzzy logic and probability theory. In fact, probability theory is not even truth-functional and is actually at least somewhat context-aware, since it requires the user to be mindful of the context in which events occur and what dependencies exist between them (e.g. "the probability of B given that A has already occurred" etc). By comparison, fuzzy logic isn't context-aware, but luckily doesn't really need to be since it deals with vagueness instead of with real uncertainty[17]. Thus, probability theory and fuzzy logic are both safe for use, even though many other multi-valued logics are not.

Though fuzzy logic is truth-functional, and hence lacks real context awareness, it still has lots of redeeming value due to the fact that the values it returns are still accurate even despite the lack of context awareness. This retained accuracy comes from the stable nature of the min-max method combined with the fact that fuzzy logic models vagueness instead of uncertainty. However, in contrast, the 3-valued logics attempt to model uncertainty in a way that is far too crude and heavy-handed, and as a result suffer a heavy penalty for their lack of context awareness.

For instance, under any of the 3-valued logics $A \wedge \neg A$ will return U when A is U, which doesn't really actually make much sense. Such an expression should return F, since a contradictory form must always be false regardless of whether the participants' values are unknown. Fuzzy logic might seem to behave similarly at first glance, but the difference in interpretation is significant. In fuzzy logic, under the min-max method, if $A = 0.50$ then $A \wedge \neg A = 0.50$. This means that the degree to which both A and $\neg A$ can be considered simultaneously true is 50%.

This result is in fact exactly what you would expect and is conceptually correct, even despite the lack of context awareness. Similarly, let's consider disjunction: If $A = 0.60$ then $A \vee \neg A = 0.60$ because at least one of A or $\neg A$ can be considered 60% true, which is exactly how we intuitively would want this expression to behave under fuzzy logic. For example, we could use $A \vee \neg A \geq 0.90$ as a threshold condition to express that we want at least one of the two possibilities to be very true, i.e. that we want to exclude ambivalence, and this could be a useful criteria to specify under certain circumstances.

By the way, the reason why I don't recommend the use of an analog of material implication for fuzzy logic (e.g. as in $A \rightarrow B \leftrightarrow \neg A \vee B$) is because any artificial intelligence evaluating such expressions would be constantly led to believe that numerous different incorrect inferences are valid. If an artificial intelligence evaluates an expression of the form $A \rightarrow B$ and A is false, then it would not be valid for the AI to add this implication to its understood rules of inference, but that's exactly what having an analog of material implication would push the AI towards doing.

Other kinds of programs also often wouldn't react well to the presence of such an operator either. There's a reason why almost no programming languages have material implication as a built-in operation. Material implication is often a misleading and counterproductive way of thinking of implication for most real-world applications. If an implication operator doesn't at least express valid derivability of some kind, then it cannot possibly even be a correct model of real implication to begin with.

Like the 3-valued logics, standard classical logic is also not context-aware, and this is part of what causes nonsensical behavior related to the material implication operator. However, at least standard classical logic doesn't attempt to incorrectly define an "unknown" value, which it would be incapable of actually modeling due to the inherent limitations of being truth-functional. In contrast, the 3-valued logics just recklessly run

---

[17] e.g. When we say that something is "somewhat true" in fuzzy logic we are saying that we are *certain* that it is "somewhat true". There is no uncertainty about it. If a fuzzy predicate is "somewhat true" of an object, then it simply means that the relevant object *somewhat* possesses the indicated property.

face first into that problem by attempting to define unknown values regardless. Anyway though, since a lack of context awareness has been the norm for our discussion so far, let's mix things up to create a point of contrast by having our next section be about a branch of logic that is all about context awareness: relevance logic.

4.5 Relevance logic

In standard classical logic, you may have noticed that the condition and consequence of an implication need not have any relationship to each other whatsoever. This is caused by the truth-functional nature of standard classical logic, which defines all logical expressions as merely being a function of whatever the truth values of the corresponding inputs are. In this way, standard classical logic has no context awareness.

However, thinking about it from a principled standpoint, shouldn't the conditions of implications have some kind of genuine causal relationship to their consequences? Could we really even call it implication if this weren't true? If A truly does imply B, then shouldn't that mean that A should be sufficient to conclude B, and hence that A must be at least in some way related to B?

This discrepancy is the inspiration for relevance logic. The goal of relevance logic is to formalize what it means for a condition to be relevant to a consequence, and to subsequently use this knowledge to disallow the use of implications wherever no genuine relationship between the condition and the consequence exists. For example, consider the statement "If $2 + 2 = 4$ then dogs are mammals.". It has two parts: the condition "$2 + 2 = 4$" and the consequence "dogs are mammals", both of which are true.

In standard classical logic, any implication with both a true condition and a true consequence must be true, and hence standard classical logic would conclude that $2 + 2 = 4$ does indeed imply that dogs are mammals. Think about that though. Does that really make any sense? Does $2 + 2 = 4$ really imply that dogs are mammals? Classical logicians may be accustomed to thinking that it does, but if you think from first principles instead of merely from what you've been trained to think, then it seems like there is clearly no genuine implication relationship between the fact that $2 + 2 = 4$ and the fact that dogs are mammals.

Would a detective ever say "Aha! $2 + 2 = 4$ and therefore dogs must be mammals. Case closed!"? I think not. It doesn't seem logical in the slightest. No sane real-world detective would think like that, and indeed I doubt that even a mystery novelist (i.e. a writer of fiction) could get away with it. Why then do we allow classical logicians to do it, considering that even *writers of fiction* can't get away with it without losing all credibility? It really makes you think, doesn't it? Material implication is bonkers in all kinds of weird ways really. The fact that having a truth-functional implication operator seems so convenient on the surface doesn't make its behavior any less absurd.

So, naturally, some people think that the way to fix this problem is to find a way to require that every condition be relevant to its consequence before an implication can ever be considered to be valid. Thus, we have the idea of relevance logic. Here's the tricky part though: How does one define what relevance means? How do we devise a mechanical process that is capable of discerning whether or not a condition bears a genuine relationship to its consequence? It may not be immediately obvious how to do this. To make matters even more difficult, multiple different methods for seemingly accomplishing it may exist, each with conflicting points of nuance. What are some of these different ways of attempting to create a relevance logic? Which systems seem to satisfy the relevance criteria in the most conceptually correct way?

As it happens, unified logic will be one such system of logic that in at least one sense satisfies the relevance criteria. You'll see why later, once we get to that part. In fact, it seems to me that unified logic satisfies the relevance criteria quite well. Indeed, it seems to do so with perhaps little to no negative trade-offs, while still being easy to understand and easy to use. However, what about the *existing* methods for implementing relevance logic though? What about the *traditional* approaches?

Well, unfortunately, it appears that there are conflicting accounts of how best to implement relevance logic in the literature currently. Not only that, but it seems that the existing systems for relevance logic also appear to have significant weaknesses, wherein in exchange for satisfying the relevance criteria each system must give up at least one of its other intuitive logical properties, e.g. such as being forced to abandon the law of disjunctive syllogism (i.e. to abandon $\neg A \wedge (A \vee B) \to B$).

However, regardless of the specific details of the various conflicting systems for how to do relevance logic, one really commonly mentioned principle in the relevance logic literature (as it stands when this book was originally written) seems to be the concept of "variable sharing". The idea behind variable sharing is that only when the condition and consequence of an implication share at least one variable in common can the implication be considered potentially relevant and hence potentially valid. That's the idea anyway. It may not even be true.

However, keep in mind that the variable sharing criteria is considered to be a *necessary but not sufficient* criteria for relevance. To say that this criteria is necessary but not sufficient means that every relevant implication must share variables, but not all implications that share variables are relevant. The idea is that if an implication does not share at least one concept between the condition and consequence, then there's supposedly no way the condition and consequence could be relevant to each other, and hence in that case we can eliminate the possibility of the implication being relevant.

However, even if the condition and consequence *do* share variables in common then they could still be unrelated from the standpoint of inference. Another shortcoming of variable sharing is that sometimes variables can mask underlying hidden relationships and shared objects. Variables can easily have different names while still referring to the same thing. And, even when variables don't refer to the same thing, they can still implicitly refer to related things that are logically dependent upon each other.

Thus, it isn't safe to assume that A and B are unrelated concepts just because they have different names. Variables can also mask underlying shared objects in the domain, shared objects which might only become apparent if you used quantifiers and predicates bound to those objects, thereby potentially revealing more hidden relationships, rather than merely if you used unquantified atomic propositions, whose contents may be more opaque.

Simply applying the principle of variable sharing is one way to approach relevance logic, in that it gives you a partial implementation of relevance constraints, one that may eliminate *some* implications that lack relevance. However, for more full-fledged implementations it appears that other techniques may be used. One of the more common techniques for implementing relevance logic variants in the literature appears to be to utilize some form of the concept of "possible worlds" borrowed from modal logic and "Kripke semantics" that involves introducing "inconsistent worlds" and "non-normal worlds". These worlds supposedly allow for contradictions and random truth assignments to be treated as true within them. For that reason, since asserting contradictions as true isn't compatible with good reasoning generally, this approach to relevance logic (though apparently popular in some circles, maybe) therefore strikes me as probably fundamentally ill-founded and wrong.

That being said, I admittedly have little to no familiarity with Kripke semantics and the related literature. I tried to learn some of it a few times but it seemed conceptually confused and probably mostly a waste of time so I abandoned it. Besides, it is probably actually irrelevant to the thrust of the material in this book, as you will see. We will be taking an entirely different approach to implementing the concept of relevance once we get to the part where we create unified logic. In stark contrast though, unified logic will have no need whatsoever for the counterintuitive convolutions that these more traditional methods like Kripke semantics appear to suffer from.

Consequently, don't worry if you don't understand any of my comments here about "possible worlds" and "Kripke semantics" and such. It ultimately won't end up being important for us in this book. Frankly, a lot of the traditional methods in the literature for non-classical logic seem far too tedious to be worth bothering with, even if they did work correctly, which I actually find rather doubtful in some cases.

Tedious and counterintuitive systems typically never see wide adoption. That's just human nature. Not only that, but it's also how it *should* be. If a system can't be explained with clarity and ease, then people will have little reason to trust it. In this respect, I think you will find that unified logic will be more intuitive to work with than most of the traditional non-classical logics in the literature, thankfully.

Anyway, we've basically got what we came here for now. Understanding the core motivating principle of relevance logic, i.e. the notion of a condition being *relevant* to its consequence, was our real goal here. As such, I won't be covering the details of actually *performing* traditional relevance logic here. In fact, I don't even know how to, so I couldn't really teach it to you even if I wanted to. I mostly just know some overview information about it. However, be that as it may, it is of no real consequence for the goals we are pursuing in this book. Having an overview of the purpose of relevance logic (and a rough sketch of why the current systems for it seem to probably be inadequate) is more than sufficient for our purposes in this book.

If you want to learn more about how systems of relevance logic are traditionally implemented then you will have to look it up yourself[18]. If you ask me though, it's not worth the pain and you should just stick around until after we've created unified logic. When that time comes, I'll demonstrate to you that unified logic does indeed

---

[18] By the way, relevance logic is sometimes also referred to as "relevant logic", but I prefer the term "relevance logic" because it seems less ambiguous to me. For example, someone could say "make sure you use the relevant system of logic" and in doing so may *not* be referring to relevance logic, thus creating ambiguity, and hence my reason for preferring the term "relevance logic" instead.

satisfy the core motivations of relevance logic, but without nearly as much cruft and counterintuitive results, compared to what traditional systems of relevance logic seem to suffer from.

In any case though, it seemed fairly difficult to find any truly clearly explained tutorials on how the traditional approach to relevance logic actually works. Personally, I just gave up after a certain point, as a matter of pragmatism. It's outside the scope of what I care about for the purpose I'm trying to fill. This is unfortunately a common pattern, not just with the literature on relevance logic but with much of the literature of *many* of the non-classical logics (and indeed with much of mathematics in general, honestly).

So much of the literature about the non-classical logics is written in a highly academic, pedantic, convoluted, and opaque way. Yet, logic is something so fundamental and essential that its meaning must be *crystal clear* if it is to be useful at all. As such, the idea that any of these convoluted traditional systems of logic could ever become widely adopted given the currently obscure state of the literature strikes me as unlikely. Clarity is king. You can't expect a system to gain widespread use without a very high degree of clarity and accessibility. Anyway though, time to move on to the next section. Since I just recently mentioned modal logic in this section, in connection with my tangential remarks about "Kripke semantics" and such, it seems fitting that modal logic should be our next section.

4.6 Modal logic

Some truths are necessary, whereas other truths are merely possible. Capturing this distinction and expressing it formally is the most common application for modal logic. However, it is certainly not the only one. Modal logic is actually a family of related logics that all have similar structural characteristics. All modal logics focus on qualifying statements in some way with respect to some "mode" of being, or at least with respect to some specialized way of reasoning. More precisely, all modal logics qualify statements with respect to some sense in which the statements may be true either in part or in total or not at all along some situational dimension. This will make more sense once you've seen examples.

Some of the most common modal logics include: (1) alethic logic, (2) temporal logic, (3) deontic logic, (4) epistemic logic, and (5) doxastic logic. Alethic logic is the modal logic of necessity and possibility, but it is often referred to simply as "modal logic" due to its ubiquity. As such, you need to be careful when interpreting the phrase "modal logic", as it has both a narrow sense (specifically alethic logic) and a broader sense (the entire family of logics). Temporal logic is the modal logic of qualifying statements with a temporal tense (i.e. past, present, or future truth). Deontic logic is the modal logic of permission, obligation, and restriction. Epistemic logic is the modal logic of knowledge. Finally, doxastic logic is the modal logic of beliefs. Epistemic logic and doxastic logic are at least somewhat closely related.

We will primarily only be concerning ourselves with alethic logic in this section. If I mention the other modal logics at all, then it will only be in passing, in order to make some kind of tangential point perhaps. As such, if you want to learn more about the other modal logics then you will have to look them up yourself. Furthermore, even though our focus will be on alethic logic, I am certainly no expert on it. Indeed, I only have a basic familiarity with the underlying principles, and I am only acquainted with a few of the more trivial rules. That's ok though. Our only real purpose here is to extract the core motivating principles of modal logic and then move on. Unified logic will satisfy the same motivating principles as modal logic, but arguably with less conceptual mud than the traditional modal logics.

Anyway, let's get started on our discussion of alethic logic. Naturally, since alethic logic is all about necessity and possibility, our first step should be figuring out what exactly it means for something to be necessary versus to be possible. The basic idea is not hard to grasp, although some of the points of nuance may give you pause.

Essentially, to say that something is a necessary truth means that it inherently must be so or is in some sense inevitable, whereas to say that something is a possible truth merely means that it could be true or just happens to be true but in principle doesn't have to be. For example, in the real world, $2 + 2 = 4$ is a necessary truth whereas the fact that some specific lamp is sitting on some specific table may be considered as merely a possible truth, albeit one that happens to be true. The lamp could be moved somewhere else, so in that sense its presence at the current table on which it sits can be considered merely a possibility, whereas there doesn't seem to be any coherent way to make $2 + 2 = 4$ no longer true in the real world.

There's an obvious philosophical objection to this entire distinction between necessity and possibility though, and that's the conflicting worldviews of determinism versus free will. Determinism is the idea that everything that was, is, or ever will be has been predetermined by causality, and hence that everything that becomes true necessarily must have come to pass in exactly that way. Thus, if determinism is true, then one can argue that all things that are true are necessarily true, and therefore that in a certain sense the concept of "possibility" is nothing more than a figment of the imagination.

The concept of free will though, in contrast, argues that events are not merely predetermined, but instead that we can genuinely change them by exercising our free will to change the future. Thus, people who believe in free will also believe that possibilities are real things that we really can choose between, rather than merely figments of our imagination caused by limited access to information.

Regardless of whether determinism or free will is the correct way of understanding existence and choice, it is still nonetheless clear that we humans are inevitably bound to the limitations of possessing incomplete information about the world around us. Therefore, regardless of whether determinism or free will is true, it is still useful for us as human beings to consider multiple different possibilities so that we can plan our lives as advantageously as possible. People sometimes like to say that "Anything is possible.". However, in reality, as appealing as this idea may be, it is very much a false sentiment. Many things are not possible. A wise person should cultivate a balanced perspective.

Indeed, some people take this idea too far. Some people find the idea of determinism crushingly depressing and use it as an excuse to give up on their ambitions and dreams prematurely or to hold self-limiting beliefs about what they are capable of. However, such an attitude is always guaranteed to be counterproductive *regardless* of whether determinism or free will is true. Even determinism itself, if true, actually does nothing to imply that you should resign yourself to your fate. It is always still in your best interest to act as if you *do* have a choice in determining your future, relentlessly striving to maximize your own well-being and life outcomes, regardless of whether everything is ultimately predetermined or not.

For that reason, you should never let the philosophical dilemma between determinism and free will actually bother you. Simply live life to the fullest regardless. Self-limiting beliefs are generally not helpful. You can't test your boundaries if you never try to push beyond them. Dream big, but do so in a controlled and emotionally stable way by embracing objectivity, realistic pragmatism, confidence, creativity, a perpetually fresh state of mind, and a willingness to appreciate all the little nuances of life. Neither make your life into a self-fulfilling prophecy of demoralizing pessimism, nor into a delusional quagmire of unsustainable optimism. Seeking an adaptable attitude and peace of mind is realistic and healthy, whereas seeking endless thrill and elation is not. A balanced perspective is generally best.

Anyway though, putting that tangent aside, let's return our focus to alethic logic and our discussion of necessity and possibility. Generally, in modal logic, people like to understand the concepts of necessity and possibility by thinking in terms of "possible worlds". Basically, the idea is that you imagine that there exists some set of "possible worlds", each of which represents one conceivable way that the world could be.

You then identify some property of interest, as expressed in some statement about the world, and ask yourself in which of the possible worlds that statement is true. If the statement is true in *all* of the possible worlds (e.g. $2 + 2 = 4$) then you say that it is necessarily true. In contrast, if the statement is only true in *some* of the possible worlds (e.g. some specific lamp is on some specific table) then you say that it is merely possibly true.

Notice that this is very similar to the concepts of universal and existential quantification from predicate logic. Universal quantification makes a statement about *all* members of a set, in much the same way that modal necessity makes a statement about *all* possible worlds. Likewise, existential quantification makes a statement about *some* members of a set, in much the same way that modal possibility makes a statement about *some* possible worlds. The concepts of necessity and possibility also have close relatives in set theory itself (specifically subset relations and intersection relations), but I'll have more to say about that later, when we discuss unified logic.

Anyway though, as alethic logic focuses on the distinction between necessity and possibility, you might be tempted to think that all statements posed in terms of alethic logic must be qualified by either necessity or possibility. However, you would be wrong. Really, there are actually three different fundamental types of qualified statements in alethic logic: currently true, possibly true, and necessarily true. Modal logic treats these qualifiers as unary operators, prefixed to the expressions to which they apply.

Necessity is indicated by a $\Box$ symbol and possibility is indicated by a $\Diamond$ symbol. In contrast though, if you merely want to say that a statement is currently true, and leave the matter of whether it is necessary or possible unstated, then you simply don't attach any prefix symbol at all. Thus, if you have some statement X, then $\Box X$ and $\Diamond X$ would usually be read as "necessarily X" and "possibly X" respectively. In contrast though, when X stands on its own, you can read it as "currently X", although this practice is not standard and people seem to more often just read it as "X".

All currently true statements are possibly true, but not all possibly true statements are currently true. Some currently true statements are also necessary, but some are not. All necessarily true statements are also currently true though, regardless of what frame of time or possible world you examine. Some possible truths are also necessary truths. Possibility doesn't exclude necessity, just as inclusive disjunction (e.g. $A \lor B$) doesn't exclude the possibility of both participants being simultaneously true.

Knowing that something is possible is *not* enough to conclude that its negation is also possible. However, in addition to the primary modal qualifiers of alethic logic, there are also two compound qualifiers made possible by the use of logical negations: impossibility and contingency. Thus, in practice, there are actually (in a sense) five common ways of qualifying a statement in alethic logic. Suppose X is a statement. The five common ways of qualifying statements in alethic logic, as applied to X, would look like this:

qualifier	expression
current	X
necessary	$\Box X$
possible	$\Diamond X$
contingent	$\Diamond X \land \Diamond \neg X$
	(alternatively: $\neg \Box X \land \neg \Box \neg X$)
impossible	$\neg \Diamond X$
	(alternatively: $\Box \neg X$)

Notice that contingency behaves the same as possibility except that it excludes the possibility of necessity. In other words, contingency is to possibility what exclusive disjunction (XOR) is to inclusive disjunction (OR). As you may have guessed by now based on these definitions, an analog of De Morgan's laws does indeed apply to modal qualifiers. This is much like how we were able to extend De Morgan's laws to also apply to universal and existential quantifiers in the section on predicate logic.

This is not surprising really if you think about it. Conceptually, the modal necessity and modal possibility qualifiers amount to little more than specialized domain-specific universal and existential quantifiers, so it isn't surprising that they end up behaving remarkably similarly. Suppose X is a statement. Here's what the modal versions of De Morgan's laws look like when applied to X:

$$\text{necessity negation:} \quad \neg \Box X \leftrightarrow \Diamond \neg X$$
$$\text{possibility negation:} \quad \neg \Diamond X \leftrightarrow \Box \neg X$$

As such, it is also possible to define necessity and possibility in terms of each other. Hence, you technically don't need to define both qualifiers, since you can always express the other qualifier as long as you have at least one of the qualifiers defined and also have logical negation available. You'll also want to define conjunctions or disjunctions though, so that you'll be able to express contingencies as well as possibilities. Anyway, here's how you define necessity and possibility in terms of each other via negation:

$$\text{necessity in terms of possibility:} \quad \Box X \leftrightarrow \neg \Diamond \neg X$$
$$\text{possibility in terms of necessity:} \quad \Diamond X \leftrightarrow \neg \Box \neg X$$

However, these alternative equivalent forms are more of a mouthful than simply using necessity and possibility directly, and it's much harder to intuitively grasp the meaning of "not possibly not X" and "not necessarily not X" than it is to intuitively grasp "necessarily X" and "possibly X". Oh, and by the way, make sure you don't get the concept of "sufficient versus necessary conditions" from the theory of implications confused with the concept of "necessity versus possibility" from modal logic. To say that B is necessary for A means that A implies B, whereas saying that B is necessarily true is a universal quantification over possible worlds.

If you find this distinction between necessary conditions and necessary truths confusing (which would be understandable), then you can try replacing the word "necessarily" with "inherently" and that may help you keep the distinction between the two concepts clear in your head. Thus, instead of reading $\Box X$ as "necessarily X" you could instead read it as "inherently X". This also makes it easier to safely use both the concept of necessary conditions and the concept of modal necessity in the same discussion, which otherwise might be confusing if you use the same term ("necessary") to refer to both.

On the other hand though, there's a reason why the same word ("necessary") is used for both. There's a strong underlying connection between the concept of a necessary condition and a necessary modal truth. In particular, when we say that B is a necessary truth in modal logic, what we are really saying is that the set of all possible worlds is a subset of the set of all conceivable worlds where B is true. This has the structure of $A \subseteq B$, which also perfectly mirrors $A \rightarrow B$.

Combine this fact with the fact that $A \rightarrow B$ can be read as "B is necessary for A" (or as "A is sufficient for B") and this is probably why the same word ("necessary") appears in both the sufficient versus necessary condition distinction and in modal logic's qualifier names. Thus, you can either choose to illustrate this underlying connection by continuing to use the word "necessary" in both cases, or you can choose to avoid any potential ambiguity by using the word "inherent" instead of "necessary". Pick your poison. They each seem to have pros and cons.

This specific system of alethic logic that we've been discussing here is just the most common variant of modal logic. This variant is sometimes also referred to as classical modal logic. Other variants of modal logic exist as well, each with slightly differing notions of which rules of inference are permitted and which are not. Each variation is traditionally symbolized by a specific cryptic letter, sometimes with a subscripted number attached.

These cryptic names originate from historical texts on modal logic and will inevitably seem quite odd and arbitrary to the uninitiated (including me). Frankly, this traditional naming system for these variants of modal logic seems contrived and uninspired. It would probably benefit greatly from some reorganization and refactoring.

Some common examples of these cryptically named variants of modal logic include: K, D, T, S_4, and S_5.

Besides just these cryptically named technical variants of modal logic, there are also of course the variants that embody different *interpretations* of "mode" than alethic logic does: temporal logic, deontic logic, epistemic logic, and doxastic logic. I mentioned these near the beginning of this section (see page 352). From what I've read, some of the cryptically named variants are related to some of the different interpretations of mode, but I am not familiar enough with modal logic to know the extent of this relationship. It doesn't really matter for our purposes. However, you will notice in all cases that the same underlying pattern of behavior (i.e. behavior analogous to universal and existential quantification) is essentially always there in some way, regardless of which variant or interpretation of modal logic you are studying.

In any case, that pretty much sums up our discussion of modal logic. If you want to learn more, then just look it up. Our purpose here was simply to extract the core motivations and basic sense for what modal logic is. It's time to move on now. What non-classical logic should we explore next? Well, how about a logic whose purpose is to try to address the explosively ill-behaved nature of the material conditional? After all, that has always been one of our leading causes of bafflement throughout our explorations of logic.

It's time we gave it some more thought. Classical logicians may think that falsehoods should imply anything, but an independently minded observer, one thinking from first principles, would likely disagree. It really doesn't make much sense that falsehoods should imply anything and everything. We'd like to be able to work with false premises without accidentally summoning the apocalypse every time we attempt to do so. That brings us to our next branch of non-classical logic: paraconsistent logic.

4.7 Paraconsistent logic

Any system of logic in which the principle of explosion ("from falsehood, anything follows") does not hold can be considered to be a paraconsistent logic. Paraconsistent logics can roughly be divided into two camps: weak and strong. This classification scheme is not a value judgment (strong doesn't mean superior here), but rather it is a measure of how extreme the corresponding paraconsistent philosophy is.

Weak paraconsistent logics are designed to invalidate the principle of explosion, but strive to do so in a way that otherwise leaves the traditional principles of reasoning as intact as possible. Strong paraconsistent logics, in stark contrast, accomplish their immunity to the principle of explosion by outright allowing for the existence of "true contradictions", i.e. by allowing some statements to be considered both true and false simultaneously.

In other words, weak paraconsistent logics are rooted in a desire to merely stop theories from degenerating into nonsense from the presence of a single contradiction, whereas strong paraconsistent logics are rooted in dialetheism (i.e. in belief in "true contradictions", a very extreme and doubtful philosophy). If you ask me, only the weak forms of paraconsistent logic are respectable. Dialetheism in contrast is essentially a self-defeating and incoherent worldview.

By allowing statements to be both true and false at the same time, strong paraconsistent logics (i.e. formal variants of dialetheism) essentially render the entire concept of negation hollow and not even representative of what it means to negate something. The *definition* of falsehood is the absence of truth, and hence any worldview that allows something to be both truth and false at the same time and in the same sense cannot possibly be reasonable.

Anyway, I've mentioned the principle of explosion before, way back in the classical logic section (see page 111, the first time we ever discussed it). However, since the dangerous behavior of the principle of explosion is the primary inspiration for paraconsistent logic, it would be wise to discuss the principle of explosion again here in greater depth. You'll need to understand the principle of explosion if you have any hope of ever understanding the underlying motivations for creating a paraconsistent logic. It is necessary for context. So, naturally, it's what we'll be discussing next.

The principle of explosion is pretty easy to understand. It is very closely related to the paradox of entailment[19]. In fact, both the principle of explosion and the paradox of entailment express the idea that "from falsehood, anything follows" in some sense. However, the principle of explosion is typically exemplified by a sequence of formal reasoning steps through which any arbitrary statement can be "proven", whereas the paradox of entailment is typically exemplified by the way material implication reacts to the presence of false conditions.

The principle of explosion can apply to more systems of logic than just classical logic though, whereas the paradox of entailment is most often concerned specifically with classical logic and material implication. However, both terms are sometimes used synonymously. The distinction is a bit sloppy and can vary from source to source. Conceptually though, the principle of explosion and the paradox of entailment are essentially the same idea, just expressed a little differently. Thus, the distinction often doesn't really matter that much, and consequentially you don't always need to worry about it too much. Anyway, here's how the principle of explosion is usually presented:

step #	form	rule of inference (justification)
1	$A \land \neg A$	assumption
2	A	conjunction elimination (from 1)
3	$A \lor B$	disjunction introduction (from 2)
4	$\neg A$	conjunction elimination (from 1)
5	B	disjunctive syllogism (from 3 and 4)

This argument is the principle of explosion in action. The sequence of steps in this argument is essentially a sequence of applications of meta implications. We begin with a set of forms consisting only of $\{A \land \neg A\}$. This is our initial "rule set" or "theory", i.e. the set of forms that we have so far determined (or assumed) to be "true". We then begin

---
[19] The paradox of entailment was first mentioned on page 110.

applying various meta implication rules that we know to be valid in classical logic to this set.

First we use conjunction elimination (i.e. $\{\alpha \wedge \beta\} \prec \{\alpha\}$ or $\{\alpha \wedge \beta\} \prec \{\beta\}$), which allows us to add A to our set, yielding $\{A \wedge \neg A, A\}$. Next we apply disjunction introduction (i.e. $\{\alpha\} \prec \{\alpha \vee \beta\}$ or $\{\alpha\} \prec \{\beta \vee \alpha\}$), which allows us to add $A \vee B$ to our set, yielding $\{A \wedge \neg A, A, A \vee B\}$. Then we apply conjunction elimination a second time, adding $\neg A$ to our set and thus yielding $\{A \wedge \neg A, A, A \vee B, \neg A\}$. Finally, we combine $A \vee B$ and $\neg A$ via disjunctive syllogism (i.e. $\{\alpha \vee \beta, \neg \alpha\} \prec \{\beta\}$ or $\{\alpha \vee \beta, \neg \beta\} \prec \{\alpha\}$) to conclude that B must be true, yielding $\{A \wedge \neg A, A, A \vee B, \neg A, B\}$.

However, notice that B can be any arbitrary statement whatsoever and yet this sequence of inferences will still always work regardless. Thus, in classical logic, if a contradiction is assumed to be true then *all* conceivable statements must also be true as a consequence (including their own negations), thereby rendering any theory that contains even a single assumed contradiction immediately nonsensical and incoherent. It is common to say that such a theory is "trivial", but I very much prefer to use the terms "incoherent" or "explosive" instead.

One could say that a very simple theory is "trivial" and not mean to say that it is explosive. This is a very common manner of speaking. Thus, the common practice of referring to explosive theories as being "trivial" is actually a very bad terminology choice, despite how popular it is. Using the terms "incoherent" or "explosive" is far less ambiguous. Thus, you shouldn't use the phrase "trivial theory" to refer to such cases if you care about clarity. Instead, you could say "incoherent theory" or "explosive theory". I personally prefer "explosive" most though. Popularity does not imply logical merit. Don't be burdened by the weight of mindless adherence to tradition. Let your mind fly free.

Allowing theories to degenerate into nonsense the moment a single contradictory error has crept into the system is clearly counterproductive. We would instead ideally want our system of logic to be more error resistant than that. Not only that, but from a principled standpoint it doesn't really even make much sense that one would ever be able to conclude any arbitrary unrelated statement B merely from a contradiction $A \wedge \neg A$. Indeed, if A and B have no relevance to each other, then deducing B from $A \wedge \neg A$ would seem very arbitrary and irrational. It doesn't really seem logical at all.

Just because an idea was beaten into you in school doesn't mean you should believe it. Sometimes these ideas we are taught in school get so ingrained in our minds though, that we sometimes lose all conscious awareness that the ideas are actually just preconceptions, and subsequently we fail to realize that things could be different. Paraconsistent logic is inspired by the idea that theories should not explode from the mere presence of a single contradiction. The question is though, how do we actually implement this paraconsistent philosophy?

Well, the most common approach seems to be to somehow block one or more of the rules of inference that allow the principle of explosion to operate. Look at the line by line argument expressing the reasoning behind the principle of explosion above. What would happen if we no longer held some of the rules of inference that justify those steps as valid? This could stop the user from ever completing the argument, thereby having the net effect of preventing explosion from occurring. In this respect, the most common approach to implementing paraconsistent logic seems to perhaps be to deny the validity of either disjunction introduction (i.e. $\{\alpha\} \prec \{\alpha \vee \beta\}$ and $\{\alpha\} \prec \{\beta \vee \alpha\}$) or disjunctive syllogism (i.e. $\{\alpha \vee \beta, \neg\alpha\} \prec \{\beta\}$), or both.

This does indeed block the principle of explosion. However, it also comes at a high cost. Disjunction introduction and disjunctive syllogism are both extremely intuitive, and it seems very odd and perhaps even unnatural to deny their validity. Thus we are faced with a trade-off. We may block the counterintuitive and dangerous principle of explosion by denying disjunction introduction or disjunctive syllogism, but only at the cost of now counterintuitively no longer being able to apply disjunction introduction or disjunctive syllogism despite how obviously reasonable they seem. Such trade-offs are common when working with non-classical logics.

You might be tempted to think that another way of blocking the principle of explosion would be to deny conjunction elimination, as that is also one of the steps used in the argument for the principle of explosion, but you'd be wrong and you should be careful about thinking like that. The reason I say this is because there is actually a second (perhaps less commonly presented) way of constructing the argument for the principle of explosion, one that never uses a conjunction operator. It looks like this:

step #	form	rule of inference (justification)
1	A	assumption
2	$\neg A$	assumption
3	$A \vee B$	disjunction introduction (from 1)
4	B	disjunctive syllogism (from 2 and 3)

This is a bit shorter, and probably even a bit clearer, than the usual way that the principle of explosion is presented. It shows the underlying structure of the principle of explosion a bit more concisely and takes one less step. More importantly though, since this argument never uses conjunction (i.e. $\wedge$), it would still be a valid expression of the principle of explosion *even if you denied conjunction elimination*.

That's why it would be incorrect to say that denying conjunction elimination would block the principle of explosion. It actually wouldn't. This is yet another example of the need for cautious thinking when working with logic. Blocking a single step in an argument won't necessarily block the entire argument, although sometimes it will. It depends on how fundamental the step is to the integrity of the argument.

By trimming the argument down like this, and eliminating unnecessary steps, we increase the probability that the remaining steps are fundamentally essential to the argument. Thus, on this basis, it appears that disjunction introduction and disjunctive syllogism are probably essential to making the principle of explosion work, at least in classical logic. Indeed, this may be true in classical logic, but it won't be true in unified logic. Unified logic operates on different foundational principles than classical logic, and thereby will be able to be paraconsistent without ever denying disjunction introduction or disjunctive syllogism.

If we didn't find a way to block the principle of explosion, then statements such as "The fact that water is wet and water is dry implies that gravity is a force of repulsion."

would have to be treated as true, which just doesn't seem like good deductive reasoning. As such, we would like to be able to deny the principle of explosion without having to make any major negative trade-offs by doing so. Unfortunately though, all the existing paraconsistent logics in the literature seem to come with hefty negative trade-offs.

Like most of the traditional non-classical logics, these paraconsistent systems seem to have counterintuitive and tedious properties that generally make them not worth the effort to actually use. In contrast, a more healthy and natural balance of logical properties with be built into unified logic, as you will see once we create it. In the meantime though, we'll need to continue our survey of the non-classical logics so that we can properly understand the background and motivations underlying each of them. Each branch of non-classical logic essentially illuminates a different facet of the hidden flaws lurking in classical logic. They each provide clues about what is truly wrong.

By exploring and understanding these branches, we gradually will come to better understand what the real problem with classical logic is. The better we understand something, the better we can solve it. Problems that seem impossible when not clearly stated or not clearly understood often become very easy to solve once all of the conceptual mud involved has been eliminated. The difficulty involved in solving many problems is often actually an illusion, an illusion originating from a simple lack of clarity. A lack of clarity often makes even the simplest problems seem prohibitively complex.

Anyway though, that pretty much wraps up our discussion of paraconsistent logic. We got what we came for. As always, feel free to research more about paraconsistent logic if you feel like it though. It's time to move on to the next branch of non-classical logic. Before we do though, I want to make sure that you're aware that all of the non-classical logics are not necessarily exclusive categories. Some logics will belong to multiple types of non-classical logic. For example, some variants of paraconsistent logic are also relevance logics. Some are also multi-valued logics.

Whenever someone says that something belongs to a certain category, you should never assume that membership in that category is exclusive unless you have good reason to do so. This is a common mistake among inexperienced thinkers. It's important to remember that information that isn't explicitly stated or evident can't really safely be assumed generally. You need to understand things exactly as they are and not overreach with your conclusions. This is exactly contrary to the way people often habitually behave in social settings, where you commonly see people gossiping and overreaching wildly based on even the slightest and most ludicrously insignificant and unreliable pieces of "information" about other people.

Thus, logical thought often very much goes against the grain of social conditioning, social conditioning which unfortunately is often very toxic and counterproductive in nature. You'll need to be mindful of this for that reason, to make sure that the presumptuous thinking patterns frequently encountered in your social conditioning don't bleed over into your pursuit of truth, where such ways of thinking would often be extremely damaging and misleading. Truth is a timeless monolith, an unassailable monument which stands perpetually unmoved by the idle whims of society. Just as a magnetic compass is compelled to point towards the north, so too is a timeless mind compelled to point towards the truth.

Anyway though, we need to decide what branch of non-classical logic to cover next.

What should we pick? Well, as it happens, some of the more extreme forms of paraconsistent logic (e.g. dialetheism) are tantamount to denying the law of non-contradiction, i.e. to denying that $A \land \neg A$ is always only false. Coincidentally, one of the subbranches of constructive logic (specifically intuitionistic logic) can be viewed as being tantamount to denying the law of the excluded middle, i.e. to denying that $A \lor \neg A$ is always only true. The law of the excluded middle and the law of non-contradiction are a complementary pair and are very closely related to each other, although not the same. For that reason, let us use this trivia as an excuse to segue into constructive logic next.

4.8 Constructive logic

Constructivism is a general mathematical philosophy, one that holds that the only way to prove that an object exists is to actually construct (or find) the object, rather than to merely "prove" the object's existence indirectly through methods such as proof by contradiction. Any proof satisfying this philosophy is referred to as a constructive proof, and any proof which is not a constructive proof is called a non-constructive proof.

Constructivism is somewhat outside the norm. Most logicians and mathematicians will accept non-constructive proofs as valid, although many find constructive proofs to be more satisfying and more illuminating regardless. A constructive logic is simply any system of logic which adheres to some kind of philosophy of constructivism when deciding which proofs are considered valid and which are not.

People who adhere to the philosophy of constructivism are called constructivists. Constructivism is a more conservative and cautious philosophy than the philosophy implicitly embodied by classical logic. The closest analog to constructivism outside of mathematics, found in everyday life, is the informal attitude of "I'll believe it when I see it", which one often encounters among skeptical people.

To such people, it is often not enough to merely make an argument that seems to prove a certain point, but rather such people require actual tangible evidence that they can reach out and touch before they will believe that a claim has truly been proven. Only tangible constructions, such as specific concrete examples and airtight algorithmic processes, rather than merely indirect proofs, are considered to be real evidence by constructivists.

To be clear though, constructivists don't necessarily require something to be constructible in the real world for it to still nonetheless be considered constructively proven. If a well-defined method for iteratively building towards constructing any arbitrary variation of a type of object is given, then a constructivist would usually consider such a method to be a perfectly valid proof of the existence of all such variations of that object, even if not all such objects are practical to actually construct.

Constructivists generally accept generative processes (e.g. applications of the principle of mathematical induction, computer programs, arbitrarily long sequences of transformations of forms, etc) as proofs. They just don't accept indirect proofs, such as proof by contradiction. Constructivists also tend to be very strict about avoiding "handwaving" and tend to treat underspecified or outlandish claims with higher than usual suspicion.

Constructivists also tend to favor computable objects over intractable or purely theoretical ones. Constructivists are more philosophically aligned with computer science and real-world practical concerns than some other types of logicians and mathematicians are. However, be aware that there are multiple different variations of constructivism, so it should not be considered as one monolithic viewpoint.

Some types of constructivists (e.g. finitists and ultra-finitists) restrict the use of the concept of infinity. For example, finitists and ultra-finitists both believe that an object does not exist unless it can be reached in a finite number of steps, via some process that builds upon some pre-existing finite object. Regular finitists still consider infinity to exist in a certain sense, but treat it strictly. They allow for "potential infinity" but not "complete infinity".

Ultra-finitists though take it one step further and theorize that infinity may not even exist at all, due to there perhaps being a largest possible number that our universe is capable of representing. The reason is because there is only a finite amount of observable matter and energy in existence (as far as we are aware) with which one could attempt to represent the greatest number, and hence in that sense infinity may not really exist at all.

However, most constructivists are usually fine with virtual existence. An object doesn't need to exist in our universe for it to still be considered logically coherent as something that could hypothetically exist in some other system or alternate universe. Logic and mathematics are concerned with the study of *all* conceivable logically coherent universes. Sometimes that coincides with the real world and sometimes it doesn't. Just because logic and mathematics allow for the exploration of completely arbitrary logically coherent systems doesn't imply that all ways of proving things within those systems are equally compelling though. The subject matter of each system may be arbitrary, but the logical coherence and credibility of specific claims within each system are definitely not.

A completely arbitrary and imaginary logical system, one with no connection to physical reality, can still nonetheless have an abundance of constructive proofs pertaining to its content. Constructive provability and correlation with reality are two totally independent aspects of a system that have little to nothing to do with each other, and it is important to understand this. The question in proving the existence of an object constructively is *not* "Is this object a part of physical reality?" but rather it is "Can a completely constructed tangible example of this object be built within the system being studied, with nothing at all left to the imagination?". If the answer to the later question is yes, then the object is indeed constructively provable.

The most widely known form of constructive logic is a system of logic known as intuitionistic logic. Intuitionistic logic's primary characteristic is that it generally forbids the use of proofs by contradiction. It does this in order to force the user to use direct proofs instead of indirect proofs when attempting to determine whether or not a given object exists. Intuitionistic logic is similar to classical logic in most respects, except mainly for the fact that it does *not* hold the law of the excluded middle ($A \vee \neg A \to T$) or the law of double negation elimination ($\neg \neg A \to A$) to be valid. The law of non-contradiction ($A \wedge \neg A \to F$) is still held as valid though, and a constructively verified contradiction is still sufficient to dismiss a claim as being false.

However, while the law of double negation elimination ($\neg\neg A \to A$) may be invalid in intuitionistic logic, the law of double negation introduction ($A \to \neg\neg A$) is still valid. Weird right? This is yet another example of how the traditional non-classical logics tend to require you to make some very counterintuitive sacrifices. You can gain a specific non-classical property that you would like to have, but only at the cost of losing an already existing intuitive property of classical logic that you would rather not lose.

Working with the traditional non-classical logics often feels outlandish. It feels as if no matter what you do you always seem to end up taking things a bit too far. You always seem to end up with a bunch of unintended consequences. It's like if your solution to end all human suffering was to kill all humans or if your solution to stop someone from ever stealing again was to chop off their hands. Technically it solves the problem, but unfortunately it also creates new problems. Alas, these new problems are often far worse than the original ones.

The properties that you lose by adopting a non-classical logic are often at least as intuitively important (if not more so) as the non-classical properties you desire to gain. This is a large part of why classical logic still so overwhelmingly dominates the field of logic and mathematics, despite the existence of all these alternatives. You very often lose more by abandoning classical logic than you gain. This is true for most of the traditional non-classical logics at least. However, that being said, unified logic will generally not require such trade-offs and yet it will indeed still be a non-classical logic. Transformative logic also does not require such trade-offs either, although admittedly only by virtue of making so few assumptions[20].

Another way that intuitionistic logic differs from classical logic, besides just not holding the law of the excluded middle ($A \vee \neg A \to T$) and the law of double negation elimination ($\neg\neg A \to A$) as valid, is that expressions in intuitionistic logic do not always have well-defined truth values. In intuitionistic logic, some expressions are clearly true and some expressions are clearly false. However, some expressions are also indeterminate and have no truth value whatsoever, not even some kind of third "unknown" truth value such as is found in certain multi-valued logics such as Kleene's 3-valued logic.

This implies that intuitionistic logic is not a truth-functional logic. Intuitionistic logic fundamentally cannot ever be distilled down to a bunch of truth tables. It is inherently always more context-dependent than that. Truth values in intuitionistic logic are more akin to specific individual proofs of derivability than to the simplistic truth values you would use in classical logic or in other truth-functional logics.

This is a good example of why it was necessary to create an ultra-minimalistic framework like transformative logic, one that makes little to no assumptions about the nature of truth. Many different possible systems of logic just fundamentally cannot be implemented using a truth-functional approach. Truth-functional logics are just a small subset of all possible logics, and trying to fit everything about an arbitrary logic's intended behavior onto a bunch of truth tables imposes some truly crippling limitations on what it is actually possible for you to do.

A principled and uninhibited exploration of the underlying philosophy of logic and

---

[20] Don't confuse transformative logic itself with the specific systems of logic expressed in terms of it. Transformative logic itself has very few constraints, but specific transformative rule sets are often very heavily constrained.

mathematics, such that we are emboldened to compare and contrast all of the different possible philosophical viewpoints and variations on how a system of logic could conceivably behave, would be totally impossible without some way of ridding ourselves of the very narrow-minded assumptions embodied by classical logic and other truth-functional logics. That's what transformative logic empowers us to do.

So, as you could imagine, defining some rules of inferences and being careful to exclude the law of the excluded middle ($A \vee \neg A \rightarrow T$) and the law of double negation elimination ($\neg \neg A \rightarrow A$) would be one way to go about implementing at least some parts of intuitionistic logic. However, is there a way to evaluate expressions in intuitionistic logic without reframing everything purely as chains of manipulations of uninterpreted forms though? Is there a way to plug in "values" into intuitionistic logic expressions and get back a result, without actually going so far as to use truth tables?

Well, from what I've read, there is at least one system for doing something like this. It's called "Heyting algebra" apparently. It defines intuitionistic logic in terms of "semantics" (i.e. interpreted values) instead of in terms of "syntax" (i.e. uninterpreted symbols). I don't really know anything about Heyting algebra besides its existence though, so I won't be explaining it here. I just thought I should mention it for reference. My instinct at first glance though is that it seems convoluted and perhaps not worth the time investment. I could be wrong though.

Anyway though, that pretty much covers what we want to know about constructive logic for the purposes of this book. If you want to know more though, then you can of course just look it up. I know this is a sparse treatment of the subject, but that's okay. We mostly only care about getting basic overviews of the foundational principles and motivations of each of the major branches of non-classical logic, rather than actually learning each one in-depth. Plus, I don't really know much about some of these branches of logic anyway, besides just knowing a general overview of them and what the underlying motivations are, and thus I wouldn't be able to cover them well anyway, at least not without investing a lot more time into learning them, more than would be worthwhile right now probably.

Constructive logic has the unusual trait of guaranteeing the existence of an algorithm capable of generating each object corresponding to a proof. This has made it be of some interest in computer science. Be that as it may, the counterintuitive properties of constructive logic's most typical manifestations (e.g. intuitionistic logic, especially its lack of the law of the excluded middle and the law of double negation elimination) have prevented it from becoming truly popular.

However, constructive logic is nonetheless one of the most widely known and widely explored of the non-classical logics. There's just one more branch of non-classical logic I want to cover before we'll be ready to move on to creating unified logic itself. Unlike constructive logic though, this one is one of the more obscure branches of non-classical logic and is not very widely known. It's name is *connexive logic* and it will be the subject of the next section.

4.9 Connexive logic

The ideas underlying connexive logic actually originate in antiquity and date all the way back to Aristotle's time, i.e. to the very beginning of the creation of the first system of formal logic ever invented in human history. Back then, the term "connexive logic" did not exist. I tried to figure out the exact origin of the term, but the information I found online was somewhat conflicting and unclear. Here are the two main possible origins of the term, as far as I can see:

1. **According to various online encyclopedias and other pages**: The Greek and Roman philosopher Sextus Empiricus (who lived roughly between the 1st and 2nd centuries CE) may have used some Greek or Latin equivalent of the word "connexion" or "connection" to refer to logics with connexive properties. Perhaps the original word was Latin and was specifically "coniunctionem"? That seems like a fairly likely possibility, but I'm just guessing. I don't have the original source and it isn't worth my time to bother looking for it.

   The term supposedly appears in a translation of some of Sextus Empiricus's work by "Kneale and Kneale". The "Kneale and Kneale" here most likely refers to a husband and wife pair: William Kneale and Martha Kneale, who are most well-known for writing a book entitled "The Development of Logic". The specific passage from Sextus Empiricus is this: "And those who introduce the notion of connexion [connection?] say that a conditional is sound when the contradictory of its consequent is incompatible with its antecedent."

2. **According to sites.google.com/site/connexivelogic**: The term "connexive logic" was perhaps coined much more recently, supposedly by Storrs McCall, a philosopher who was still alive at the time this book was written. The specific passage on the website that indicates this is this one: "More specifically, Angell aimed at devising a formal system that realizes what he calls the principle of subjunctive contrariety, i.e. the principle that 'If p were true then q would be true' and 'If p were true then q would be false' are incompatible. Based on this work of Angell, McCall introduced the terminology 'connexive logic', and studied extensively the formal system presented by Angell."

Anyway though, putting trivia about the origin of the term aside, let's continue our discussion. Indeed, even though the term "connexive logic" came much later, the first documented uses of connexive principles actually came from Aristotle. Aristotle's own system of logic assumed connexive principles to be true, as evinced by certain statements that he made in his work.

Thus, since Aristotle's syllogistic logic dominated the field of logic for more than a thousand years, the principles of connexive logic were likewise implicitly assumed to be true by most people who studied logic. It was only with the advent of classical logic (and its inherently truth-functional ideas about how reasoning should work) that connexive principles were abandoned and indeed were nearly forgotten entirely by most logicians and mathematicians.

The principles of connexive logic may have got their start with Aristotle's syllogistic logic, but that is not the only place from which they have drawn inspiration. In fact, as understood today, the basic underlying principles of connexive logic actually originate from a combination of both Aristotle's work and the work of a much later philosopher named Boethius. Boethius lived during the time of the Roman Empire, his life spanning from approximately 480 CE to 524 CE. His work would end up becoming popular for much of the Middle Ages. Boethius's studies and work included some logic and mathematics, in addition to some much more hand-wavy and superficial philosophical and religious material.

The defining characteristic of a connexive logic is that it adheres to at least one of two specific pairs of rules. These two pairs of rules are known as Aristotle's theses and Boethius's theses. The word "theses" is the plural form of "thesis", which means a premise or theory that a person is claiming to be true and is offering for consideration as a potential basis for further reasoning. In other words, a "thesis" is just an axiom, and "theses" just means multiple axioms. The best way to understand connexive logic is to understand these two pairs of theses. It's best to just dive in. Few things explain connexive principles as clearly and concisely as the two pairs of connexive theses themselves do. Here they are:

Aristotle's Theses:

1. $\neg(A \to \neg A)$

2. $\neg(\neg A \to A)$

Boethius's Theses:

1. $(A \to B) \to \neg(A \to \neg B)$

2. $(A \to \neg B) \to \neg(A \to B)$

Translated into plain English, Aristotle's theses together say "No statement can imply its own negation." and Boethius's theses together say "No statement can imply both some other statement and that other statement's negation at the same time.". These rules seem like something that should obviously be true right? Indeed, one would think that any sane logic would obey these rules. That's why it's so shocking that classical logic actually *doesn't* obey these rules. Let that sink in for a moment. Classical logic allows for the possibility of a statement implying its own negation. Weird, right?

In fact, it's not just weird. If you really stop and think about it, it is practically tantamount to a violation of the law of non-contradiction. The law of non-contradiction says that it is impossible for both a statement and its own negation to be true simultaneously. However, by allowing statements to potentially imply their own negations aren't you essentially violating that principle? After all, the definition of falsehood is the absence of truth. Likewise, truth is the absence of falsehood. How then could the truth of a statement ever imply its own falsehood? The very idea of it is absurd on its face.

One thing you might be thinking at this point though is "Why not just add these into classical logic as axioms? Wouldn't that solve the problem?". Well, unfortunately, no. It definitely would not. Not only would it not solve the problem, but it would be catastrophic if you attempted to force it. To understand why, one need only translate the material implication operator into its classically equivalent form as a disjunction (via $A \to B \leftrightarrow \neg A \vee B$) in each of the expressions for the connexive theses and then simplify. If you understand the nature of axioms, then you will immediately see a rather alarming problem. Here, I'll show you:

Aristotle's Theses:	**Boethius's Theses:**
1. $\neg(A \to \neg A)$ $\twoheadrightarrow \neg(\neg A \lor \neg A)$ $\twoheadrightarrow A \land A$ $\twoheadrightarrow A$	1. $(A \to B) \to \neg(A \to \neg B)$ $\twoheadrightarrow \neg(\neg A \lor B) \lor \neg(\neg A \lor \neg B)$ $\twoheadrightarrow (A \land \neg B) \lor (A \land B)$ $\twoheadrightarrow A \land (\neg B \lor B)$ $\twoheadrightarrow A \land T$ $\twoheadrightarrow A$
2. $\neg(\neg A \to A)$ $\twoheadrightarrow \neg(A \lor A)$ $\twoheadrightarrow \neg A \land \neg A$ $\twoheadrightarrow \neg A$	2. $(A \to \neg B) \to \neg(A \to B)$ $\twoheadrightarrow \neg(\neg A \lor \neg B) \lor \neg(\neg A \lor B)$ $\twoheadrightarrow (A \land B) \lor (A \land \neg B)$ $\twoheadrightarrow A \land (B \lor \neg B)$ $\twoheadrightarrow A \land T$ $\twoheadrightarrow A$

Notice that all of these expressions transform into either A or $\neg A$ when simplified using the rules of classical logic. Think about what that implies about what would happen if we attempted to assert even a single one of these expressions as a new axiom in classical logic. The symbol A here stands for *any* arbitrary statement. Asserting an expression as an axiom in classical logic implies that you are saying that the indicated expression is *always true*.

Combine those two facts together and the consequence is that adding any of Aristotle's theses or Boethius's theses to classical logic as axioms would be equivalent to asserting *all* conceivable statements to always be true. This would clearly instantly render our entire system of reasoning in classical logic totally worthless. A system of logic where everything is always true has no value. It would be incapable of distinguishing between anything and hence also incapable of reaching any meaningful conclusions.

Thus, as you can see, it is logically impossible to safely add connexive properties to classical logic, despite the fact that doing so would seem to make so much sense whereas not doing so would seem so absurd. This is quite a problem. If we *do* add the connexive theses to classical logic, then everything will be treated as true and it will all therefore degenerate into nonsense.

On the other hand though, if we *don't* add the connexive theses to classical logic, then we are essentially allowing the law of non-contradiction to be indirectly violated at least some of the time by either allowing statements to sometimes imply their own negations (in violation of Aristotle's theses) or else by allowing statements to sometimes imply some other contradictory pair of statements (in violation of Boethius's theses). We're screwed either way. It is impossible to fix this problem under the constraints imposed by classical logic. This strikes me as probably being "smoking gun" evidence that classical logic can't possibly really be the most correct system of logic.

In addition to being evidence of the inadequacy of classical logic, this is also evidence that the classical disjunction definition of implication ($A \to B \leftrightarrow \neg A \lor B$) is probably fundamentally misguided and wrong. It inevitably leads to incorrect inferences at least some of the time. A correct implementation of the concept of implication inherently requires context sensitivity, and no truth-functional logic can ever provide this.

The only reason truth-functional logics are able to thoughtlessly return truth values every time you feed them some inputs is because they ignore context. Contextual awareness of the specific relationships involved in every possible scenario can't possibly be crammed into any kind of truth table. There's just nowhere to encode the information. Crude labels like "true", "false", and "unknown" are simply not enough.

Nonetheless, based on the traditional literature, it appears that some connexive logics *do* try to use truth tables. Generally speaking, when you read the existing literature on connexive logic, you will probably notice that the same problem that appears for many other non-classical logics also appears here. The various prototypes for connexive logic that people have suggested in the literature seem to force you to lose at least some other important or intuitive properties in order to gain the connexive properties that you desire. The end result is a quirky system of reasoning that still doesn't feel correct and which suffers from counterintuitive behavior. Unified logic won't have this problem though, despite the fact that it will still satisfy all the connexive theses.

Connexive logics span a spectrum from weak to strong, with the "weak" logics satisfying less connexive properties and the "strong" logics satisfying more connexive properties. On the weak side, a connexive logic might only satisfy one of Aristotle's theses. In contrast, on the strong side, a connexive logic might satisfy all of both Aristotle's theses and Boethius's theses. Thus, for example, in a strong connexive logic, not only are $A \to \neg A$ and $\neg A \to A$ not allowed as inferences, but they are also never true for any conceivable substitution of A, and similarly for Boethius's theses.

As for the traditional literature on connexive logics, I will not be covering it here, nor do I really know much about it. It would probably be a waste of time. Like the vast majority of the traditional non-classical logics, the existing connexive logics in the literature seem to be esoteric and unnatural and seem to not really be worth the time investment. Besides, I am not proficient enough in them to teach them even if I wanted to. I just know enough of a vague overview to make an educated estimate that learning them would probably not be worth my time for what my goals here are. You are of course free to look up more information on your own however, as always.

Another thing I want to point out though, is that Aristotle's theses are kind of a special case of Boethius's theses. Aristotle's theses say that no statement can imply its own negation, whereas Boethius's theses say that no statement can imply two self-contradictory statements. The statement B in Boethius's theses could be any arbitrary statement, including even A itself. As such, if we substitute A to replace B in Boethius's first thesis and then simplify then Boethius's first thesis becomes the same as Aristotle's first thesis. Observe:

Boethius's Theses:

1. $(A \to B) \to \neg(A \to \neg B)$
$\to (A \to A) \to \neg(A \to \neg A)$
$\to (\neg A \lor A) \to \neg(A \to \neg A)$
$\to T \to \neg(A \to \neg A)$

$$\twoheadrightarrow F \vee \neg(A \to \neg A)$$
$$\twoheadrightarrow \neg(A \to \neg A)$$

However, if we apply this same approach to Boethius's second thesis using the laws of classical logic then it won't end up being the same as Aristotle's second thesis. There's a bit of an asymmetry here when you're dealing with classical logic. Be that as it may, if we put aside the quirks of classical logic then Aristotle's theses are still nonetheless special cases of Boethius's theses from a conceptual standpoint. Boethius's theses say that a statement A cannot imply any contradiction B and $\neg B$. The case where the statement A is contradicting *itself* is clearly just a special case of that.

If that's not enough evidence for you, then consider this: Aristotle's second thesis $\neg(\neg A \to A)$ can be turned into Aristotle's first thesis $\neg(A \to \neg A)$ by simply observing that because A represents any arbitrary statement we can simply substitute $\neg A$ as a replacement for A in the expression, which has the net effect of transforming $\neg(\neg A \to A)$ into $\neg(A \to \neg A)$. Here's what the transformation path for this would look like:

$$\neg(A \to \neg A) \twoheadrightarrow \neg(\neg A \to \neg\neg A) \twoheadrightarrow \neg(\neg A \to A)$$

Consequently, taking all these considerations into account, it seems that any connexive logic that satisfies Boethius's theses should generally also satisfy Aristotle's theses. Perhaps there might be connexive logics where this is not true, but I don't see why such systems would exist. If a connexive logic satisfied Boethius's theses but not also Aristotle's theses then it would probably have to be a very weird system indeed, and hence it would be unlikely to be useful for reasoning.

Care must be taken when designing a connexive logic to ensure that it is not explosive, i.e. that it does not imply that all conceivable statements are true. This may be an easy mistake to make if your approach to connexive logic is truth-functional. I seem to recall reading that at least one connexive logic is also a paraconsistent logic. This makes sense considering that it is so important to avoid explosion in a connexive logic, and paraconsistent logics are precisely the logics where the principle of explosion does not hold.

The connexive theses really do intuitively seem like they should hold, which is why it is so weird that many people aren't even aware of the existence of connexive logic at all. Sometimes it is amazing how easily an entire society can overlook even a very simple idea and completely fail to realize it. On the other hand though, new ideas often spread surprisingly easily throughout society, resulting in stunning new levels of prosperity in the blink of an eye. Social momentum is a powerful thing, but it is also a blade that cuts both ways. Collective intelligence has always been both a blessing and a curse for humanity.

Oh, and by the way, as a minor point of trivia, the word "connexive" in "connexive logic" appears to just be an archaic British spelling variant of the word "connective". Similarly, "connexion" is just a spelling variant of "connection". The term "connexive logic" was thus probably chosen to emphasize the importance of the "connection" between the condition and consequence in an implication, and hence the term "connexive logic" was born. I just thought I'd mention that, because otherwise the reason for choosing the term may have been unclear.

Anyway though, that pretty much covers everything I want to say about connexive logic here. I'd say that it's time to move on to the next branch of non-classical logic, but we are actually done with that for now. We've covered all of the branches of non-classical logics that I felt were prudent to cover. There are also some other more miscellaneous branches of non-classical logic, but I deemed them not important enough for inclusion. We now have the necessary background to properly appreciate the motivations behind unified logic. All we needed was a rough overview.

The non-classical logics each provide us with insight into a specific conceptual flaw in classical logic and get us thinking about possible alternatives. However, we always seem to have to make trade-offs. We always seem to have to give up at least one important property in order to gain a new property that we want. Unified logic strives to overcome this problem by giving us a more balanced logic that satisfies the core motivations of all of these different branches of logic simultaneously, while still remaining highly intuitive. So, without further ado, it is now officially time to create unified logic. Doing so will be the subject of the next chapter, the apex and namesake of this book.

Chapter 5

What is unified logic?

5.1 How is it related to the other logics?

Sometimes the state of human knowledge has an odd tendency to circle back upon itself. Sometimes we start off with an assumption that we instinctively feel should be true, but then later realize that this assumption does not seem to be fully adequate for what we have in mind. We then create a new system based on different assumptions, but it is only after exploring that new system in greater detail that we eventually come to realize that the original assumption that we were first inspired by can actually be modified to work, at which point we may return to it. We explore different possibilities, eventually coming back to our original thoughts, but this time armed with a deeper and more insightful perspective. In this way, the journey of discovery makes us stronger, even if we ultimately don't end up moving very far from our original ideas in the end. Unified logic is kind of like that.

You see, unified logic is essentially just an extension, modification, and reinterpretation of set theory. Set theory in turn is just a more highly evolved and refined descendant of syllogistic logic, which in turn is the first system of formal logic ever devised in human history. It is in this sense that our exploration of unified logic will feel sort of like "coming full circle" and acquiring deeper insights into what was really going on the whole time. We will come to see more clearly what precisely it was that really went wrong in the creation of classical logic and how to fix it, and by so doing we will end up returning partially to an older perspective, but this time armed with greater insight and wisdom.

Do you remember our discussion of how the empty set ($\emptyset$) and the universal set ($\mathcal{U}$) could mimic the behavior of false (F) and true (T) respectively, allowing set theory to emulate expressions from classical logic? Do you remember how I hinted that this would become important later? Well, that time is almost upon us. Earlier in the book, we looked at a large table of remarkably similar laws found in both classical logic and set theory (see page 147) and then segued into a discussion of the correlations between the empty set ($\emptyset$) and false (F) and between the universal set ($\mathcal{U}$) and true (T). This remarkable correspondence was further demonstrated in a table on page 149 for the case of disjunction and union. You might have already been getting suspicious that unified logic would end up being closely related to set theory if you were paying close attention to the trajectory of our discussions around the time we discussed these two tables.

Indeed, going back even further, all the way back to the first chapter, you may remember our discussion of the concept of "existence-based truth", where we explored some thought experiments that demonstrated the apparent conceptual superiority of existence-based truth over "statement-based truth". As you may recall, existence-based truth inherently relies upon the concept of sets and uses the non-emptiness or emptiness of sets of objects to represent whether or not something is considered true. For reference, our first official definition of existence-based truth occurred on page 27 and was followed shortly thereafter by our first official definition of statement-based truth. Thus, even as far back as the first chapter of this book you may have already been becoming suspicious that unified logic would end up being closely related to sets.

This idea of truth as being the same thing as existence, as being the same thing as the non-emptiness of sets, is pretty straightforward and easy to grasp in itself. However, the implementation details are where things get murky. How do we represent truths in terms of sets exactly? There are many statements in logic and mathematics that would seem to not really be about sets. Isn't set theory only capable of talking about sets though? Is there a way to consistently translate general ideas into terms of set theory? These are some of the kinds of questions we should keep in mind as we construct unified logic.

Indeed, the perception of set theory as only being able to talk about "set-like" concepts is probably part of what has caused many people to fail to connect the dots necessary to realize that an idea like unified logic might be the solution to many of the problems and shortcomings that plague modern classical logic. As it turns out though, set theory isn't as limited as people think it is. It can express some ideas that many people don't realize that it can.

You see, modern set theory is actually missing several of its most basic operators and most fundamental concepts. Some parts of set theory were never actually fully and correctly defined, and thus a vast swath of set theory's potential was completely overlooked. As hard to believe as that may sound, it is true. You will see for yourself once I show you the evidence for it. Even natural language itself, the very words you are reading right now on this page, has structure in common with unified logic and is thus partial evidence in favor of it. This will become more obvious once you see what I'm talking about later on.

Unified logic may cast a wide net, but we're not going to try to absorb all of it at once. That would be too overwhelming probably. Instead, we will be splitting our discussion of unified logic into two parts. The first part will cover only a more crude subset of unified logic, a subset that feels very much like the standard set theory that you may have learned in school except with the addition of a few new operators and a few new ways of thinking about things. We will demonstrate that this subset of unified logic satisfies all of the major motivations underlying the non-classical logics that we previously discussed, at least for what it is capable of expressing.

The second part is were things will get really weird though. That's when we'll expand beyond the domain of standard set theory and open your eyes to some possibilities that would otherwise seem absurd if not for the evidence right in front of your eyes. How seemingly absurd am I talking about here? Well, let me put it this way: It will enable us to literally divide by zero. I'm not talking about some kind of mere label like "not a number" or "undefined". I'm talking about a genuine answer to how to divide by zero that is meaningful and has a well-defined and non-trivial behavior when combined with other expressions.

However, if you think merely dividing by zero is weird, then you haven't seen anything. There are things that unified logic is capable of that arguably make division by zero look totally tame by comparison. For instance, have you ever applied a number to another number *as an operation* before? Well, you're about to learn how to once we get to that section. It'll open your mind to some fresh new ways of thinking.

You will literally be able to "four the number three" in the same sense that you might "negate the number three". You may never see math the same way again. Exciting stuff awaits us. These ideas will be accompanied by supporting evidence and concrete examples. I'm not even messing with you. I'm dead serious. I know some of these ideas may sound crazy to some people, but just humor me for now. Once you see how these ideas will work, it will all make a lot more sense.

Let's not get ahead of ourselves yet though. We first need to understand the simpler subset of unified logic, and then build up to the more advanced stuff gradually. The first part of unified logic, the part that feels pretty much the same as regular old set theory but just with some new operators and new perspective is what we will call **primitive unified logic**. In contrast, the more complete version of unified logic, the one that will make your brain explode and allow us to divide by zero (among many other things) is what we will call **relational unified logic**. Let's get started. It's time to learn primitive unified logic.

5.2 Introduction to primitive unified logic

You may have noticed that I occasionally distinguish between two different types of expressions in formal languages: value expressions and statement expressions. It has probably always been pretty clear from context what these terms have referred to, but I think it's worth discussing the nature of this distinction in greater detail here. Primitive unified logic will be adding on both some new value expressions and some new statement expressions to our set theory vocabulary, so it will probably be worthwhile to make sure that we have a clear understanding of what the difference between the two types is.

We actually did already define the two types way back on page 131, briefly and in passing, but let's review the concepts again anyway, for clarity. For those with a computer science background, a good analogy for the difference between the two types would be that value expressions are kind of like functions that have return values whereas statement expressions are more like procedures that don't return any values (a.k.a. "void functions").

For those of you without a computer science background though, I will explain the distinction more directly so that you can understand. Traditional mathematics literature is often kind of sloppy about the distinction between the two types, even though the distinction is often very important. In fact, much of the mathematics literature would never compile as a computer program, due to the widespread lack of precision in how types are treated in the math literature.

Anyway though, let's start with value expressions. Value expressions are expressions that can be treated as representing objects. Value expressions can be chained together into larger expressions very easily. One thing that value expressions never do though is make claims. A value expression is not an assertion, but rather it is merely a description of some object, often phrased in terms of some other object(s). Value expressions are kind of like nouns: they represent things, but don't make statements about those things. For example, "2 + 2" is a value expression. It does not make any claims. It merely represents an object. It is composed of objects (two numbers and one operation in this case) but can also be thought of as an object in and of itself, treated as a whole, independently of its parts.

Statement expressions are a bit different. Statement expressions make claims, but cannot be used as values (at least not without abusing notation). This makes statement expressions sort of the opposite or the complementary counterpart of value expressions. Admittedly, sometimes statement expressions can be interpreted as returning truth values corresponding to whether or not the indicated statement is true, but this is not always the case and is sort of an abuse of the concept of a statement expression in some ways. Statement expressions are like complete sentences: they make assertions about the state of a system.

Whereas value expressions always simply return what they represent and generally never clash with each other, statement expressions sometimes express things that are contrary to what other statement expressions say or that are otherwise contrary to the reality of the system. Thus, statement expressions have an additional confounding factor that value expressions lack, and for this reason it is often critical when working with statement expressions to make sure no false statements creep into the reasoning process.

For example, "2 + 2 = 4" is a statement expression. It makes a claim, but it cannot be embedded inside another expression as a value without abusing notation. Strictly speaking, for instance, you cannot afterwards add on 5 to this by saying "(2+2 = 4)+5". The expression "2+2 = 4" is an assertion about a relationship. It does not have a return value, at least not in the traditional sense of what it means to have a return value, and thus it cannot be chained together with "+ 5". Attempting to do so would be analogous to saying something like "Bob went to the store and Jane." in English. Indeed, in English we would say that this "sentence" isn't "grammatically correct", but in the jargon of formal logic, mathematics, and computer science we would instead often say that such an expression isn't "syntactically correct", i.e. that it has "bad syntax"[1].

This distinction between value expressions and statement expressions is important to understand. The sloppy treatment it often receives in classical logic (and also in other parts of the mathematical literature unfortunately) invites confusion. Probably one of

---

[1] However, syntax in logic, mathematics, and computer science tends to be much stricter than grammar in natural languages. If you think grammar-obsessed writing teachers are strict then you should try talking to a compiler sometime. Compilers make grammar nazis look like saints. If even one semicolon is out of place then a computer program very often won't even work.

the most glaring examples of this potential confusion is the way material implication is used and interpreted. If I write $A \rightarrow B$ on a sheet of paper in classical logic, then what can the reader conclude that I am trying to say? Am I saying that A in fact does materially imply B? Or, am I not actually asserting anything at all and am I instead merely treating $A \rightarrow B$ as an expression to be evaluated to determine whether it is true or false? In classical logic, you can't always be sure which is the intent.

Indeed, this problem is not even unique to material implication. *All* expressions in classical logic are ambiguous in this respect. For example, is $A \wedge B$ an assertion that both A and B are true, or is it merely an inert truth-functional expression to be evaluated? Similarly, to take another example, does simply just writing the name of any arbitrary boolean variable X indicate an assertion that X is in fact true, or does it instead merely indicate the truth value of X, whatever it may be? Without context, it's unclear what the intent even is. That's a problem. It's kind of crippling to not even be able to clearly distinguish between when someone is making an assertion versus when they are merely referring to some described object. This is something that should definitely be communicated more clearly.

Contrary to its reputation as the pinnacle of precision, traditional mathematical vocabulary is actually rife with such ambiguities and hidden assumptions, quite unlike modern computer science by the way, which in contrast is almost universally rigorous and unambiguous by necessity. Computers have little to no tolerance for ambiguity at the machine code level, so this is not really that surprising if you think about it. Mathematics needs to become more precise and more unambiguous. It needs to become more like computer science. Ambiguity and hidden assumptions are poison to scientific progress. In contrast, precision and clarity tend to lead to fertile scientific ground. Scientific endeavors tend to stagnate in the absence of clear thinking.

Anyway though, I think it's almost time to start defining the new operators of which primitive unified logic is comprised. Some of these operators will be value expressions and others will be statement expressions. Let's start with the value expression operators first. As you may recall, unified logic is essentially just an extension, modification, and reinterpretation of set theory. So, naturally, these new operators will all be related to sets.

Before we do that though, let's briefly refresh our memory as to what the already existing set operations we have defined are. Some of the operators have multiple names and multiple symbols in use. For now, we are only interested in the operators that are value expressions, and specifically only those that return sets and not any other type of object. Here's a table of the set theory operators we have previously covered that satisfy those criteria:

Operator Name(s)	Example
not / negation / complement	$\neg A$
and / intersection	$A \cap B$
or / union	$A \cup B$
set difference / set removal	$A - B$
set xor	$A \veebar B$
set nand	$A \uparrow B$

set nor	$A \downarrow B$
cartesian product	$A \times B$
binary pairing	$A \mathbin{\emptyset} B$
concatenation	$A \mathbin{\multimap} B$
power set	$\text{PowerSet}(A)$

For reference, the comprehensive list of all of the set theory operators and relationships we previously covered way back in the original set theory section, before we ever even talked about transformative logic, can be found on page 145. Feel free to look at the old table there if you're interested, but I'll be walking through all of the relevant stuff item by item here as we go along, so you won't really need to. It's all going to be pretty straightforward really. Reviewing some of the old set theory content may help you start thinking more in terms of set theory again though, which may help you with acquiring a proper understanding of unified logic, given that unified logic is based on set theory.

The cartesian product[2] ($\times$), binary pairing ($\emptyset$), concatenation ($\multimap$), and power set (PowerSet) operators here are of little to no concern to us in the following discussion. I have only listed them here for completeness sake since they are indeed set theory value expressions. The set difference ($-$), set xor ($\uplus$), set nor ($\downarrow$), and set nand ($\uparrow$) are also redundant, since they can be built up out of the more basic set operators (negation, intersection, and union), but are sometimes nice shorthands despite admittedly being a bit esoteric. The primary operators of set theory are of course the negation ($\neg$), intersection ($\cap$), and union ($\cup$) operators, which occur in great abundance in countless different contexts.

Do you remember the table of remarkably similar laws found in both classical logic and set theory (on page 147) that I mentioned earlier? It was followed by another table (on page 149) showing how $\emptyset$ corresponds to F and $\mathcal{U}$ corresponds to T in the context of classical logic's disjunction operator versus set theory's union operator. I also mentioned that the same pattern held true for all the other corresponding operators between classical logic and set theory as well. Well, shortly after that second table, in a paragraph on page 149, I hinted that an interesting hidden relationship between classical logic and set theory existed, and that it had something to do with implication. I said that it would become important later. That time has finally come. Here's the specific paragraph on page 149 that I'm referring to, where I originally alluded to this:

> "You will notice however that one of the basic operators of classical logic has been left out of this comparison: material implication. The reason for that is because implication is special and deserves its own in-depth treatment, which will come later in the book. Understanding that a connection between implication and set theory exists, and fleshing it out properly, is part of the key to revealing the hidden secrets that enable the creation of unified logic."

---

[2] Remember, the "cartesian product" operator is nothing more than an ambiguous symbol found in the traditional math literature that actually always refers to either binary pairing or concatenation. It is redundant. You'll see it sometimes though, in which case you'll need to know the symbol for it so that you can read it.

Here's the secret: Set theory actually has its own implication operator, it's just that when set theory was originally invented everybody apparently failed to realize it. Negation, intersection, and union are *not* the only fundamental operators of set theory. There is a fourth fundamental operator, one that expresses the concept of implication *without* succumbing to the conceptual problems that plague classical logic's material implication. The idea is simple, but its impact is large.

Once you add this missing implication operator into set theory, then the list of uncanny similarities between the laws of classical logic and set theory (see the table on page 147) expands even more, at which point it becomes blindingly obvious that set theory is definitely a full-blown branch of logic, rather than merely a separate thing, if ever there was any doubt before.

Set theory has analogs not just of the negation, disjunction, conjunction, and equivalence laws that classical logic has, but also of the implication laws of classical logic as well. Not everything is the same though. Some of the laws are different. Set theory's version of the implication operator seems to have far fewer conceptual flaws, if any. Set theory's version has more logical and conceptual integrity, and also better correlation with people's instincts about the way implication should behave and how it relates to natural language.

Actually, interestingly, it's not just one implication operator that set theory is missing, but two. You see, set theory's version of implication actually comes in two different variants whose behaviors differ slightly from each other. These two versions of implication will henceforth be referred to as *sub implication* and *super implication*, and together they form a more general type of operation that we will call *unified implication*.

In other words, sub implication and super implication are both specific types of unified implication. Admittedly though, whether or not you want to think of sub implication and super implication as two distinct operators or as just different manifestations of the same operator (unified implication) is somewhat a matter of taste. Anyway though, here's our new formal definitions for them:

Definition 118. *Let* $\mapsto$ *be the symbol we use to represent* **sub implication**. *Given any two arbitrary sets A and B, let the expression* $A \mapsto B$ *be read as "A sub B", where A acts as the condition and B acts as the consequence. Sub implication is a value expression, not a statement expression, and returns a set. As is typical for implication operators, the symbol may be written facing either direction according to the user's tastes.*

Thus, the expressions $A \mapsto B$ *and* $B \mapsfrom A$ *both mean the exact same thing, just written different ways. The expression* $B \mapsfrom A$ *is read as "B sub by A". Notice the word "by" is inserted to indicate that the operand order is reversed. In other words, when reading from left to right,* $\mapsto$ *is read as "sub", whereas* $\mapsfrom$ *is read as "sub by". Evaluation of sub implication expressions proceeds as follows:*

1. *If at least one of A or B is the empty set, then return the empty set.*

2. *Otherwise, if A is a subset of B then return A, else return the empty set.*

Definition 119. *Let* $\leftmapsto$ *be the symbol we use to represent* **super implication**. *Given any two arbitrary sets A and B, let the expression* $B \leftmapsto A$ *be read as "B super A" (notice the* order), where A acts as the condition and B acts as the consequence. *Super implication* is a value expression, not a statement expression, and returns a set. As is typical for implication operators, the symbol may be written facing either direction according to the user's tastes.

Thus, the expressions $B \mapsfrom A$ and $A \rightarrowtail B$ both mean the exact same thing, just written different ways. The expression $A \rightarrowtail B$ is read as "A super by B". Notice the word "by" is inserted to indicate that the operand order is reversed. In other words, when reading from left to right, $\mapsfrom$ is read as "super", whereas $\rightarrowtail$ is read as "super by". Notice that this is the reverse arrow direction order of how sub implications are read. Evaluation of super implication expressions proceeds as follows:

1. If at least one of A or B is the empty set, then return the empty set.

2. Otherwise, if B is a superset of A then return B, else return the empty set.

Definition 120. *Together, sub implication and super implication are instances of a broader type of implication named* **unified implication**. *These two variants of unified implication can either be thought of as separate operators or as just instances of the same operator: unified implication. In fact, the two separate algorithms for evaluating sub implications and super implications can easily be generalized into just a single algorithm. Suppose we have an expression of the form $A \mapsto B$ or $A \rightarrowtail B$ (etc). Here's what the more general unified implication algorithm would look like then:*

1. If at least one of A or B is the empty set, then return the empty set.

2. Otherwise, if A is a subset of B (or equivalently: if B is a superset of A) then return whichever set is on the side with the vertical line adjacent to the arrow symbol, else return the empty set.

Definition 121. *In addition to the notation above, an alternative notation for writing unified implications will also be defined, so that the user may choose according to their preference. The alternative notation will be called* **implicit unified implication notation** *and will use essentially the same symbols as before ($\mapsto$ and $\rightarrowtail$, etc), except for that the symbols will lack the vertical lines and will be read differently. In particular, implicit unified implication notation eliminates the need for the vertical line that normally distinguishes sub implication from super implication by instead using only the direction the arrow faces to encode that information.*

Thus, in implicit unified implication notation, $A \rightarrow B$ always represents "A sub B" (i.e. sub implication, $A \mapsto B$) and $B \leftarrow A$ always represents "B super A" (i.e. super implication, $B \mapsfrom A$). In other words, in implicit unified implication notation, the set on the left is always the set returned, if any non-empty set is returned at all. The upside of this notation is that it takes less effort to write, has fewer moving parts, and flows naturally. The downside of this notation is that it is perhaps less explicit and it is also no longer possible to write the arrow facing two different directions for both sub implication and super implication.

For the sake of completeness and to aid in communication, the other notation for unified implication, the one that does use vertical lines (e.g. as in $\mapsto$ or $\rightarrowtail$), will be referred to as **explicit unified implication notation** in contrast.

Definition 122. *Much like other implication operators found in other branches of logic, unified implication can also be bidirectional, meaning that both directions of implication (e.g. $A \mapsto B$ and $A \mathbin{\leftarrowtail} B$) could be occurring at the same time. In this case, the distinction between sub implication and super implication becomes irrelevant, since the same set would be returned either way if any non-empty set is returned at all.*

There is therefore never any reason to use vertical lines for the symbol we choose to represent it. We call a bidirectional unified implication **unified equivalence**, *and use the symbol $\rightharpoonup$ to represent it. If A and B are two arbitrary sets, then the unified equivalence of them would thus be written as $A \rightharpoonup B$. The expression $A \rightharpoonup B$ may be read as "A equiv B" for short. As you would expect, here is what the algorithm for evaluating unified equivalences looks like:*

1. *If at least one of A or B is the empty set, then return the empty set.*

2. *Otherwise, if both sets are subsets of each other (i.e. if both sets are the same set) then return the set, else return the empty set.*

Notice that in explicit unified implication notation the vertical line is always on the side of the return set, thus providing a very easy way to determine whether you are looking at a sub implication or a super implication at a glance. If the vertical line is on the tail end of the arrow then it's a sub implication, whereas if the vertical line is on the head end of the arrow then it's a super implication. Otherwise, if the arrow lacks any vertical line at all, then the set on the left is always the set that is returned, if any non-empty set is returned at all.

This system makes it easier to remember how to work with unified implications than if the chosen symbols for each variant of the operators had been chosen more arbitrarily. The unified implication arrow symbol ($\rightharpoonup$, etc) also has the advantage that it can be written with a single stroke of the pen and is visually distinct from the material implication symbol ($\rightarrow$).

Already, by reading how unified implication is evaluated, you might have noticed that unified implication possesses a rather unusual logical property that no other implication operator that we have previously discussed has ever managed to have. Do you see it? It's easier to see if you keep in mind the basic principles from chapter 1, where we discussed existence-based truth (see page 27) and how in an existence-based truth framework emptiness is the same thing as falsehood, a.k.a. the principle of "void falsehood" (see page 58).

What is this special property of unified implication that I've been hinting at here? Well, look at how unified implication reacts to the presence of empty sets in its operands and you'll soon see. The property of which I speak is *vacuous falsehood*. In other words, whereas in classical logic false premises imply anything, in unified logic false premises imply nothing. It's exactly the opposite of vacuous truth. I actually mentioned that vacuous falsehood would hold in unified logic once before, way back in the section on predicate logic on page 311.

Do you remember how vacuous truth was inescapable in classical logic, how the nature of the truth tables for material implication forced us to define all of the cases where the condition was false to always return true? Let me refresh your memory. Look at this relevant paragraph from page 111 from the section on the paradoxes of material implication, where I explained why it's logically impossible to prevent falsehood from implying everything in classical logic:

> "If you tried to define material implication any other way in classical logic then you'd either (1) cause implication to become the same thing as equivalence, (2) cause implication to be determined entirely by the truth value of the consequence, or (3) cause implication to become the same thing as conjunction. None of these outcomes is acceptable. Try changing the truth value of $A \to B$ for the cases where A is held false and B is allowed to vary and you'll see what I mean. Thus, classical logic must accept the paradox of entailment as a law, despite the seemingly nonsensical consequences of doing so."

Well, it turns out that in set theory not only is vacuous truth not inescapable, but it is easily avoided. This is a big deal. Vacuous falsehood will enable us to escape from a lot of the bad properties of classical logic. It seems that logicians have apparently always thought that vacuous falsehood was impossible to implement in a logically coherent and useful way, seeing as the idea of doing so does not appear anywhere in the literature as far as I can tell. However, unified implication proves them wrong. Once you start to think in terms of sets and existence-based truth, everything starts making a lot more sense and many problems in the philosophy of the foundations of logic disappear. Let's make a formal definition for vacuous falsehood, as well as for some other related terms:

Definition 123. *In unified logic, we use an existence-based concept of truth, rather than a statement-based concept of truth. As such, in unified logic, all non-empty sets are thought of as being "true" and all empty sets are thought of as being "false". As a consequence of this, the way that unified implication is defined causes it to have the property of* **vacuous falsehood**. *Vacuous falsehood means that an implication with a false condition is always false, i.e. that nothing follows from false premises.*

Or, considering that falsehood and nothingness and emptiness are all the same thing from an existence-based truth perspective, one could also perhaps equivalently say that vacuous falsehood means that "nothingness implies nothingness", i.e. that empty sets cannot have any properties other than those of emptiness itself. Notice how this differs from the classical way of thinking.

In classical logic, "All of the eggs in the basket are red." and "All of the eggs in the basket are blue." would both be considered true statements given an empty basket containing no eggs. However, in stark contrast though, in unified logic both statements would be considered false, because no non-existent thing could ever have any property other than being non-existent and hence certainly could never be red or blue. Notice how much more rational and natural the unified perspective is for this example compared to the classical perspective.

Definition 124. *One point of confusion here might be over the relationship between "vacuous truth" and "the paradox of entailment" and their analogs in unified logic. In classical logic, vacuous truth and the paradox of entailment are two different things* with similar natures. Vacuous truth concerns itself with the truth of statements quantified over empty domains, whereas the paradox of entailment concerns itself with the consequences of the truth-functional nature of material implication. In other words, in classical logic, vacuous truth applies to sets, whereas the paradox of entailment applies to binary truth values.

However, in stark contrast, in unified logic everything is always thought of in terms of sets, so this distinction no longer seems to necessarily make sense in the context of unified logic. Vacuous truth and the paradox of entailment are subtly different things in classical logic, but the analogous concepts in unified logic do not really seem necessarily distinct. However, if I were to name the analogous counterpart of the paradox of entailment for unified logic then I would call it **integrity of entailment**.

Let's mix things up a bit to force the distinction to be meaningful again though. Let's define it like this: An implication operator has integrity of entailment if when given false premises it does not imply everything. Do you see the subtle distinction here? Vacuous falsehood says that any implication with false premises is always false. In contrast though, integrity of entailment merely says that an implication with false premises cannot imply everything. Given false premises, an implication with integrity of entailment may imply some things, or it may imply nothing, but it can never imply everything.

All implication operators that have vacuous falsehood also have integrity of entailment, but not all implication operators that have integrity of entailment necessarily also have vacuous falsehood (if any such operators exist at all). In other words, integrity of entailment is (nearly?) a synonym for paraconsistency (see page 357). This would seem like a pointless term to define then perhaps, seeing as it is apparently (nearly?) synonymous with paraconsistency, but we're also going to define another related term that is certainly not the same thing as paraconsistency. This is the subject of our next definition.

Definition 125. *In addition to potential problems related to the behavior of the condition of an implication (e.g. the paradox of entailment), there is also the behavior of the consequence to consider. If you recall, in classical logic, besides just being vulnerable to the paradox of entailment, the material implication operator was also vulnerable to a problem called the paradox of consequence (see page 111). The paradox of consequence says that any material implication with a true consequence will always be evaluated as true, even though this may not make sense.*

Unified implication does not have this problem. Therefore, let's define a corresponding new term, like so: An implication operator has **integrity of consequence** *if when given a true consequence it does* not *always evaluate as true. In other words, an implication operator has integrity of consequence precisely if the paradox of consequence does not apply to it. This is sort of similar to paraconsistency, but is different in that it concerns itself with the behavior of the consequence instead of with the behavior of the condition.*

Definition 126. *In classical logic, given any two contradictory premises that have been accepted as simultaneously true it becomes possible to construct an argument "proving" every conceivable statement. This is known as the principle of explosion, and any* *theory (i.e. any set of statements) that it can be applied to is said to be "trivial" or "explosive". Explosive theories are useless, due to the fact that they cannot make any meaningful distinctions between anything.*

The principle of explosion differs from vacuous truth and the paradox of entailment subtly in that the principle of explosion is a way of constructing arguments from contradictions, *whereas vacuous truth and the paradox of entailment are immediate consequences of assuming empty or false premises respectively. Review the tables on pages 358 and/or 360 if you want to be reminded of what the structure of an argument that uses the principle of explosion looks like.*

*Unified logic has its own analogous counterpart of the principle of explosion, which we will refer to as the **principle of implosion**. As you would expect, the principle of implosion behaves in essentially the opposite way as the principle of explosion. For example, two contradictory sets will always result in the empty set when intersected with each other, and hence any unified implication whose condition is a contradiction will always result in the empty set, i.e. will always imply nothing.*

In this way, attempts to base arguments on contradictory premises in unified logic will tend to spontaneously collapse and be rendered sterile. Any other arguments touched by the same contradictory premises will also tend to collapse upon themselves. Thus, unified logic is naturally resistant to contradictions in that it tends to spontaneously quarantine them via implosion, whereas classical logic is hypersensitive to contradictions and becomes explosive the moment even a single contradiction appears.

That sure was a lot of definitions clustered together wasn't it? Sorry about that, but it seemed probably best to just get them all out of the way at once due to how closely related these terms are to each other. It's a bit overwhelming maybe, considering the somewhat subtle nature of the distinctions between some of these terms, but it's probably better to just go ahead and clear the air by just diving in. Vacuous falsehood is the term you'll probably hear me using most often to refer to how unified implication reacts to empty sets, but it's still worthwhile to define the other terms anyway, so that you can have a proper appreciation for some of the subtleties and nuances involved.

I'm not totally sure about some of these more subtle distinctions and details, but that's ok. Life is a work in progress. You have to build tentative structures sometimes. Sure, sometimes you'll be wrong, but you couldn't really make progress any other way. You'd be too crippled by analysis paralysis otherwise. Life is like a road that's always under construction. We don't know what the final journey will look like, but we'll get there eventually anyway.

Anyway though, I wanted to point out at least one of unified implication's interesting properties before we continued our discussion of the broader context. Vacuous falsehood was a good pick for that due to the fact that its applicability is relatively readily apparent from just reading the definitions of the various unified implication variants we defined. That being said, it would probably be smart to shift gears and explore some actual simple concrete examples now. After that, we'll also want to discuss how unified implication relates to some other possible ways of defining implication and the ambiguities involved in getting a sense for what an "implication" operator should really mean.

More on that later though, let's go ahead and do some simple concrete examples now. Nothing tests a person's understanding of a concept better than concrete examples. We'll start with some abstract mathematical examples, so that we can focus on verifying our understanding of the raw mechanics first. Then, shortly afterwards, we'll also consider the broader philosophical connotation of these operators in order to give you a better sense for what they really mean on a deeper level.

For the examples below, I'm going to be using equal signs to indicate what each unified logic expression would return if evaluated. We could alternatively use transformative implication arrows to indicate this, but then we'd have to also use quotation delimiters sometimes to work around the way transformative logic treats set notation. Anyway though, here's our first set of examples:

1. $\{a, b, c\} \mapsto \{a, b, c, d\} = \{a, b, c\}$

2. $\{a, b, c\} \rightarrowtail \{a, b, c, d\} = \{a, b, c, d\}$

3. $\emptyset \rightharpoonup \{\alpha, \beta, \gamma\} = \emptyset$

4. $\{\alpha, \beta, \gamma\} \leftharpoonup \emptyset = \emptyset$

5. $\{x, y\} \rightharpoonup \{x, y\} = \{x, y\}$

6. $\text{Int} \rightarrowtail \text{Even} = \text{Even}$

7. $\{\} \rightharpoonup \{60, 12, 2, 30, 12, 1\} = \{\}$

8. $\emptyset \mapsto \emptyset = \emptyset$

9. $\{p, q, r, s\} \rightarrowtail \emptyset = \emptyset$

10. $\bigl(\{1, 2\} \mapsto \{1, 2, 3\}\bigr) \mapsto \{1, 2, 3, 4\} = \{1, 2\}$

11. $\bigl(\{1, 2\} \rightarrowtail \{1, 2, 3\}\bigr) \rightarrowtail \{1, 2, 3, 4\} = \{1, 2, 3, 4\}$

12. $\bigl(\{x : x = k^2 \land k \in \text{Int}\} \rightharpoonup \text{Cont}\bigr) \rightharpoonup \text{Ori} = \{x : x = k^2 \land k \in \text{Int}\}$

As a reminder, $\emptyset$ and $\{\}$ mean the same thing. They both represent the empty set. I put a few instances of both of them in the example list just to test and refresh your memory a bit. Most of the examples in the list use generic letters or numbers. However, as you can see, I've also included a few examples that use common predefined sets such as the set of all integers (Int), the set of all continuum numbers (Cont), and the set of all orientational numbers (Ori). Finally, the last example in the list uses set builder notation (see page 130) to serve as a reminder that sets can be specified multiple different ways, as is common in set theory. We probably won't be using much set builder notation in our discussion of unified logic, but I just felt like mentioning it again anyway.

I've chosen a fairly random and diverse set of examples here in order to make sure that you don't have any obvious holes in your understanding of how these operators work. That being said, these operators are really quite simple and it shouldn't be that hard to understand what they do. In contrast though, what is likely far harder to grasp is why exactly these specific operators would be chosen and what their real philosophical impact is. Luckily, this becomes much clearer once you consider real-life sets outside of the realm of abstract logic and mathematics.

5.3 Unified implication and its relation to language

In particular, as you will see, the relationship of the unified implication operators to natural language is quite surprising and illuminating. So far, some of you might be thinking that unified implication feels a bit artificial and may be wondering what the point is. However, once you see the relationship unified implication has to natural language, and it really clicks, then you won't have any doubt left that unified implication is a legitimate operator. Indeed, not only is unified implication legitimate, but you use the operator every day without even consciously realizing it.

Specifically: Many of the sentences you frequently construct when you communicate using natural language would actually be impossible to construct without unified implication. In fact, it seems to me that *all* natural languages probably have unified implications hidden in them at least somewhere. It appears that unified implication may be an inherent part of the structure of all languages, a critical concept hidden in plain sight, part of the bridge between formal logic and linguistics.

Some of the most basic parts of speech in natural languages are actually just unified implications in disguise. Not only that, but many sentences in natural languages are actually composed almost entirely of unified implications and sets. Once you start to see the parallels, and you learn what the natural language equivalents of each of the variants of unified implication are, then the conceptual essence of what unified implication really is will become substantially more apparent.

Let's start with sub implication. What is the natural language equivalent of sub implication? Take a moment to think about it if you'd like to see if you can figure it out on your own. If not though, don't worry. The relationship is fairly subtle, so it is understandable if you have trouble finding it. Once you see the connection though, the relationship will become much clearer and much more obvious. Anyway, are you ready? Here's the answer: Sub implication is roughly the formal logic equivalent of the words "the" or "of" in natural language. There are also some other words that are related to sub implication, but we'll get to that later.

Think of it like this maybe: The words "the" and "of" are two of the most pure analogs of sub implication that exist in natural language. These two words express almost nothing except for pure sub implication[3], whereas certain other words also implicitly involve sub implication but are more "muddled" and have other concepts mixed in.

Does this correspondence between sub implication and these two words ("the' and "of") seem weird to you? Don't worry, that's very understandable. I'll show you some

---

[3]Well, actually, I wonder if maybe "of" is impure. The word "of" has the connotation of "related to", "coming from", or "made partially or entirely from", whereas "the" more strictly has only the connotation of "member of the type". Mostly they seem to behave similarly though. I'm not sure how this point of nuance should be handled in the system, but my current approach is to treat "the" and "of" as essentially the same (i.e. as sub implication). I could be wrong.

examples and then you'll gradually see what I'm talking about better as we continue our discussion. For example, consider the following list of natural language expressions that all contain the words "the" or "of" somewhere and think about how sub implication might relate to them:

1. Jupiter the planet

2. Bob Ross the painter

3. citizens of Germany

4. product of China

5. a sword of gold

6. people of the world

7. orange the fruit

8. orange the color

9. Griswold the Blacksmith[4]

10. Frodo of the Shire[5]

11. Dracula the Vampire[6]

12. Brienne of Tarth[7]

13. Intel and Google the corporations

14. 2, 3, 5, and 7 the prime numbers

15. copper, silver, and gold the metals

16. the flavor of the food

17. the dangers of chainsaws

18. the risks and opportunities of entrepreneurship

---

[4] Griswold is a character from the classic 1996 video game *Diablo*. He is the only blacksmith residing in Tristram, which is the town you start in. You can buy new weaponry and armor from him, or pay him to repair your existing equipment.

[5] Frodo is the main character of the influential genre-creating fantasy novel series *The Lord of the Rings*. He is a member of a race of short humanoids called Hobbits. His hometown is known as the "the Shire".

[6] *Dracula* is the name of a famous novel which popularized the concept of vampires. The character named Dracula in it is one such vampire.

[7] Brienne of Tarth is a character from the popular fantasy novel series *A Game of Thrones*. Tarth is her homeland, the island where she grew up.

Did you notice how the entity on the left side is always a subset of the entity on the right side in some sense? Did you also notice how all of these expressions are *value expressions* and that they all return the entity on the left side (the subset) after being evaluated? If these weren't value expressions then you wouldn't be able to say things like "The knight vanquished the demon by impaling it with a sword of gold." because then you wouldn't be able to chain "a sword of gold" into the surrounding expression (etc).

Notice also that it is not gold of just any kind that the knight is impaling the demon with in the previous example, but rather it is a *sword* made of gold. All of these expressions in the list clearly focus on the subset (the entity on the left) and not on the superset (the entity on the right). That's why these expressions are all sub implications and certainly not super implications or unified equivalences.

Notice also that all of the participating entities are sets, regardless of which side of "the" or "of" they are on. Sometimes the entity is a set of only one object, but other times the entity is a set of multiple objects. Take, for example, the expression "citizens of Germany". On its own, the "citizens" part of this expression refers to some specific unknown set of citizens, but when you combine it with "of Germany" our minds automatically make the connection that the phrase must be talking about specifically the citizens of Germany as opposed to some other set of citizens.

In other words, our minds automatically search for suitable pairs of sets that would make the expressions not return empty sets, and then use the most likely of such pairs in order to interpret what the expressions should mean. If we can't find any such set, then the set our mind retrieves will end up being empty, at least until such time as we become aware of two sets which work for the expression. Notice how closely this mirrors our definitions for unified implication. Except for the "searching our minds for the most likely pair of sets" aspect, it is essentially the same.

However, a critical realization you must have in order to see what is really going on in these natural language expressions is that we humans use a lot of shorthand and implicit context when we express ourselves. Sometimes there's hidden structure that has been omitted for the sake of brevity and efficiency. This is reasonable considering how laborious and long-winded even brief expressions in natural language can sometimes be.

The use of hidden structure is an effective optimization in this respect. For example, did you notice that some of the expressions in the list above included some extra instances of "the" (e.g. "people of *the* world", "*the* flavor of *the* food", etc) that maybe don't appear to you to be behaving as sub implications? Did this confuse you?

Well, it turns out that those extra instances of "the" in these expressions actually *are* sub implications. It's just that you can't see it as easily because they have some hidden structure. You see, the word "the" sometimes actually refers to an implicit object that is known to the participants in the conversation from context[8]. For example, the expressions "citizens of Germany" and "product of China" only have one sub implication

---

[8] Whether the context occurs in the past, present, or future doesn't matter. For example, someone could say "Go to my room and get *the* coat hanging near *the* window." and the person retrieving the coat could still understand this context even though it occurs in the future. The sets will be filled in by the mind once the person arrives at the room and sees the items.

in them, but in contrast "Frodo of the Shire" actually has two and "the flavor of the food" actually has three. These later two expressions have hidden structure. The extra instances of "the" floating around in the text are what give it away.

When we say "the Shire" or "the flavor" or "the food" we are referring to some *specific* instances of each of those things, as opposed to the more general concept of "shires" or "flavors" or "food". For example, if we just said "the flavor of food" (instead of "the flavor of *the* food") then it might be interpreted as meaning all of the flavors of all foods, i.e. as referring to the concept of foods having flavors in the most generalized possible sense, which would probably not be what we want in this case[9].

Thus, the real structure of these expressions must actually be something along the lines of "X the Shire", "Y the flavor", and "Z the food". The hidden variables X, Y, and Z here serve to indicate that it is some specific subset or instance of each concept that we are referring to, as opposed to the more general sense. X is a specific Shire[10], Y is a specific set of flavors, and Z is a specific set of food.

Here, let's translate "Frodo of the Shire" and "the flavor of the food" into unified logic to make the point clearer. Let F be the set containing only Frodo, let S be the set of all shires (not just the Shire from *The Lord of the Rings*), let L be the set of all flavors, and let D be the set of all foods. Let X, Y, and Z serve the purpose previously mentioned (i.e. placeholder variables to provide specificity). Then, the translation into unified logic (revealing all the hidden structure) looks like this:

1. Frodo of the Shire: $F \rightharpoonup (X \rightharpoonup S)$

2. the flavor of the food: $(Y \rightharpoonup L) \rightharpoonup (Z \rightharpoonup D)$

In these expressions, $(X \rightharpoonup S)$ means specifically the Shire from the *Lord of the Rings* (not just any shire), $(Y \rightharpoonup L)$ means the specific set of flavors being experienced, and $(Z \rightharpoonup D)$ means the specific set of food being consumed. The word "the" here is simply serving to bind to specific instances (specific subsets) of these more general concepts (shires, flavors, and foods) in order to disambiguate them via implicit context. This is a efficient way of eliminating some verbiage. It is part of how natural languages are able to use variables without having to actually explicitly name all of them.

Use of "the" as a prefix to an object usually indicates that some hidden variables are in play somewhere. Have you ever thought it's weird that natural languages can be so expressive without using many named variables? Well, now you know part of why that probably is. It's a trick based on simply chaining unified implications together, threading them together in such a way as to eliminate the need for too many awkward explicitly named variable introductions by allowing large amounts of specificity to be determined by context, a clever trick which natural languages exploit quite often.

---

[9] Story time: Bob and John sat at the table, busily eating their breakfast. "Hey Bob, how do you like the flavor of food?", asked John. Eyes glazed over with impatience and early morning irritability, Bob stared blankly back at John for a while until finally replying: "Yes John… after all, it would certainly suck if food never had any flavor. Now shut up and go learn better grammar."

[10] The word "shire" is apparently sometimes used as a synonym for "county", especially in England. It means a regional government, especially one for a rural region. Thus, there could be multiple different "shires" and hence the potential need for disambiguation.

Fun fact: By successfully connecting the concept of sub implication to the concept of the word "the", we are now in a position to give the word "the" a formal definition. Have you ever tried defining "the" before? Ask someone to define "the" for you without using a dictionary. Lots of people will be baffled by the task.

The dictionary definition for "the" (according to Google at the time this book was written) is (1) "denoting one or more people or things already mentioned or assumed to be common knowledge" or (2) "used to point forward to a following qualifying or defining clause or phrase". These definitions are good enough for informal purposes. In fact, notice that these definitions also seem to have interesting correlations to unified implication. However, here's a more formal definition:

> **the**: sub implication as it appears in natural language, such that it also supports the optional use of a hidden variable in the subset operand

In addition to the ordering convention used for the list of examples above (on page 387), the opposite ordering convention also sometimes occurs in natural language. This is analogous to reversing the direction of the sub implication arrow symbol (i.e. from $\mapsto$ to $\mathrel{\rotatebox[origin=c]{180}{$\mapsto$}}$), as can happen when you use explicit unified implication notation. The word order in the natural language analog isn't necessarily the same as the symbol order in unified logic though. Here's some examples of reverse ordered sub implications in natural language:

1. the color orange

2. the classic 1996 video game *Diablo*

3. the corporations Intel and Google

4. my uncle Gary

5. Olivia's pet cat Shadow

6. the famous game programmer John Carmack

Let's take a moment to analyze a few of these examples so that we can see what their structure is more precisely. Consider, for example, the phrase "the classic 1996 video game *Diablo*". Do you see which part is the superset and which part is the subset? Here's the answer: The superset is the set of all classic 1996 video games, whereas the subset is the set containing only *Diablo*. Let C be the set of all classic 1996 video games and let D be the set containing only *Diablo*.

Then, the phrase "the classic 1996 video game *Diablo*" can be translated into unified logic as $C \mathrel{\rotatebox[origin=c]{180}{$\mapsto$}} D$. The meaning is the exact same as the natural language phrase. Notice however that "the" prefixes the two operands in English, whereas in unified logic the corresponding sub implication is an infix operator. Notice also that "*Diablo* the classic 1996 video game" has the exact same meaning as "the classic 1996 video game *Diablo*", and would translate into unified logic as $D \mapsto C$, just as expected.

As you can see, most of the examples I gave in the list above adhere to a format of "the *B A*" where *B* is the superset and *A* is one of its subsets. In unified logic this becomes $B \rightharpoonup A$. This is one of the more obvious forms of reversed order sub implication that occur in natural language. However, other forms may occur as well, and a few of the examples I selected demonstrate that.

Take, for example, the phrase "Olivia's pet cat Shadow". Isn't Shadow a subset of the set of all of Olivia's pet cats? After all, for all we know, Olivia might have multiple pet cats and not just one. Similarly, isn't it possible that I have multiple uncles and Gary is just one of them, and hence that "my uncle Gary" can likewise be considered as containing a hidden sub implication?

Thus, as you can see, expressions of the form "*B A*", where *A* and *B* are both nouns, can *also* at least sometimes be instances of reversed order sub implication[11]. Notice that "the *B*" and "*A*" from the other form we discussed ("the *B A*") also both constitute nouns. For example, "the color" and "orange" are both perfectly good independent nouns. Splitting up "the color orange" in this way does not create grammatical problems. This is typical of reversed order sub implications in English.

However, beware that many phrases of the form "the *B A*" and "*B A*" are *not* sub implications of the form $B \rightharpoonup A$[12]. For example, consider the phrase "the colorful shirt". Are "the colorful" and "shirt" *both* perfectly good independent nouns? Nope, not really. The phrase "the colorful" is not a noun, at least not without stretching the semantics of it quite a bit. You wouldn't normally say "I went to the store and bought some colorfuls." would you? I think not. Even more revealingly though, watch what happens if you assume that "the colorful shirt" is a sub implication and then try to reverse the operand order. Reversing "the colorful shirt" turns it into "shirt the colorful". Sounds weird right?

That's because "colorful shirt" is actually an *intersection* of two sets (the set of all colorful things and the set of all shirts), and the "the" in "*the* colorful shirt" is actually just indicative of a hidden variable providing specificity. As such, let C be the set of all colorful things, let S be the set of all shirts, and let X be some specific shirt. Then, "the colorful shirt" would actually translate into unified logic as $X \rightharpoonup (C \cap S)$.

Translating it as $C \rightharpoonup S$ would be wrong. Not all shirts are colorful. Some are monochrome. The set of all colorful things and the set of all shirts merely intersect. Neither is a subset of the other. You need specificity to escape that fact, and you can't get specificity if you've already consumed the sub implication's subset operand by translating the phrase into $C \rightharpoonup S$. Hence $X \rightharpoonup (C \cap S)$ must be the correct translation.

You need to be careful how you read things, and make sure you are not confusing intersection with sub implication. The intersection and the sub implication of a pair of sets are often both simultaneously true, so that's part of why this is an easy mistake to make. Generally speaking, adjectives tend to be participants in intersections of sets,

---

[11] Parenthetical expressions can also sometimes be sub implications. Notice how "Olivia's pet cat Shadow", "Shadow (Olivia's pet cat)", and "Olivia's pet cat (Shadow)" all mean the same thing and imply that Shadow is a member of the set of Olivia's pet cats. Notice also that the focus (the return value) of each expression is Shadow, and not all of Olivia's pet cats, and that therefore these expressions must indeed be sub implications and not super implications.

[12] Phrases of the form "the *B A*" *will* always contain a sub implication corresponding to at least the word "the", but not always in the form of $B \rightharpoonup A$ or one of its equivalents.

not implications. Nouns can go either way though, since sometimes nouns are used like adjectives and sometimes not. For example, notice that the phrase "classic 1996 video game" in "the classic 1996 video game *Diablo*" can be decomposed into an intersection of sets.

Allow me to demonstrate: Let L be the set of all classic things, let Y_{1996} be the set of all things that happened in 1996, let V be the set of all video games, and let D be the set containing only *Diablo*. Then, the phrase "the classic 1996 video game *Diablo*" can be translated into unified logic as $(L \cap Y_{1996} \cap V) \rightarrowtail D$. Notice that previously we wrote this as $C \rightarrowtail D$, where C was the set of all classic 1996 video games.

In other words, $L \cap Y_{1996} \cap V = C$. Notice that "1996" and "video game" are both nouns, even though they are acting like adjectives and are participating in an intersection of sets. Natural language can be tricky and nuanced sometimes. Pay more attention to the actual meaning of phrases than to the superficial structure and quirks. Natural languages often take shortcuts that create subtle ambiguities. You'll need to pay close attention to any implicit information that may be present.

Anyway, while we're still talking about the distinction between sub implications and intersections, let's go ahead and briefly look at some examples of pure intersections. Doing so will help you form the distinction between sub implications and intersections more clearly in your mind, so that you can translate natural language expressions into unified logic expressions more accurately. This time there won't be any unified implications at all, only intersections of sets. I want you to see what that looks like. I'm also going to throw in a few unions of sets just to mess with you. See if you can spot them. Here we go:

1. autumn leaves

2. haunted houses

3. big fluffy bunny rabbits

4. sweet candy and sour grapes

5. bread, oats, and potatoes

6. sweet and salty popcorn

Let's look at the easier examples first. Consider, for example, the phrase "autumn leaves". What is the structure of this phrase, when translated into unified logic? Well, basically it's just a simple intersection of "autumn" and "leaves". Let A be the set of all things associated with autumn and let L be the set of all leaves. Then, the phrase "autumn leaves" can be translated into unified logic as $A \cap L$.

Easy right? The next two items, "haunted houses" and "big fluffy bunny rabbits", translate similarly. The only difference is that "big fluffy bunny rabbits" requires three sets instead of two. As such, let B be the set of all big things, let F be the set of all fluffy things, and let R be the set of all bunny rabbits. Then, the phrase "big fluffy bunny rabbits" can be translated into unified logic as $B \cap F \cap R$. Similarly, if N was the set of all haunted things and S was the set of all houses then $N \cap S$ would be the set of all haunted houses.

The last three items in the list are where things get maybe a little bit tricky. The main problem for interpretation here is that the word "and" in English means different things in different contexts. Sometimes English "and" means intersection, and sometimes it means union. Can you see which is which in the last three items in the list above? I'll explain all three of them, since they each demonstrate different contexts where English "and" is used.

For item #4, we have "sweet candy and sour grapes". As such, let W be the set of all sweet things, let C be the set of all candy, let R be the set of all sour things, and let G be the set of all grapes. Do you notice how "sweet candy and sour grapes" has a split structure and two different kinds of components? Well, that's because the phrase actually translates into $(W \cap C) \cup (R \cap G)$. The "and" in this case is actually a union, despite the fact that we normally associate "and" with intersection in logic.

The real intersections in this expression are actually the adjective-noun pairs on both sides. It's kind of weird that the English "and" here is actually a set theory "or" operator (a.k.a. a union) in disguise, right? That's just the way English is though. Logical connectives are used a bit inconsistently in some natural languages. English "and" serves as both a constraint specification mechanism and a listing mechanism, and that is the origin of this quirk.

For item #5, we have "bread, oats, and potatoes". In this case, the expression uses only one type of operator. Do you see which one? Try performing intersections or unions on the sets and you'll soon see. Let B be the set of all bread, let O be the set of all oats, and let P be the set of all potatoes. Then, the phrase "bread, oats, and potatoes" translates into unified logic as $B \cup O \cup P$.

Lastly, for item #6, we have "sweet and salty popcorn". In this case, the "and" actually *is* being used as an intersection. Specifically, it is being used as an intersection of adjectives attached to a noun. Let W be the set of all sweet things, let L be the set of all salty things, and let P be the set of all popcorn. Then, the phrase "sweet and salty popcorn" translates into unified logic as $W \cap L \cap P$.

As you can see, it is just a chain of intersections, and indeed the "and" in it is arguably redundant. The phrase "sweet salty popcorn" would mean the same thing, but I guess us English speakers like the cadence of "sweet and salty popcorn" more apparently, since it sounds slightly more natural than "sweet salty popcorn" for some reason.

On the other hand though, the prior list of examples aside, consider the phrase "red and blue glass". How would it translate? Well, let R be the set of all red things, let B be the set of all blue things, and let G be the set of all glass things. Then, the phrase "red and blue glass" could translate to either $(R \cup B) \cap G$ or $R \cap B \cap G$ depending on the interpretation.

In the case of $(R \cup B) \cap G$, this would be equivalent to $(R \cap G) \cup (B \cap G)$ by the law of distributivity and would mean a mixed set of red glass and blue glass, not requiring any piece of glass to be both red and blue but perhaps permitting it[13]. In contrast though,

---

[13] This depends on how loosely or strictly you interpret the adjectives. A "red object" could mean an object that is either partly red or entirely red, depending on interpretation. Natural language is often sloppy on this point. For our glass example here, we are

in the case of $R \cap B \cap G$, this would mean a set of glass where *all* of the pieces of glass are required to be both red and blue in some sense (e.g. as a conglomerate of separate chunks of red and blue glass fused together, or as purple glass, or in a strict exclusive sense that would yield an empty set, etc). The translation of natural language into formal language often depends on the implicit intended meaning of the specific phrase. Be careful.

Anyway, adjectives aren't the only things in natural language that translate into intersections of sets. Many prepositional phrases also do. To refresh your memory of grammar class, remember that prepositions are words like "in", "on", "above", "under", "at", "by", "near", and so on. Do not confuse pr**e**positions (a type of grammatical object) with pr**o**positions (statements with truth values). Prepositions are a category of words, not statements. The close similarity of the spellings of the two terms ("preposition" and "proposition") is unfortunate.

Essentially, prepositions modify the set they are attached to in order to focus in on part of that set in some specific way. In this way, prepositional phrases allow you to communicate what set of objects you are talking about more specifically and also to describe how that set relates to other objects more flexibly. Prepositional phrases have more of a relational connotation, whereas adjectives have more of a qualitative connotation, but both serve a similar function in that they are used to constrain sets with additional criteria. Here's some examples of prepositional phrases for you to consider, in light of these thoughts:

1. the power outlet under the desk

2. the lamp near the door

3. lunch for tomorrow

4. the mailbox at the train station

5. the research paper by Grigori Perelman

6. a letter from grandma

7. the house on Water Lane in Lincolnshire, England

8. books about mathematics

9. restaurants near museums

10. the bookstore between the bakery and the art supply store

Let's formalize a few of these so that we can see what the underlying structure is more clearly. For example, consider the phrase "the power outlet under the desk". Let X be the specific power outlet we have in mind, let P be the set of all power outlets, and let U be the set of all things under the desk. Then, the phrase "the power outlet

---
assuming the looser interpretation, i.e. that the adjective need only apply partially and not necessarily entirely.

under the desk" could be translated into unified logic as $(X \rightharpoonup P) \cap U$. Notice how "the power outlet" has a sub implication in it to provide specificity, but the overall phrase "the power outlet under the desk" itself is actually an intersection of sets. This kind of phrasing can sometimes be slightly misleading. The "the" in front can make you think that the phrase should maybe be considered a sub implication, but this is an illusion in this case. A power outlet is not (in general) a kind of "thing under a desk". Only *some* power outlets are members of the "under the desk" set.

Next, let's formalize the phrase "the house on Water Lane in Lincolnshire, England". Let X be the specific house we have in mind, let H be the set of all houses, let W be the set of all things on Water Lane, and let L be the set of all things in Lincolnshire, England. Then, the phrase "the house on Water Lane in Lincolnshire, England" translates into unified logic as $(X \rightharpoonup H) \cap W \cap L$. Notice that this time we have one more set involved than we did before. That's because "on Water Lane" and "in Lincolnshire, England" are both perfectly good independent prepositional phrases. Prepositional phrases often work together like this, forming large chains of relational constraints designed to precisely pin down what exactly it is you are talking about, in order to prevent miscommunication or ambiguity.

Finally, let's do one last example from this list. Consider the phrase "books about mathematics". Notice that this time, unlike the previous two examples we formalized, there is no "the" prefixing the phrase. How will this change the structure of the formalized version? Well, it will change it exactly as you'd expect. It will simply cause the first set to not be a participant in a sub implication, but rather to merely be a broad category of things, i.e. to be a general set rather than a specific one. The result is very simple: Let B be the set of all books and let M be the set of all things that are closely related to mathematics. Then, the phrase "books about mathematics" can be translated into unified logic as $B \cap M$. Easy right?

Intersections, unions, and sub implications of sets occur in great abundance in natural language in various different forms. Intersections and unions are their own separate thing and cover a pretty broad spectrum of language phenomenon, from adjectives to prepositional phrases (and perhaps other things too). However, sub implication certainly isn't the only form of unified implication that shows up in natural language.

Both super implication and unified equivalences also show up as well. As such, I think it's time we talk about the correspondence between natural language and super implication now some. Doing so will help flesh out our understanding of the nature of the complementary relationships between sub implication and super implication and will help us appreciate the nuanced differences between how they feel in expressions.

Recall that the only real difference between sub implication and super implication is which of the two participant sets they return. If the condition set is a subset of the consequence set then sub implication will return the condition set, whereas super implication will return the consequence set. The difference is all in the return value.

Therefore, all we have to do to find analogs of super implication in natural language is to find expressions that focus on a superset while simultaneously alluding to the existence of some valid subset of that superset. Can you think of any expressions in natural language that fit that description, i.e. that *give examples* of members of a set but without shifting the grammatical focus to such members and instead keeping the focus on the encompassing set?

Want another hint? Here you go: The key word here is *examples*. Here's the answer: Super implication is the formal logic equivalent of the phrases "e.g.", "for example", "for instance", "such as", "including", and "which includes" (and any other strong synonym of these phrases) from natural language[14]. Makes sense if you think about it, right?

When you talk about a set, but you include a few examples of that set for reference, you often still want to keep the grammatical focus on the original set. The examples you give at such times essentially probe the readers' understanding of the set, to make sure they're on the same page as you about what kinds of things are in that set, thus providing additional clarity and refreshing the readers' memory and state of mind. Some concrete examples will make this clearer. Here's a list for you to consider, in light of super implication:

1. polygons, such as triangles

2. blacksmiths (e.g. Griswold from *Diablo*)

3. prime numbers (for example: 2, 3, 5, and 7)

4. game engines (e.g. Unity, Unreal Engine, Godot, and RPG Maker)

5. the library's catalog of books, which includes many science, engineering, art, and history books

6. animals, including polar bears and honey badgers

7. economically prosperous professions, for instance computer science, electrical engineering, and medicine

8. corporations such as Intel and Google

See how the focus is on the superset this time, but the expressions are still implicitly testing whether or not the subset is indeed a member of the superset? That's super implication for sure. It's very characteristic of the behavior of the operator. Notice how the sets of examples (the subsets) actually sort of disappear into the superset, and that it is still the superset we are really talking about. The sets of examples are just there to validate the reader's understanding of the superset and to refresh their memory. The return value of the overall expression is very clearly the superset. This is the opposite behavior of sub implication. However, not all appearances of theses phrases in natural language necessarily have straightforward translations into terms of super implications, so be a bit careful.

Let's formalize a few of these, just for practice. Consider, for example, the phrase "polygons, such as triangles". To refresh your memory of geometry, recall that a polygon is any 2-dimensional closed shape whose boundary consists entirely of some ar-

---

[14]Often these phrases occur inside comma delimited segments of sentences or inside parentheses, but not always.

bitrary number of flat sides[15]. Similarly, a triangle is any 2-dimensional closed shape with exactly 3 flat sides. Let P be the set of all polygons and let T be the set of all triangles. Then, the phrase "polygons, such as triangles" translates into unified logic as $P \leftarrow T$. The expression $P \leftarrowtail T$ would also be equivalent of course, as would $T \rightarrowtail P$ (if you ignore the superficial ordering correspondence).

Next, let's do the game engine example. Suppose that G is the set of all game engines and that E is the specific set of game engines {Unity, Unreal Engine, Godot, RPG Maker}. In that case, the phrase "game engines (e.g. Unity, Unreal Engine, Godot, and RPG Maker)" would translate as $G \leftarrow E$, just as you'd expect. By the way: Game development is one of the most powerful and fulfilling creative mediums ever invented. You can create your own little virtual worlds filled with whatever you want. Your world can be passive and contemplative, or it can be fun and gameplay-oriented. It can be whatever you want. The sky is the limit. With sufficient skill, you can turn your wildest dreams into reality. Why not give it a try?

As an aspiring game developer myself, who has worked in the industry, I personally currently recommend Unity as the best overall game engine to use for most people. It is both powerful and easy to use. The user interface is clean and free of noise. It seldom misleads you. It is the opposite of overwhelming. Unity does require programming experience to use effectively though, so expect to do a lot of coding.

However, if that's too hard for you, I hear RPG Maker is relatively beginner friendly, has lots of pre-made content to play with, and nominally doesn't require programming experience. Be that as it may, RPG Maker is also extremely limited in terms of the kinds of games you can make with it. Perhaps try some of the other engines too. Unreal Engine and Godot are probably Unity's main competitors. Each engine has different pros and cons, but I still recommend Unity the most, both for beginners and experts. Programming is not even remotely as difficult as many people think it is. With enough patience and persistence, anyone can learn it. Don't be intimated.

Truly powerful game engines *always* require programming experience[16]. Only programming languages are capable of fully describing completely arbitrary concepts, and this is why learning to program is a necessity if you want to push any game design boundaries. Creating new gameplay mechanics fundamentally requires programming. Anything is better than nothing though, so don't tie yourself in knots trying to decide between all the different game engines out there. Avoid analysis paralysis. Make your

---

[15] For the curious: The correct term for an *n*-dimensional "polygon", i.e. for any "polygon-like" (i.e. flat-sided) shape in any number of dimensions is a *polytope*. The 3-dimensional case is called a *polyhedron*. Thus, we have polygon (2-dimensional), polyhedron (3-dimensional), and polytope (*n*-dimensional). In the 3D modeling industry though, arbitrary geometric objects are usually referred to simply as *meshes* instead of as polyhedra.

[16] By the way, visual scripting (which is often advertised disingenuously as "not requiring you to learn programming") still 100% definitely requires you to learn programming. It just looks fancier is all. If anything, visual scripting is actually *more* awkward and difficult to work with than regular text-based programming is. It's arguably an instance of an anti-pattern called the Inner Platform Effect (look it up). The supposed user friendliness of such systems is mostly just a superficial illusion. The only truly helpful part of such systems is that some of them have built-in continuous type checking that can prevent you from connecting things together incorrectly. It thus can eliminate some syntax problems. Otherwise, such systems are usually just a hindrance. The "ease of use" for "non-programmers" is mostly just a placebo effect. It's still programming. Programming is already fairly easy. Intimidation is the main thing that drives people away. Visual scripting mostly just gets "non-programmers" to lower their guard, bypassing the "non-programmer's" toxic self-limiting beliefs.

tasks small and actionable. Every day, ask yourself what's one little thing you can do to move forward or to learn more about the systems available to you and then just do it. Don't overthink it too much.

Just get the ball rolling enough that your creative spirit is ignited and then see where it goes. Forget about self-limiting beliefs. You might be surprised what you are capable of. Almost everyone who says they can't do art, or can't do music, or can't do math, or can't program (etc) is wrong about it. That's all just in your head. Believe me, I've been there. It's not really that hard to acquire basic proficiency in any of these skills, especially if you use lots of good sources of information (e.g. tutorials, documentation, examples, books, etc) and do plenty of motivated self-study and real-world practice. It does require patience and a healthy work ethic though.

True mastery is a somewhat different matter perhaps, but you'd be surprised just how quickly you can learn 80% of the most valuable fundamentals of any skill. Sometimes you can acquire basic proficiency within even just a few weeks of practice, if you work efficiently enough. It is very liberating. Just try it some time. Stop doubting yourself. Unleashing your creative spirit will make your life feel much more fulfilling and satisfying. There is no joy more lasting than the joy of creation.

By the way, the fact that you can learn 80% (or some other high percentage) of a skill in a fairly short period of time is not merely some isolated phenomenon specific only to learning new skills. It is actually more general than that. In particular: This example is just one of countless examples of the Pareto Principle in action, which broadly speaking says that often 80% of the results of something come from 20% of the efforts (or some other similar high ratio, e.g. 90:10, 70:30, etc).

The Pareto Principle is a useful rule of thumb and is worth considering. It is valid surprisingly often. It may sound arbitrary, and sort of is, but when you cut away the more arbitrary aspects of it and focus on the real fundamental underlying essence of it though, the Pareto Principle generalizes easily: It is simply the observation that exponential distributions ("power laws") are extremely common in both natural and artificial systems, especially in any system that has any form of feedback or self-amplifying assets.

This principle is the same reason why the more wealth you have the easier it becomes to acquire even more. The more resources you have available to you the easier it is to invest those resources to generate even more assets. It is an example of exponential growth in the real world. Just as placing a microphone near a speaker creates an exponential growth of sound that blasts the speakers, so too do countless other natural and artificial systems experience a threshold effect where explosive exponential growth occurs easily under the right conditions.

Exponential growth is so powerful that once even a small amount of it starts snowballing it can rapidly generate explosive abundance in the blink of an eye. Being aware of this fact and leveraging it to your advantage can be extremely effective. In fact, awareness of this (whether conscious or subconscious) is part of the secret sauce to what typically makes the difference between weak economic actors and strong ones, i.e. between poverty and wealth.

Wealthy people tend to know how to take advantage of these kinds of exponential growth effects and often have a much better sense of proportionality and impact with respect to their economic choices than most poor people do. Wealthy people tend to think in terms of investment, prioritization, and economic scalability (e.g. creating permanent passive income through product creation and IP ownership, understanding that "you have to spend money to make money", etc), whereas poor people tend to pursue wealth in extremely inefficient and unscalable ways (e.g. selling their valuable creative work, energy, and time to others for one-time payments or hourly wages, etc).

Anyway, we got sidetracked on a tangent there for a moment. I love these tangents though because they tend to generate a lot of unexpected serendipitous value. Structuring your thoughts too rigidly and not allowing your mind to flow where it naturally wants to flow, in contrast, tends to suppress serendipitous value and reduce its frequency and depth. Let your mind flow where it wants to flow, just as long as you keep the flow within reasonable constraints that account for productivity. In this way, you can often uncover hidden wisdom.

When you listen to the lay of the land, when you follow its contours and shape to find the paths of least resistance and the points of highest visibility, it will often tell you where you should go. It will whisper secrets to you, hints of where the highest-value insights may lay hidden. No unexplored path ever came equipped with pavement. Yet, all you have to do is look around you to see such paths in abundance, beckoning to anyone who is willing to quiet their mind long enough to hear the call, locked gateways that open only to those who can spare a patient moment to live above the meaningless hustle and bustle of everyday tedium and norms.

Anyway, we got sidetracked on a tangent there for a moment (*deja vu*, heh). Let's return our thoughts to super implication now. I want to do one last example from the most recent list before we move on. The example I have in mind this time is the phrase "the library's catalog of books, which includes many science, engineering, art, and history books". There are perhaps multiple ways to formalize this phrase, but I'll show you the first way that popped into my mind. As usual, we'll need to name some sets first.

As such, let X be the set of all books held at the specific library we have in mind, let L be the set of all books that are held by at least one library in the world (i.e. all library books), let B be the set of all books (whether held in any library or not), let S be the set of all science books in X, let E be the set of all engineering books in X, let A be the set of all art books in X, and let H be the set of all history books in X. Then, the phrase "the library's catalog of books, which includes many science, engineering, art, and history books" may be translated into unified logic as:

$$((X \rightharpoonup L) \rightharpoonup B) \leftharpoonup (S \cup E \cup A \cup H)$$

The expression is a fair bit bigger this time, isn't it? Those subtle differences in phrasing can have a significant impact on what the closest translation looks like. Notice that this time the example actually includes both sub implications and super implications, as well as some unions too. I wanted to show you this example so that you could see what it might look like when sub implications and super implications are mixed together in phrases, so that you are better able to detect them when they occur.

Of course, you can simplify expressions like this by just not caring about some of the subtler underlying structure. For example, you could just work in terms of two sets

S_1 and S_2, such that $S_1 = (X \rightarrow L) \rightarrow B$ and $S_2 = S \cup E \cup A \cup H$ under the hood, but never actually formalize or describe the underlying structure of S_1 or S_2. In that case, the expression would then just become $S_1 \leftarrow S_2$. How much detail you want to include is somewhat a matter of taste. It depends on what seems optimal for the purpose you are trying to fulfill. There's often multiple ways of doing things.

Anyway, just as with sub implication, super implication can also sometimes be phrased in reversed order in natural language. In the above examples, the superset was always on the left and the subset was always on the right (resembling $S_1 \mapsto S_2$ or equivalently $S_1 \leftarrow S_2$). However, let's do some examples where that trend is reversed, i.e. where the superset is on the right and the subset is on the left (i.e. $S_2 \rightarrowtail S_1$). Phrases with super implication in reversed order are perhaps a bit less common (at least in English), but they do still exist. Here's some examples:

1. catfish, indeed all fish

2. biology, like all sciences

3. Adobe, Autodesk, and Pixologic, indeed all software companies who create art tools

4. fried chicken, in fact all fried food in general

5. brick-and-mortar bookselling, as with all fading markets

6. role-playing games, well all games really

7. tetrahedrons, indeed polyhedra in general

8. foxes, as members of the Canidae family

9. foxes, as canids

There may exist other phrases that also correspond to reversed order super implication, but these were the ones that came to mind as I wrote this. As you can see, phrases of the form "X, indeed all Y", "X, like all Y", "X, in fact all Y", "X, as with all Y", "X, well all Y really", "X, indeed Y in general", and "X, as Y" all seem to generally correspond with reversed order super implication. The most concise correspondence here seems to be the word "as". It is like the reversed order version of "e.g." sort of.

However, as you may have noticed, the grammatical focus of these expressions is a bit more hazy than for the other correspondences between natural language and the unified implication variants that we have previously discussed. The focus still sort of feels like it is on the subset, yet logically speaking it is fairly clear that whatever we say next will apply to the entire *superset* and not just to the specific subset we have mentioned. Thus, the return value of these expressions is indeed the superset, but it could still sometimes feel like we are "talking about" the subset in some sense. As such, it's best to think more about the actual underlying logical structure of these phrases, rather than merely thinking about "how they feel" or "how they sound". It's more foolproof that way.

Let me use one of these examples in a complete sentence so that the fact that it does indeed correspond to super implication becomes more conceptually clear, if it isn't already. For example, consider the sentence "Catfish, indeed all fish, are creatures that live in the water.". Do you see how the assertion "are creatures that live in the water" attached to "catfish, indeed all fish" here is actually making a statement about *all* fish and not just about catfish, and that therefore "catfish, indeed all fish" must be a super implication?

What does this look like formalized in symbols? Well, let's see. Suppose that C is the set of all catfish and that F is the set of all fish. Then, the most natural translation of "catfish, indeed all fish" into unified logic would be $C \rightharpoonup F$. Likewise, let's formalize "biology, like all sciences". Suppose that B is a set including only biology and that S is a set including all branches of science. Then, the translation to unified logic would be $B \rightharpoonup S$, as expected.

Finally, let's do an example that has slightly more complicated internal structure. Let's formalize "Adobe, Autodesk, and Pixologic, indeed all software companies who create art tools". Do you see why this example has a bit more internal structure than some of the other examples of reversed super implication? It's because we have a list of multiple items for the subset and also a preposition attached to the super set (specifically "who create art tools").

Let P be the set {Adobe, Autodesk, Pixologic}, let S be the set of all software companies, and let A be the set of all creators of art tools. Then, the phrase translates into unified logic as $P \rightharpoonup (S \cap A)$. Alternatively, we could choose to not bother naming the set {Adobe, Autodesk, Pixologic} as P, in which case we would simply write this:

$$\{\text{Adobe, Autodesk, Pixologic}\} \rightharpoonup (S \cap A)$$

Notice how $S \cap A$ means "software companies who create art tools". It would be wrong to say something like $S \rightarrow A$, since not all software companies are creators of art tools. I just thought I'd point that out to you again, as another reminder, to make sure that the distinction between unified implications and set intersections sinks in properly. Also, remember that intersection combines constraints whereas union allows any object satisfying either constraint in.

Thus, if we instead did a union, as in $S \cup A$, then it would actually correspond to "software companies or creators of art tools", rather than corresponding to "software companies who create art tools", and thus the overall expression would instead need to be something like "Adobe, Autodesk, and Pixologic, indeed all software companies or creators of art tools" in that case.

By the way, it's no accident that I designed implicit unified implication notation (i.e. the $\rightarrow$ and $\leftarrow$ operators) to use the operand ordering that I did, such that the subset of sub implication and the superset of super implication are both always on the left. I did it in full knowledge of the fact that expressions like "X the Y" and "X e.g. Y" tend to be a bit more natural in English than the reversed order variants sometimes are.

Having the left operand always be the return set is also very easy to remember, so that was another contributing reason for the choice. I chose the notations to be as natural as possible, basically. However, be that as it may, I wonder if other natural languages (i.e. languages other than English) have different optimal orderings. I don't know enough about other languages to discern the answer to that though.

Anyway, that covers super implication. It's now time to move on to the final variant of unified implication: unified equivalence. It too has a strong correlation with natural language. Can you guess what it is? Perhaps, now that you have seen multiple examples of finding these correlations between unified implication variants and natural language, you may find it easier this time. Take a moment to try to find the correlation, if you feel like it.

The answer is this: Unified equivalence is the formal logic equivalent of the natural language acronym "a.k.a." and its equivalent phrase "also known as", and sometimes "i.e." and other similar phrases too. Unified equivalence also closely corresponds with the word "equivalently", which is more formal-leaning than "a.k.a." but is still a natural language word arguably. As such, consider the following examples in light of this insight:

1. Samuel Clemens, also known as Mark Twain

2. electrical engineers, also known as "double Es"

3. K9s (i.e. trained police dogs)

4. programmers, a.k.a. coders, a.k.a. software engineers, a.k.a. software developers, a.k.a. computer scientists, a.k.a. hackers

5. police, also known as cops, also known as officers of the law

6. three-sided polygons (equivalently: triangles)

7. the third smallest prime number, a.k.a. 5

8. the value of $2 + 4$ a.k.a. 6

9. the value of $\sqrt[12]{2}$, a.k.a. approximately 1.0595, a.k.a. the frequency multiplier per semitone in 12-tone equal temperament according to the rules of music theory, a.k.a. what you have to multiply the frequency of any given note on a standard piano by in order to determine the correct frequency of the next note on the piano

10. all complex numbers, or equivalently all orientational numbers, or equivalently all oris

Let's formalize a few of these to get a better sense for them. Consider, for example, the phrase "Samuel Clemens, also known as Mark Twain". Let C be the set containing only Samuel Clemens and let T be the set containing only Mark Twain. Then, the phrase "Samuel Clemens, also known as Mark Twain" translates into unified logic as $C \rightharpoonup T$. Likewise, let's translate "three-sided polygons (equivalently: triangles)". As such, let P be the set of all three-sided polygons and let T be the set of all triangles. Then, the phrase "three-sided polygons (equivalently: triangles)" translates into unified logic as $P \rightharpoonup T$.

Let's try one of the bigger examples now. Let's try translating the phrase "programmers, a.k.a. coders, a.k.a. software engineers, a.k.a. software developers, a.k.a. computer scientists, a.k.a. hackers". Suppose that P is the set of all programmers, that C is the set of all coders, that E is the set of all software engineers, that D is the set of all software developers, that S is the set of all computer scientists, and that H is the set of all hackers. Then, in the most general possible sense of the meaning of each of these terms, these sets all refer to the same set, and the phrase translates into unified logic as follows:

$$P \rightleftharpoons C \rightleftharpoons E \rightleftharpoons D \rightleftharpoons S \rightleftharpoons H$$

Notice that I said "in the most general possible sense of the meaning of each of these terms". Those of you who are familiar with these terms and understand the nuances of each of them may feel inclined to claim that these are not all exactly the same sets really. Under a certain set of arbitrary definitions, you could be considered to be correct in such an objection of course. For example, the term "hackers" has the connotation of a specific subset of programmers: those who understand the more shadowy aspects of network programming and software exploitation, i.e. the kind of programmers who write computer viruses, who engage in cybercrime, or who work for government spy agencies (etc).

Similarly, the term "computer scientists" has the connotation of specifically programmers who do research on the foundations of computer science, i.e. programmers who research new techniques for programming, who delve deeply into the theory of computing, and who often work at universities and other kinds of research institutes. The terms "programmers" and "software developers" are very neutral in tone, with no particular connotation in any direction, except perhaps as they stand in contrast to other more specific kinds of programmers. The term "coders" is more of a slang term, and sounds a bit more informal, but still has mostly a neutral connotation and would usually be treated as referring all programmers.

Finally, the term "software engineer" is also somewhat neutral, but is sometimes used with the intent of (falsely) making the bearer of the title sound more skilled than someone who is "just a programmer". The idea is that supposedly a "software engineer" is stricter and more principled in how they manage software production, but this is not generally true in reality. The term is somewhat of a propaganda piece.

In an academic context, "software engineering" is associated with software management and planning, but many of the associated ideas (e.g. UML, fad software methodologies, etc) are often actually very out-of-touch with what really matters for producing high quality software. The term thus has an air of pretentious false superiority sometimes, e.g. like when a programmer who hasn't actually programmed much in a long time nonetheless feels compelled to meddle in other programmers' work in overbearing, out-of-touch, and counterproductive ways.

Purely from a descriptive standpoint though, the term "software engineer" is very accurate (i.e. *all* programmers *do* engineer software, obviously). It's only the efforts of a certain subset of programmers to elevate themselves artificially above other programmers via the label (without real merit) that has given the term a bit of a pretentious connotation. Otherwise though, it is a perfectly fine term[17]. Ultimately, it's best to treat "programmer" and "software engineer" as completely synonymous and to never treat either one as being superior to the other. The distinction is mostly just social posturing.

There is also another term for programmers, not listed in the example, which is "code monkeys". However, the term "code monkeys" refers specifically to low quality or unskilled programmers who tend to just slap together a bunch of code haphazardly to get it working, or otherwise to programmers who work on very trivial and crude code that requires little to no thought. Thus, the term "code monkeys" would be unlikely to be used to refer to all programmers, and that is why it was not included in the unified equivalence example above. Language is nuanced and often the same words are used in multiple different ways.

All of the terms in the unified equivalence example (i.e. programmer, coder, software engineer, software developer, computer scientist, and hacker) can be used either in the specific sense (with the specific connotations described above) or in the more general sense (where all of the terms refer to all programmers, without any distinction). We must choose which sense we mean and we must communicate it clearly. We must pick a single definition for each term we use when working with logic, otherwise we would be guilty of the fallacy of equivocation and any argument we used the terms in thereafter would be rendered ambiguous and untrustworthy.

If we use these programmer terms in the most general sense, then the unified equivalence would end up returning a non-empty set, since all of the sets would be the same set in that case. In contrast though, if we accounted for the more specific and more nuanced connotation of each of the terms, in such a way as to render at least one of the sets as less than the set of all programmers, then the unified equivalence would return the empty set. Notice also that this example demonstrates that it is perfectly acceptable to chain a whole bunch of unified equivalences together, both in unified logic and in natural language. This is just as expected considering the operator is a value expression, and composition is something all value expressions tend to naturally do.

Anyway, one interesting thing about unified equivalence is that it allows us to legitimately inline equivalences inside of surrounding value expressions. Do you remember how I mentioned once before (on page 376) that expressions like "$(2 + 2 = 4) + 5$" are technically invalid, because $2 + 2 = 4$ has no return value suitable for having 5 added to it? Well, unified equivalence is a form of equivalence that does *not* suffer from any such problems.

For example, suppose we had three arbitrary sets A, B, and C. If we then said something like $(A = B) \cap C$ then this would be invalid of course, since you can't intersect an equation with a set. However, notice that this would work perfectly fine with unified equivalence. The expression $(A \leftrightharpoons B) \cap C$ would be a perfectly valid expression.

---

[17] Perhaps another term for programming should be "the field of many names", as apparently we can't seem to stop ourselves from creating new synonyms for our profession. How many terms will there be by the year 3000 CE I wonder? Perhaps 256? Why wait to find out though? Let's get started. Useless new programmer synonyms here we come! Here's a preview: compiler whisperer, professional computer kicker, bugsmith, Skynet soother, MS Paint artist, undocumented computer system designer, software overengineer, user frustration specialist, square wheel innovator, unexpected computational behavior therapist, technical buzzword fountain, software paradigm cultist, code smeller, error emissary, algorithmic slacker, …

If *A* and *B* were the same set, then that set would intersect with *C* and return non-empty, else it would return an empty set to indicate no match to the criteria. Voila! We now have equivalences as value expressions. Thus, in unified logic, you can put equivalences inside of value expressions and still continue working, unlike in traditional math. Cool trick right? Unified implications and unified equivalences significantly increase your ability to seamlessly express certain common concepts. It's quite a vocabulary boost.

Anyway, that covers unified equivalence. We have now amassed quite a good variety of interesting correlations between all the unified implication variants and natural language. These correlations with natural language have obvious direct value, but in addition to that they also provide us with new alternative names for the unified implication variants.

As you may recall, when we originally defined the unified implication variants (on pages 379 through 381), I defined some official readings for those symbols, i.e. ways in which the unified implication arrow symbols could be read off the page and spoken aloud. Now that we have identified these additional correlations between natural language and unified logic though, we have some new synonyms for these operators. As such, here's a table of some formal and natural readings for the various different unified implication symbols, for ease of reference and comparison:

Symbolic Expressions	Formal Reading	Natural Readings
$X \mapsto Y$ $X \rightharpoonup Y$	X sub Y	X the Y, X of Y
$X \rightharpoondown Y$	X sub by Y	the X Y
$X \mapsfrom Y$ $X \leftharpoonup Y$	X super Y	X e.g. Y
$X \leftharpoondown Y$	X super by Y	X as Y
$X \rightleftharpoons Y$	X equiv Y	X a.k.a. Y, X i.e. Y

Thus, as you can see, these natural readings can be used as alternative names for the corresponding formal operators. For example, you could say the "the" operator instead of saying the "sub implication" or "sub" operator (etc). Notice that the table above is not comprehensive. I did not include all of the possible readings and correlations. I only picked the most concise and prominent ones. Doing so felt cleaner. It helps the relationships "pop" a bit more.

Natural languages tend to evolve in such a way that the most fundamental and common words are very short, whereas the less common words are much longer. Notice that this is exactly the case here. All of the unified implication variants have very short corresponding words in natural language. I do not think this is a coincidence, considering that these operators all occur in great abundance, hidden throughout natural language in various different places and forms, ingrained within the underlying structure of language itself.

It's fascinating how easily so many expressions from natural language can be translated into simple terms of sets, isn't it? We can construct some very natural expressions in logic now, formal expressions which flow and combine much like natural language does. These correlations between natural language and unified logic empower us to express ourselves much more effectively. However, this is hardly the end of it.

As you may recall, I mentioned earlier (on page 375) that we would be defining both some value expressions and some statement expressions for primitive unified logic. Well, we have just finished with the value expressions, but now it's time for the statement expressions. The statement expressions will expand our vocabulary to include not just descriptions of sets (as it has so far), but also complete sentences that make claims. Such statement expressions will be known as *existential statements* in unified logic, and they are the subject of the next section.

5.4 Existential statements and their relation to language

Think back to our discussion of existence-based truth, and how it compared to traditional statement-based truth. In existence-based truth, truth is a property of sets. Sets that are not empty have the property of being "true", whereas sets that are empty have the property of being "false". In contrast, in traditional statement-based truth, truth and falsehood are their own separate objects, and are merely labels that we attach to statements.

Clearly, given any arbitrary well-defined set expression in unified logic, the set will either be non-empty or empty, and this will determine whether it is true or false. This is all well and good, but the problem is that it does not yet allow us to actually assert a claim, but rather it only allows us to evaluate a description of a set to see what it ends up being, i.e. to see whether the described set is non-empty or empty.

The plus side of value expressions is that we never need to worry about conflicting statements, since there aren't any statements that could possibly conflict in the first place. However, from a practical standpoint, we nonetheless need a way to make assertions. Statements are a crucial component of effective communication. Merely having value expressions is insufficient, because if you only had value expressions then you would only be able to describe things.

You would never actually be able to make claims about those things, if all you had available to you was value expressions. This is where existential statements come in. Existential statements are a way of making statements that still adhere to the rules of existence-based truth, but allow us to make claims. They straddle the line between between existence-based truth and statement-based truth, essentially, giving us what we need to communicate properly.

Fundamentally, value expressions and statement expressions are both just different kinds of arbitrary forms. They are simply interpreted however we want them to be. Both types of expressions are subject to different sets of transformation rules, just as you might describe via transformative logic or some other suitable system of reasoning. As such, what we really need in order to implement existential statements is simply a way of modifying any given form whose value is a set so that the new modified form instead represents an assertion that the set is either empty or non-empty, according to our intent. This is not hard to do. All we really need is some new symbols and notation to represent it, i.e. some new operators. Let's define the first, most straightforward, and most general-purpose of such operators now:

Definition 127. *In unified logic, in addition to needing a way to describe sets, we also need a way to make assertions. As a system that uses existence-based truth, unified logic conceives of truth and falsehood as being something that depends on whether or not any given set is non-empty or empty. Thus, in order to express this, we will need an operator whose role is to mark whatever set expression it is attached to as either non-empty or empty according to our intent.*

*To mark a set as non-empty, we will prefix it with the symbol $\exists$, adding parentheses if necessary. We will call this operator the **existential truth operator**, or simply the **exis operator** for short. This operator uses the same symbol as the existential quantification ("there exists") operator from predicate logic, and is similar in many respects, but behaves slightly differently and thus is not technically the same operator. The exis operator accepts only a single set as input and is not a quantifier. Using the same symbol for both existential truth and existential quantification is useful as a memory aid. The difference between their roles does not seem great enough to justify creating a different symbol for each. They feel nearly synonymous in practice.*

Most often, we will read the symbol $\exists$ as "exis". One may be tempted to read an existential truth $\exists$ as "there exists", but doing so seems unwise because it could potentially confuse people about whether we mean existential quantification or existential truth. Thus, the standard policy will be to read $\exists$ only as "exis" in the context of existential truth. For example, the expression $\exists A$ can be read as "exis A" and would mean that the set A is non-empty (equivalently: that it is true, in the existential sense). Similarly, the expression $\exists(A \cap B)$ can be read as "exis A and B" or "exis A intersect B" and would mean that the set $A \cap B$ is non-empty. Notice that we added parentheses to $\exists(A \cap B)$ to avoid ambiguity.

*Similarly, in addition to having a way of asserting that a set is non-empty, we should also have a way of asserting that it is empty. To do this, i.e. to mark a set as empty, we may prefix it with the symbol $\nexists$, again adding parentheses if necessary. We will call this operator the **existential falsehood operator**, or simply the **not exis operator** for short. Besides asserting that a set is empty instead of non-empty, this operator will behave the same as the exis operator. For example, the expression $\nexists(A \cup B)$ can be read as "not exis A or B" or "not exis A union B" and would mean that the set $A \cup B$ is empty. This would be true whenever both A and B are empty.*

Alternatively, instead of using $\exists$ and $\nexists$, you can choose to use equations, in which case you will need to use the equals sign operators ($=$ and $\neq$) and a symbol for the empty set (either $\emptyset$ or $\{\ \}$, according to your preference). In this case, any time you wish to assert the existential truth of a set, you will need to append "$\neq \emptyset$" or "$\neq \{\ \}$" to the set expression to transform it into an equation. Similarly, as you would expect, asserting existential falsehood requires instead appending "$= \emptyset$" or "$= \{\ \}$". Thus, for example, the expression $A \rightarrowtail B \neq \emptyset$ would be an assertion that $A \rightarrowtail B$ does in fact exist, i.e. that A is indeed a subset of B. Likewise, the expression $(A \cup B) \cap C = \emptyset$ would be an assertion that $(A \cup B) \cap C$ does not exist, i.e. that the set is empty.

You might find the equation form easier to apply set theory laws to sometimes, in order to change the form of the equation to solve for something. However, notice that us-

ing the equals sign causes the associated negations to be reversed in a certain sense, in that assertions of truth require a negative equals sign, whereas assertions of falsehood require a positive equals sign.

In this respect, the exis operators arguably feel a bit more like direct claims of truth or falsehood, and hence may feel slightly more atomic. The choice is ultimately subjective though. Pick whichever method you prefer. I often like to provide multiple options for things in the interest of diversity. Natural selection processes will tend to refine the system over time, reducing the prevalence of inferior choices over time. One often cannot easily predict what approach is best, so it is often wise to simply provide multiple options and then let natural selection run its course.

Pretty straightforward right? It's really not a complicated operator. It's about as simple as they come. However, despite this simplicity, the exis operator is nonetheless quite expressive. Why? The reason is because the exis operator takes any arbitrary set as input, thereby permitting the expression of the existential truth of any set that is describable with set theory and unified implication. This indirect generality is the true reason why the exis operator is so expressive despite its simplicity.

Furthermore, just as unified implication has strong correlations with natural language, so too do existential statements. What any given existential statement corresponds to in natural language depends on what the set expression it is attached to is. Also, just as before, there are often multiple natural language constructs that correspond to a single unified logic operator. You might be surprised what some of the connections are. Set theory operators correspond to a lot more natural language expressions than you might think. Let's do some examples. I'll list a bunch of natural language expressions that have equivalents in terms of unified logic. All of the examples I give will include the exis operator somewhere. Try to guess how the expressions will translate into unified logic, if you'd like to. Here's the list:

1. Biologists are scientists.

2. Some programmers are also mathematicians.

3. The house on Water Lane in Lincolnshire, England, is Isaac Newton's home.

4. Green and blue are colors.

5. Some native English speakers are possibly Australian.

6. Nebulas exist.

7. Alpha Centauri is a solar system.

8. Currently enrolled college students are possibly future graduates.

9. Many engineers are electrical engineers.

10. Chocolate is a type of candy.

11. Painting and sculpting are both art.

12. Either Santa Claus or reindeer exist, or both.

Did you notice how these examples are all *complete sentences* this time? Well, that's because unified logic is actually able to completely (or nearly completely) formalize a large subset of natural language sentences. Admittedly, there are many natural language sentences for which I have no idea how you'd translate some of the information in them into unified logic, but nonetheless the subset that *is* translatable is fairly large and expressive.

Anyway though, did you figure out how to translate some of these expressions into unified logic? Well, regardless, let's see how to now. We'll start with one of the simplest examples. Consider, for example, the expression "Nebulas exist.". How do we translate this into unified logic? The answer is easy. Let N be the set of all nebulas. Then, the expression "Nebulas exist." translates into unified logic as $\exists N$. If indeed this is the case (and it is), then the statement is true.

Remember, however, that expressions wrapped in $\exists$ are not necessarily true. Such expressions are just arbitrary forms, just as arbitrary as forms we might manipulate using transformative logic, and as such are open to context and interpretation to an extent. Maybe we are talking about a virtual world where nebulas actually don't exist. Maybe not. Who knows really? It depends on context and interpretation.

The transformation rules for value expressions and statement expressions are different, but the goal in either case is the same: to conserve the properties and logical consistency of whatever is being described. Existential statements are just forms whose interpretation and rules are designed to adhere to the standards of existence-based truth, rather than to the relatively naive standards embodied by traditional classical truth.

Anyway though, let's move on to the next example from the list. Consider, for example, the expression "Biologists are scientists.". Do you see what the underlying operator in the set expression that the exis operator is wrapping here has to be? It needs to be either sub implication or super implication, since this sentence is expressing a subset relationship. Which one though? Sub implication or super implication?

The answer is that it doesn't matter in this case. The distinction between sub implication and super implication often becomes irrelevant at the top level of an expression in the context of existential statements. Sub implication and super implication *can* make a difference in the outcomes internally in a set expression, but at the top level, as long as the resulting set is not empty, the distinction tends to often not matter.

As such, both of the two expressions $\exists(B \mapsto S)$ and $\exists(B \rightharpoonup S)$ would be usable translations of "Biologists are scientists.", given that B stands for the set of all biologists and S stands for the set of all scientists. The other equivalent ways of writing these (i.e. $\exists(B \rightarrow S)$, $\exists(S \leftarrow B)$, $\exists(S \leftharpoonup B)$ and $\exists(S \mapsfrom B)$) would also work of course, but some feel more natural than others. I personally prefer $\exists(B \mapsto S)$ and $\exists(B \rightarrow S)$ the most, because the sentence feels like it focuses more on "biologists" than on "scientists", and also because $\exists(B \mapsto S)$ and $\exists(B \rightarrow S)$ have the same ordering as the natural language sentence. All of these variants are equally true though, and all of the fundamental elements of the structure of the sentence are intact in every case given, so the preference here is somewhat superficial.

Next let's evaluate the expression "Some programmers are also mathematicians.". What does it translate into? What operator do we know from set theory that is capable of expressing partial overlaps between sets? The answer is of course intersection, i.e. set theory's version of a "logical and". Notice that there's no reason to think there's any unified implications here though. As such, let's begin the translation. Suppose P is the set of all programmers and suppose M is the set of all mathematicians. Then, the expression "Some programmers are also mathematicians." translates into unified logic as $\exists(P \cap M)$. If you think about it, this is a pretty clear correspondence. Clearly, if $P \cap M$ is non-empty then this implies that some programmers are also mathematicians.

Next let's do one of the big ones. Let's do "The house on Water Lane in Lincolnshire, England, is Isaac Newton's home.". Did you catch the reference to the example list on page 394, where we used "the house on Water Lane in Lincolnshire, England" as an example of prepositions translating into set intersections? Well, let's extend that example to this case involving a complete sentence now. Suppose, as before, that X is the specific house we have in mind, that H is the set of all houses, that W is the set of all things on Water Lane, and that L is the set of all things in Lincolnshire, England. Furthermore, let I be the set of all of Isaac Newton's homes[18]. Then, one possible translation of the phrase is as follows:

$$\exists([(X \rightarrowtail H) \cap W \cap L] \rightarrowtail I)$$

Interesting how such basic operators can capture the essence of such a complex sentence so easily, isn't it? As you can see, I've used some square brackets (i.e. the [ ] symbols) here to make the parenthetical nesting easier to read. You can often reduce the parenthetical nesting level by one pair of parentheses by using equation notation instead of exis operator notation though. In that case, the expression would instead look like this:

$$[(X \rightarrowtail H) \cap W \cap L] \rightarrowtail I \neq \emptyset$$

Next let's do one that has slightly tricky wording. Let's do "Painting and sculpting are both art.". Do you see why the wording of this one is maybe a bit tricky? Want a hint? Look at the word "and" here and remember that I mentioned before that "and" in English can act like either a conjunction or a disjunction depending on the context it is placed in. Which one do you think it is here? The answer is actually disjunction, more specifically set union in this case.

Having determined this, we're now ready to translate the sentence. As such, let P be the set of all things associated with painting, let S be the set of all things associated with sculpting, and let A be the set of all things associated with art. Then, the sentence "Painting and sculpting are both art." may be translated into unified logic as $\exists((P \cup S) \rightarrowtail A)$. There are of course also some other admissible translations (e.g. using $\mapsto$ or $\rightarrowtail$, etc), but that doesn't matter much.

Let's do another one that involves union, but in a different way. Consider the sentence "Either Santa Claus or reindeer exist, or both.". Which operations from set theory

---

[18] People can have multiple homes. I'm not saying Isaac Newton did, I'm just saying that people can, so we might as well frame it as generally as possible.

or unified logic does this expression decompose into? Well, let's find out. Let S be the set containing only Santa Claus (if anything)[19] and let R be the set of all reindeer. Then, the sentence "Either Santa Claus or reindeer exist, or both." translates into unified logic as $\exists(S \cup R)$. Notice that this time there is no need for any sub implications, since nothing is being claimed to be a subset of anything else.

Finally, let's translate an example that demonstrates a point of nuance that unified logic will fail to completely capture. Let's translate the sentence "Many engineers are electrical engineers.". Suppose that E is the set of all engineers and that L is the set of all electrical engineers. Clearly, not all engineers are electrical engineers. Therefore, this sentence will translate as $\exists(E \cap L)$. Do you see why this translation looses some of the nuance of the original natural language sentence?

The original sentence said that *many* engineers are electrical engineers, i.e. that a large quantity of them are. However, our translation only really says that *at least one* engineer is an electrical engineer. It doesn't capture the magnitude of the quantity at all. This is partly because how much "many" is exactly is subjective and is hard to quantify precisely. However, mostly it is just a limitation of the system. Ideally, you'd want to be able to express all conceivable points of nuance in any translation you perform, but doing so doesn't really matter much to us for the purposes of this book so we aren't going to bother with it.

There's one more thing I want to point out for this example though. Did you notice that the set of electrical engineers is a subset of the set of all engineers, and yet we are using intersection instead of unified implication in the expression? That's no accident. Translating it as a unified implication, as in $\exists(E \mapsfrom L)$ for example, would be *wrong*. This statement would be true, but just because a translation would be true doesn't make it a correct translation.

A correct translation needs to actually correspond to the structure of the original source material being translated as well as possible. If I said that many X are Y without telling you what the sets X and Y actually are, would you be able to assume that Y is a subset of X? Certainly not. There's nothing in the structure of the statement that actually indicates that. We can't cheat when translating. We have to honor the structure as given.

We can't change it based on extra information that we just happen to know by coincidence. However, if instead of "Many engineers are electrical engineers." we had "Engineers include electrical engineers as a subtype." then *that* would indeed correctly translate as $\exists(E \mapsfrom L)$. Similarly, if we had "All electrical engineers are engineers." then this would translate most naturally as $\exists(L \mapsto E)$.

Anyway, that covers our introduction to the exis operator. Now it's time for us to talk about the other existential operators. Yes, there are others, but they are all technically derived from the exis operator and are redundant. You will see what I mean. Nonetheless, despite the redundancy of the remaining existential operators, they serve a useful purpose. Mostly, they are useful for two things: (1) shorthand and (2) giving

---

[19]This set will only be non-empty if Santa Claus is indeed considered to exist in the system under consideration (e.g. real life vs fiction vs simulation etc). Otherwise, if he does not exist in the system under consideration, then this specification of the set S would return the empty set since no real object would match the criteria.

names to certain common concepts, to make the relationships held by those concepts easier to discuss, think about, and illuminate. This time we're just going to define them all at once in rapid fire, and then discuss them all together afterwards. Here's the new set of definitions:

Definition 128. *When we wish to say that two sets A and B have a non-empty intersection, we may write A ⊓ B. We refer to ⊓ as the* **existential intersection** *operator, or* **exis and** *for short. The expression A ⊓ B has exactly the same effect as writing ∃(A ∩ B) or A ∩ B ≠ ∅.*

Additionally, we also define the symbol ◇ to mean the same thing as ⊓, but in contrast we prefer to read it as "possibly" and to refer to it as the **possibility** *operator. The reasons why this redundant operator symbol was created will be explained later. This diamond symbol will also have several related visually similar variants, which will allow it to be more expressive and versatile than the ⊓ symbol. Thus, we will often tend to prefer ◇ and its variants over ⊓. We define both regardless though, mostly in order to illuminate certain relationships, as you will see later. Anyway though, in total there are four variants, which are defined and named as follows:*

1. *The* **possibility** *operator, which is written as A ◇ B and means that we know that the intersection of A and B is non-empty.*

2. *The* **left-proper possibility** *operator, which is written as A ◆ B and means that we know both that the intersection of A and B is non-empty and that A is not a subset of B.*

3. *The* **right-proper possibility** *operator, which is written as A ◆ B and means that we know both that the intersection of A and B is non-empty and that B is not a subset of A.*

4. *The* **proper possibility** *operator (a.k.a. the* **contingency** *operator), which is written as A ◆ B and means that we know both that the intersection of A and B is non-empty and that neither A nor B is a subset of the other.*

Definition 129. *When we wish to say that two sets A and B have a non-empty union, we may write A ⊔ B. We refer to ⊔ as the* **existential union** *operator, or* **exis or** *for short. The expression A ⊔ B has exactly the same effect as writing ∃(A ∪ B) or A ∪ B ≠ ∅.*

Definition 130. *When we wish to say that two sets A and B are both non-empty and that A is a subset of B, then we may write A ⊑ B. We refer to ⊑ as the* **existential subset** *operator, or* **exis sub** *for short. The expression A ⊑ B has exactly the same effect as writing ∃(A ⇀ B) or A ⇀ B ≠ ∅. Any other similar expression wrapped in ∃ (i.e. any other expression that uses a different unified implication variant, where A is the subset and B is the superset) would also work, since which set is returned by the unified implication wouldn't matter for determining whether the resulting set is merely non-empty or not.*

In addition, we also define the related symbol ⊏ and refer to it as the **existential proper subset** *operator, or* **exis proper sub** *for short. This operator behaves the same* as ⊑, except for that it also specifies that the two sets are not equal to each other. The relationship between ⊑ and ⊏ is extremely similar to the relationship between ⊆ and ⊂ from standard set theory. The relationship differs only in that ⊑ and ⊏ are existential, whereas ⊆ and ⊂ are non-existential. The difference is that ⊑ and ⊏ can only be true if both sets exist (i.e. if both sets are non-empty), whereas ⊆ and ⊂ can sometimes be true even if one or more of the sets are empty.

Definition 131. *When we wish to say that two sets A and B are both non-empty and that B is a superset of A, then we may write $B \sqsupseteq A$. We refer to $\sqsupseteq$ as the **existential superset** operator, or **exis super** for short. The expression $B \sqsupseteq A$ has exactly the same effect as writing $\exists(B \leftarrow A)$ or $B \leftarrow A \neq \emptyset$. Any other similar expression wrapped in $\exists$ (i.e. any other expression that uses a different unified implication variant, where B is the superset and A is the subset) would also work, since which set is returned by the unified implication wouldn't matter for determining whether the resulting set is merely non-empty or not.*

*In addition, we also define the related symbol $\sqsupset$ and refer to it as the **existential proper superset** operator, or **exis proper super** for short. This operator behaves the same as $\sqsupseteq$, except for that it also specifies that the two sets are not equal to each other. The relationship between $\sqsupseteq$ and $\sqsupset$ is extremely similar to the relationship between $\supseteq$ and $\supset$ from standard set theory. The relationship differs only in that $\sqsupseteq$ and $\sqsupset$ are existential, whereas $\supseteq$ and $\supset$ are non-existential. The difference is that $\sqsupseteq$ and $\sqsupset$ can only be true if both sets exist (i.e. if both sets are non-empty), whereas $\supseteq$ and $\supset$ can sometimes be true even if one or more of the sets are empty.*

Definition 132. *Just as unified implication can be applied in both directions, so too can existential subset and superset relations. When it is simultaneously true that $A \sqsubseteq B$ and $A \sqsupseteq B$, then we know that A and B must be the same set, but we also know something more: we know that the set cannot be empty. We will refer to this as **existential equivalence**, or **exis equiv** for short.*

Existential equivalence is represented by the symbol ⊨⊨, i.e. by an equals sign with two vertical lines placed on its sides. Existential equivalence is the statement form of unified equivalence. As such, an expression of the form $A ⊨⊨ B$ has the same meaning as $\exists(A \leftarrow B)$ or $A \leftarrow B \neq \emptyset$. It is probably best to read $A ⊨⊨ B$ as "A exis equiv B", seeing as "existential equivalence" is quite a mouthful compared to "exis equiv".

Definition 133. *Sometimes it is useful to be able to concisely declare that two sets share nothing in common, i.e. that their intersection is empty. To do this, we will simply write $A \boxtimes B$. We will refer to $\boxtimes$ as the **mutual exclusion** operator, or **mutex** for short. We may read an expression of the form $A \boxtimes B$ as "A mutex B" or "A does not intersect B", and it is equivalent to $\nexists(A \cap B)$ or $A \cap B = \emptyset$. It is essentially a negated existential intersection.*

Alternatively, instead of using $\boxtimes$, the mutual exclusion operator may be written as $\not\cap$ or $\obslash$, as in $A \not\cap B$ or $A \obslash B$, since the mutex operator is logically equivalent to the negation of the existential intersection operator, and drawing a slash through symbols is often used to indicate logical negation. However, the $\not\cap$ symbol tends to not look as aesthetically appealing as $\boxtimes$ or $\obslash$ arguably, so I prefer to use $\boxtimes$ or $\obslash$ instead.

413

By the way, the × symbol in the box in ⊠ also serves as a memory aid: you can think of it as standing for "mutually *ex*clusive". The × symbol is also used for multiplication, which has a special relationship to logical conjunction. On the other hand though, you may find the logical negation convention employed in ⊘ to be more memorable. Use whichever symbol you prefer. Oh, and by the way, The ⊘ symbol may be read as "impossibly", in addition to being readable as "mutex".

Definition 134. *Mutual exclusion is a commonly used concept, and for good reason, but it also has a related cousin whose existence is perhaps seldom (if ever) acknowledged. This related cousin differs only in that the intersection in the definition has been replaced with a union. As such, to fill this role, we may write $A \boxplus B$ or $A \mathbin{\underline{\cup}} B$ (whichever we prefer). The expression $A \boxplus B$ is equivalent to $\not\exists(A \cup B)$ or $A \cup B = \emptyset$, and thus $\boxplus$ has the net effect of asserting that A and B both don't exist. We refer to this operator as the* **mutual non-existence** *operator, or* **munex** *for short.*

Alternatively, the operator could also be referred to as a negated existential union, of course, since that's what it is. Notice that the $\boxplus$ symbol contains a +, which is used for addition, which incidentally also happens to have a special relationship to logical disjunction. I did this intentionally as a memory aid. The symbols for both mutex and munex were chosen specifically with these relationships in mind.

Definition 135. *Given two sets A and B, to say that A includes more than just B we may write $A \boxminus B$. We refer to $\boxminus$ as the* **set survival** *or* **existential set difference** *operator. We may read $\boxminus$ as "survives" if we wish, and thus $A \boxminus B$ may be read as "A survives B". Saying $A \boxminus B$ has the exact same effect as saying $\exists(A - B)$ or $A - B \neq \emptyset$.*

As you may recall, the $-$ operator in $A - B$ represents the set difference (a.k.a. set removal) operator from set theory. An expression of the form $A - B$ returns the set of everything in A minus everything in B. As you can see, the $\boxminus$ symbol includes a "$-$" symbol inside the square. This was done to make it easier to remember the close relationship the set survival operator has with the set difference (a.k.a. set removal) operator.

Definition 136. *If we wish to state that the set nand of two sets A and B is non-empty, then we may write $A \mathbin{\top} B$. We will refer to $\top$ as the* **existential nand** *operator, or* **exis nand** *for short. Writing $A \mathbin{\top} B$ has the exact same effect as writing $\exists(A \uparrow B)$ or $A \uparrow B \neq \emptyset$. Be careful not to confuse existential nand with mutual exclusion. Existential nand and mutual exclusion represent two different senses in which an existential intersection can be negated, and thus they can potentially be confused if you aren't paying attention.*

Definition 137. *If we wish to state that the set nor of two sets A and B is non-empty, then we may write $A \mathbin{\bot} B$. We will refer to $\bot$ as the* **existential nor** *operator, or* **exis nor** *for short. Writing $A \mathbin{\bot} B$ has the exact same effect as writing $\exists(A \downarrow B)$ or $A \downarrow B \neq \emptyset$. Be careful not to confuse existential nor with mutual non-existence. Existential nor and mutual non-existence represent two different senses in which an existential union can be negated, and thus they can potentially be confused if you aren't paying attention.*

Alright, so that covers all of our existential operators. See why I said that they're all redundant? They're all just alternative representations of various unified logic value expressions that have been asserted to be either non-empty or empty. Notice also that most of the existential operators use symbols that are just squared off versions of their more curvy value expression counterparts. This was done intentionally of course, in order to make most of the operators a lot easier to remember and to give the resulting notation a more consistent and aesthetically pleasing look and feel.

Admittedly, a bunch of these operators are rather esoteric and not that likely to be useful, but I decided to define them anyway. The existential nor (⊥⊥) and existential nand (⊤) are probably the most extreme cases of that, i.e. of existential operators that are a bit too esoteric to have much chance of ever being widely used. You're probably better off decomposing them into their more fundamental components, such as by writing $\exists(\neg(A \cup B))$ or $\neg(A \cup B) \neq \varnothing$ instead of using obscure symbols such as $A \perp\!\!\!\perp B$. Still, it's nice to have a complete vocabulary and $A \perp\!\!\!\perp B$ is a lot more concise. Plus, maybe it has some especially convenient use that I'm not aware of.

In fact, pretty much all of these existential operators are perhaps best written in terms of their more fundamental components. Reducing the number of redundant symbols you have to differentiate from each other tends to make logical and mathematical expressions easier to work with, although not always. Specialized composite symbols obscure the relationships between the underlying concepts to an extent, so working with a minimum of symbols can often make it easier to see relationships and allowable transformations. Extraneous symbols can be seen as a form of noise in a certain sense, i.e. as unnecessary cognitive load essentially.

On the other hand though, defining all these more specialized operators gives us some useful names and symbols for talking about things. Also, decomposing everything into the most fundamental components, while more uniform and consistent from a conceptual standpoint, does on the other hand sometimes feels a bit too atomic and overwhelming, much like programming everything in assembly language would in computer science. Defining these composite operators is still worth doing in this case, I think. The worst that can happen is that the symbols might just not end up being widely used.

One thing you may be wondering though, is how to deal with multiple instances of these symbols in a single expression, i.e. composite expressions made up of multiple existential operators, such as $(A \sqcap B) \sqcap C$ for example. Translated literally, as we have defined things so far, this would be problematic. The expression $(A \sqcap B) \sqcap C$ would translate into something like $\exists(\exists(A \cap B) \cap C)$, which wouldn't have any coherent meaning. You can't intersect the statement $\exists(A \cap B)$ with the set C. Their types are incompatible. However, we will define a convention for how to interpret such expressions regardless, so that the use of multiple existential operators in the same expression will become possible. Doing so will make the notation more convenient and expressive.

The rule is simple: whenever you encounter an expression that uses multiple existential operators, convert all such operators to their corresponding value expression forms and then either wrap the entire expression in a basic existential assertion (i.e. $\exists$ or $\nexists$, depending) or else turn it into an equation, whichever you prefer. Doing so will make the expression's meaning unambiguous, assuming adequate parentheses are given. Seeing examples will make what I mean here clearer. Here is a table of some examples of such conversions:

Multi Exis Ops	One Exis Op	As Equation
$(A \sqcap B) \sqcap C$	$\exists((A \cap B) \cap C)$	$(A \cap B) \cap C \neq \emptyset$
$(A \sqsubseteq B) \sqsubseteq C$	$\exists((A \rightarrowtail B) \rightarrowtail C)$	$(A \rightarrowtail B) \rightarrowtail C \neq \emptyset$
$A \sqsupseteq (B \sqcup\!\!\!\downarrow C)$	$\exists(A \leftarrowtail (B \downarrow C))$	$A \leftarrowtail (B \downarrow C) \neq \emptyset$
$(A \sqcup B) \boxminus C$	$\exists((A \cup B) - C)$	$(A \cup B) - C \neq \emptyset$
$A \boxtimes (B \bowtie C)$	$\not\exists(A \cap (B \uplus C))$	$A \cap (B \uplus C) = \emptyset$
$(A \mathbin{\top} B) \boxplus C$	$\not\exists((A \uparrow B) \cup C)$	$(A \uparrow B) \cup C = \emptyset$

There is one wrinkle I need to point out though. Most of the conversions are obvious, but the $\sqsubseteq$ and $\sqsupseteq$ operators have to be treated specially. The rule for them is this: Instances of $\sqsubseteq$ are always converted into $\rightarrowtail$ (or equivalently $\mapsto$), and instances of $\sqsupseteq$ are always converted into $\leftarrowtail$ (or equivalently $\mapsfrom$).

Technically, there are other possible ways of converting $\sqsubseteq$ or $\sqsupseteq$, due to the fact that other variants of the unified implication arrows also exist. However, troubling ourselves over such cases is not worth it, and thus we will simply ignore such cases. Which unified implication a $\sqsubseteq$ or $\sqsupseteq$ converts into could impact its behavior inside the expression, and hence could also impact the final result. That is why I have chosen to standardize the conversions of $\sqsubseteq$ and $\sqsupseteq$ in this way. It eliminates this problem and forces people to use an unambiguous translation.

Most of the existential operators were included at least once in the example list above. The only exceptions were the proper variants of all the operators that have proper variants. The proper variants of those operators (i.e. ◇, ◈, ◆, ⊏, ⊐) cannot yet be expressed this way in their equivalent forms using only one $\exists$ or $\not\exists$, but don't worry about that for now. You can just break those up into multiple separate statements, using transformative logic and such. Those operator variants are not that important for our purposes here anyway though.

By the way, I personally find it quite odd that so many of these basic operators are missing from the vocabulary of logic and mathematics. You would think that with so many logicians and mathematicians (and also people in related fields like computer science and engineering) that more of these basic operators would have been defined by someone by now, but apparently not. This is actually a very interesting point to me. It shows how easily an entire society can overlook even the most basic of concepts and yet still be totally oblivious to it. It shows the reality-bending power of preconceptions.

It shows how incredibly easily even a small amount of social conformity and presumption can blind people to even the simplest of things. I imagine that much of the blindness involved in these kinds of preconceptions comes from the common incorrect belief people have that "If the founders of this field of study haven't discovered it yet, then it must not exist or must not matter.". People are sometimes far too quick to assume that the founders or predecessors of a field of study have already thought of everything. Simple ideas are often the most fleeting. They drift along like frail little feathers carried on the wind. You have to be paying close attention to catch them.

You shouldn't assume that the basics of any given field of study are necessarily already complete. To do so would essentially be a fallacy of authority and conformity, an overeagerness to believe in whatever the current social order happens to be. It comes from the same place as a person's desire to fit in, to not be subjected to any uncomfortable attention from others that may threaten the person's place in a social group. It is often far easier to adhere to the social norms of a field of study, to toe the party line so to speak, than it is to pose original thoughts and to thereby put yourself at risk of ridicule.

However, if the full truth is to be discovered, then we must not succumb to such base impulses and overblown fears. Truth is a selfless pursuit. It does not feel fear nor long for belonging. It simply is. You cannot threaten the ego of truth, for it has no ego to begin with. Likewise, you cannot compel the truth to "fit in", for it already "fits in" with everything that exists, because it *is* everything. In stark contrast though, the forces of conformity often strive to deceive us into believing that conformity is merely some form of innocent solidarity, merely a sacrifice for the good of the team. But this surface-deep feeling of comfort that conformity provides us is often just an illusion.

Conformity is not selfless sacrifice, but rather it is merely just another form of lazily looking after one's own self-interest in preference to putting forth whatever effort may be necessary to honor what is right and true and good in this world. Conformity is merely cowardice. Truth, in contrast, is governed by evidence, not norms. To find the truth we must be brave enough to question our own assumptions. We must be willing to be wrong, if we want to be right.

Anyway, we got off on a tangent there for a bit. Let's get back on track now. As you may recall, on page 408 I gave a list of examples of complete sentences from natural language that could be successfully translated into terms of existential statements. I hinted that some of the existential operators would have a close correlation with certain words or phrases from natural language, just as the same was true of value expression operators such as unified implication, intersection, and union. Well, having all these extra existential operators defined and named now makes it a bit easier to talk about some of those correlations.

Some of the correlations should already be fairly apparent just from looking at the example list on 408. Regardless though, there are still some points of nuance that merit further discussion. Many of these operators correspond to multiple different natural words or phrases. Also, some of them only correspond to natural language phrases under the correct context, i.e. when another suitable natural language word or phrase is present or absent somewhere in the sentence, depending on each specific case.

Let's look at some specific examples. Consider, for example, the sentence "Some programmers are also mathematicians.". Using just the exis operator, we would translate this sentence into something like $\exists(P \cap M)$, where P is the set of all programmers and M is the set of all mathematicians. However, using our new shorthand exis operators, we could also translate it into either $P \sqcap M$ or $P \diamond M$, depending on our preferred notation.

On the other hand though, consider the sentence "Biologists are scientists.". In this case, the basic exis operator translation would be $\exists(B \rightarrow S)$, and the shorthand exis operator translation would be $B \sqsubseteq S$, where B is the set of all biologists and S is the set of all scientists. Notice, however, that both "Some programmers are also mathematicians." and "Biologists are scientists." use the word "are" between the two sets. What then is the distinguishing factor that makes the difference in how these sentences translate? It is of course the presence or absence of the word "some", or of any other sufficiently similar word. Notice that all of the following sentences are just different ways of stating that the intersection of two specific sets is non-empty:

1. Some programmers are also mathematicians.
2. Some native English speakers are possibly Australian.
3. Currently enrolled college students are possibly future graduates.
4. Many engineers are electrical engineers.
5. Some Europeans are Greek.
6. A reptile is possibly a snake.
7. Few English majors are also computer science majors.
8. Most computer engineers are proficient in at least the basics of programming.
9. Tons of novelists are participants in creative writing communities.
10. Almost no priests are scientists.

All of these sentences can be captured as $A \sqcap B$ (or equivalently $A \diamond B$), where A and B are the relevant sets. The only lost detail is the magnitude of the intersection, i.e. the rough sense of what *proportion* of the first of the sets is enclosed inside the second, as indicated by such words as "many", "few", "most", "tons of", "almost no", or any other similar word or phrase. Putting aside the sense of proportion of the intersection though, the most fundamental words corresponding to existential intersection here seem to be the words "some" and "possibly". In contrast though, let's look at some sentences that otherwise look similar, but do not contain any form of "some" or "possibly":

1. Biologists are scientists.
2. Green and blue are colors.
3. Alpha Centauri is a solar system.
4. Chocolate is a type of candy.
5. Painting and sculpting are both art.
6. All electrical engineers are engineers.
7. All Greeks are Europeans.
8. Snakes are reptiles.

All of these sentences, in contrast to the prior set of sentences, can be captured as $A \sqsubseteq B$, where A and B are the relevant sets. Clearly then, the default behavior of "is" and "are" is to behave like an existential subset statement. As such, as you can see, we have now discovered two of the correspondences between natural language and existential statements: namely that "some" and "possibly" (and other similar words and phrases) correspond to $\sqcap$ (or equivalently $\Diamond$), whereas "is" and "are" (without "some" or "possibly", but optionally with "all") correspond to $\sqsubseteq$.

In contrast though, notice that I did not include any statements of the form "Not all X are Y." in the list of examples of existential intersection statements. That's because saying that not all X are Y doesn't actually imply that at least some X are Y (although some might think it would), but rather it only implies that we know that X is not a subset of Y. You need to be careful when reading statements to make sure that you don't fall into these kinds of subtle interpretation traps.

A sentence of the form "Not all X are Y." might seem at first glance like it could translate into $X \sqcap Y$ (or equivalently $X \Diamond Y$), but in reality the correct translation would be something like $\not\exists(X \rightharpoonup Y)$ or $X \rightharpoonup Y = \emptyset$ or $X \not\sqsubseteq Y$. Indeed, given only that we know that not all X are Y, it would still be possible that X or Y (or both) is empty, or even just that X and Y are both non-empty but X doesn't intersect with Y in any way, and hence $X \sqcap Y$ (or equivalently $X \Diamond Y$) would not be a safe inference, at least not without possessing additional information.

Anyway though, besides just serving to clarify when to use $\sqcap$ and when not to, this discussion of what "not all" actually translates into has provided us with one more successfully identified correspondence between natural language and the existential statements of unified logic. This brings the total number of natural language correspondences we have identified for existential operators up to three: $\sqcap$, $\sqsubseteq$, and $\not\sqsubseteq$. For some of the other existential operators though, I actually wasn't able to identify any reliable correspondences to natural language. However, there are still a couple more correspondences I was able to successfully identify.

The first of these is the correspondence in natural language to the mutex operator (a.k.a. $\boxtimes$). In particular, the mutex operator corresponds to phrases of the form "No X is/are Y" or "X is/are never Y", or any other synonymous phrasing. Thus, sentences such as "No cats are reptiles." or "Bob is never late." would be valid examples of natural language sentences corresponding to the mutex operator. Notice that this is definitely not the same as the "Not all X are Y." case that we were discussing earlier. Just because both phrases have negations in them (e.g. "not", "no", "never") doesn't mean you should assume they have the same structure. Be careful. Patience is essential in logic.

The second of the additional correspondences of existential statements to natural language I was able to identify is that of the existential equivalence operator. Initially, this correspondence will not surprise you. However, there is a specific point of nuance about it that you may not have expected or ever thought about before, one which is quite instructive and illuminating. More on that point in just a moment though. First consider the basic idea: The existential equivalence operator corresponds to phrases of the form "X equals Y" or "X is/are the same thing as Y", or any other synonymous phrasing. *Whoop dee doo* though, right? It seems blatantly obvious, doesn't it? Why am I even bothering telling you about it then?

Well, not so fast. Do you remember how existential equivalence is not actually the same as regular equivalence? Specifically, as you may recall, existential equivalence also requires that the two sets in the equivalence *must not be empty* in order for the statement to be considered true, else it would actually be considered false. This has a few interesting consequences. It might sound initially like this would be bad behavior for the operator. After all, shouldn't two empty sets always be considered equal to each other? That's what a lot of people's intuition would say.

However, there are actually some really strong reasons why not letting two empty sets be considered equal to each other might sometimes (under the right circumstances) actually be better. In particular, it has to do with the fact that different nonsensical sets, each described in terms of different properties, arguably shouldn't be considered equivalent to each other. A set with self-contradictory properties, or even a set which merely has no members, will return an empty set when evaluated. Yet, it would often be strange to say the sets are equal, considering they are described by different properties and may in fact even have no conceptual relationship to each other whatsoever (other than being empty).

Consider, for example, the set of all dry water and the set of all unicorn wizards. Neither of these sets exist in real life, as far as all evidence indicates[20]. This implies that both sets are empty sets. Yet, does that mean we should treat them as equal? Not necessarily. Certainly in one sense they are equal, i.e. in the non-existential sense of equality. However, in a different sense though, i.e. in the existential sense of equality, they are not equal.

These sets are described by different properties, and indeed few people would say that dry water is conceptually the same as unicorn wizards, regardless of the fact that neither exists. Existential equivalence captures this point of nuance in how it behaves. Natural language doesn't treat all empty sets as equivalent, and neither does existential equivalence. Thus, existential equivalence is perhaps actually closer to natural language's version of the concept of equivalence than the more traditional non-existential concept of equivalence from mathematics is, in a sense.

On the other hand though, what about the case where we are merely talking about two empty sets directly, rather than describing them indirectly through some set of unsatisfiable properties? Does existential equality make much sense then? Does the fact that $\emptyset \mathrel{\vDash} \emptyset$ is false make sense? Well, if you think of it in terms of the value expression version of $\mathrel{\vDash}$, i.e. in terms of unified equivalence ($\leftrightharpoons$), then yeah it kind of does make sense.

As you may recall, unified equivalence behaves like the natural language expression "a.k.a.", and indeed a unified equivalence of empty sets (i.e. $\emptyset \leftrightharpoons \emptyset$) returns exactly what you'd expect under this semantics. The empty set is indeed *also known as* the empty set, so having it return itself is still exactly what we want, regardless of the fact that the empty set also represents falsehood.

In fact, in unified logic, emptiness is the essence of falsehood itself. Therefore, if

---

[20] If the reader knows of any unicorn wizards they would like me to meet, then by all means send me the evidence. As for dry water though, it is a self-contradictory concept and therefore is logically impossible and hence can be dismissed immediately without even considering whether any physical evidence exists or not.

you wrap ∅ ↺ ∅ in an existential statement, as in ∅ ⊨ ∅ (a.k.a. ∃(∅ ↺ ∅)), then you shouldn't be surprised that it evaluates as false, since saying it evaluates as false is actually the same as saying that it evaluates as exactly the same thing as what it already is: emptiness. Thus, this behavior is not as odd as it may seem at first glance. It just feels that way because we're so used to thinking in non-existential terms that it blinds us to alternative ways of thinking.

Saying that ∅ is not existentially equivalent to ∅ in no way denies that both sets are still equally empty. Rather, it only denies that they have any equivalent *existential* properties. An object can't have any substantive properties if it doesn't even exist. Only things that exist can have real properties. Non-existent things can only ever possess the property of being empty, i.e. of *being nothing*, and nothingness is the same thing as falsehood in unified logic.

In this sense, existential equivalence blocks all non-existent sets from ever implying that they have any property greater than emptiness, and this even includes blocking any implications from ever having any of the nonsensical or non-existent properties that any other descriptions of empty sets may hypothetically pose.

Anyway, that basically covers all of the correspondences between natural language and existential operators that I will be identifying in this book. We should make a summary table now, so that we can get a better overview of what all the correspondences we've discussed here are. Combined with the table on page 405, which in contrast describes the correspondences between some of the value expression operators and their own natural language analogs, these two tables will provide a good quick overview of how primitive unified logic and natural language relate to each other. Anyway though, here's our new table of correspondences for the existential operators:

Symbols	Formal Reading	Natural Readings	
$X \boxtimes Y$	X mutex Y	No X is/are Y.	X is/are never Y.
$X \sqcap Y$	X exis and Y	Some X is/are Y.	X is/are possibly Y.
$X \sqsubseteq Y$	X exis sub Y	All X is/are Y.	X is/are Y.
$X \vDash Y$	X exis equiv Y	X equals Y.	X is/are the same thing as Y.
$X \not\sqsubseteq Y$	X not exis sub Y	Not all X are Y.	X is/are not a type of Y.

By the way, another way of expressing $X \vDash Y$ in natural language would be to say "X is/are Y and Y is/are X.", just as one would expect considering that equivalence is really just a bidirectional implication or subset relation, i.e. is really just the conjunction of both possible directions of valid inference. Notice that it would be incorrect to think that "X is/are Y." by itself would be enough to indicate equivalence, despite the fact that many people often do so and mistakenly treat "is/are" as a synonym for "equals". In reality, "is/are" only consistently indicates a one-directional implication. A two-directional implication might also exist, by coincidence, but you'd need more information in order to be able to actually assume that.

Those of you who have studied Aristotle's syllogistic logic before may have noticed that many of these phrases in the table above are identical to phrases found on the

"Square of Opposition"[21] from syllogistic logic. This is no accident of course, since syllogistic logic is just an ancient ancestor of modern set theory, so it's not surprising that there are parallels. After all, unified logic is really just an extension, modification, and reinterpretation of set theory, so it's only natural that this would happen.

However, related to that same point though, you may have noticed that this table appears to be "missing" the case from the Square of Opposition expressed by the phrase "Some X are not Y.". Notice that the difference is that the Y set is negated this time, instead of positive. However, properly understood, this is not actually a real difference. The case expressed by the phrase "Some X is/are Y." in the Square of Opposition actually covers both cases, i.e. both the case where we have just Y and also the case where we have $\neg Y$.

The reason is simple. The set Y can be substituted for any set, including negated sets as well of course. Thus, we could easily just substitute A into X and $\neg B$ into Y and then we'd have "Some A are $\neg B$.", thereby covering both cases. The names of the set variables are superficial. We can mess with the form of the X and Y expressions as much as we want, by simply substituting in any arbitrary set expressions that we feel like using. It's only when we've started to actually make claims about real sets that we have to be more careful.

These correspondences between the operators of unified logic and the words and phrases of natural language seem important enough to merit having a formal term to refer to the entire group of all such correspondences. Providing this term will make it easier for people to talk about these relationships potentially, so it seems worth doing. As such, here's our new formal term to refer to it:

Definition 138. *There are many interesting and uncanny correspondences between the operators of unified logic and the words and phrases of natural language. Many words and phrases that occur in natural language can be formalized into terms of unified logic expressions. We may refer to this high-level relationship between natural language and unified logic as the **natural-unified language correspondence**.*

Much of natural language is actually composed of nothing more than just a bunch of claims that various different specific sets are non-empty (a.k.a. true) or empty (a.k.a. false). Indeed, many chains of sentences in natural language can actually be thought of as simply being big lists of equations (or existential statements) about sets, in disguise. In this way, people communicate classification information about objects from their own minds to the minds of others simply by describing how those objects relate to each other as sets. However, as striking as this relationship may be, there are still many natural language expressions that seem to have no translation into terms of unified logic, at least not as have yet been discovered.

Beyond this natural-unified language correspondence though, interesting as it may be, there is also an even broader sense in which logic, language, and our universe are all connected together. Do you see the parallels? Think back, for example, to the concept of logical naturalism that we defined on page 105. Logical naturalism is the idea that

---

[21] Look it up if you're curious. I'm not going to bother explaining what the Square of Opposition is in this book.

logic is not merely something we do in our heads, but rather it is also part of the structure of the universe itself.

One can see the strikingly close similarities between (on the one hand) natural laws like the law of conservation of mass and energy and (on the other hand) logical laws like the principle of property conservation (i.e. the basis of all proofs and operations in logic and math) as evidence that the universe itself is interwoven with the structure of logic, i.e. that logic is part of the substance of the universe rather than merely something apart from it.

This also makes even more sense when you observe how uncannily and consistently effective logic and mathematics are at modeling our universe. I suspect that this is no mere coincidence. I suspect that our universe itself is in some sense *made of* logic, at least partially. Our universe is just one such possible world, and logic enables you to simulate others as well[22].

Yet, even that is not the end of the interesting correlations here. Consider, for a moment, not just the correspondence between natural language and logic and the universe, but also the origin and underlying principles upon which natural language in particular appears to be based. How is it that we come to possess natural languages? What is the fundamental operating principle underlying the creation and existence of natural languages? I suspect that a lot of it (perhaps even all of it?) is actually just set identification, set classification, and applications of the principle of property conservation.

Even we ourselves as living creatures are a part of this all-permeating logical fabric of the universe. As such, it would be unsurprising if our own natural languages are actually fundamentally comprised of these same underlying logical principles as the universe itself is. The nature of language is probably in harmony with the nature of our universe. How else would you do it? After all, you'd want a language to be able to express anything about whatever world it finds itself in, to be adaptable and flexible, and how better to do that than to match the structure of language to the structure of all conceivable universes (i.e. to the structure of logic itself, which seems to be able to simulate any conceivable universe).

Indeed, not all animals may be able to speak and write as we humans do, but all animals *are* able to identify and classify sets and to react to invariants in their environment. They wouldn't be able to hunt or to avoid poisonous food (etc) if that wasn't true. They could only do so if they had some capacity for reasoning with sets, whether subconscious or conscious. Thus, surely they do. A creature can't keep separate things separate in its mind unless it can classify things into different sets so that it can react differently to each of those sets.

Furthermore, look at the structure of much of natural language, as evinced by our own discussion of the natural-unified language correspondence here. The structure of much of natural language appears to just be expressions of relationships among sets. I think this is a large part (perhaps even all?) of the origin of natural languages. Natural languages arise from our natural ability to think in terms of sets, which is a gift we have because we are part of this universe, a universe which is itself a living and breathing logical system filled with sets and property conservation rules. In this sense then, one

---

[22] This is especially true when logic and mathematics are combined with the enormously enabling power of computer science.

could even argue that surely all animals and living creatures must also in some sense "have language". It wouldn't really be unique to us. It would just be expressed in different ways, sometimes only within the creature's own mind for example.

After all, if language is just reasoning with sets and with properties of an environment, and all animals have to be able to identity sets in order to survive (and they do), then it would naturally follow as a consequence that all animals and living creatures actually *do* have language in that sense. Logic is the universal language. All living creatures have at least some form of logic, otherwise they couldn't react differently to different things in their environment, and therefore all animals in some sense probably have some kind of language capacity, even if only in their own mind.

They may not be able to express it like we do, via speech or writing, but there is good reason to believe they *do* have it, since it is apparently perhaps nothing more than a manifestation of these inherent logical operating principles of our universe, principles shared by anything and everything that exists in our universe. Cool thought, right? It really makes you pause before making overreaching or arrogant assumptions about the frame of mind of the other living creatures that we share our world with. They may be vastly more aware and sentient than human beings currently give them credit for.

Look also at the way that even extremely simple sets of logical rules can easily create a massive cornucopia of diverse effects and spontaneously emerging beauty and complexity. You certainly don't need any kind of artificial human-like or god-like "intelligent designer" to accomplish any of it. Such centralized control and sentience is not even remotely necessary. The idea that a designer would be necessary becomes utterly laughable once you have enough experience studying emergent systems. Mindless emergent systems probably have a far greater capacity for creating spontaneous beauty than even human designers do. Look at things like L-systems, John Conway's Game of Life, and biological evolution for example (all are specific types of emergent systems). Look at any kind of emergent system really.

And, as you study these kinds of systems, notice how incredibly easily even an extremely crude set of logical transformation rules and initial forms is able to generate such a stunning array of unexpected effects. Tiny rule sets can generate everything from realistic models of real-life plants (e.g. via L-systems) to living simulations capable of full computational generality and emulation of any conceivable artificial intelligence program (e.g. via John Conway's Game of Life) and beyond. It's easy. Yet, all of this can be captured by the seemingly quaint vocabulary of the basic principles of logic[23].

This is yet more evidence of the deep connection between logic, language, and our universe. This is yet more evidence that logic is not merely a separate thing from our universe, but indeed may actually be a fundamental part of the substance of our universe in some sense. All the complexity of our world may just emerge spontaneously from this logical substrate, including even our natural languages, the way we think, and the emotions we feel.

---

[23] However, computer programming may also be necessary (on a case by case basis) to overcome the ambiguities, imprecision, and computational limitations that often plague traditional hand-written mathematics and logic. The fundamental point still stands though. Computers are just better at guaranteeing strict interpretations of everything and at making tasks much more computationally tractable. In principle though, anything a computer does could also be done by hand, equally precisely, with sufficient time and with extreme mental discipline.

That's what the evidence seems to suggest. Spontaneous emergence of complexity is the most natural explanation for the way everything is. It would be the simplest and most parsimonious explanation for existence as we know it, the theory with the least unnecessary or unjustified assumptions, the one that fits best. The exact details are often unclear, since there are so many different variables involved, but the general idea of emergent phenomena seems to be by far the most likely basis for the way the universe functions. Anyway though, putting that tangent aside, let's define a new formal term for the idea that logic, language, and the universe are extremely intimately connected and interwoven with each other like this:

Definition 139. *We may refer to the idea that our universe is structured in such a way that logic, language, and the physical substance of universe itself are all really just different manifestations of the same thing (i.e. of logic itself in some sense) as the* **Logical Language Universe Theory**, *or the* **LLUT** *for short. The Logical Language Universe Theory is one specific type of logical naturalism, specifically one where language itself is also considered to be a close part of the relationship. However, one could imagine other viewpoints that also adhere to logical naturalism, yet do not adhere to this idea that language is intimately connected to the underlying logical properties of the universe.*

Oh, and by the way, there is also an idea from linguistics that there exists some kind of "Universal Grammar" or "Innate Language" which all people possess. Personally, I suspect that logic is actually exactly this "Universal Grammar" or "Innate Language" of which these linguists theorize. We may not yet be at a point where every aspect of natural language can be formalized in terms of logic, but I suspect that one day humanity will eventually achieve it.

When that day comes, humanity will perhaps have finally discovered the true form of logic, insofar as human beings are capable of understanding it under the constraints of our limited minds[24]. I don't think unified logic or transformative logic are likely to be the true form of logic and language, but I do think they will help us move a little bit closer to that goal and that one day someone will finish the job and truly bridge the gap completely. What a glorious day that will be for the study of logic and language, when it finally comes.

Anyway, it is now time to return once again to the meat of our discussion. We've now got both the value expression and statement expression operators covered for primitive unified logic. However, you may have noticed that we've been focusing *a lot* on finding correspondences between natural language and the operators of primitive unified logic, yet we've been focusing *very little* on actual implications in the more conventional sense of what one normally does in logic. That's because it was easier (and also more natural)

---

[24] After all, it isn't necessarily a given that we as humans are even capable of fully understanding the universe. Our brains may inherently not be up to the task, for all we know. Other animals aren't capable of certain tasks, so it would be unsurprising if the same is true of us in some way that we are unaware of, yet that other creatures in the universe might be aware of. We shouldn't assume we're necessarily on the top of the ladder of sentience and self-awareness. To think so would be unjustified and irrational arrogance. We are evolutionarily young. The evidence suggests that our species only very recenty became this intelligent. It is therefore unlikely that we have already climbed to the apex of sentience.

to talk about the correspondences with natural language at the same time as we were defining the operators.

By discussing both the formal definition of each operator and each operator's correspondences to natural language at the same time, it is easier to get a good intuitive grasp on what each of the operators really are and what they really do and express. Now though, it's finally time to turn our attention to the more conventional sense of what these logical operators are used for, i.e. to deduction and inference and so on, as it pertains to unified logic.

It will also soon be time to start exploring in greater detail how unified logic actually relates to the other non-classical logics, especially with respect to why it is called *unified* logic: i.e. with respect to why it seems to satisfy the primary motivating principles of all of the most important branches of non-classical logic simultaneously. However, first we should talk more about the basics of reasoning with primitive unified logic. Let's begin.

5.5 Reasoning with primitive unified logic

Do you remember learning about material implication in school? Was that something that your education covered? Even if your education wasn't heavy on math and science, there's a good chance you did at least learn about the truth tables for the various classical logic operators at some point, including material implication probably. Maybe you don't remember though, or maybe it just wasn't included in the curriculum you were exposed to. Regardless though, if you're like many people, you may have instinctively felt that something was off about material implication when it was taught to you. It may have felt a bit too arbitrary and a bit too forced, like it didn't really model implication in a substantive way, like something was being left out.

Let's talk about that some. It's relevant to understanding how reasoning with unified logic differs from classical logic. Let's start by briefly refreshing our memory of material implication. Material implication, as you may recall, is a value expression (i.e. is *not* an attempted assertion of a fact) and is represented by the symbol $\rightarrow$. It returns a binary truth value (i.e. only true or false). A material implication of the form $A \rightarrow B$ is treated as being logically equivalent to $\neg A \vee B$. In other words, material implication is defined as being true whenever either the condition is false or the consequence is true (or both), or otherwise false. Material implication's truth table looks like this:

A	B	$A \rightarrow B$
F	F	T
F	T	T
T	F	F
T	T	T

The first thing about this definition that will likely seem odd to many students is that the cases where the condition is false are always true. Why would implications with false premises always be true? That seems pretty odd, doesn't it? Furthermore, many students are likely to be confused (either consciously or subconsciously) by the fact that this operator is not actually making a claim, but rather it is merely a truth function that is being evaluated. Why treat it as being a complete model of implication when (interpreted strictly) it isn't even capable of making a claim?

To make matters even more confusing, the *sense* of what an expression of the form $A \to B$ actually means is typically not adequately explained to the student. The operator is introduced as "the definition of implication", the corresponding truth table is given, and that's pretty much it usually. If the student is lucky, some reasoning for why material implication is the only usable way of defining a truth-functional implication operator within the constraints of classical logic will be given, but even then the explanation will often still feel somehow unsatisfying.

Indeed, the way material implication is typically introduced to people is actually *actively misleading*. Seldom is the student informed of the actual underlying spirit and operating principle of the operator, which is simply this: **Evaluation of $A \to B$ is based on determining whether the truth values of A and B would *not disprove* that A implies B.** Notice the wording. Notice that nowhere in this sentence does it ever say that a true material implication would indicate that the *implication* is true. To think that a true material implication indicates that the *implication* is true is actually *wrong*. A true material implication only indicates that the given truth values of A and B would *not disprove* the *possibility* that A implies B.

Just because an implication is not disproven does not imply that it is true. Not being disproven and being true are not the same thing, yet material implication is very often taught (unwittingly) as if not being disproven and being true *are* the same thing. The literature on material implication is thus riddled from top to bottom with *equivocation*, i.e. with treating two different things (in this case: not being disproven vs being true) as if they are the same thing when they are not.

The way material implication is interpreted in classical logic thus rests upon a falsehood. Material implication does not even model real implication, despite the fact that the way material implication is phrased and discussed makes it *sound* like it does. This is the source of a large part of the confusion surrounding material implication, but it is only part of it. There are other sources of confusion involved as well.

Material implication only tests for a single way of potentially disproving the idea that A implies B. There are many other ways that the idea that A implies B could be proven wrong though, even if the corresponding material implication evaluates as true. Really, material implication would be more accurately named as something else. A better name would actually be something like "**possible implication**".

In fact, I think that "possible implication" is so clearly superior to "material implication" as a terminology choice that in the long term people should actually stop using "material implication" entirely and should instead favor "possible implication". Doing so will help prevent future generations from being confused by what this operator really represents. Names have a *huge* impact on accessibility and understandability. A badly named concept will seldom be used correctly by the intended audience, or may even be ignored entirely due to how incomprehensible its purpose is, whereas a well-named concept can often be understood nearly instantaneously and is vastly more likely to be widely adopted.

Terminology choices are not superficial. How well terminology is chosen can easily determine whether a concept lives or dies. A small change in wording can easily turn something incomprehensible into something blatantly obvious. Names should be chosen with the utmost care, and should be as descriptive and unambiguous as possible, within the constraints of reasonable brevity.

The tradition of sometimes disregarding such "trivial" details in mathematics as unimportant is poison to progress. The more difficult the true essence of a concept is to understand, the fewer potential uses of that concept will become apparent, and thus the slower the pace of research will inevitably go. Foundational concepts should be given the most perfectly clear names possible. Even small flaws in how foundational concepts have been named will tend to amplify a thousand fold at the higher-levels, spreading confusion, misleading information, additional conceptual errors, wasted time, and needless difficulty throughout the entire system like a plague.

When concepts are poorly named, things that should take mere minutes to master can easily end up taking months or years instead. In such cases, the student's mind must constantly fight against the poor terminology until the student's mind finally somehow manages to build up a correct mental model. Even after that though, the student will probably find it difficult to explain the concept well to anyone else, and thus the cycle of ignorance will continue. This price is paid repeatedly by every single student who ever tries to learn the poorly described concept, thereby causing such "small" naming flaws to actually have unimaginably vast human costs in terms of wasted time when spread out over the long-term history and future of humanity.

Literal entire lifespans worth of human time are wasted by such seemingly small details when you actually sum it all up over many people. How is that much different, substantively speaking, than actually killing off the same number of real people as the number of wasted lifespans worth of time? It sort of isn't. It's just that most people don't consciously realize this. Perpetually wasting people's time is thus, in effect and in a certain cumulative sense, indirect murder. Life is short and time is precious. A person's time *is* their life. Wasting one person's lifespan worth of time thus has the same net effect as wasting one person's entire life. It's just that this loss is distributed over multiple people, thereby making it harder to notice the impact.

Anyway though, as I was saying, not only is the name of material implication poorly chosen, but material implication is also not actually even a real implication itself, but rather it merely tests for the *possibility* that an implication it corresponds to could be true. The alternative name I have given here ("possible implication") also clearly helps to demonstrate *why* the operator is so subtly and insidiously confusing to people.

Once you reveal the underlying meaning of the material implication operator by giving it a more accurately descriptive name, the true nature of it becomes much more readily apparent, and mere "possible implication" is certainly not what you would expect if you were thinking of material implication as being a complete representation of implication. Material implication is partial implication at best. It is inherently incomplete.

Subtle and unexpected differences between how something is described and what that something actually is in reality are the worst kinds of differences you can possibly have in a system if you care about clarity of communication and ease of use. This is just as true of all forms of communication as it so famously is in computer programming. Subtle and unexpected differences are the most difficult possible kind of information for a human mind to manage and detect. This is likely a big part of why the true nature of material implication is currently so seldom understood.

It is very tempting to think that a true material implication would indicate that the *implication* is true, but to do so would be wrong. Thus, the way material implication is typically described in the current literature absolutely *begs* for widespread misunderstanding and misuse of the operator. At best, a material implication returning true merely reduces the *chances* that the implication is false. Material implication only tests *one* of many different possible ways that an implication could be false.

It is only with this understanding in mind that the material implication operator can ever be used correctly and safely by anyone. Material implication is nothing more than a *single* computationally inexpensive but *frequently inconclusive* test of the potential truth of an implication. Only in the case where a material implication is *false* is the result of a material implication conclusive.

A false material implication does indeed indicate that the implication being evaluated cannot possibly be true (i.e. that it is *impossible*[25]), but a true material implication in contrast actually tells us very little information. A true material implication only tells us that the implication could *possibly* be true, i.e. that it is either true or false but we don't know which. In classical logic though, saying that something is either true or false provides essentially no meaningful information, since that's already true of all classical truth values regardless of what they represent anyway.

Why then, if the true essence of material implication is so clear, is the material implication operator so often introduced in such an opaque way, i.e. in a way that does not clearly convey any of this information? Why is it not explained unambiguously to students that material implication is *not* really representative of real implication? The true nature of this operator is not that hard to see, yet few seem to truly grasp it. Why was this not realized sooner? Why are the standard explanations of material implication so vapid and so lacking in any of the real conceptual essence of what the operator actually is?

The answer, I think, is perhaps mostly the toxic culture of obscurantism that permeates some parts of the mathematics community and literature. Conceptual correctness and substantive expression of the true essence of things is too often ignored or dismissed as "trivial", whereas convoluted verbiage and obfuscation is embraced widely without the slightest hesitation. Concepts are defined and proven, but never truly grasped. The foundational principles and intuitions embodied by ideas are cast aside carelessly, as if not of value.

Clear communication of ideas is neglected and treated as optional. The limitations of the human mind are not acknowledged. Advocating for clarity and accessibility often makes you a target for being treated as inferior, as if wanting to reduce the cognitive load of ideas must surely indicate that you have an inferior mind[26], rather than merely

---

[25] Notice the presence of the word "possible" in "impossible" here. It is no mere coincidence. The word "impossible" literally means "not possible". Therefore, the logical negation of "impossible" cannot possibly be the value "true". Rather, the logical negation of "impossible" must be merely "possible".

[26] "You must just not be able to handle the complexity, unlike us enlightened few, who know that the true purpose of commu-

indicating that you are *not insane* and that you understand the enormous value and enabling power of ease of use and genuine understanding, as it does in reality.

The true essence of concepts is all too often just swept under the rugs, while the most petty details, technicalities, and fringe concerns take center stage and are treated as the only matters of any importance. Motivation is too often absent, and complexity for complexity's sake is too often glorified. The absence of real-world value ("purity") is seen as a badge of honor, whereas work that involves well-motivated principles that add genuine value to the lives of fellow human beings is seen as grunt work, is seen as something that only the filthy peasants participate in.

And yet, value creation is the essence of life. A profession that does not create value has no real purpose and hence no real reason to exist. The mathematics community seems unique in this respect, in that it has become one of the only communities where a culture of *deliberately sneering* at value creation has managed to somehow take root and proliferate. Without value creation though, all interest in a field of study will inevitably evaporate over time and fewer and fewer people will invest time in learning it.

Thus, in some ways, mathematics has become one of the only fields in existence that seems to actively work against its own value proposition, by increasingly failing to escape from the intellectual tarpit of obscurantism, and thereby over time perhaps becoming less and less attractive to newcomers and less and less valuable in the eyes of outsiders. As such, in these respects "purity" in mathematics has in many ways become akin to a plague, not something to be glorified. The popular mentality of merely loving complexity for complexity's sake, with little to no concern for actually creating tangible value for anyone, nor for adequately investing in genuine clarity of communication, is poison to mathematics' ability to sustain long term traction and to gain more widespread appreciation by society.

Astoundingly, the origins of mathematical ideas are often erased from history by their own authors. The thought processes, insights, and intuitions that led to those ideas being derived are often intentionally left undocumented, in the hope that the work it took to get there will be forgotten, leaving it "pure", lest any of the "trivial" underlying thoughts and insights ever come to light and thereby besmirch the creator's tiny and fragile ego. Clear thoughts are so much easier to criticize, after all. What better way to protect yourself from any criticism than by scouring away the underlying essence of what you are talking about, so that it is no longer readily apparent to anybody?

The thought processes, insights, and intuitions that led to these ideas are sometimes later rediscovered and reverse engineered, but often only at great human cost[27], if ever. Immense amounts of time and productive energy are wasted needlessly. Furthermore, the damaging effects of this behavior compound over time. The less information is available about the true origin of ideas, the less well understood those ideas will inevitably become, causing even more misunderstandings to proliferate.

---

nication is not to convey useful information but rather to make ourselves look much smarter than we actually are. We are absolutely terrified of anybody ever realizing how uninspired and aimless most of our ideas actually are, so we therefore subconsciously try to undermine clear communication at every possible opportunity." — Sincerely, Academia

[27] Some parts of this book are actually examples of this. It takes a lot of time and effort to unravel and undo conceptual misunderstandings after those misunderstandings have already spread throughout society.

The net result is that many concepts in mathematics are "understood" only as pale and pathetic shadows of their true selves, thereby resulting in a brittle, inflexible, and unenlightening framework of thought that leaves much to be desired. The untapped potential here is enormous. Math content is all too often drained of its substance and essence, leaving nothing but a hollow husk of seemingly meaningless instructions that provide little to no insight or intuition whatsoever into the underlying principles. Yet, it is only by firmly grasping the underlying principles of ideas that the true power and expressiveness of those ideas can ever be unlocked.

Mathematics is the language of magic. It is the means by which any arbitrary conceivable universe can be described and made real by simulation (when combined with computer science). Yet, this immense power can only be wielded effectively if one possesses a deep and genuine understanding of the underlying concepts in play. Shallow rote memorization is not enough. Math content drained of its substance and essence is indeed a very sad and boring affair. It's no surprise that so many people hate it.

For most people, math is only enjoyable if you ignore all the meaningless convoluted crap (e.g. most superficial technical details, most fringe concerns, and most "advanced research") and instead only focus on the stuff that has real expressive value and strong conceptual clarity. If you don't provide any context or motivation to support an idea, then you shouldn't expect anybody to care. Why should they? People generally only care about what adds value to their lives. That is the most rational and optimal way to live. Human life is short and precious, so why would people want to waste what precious little time they have to live on unmotivated ideas that are unlikely to return anything of value? Hostility to wasted time is *rational*, and thus it should come as no surprise.

This common distaste for much of mathematics is not necessarily some kind of bias against math itself, but rather it is often simply how things *should* be. Many of these people expressing distaste for mathematics are actually responding in an entirely healthy way to what they've so far seen and experienced in their lives regarding mathematics. People are constantly making subconscious tweaks to their value judgments of all the things they have experienced in life so far. They then subsequently subconsciously make use of this perpetually evolving value estimate to guide all their decisions and attitudes in the present moment.

Consistently failing to provide any motivation for something *inevitably* leads to people's subconscious value judgment of that thing decaying over time. Unfortunately, the current mathematics culture seems geared specifically to making this subconscious value judgment decay happen *as rapidly as possible* in the eyes of most people. Typically, only people who *already understand* the value of mathematics from some other context (e.g. aspiring game programmers) will see the huge value in studying it.

Those are the kind of people most likely to continue being highly motivated to study math even despite lacking any clear immediate display of value for any given specific mathematical idea. Only things that demonstrate *value* should ever be expected to be interesting to people. Consequently, if you want more people to care about math then you will have to *show them its value*. In contrast though, fruitless abstractions with no connection to anything of value will often just turn people off.

This widespread lack of focus on conceptual correctness, clarity of communication, and genuine value creation is bad enough on its own. To make matters worse though, people are far too slow to aggressively question assumptions and far too quick to trust in authority figures. Too many people think "Oh, well I've heard that these concepts have been explored by a lot of experts, so I bet there's nothing here that anyone could have possibly missed." and don't realize the extreme extent to which mass hysteria and delusion over even the simplest things has been the norm throughout human history. It is no surprise then that the community of mathematics, like all communities, could gradually become completely blind to even some of the most basic ideas and alternative ways of thinking. Material implication's widespread misinterpretation is just one of many such examples.

Whole societies regularly suffer from mass delusions. It can happen pretty easily actually. It's not even a rare occurrence. People need to learn to be more skeptical of "the wisdom of our ancestors" and such. Far more often than people realize, "treasured beliefs" and norms are actually based on nothing more than quirks of history and have no real substance whatsoever. Sometimes ideas survive purely based on momentum and self-reinforcing feedback mechanisms. Popularity does not reliably imply merit.

Social norms are a flimsy foundation for beliefs at best. That's just as true of specialized professional communities and their norms as it is of the uneducated masses. It sure is fun to pretend like us hard science people and engineers are totally immune to all that though, isn't it? It's a nice big juicy ego massage, and also false. Human nature is human nature. One must be vigilant against ignorance, lest it inevitably return and take root over time, spreading over everything it touches, suffocating value creation and extinguishing human progress everywhere it goes.

Anyway though, I got off on one of my crazy out-of-control tangents again there. Let's turn our attention back to material implication now, and our illuminating discussion of what its real essence is. Doing so will help us frame our thoughts for how reasoning will work in unified logic, as it stands in contrast to classical logic. As you may recall, we were discussing how material implication does not really model real implication and that it would be more accurately named something like "possible implication" instead, due to the fact that it only actually tests whether the implication is *not disproved* by the given input values.

Well, this problem is already quite alarming by itself, but there is also yet another problem: namely that there are multiple different interpretations for what an "implication" operator should even mean in the first place. Indeed, we have already met some of those other interpretations. Transformative implication, meta implication, unified implication, and existential subset and superset relations are all examples of some possible interpretations of "implication". Each is different from the others. Which (if any) is best?

The answer is none of them. Each one serves its own purpose and has its own pros and cons. Which one you want to use depends on the context and on what you want to accomplish. Do you want to *evaluate* whether an implication is true or *assert* that it is? The former requires a value expression whereas the later requires a statement expression. These two types of expressions are treated differently on the surface, but are both really just different ways of interpreting arbitrary forms, just different sets of transformation rules applicable to some corresponding sets of symbols. Even a degenerate operator like material implication can still have some beneficial use, as long as you remember what it really is, which is merely "possible implication".

Let's do a little thought exercise. Let's forget for a moment what the symbol $\Rightarrow$ is usually used to represent in the literature, and instead think of $\Rightarrow$ as merely representing "implication" in some vague and undetermined sense. The $\Rightarrow$ symbol is a good choice for this role, since it is sometimes used to indicate "formal implication" without actually adequately explaining what that really means. What then are some of the possible things we could mean when we say $A \Rightarrow B$?

Let's start with one of the most obvious and natural interpretations, an interpretation close to what many people intuitively mean by "implication". Let's suppose that we want $A \Rightarrow B$ to indicate that if we already have A then we can conclude B. In this interpretation of $A \Rightarrow B$, what we really want to be able to do is to assert $A \Rightarrow B$ as fact. We aren't *testing* whether this rule is true, but rather we are *creating* this rule, so that it embodies the relationship we wish to express. We are creating a new admissible rule of inference and *forcing* it to be allowed.

Perhaps B following from A is simply a fact that we have observed, i.e. is simply a law of nature or else of some hypothetical abstract system that we are studying. We now simply want to formalize this observation of fact as a rule. Alternatively, perhaps the rule we have in mind is just something we want to introduce into a system that we are arbitrarily designing and exploring. Either way is fine. Regardless though, to accomplish this goal we must actually perform rule generation.

There is nothing to evaluate. We are actually creating a new axiom in some sense, a new foundational assumption. Specifically, we have some existing set of things that have already been established in the system somehow, to which we now wish to add another set of things as a consequence. Clearly then, under this interpretation of "implication", $A \Rightarrow B$ would actually be meta implication, i.e. would actually be $\{A\} \prec \{B\}$ in disguise.

Naturally, this also works for any other arbitrary forms too of course, besides just A and B. For instance, if we observed in some system that whenever we have $\neg Z$ then we may also infer R (with no exceptions and with no missing cases), then we could safely add $\{\neg Z\} \prec \{R\}$ as a new rule about the system we are studying. Notice that this need not even require truth values. We are merely saying that to a set that includes $\neg Z$ we may add R. We don't even need to know what $\neg Z$ and R actually mean. The $\neg$ symbol could potentially not even represent negation, for all we know. We could just as easily say $\{\lambda \odot \delta\} \prec \{\beta \theta \odot\}$ and it could be just as valid[28].

What's another possible interpretation of $A \Rightarrow B$, besides meta implication? Well, it could be that we want to say that whenever something belongs to category A then it must also belong to category B, such as if all objects that have the properties represented by A also have the properties represented by B. This is a common occurrence. It allows us to infer more general properties of objects than just the initial specific set of information that was given. What we are really talking about here is sets and how they stand in

---

[28] Perhaps $\beta\theta\odot$ is a really terrible fraternity or sorority that follows $\lambda\odot\delta$ around everywhere it goes, like some kind of envious groupie. On the other hand though, the entire concept of fraternities or sororities is already pretty terrible regardless, so maybe this possibility is not even worth mentioning.

relation to each other. In this case then, $A \Rightarrow B$ would actually be an existential subset expression, i.e. would actually be $A \sqsubseteq B$.

Subset and superset relations can be thought of as a type of implication relationship. Membership in the subset *implies* membership in all of its supersets as well, and indeed this knowledge can be used to make numerous different kinds of logical deductions. Deductions from subset and superset relations are not the only possibilities though. Even if neither of two sets is a subset of the other, there are still some things that can sometimes be deduced merely from knowing whether or not the two sets intersect, although admittedly less than what is admissible when given a subset or superset relation.

What else? What other interpretations of $A \Rightarrow B$ are there? Well, there is also the case where we wish to say that something of one form may be transformed into something of another form without losing any of the essential qualities that we care about. For example, we may wish to say that $2 + 2$ may be transformed into 4 without changing its numeric value, i.e. that the form $2 + 2$ *implies* the value 4. In this case, $A \Rightarrow B$ would really just be transformative implication, i.e. would really just be $A \rightarrow B$.

Transformations like this don't have to represent conceptually equivalent forms like $2+2$ and 4 though. The permitted transformations can be *totally arbitrary*. It all depends on what properties *you* want to express and conserve in your system. The forms can evolve and mutate in whatever way you see fit. As long as the transformations are still suitable for the system you have in mind, i.e. are still an accurate representation of what you are trying to model, then anything is permissible. The sky is the limit.

In the extreme case, the only thing being conserved may even just be "adherence to the given set of transformation rules". The rules don't need to be conventionally intuitive in any sense, as long as the rules are at least logically consistent and coherent. Indeed, the rules only need to be consistent and coherent in the sense of being well-defined in terms of transformative logic. One should not even assume that conventional concepts such as "contradictions" and "tautologies" (nor even truth values of any kind) will even have any effect on what is allowed or not.

Consistency and coherence here should thus be interpreted in the broadest possible sense. Not all sets of transformation rules will even model contradictions or tautologies (nor even truth values of any kind). In fact, most won't. It's all just arbitrary manipulation of forms. In fact, interestingly, since paradoxes often depend upon the existence of truth values and contradictions, many transformation rule sets are therefore not even capable of containing paradoxes.

A paradox in a rule set that works purely in terms of uninterpreted forms would be a non-sequitur, in a certain sense, if you really stop and think about it. It doesn't actually make much sense. In fact, arguably, paradoxes are *still* non-sequiturs even in systems that *do* have truth values and contradictions, since even then you could still just reframe those "truth values" as uninterpreted forms instead, and thereby still retain the ability to perform arbitrary deductions and transformations even in the presence of such "paradoxes".

Anyway though, what if instead of the above interpretations of implication we actually wanted $A \Rightarrow B$ to test whether A is a member of B, but to not actually assert this relationship as if it were fact, but rather we merely want to *evaluate* it? What if we essentially want *implicit inline type checking* of objects? We would need some way to *imply* that A *may* be a subset of B, but without actually asserting that this is necessarily true.

We certainly wouldn't want to add this as a new axiom. That would ruin the whole point of what we are trying to do here. We are *not* assuming that A is an existential subset of B. We are *not* creating a new rule that would allow us to automatically conclude that all A are B by definition, but rather we are merely *evaluating* A in terms of whether or not it is a subset of B.

In this case then, we would want $A \Rightarrow B$ to act the same way as unified implication, i.e. to act the same way as $A \rightharpoonup B$ (or whichever of its other variants is appropriate in the context). By doing so, we would thereby be empowered to describe sets in a much more seamless way. We would be enabled to chain sets together arbitrarily in any way that embodies the relationships we think those sets *may* have to each other. Indeed, natural language exploits this trick frequently and it is quite useful.

Finally, we may just want a computationally inexpensive way to test whether we can quickly eliminate the possibility that A implies B. If we find that A is true at the same time that B is false, then surely A cannot imply B, for if A *did* imply B then B would always be true whenever A is true, but this cannot be the case if at any moment we find that A is true and B is false simultaneously.

On the other hand though, any other possible combination of truth values for A and B would not permit us to determine whether A implies B or not. In the absence of additional information, the possibility would merely be left open. Thus, as you would expect, this interpretation of $A \Rightarrow B$ would of course be the case where $A \Rightarrow B$ would represent material implication (a.k.a. "possible implication"), i.e. $A \to B$.

Thus, we have discovered that "implication" can have at least five different possible meanings. Four of the possible meanings (transformative implication, meta implication, unified implication, and existential subset/superset relations) are more fundamental and truly capture the essence of some specific aspect of implication as we naturally think of it. In contrast though, the fifth possible meaning we have described here, i.e. material implication (a.k.a. "possible implication"), is actually just an auxiliary test function that only partially corresponds to any form of real implication. Material implication is thus a fraud, in a certain sense. It does not truly model implication. Material implication is the pyrite[29] of logic. Material implication is to real implication what fool's gold is to real gold.

That basically covers the main interpretations of implication. Each has its own use. It's crazy how these distinctions seem to be basically just ignored or overlooked in the mathematics literature though, and are not even remotely adequately explained, isn't it? Unfortunately, that seems to be pretty typical of the existing mathematical climate and culture. True conceptual clarity is scarcely available, despite how essential it is.

It reminds me, for instance, of how mathematicians *still* regularly shrug off most suggestions that they should start thinking and working more like computer scientists. Many mathematicians seem to totally lack the imagination necessary to truly under-

---

[29]Pyrite is a mineral. Pyrite is also known as "fool's gold", due to the way it superficially resembles gold in terms of color and appears metallic. Pyrite tends to form cube-shaped crystals though, unlike gold, and thus the difference is easily spotted by a trained eye.

stand how *extremely error prone* the human mind really is, which is something that few things other than interacting with a compiler will ever truly teach you. This attitude is essentially rooted in false complacency. Formal math education isn't really adequate for truly understanding how to think precisely. Computer science is actually far more precise and far more rigorous than math.

Most mathematicians are wrong about how precise they think their own minds and reasoning systems are, and yet they don't even realize it. What most people would pass as "obviously perfect mathematical reasoning", a compiler would probably instantly tell you actually contains more than 20 errors. That's how it tends to go. Most humans are utterly oblivious to how often they are really wrong. I have never encountered any tool more effective at teaching people that lesson than having them try to interact with a compiler.

To this day, I still remember how learning to program utterly blew away all delusions I had about the precision of my own reasoning. Specifically, my programming experience is in general-purpose programming languages, *not* in automated mathematical proof systems. However, the same principle will still apply regardless. Learning to program is an eye opening and life changing experience. It truly awakens you to the extraordinary weakness of human nature. It causes the necessity of self-discipline and precision to become irrefutably clear to you. Even if you don't end up using programming in your daily life, it will still change the way you think. To err is human. Interacting with a compiler very quickly makes you realize what a incredibly foolish human being you truly are.

Indeed, it may even be wise for mathematicians to actually start compiling *all* their work using computer assistance eventually, instead of merely continuing to only work by hand. Instead though, mathematicians continue to abuse their own notation and concepts severely, thereby overloading their framework of thought with lots of implicit assumptions and subtle inconsistencies. Truly basic and foundational ideas may be safe to explore by hand, if great care is taken, but I have a lot of doubt about higher-level research. In fact, I'd guess that huge swaths of higher-level research mathematics are actually just incoherent nonsense in disguise. It wouldn't even surprise me in the slightest if more than half of all higher-level research mathematics one day turned out to be nothing more than meaningless noise.

If mathematicians weren't so quick to make implicit assumptions, and were much more strict about the conceptual and computational meaning of their work, and really took the time to truly understand the underlying natural principles and intuitions captured by each concept, then maybe they would have realized by now that even such a basic concept as "implication" was so highly ambiguous and so poorly explained.

Sure, there may be some mathematicians who are good at explaining the underlying essence of concepts clearly, but it seems like far fewer of them exist than there should be. So much of the literature seems so bogged down in convolution and poor communication that it often feels as if perhaps even the majority of concepts in math are never truly explained clearly *anywhere*. The underlying essence of concepts is far too often either poorly communicated or not communicated at all. How can we trust an idea though, if even the author of that idea cannot explain it clearly? If a concept cannot be communicated clearly, then it should not be trusted.

Clarity should be treated as a prerequisite to credibility. It should not be optional. New ideas should usually not be accepted until they have been explained clearly. Even if a convoluted proof *looks* correct, it should still usually only be considered acceptable if it has been explained with a truly high degree of conceptual clarity and accessibility. Mathematicians (and society in general really) should stop being so easily impressed by overcomplicated crap.

If some ideas are explained in an overly convoluted and opaque way, then the default assumption should be that those ideas cannot yet be trusted (if ever), rather than merely unjustifiably and prematurely assuming that the creator of those ideas "must just be super smart". Obscurantism should be viewed as a sign of low credibility instead of as a badge of prestige. Complexity is not inherently virtuous. One should not be impressed by it, unless it cannot be greatly simplified and is accompanied by sufficient value.

People should be very careful not to accept something just to "not look dumb". Intellectual intimidation, pressure to conform to pre-existing time-honored ideas, and social posturing is a pervasive problem in not just society as a whole, but also in the current culture of mathematics. Convoluted content is too often viewed as impressive, when in reality it should probably be viewed as a huge red flag.

Human beings cannot generally be trusted to think that they truly understand convoluted or opaque work. The human mind is far too error prone for such trust to ever be safe. When complication is unavoidable though, extra care must be taken to be as rigorous as possible, to attack the problem from multiple different angles and perspectives, to consider other possibilities (i.e. to "play devil's advocate"), to use computer assistance, and to still explain each individual component as clearly as possible, even if the complete system as a whole still remains hard to understand.

Ideas that aren't fully understood still have value, and are of course still worth exploring, but clarity should nonetheless always be one of the central concerns of all good intellectual work. Clarity should be maximized as much as possible, within reasonable time and resource constraints. You can't have maximum impact without maximum clarity. It is pretty much impossible. Lack of clarity is perhaps the biggest limiting factor in intellectual progress. It is in some sense the main bottleneck. Clarity is the lens through which all information must eventually pass in order to be truly understood, and thus its absence is inherently prohibitive.

The smart way to overcome complexity is usually to simplify it, or at least (failing that) to compartmentalize it. There's a maximum limit to how much the human mind can stretch itself to navigate complexity. It is usually far better to make your problems fit your mind than to try to make your mind fit your problems. Clarity scales, whereas sheer masochistic persistence does not. Hard work is often not nearly as effective as smart work.

Always remember that creating value should be your focus. Nobody ultimately cares how hard you worked to create something. They only care about the *value* of what you have created[30]. Have the humility to accept the limitations of your human mind, rather than arrogantly believing yourself to be able to swallow any amount of

---

[30] By the way: True fairness is compensating people in proportion to the value they produce, not in proportion to how hard they work. Mere hard work has no inherent merit of its own whatsoever. For all we know, you could be unwittingly working very hard to make the world a much *worse* place. Only value creation merits compensation generally. Safety nets and support structures

complexity whole. A love of complexity for complexity's sake is the hallmark of idiocy, not of intelligence. Toxic elitism helps nobody.

Fake intelligence focuses on deliberately obfuscating concepts and delighting in complexity, whereas real intelligence focuses on maximizing value creation while minimizing the burden of thought. To outsiders, fake intelligence and real intelligence may appear similar at first glance, but in reality their true natures are exactly opposite. Fake intelligence is like darkness. It obscures and conceals. It feels like mysticism. Real intelligence though is like light. It illuminates and reveals. It feels like science or art.

Fake intelligence feels vague and noisy and shadowy, whereas real intelligence feels crystal clear and quiet and plain as day. Fake intelligence is attracted to complexity, whereas real intelligence is attracted to clarity. Clarity empowers, whereas complexity cripples. Thus, naturally, a wise person, a person who genuinely cares about creating real value for their fellow human beings, will choose clarity every time, whenever it is possible and the means to do so are apparent.

Anyway though, let's turn our thoughts back to our discussion of the various different interpretations of what "implication" should mean. The fact that these different possible interpretations of "implication" are never actually clearly explained to students as conceptually distinct entities, combined with the fact that the *sense* (the conceptual essence) of each of these operators is *also* never truly explained in a crystal clear way, is perhaps the real fundamental reason why students often feel somehow unsettled when "implication" is introduced to them in logic class.

Students' instincts that there is something off about material implication, that it feels strangely arbitrary, is thus actually entirely correct. It is the teachers, who force the concept of "implication" as just being "defined that way" and as just being "something you have to accept", who do not actually understand the fundamental underlying essence of what they are even talking about, who are actually the ones in the wrong here. I don't really blame them though.

The teachers never understood the concepts either. It's the blind leading the blind. It's simply the result of a long history in mathematics of disregarding genuine understanding and instead trusting far too much in unmotivated technical details and in the authority of the past. It's the result of gradually losing all sense of natural principles and justified intuition, and instead thinking only in blindly mechanical and presumptuous ways. It's the loss of conceptual clarity. It's the loss of the ability to feel the heartbeat of reason, to sense the primordial landscape of all conceivable logically coherent universes, to see with fresh and newborn eyes.

Now that we've disambiguated the concept of implication though, we finally have some real conceptual clarity. We finally can see implication for what it truly is, without feeling so unsettled by it. Of course, it is still possible that there are other important forms of implication that I have overlooked, or that there are some aspects of it that I have in some way otherwise neglected. No human is immune to error of course. We all have an abundant capacity for stupidity. No amount of scrutiny can ever truly eliminate the possibility of mistakes being made[31].

---

are usually the only exception to this rule.

[31] By the way, this is also part of why it's so important to design robust and error-tolerant systems. Pretty much all systems

It is impossible for a person to be sure that they have eliminated all of their own biases, since any frame of mind used to consider whether this is true could itself be biased. However, the point is that we are in a much better place now, with far less conceptual mud involved. The situation is no longer so murky that it makes us feel so uneasy. We've talked about a lot of stuff though, so let's take a moment to gather together a summary table, to list and compare the various different interpretations of implication all in one place. Here's the table:

Concept	Symbols	Type	Domain
transformative implication	$A \rightarrow B$	statement	forms
meta implication	$\{A\} \prec \{B\}$	statement	forms
unified implication	$A \rightharpoonup B$ (etc)	value	sets
existential subset relation	$A \sqsubseteq B$	statement	sets
material/possible implication	$A \rightarrow B$	value	classical truth

A few of you may be wondering where formal implication ($\Rightarrow$), syntactic consequence ($\vdash$), and semantic consequence ($\vDash$) are in this table. Well, I intentionally didn't include them. The formal implication operator ($\Rightarrow$) is ambiguous, but is most often just used to represent material/possible implication, depending on the author. We hardly discussed syntactic consequence and semantic consequence during this book, but I will nonetheless discuss them again briefly here anyway for the sake of completeness.

As for syntactic consequence, it is essentially the same basic idea as meta implication, in some sense, but just with different notation conventions and perhaps more conceptual mud in some respects. As for semantic consequence though, it is also somewhat similar to meta implication maybe, but is based on evaluating all possible truth values of a system to see if an implication holds, instead of being based purely on manipulating symbols as syntactic consequence is.

In my view though, implicitly attaching truth values to the uninterpreted forms involved in meta implication in this manner for semantic consequence would perhaps be a pointless and over-complicating decision. Thus, it seems to me that maybe semantic consequence would serve no real purpose above what is already granted by simply having meta implication available. The extra assumptions that semantic consequence seems to make about truth values strike me as unnecessary and beside the point probably. It seems redundant. In fact, it may even be misleading or poorly defined. Thus, it too was omitted from the table above[32].

Anyway though, now that we've outlined the concepts we've covered, and framed our perspective a bit more, I think it is time we turn our attention more to the actual

---

will tend to fail at least sometimes. Defense mechanisms and fallback systems should thus be put in place in advance to account for this. No amount of brilliance is foolproof. Truly smart systems are built to be error-resistant and capable of recovery, not to be perfect. Perfection is unrealistic and almost never happens. Few things are more frightening than a system designer claiming that their system "cannot possibly fail".

[32] Part of communicating clearly is knowing when to omit noisy low-value information and redundancies, so that the reader can better focus on what is actually important. Overwhelming people with options that differ only slightly (or not at all) from each other may sometimes cause crippling analysis paralysis. Admittedly, I myself am not always the best at omitting unnecessary information (notice the length of this book etc), but nonetheless I think the principle stands true.

reasoning process involved in using unified logic. In essence, it is actually a hybrid process that exploits whatever combination of transformative implication, meta implication, unified implication, and existential statements is necessary to get the job done.

Transformative implication and meta implication are used as a framework through which to describe how the laws governing unified implication and existential statements work, and also as a framework for building up whatever arbitrary vocabulary of forms we actually want to talk about in unified logic. Indeed, remember that conventional mathematical concepts (e.g. arithmetic, algebra, calculus, etc) are just one possible thing we might want to talk about. Literally any other set of arbitrary forms with any other arbitrary set of applicable rules could be what we want to talk about.

It's possible that even seemingly fundamental rules like $2 + 2 \rightarrow 4$ may not apply to the system we want to study. In this respect, transformative logic frees us from the constraints of such conventional concepts and thereby enables us to talk about anything we want to. Unified logic then subsequently further enables us to talk about the sets of forms generated by our transformation rules, making it possible to express many additional things about those sets of forms and how they stand in relation to each other. Transformative logic and unified logic are thus synergistic in this sense. They both enhance what the other is capable of saying.

To understand what I mean here, it is important to not forget that expressions in set theory are just as much arbitrary forms as any other forms are. Set theory is itself a transformative language, just like any other transformative language that you might conceive of. Granted, set theory is special in terms of how ubiquitous it is and in terms of how its use is necessary for making transformation rules easier to specify, but it is still nonetheless ultimately just another set of transformation rules, regardless of how much more fundamental it is than most other transformation rule sets. Thus, naturally, transformative logic is the language we typically use to talk about many of the rules that apply to reasoning in set theory and unified logic. Consider, for example, the following three equivalent meta implications[33]:

$$\frac{A \sqsubseteq B \quad B \sqsubseteq C}{A \sqsubseteq C} \qquad \frac{\exists(A \rightarrow B) \quad \exists(B \rightarrow C)}{\exists(A \rightarrow C)} \qquad \frac{A \rightarrow B \neq \emptyset \quad B \rightarrow C \neq \emptyset}{A \rightarrow C \neq \emptyset}$$

These meta implications clearly just express the law of transitivity as it applies to existential subset statements. The meaning is quite clear. From the perspective of transformative logic, A, B, and C are just symbols[34]. Transformative logic doesn't know that they are sets. We, as the users of this rule set, only know that A, B, and C are sets implicitly. Transformative logic's special way of handling sets only kicks in when the set notation appears on the top level, such as when forms have been surrounded by tran-

---

[33] Remember that meta implication can be written either inline using ⊲ or vertically using a horizontal line. Regardless of how you write it though, it is still the same operator. Each format has its own pros and cons.

[34] The same is true of $\sqsubseteq$, $\exists$, $\rightarrow$, $\neq$, $\emptyset$, and parentheses. They're all just arbitrary uninterpreted symbols from transformative logic's perspective, as usual. Only a small handful of symbols are ever given special treatment by transformative logic. As a reminder though, you can force all symbols to be treated as literals by using the ⌜ ⌟ quotation delimiters.

sience delimiters or { }. For example, recall that $\{X, Y\} \to Z$ would actually have the same meaning as writing both $X \to Z$ and $Y \to Z$.

Notice however that those mechanics are not applicable to the above meta implication rules though, since those rules are not visibly using sets and thus will be assumed by transformative logic to be referring to nothing more than uninterpreted forms. Writing something like $A \to B$ in transformative logic will *not* be interpreted as if A and B are sets, not even if A and B *are* sets.

In contrast though, writing something like $\cdot | A | \cdot \to \cdot | B | \cdot$ (or some other appropriate choice of transience delimiters) would indeed pull out the values of the sets. Thus, $\cdot | A | \cdot \to \cdot | B | \cdot$ would indeed indicate that any of the forms in set A could be transformed into any of the forms in set B, subject to the constraints and nuance of which specific type of transience delimiter has been selected (trans-closed in this case, but there are other possibilities).

You'll need to understand how the conventions that transformative logic adheres to work if you're going to be able to properly understand how to talk about the properties of unified logic on the meta level. Keep in mind though that the rules that govern unified logic itself are not generally the same thing as the rules that govern the subject matter you are investigating. A rule like $\{A \sqsubseteq B, B \sqsubseteq C\} \prec \{A \sqsubseteq C\}$ is a rule about unified logic itself, whereas a rule like $2 + 2 \to 4$ is a rule about the subject matter, i.e. about some specific arbitrary system you are exploring or creating. Also, remember that transformative logic can be used to produce sets of forms, and those sets of forms can then subsequently be used as input into unified logic expressions, which can then again be processed by transformative logic. There's a lot of potential for crossover.

Anyway though, one of the main differences between how reasoning with unified logic works compared to how reasoning with classical logic works is that unified logic has a much more nuanced and sensitive way of dealing with conjunctions and disjunctions than classical logic has. Classical logic has only a single way of framing conjunctions and disjunctions, whereas unified logic actually has multiple. Each individual proposition in classical logic always stands for a specific independent statement about the universe as a whole. Even seemingly narrowly scoped statements like "My notebook is red." are actually statements about the entire universe in disguise in classical logic. For example, "My notebook is red." actually means "There exists a universe such that my notebook is red and this is that universe." when read with a more careful eye.

As such, whether or not two given classical propositions are related to each other does not change what the resulting truth value of their conjunction or disjunction will be. Only the truth values themselves have any impact. The relationships the participants have to each other are ignored. However, this is not true of unified logic. Unified logic is partially context-aware, in the sense that what sets are participating in a conjunction or disjunction in unified logic can change the result's truth value[35]. In unified logic, all the participants in a conjunction could be true and yet their conjunction could still be false[36]. This could never happen in classical logic, and the independently asserted

---

[35] In this respect, unified logic perhaps qualifies as a "non-monotonic logic" (look it up). However, non-monotonic logic is not something I am sufficiently familiar with to actually make this determination. There is possibly a connection though.

[36] Remember that in unified logic all non-empty sets are considered "true", whereas all empty sets are considered "false". We

nature of all classical propositions is why. The participants in expressions in unified logic are *not* necessarily independent, unlike in classical logic, where they always are.

This might sound a bit odd, but in practice it is actually very straightforward and easy to understand. The way you express conjunctions and disjunctions in unified logic is actually very simple. It is essentially exactly what you would expect. Conjunctions are expressed using set intersections, and likewise disjunctions are expressed using set unions. Other logical expressions are also expressed exactly how you'd expect them to be. The *real* difference is that unified logic distinguishes between two different forms of expressions that classical logic fails to differentiate between. You will understand what I mean soon. Let's start with the more obvious case first.

The first way of framing a conjunction or disjunction in unified logic is to simply place all of the involved sets in corresponding intersections or unions. This is nothing new really. Such expressions occur abundantly in set theory. For example, if we wanted to evaluate whether or not any object that satisfies the properties embodied by sets A, B, and C simultaneously must also satisfy the properties of set D as a consequence, then we could write this:

$$A \cap B \cap C \rightarrow D$$

This is unified logic's equivalent of evaluating whether "if A and B and C then D" is true or not. If this expression returns a non-empty set, then it must be true and one can therefore subsequently conclude that any object that is a member of all three sets A, B, and C is also a member of set D. If, on the other hand though, this expression returns the empty set, then it must be false and one can therefore subsequently conclude that it is *not* safe to assume that any object that is a member of all three sets A, B, and C is also a member of set D. Notice that this is a value expression. If instead we wanted to actually *assert* that A, B, and C did indeed together imply D, then in that case we would instead want to write this (or any other equivalent form):

$$\exists (A \cap B \cap C \rightarrow D)$$

How the value expression version and the statement expression version are used is somewhat different. The value expression version requires that you know the full contents of A, B, C, and D and that you evaluate whether the described relationship between them really does hold true or not, after which you can subsequently make use of the knowledge you gain from determining this.

In contrast though, even if you didn't know the full contents of A, B, C, and D, you could still *assert* that this relationship does hold true for them, as an additional axiom (i.e. an additional unquestioned assumption) in your system of reasoning, and subsequently combine that fact with other facts in order to deduce additional existential statements from it.

For instance, if you also knew that $\exists (X \rightarrow A \cap B \cap C)$ was true then you could also conclude that $\exists (X \rightarrow D)$ must be true as well (via transitivity), regardless of whether you knew the full contents of these sets or not. Both the value expression version and the statement expression version have their uses.

---

always work in terms of the actual sets in unified logic, not their truth values. Truth is merely a *property* of sets in unified logic, rather than being the entire value itself as it is in classical logic.

How you use them depends on the context and on what you want to accomplish. In fact, in the case of the value expression version, sometimes you don't even *care* whether the expression actually is true or not, but rather you are merely trying to describe some hypothetical object, which may or may not exist as you have described it, whose value may or may not be used for something else later on, according to your needs. Sometimes you care far more about the set itself than you do about its truth value. There are lots of different possibilities.

One thing you may have noticed here though, is that these set expressions seem limited in what they are actually able to say. Doesn't doing things this way inherently restrict us to only be able to talk about specifically *set relationships*? How can we claim to have full generality with such a system? Sure, $A \cap B \cap C$ works if we only care about *sets*, but what about if we wanted to assert three different independent *statements*? How could set expressions ever hope to capture something like that? These are all good questions. You would be wise to ask them. It is very understandable if you are mystified as to how we are supposed to somehow use set expressions to capture any conceivable logical statement. It perhaps just doesn't look possible at first glance.

The trick though is that it *becomes* possible once you realize that there is actually an easy way to reframe sets to act like independent statements, without even leaving the confines of set theory and unified logic. If you were paying close attention to an example I gave earlier, then you might have potentially even realized how to do it yourself by now. Do you remember the little side comment I made earlier (on page 441) about "My notebook is red." actually being equivalent to "There exists a universe such that my notebook is red and this is that universe." when read with a more careful eye? This side comment actually has interesting implications, in a sense. It hints at the secret of how to bridge the gap between sets and independent statements. Properly understood, it tells us how to frame independent statements *as sets*.

All you need to do to make it work is to frame the statement you wish to make in terms of whether or not a corresponding set of objects or relationships exists in the current universe or not, such that the constraints imposed upon that set mirror the constraints embodied by the statement you wish to make about that universe. Once you have chosen a suitable set for this purpose, you just need to express that the set is part of the current universe, i.e. that it is a subset of the universal set ($\mathcal{U}$), and write that down in terms of the language of unified logic. For example, if X was our set, and we wanted to frame it as an independent statement instead of as a normal set, then we could write it like so:

$$\mathcal{U} \leftarrow X$$

Notice how evaluation of $\mathcal{U} \leftarrow X$ behaves differently than if we had just written X alone. The expression $\mathcal{U} \leftarrow X$ will always return either the universal set ($\mathcal{U}$) or the empty set ($\emptyset$). When does it return each one though? Well, naturally, it returns $\mathcal{U}$ whenever X is non-empty, whereas it otherwise returns $\emptyset$. Think about what that implies though. Do you remember those tables I showed you a long time ago on pages 147 and 149, the ones where I showed you that $\mathcal{U}$ and $\emptyset$ in set theory correspond almost perfectly to T and F in classical logic, respectively? Well, don't you think it's interesting that putting

X in the form $\mathcal{U} \leftarrow X$ causes it to instantly start behaving like a binary truth value that closely mirrors how a classical truth value would work?

Indeed, this is no mere coincidence. This is in fact the true form of independent statements. It only becomes apparent when you start thinking in terms of sets. Independent statements are just sets that have been reframed in terms of how they stand in relation to the universal set, which are then evaluated in light of that context. There is hidden structure in what an independent statement really is, and this is that structure. This way of framing sets is one of the central aspects of how unified logic and classical logic relate to each other. The true nature of the relationship between unified logic and classical logic cannot be fully grasped without understanding it. Therefore, in order to make this concept easier to refer to and to think about, it is now time that we define a corresponding formal term for it:

Definition 140. *When a set X is placed in a super implication as the subset operand, with the universal set ($\mathcal{U}$) as the superset operand, then we say that X is in **universal evaluation form**, or **unival form** for short. Unival form allows independent statements to be framed as sets. It causes sets to behave similarly to classical truth values, with $\mathcal{U}$ acting like T and $\emptyset$ acting like F, but with some somewhat different properties in certain cases. In this book, we will usually write the unival form of X as $\mathcal{U} \leftarrow X$, but the equivalent forms $\mathcal{U} \mapsfrom X$ and $X \mapsto \mathcal{U}$ would also of course be acceptable, if you would prefer to use one of them instead.*

*For every set X, there are two different corresponding variants of its unival form: **positive unival form** and **negative unival form**. Which one you use depends on what you want to say. Positive unival form looks like $(\mathcal{U} \leftarrow X)$, and has the same effect as claiming that X is true (i.e. that X exists, i.e. that X is non-empty). In contrast though, negative unival form looks like $\neg(\mathcal{U} \leftarrow X)$, and has the same effect as claiming that X is false (i.e. that X does not exist, i.e. that X is empty).*

However, care must be taken that $\neg(\mathcal{U} \leftarrow X)$ is not confused with $(\mathcal{U} \leftarrow \neg X)$. The later is the positive unival form of $\neg X$, not the negative unival form of X. They have different meanings. $(\mathcal{U} \leftarrow X)$ and $(\mathcal{U} \leftarrow \neg X)$ are very often simultaneously true, whereas $(\mathcal{U} \leftarrow X)$ and $\neg(\mathcal{U} \leftarrow X)$ are never simultaneously true.

*Given some set X, we may refer to $(\mathcal{U} \leftarrow \neg X)$ as its **faux negative unival form**, as a warning against any potential confusion. The word "faux" here means "fake" or "imitation", hence the terminology choice. However, keep in mind that the expression $(\mathcal{U} \leftarrow \neg X)$ does still have plenty of valid uses of course. It is only when you mistake it for the negative unival form of X that it creates problems.*

Definition 141. *For the purposes of contrasting sets that are in unival form against those that aren't, one may refer to sets that aren't in unival form as being in **non-unival form**. For example, the non-unival form corresponding to the unival form $\mathcal{U} \leftarrow X$ would just be X. This definition is not that useful in and of itself. Its purpose is purely to ensure that both types of forms have names, to make it easier to talk about them. This definition is mostly only useful in the context of unival forms and related considerations, and indeed there is probably very little incentive to ever use it outside that context.*

Now that we have defined unival form though, let's take some time to actually work through some examples of how to use it. How do we actually frame statements that we wish to evaluate the truth value of using this system? Well, the most straightforward and obvious case is when we just want to evaluate whether or not certain sets exist. In such a case, all we will need to do is to directly put the relevant sets into unival form and then combine them into an appropriate logical expression.

For example, suppose we wanted to evaluate whether or not both programmers and mathematicians exist. Assume here that the universe we are considering to be the context for this claim is the real one, the one we live in, rather than some other hypothetical universe or arbitrary logical system. Let $\mathcal{U}$ be the set of all things in the real world, i.e. reality's universal set. Let P be the set of all programmers and let M be the set of all mathematicians. Then, this claim (that both programmers and mathematicians exist) could be evaluated using unival form like so:

$$(\mathcal{U} \leftarrow P) \cap (\mathcal{U} \leftarrow M)$$

Notice that the existence of P and M are treated as independent, even though they are being joined by a set intersection. Compare that with what would happen if we instead had just $P \cap M$. Imagine, for instance, that we lived in an alternate universe where both programmers and mathematicians existed, but nobody was simultaneously both a programmer and a mathematician[37]. In such a world, $P \cap M$ would return an empty set (and hence would be false) but in contrast $(\mathcal{U} \leftarrow P) \cap (\mathcal{U} \leftarrow M)$ would return $\mathcal{U}$ (and hence would be true, like all non-empty sets). In this way, unival form allows us to bypass the normally much stricter behavior of the set theory operators so that the participating sets are (in effect) considered separately instead of together.

Of course, this is not the only way you can pose independent statements though. For example, you could alternatively choose to use existential statements instead of sets in unival form. You could have some kind of working list of facts you know about the system, framed as existential statements, and you could then just add $\exists P$ and $\exists M$ to that list to indicate that you also know that P and M exist. You could then perhaps proceed to make additional inferences (if possible) by combining these facts together with other pre-existing facts in your list of existential statements by applying meta implication to the set of everything you know so far.

However, sets in unival form are value expressions, whereas existential statements are statement expressions (i.e. assertions). Thus, the two approaches may feel and act a bit different, and each may have limitations or quirks that the other does not. It's hard to say which (unival form or existential statements) is generally better (if any). I am honestly not sure. Probably both approaches have their own pros and cons.

In any case though, let's now consider the case where we want to use unival form to express something *other* than just the existence or non-existence of certain specific sets of simple objects. Consider, for example, what we'd have to do if we wanted to express that $2 + 2 = 4$ is a true relationship. The first thing you need to notice here, is

---

[37]This is false of course. Programming and mathematics are very closely related fields. Tons of people are highly proficient in both. In fact, learning programming without picking up at least a little bit of knowledge of mathematics (even if only by accident) is nearly impossible. However, just humor me anyway as a thought exercise.

that since $2 + 2 = 4$ could potentially just be some arbitrary form, whether or not it is true will depend entirely upon what universe we are currently operating in. We cannot simply assume it as a given. Some systems may not have it. It does indeed hold true in the real world, but that doesn't mean it holds true in some other arbitrary hypothetical logical world.

The second thing you need to notice here, in order to understand how to say something like this in unival form, is to realize that all relations are just arbitrary forms and thus can just as easily be treated as plain objects as any other kind of object can. You see, in set theory, people are accustomed to thinking of sets as containing only conventional objects, such as cats, numbers, books, matrices, colors, people, and so on. However, this is actually a very narrow-minded perspective. In actuality, there is nothing to stop sets from containing completely arbitrary relationships as well. Relationships are just as much objects as anything else is. The set $\{2 \cdot 3 = 6, x^6 = 60, r \in \text{Cont}\}$ is just as much a valid set of objects as $\{\text{red}, \text{green}, \text{blue}\}$ is.

Thirdly, it is important to realize that the universal set is essentially the set of all valid forms within a system. For reality, this is the set of all real objects and all real relationships between them. There's no way to know what all of the objects and relationships in the universal set of physical reality actually are without discovering them through direct evidence and experimentation. This can be quite laborious.

However, in logic and mathematics, things are usually more straightforward and flexible. In logic and mathematics, we are more often interested only in the objects and relationships that can be generated from some set of axioms and initial forms, i.e. from some sort of rule set, i.e. from some sort of transformative language. It is *those* objects and relationships that constitute our universal set in most systems we study.

Thus, naturally, different transformative languages will tend to generate different corresponding universal sets. It is these corresponding universal sets, each generated by some arbitrary transformative language, that we are really testing relationships against when we are trying to claim that they are true. Thus, it follows that in order to claim that an arbitrary relationship is true via unival form, we need only place the relationship we intend to claim into a set, and then express that set in unival form. As such, this is what it would look like if we wanted to evaluate whether or not $2 + 2 = 4$ is true in the current universe (i.e. in the current arbitrary logical system we are studying), using unival form:

$$\mathcal{U} \leftarrow \{2 + 2 = 4\}$$

As you can see, the result is quite simple and intuitive. If indeed $2 + 2 = 4$ is true in the universe we are currently considering, then this expression will return $\mathcal{U}$ and hence will be true. Otherwise, it will return $\emptyset$ and hence will be false. Furthermore, notice that we need not only claim one relationship at a time with this format. We can easily place as many relationships as we wish to claim in the same set, and then place that set in unival form. For example, the following expression would also of course be permitted:

$$\mathcal{U} \leftarrow \{2 + 2 = 4, \quad 2 + 3 = 5, \quad 2 + 4 = 6\}$$

It would have exactly the effect that you would expect it to have. If indeed $2 + 2 = 4$, $2 + 3 = 5$, and $2 + 4 = 6$ were all valid in the universe under consideration, then the expression would return $\mathcal{U}$, but otherwise would return $\emptyset$. Once nice thing about the fact that this is permitted is that it allows you to reduce the amount of redundancy in an expression in unival form. If including multiple relationships in one set like this wasn't permitted, then we could still express the same thing, but it would be more verbose and would instead look like this:

$$(\mathcal{U} \leftarrow \{2+2=4\}) \cap (\mathcal{U} \leftarrow \{2+3=5\}) \cap (\mathcal{U} \leftarrow \{2+4=6\})$$

This expression is far more verbose, but it is still valid. It is always permissible to express things this way, if you prefer doing so. For example, perhaps you want to group certain similar relationships together in the same set for better readability or clarity. It's a matter of taste and style. It's just an option available to you. Sometimes though, there are cases where separating the relationships into different sets is not optional. Sometimes the separation is required. For example, consider the following expression:

$$[\,(\mathcal{U} \leftarrow \{2+2=4\}) \cup (\mathcal{U} \leftarrow \{2+3=5\})\,] \cap (\mathcal{U} \leftarrow \{2+4=6\})$$

Notice that this expression has a different meaning, and hence cannot be wrapped up into a single unival form. It evaluates whether or not at least one of $2+2=4$ or $2+3=5$ is true and $2+4=6$ is true. This can only be expressed with multiple unival forms. There's no way to express this in a form where $2+2=4$, $2+3=5$, and $2+4=6$ are all written in the same set.

Anyway, by the way, so far we have only talked about sets consisting entirely of conventional objects (e.g. $\{2,3,4,5\}$) and sets consisting entirely of relationships (e.g. $\{2+2=4,\ 2+3=5\}$), but sets that mix both types of objects are also permitted. Regardless of whether an object is a conventional object or a relationship, saying that it is a part of the universal set is simply the same as saying that it is valid, i.e. that it does indeed exist in that universal set. Consider, for example, the following expression:

$$\mathcal{U} \leftarrow \{2,\ 3,\ 4,\ 5,\ 2+2=4,\ 2+3=5\}$$

This expression claims that 2, 3, 4, and 5 are all valid standalone objects in the system we are considering, and also that $2+2=4$ and $2+3=5$ are both valid relationships. What would happen if we left out 2, 3, 4, and 5, but allowed $2+2=4$ and $2+3=5$ to remain? How would that differ in terms of what claim the expression would represent? In what sense would the meaning of $\mathcal{U} \leftarrow \{2+2=4, 2+3=5\}$ be different from the above expression?

Well, the difference would be that we would no longer be claiming that 2, 3, 4, and 5 are necessarily valid standalone objects. We'd only be claiming that the forms $2+2=4$ and $2+3=5$ are valid. Even if this modified expression returned true, then it would not necessarily be safe to assume that 2, 3, 4, and 5 standing on their own would be valid objects in the system. It could nonetheless easily be true, but we can't just blindly assume that. A symbol can be valid in a relationship while simultaneously not being allowed to stand on its own.

Not all objects can exist outside of relationships, so it is not safe to assume that an object also exists in standalone form just because it does exist in some relationship or other form. Forms and transformation rules are very flexible concepts, so it is important that we not make any premature assumptions about how they work, otherwise we would be limiting what we are capable of expressing with them. Let's define official terms for these different types of forms, just in case it becomes useful at some point:

Definition 142. *When a form exists as a valid object in a system, without having to always be a part of some other form, then we may say that that form is an **independent form**. Whether or not a given form is an independent form depends on what the current system being studied is and what transformation rules it has. It does not make sense to talk about whether a form is independent or not unless there is a specific system of rules being considered. Otherwise, the form would just be a bunch of meaningless symbols and the concept of being independent or not wouldn't make any sense.*

Definition 143. *When a form only ever exists in a system as part of another form, and thus cannot stand on its own in that system as a valid object, then we may say that that form is a **dependent form**. This is the opposite of an independent form. These terms are mutually exclusive. No independent form is a dependent form and no dependent form is an independent form. However, independent forms may or may not also appear as part of other forms, in addition to being capable of standing on their own.*

Definition 144. *If a form exists only in isolation, and never as part of any other form, then we may say that that form is an **isolated form**. Isolated forms are always also independent forms, but most independent forms are not isolated forms. Isolated forms are never dependent forms, of course, since independence and dependence are mutually exclusive concepts.*

Notice that I added on this third definition (the definition for isolated forms), even though we didn't talk about that case earlier. I did that for the sake of completion, since it can't hurt to have a more complete vocabulary for expressing ourselves. The more words we have, the more avenues for our thoughts open up to us. Words are empowering. Human intelligence has always been strongly dependent upon the power of language. We wouldn't be very smart without it. Intelligence doesn't scale well without language.

Next, let's talk about how to read expressions in unival form in a natural way. One of the most important things to remember when reading unival expressions is that they are value expressions, not statement expressions. Expressions in unival form represent claims that are being evaluated, not assertions that are being forced into the rule set. As such, an expression of the form $\mathcal{U} \leftarrow X$ shouldn't really be read as an assertion that X does exist in the universe, but rather it should be read as being more tentative. For example, consider the following expression:

$$\neg(\mathcal{U} \leftarrow X) \cap (\mathcal{U} \leftarrow Y)$$

This expression tests whether or not it is true that X is false and Y is true. It does not force the claim to be true. It is like saying "evaluate whether this is a universe such that X is false and Y is true", rather than being like actually asserting that "this is a universe such that X is false and Y is true". I know we talked about this already before, but it seems worth repeating here to make sure it really sinks in.

Value expressions aren't assertions, even though they often sort of read like assertions in a certain light, because they *do* often represent claims that are being tested. This is treated sloppily in some parts of the literature, but it is actually a very important distinction. Values are not assertions. Describing a possibility isn't the same as asserting it to be true. We can describe objects without saying they actually exist. I can *describe* a fire breathing unicorn without necessarily implying that I think it actually exists. The same is true of logic.

The way that unival forms behave is remarkably similar to how expressions in classical logic behave. Being aware of unival form and how to use it does indeed remove some restrictions on what you would otherwise be able to say with sets. Specifically, unival form allows you to express independent statements without ever leaving the language of sets, but this ability comes at a cost. What you gain in generality by using unival form, you lose in inferential strength. What you can infer based on expressions in unival form is often far weaker than what you can infer based on expressions that are not in unival form. This is exactly as it should be.

To understand why, it is instructive to consider the case where an expression in unival form is used as the condition of another unified implication. This is the same as evaluating whether the expression in unival form *implies* some other set as its superset. Under what circumstances could such an expression ever be true? Not many, actually. Think about it. Expressions in unival form always return either $\mathcal{U}$ or $\emptyset$, but what would $\mathcal{U}$ or $\emptyset$ do if given as the condition of a unified implication? An example will make this clearer. Consider, for example, the following expression:

$$((\mathcal{U} \leftarrow A) \cap (\mathcal{U} \leftarrow B)) \rightarrow C$$

Under what circumstances could this expression return a non-empty set? Clearly, if it's going to return a non-empty set, then we will first need the condition side of the expression (the left side here) to return a non-empty set. That will only happen when both A and B are non-empty sets, in which case the condition expression will become $\mathcal{U}$. What then must C be in order for $\mathcal{U} \rightarrow C$ to be non-empty though? The answer is that C must also be $\mathcal{U}$.

However, think about what that implies about the inferential strength of unival form expressions. It implies that unival form expressions can at best only imply their own memberships in the universal set. And what, exactly, does it mean to be a member of the universal set though? Well, it merely means that the member is *true*. The universal set is, after all, nothing more than the set of all valid objects and relationships in the universe under consideration. This means that even if C *is* compatible with the condition expression, then it still won't really imply any meaningful superset relation beyond what is already known.

Being a part of the universal set says almost nothing about the properties of the members, except merely validity. This renders attempts to derive meaningful superset relations from expressions in unival form essentially pointless. It is redundant. Even if the expression in unival form in the condition were true, no additional tests of unified implications would actually provide any new information, since any unified implication with an expression in unival form as the condition set can only be true if the consequence set is also $\mathcal{U}$.

Making strong inferences would require much more highly constrained sets. The universal set is far too all-encompassing. The less constraints something has, the less you can logically infer about it. That's the cost of abstraction. More abstraction gives you more flexibility, but only at the cost of having less inferential strength. This is a fundamental trade-off involved in all logical systems. It is an unavoidable law of nature. You can't have maximal abstraction and maximal inferential strength at the same time. It is logically impossible.

In contrast though, notice what happens when you have expressions that are *not* in unival form. Expressions in non-unival form may not be able to evaluate unrelated sets independently of each other, and this does limit what you can do with them in certain respects, but what expressions in non-unival form lack in generality they make up for by having vastly more inferential strength. Unival form and non-unival form each have their own advantages and disadvantages. Consider, for example, the following non-unival expression as compared to the previous unival expression:

$$(A \cap B) \rightarrow C$$

Notice that this time the value of C is meaningful. If we can find a C which is indeed a unified implication of $A \cap B$ (i.e. where $A \cap B$ is an existential subset of C), then this could easily give us some meaningful information, upon which we could potentially base additional logical inferences. For instance, suppose that A is the set of all artists, that B is the set of all book authors, and that C is the set of all computer users. Suppose furthermore that we evaluate $(A \cap B) \rightarrow C$ and thereby determine that it is non-empty, and hence true[38].

This information is meaningful. Knowing it has a substantive impact on what we can potentially infer about other parts of the system. It tells us that we know that all people who are both artists and book authors must also be computer users. We can now combine this fact with other facts in order to potentially obtain more information than we started with. Information that has inferential strength has more than just its immediate face value. Suppose, for example, that we also knew that all computer users grew up in families with yearly incomes above $40,000. We could now infer that all people who are both artists and book authors grew up in families with yearly incomes above $40,000. In this way, we have now obtained more information than was originally given to us, and that is a big part of the whole point of logic.

In contrast though, as we have seen, expressions in unival form are not an effective way to build up additional unified implications. Additional inferences from the truth of expressions in unival form will have to come from the *content* of those expressions, rather than from more easily inferred superset relations. Interestingly, notice that while it is true that expressions in unival form *do* have remarkable similarities to how propositions in classical logic work, expressions in unival form are much more limited in what they are allowed to imply.

---

[38] This is just a hypothetical example. I'm not saying that it is actually true. In fact, it probably isn't. There is probably at least one person who is both an artist and a book author, but who is *not* also a computer user.

This is not a bad thing. In fact, classical logic overreaches in what it allows its implication operator (material implication) to do[39]. The more constrained nature of expressions in unival form is actually beneficial. It blocks the user from making potentially unsafe inferences and thereby forces the user to find a safer way of reaching their conclusions. Attempts to build up additional unified implications from independent statements will tend to fail in unival form, thereby blocking you from overreaching and obtaining the kinds of unsafe conclusions that material implication would let you have.

In a sense though, classical logic is actually always implicitly in unival form, but fails to recognize the structural limitations that this implies about what conclusions it can reach safely or not. Univaluniv form in unified logic in contrast makes those limitations more plainly apparent and prevents you from doing more with potentially unrelated independent claims than you should really be able to. Classical logic has an incomplete understanding of its own components, and thus unsurprisingly is misleading. Its model of implication is wrong.

Genuine implications require relevance. Expressions in unival form allow you to combine claims about unrelated entities together, but only at the cost of blocking you from forming any additional unified implications, due to the fact that relevance can no longer be guaranteed once you've allowed yourself to combine potentially unrelated claims. In contrast though, expressions in non-unival form can only end up true if the participants are genuinely related to each other in at least some way, such as by having some set theory based relationship to each other for example. Hence, it should come as no surprise that expressions in non-unival form can have additional meaningful unified implications, whereas expressions in unival form cannot (besides just the trivial $\mathcal{U}$ case). This is in fact exactly how things should be.

Anyway though, it is worth noting that in addition to being able to put expressions into unival form or non-unival form, you can also mix both unival form and non-unival form together in the same expression. Considering that expressions in unival form and expressions in non-unival form both return sets, this kind of mixing must of course be admissible in expressions in set theory. Indeed, in general, anything that returns a set can be given as input to set theory, just as one would expect. Anyway though, here's an example of what mixing unival form and non-unival form together might look like:

$$((\mathcal{U} \leftarrow A) \cap B) \rightarrow C$$

The main thing you need to understand when working with mixed expressions like this, in order to interpret them correctly, is how the universal set ($\mathcal{U}$) behaves in the presence of other sets in intersections and unions. Specifically, if B is some arbitrary non-empty set, then $\mathcal{U} \cap B$ will always return B and $\mathcal{U} \cup B$ will always return $\mathcal{U}$. In mixed expressions such as the one above, this behavior will determine how the different parts of the expressions will interact with each other, and thus will also determine how you should interpret and read the overall expression.

---

[39] Well, technically, it is not the material implication operator itself that is the problem, but rather it is material implication's incorrect interpretation in classical logic that is the problem. If material implication were instead reinterpreted in the conceptually correct way, i.e. as merely "possible implication", then none of its (now correspondingly much weaker) conclusions would actually constitute overreach.

Consider the above example, for instance. An expression in unival form like $(\mathcal{U} \leftarrow A)$ would normally inhibit meaningful unified implications, but the fact that it is also being intersected with B makes the expression as a whole act differently. If $(\mathcal{U} \leftarrow A) \cap B$ ends up being true (i.e. non-empty) then it must have the value of B. Thus, the expression as a whole will only return a true (i.e. non-empty) set if indeed B sub implies C. In other words, the above example is really just like evaluating whether "A is true and B is an existential subset of C" is true or not, except for that it returns the set B instead of merely a crude binary truth value like F or T (or equivalently, $\emptyset$ or $\mathcal{U}$ respectively).

Anyway though, with that said, we've now basically covered the most essential components of how to reason with unified logic. We've discussed a few different specific examples and ways of approaching the reasoning process, and these have been instructive, but it is important to remember that reasoning with unified logic is very much a hybrid process. You simply use whatever combination of transformative implications, meta implications, unified implications, and existential statements is necessary to get the job done. The exact approach you may need to use in a specific circumstance could vary quite a lot, and sometimes may not even resemble the examples we have discussed here very closely.

Sometimes you will need to lean really heavily on one technique or another. For instance, some systems will tend to mostly just require the use of transformative implication and meta implication, and may have little to no use for unified implication and existential statements. Sometimes all you really need is transformative logic, and all this stuff about set theory and unified logic may just be a waste of time in such cases. However, sometimes the properties of unified implication and existential statements are quite advantageous and will really help you to express what you have in mind. It depends on the situation and what you are trying to do.

We do not have time in this book to discuss all the possible ways of approaching problems, nor am I probably even aware of what all such approaches actually are. Indeed, considering that this book is intended primarily to blaze new trails and to illuminate the deeper insights underlying certain concepts, rather than to provide comprehensive coverage of every nook and cranny of the theory, such coverage would be far outside the scope of the intent of this book. Much of the material in this book is very much still "fresh out of the oven". In fact, many of these ideas are still fresh even to me, and even I myself doubtlessly do not fully realize what the final impact and reach of many of these ideas will actually be.

Even though I have focused so overwhelmingly on just the most basic fundamental ideas underlying the theory presented in this book, the book has still ended up consuming an enormous amount of time and energy. It has already expanded far beyond its original intended scope and size, and if I were to attempt to be comprehensive in my coverage at this point then it would derail the core thrust of the book far too much.

I often find that it's best to just take things one step at a time, letting things unfold naturally, doing whatever you can to create value in the here and now, rather than trying to always do everything up-front and all at once. Life is a work in progress. Natural selection will run its due course, as it always has. That's just as true of abstract ideas as it is of biology. We have a schedule to keep and priorities to attend to.

True perfection is maximizing the amount of value you produce per unit time in what precious little time you have in life. True perfection is pragmatism. True perfection is always keeping the big picture in mind, is always orienting your thoughts and actions towards creating the greatest possible good for both your fellow human beings and yourself. In stark contrast though, the siren call of so called "perfectionism", of getting distracted by every little low-value detail, of filling in every possible omission without respect to how important it may actually be, is the opposite of this. It is thus with great irony that what many people call "perfectionism" is in fact nothing more than poor time management skills in disguise. Perfectionism is poor time management made glorified.

Perfectionism has nothing to do with real perfection. Real perfection is proportionality and prioritization. Real perfection is nuanced and perpetually evolving adaptation to the constraints and limited resources of reality, not unrequited pursuit of some unattainable and impractical ideal. Having "flaws" does not necessarily do anything to diminish something's worth; for in the tapestry of all time stretching forward from this moment, it is only value creation which is the true measure of merit. Just as a story is not made imperfect by the presence of any single bad moment within it, and indeed may even be enriched and enlivened by it, neither is any other thing made less perfect merely for the presence of a "flaw". It is only when the full story is told, when all the pieces come together to form the complete symphony, that the real measure and value of a thing can truly be determined.

A thing is more than just a snapshot of itself at one moment of time. It is a tapestry of relationships and connections, a dynamic thread stretching forth from the very beginning to the very end of its own existence, interwoven with everything it ever touches or interacts with. A note played on a piano seldom sounds alone. It is only in the bigger picture that its true value can ever really be determined. It is only when the overall composition is held in mind, when both the melody and harmony are properly absorbed, that a thing can ever truly be accurately judged. Everything is a story in time. Even the most inanimate object has a story, even indeed despite the fact that it may have no power to act on its own.

Anyway though, putting my tangential philosophical commentary on the nature of real perfection aside, my point is that a comprehensive discussion of every law that could ever be applied to unified logic would be impractical here. It would also be fairly dry and monotonous and lengthy. I think it would be far better to instead just cover a small handful of the applicable laws, just to give a sense for what some of them are.

I will simply select a few such laws, semi-randomly, according to my whims, and then discuss them some. A more comprehensive exploration of what laws are applicable to unified logic can be performed at some later date, after the release of the book maybe. Whether that exploration is performed by me or by someone else doesn't really matter, so I might as well leave a lot of meat on the bone for other people to play with too, rather than hogging it all for myself and further delaying the publication of this book.

Let's begin. First, I'd like to point out an especially interesting point of nuance about unified logic, which is that in addition to having its own form of the law of non-contradiction (specifically $X \cap \neg X \to \emptyset$), unified logic also has a related law that has almost the opposite connotation but which is still valid. This related law may feel sort of like the opposite of non-contradiction at first glance, but it actually says something quite different though, and thus does not actually conflict with the law of non-contradiction in any way. Here it is, as a meta implication, in two equivalent forms:

$$\frac{X \neq \mathcal{U} \quad X \neq \emptyset}{\neg X \neq \emptyset} \qquad \frac{X \neq \mathcal{U} \quad \exists X}{\exists \neg X}$$

Do you see what this law is saying? It is saying that for every conceivable set X, if X is not the universal set, but X is still at least non-empty, then $\neg X$ must also be non-empty. Taking into account an existence-based view of truth, this is essentially the same as saying that as long as X is not $\mathcal{U}$ (i.e. as long as X is not the set of everything in the entire universe) then both X and $\neg X$ must always be true. In other words, X and $\neg X$ must always co-exist in the same universe, so long as neither set is empty or universal. It is logically impossible for a set to exist without its negation also existing, unless the set happens to be the universal set (whose negation, in contrast, would be the empty set of course).

Do you see how this law sort of has the opposite connotation as the law of non-contradiction, in a sense? The law of non-contradiction says that something and its negation can never be true simultaneously, whereas *this* law says that as long as a set is not the universal set and is non-empty then both it and its negation must always be simultaneously true (i.e. must co-exist). Despite this, both the law of non-contradiction and this law are simultaneously true in unified logic. This is interesting and worth consideration. As such, this law needs a name, so that we might more easily refer to it. What name would be appropriate to capture the essence of this law though? Well, here's my choice:

Definition 145. *In set theory and unified logic, whenever we know that $X \neq \mathcal{U}$ and $X \neq \emptyset$, we can conclude that $\neg X \neq \emptyset$ must be true as well. This can be expressed by the meta implication $\{X \neq \mathcal{U}, X \neq \emptyset\} \precsim \{\neg X \neq \emptyset\}$, or any other equivalent form. We will refer to this law as the **law of yin and yang**.*

In Chinese language and Taoist philosophy, "yin" means darkness and "yang" means light, roughly. These terms are commonly used as a metaphor for illustrating the complementary roles of the forces of darkness and light in the world, i.e. how darkness and light are like two sides of the same coin, such that neither could exist without the other and each serves to counteract the other to maintain balance in the world. This has roughly the connotation we are looking for. The correspondence is pretty close. The phrase "the law of yin and yang" doesn't imply contradiction, but it does imply co-existence, and thus it is a fitting choice of words for the name of this law of logic.

It may seem weird that both the law of non-contradiction and the law of yin and yang apply to the same system, considering that their connotations are so different, but nonetheless it actually does still make sense. Having both these laws helps to capture the nuances of how contradictory things can relate to each other, and does so better than only having the law of non-contradiction would. The law of non-contradiction and the law of yin and yang, considered together, essentially say that contradictory properties never exist in the same object(s) and yet contradictory properties usually *do* exist separately. If objects exist in a system which have the property embodied by set X, then, unless all objects in the system have the property X, some won't have it. Thus, X and $\neg X$ tend to co-exist, but do so separately.

This law is also closely related to why you need to be careful when using unival form in unified logic. If you accidentally use the faux negative unival form of a set X, i.e. $(\mathcal{U} \leftarrow \neg X)$, instead of the real negative unival form, i.e. $\neg(\mathcal{U} \leftarrow X)$, then you will very often receive unexpected behavior from the resulting expression, due to the fact that X and $\neg X$ usually co-exist in the same universe. Existence-based truth may be more principled and foolproof than statement-based truth[40], but you do need to make sure you understand that its underlying mechanics are a bit different than what you may be used to from classical logic.

I personally find the combination of the law of non-contradiction and the law of yin and yang to be somewhat philosophically interesting. Generally speaking, when people believe in the existence of "true contradictions" (e.g. dialetheism) they are essentially just confused. They are victims of their own imprecise and unprincipled perspective on how logic and reality actually work. Their minds lack discipline and clarity. You see, whenever someone thinks they have found an instance of a "true contradiction" it is very often actually the case that the supposedly "contradicting" properties of the object are in fact *separate* parts of that object.

Separately contradicting properties do *not* constitute a real contradiction though. It's only when you try to force incompatible properties to apply to the same part of an object that a real contradiction is born. There is also the matter of fuzzy logic[41], vagueness, and "degrees of truth", but even that too does not truly violate the law of non-contradiction when properly understood. The law of non-contradiction is, at its core, absolute, even in contexts where it looks at first glance like you might be able to avoid it. You simply cannot have an object that both has a property and does not, unless those properties apply to *separate* parts of that object. The law of yin and yang acts as yet another point of nuance in this respect, to help people realize more precisely where the boundaries between contradiction and non-contradiction really are, much like fuzzy logic also does.

Next, let's explore an example of a law that you will frequently need to use when reasoning with existential intersections. You can do more with complete implication relationships, such as existential subset relations and transformative implications (etc), than you can with mere intersections, but there are still things you can nonetheless infer based on the more incomplete information that intersection relationships provide. Here is one such law, one of the most essential, written in four equivalent forms:

$$\frac{A \Diamond B \quad B \sqsubseteq C}{A \Diamond C} \qquad \frac{A \sqcap B \quad B \sqsubseteq C}{A \sqcap C} \qquad \frac{\exists(A \cap B) \quad \exists(B \to C)}{\exists(A \cap C)} \qquad \frac{A \cap B \neq \emptyset \quad B \to C \neq \emptyset}{A \cap C \neq \emptyset}$$

---

[40] For reference: existence-based truth and statement-based truth were originally defined starting on page 27.

[41] For reference: Most of our discussion of fuzzy logic occurred in the multi-valued logic section, which began on page 313.

As you can see, this law looks a lot like the law of transitivity. The main difference is that instead of using only implication or subset relations, this law uses a mix of both subset relations and intersection relations. Otherwise, the structure is pretty similar. This is a fairly common rule to use when working with sets. This law is already widely known, so I tried to find an existing name for it, but I couldn't find one. Thus, as usual, to make the concepts we are working with easier to refer to, talk about, and think about, let's define a new term:

Definition 146. *When we know that two sets A and B have a non-empty intersection, and that B is an existential subset of C, then we can conclude that A and C must also have a non-empty intersection. We will refer to this law as the **law of intersection implication**. The law can be expressed as a meta implication of the form $\{A \diamond B, B \sqsubseteq C\} \prec \{A \diamond C\}$, or any other equivalent form.*

In contrast to this law though, notice that $\{A \diamond B, B \diamond C\} \prec \{A \diamond C\}$ (or any other equivalent form) would not be a valid rule. Given that we knew that $A \diamond B$ and $B \diamond C$ were true, then we would at least know that some A are B and that some B are C, since that is indeed what $A \diamond B$ and $B \diamond C$ literally mean. Furthermore, just as one would expect, it would also be possible that some A are C under such circumstances. However, while it may be *possible* that some A are C, we wouldn't actually *know* whether or not some A are C, based only on what we've been given so far. We'd need more information. Existential intersection relationships are not transitive.

Next, let's discuss a law of unified logic which is closely analogous to a law we've previous discussed, but which is just a bit different. Do you recall our discussion of the concept of meta modus tollens? It was a law of transformative logic, which governed in what sense something like modus tollens or contraposition could be applied to transformative implications. We defined it on page 274. Well, here's a structurally similar law about sets, which is also valid:

$$\frac{A \sqsubseteq B \qquad X \not\sqsubseteq B}{X \not\sqsubseteq A}$$

Look at meta modus tollens on page 274 and compare it to this rule. Do you see the similarities? Both rules have essentially the same structure. The only difference is that whereas meta modus tollens uses $\to$ this rule instead uses $\sqsubseteq$, and whereas meta modus tollens uses $\not\to$ this rule instead uses $\not\sqsubseteq$. Otherwise, the two rules are the same.

Both rules are closely analogous to classical modus tollens (i.e. $\{A \to B, \neg B\} \prec \{\neg A\}$), except that they have a third participant X, such that the negations of A and B are applied to a *relation* from X to A or B, instead of being applied to A or B directly. This results in both rules working in terms of derivability and non-derivability, instead of merely working in terms of direct assertion and denial as classical modus tollens does. The "$X \not\sqsubseteq$" parts of the conditions of the above rule are analogous to the "$\neg$" parts of classical modus tollens, and thinking of it that way may make it easier for you to see the nature of the connection to classical modus tollens here more clearly.

In short though, the above rule is simply a subset relation version of meta modus tollens. It says that if all members of *A* are members of *B*, and not all members of *X* are members of *B*, then not all members of *X* can be members of *A*. At least one member of *X* must not be a member of *A*, given that $A \sqsubseteq B$ and $X \not\sqsubseteq B$ are true. Otherwise, if $X \sqsubseteq A$ were true, then we could use the fact that $A \sqsubseteq B$ is true to conclude that $X \sqsubseteq B$ must also be true, but this would contradict our assumption that $X \not\sqsubseteq B$ is true, and hence $X \sqsubseteq A$ must be false. As usual, let's define a new official term for this law, to make it easier to refer to, think about, and work with:

Definition 147. *Whenever we know that $A \sqsubseteq B$ and $X \not\sqsubseteq B$ are true, we may also conclude that $X \not\sqsubseteq A$ must be true as well. We will refer to this as the law of* **existential subset modus tollens**. *It is the existential subset relation analog to meta modus tollens. It may be written as a meta implication of the form* $\{A \sqsubseteq B, X \not\sqsubseteq B\} \prec \{X \not\sqsubseteq A\}$, *or any other equivalent form.*

Oh, and while we're at it, we might as well define the analogous law for non-existential subset relations as well. After all, not all subset relation statements are existential. Regardless of the fact that existential subset relations are more aligned with the philosophy of unified logic than non-existential subset relations are, we might as well still give a formal definition for the non-existential case anyway. It works exactly as you'd expect. Here it is:

Definition 148. *Whenever we know that $A \subseteq B$ and $X \not\subseteq B$ are true, we may also conclude that $X \not\subseteq A$ must be true as well. We will refer to this as the law of* **non-existential subset modus tollens**. *It is the non-existential subset relation analog to meta modus tollens. It may be written as a meta implication of the form* $\{A \subseteq B, X \not\subseteq B\} \prec \{X \not\subseteq A\}$, *or any other equivalent form.*

This is what it would look like written in vertical meta implication notation:

$$\frac{A \subseteq B \\ X \not\subseteq B}{X \not\subseteq A}$$

While these two laws may indeed closely resemble meta modus tollens, the way they relate to contraposition is not quite the same as how meta modus tollens does. In transformative logic, meta modus tollens permits $\{A \rightarrow B, X \not\rightarrow B\} \prec \{X \not\rightarrow A\}$, which is sort of related to contraposition, but does not allow actual contraposition in the conventional sense. The rule $\{A \rightarrow B\} \prec \{\neg B \rightarrow \neg A\}$ would be more directly analogous to contraposition, if it were allowed, but it is not valid for transformative logic. Meta modus tollens is thus the closest you can get to contraposition in transformative logic. However, for set theory and unified logic, in contrast to transformative logic, direct contraposition is valid. Indeed, both the existential and non-existential variants are valid. Here are the corresponding definitions:

Definition 149. *Given that we know that $A \sqsubseteq B$ is true, we may also conclude that $\neg B \sqsubseteq \neg A$ is true. We will refer to this as the law of* **existential subset contraposition**.

This may be expressed as a meta implication of the form $\{A \sqsubseteq B\} \prec \{\neg B \sqsubseteq \neg A\}$, or any other equivalent form. Here is what it would look like when written in vertical meta implication notation:

$$\frac{A \sqsubseteq B}{\neg B \sqsubseteq \neg A}$$

Definition 150. *Given that we know that $A \subseteq B$ is true, we may also conclude that $\neg B \subseteq \neg A$ is true. We will refer to this as the law of **non-existential subset contraposition**. This may be expressed as a meta implication of the form $\{A \subseteq B\} \prec \{\neg B \subseteq \neg A\}$, or any other equivalent form. Here is what it would look like when written in vertical meta implication notation:*

$$\frac{A \subseteq B}{\neg B \subseteq \neg A}$$

There are also variants of these contraposition laws which use transformative implication instead of meta implication, but we will still usually refer to them by the same names (i.e. as simply existential or non-existential subset contraposition), since the distinction is not all that important most of the time. Indeed, often we'll be even lazier and just say "contraposition", instead of specifying the type any further. However, if you really want to distinguish each type carefully though, you can tack the word "transformative" or "meta" onto the beginning of the terms to distinguish each case precisely. Thus, the following table shows some more precise terms for each variant:

Contraposition Variant Name	Rule
transformative existential subset contraposition	$A \sqsubseteq B \to \neg B \sqsubseteq \neg A$
meta existential subset contraposition	$\{A \sqsubseteq B\} \prec \{\neg B \sqsubseteq \neg A\}$
transformative non-existential subset contraposition	$A \subseteq B \to \neg B \subseteq \neg A$
meta non-existential subset contraposition	$\{A \subseteq B\} \prec \{\neg B \subseteq \neg A\}$
transformative classical contraposition	$A \to B \to \neg B \to \neg A$
meta classical contraposition	$\{A \to B\} \prec \{\neg B \to \neg A\}$

As you can see, there are a fair number of them, but the spirit of what they are doing is essentially the same. You may notice though that this list of contraposition variants does not include any modus tollens rules, even though modus tollens rules also have a similar spirit to what contraposition does. That's because modus tollens rules are a bit more distinct from contraposition, despite the similar reasoning. I think it would be wise to list a table of the modus tollens variants here too, for completeness. However, before we do that, there's one more variant of modus tollens for subset relations that I didn't mention yet. Here it is:

Definition 151. *Given that we know that $A \subseteq B$ and $B = \emptyset$ are true, we may also conclude that $A = \emptyset$ is true. We will refer to this as the law of **simplistic subset modus tollens**. This may be expressed as a meta implication of the form $\{A \subseteq B, B =$*

∅} ⊰ {A = ∅}, *or any other equivalent form. Here is what it would look like when written in vertical meta implication notation:*

$$\frac{A \subseteq B \quad B = \emptyset}{A = \emptyset}$$

This version of modus tollens is not as powerful or as general as the existential and non-existential subset modus tollens rules that involve the extra X variable, hence the qualifier "simplistic" here. Another problem with this approach to subset modus tollens, compared to the approach that uses the extra X variable, is that this approach only works for the non-existential subset case.

If you tried to make this same argument using an existential subset relation between A and B, instead of a non-existential one, then you would have to admit self-contradictory premises by doing so, which wouldn't be allowed. You'd have to claim both $A \sqsubseteq B$ and $B = \emptyset$ as required conditions of the rule, but that would be invalid because $A \sqsubseteq B$ would require that both A and B are non-empty, while simultaneously $B = \emptyset$ would say that B is empty, which would be self-contradictory and hence nonsensical. Anyway though, now that we've got that out of the way, time for that table of the modus tollens variants that I mentioned earlier:

Modus Tollens Variant Name	Rule
existential subset modus tollens	$\{A \sqsubseteq B, X \not\sqsubseteq B\} \prec \{X \not\sqsubseteq A\}$
non-existential subset modus tollens	$\{A \subseteq B, X \not\subseteq B\} \prec \{X \not\subseteq A\}$
simplistic subset modus tollens	$\{A \subseteq B, B = \emptyset\} \prec \{A = \emptyset\}$
classical modus tollens	$\{A \to B, \neg B\} \prec \{\neg A\}$
meta modus tollens	$\{A \to B, X \not\to B\} \prec \{X \not\to A\}$

Do you see how contraposition and modus tollens differ from each other more precisely now? They have different structure, but similar underlying reasoning. These lists of variations of contraposition and modus tollens maybe look a bit overwhelming at first perhaps, but the reasoning involved in each of them is very predictable and unsurprising, so in practice accounting for all of them isn't really that difficult.

Anyway though, let's do one more example of a law of unified logic before we move on to the next section. Let's do another example that helps us compare and contrast unified logic and classical logic a bit more. Specifically, let's consider the nature of unified logic's equivalent of the law of disjunctive syllogism. Classical disjunctive syllogism has the form $\{A \lor B, \neg A\} \prec \{B\}$. Admittedly, I've only mentioned disjunctive syllogism briefly and in passing, prior to this point in the book. However, notice that I've also previously mentioned that material implication has an equivalent form as a disjunction. Specifically, $A \to B \leftrightarrow \neg A \lor B$ is true in classical logic[42].

---

[42] See the paragraph immediately following the definition of the paradox of manifest denial on page 114 to see an example of somewhere I've previously mentioned this equivalence. There are also several other places where I've mentioned it in this book. It is also very widely known among logicians.

Do you see that material implication and disjunctive syllogism are thus closely related in classical logic? Indeed, in classical logic, one can think of applications of material implication as being essentially the same thing as applications of disjunctive syllogism. Given that $\neg A \vee B$ and A are true, one can use disjunctive syllogism to conclude that therefore B must also be true. Notice that this is not really any different from being given that $A \rightarrow B$ and A are true and then subsequently using this fact to conclude that B must also be true. In classical logic, these two inferences are fundamentally the same.

However, this is not true of unified logic. Implication and disjunctive syllogism are two very different things in unified logic, even though both are nonetheless valid in it. This is an important point of nuance. Classical logic steamrolls over the distinction between implication and disjunctive syllogism, whereas unified logic does not. In classical logic, implication and disjunction syllogism seem like pretty much the same thing, whereas in unified logic they are clearly quite different. As such, here's our definition for the unified logic version of disjunctive syllogism:

Definition 152. *Given that we know that $A \sqcup B$ and $\not\exists A$ are true, we may also conclude that $\exists B$ must be true. We will refer to this as the law of **existential disjunctive syllogism**. It is the set theory and unified logic version of disjunctive syllogism. It may be written in inline meta implication notation as $\{A \sqcup B, \not\exists A\} \prec \{\exists B\}$, or any other equivalent form. Also, here is what it would look like when written in vertical meta implication form, in three equivalent forms, so that you may choose one according to your preference:*

$$\frac{A \sqcup B \qquad \not\exists A}{\exists B} \qquad \frac{\exists(A \cup B) \qquad \not\exists A}{\exists B} \qquad \frac{A \cup B \neq \emptyset \qquad A = \emptyset}{B \neq \emptyset}$$

Notice that $A \sqsubseteq B$ and $\neg A \sqcup B$ would not be equivalent. Unlike in the analogous case in classical logic, you cannot transform one into the other. Indeed, even if we were given that $\neg A \sqcup B$ and $\not\exists \neg A$ were true, and hence used it to correctly conclude that $\exists B$ must also be true, then this still wouldn't do anything to indicate whether or not $A \sqsubseteq B$ is true. The expression $A \sqsubseteq B$ can only be true if A is in fact an existential subset of B. The inference from $\neg A \sqcup B$ and $\not\exists \neg A$ to $\exists B$ though, in contrast, works regardless of whether $\neg A$ and B have any relationship to each other whatsoever.

Thus, existential subset statements (i.e. set theory's version of implication statements, essentially) can't possibly be translated into terms of existential disjunctions in unified logic, unlike in classical logic where the analogous argument (i.e. $A \rightarrow B \leftrightarrow \neg A \vee B$) would work. This is interesting. It shows that disjunctive syllogism truly is distinct from implication, which is an important point of nuance about how logic actually works, but one that the overwhelming dominance of classical logic has unfortunately made a lot of people blind to.

We've been discussing a lot of laws about statement expressions in unified logic here, but I want to remind you that there are also numerous laws that govern value expressions in unified logic as well. Strictly speaking, meta implications should typically be thought about in terms of statement expressions rather than in terms of value expressions, since value expressions do not actually assert anything to be true and thus the inclusion of value expressions in the set of valid forms via meta implication doesn't necessarily actually say anything.

Classical logic is sloppy on this point. For example, classical logicians are accustomed to applying the meta implication $\{A \rightarrow B, A\} \prec \{B\}$ (classical modus ponens), but failing to realize that technically the expressions $A \rightarrow B$, A, and B don't actually even necessarily say anything, since they are just value expressions and thus are not technically even making assertions about the truth or falsehood of anything. However, that being said, this sloppy convention can arguably be a nice convenience in some ways, and still feels fairly natural. It's a bit debatable.

Basically, when you use value expressions like this in classical logic, you are actually just abusing your notation and implicitly asserting that the indicated value expressions are not merely valid (i.e. not merely derivable from the current rule set and initial forms) but are also *true*. Technically, value expressions could be true or false, and you don't necessarily know which. Thus, in a sense, classical modus ponens would maybe be more accurately written as $\{A \rightarrow B = T, A = T\} \prec \{B = T\}$, to make it clear that you are claiming that those expressions are actually true, but nonetheless classical logicians tend to use $\{A \rightarrow B, A\} \prec \{B\}$ anyway, since it is less verbose and more convenient, even though it is arguably less precise.

That being said, putting the weird conventions of classical logic aside, most laws about value expressions actually tend to take the form of transformative implications, not meta implications. This makes perfect sense of course, when you consider what value expressions are supposed to represent. One does not generally consider a whole bunch of value expressions standing separate from each other, as if each one made a claim (they don't), but rather one tends to consider them in light of how they might fit together with other value expressions and statement expressions in which they are embedded. For that way of thinking to work properly, you generally need to be able to transform the value expressions in-place, which is something that transformative implication seems to be better at doing than meta implication.

Here's a very incomplete list of a few random value expression laws for unified logic, so that you can get a better sense for how value expression laws tend to feel and behave compared to statement expressions laws:

Value Expression Law Name	Rule
order independence of union	$A \cup B \leftrightarrow B \cup A$
order independence of intersection	$A \cap B \leftrightarrow B \cap A$
vacuous falsehood of condition	$\emptyset \rightarrow B \leftrightarrow \emptyset$
vacuous falsehood of consequence	$A \rightarrow \emptyset \leftrightarrow \emptyset$
union negation (De Morgan's laws)	$\neg(A \cup B) \leftrightarrow \neg A \cap \neg B$
intersection negation (De Morgan's laws)	$\neg(A \cap B) \leftrightarrow \neg A \cup \neg B$
dist. of sub impl. over conseq. intersec.[43]	$A \rightarrow (B \cap C) \leftrightarrow (A \rightarrow B) \cap (A \rightarrow C)$
identity element of union	$A \cup \emptyset \leftrightarrow A$
identity element of intersection	$A \cap \mathcal{U} \leftrightarrow A$

the law of identity (for sets)	$A \leftharpoonup A \leftrightarrow A$
the law of the excluded middle (for sets)	$A \cup \neg A \leftrightarrow \mathcal{U}$
the law of non-contradiction (for sets)	$A \cap \neg A \leftrightarrow \varnothing$

Notice that value expression laws need to conserve the *entire value* of the expressions they govern, under all possible inputs. Conserving only truth (i.e. in unified logic, the empty or non-empty status of the sets in the chain of reasoning) is insufficient here. This is an important pitfall to be aware of. You need to really think about what you are actually doing. You need to always be mindful of *what* you are trying to conserve when you work with transformative logic.

Many of these value expression laws above also have statement expression analogs. The fact that there are so many subtly different variations of many of these laws may make giving all of them precise and descriptive names a bit of a nuisance. Indeed, it wouldn't really surprise me if I've accidentally created some naming conflicts somewhere, as consequences of my choices of terminology in this book. However, that's just the nature of the beast. Progress is an iterative process.

In order to not slow progress down to a crawl, it is often necessary to let oneself be open to the possibility of sometimes making mistakes. All one can do is to try to make a "best estimate" of what the ideal might be. It would be too prohibitive to try to think of every possibility at once. That would not be realistic and would be a massive waste of time. It would cripple us with analysis paralysis.

People can collectively refine the ideas and terminology over time, whenever conflicts inevitably do appear. Natural selection always finds a way. Iterative improvement and evolution is the way of the world. It is the optimal workflow, if you want to maximize the amount of value you produce per unit time. Better initial choices do save a lot of time down the road, so it is indeed best to be careful, but there's only so much that one can predict within a reasonable span of time and under the pressure of resource constraints. Regardless, I think you get the point though.

Anyway, that concludes our current discussion of our small sample set of various different laws of unified logic. There are surely many more laws applicable to unified logic than just the ones we have covered above. There is plenty of room for exploration and discovery. I leave it to the interested reader to discover more of the laws on their own and to potentially publish those discoveries in their own work if they feel so inclined.

A comprehensive treatment of every law of all of the systems of reasoning I have devised in this book would be far beyond the scope of my plans for this book. It would cause huge delays in the book's release. Thus, it is better for me to limit my coverage of the details and to focus instead on expressing the core conceptual essence of this new avenue of thought as best I can. There's also no reason for me to hog all of the ideas for myself when there are so many other intelligent and highly capable people out there. Truth belongs to everyone.

Humanity's collective knowledge has a life of its own, like a living creature, like a hive mind. The more people have a chance to think about an idea, the more that idea has

---

[43] abbreviation for: "distribution of sub implication over consequence intersection"

a chance to grow into something truly great, something with impact. The economies of scale necessary for human progress can only occur when ideas are sufficiently widely available that many people are exposed to those ideas and are given abundant opportunities to contribute. The rate of human progress seems to be strongly proportional to how widely accessible and clearly expressed the corresponding ideas are. Public accessibility and ease of use are essential.

Anyway though, it is now time for us to transition from discussing the basics of primitive unified logic to discussing how unified logic relates to the other branches of logic, especially how unified logic manages to satisfy most of the core motivating principles and criteria of most of the major branches of non-classical logic simultaneously. After that, we will finally transition into the more advanced form of unified logic: relational unified logic. However, more on that later. For now, as I said, it is time to talk about how unified logic relates to the various other branches of logic.

5.6 Unified logic's relationship to the other branches of logic

5.6.1 Unified logic's relationship to set theory

Let's begin with the simplest and most obvious case. Specifically, let's take a moment to consider how unified logic relates to set theory. Set theory is a branch of logic just like any other (even though it is not always thought of as such by some people), and not only that but it is also a *non-classical* branch of logic. Non-classical logics are of interest to us primarily in that they seem to identify potential conceptual shortcomings in classical logic, i.e. desirable properties that one would like a logic to have, but which classical logic lacks. Each branch of logic represents a different philosophy on the nuances of how correct reasoning should work, or if not that then each branch at least serves a certain niche of subject matter with regards to a well-defined reasoning process.

Naturally then, the philosophies of the various branches of logic differ from each other in a multitude of ways. Unsurprisingly though, some branches are quite closely related to each other whereas others are far more distantly related. Some branches of logic are quite narrowly scoped, like isolated little islands, whereas others can cover quite a broad domain and can intersect with other logics' domains to a very great degree. Unified logic is among the broadest scoped of these logics perhaps. It can stretch its vines far and wide across the logical landscape, especially when combined with transformative logic. However, this does not imply that unified logic does not have close relatives. Indeed, among all of the branches that unified logic is related to, it is clearly most closely related to set theory.

This should come as no surprise of course. After all, if you think back to when we first defined unified logic, I did mention that unified logic is essentially just an extension, modification, and reinterpretation of set theory (see page 373). It is thus only natural that unified logic and set theory are extremely closely related. Indeed, primitive unified logic in particular is essentially just set theory extended with the unified implication operators and some related material, thereby giving set theory access to conditional operators in a way that it previously did not have. In contrast though, relational unified logic does not quite make the same foundational assumptions as traditional set theory does[44], and thus is a bit more distantly related, but we'll get to that later.

In any case though, the fact that unified logic is mostly just a generalized version of set theory clearly implies that unified logic will satisfy the same basic motivating principles and criteria as set theory does. The part of unified logic that unified logic shares with set theory is mostly left untouched, and the net result of this is that unified logic can generally express all the same things that set theory can, plus more. Any nuanced differences in the foundations that might otherwise exist would not be enough to disqualify unified logic from still satisfying the same underlying spirit and motivating principles as set theory.

Thus, as the connection between unified logic and set theory is so obvious, I will now consider the relationship between them to be adequately illuminated, and hence will now move on to discussing unified logic's relationship to the other branches of logic. You may refer back to the section on set theory on page 123 if you would like to review set theory in greater depth, but I don't think it is likely to be necessary, so it probably wouldn't be a very good use of your time.

Oh, I will however mention one other piece of low hanging fruit though, which is related. Specifically, I'd like to point out that since syllogistic logic is just the ancient (weaker) ancestor of set theory, and we have already deduced that unified logic satisfies the basic principles of set theory, then it also follows that unified logic likewise satisfies the basic principles of syllogistic logic, just as set theory does. Thus, we have now finished discussing how unified logic relates to both set theory and syllogistic logic. These are the most closely related branches of logic to unified logic, so it is not at all surprising that these relationships are so easy to see. That's two branches down, but a whole bunch more to go.

Of the remaining branches, one of them will not really be covered much at all. Its potential relationships to unified logic will mostly be omitted. Can you guess which branch? It's not too difficult to guess. Here's a hint though: Which branch among all those we discussed is probably the most difficult to analyze formally? The answer is informal logic. The difficulty of approaching informal logic from a formal standpoint is even evident in its name: "*informal* logic", a phrase which literally means "*not formalized* logic".

5.6.2 Unified logic's relationship to informal logic

Informal logic is too vague and diverse for its relationship to unified logic to be easily analyzed. No doubt such a relationship exists in some form, but I have decided that attempting to address it further is not worth our time right now. My attempts to discover clear and unambiguous connections between unified logic and informal logic were not very fruitful. I do think there are connections there, and that its maybe worth doing some

---

[44] Indeed, relational unified logic actually makes some fairly radical changes to what it interprets a set as being, relative to traditional set theory. Much of the same spirit and behavior is there, but nonetheless relational unified logic definitely doesn't rest on the same foundation as traditional set theory.

day, whether by me or someone else, but for the time being it doesn't seem important enough to bother sinking much more time into it.

On the other hand though, unified logic's relationship to natural language (which some might consider "informal" in a certain sense) is quite clearly significant. Think back to our discussion of unified implication and its relationship to many common words and structures that are employed in natural language in great abundance (e.g. "the", "e.g.", "a.k.a.", etc). See the section on page 386 if you want to refresh your memory on the striking relationship between unified logic and natural language. However, unified logic's relationship to natural language won't really be important for our upcoming discussion of the other branches of logic, so don't worry about it too much. You don't need to remember it to understand the stuff we are about to talk about.

Anyway though, with that out of the way, we now have set theory, syllogistic logic, and informal logic done. That's three branches down. We have also reminded ourselves that unified logic has a special relationship to natural language, in that it embodies a special kind of "linguistic logic" or "natural logic" in a certain sense, if you will. However, I'm not sure if I should count this relationship of unified logic to natural language as being indicative of the existence of another "branch", seeing as there apparently isn't really an existing branch of non-classical logic for "the ability to express natural language directly" as far as I'm aware, and also seeing as unified logic can only express a subset of natural language, and probably not all of it.

5.6.3 Unified logic's relationship to predicate logic

Putting all that aside though, I'd like to now start going through the other branches of logic, roughly in the same order as which they occurred in the chapter on non-classical logic[45]. As such, let's start with predicate logic now. To refresh your memory, predicate logic is primarily about two things: (1) predicates and (2) quantifiers.

Predicates are functions that return truth values, which are often used to test input objects for whether or not some property embodied by the predicate applies to them. Quantifiers express "how much" of a set a predicate applies to in a statement. The two most common quantifiers are existential quantification (a.k.a. "for some" or "there exists") and universal quantification (a.k.a. "for all"). Both predicates and quantifiers are necessary in order for predicate logic to work. If either didn't exist then predicate logic wouldn't be possible. You can't quantify[46] an expression that doesn't contain any predicates.

A quantified statement could be a value expression (in which case it would return a truth value and would not actually assert a claim), or alternatively it could be an assertion of a fact (i.e. an axiom, i.e. an assumption) to be treated as a given and used for future reasoning. Like classical logic, predicate logic shares this sloppy convention of using the same expression to represent either a value expression or a statement expression depending on the implicit context.

---

[45]The chapter on non-classical logic began on page 291. However, if you want to refresh your memory of what order we covered the various branches of non-classical logic in, then looking at the table of contents (pages 3 to 8) would be easiest.

[46]"quantify" in the sense of predicate logic, not in the more general sense of "quantify" as "formalizing something or translating it into terms of numbers and/or the language of mathematics or science"

Indeed, predicate logic is most often used as an extension of classical logic, and so it often uses classical truth values. However, other variations of predicate logic besides the classical case are possible as well, such as fuzzy predicate logic for example, where the predicates return fuzzy truth values (typically represented as continuum numbers between 0 and 1) instead of classical truth values.

Anyway though, the section on predicate logic is located on page 301. You can look at that section if you'd like a more complete overview of what predicate logic is. However, reviewing that section would probably be overkill and is not really necessary here. I'll give a few examples of quantified expressions here to refresh you memory, and that will probably be sufficient. Let's do one existentially quantified expression and one universally quantified expression. Here we go:

1. Suppose we wanted to express "There is at least one book that is available in English, Japanese, and German.". Well, suppose that B is the set of all books, that $E(x)$ is a predicate for whether x is available in English, that $J(x)$ is a predicate for whether x is available in Japanese, and that $G(x)$ is a predicate for whether x is available in German. Let x represent a generic object. Then this claim may be represented in predicate logic by the following expression:

$$\exists (x \in B)(E(x) \land J(x) \land G(x))$$

2. Suppose we wanted to express "All cats are either black or not black.". Well, suppose that C is the set of all cats and that $B(x)$ is a predicate for whether x is black. Let x represent a generic object. Then this claim may be represented in predicate logic by the following expression:

$$\forall (x \in C)(B(x) \lor \neg B(x))$$

However, remember that the way these expressions have been written above are not the only ways of writing these expressions. For example, the predicate function notation is not always necessary. The predicate functions can often be rephrased in terms of membership in sets instead. Here's what it could look like if we did that:

1. $\exists (x \in B)((x \in E) \land (x \in J) \land (x \in G))$

2. $\forall (x \in C)((x \in B) \lor (x \in \neg B))$

Notice that this time E, J, and G (in the first example) and B (in the second example) are sets instead of predicate functions. The sets are conceptually collections of objects, whereas the predicate functions are conceptually functions that return truth values, but the net effect of the overall quantified expressions here is nonetheless exactly the same as before.

The reason I'm pointing this out is because I want you to realize how closely related predicates are to sets. The closeness of this relationship is critically important to understanding how unified logic and predicate logic are related to each other. Indeed, many predicates can be rephrased entirely in terms of set expressions. This is the key factor that enables you to build a bridge between unified logic and predicate logic, such that unified logic ends up covering a lot of the same territory as predicate logic.

You see, the existential quantifier and the universal quantifier each correspond closely to a similar concept from unified logic. Due to the ambiguity in how existential and universal quantifiers are interpreted in predicate logic though (i.e. as either value expressions or statement expressions depending on the context), each quantifier actually corresponds to *two* different things in unified logic. However, the conceptual essence of each of those pairs of things is still much the same. Can you think of what the correspondences might be? Well, as a huge hint, here's what the above examples of quantified expressions could look like when translated into unified logic:

As Value Expressions:

1. $B \cap E \cap J \cap G$

2. $C \rightharpoonup (B \cup \neg B)$

As Statement Expressions:

1. $B \cap E \cap J \cap G \neq \varnothing$

2. $C \rightharpoonup (B \cup \neg B) \neq \varnothing$

Of course, there are other ways to write these expressions. You could change the order of the operands in either expression whenever the result would still be equivalent. You could also use any of the other alternative notation systems that I've defined in this book. For example, you could write $\exists (B \cap E \cap J \cap G)$[47] or $B \sqcap E \sqcap J \sqcap G$ instead of $B \cap E \cap J \cap G \neq \varnothing$ and it would mean the same thing. Similarly, you could write $\exists (C \rightharpoonup (B \cup \neg B))$ or $C \sqsubseteq (B \sqcup \neg B)$ (or even mixed variations such as $C \sqsubseteq (B \cup \neg B)$) instead of $C \rightharpoonup (B \cup \neg B) \neq \varnothing$ and it would likewise still mean the same thing.

As you can see though, existential quantification corresponds closely to intersection or existential intersection, and universal quantification corresponds closely to unified implication or existential subset relations. If you really think about it, these operators from unified logic express the same fundamental concepts as the quantifiers of predicate logic do, but do so in a different form.

Existentially quantified expressions are all about testing or asserting that at least one object exists in one or more specified sets, but if you think about it this is really the same thing as testing or asserting an intersection of sets. Similarly, universally quantified expressions are all about testing or asserting that all of some set of objects also belong in some other set, but if you think about it this is really the same thing as testing or asserting a subset relation among sets. Thus, we have this close relationship between unified logic and predicate logic.

Indeed, I wonder if predicate logic's existence is itself in some sense evidence that a set-based way of thinking of things may be more fundamental than a classical way of thinking of things. Think about it. Classical logic, when restricted to the use of only atomic propositions and not allowed to use quantifiers, is quite limited. In fact, classical logic without quantifiers is quite poor in its expressive capabilities, and users of it very quickly realized this and were forced by necessity to extend it with predicate logic in order to make it more useful.

However, we have seen from this discussion that predicate logic itself is closely related to set theory. Indeed, predicate logic can sort of be thought of as a way of forcing

---

[47] Notice that $\exists$ is the "exis operator" here, not the existential quantifier, even though the symbol looks identical.

classical logic to be able to talk about sets. Why does that need to happen though? If classical logic were more fundamental as a basis for logic than set theory, then why does classical logic need to be extended with the power of predicate logic, lest it be immediately crippled and rendered incapable of expressing more advanced concepts?

That bodes badly for the idea that classical logic could ever be the fundamental essence of logic. In a sense then, predicate logic is kind of like a poor man's set theory, at least in part if not in whole. The need for an ability to express set-like concepts arises so quickly when working with logic that the absence of this ability in classical logic causes classical logic to almost immediately manifest a need to be extended with predicate logic, almost as if compelled by a symptom of a disease, almost as if classical logic has a fever and can only get relief from that fever by extending itself with predicate logic, i.e. by making itself a bit more like set theory via enabling itself to reason about collections of objects in at least some way.

Classical logic seems incapable of resisting the pull of set theory in light of this, seeing as classical logic so desperately requires the assistance of predicate logic in order to become more fully expressive. It feels almost as if nature itself is trying to guide our hands as logicians closer towards the truth, as if pulled in by an inescapable gravity towards something more like set theory as the proper foundation.

However, there is one important catch here, apparently. Specifically, there are many things that require the ability to use dummy variables as inputs so that you can more precisely specify the nature of the relationships you wish to talk about. Consider, for example, a more complicated quantified expression that includes multiple input variables in some of the predicates. How would we represent such an expression entirely in terms of a set expression, in the most general possible case? How do we handle complicated predicates involving multiple input variables, where some of the variables are cross-referenced in multiple different parts of the quantified expression and stand in some specific relation to each other? Can pure set expressions capture such scenarios in all possible cases or not?

There may actually be a way, in some such cases, to capture these nuances using some of the concepts from relational unified logic, but we haven't got to that part yet. For the time being though, my point is that not all predicate logic expressions can necessarily be easily translated into terms of simple set expressions. Indeed, even with what relational unified logic will give us, we may still yet have need for the kind of variable usage that predicate logic employs under certain scenarios. However, that is a subject for another day. It is outside of the scope of what I plan on considering in this book. It is maybe worth while to investigate in greater depth at a later date however, in principle.

The core point here though, of talking about these close relationships between unified logic and predicate logic, was to demonstrate to you that unified logic successfully covers a large chunk of the territory of predicate logic. There are many predicate logic expressions that can be easily expressed entirely in the language of unified logic. Moreover, even regardless of the cases where the translation is not apparent, it is clear that existential quantification is very closely related to intersection and that universal quantification is very closely related to subset relations.

Unified logic's vocabulary clearly overlaps with the underlying spirit and motivating principles of predicate logic, even if perhaps only partially. That happens to be enough for our purposes here. Thus, we will now consider the relationship between unified logic and predicate logic to have been adequately expressed (for the time being), and will now move on to the other branches.

5.6.4 Unified logic's relationship to multi-valued logic

Next up is multi-valued logic. As such, let's now take a moment to refresh our memory of what multi-valued logic actually is. There are multiple forms of it. It's not just one system of logic. The section on multi-valued logic began on page 313. You can reread that section if you'd like a more in-depth review of multi-valued logic, but it probably won't be necessary. The coverage I will give here to refresh your memory will likely be sufficient.

Simply put, multi-valued logic is any logic which has more truth values than classical logic has, i.e. which has more than just true or false. The most rudimentary multi-valued logics are those which introduce a third value to stand for "unknown" in some sense. Such logics were among the first multi-valued logics to be conceived. The idea they employ, i.e. that perhaps there should be a third truth value to represent unknowns, is a relatively obvious one. It is an idea that could easily occur to pretty much anyone who is trying to conceive of their own alternative to classical logic.

Introducing new truth values is indeed one of the most obvious moves that a person could think to make in order to create their own system of reasoning. However, unfortunately, these kinds of systems are often quite limited and don't actually tend to work very well. Systems of 3-valued logic tend to suffer from many conceptual holes and problems. Even classical logic is often a better choice. The "unknown" truth value in these 3-valued logics is pretty much doomed to not handle all of the possible cases and contexts correctly.

No truth-functional logic can ever be truly context-sensitive[48]. Only systems of logic based on deeper principles, such as derivations of arbitrary forms (e.g. transformative logic) or sets (e.g. unified logic) seem to be able to do that. Not only that, but there is also lots of ambiguity in what the third truth value (the "unknown" value) actually should mean and how it should behave. Just because the concept of an "unknown" seems intuitively obvious to us as humans doesn't mean it is actually that easy to formally model in a way that can handle all of the relevant use cases.

Intuition can be deceptive. Just because we can form a thought in our mind doesn't make that thought actually coherent. Just because something "makes sense" in our mind doesn't mean that it actually makes sense in reality. You have to test your assumptions and fill in all the details. Only then can you truly determine whether an idea is genuinely coherent or not. Only then can you truly be sure that you understand how it will actually

---

[48]The reason is because distilling everything down to just truth table lookups inherently requires that you ignore all of the other contextual details during expression evaluation. If not, then you wouldn't be able to safely use the truth table as a lookup mechanism, because you might then miss contextual info that could change the answer. Thus, the trade-off is unavoidable. You can have easy truth table lookups, or you can have context sensitivity, but you can never have both. However, be aware that truth-functional content wrapped up inside a more general-purpose computational system, such as a computer, actually *can* be context-sensitive, but only by virtue of the extra mechanisms such a system may provide.

behave in reality. Only then will you be able to test whether or not it contains any conceptual holes.

Indeed, many different philosophies on how this third truth value representing some kind of "unknown" should work have been proposed over the years, but all of them have pros and cons and consistently fail to be comprehensive. Creating a perfect 3-valued logic is essentially impossible. Among the various 3-valued logics that introduce a third truth value for "unknown" that we have discussed in this book, we specifically covered Kleene's 3-valued logic, Bochvar's 3-valued logic, and Łukasiewicz's 3-valued logic[49].

These are three of the most famous 3-valued logics by far, and it would be fairly easy for someone to independently reinvent them by accident. I also briefly mentioned the existence of a 4-valued logic called Belnap's logic, but I did not discuss it in detail. Indeed, 4-valued logic is not the limit either. One may choose to add on as many additional truth values to a system of multi-valued logic as one wishes to. As long as the number of additional truth values is finite, then we say that any such system is a *finitely* multi-valued logic.

However, finitely multi-valued logics are certainly not the limit of what multi-valued logic is capable of. Besides the finitely multi-valued logics, there are also the *infinitely* multi-valued logics. Among the most prominent of such infinitely multi-valued logics are two especially characteristic systems: fuzzy logic and probabilistic logic. We talked about fuzzy logic quite extensively in the section on multi-valued logic, beginning on page 319 at the paragraph that starts with "Let's talk about fuzzy logic first.".

Fuzzy logic is quite a useful logic, when used in the right context, so our lengthy discussion of it was ultimately worth it, even despite how tangential it was to the primary thrust of the book. Probabilistic logic is also useful of course, as anyone who has ever worked much with probability theory and statistics can attest to, but we only ever discussed it briefly, and certainly not with sufficient detail to actually put it into practice.

Fuzzy logic and probabilistic logic both like to use continuum numbers between 0 and 1 as their truth values, generally. However, this is certainly not the only way of creating truth values for a multi-valued logic. Multi-valued logic requires only that there are more truth values than just the conventional two classical truth values true and false. Indeed, when you consider set theory and unified logic in light of an existence-based perspective on truth, it becomes clear that set theory and unified logic are in fact *also* infinitely multi-valued logics. As such, since unified logic is already very clearly a multi-valued logic on these grounds, it of course therefore automatically fulfills the corresponding underlying principles of multi-valued logic.

We could thus declare our discussion of the relationship between unified logic and multi-valued logic to be concluded, seeing as we have now satisfied the primary goal already. However, we will nonetheless continue discussing it for just a bit longer. I still have some things I'd like to say to further illustrate the relationship. Besides the usual sense in which we treat sets in unified logic as truth values, there is also at least one other perspective from which unified logic can act as a multi-valued logic, but in a somewhat different sense. It is worth considering. As always, we want to broaden our

---

[49] The corresponding truth tables for these logics are located on pages 314 (Kleene), 316 (Bochvar), and 317 (Łukasiewicz).

understanding of the nuances of the fundamental concepts in play as much as possible, so that we can truly understand and utilize the full potential of the system.

First, let's discuss the more conventional sense of unified logic being a multi-valued logic, so that we can get a better grasp of what this actually means in practice. For example, what exactly constitutes a unique truth value for unified logic when we interpret it as a multi-valued logic? Well, it actually turns out that every unique set in unified logic is also a unique truth value. This makes sense if you consider what it would mean for something to be a unique truth value. A unique truth value would be a truth value that behaves differently from at least one other truth value in at least one context. A unique set is always guaranteed to satisfy this criteria.

If a set is unique then we can be sure that there is at least one context where its behavior in a logical operation could differ from some other set's behavior, because as a unique set we know that it must differ by at least one element from all others sets. As such, if we had three sets X, A, and B, and we knew that A and B were distinct, and we evaluated $X \cap A$ and $X \cap B$, then we would know that the results of $X \cap A$ and $X \cap B$ could *potentially* differ depending on the values of X, A, and B. In general though, there must be *some* conceivable operation where A or B would cause a different result to be returned when substituted into a certain spot in an expression.

Furthermore, there are of course an infinite number of conceivable sets, each of which can be interpreted as truth values in light of this, and hence unified logic must surely be an infinitely multi-valued logic, just as previously mentioned. Each truth value (i.e. each set) behaves differently according to its contents, which could contain any arbitrary collection of objects. Thus, unified logic is really quite a flexible multi-valued logic and indeed is capable of being used in lots of different completely logically distinct contexts.

This way that sets react differently depending on what other sets they are combined with in an expression is what gives unified logic its ability to be context-sensitive. Contrast this to fuzzy logic though, for example, which is infinitely multi-valued but *not* context-sensitive. That being said, it is important to not misunderstand what "context-sensitive" really means for a logic. Classical logic and fuzzy logic are both not context-sensitive, but both can still react to the conditions of their environment when evaluated (for example, especially inside a computer program).

One might call that "context sensitivity" in a certain sense, but that is not the sense we are using here. The sense we are using here is that of allowing the outcome of an operation to change based on the structure and content of its own operands' logical relations to each other, rather than merely evaluating everything blindly in a way that steamrolls over how the operands might relate to each other or how the logical expression being evaluated was framed.

Anyway, that covers the conventional sense of unified logic as a multi-valued logic. Next, I'd like to discuss a different sense in which unified logic can be used as a multi-valued logic, one in which you interpret the contents of sets in a different way yet still perform all the operations the same way as before. In particular, unified logic can also be used as a kind of "logic of partial unknowns" by which you can combine different pieces of incomplete information in order to form a more precise understanding of the range of some unknown value.

How does it work? Well, the key is to change how you interpret the sets. Specifically, instead of thinking of each set as a collection of a bunch of objects that you already have, you can think of each set as representing uncertainty as to what some corresponding value is, such that each of the members of a set are *possibilities* for what the corresponding real value *might* be. Suppose, for example, that there is some unknown value v that we want more precise information about. Suppose furthermore that we learn somehow from one source that v could be either a, b, or c. Let's represent this information as the set $A = \{a, b, c\}$. Likewise, suppose some other source tells us that v could be either b, c, or d. Let's represent this information as the set $B = \{b, c, d\}$.

How might we make use of this information? The answer is pretty straightforward really. We simply combine the constraints of A and B in order to gain more precise information about v. Specifically, we know that v must be a member of both A and B, and thus we may conclude that v must be a member of $A \cap B$. A little bit of evaluation easily shows that $A \cap B = \{b, c\}$, and hence that we now know that v must be either b or c. This reveals more information than what we directly started with.

This is the kind of thing I was talking about when I said that unified logic could be used as a kind of "logic of partial unknowns". There are of course other operators that you can use like this besides just set intersection ($\cap$), each with its own distinct interpretations in terms of this "logic of partial unknowns", but that's outside the scope of what I wanted to talk about here. The simple demonstration above is sufficient. I just wanted you to be aware that this way of thinking about set expressions existed, so that it would get the gears in your head turning a bit more. The more we understand the nuances of what a system can express the better. Anyway, with that said, we now conclude our discussion of the connection between unified logic and multi-valued logic.

5.6.5 Unified logic's relationship to relevance logic

Next, I'd like to talk about relevance logic. Our original discussion of relevance logic began on page 349. Refer to that section if you would like a more extensive review of what relevance logic is than what I will say here. However, in the case of relevance logic, even our original discussion of it was not very detailed and lasted only a few pages, owing to the fact that understanding the details of how to actually perform traditional relevance logic was unimportant and out of scope for our purposes in this book. I mentioned that unified logic would be a relevance logic, and I also implied that it would probably be easier to understand than whatever the traditional systems of relevance logic are. Therefore, mostly in this section we will just be talking about how unified logic is a relevance logic and what that means, more so than comparing it to any of the traditional relevance logic systems.

First though, before we talk about that, let's briefly refresh our memory as to what relevance logic is. The core motivating principle of relevance logic is that a condition ought to be genuinely related to its consequence in an implication in order for that implication to have any chance of being true. After all, if you think about it, how could a condition being true imply that the consequence is true if the consequence is not in fact related to the condition in some way. From a principled standpoint, it seems like an implication should require that there is something about the nature of the condition being true which inherently implies that the consequence must also be true.

Suppose, for example, that A represented the statement "Semiconductors are the basis for computer technology." and that B represented the statement "The distribution of wealth among participants in a market very often resembles an exponential curve."[50]. Both of these statements are true in real life. Does A imply B? A classical logician would say yes. Does that actually make much sense though? In classical logic, any implication whose condition is false or whose consequence is true will always be true. Thus, as a subset of such cases, any implication with both a true condition and a true consequence will always be considered true in classical logic, regardless of content. This example with A and B is just one such case.

However, it is clear that A and B have no real logical connection to each other. At best, they are extremely distantly related (e.g. computer tech and exponential distributions are both related to mathematics and both have a big impact on the economy, etc), but regardless even of whether we admit such distant relationships into consideration, there is clearly not a causal relationship between A and B. The truth of A does not cause B to be true. The two statements A and B are *irrelevant* to each other, or at least overwhelmingly irrelevant if not entirely so. One might then be inspired to ask how one might avoid this anomalous behavior.

To do so requires the creation of a system of relevance logic. The motivating principle here is very easy to understand. The tricky part though, is to define what exactly relevance actually is. Defining this in the most general possible sense would be quite a hefty task. Indeed, we will not attempt to do so here. Instead, we will simply focus our energies on understanding in what sense unified logic is a relevance logic and how we can frame at least some statements in a relevance-preserving way within its framework of reasoning.

The key enabling factor in what makes unified logic a relevance logic is the fact that unified logic does everything in terms of either (1) direct derivations of forms from arbitrary foundational assumptions via transformative logic or (2) reasoning about relationships among sets. As for derivations of forms via transformative logic, those are simply governed by whatever arbitrary rules apply to the specific logical world you are working with. Such rules are admittedly important, but are also not necessarily applicable outside of their own narrow contexts.

In contrast, relationships among sets are typically more universal and hence more reliable. As such, I'd like to focus more on how the set-based mechanisms of unified logic are capable of expressing relevance here. The case involving transformation rules

---

[50]The reason for this is perhaps mostly self-reinforcing feedback effects and viral growth. The wealthier someone is the easier it is for them to acquire even more wealth, since they have more resources to invest and to profit from, and more free time as well. Also, people overwhelmingly tend to search for the "best" or "most popular" instance of any item whenever they shop. People seldom intentionally search for poor, mediocre, or average products. Likewise, in dating, most people invest most of their attention on the most attractive people, regardless of their own assets. Also, in business, if a product suddenly becomes successful then it will often be in part because of viral growth, which inherently is an exponential form of growth. For example, imagine every 1 new customer eventually tells 2 more people about a product... That's pretty much an instant recipe for explosive exponential growth. Either way though, whether through existing wealth or new wealth, markets tend to trend roughly towards exponential distributions or other similar distributions. Power laws are very common, both in nature and in human society.

being applied to arbitrary forms is kind of a different beast entirely. Transformations rules express the fundamental properties of the system being studied, and hence transformation rules are automatically relevant to each other in some sense. Thus, there also isn't as much to say about them here, which is another reason why we'll be focusing on the relevance properties of set-based reasoning instead.

Let's begin. Take a moment to contemplate the nature of set-based reasoning. In what way might it be capable of preserving relevance? What does it have that classical logic doesn't that might enable it to do so? Think about what relevance really requires. It requires that there be some sort of genuine connection between the condition and consequence of some implication. Classical logic doesn't have this. Classical logic doesn't care about the contents of the participants in an implication.

What about set theory and unified logic though? Do set operations care about the contents of their operands? The answer, of course, is yes. In fact, not only do set operations care about the contents of their operands, but one could argue that for set operations the contents of their operands are *everything*. The outcome of any given set operation is determined entirely by how the contents of each participating set relate to each other set in the given logical expression. Thus, a set expression's result cannot return an non-empty set (a.k.a. any true value[51]) unless some kind of genuine relationship exists between the participating sets in some sense. Thus, unified logic always guarantees at least some kind of connection between every participant in a set expression, and hence always guarantees at least some degree of relevance.

Do you remember what the unified logic equivalent of implication was? It was unified implication and existential subset relations. Under unified logic, unified implications are the value expression version of implication and existential subset relations are the statement expression version of implication. In both cases, the implication can only be true if the condition is a subset of the consequence. This does indeed guarantee at least some kind of relevance.

Suppose, for example, that A is the set of all oak trees and B is the set of all photosynthetic organisms[52]. Given these variable assignments, would the expression $A \rightarrow B$ be non-empty (i.e. true)? Yep, it sure would. Oak trees are plants and all plants are photosynthetic organisms, so it would indeed return a non-empty set. Specifically, in this case, it would return A, i.e. it would return the set of all oak trees. Similarly, it would be fair to assert that $A \sqsubseteq B$ is true, given that it indeed is.

Notice that A and B are genuinely related to each other here, and hence genuinely relevant. This is not merely some superficial truth table evaluation like you'd see in classical logic. On the contrary, it is an evaluation of whether a *real* relationship holds true between two different things, even if only that of a subset relation. Admittedly, sometimes subset relations may be arbitrary and contrived. However, the wonderful thing about subset relations is that in most real-world use cases a subset relation will tend to imply that the subset shares certain special properties in common with the superset.

---

[51] Remember that under an existence-based truth perspective all non-empty sets are considered true and all empty sets are considered false. Remember also that unified logic adheres to this philosophy.

[52] Photosynthetic organisms are any organisms that get their energy from processing sunlight to synthesize the chemicals they need. Plants are the most obvious and widely known example, but algae and cyanobacteria also perform photosynthesis.

Our example here with *A* as the set of all oak trees and *B* as the set of all photosynthetic organisms is one such example. The set *A* shares special properties in common with the set *B*, in the sense that if we are given any member of the set *A* then we may automatically conclude that it also must have the properties of a member of the set *B*. Indeed, as long as the sets you use each require all their members to meet specific properties, rather than just being composed of a random hodgepodge of unrelated objects, then this kind of reasoning can be exploited and will work. It is in this way that you will be able to apply implications in a truly relevant way via set-based reasoning.

One potential catch here though is that sets in unival form[53] won't say much, and thus independent statements expressed in terms of sets may be difficult to work with in this way. Membership in the universal set does not necessarily imply any special property other than mere existence. This is unsurprising though, considering that the universal set is simply the set of all things that exist in a given logical universe.

As such, in order to exploit the relevance properties of unified logic's set-based operations, you will typically need to find a way to frame the problem in terms of non-unival forms instead of in terms of unival forms. This may sometimes be difficult, or perhaps even outright impossible in some cases. The use of transformative logic (or else some similarly arbitrary and flexible system) may be the only recourse in such cases. In any case though, with that said, I've now covered everything I wanted to say about the connection between unified logic and relevance logic. As such, it is now time to proceed to the analysis of the next branch of logic.

5.6.6 Unified logic's relationship to modal logic

This time, let's talk about modal logic. This'll be an easy one. The relationships between unified logic and modal logic will actually be pretty structurally similar to the relationships between unified logic and predicate logic. You'll see what I mean later. Obviously, the connection between unified logic and modal logic will be different in certain respects. However, predicate logic and modal logic actually do have a rather close relationship, one that you can readily observe if you look at them with the proper perspective. The difference won't be as large as you might expect at first glance.

Before we talk more about that though, we should take some time to briefly review what modal logic is, so that we can refresh our memory. A fresh memory is better able to see connections between concepts than a stale one. As such, let's begin. If you would like to review the original section on modal logic, feel free to refer to page 352. Otherwise, if a more brief review seems adequate enough to you, then simply continue reading here.

First, it is important to remember that modal logic is not a single system of logic, but rather it is a family of logics with certain similar traits in common. However, be that as it may, one specific system of modal logic is overwhelmingly the most popular and the most common in the literature. Often it is actually this specific system that people are referring to when they say "modal logic" without qualification. This type of modal logic (the most widespread) is known as alethic logic. Some other types of modal logic

---

[53] See page 444 if you wish to review the definition of unival form.

include temporal logic, deontic logic, epistemic logic, and doxastic logic, but this is not an exhaustive list apparently.

All modal logics have a special set of qualifiers (a.k.a. "modes"[54]) which they may apply to statements. The study of how these qualifiers behave when attached to statements is the primary concern of all modal logics. Alethic logic in particular studies necessity, possibility, contingency, and impossibility as its modes. However, alethic logic typically expresses contingency and impossibility indirectly, by writing them in their equivalent forms in terms of necessity, possibility, and logical negation.

These modal qualifiers are represented as unary operators that are attached to logical expressions. Specifically, in the case of alethic logic, necessity is represented as $\Box$ and possibility is represented as $\Diamond$. The operators from classical logic are also used. For example, $\neg$ is used as the symbol for logical negation. These various operators from classical logic usually behave the same way in modal logic as they do in classical logic. Alethic logic is usually interpreted in a classical light.

If, for example, we wanted to say that $A \vee \neg A$ is necessarily true and B is merely possibly true, then we could write $\Box(A \vee \neg A) \wedge \Diamond B$. As with classical logic, this could be interpreted as either a value expression or a statement expression depending on the context. It could just be an expression we want to evaluate to determine whether it is true or false, or it could be a claim that in fact it is actually known to be true and hence should be assumed as such for the purposes of inferring additional claims.

These modal qualifiers are often interpreted in light of the concept of "possible worlds", which is to say that necessity and possibility are thought about in terms of whether the statement they are attached to is true or not in some set of hypothetical alternative circumstances under consideration. Thus, for example, necessity is taken as meaning that something is true in *all* possible worlds, whereas possibility is taken as meaning that something is true in only *some* possible worlds. Notice the parallels here between universal quantification and necessity and between existential quantification and possibility.

Part of the whole premise and way of thinking underlying modal logic is that these special qualifiers attached to logical expressions are somehow special logical objects in their own right. There's this idea that each qualifier has its own unique properties that require a corresponding unique treatment and study. However, in reality, as much as modal logic tries to make it look like these special qualifiers are unique concepts, they actually probably aren't. In fact, when properly understood, modal qualifiers aren't even real unary operators. The "possible worlds" mechanic actually is essentially a hidden implicit operand and obscures the true nature of what the modal qualifiers actually are. The unary representation of modal operators is actually a lie, contrary to what much of the literature of modal logic may lead you to believe.

Indeed, once more properly understood, all of the modal qualifiers can be refactored to eradicate the superficial distinctions between them and thereby to reframe the corresponding concepts on more solid ground, in a more unified, streamlined, and easier to reason about way. This brings us to the connection between unified logic and modal logic.

---

[54]This is where the name "modal" is derived from, by the way.

Can you guess what the connection is? Hints have been dropped in a few places in the book, which you may have noticed if you were paying close attention. In any case though, here's the answer: Modal necessity is unified implication or existential subset relations, and modal possibility is intersection or existential intersection. Not only that, but even outside of alethic logic, all of the modal qualifiers are roughly equivalent to either (1) unified implication or existential subset relations or (2) intersection or existential intersection. All modal qualifiers are in a sense actually set theory or unified logic in disguise. Modal qualifiers are just special cases of set operators, which have been confused in such a way as to create a bunch of false distinctions.

Once we unify all of this by reframing it from the perspective of set theory and unified logic, all of the modal logics seem to become instantly irrelevant and obsolete. There seems to be nothing that modal logic can express that set theory and unified logic can't express in a more flexible, generalized, and conceptually correct way. This is at least my impression at this point, in any case. I fairly strongly suspect it's true. Further discussion of this idea, with the aid of some more concrete examples, will likely make my point here much clearer.

Consider, for example, how the concept of necessity and possible worlds might be reframed in terms of sets. Suppose, for instance, that we wanted to say that all ferns are necessarily green. Forget about whether this is actually true for a moment, and just assume that it is. Let's just assume that all ferns are indeed green, for argument sake. How would we say this in modal logic? Well, if we let G represent just the statement "Ferns are green." then we could say that ferns are *necessarily* green in modal logic by writing $\Box G$ and indeed that would be a fair representation.

How could we reframe this as a statement about sets though? Well, suppose that we let G instead stand for the set of all possible worlds where ferns are green and that we let P stand for the set of all possible worlds. Then, $\Box G$ can be translated into terms of sets as either $P \rightharpoonup G$ or $P \sqsubseteq G$, depending on whether we want to evaluate it (as in $P \rightharpoonup G$) or assert it (as in $P \sqsubseteq G$). Think about what these expressions say. These expressions essentially attempt to claim that all possible worlds are a subset of all possible worlds where ferns are green. If true, then that would imply that all possible worlds have green ferns, which is indeed the exact same thing as saying that ferns are necessarily green.

Similarly, suppose we wanted to say that it is possible to create an air-cooled microchip that is twice as fast as any currently existing air-cooled microchip. Let us represent the statement that such a microchip exists as M. To express that it is possible for such a microchip to exist then, we could write $\Diamond M$ in modal logic. This much is obvious at least. How could we express this in terms of sets though?

Well, the answer is still very simple. Do you see what it is? All we need to do is to reframe the possibility in terms of intersections of sets. As such, suppose that instead of being a statement M was actually the set of all possible worlds where microchips twice as fast as the current ones are possible to create. Suppose furthermore that P is the set of all possible worlds, regardless of other criteria, as with the previous example. Then, $\Diamond M$ can be translated into terms of sets as either $P \cap M$ or $P \Diamond M$, depending on whether we want to evaluate it (as in $P \cap M$) or assert it (as in $P \Diamond M$). Notice that these expressions can only be true if indeed there is at least one possible world where these microchips do exist.

Does one of these symbols look familiar to you? Did you notice how the modal possibility operator and the unified logic possibility operator (a.k.a. the existential intersection operator) use the same symbol and even share a synonymous name? This is no accident. I made this decision intentionally in anticipation of this moment, in light of the essence of what existential intersections fundamentally represent.

You see, possibility is actually an at least binary operator, meaning it takes at least two operands to be well-defined. The same is true of necessity. Indeed, unified logic's intersection ($\cap$) and possibility ($\Diamond$)[55] operators seem to me to be the real form of the concept of possibility. Similarly, unified logic's unified implication ($\mapsto$ and $\rightharpoonup$, etc) and existential subset relations ($\sqsubseteq$) seem to me to be the real form of the concept of necessity. They seem to be the hidden fundamental essence of what underlies the whole "possible worlds" and unary modal qualifiers facade of modal logic.

This isn't merely a matter of being synonymous though. It's not just that this is some alternative way of expressing the same thing. These set-based operators are actually fundamentally more general than their modal equivalents. The alethic modal operators essentially just hard-code one of the operands to always be the set of all possible worlds, but notice there is no reason to think that you have to always do it that way in set theory and unified logic. Indeed, you can choose whatever arbitrary sets you want and they will always be permissible as operands to these set operators.

This is also why I suspect these correlations probably make *all* of the modal logics obsolete, rather than only alethic logic. All the modal logics really seem to do is to hard-code one of the set operands to always be a specific type of set, and that is where the various types of modal logic actually come from. Specifically: Alethic logic hard-codes one operand to always be a set of possible worlds. Temporal logic hard-codes one operand to always be a set related to the past, present, or future. Deontic logic hard-codes one operand to always be a set of social rules of some kind (e.g. what is permissible or forbidden, etc). Epistemic logic hard-codes one operand to always be a set related to knowledge. Finally, doxastic logic hard-codes one operand to always be a set related to beliefs. Admittedly, I'm only proficient in some of the basics of alethic logic, but nonetheless, based on what I've read from overviews of these various branches of modal logic, it sounds to me like this is what is probably really going on.

I suspect it's all the same stuff in disguise, once you start to think of it purely in terms of sets. You can even apply these ideas to others types of sets besides just the ones we've talked about here, thereby in effect creating new analogs of "modal logic" without much effort. You just need to manage the sets properly. In fact, you don't really need to restrict the sets at all. It is perfectly natural to think of intersections and subsets in terms of possibilities and necessities, even without restricting your process to some specific kind of qualifier as you would in modal logic.

For example, suppose we wanted to say that women are necessarily female. Suppose furthermore that W is the set of all women and that F is the set of all female organisms. Then, all we'd have to do is to write either $W \rightharpoonup F$ or $W \sqsubseteq F$, with the choice of which one depending on whether we want to evaluate the claim ($W \rightharpoonup F$) or assert the claim

---

[55] a.k.a. existential intersection ($\sqcap$)

($W \sqsubseteq F$). That'd do the trick. Notice that there aren't any "possible worlds" involved here, and yet this scenario is still very clearly a valid instance of the concept of necessity.

Similarly, suppose we wanted to say that shipments of corn are possibly rotten. As such, let C be the set of all corn shipments and let R be the set of all rotten things. We could represent this claim by either $C \cap R$ or $C \Diamond R$, depending on whether we want to evaluate the claim ($C \cap R$) or assert the claim ($C \Diamond R$), just as before. Notice once again that there are no "possible worlds" or anything of that sort involved here. Yet, this scenario is still very clearly a valid instance of the concept of possibility.

Clearly then, contrary to what some of the literature of modal logic might lead you to believe, necessity and possibility are far broader concepts than just selections upon possible worlds. They originate from a much more fundamental concept: relationships between sets. Admittedly, there is debate in the modal logic literature about what exactly the logical properties of the modal operators should be, but I suspect that this is probably mostly just conceptual confusion. Set theory seems likely to be a more reliable foundation for working with necessity and possibility. Modal logic, in contrast, seems riddled with unresolved and aimless controversies about which exact properties should apply to the modal operators. Set theory, in contrast, already has all those properties firmly decided, with little to no room for controversy.

Modal logic originates more from philosophical logic than from mathematical logic, and philosophers do have a fairly strong tendency to confuse ambiguity with profundity. Many "deep philosophical debates" are nothing more than the product of using poorly defined terms and creating false distinctions where such distinctions do not actually exist or matter. Much of the "work" of modern philosophy has little to no substance or merit once you strip it of all of its vagueness and pretense.

Basically, truly good philosophy ends up becoming science instead. Thus, most of the remaining "unresolved deep questions" in philosophy tend to often be meritless, confused, or pragmatically irrelevant, rather than genuinely meaningful or useful. It's a natural selection process. Good philosophy becomes science. The remainder of philosophy therefore trends towards nonsense, untestable claims, pseudointellectualism, or conceptual confusion[56].

Anyway, tangents aside, it should now be clear why the diamond symbol ($\Diamond$) was chosen as one of the allowed ways of representing possibility (a.k.a. existential intersection) in unified logic. It was in anticipation of the close relationship between modal possibility (which uses the same diamond symbol) and unified logic's version of possibility. In addition though, remember that we also have the partially and completely filled in diamond variants in unified logic to represent additional constraints on the possibility (a.k.a existential intersection).

In total, as originally defined on page 412, we have four versions of the diamond symbol available for use in unified logic: (1) possibility ($\Diamond$), (2) left-proper possibility (◆), (3) right-proper possibility (◆), and (4) proper possibility (a.k.a. contingency) (◆). When a side of the possibility diamond has been filled in black then it means that

---

[56] However, keep in mind though that improving your "philosophy of life" is not at all the same thing as studying academic philosophy. Improving your life philosophy is almost always a good thing, whereas studying academic philosophy is almost always nothing but a waste of time. Life philosophy (e.g. Stoicism) and academic philosophy (e.g. big meaningless questions) are not the same thing. Don't equivocate them.

the set on that side is not a subset of the set on the other side. This applies to both sides simultaneously.

Thus, if we say $A \Diamond B$ then it means that we know that A and B have a non-empty intersection and also that A is not a subset of B. Similarly, $A \blacklozenge B$ means that we know that A and B have a non-empty intersection and that neither set is a subset of the other, i.e. that membership in neither set implies membership in the other. This makes the ◆ diamond useful for conveniently expressing possibility in the stricter sense of knowing that the relationship is possible but not necessary in either direction (a.k.a. contingency).

Contingency is a useful variation of possibility, so it is nice to have a separate symbol for it like this. We don't reuse the modal necessity symbol (□) in unified logic though, because (1) we already have existing symbols for unified implication and existential subset relations and (2) □ is convenient to keep reserved as a generic symbol for use in transformative logic (e.g. as a placeholder for "some operator here").

Anyway though, that pretty much covers our discussion of the overlap between unified logic and modal logic. There are also other things you can express with sets that probably have a relationship to modal logic in some sense, other than what I have discussed here. However, the prior discussion suits our purposes well enough, given that the intent is mostly just to illustrate that a strong connection does indeed exist between unified logic and modal logic. Indeed, it is clear that unified logic can express much of what modal logic can, if not more, and also that set theory and unified logic seem to rest on a stronger foundation than modal logic does. As such, with modal logic now having been adequately discussed, it is time to move on to the next branch.

5.6.7 Unified logic's relationship to paraconsistent logic

This brings our discussion now to paraconsistent logic. Our main discussion of paraconsistent logic started on page 357. As usual, if you would like a more in-depth review of what paraconsistent logic is, then you can refer to that section. Otherwise, if a brief review seems sufficient to you, then simply continue reading from here. You could also look up additional info online if you feel like it, of course.

Anyway though, let's get started. Naturally, the first thing we'll want want to refresh our memory on is our understanding of what the core motivating principle of paraconsistent logic is. Do you remember what it is? It's the idea of wanting to have a logic where the principle of explosion does not hold[57]. In other words, a paraconsistent logic is a logic where the idea that "from falsehood, anything follows" is false. It means we want a logic that is *not* explosive. It means we don't want the existence of a single contradictory premise to cause all conceivable statements to instantly become true.

Classical logic, in contrast to paraconsistent logic, is very explosive. Some non-classical logics are also explosive, but that's not our concern here. Classical logic is a sufficient example. We only really need to understand one non-paraconsistent logic, for

---

[57] A paraconsistent logic also doesn't want the paradox of entailment to hold. The principle of explosion and the paradox of entailment are very similar and mostly mean the same thing, but the principle of explosion is perhaps slightly more general. Thus, I will primarily refer to the principle of explosion in this section. However, you should remember that the paradox of entailment is also included in that implicitly.

point of comparison, to properly frame our understanding of the nature of paraconsistent logic. Here's an example of a proof of the principle of explosion for classical logic, to refresh your memory on how it works, taken from page 360 of the main paraconsistent logic section:

step #	form	rule of inference (justification)
1	A	assumption
2	$\neg A$	assumption
3	$A \vee B$	disjunction introduction (from 1)
4	B	disjunctive syllogism (from 2 and 3)

As you may recall, I mentioned in the main paraconsistent logic section that a common way of creating a paraconsistent logic is to disallow disjunction introduction or disjunctive syllogism (or both) in order to make the above chain of reasoning fail. By making this argument impossible to construct, one thereby makes it impossible for the principle of explosion to be applied and hence prevents explosive behavior within the confines of the reasoning system. This is a typical example of the kind of technique that a traditional paraconsistent logic may use in order to block the principle of explosion.

According to some sources in the literature of paraconsistent logic, achieving paraconsistency inherently requires that you abandon at least one of the laws of disjunction introduction or disjunctive syllogism. However, contrary to what the literature of paraconsistent logic says, this is actually false. A paraconsistent logic that abandons neither the law of disjunction introduction nor the law of disjunctive syllogism actually *can* be constructed. One such system of logic is unified logic.

All you need is the proper foundation and it can easily work. First, let's take a moment to prove to ourselves that indeed disjunction introduction and disjunctive syllogism are both valid in unified logic. Unified logic has a different definition of what truth and falsehood mean, one based on sets instead of merely simplistic truth labels. As such, arguably, in unified logic the corresponding laws should actually be called "union introduction" and "union syllogism". However, the spirit and net effect is still the same.

Unified logic can emulate the same behavior as F and T directly by using $\emptyset$ and $\mathcal{U}$ if need be. But, even regardless of that, unified logic's union operator is still a perfectly good substitute for disjunction in reasoning involving disjunction-like structure. Sure, the way you think about things and frame problems is sometimes a bit different, since you do have to think in terms of sets, but it is still nonetheless essentially filling the same niche and need.

Let's begin with disjunction introduction. Classical disjunction introduction works by applying a meta implication of the form $\{\alpha\} \prec \{\alpha \vee \beta\}$ or $\{\alpha\} \prec \{\beta \vee \alpha\}$[58]. Let's

---

[58] Remember that classical logic expressions change whether they are value expressions or statement expressions based on the context. In proofs, each asserted classical expression effectively has an implicit "$= T$" attached to the end of it. Thus, for example, assuming α in a classical proof actually means assuming $\alpha = T$ in reality. Similarly, assuming $\alpha \wedge \beta$ would actually mean you are assuming $\alpha \wedge \beta = T$. It's implicit, unlike in unified logic. Blame the sloppy conventions of classical logic for it.

think about the actual underlying justification though. The reasoning for allowing this is simple. It goes like this:

1. Suppose we know that α is true.

2. We know that any classical disjunction ($\vee$) where at least one participant is true will also end up being true, regardless of the truth values of the other participants.

3. Therefore, we know that any other truth value β we attach to α via disjunction ($\vee$) will have no effect on the truth value of the overall expression.

4. Therefore, given that α is true we may also conclude that $\alpha \vee \beta$ is true, regardless of the value of β or its relation to α.

For the set-based version of disjunction introduction (a.k.a. "union introduction") the rule would be $\{\alpha \neq \emptyset\} \prec \{\alpha \cup \beta \neq \emptyset\}$ or $\{\alpha \neq \emptyset\} \prec \{\beta \cup \alpha \neq \emptyset\}$. Does the same analogous reasoning work for this set union operator? Sure, of course it does. Watch:

1. Suppose we know that α is a non-empty set.

2. We know that any set union ($\cup$) where at least one participant is non-empty will also end up being non-empty, regardless of whether the other participants are non-empty or not.

3. Therefore, we know that any other set β we attach to α via union ($\cup$) will have no effect on the emptiness or non-emptiness of the overall expression.

4. Therefore, given that α is non-empty we may also conclude that $\alpha \cup \beta$ is non-empty, regardless of the content of β or its relation to α.

Next, let's talk about disjunctive syllogism. Classical disjunctive syllogism works by applying a meta implication of the form $\{\alpha \vee \beta, \neg \alpha\} \prec \{\beta\}$ or $\{\alpha \vee \beta, \neg \beta\} \prec \{\alpha\}$. How does the actual underlying reasoning work though? Let's see:

1. Suppose we know that $\alpha \vee \beta$ and $\neg \alpha$ are true.

2. This means that we know that at least one of α or β is true and that α is false.

3. Therefore, by process of elimination, since α is false, β must be the one that is true.

4. Therefore, given that $\alpha \vee \beta$ and $\neg \alpha$ are true we may also conclude that β is true, regardless of the value of α and β or any relation between them.

For the set-based version of disjunctive syllogism (a.k.a. "union syllogism") the rule would be $\{\alpha \cup \beta \neq \emptyset, \alpha = \emptyset\} \prec \{\beta \neq \emptyset\}$ or $\{\alpha \cup \beta \neq \emptyset, \beta = \emptyset\} \prec \{\alpha \neq \emptyset\}$. Notice that I wrote $\alpha = \emptyset$ instead of $\neg \alpha \neq \emptyset$ and $\beta = \emptyset$ instead of $\neg \beta \neq \emptyset$. Writing

$\neg\alpha \neq \emptyset$ or $\neg\beta \neq \emptyset$ here would be wrong. It would mean something different[59]. Be careful what you are doing. Actually *think* about the meaning of what you are saying, instead of just translating each symbol mindlessly. Anyway though, does the analogous underlying reasoning for this "union syllogism" actually work? Yep, it sure does. Here, look:

1. Suppose we know that $\alpha \cup \beta \neq \emptyset$ and $\alpha = \emptyset$ are true.

2. This means that we know that at least one of α or β is non-empty and that α is empty.

3. Therefore, by process of elimination, since α is empty, β must be the one that is non-empty.

4. Therefore, given that $\alpha \cup \beta \neq \emptyset$ and $\alpha = \emptyset$ are true we may also conclude that $\beta \neq \emptyset$ is true, regardless of the contents of α and β or any relation between them.

Keep in mind that there are also of course alternative notations for the set statements above. For instance, the notation that uses $\exists$ and $\not\exists$ instead of "$\neq \emptyset$" and "$= \emptyset$" may cut down on the noise a bit. For example, you could say $\exists(\alpha \cup \beta)$ instead of $\alpha \cup \beta \neq \emptyset$ and $\not\exists\alpha$ instead of $\alpha = \emptyset$. There are also the more specialized symbols too, such as $\sqcup$ and the others. For instance, $A \sqcup B$ is shorthand for $A \cup B \neq \emptyset$, as you may recall.

Pick whichever notation you prefer. I know having all these different notations is redundant and perhaps a bit annoying to some people, but I did it so that people could choose what they prefer. My hope is that as natural selection processes gradually come into play after this book's release, the community will gradually standardize which notation is used the most based on their experiences with what tends to work the best overall. Allowing natural selection to filter and evolve ideas organically tends to work a lot better than dictating how things should be done arbitrarily. Survival of the fittest will determine which ideas and notations are the best automatically over time. That's the way it should be. Competitive diversity is almost always better than dictatorship.

Anyway though, with all that said, we have now proven that unified logic does indeed have both disjunction introduction and disjunctive syllogism as applicable laws. Well, more precisely it has "union introduction" and "union syllogism", which are strong substitutes that fill the same role. Same difference though, pretty much. As such, it is now time for the next part of our discussion, which is the part about demonstrating that unified logic is paraconsistent despite still having close analogs of both disjunction introduction and disjunctive syllogism.

Luckily, we have already done this. Do you remember what we discussed early on in the introduction to primitive unified logic that might be relevant here? In particular, I want to point out that I already mentioned that unified logic has the opposite property of the principle of explosion: the principle of implosion. It also has vacuous falsehood and integrity of entailment. These are all very closely related properties. These properties are all opposite counterparts to their classical analogs: vacuous truth vs vacuous

---

[59] Remember our discussion of the law of yin and yang and how it differs from the law of non-contradiction? That's related to this.

falsehood, principle of explosion vs principle of implosion, and paradox of entailment vs integrity of entailment.

Indeed, whereas classical logic obeys the principle that "from falsehood, anything follows", unified logic obeys the opposite principle: "from falsehood, nothing follows". Unified implication and existential subset relations (unified logic's versions of implication) both already don't suffer from the principle of explosion and its related ills. Hence, there is no need for unified logic to sacrifice its analogs of disjunction introduction and disjunctive syllogism in order to acquire immunity to the principle of explosion and hence paraconsistency. It already has paraconsistency, straight out of the box.

Thus, we are already done here. If you would like to review the reasons why unified logic and its operators have these properties in greater depth, then simply refer back to the main unified logic section. For a quick reference on where some of the relevant definitions were though, see the following list:

1. the definition of unified implication: page 380

2. the definition of vacuous falsehood: page 382

3. the definition of integrity of entailment: page 382

4. the definition of the principle of implosion: page 383

Otherwise, for other assistance in navigating to a specific section of the unified logic chapter, consider consulting the table of contents, which can be found between pages 3 and 8. Anyway though, as I was saying, we have now essentially completed our coverage of paraconsistent logic. It is now time to proceed to the next branch of logic.

5.6.8 Unified logic's relationship to constructive logic

This time we'll be talking about constructive logic. As such, if you would like a more extensive review of constructive logic then please refer to the main section on constructive logic, which can be found on page 362. However, be aware that even in the main section on constructive logic we did not really discuss the details of actually performing constructive logic, but rather we only discussed a conceptual overview of it. As such, if you want information on actually performing traditional constructive logic, then you'll need to look it up yourself somewhere.

Otherwise though, if a brief review to refresh your memory seems sufficient to you, then simply continue reading from here. Personally though, I don't think referring back to the main section on constructive logic will be necessary. We only really need to understand the most basic motivations and concepts here. That will be sufficient to illustrate the connection between unified logic and constructive logic. The details will be relatively unimportant.

Let's begin. Naturally, the first thing we should remind ourselves of is what the core motivating principle of constructive logic actually is. Specifically, we should ask ourselves what it means for a logic to be constructive. To that point, we should remind ourselves that constructive logic is part of a broader philosophy called constructivism. What's constructivism though?

Well, constructivism is essentially the idea that the only way to truly prove the existence of an object is to construct (or find) that object. Indirect proofs in contrast, such as proofs by contradiction, are considered untrustworthy. Only a fully constructed instance of the object, with nothing left to the imagination or left incomplete, is considered to be a real proof of that object's existence. It's essentially the formal equivalent of the informal everyday philosophy of "I'll believe it when I see it.".

However, constructive logic is not a single system of logic. It is actually an entire family of related logics, each of which adheres to the philosophy of constructivism in its own distinct way. Regardless of the differences between the variations though, constructive logics tend to be generative in nature, meaning that they tend to work in terms of building up objects piece by piece in such a way that an algorithm could (at least in principle) be created to actually create concrete instances of the objects. This differs from non-constructive logics, where some proofs are allowed to prove that some object satisfying some property exists, and yet still not be able to actually produce any specific example of such an object.

There seem to be multiple ways to create a constructive logic, some of which carry different trade-offs and characteristics. However, one of the most popular and widely studied systems of constructive logic is the system known as intuitionistic logic. Intuitionistic logic uses much of the same vocabulary as classical logic, but adheres to a different set of rules governing what it is allowed to do.

In particular, intuitionistic logic can be thought of as sort of like classical logic with the law of the excluded middle ($A \vee \neg A \rightarrow T$) and the law of double negation elimination ($\neg \neg A \rightarrow A$) removed. There's more to understanding the underlying philosophy and mechanics of intuitionistic logic than just that, but the lack of these two basic laws captures the essence of it fairly well. It is important to note though that intuitionistic logic is not truth-functional. There's no way to create truth tables for evaluating expressions in intuitionistic logic. Instead, you have to think in terms of derivations from forms, i.e. in terms of proofs.

This brings us to the first of the connections between unified logic and constructive logic that we will be discussing. It's a bit indirect, but it still sort of counts. As you may recall, I mentioned earlier in this book that part of the way that reasoning with unified logic works is that you use whatever combination of unified implications, existential subset relations, and transformation rules is necessary to successfully express the implications and properties that you are trying to work with.

However, notice that transformation rules are technically more closely related to transformative logic than to unified logic. I'm pointing this out because the connection I am about to make here is about transformation rules, not set-based reasoning, but I still nonetheless consider it to be part of the bigger picture of why and how unified logic has constructive properties. Indeed, it turns out that transformative logic is itself also a constructive logic.

Can you see why? The nature of transformative implication is the key. You see, transformative implication is an inherently constructive operation. All it does is to take one form and transform it into another. Sure, there are other details involved in the process, such as whether or not the form is embedded inside another form and other kinds of parsing issues, technicalities, and notational conveniences. But, at heart, a transformative implication is just a rule about constructing one form from another form based on a simple rule. Transformative implication always takes something that already exists (i.e. has already been constructed) and simply creates something new out of it. It thus always conserves constructive existence.

Each individual transformation step is foolproof and mechanical. The form begins as a finite and fully constructed object and still remains finite and fully constructed with every step thereafter. The transformation process continues however arbitrarily long the user wishes it too, but always remains well-defined at every step. This is classic constructive behavior. It's an instance of the principle of property conservation[60]. It's entirely a generative and reliable process. It's also computationally straightforward and highly suitable for computerization. A properly designed computer program that performs transformative logic can probably be made to be at least reasonably efficient, not just in theoretical terms but in real practical terms[61].

Anyway though, my point here is that since transformative logic is part of the reasoning process involved in unified logic, and transformative logic is constructive, then clearly at least that component of unified logic will be constructive at minimum. It is also of course useful to know that transformative logic is constructive, even regardless of its relation to unified logic.

That's not the only part of unified logic that is constructive though. At least some parts of the set-based components of unified logic are constructive. Specifically, all set operations whose inputs, outputs, and intermediate results are all finite sets are constructive. The reasoning is similar as for the transformative logic case. In particular, when we start with finite sets and then perform set operations upon them that all result in finite sets, then we are always guaranteed to be able to fully construct those sets. At every moment of this process we are dealing with completely constructed sets, and that's what makes it constructive.

What about the case of infinite sets though? Well, honestly I'm not sure about that case. I thought about it for a while but I couldn't really come up with anything that felt definitive enough for me to make the claim one way or the other. I just don't know whether set operations with infinite sets should be considered constructive in unified logic or not. Part of the problem is that there may be some way to work with infinite sets purely in finite terms by reframing them in terms of properties and relationships instead of infinite sets of elements, which may allow you to work with them constructively.

There's also the issue of whether or not you actually consider the infinite sets to be constructively problematic or not to begin with. I don't know really. Perhaps someone else with more experience in constructive principles can resolve this. As for me, I'm just trying to work within my limitations as best I can. There are some big gaps in my knowledge that I don't have time to fill, given the scope of what I have planned and

---

[60] Remember that? We defined the principle of property conservation on page 104. It's similar to the principle of mathematical induction, but broader.

[61] Admittedly, such a program would probably need to stay away from some of the more computationally expensive operations of transformative logic, but the more basic components of transformative logic would still probably be relatively straightforward to implement efficiently.

what my priorities are. Oh well though. Such is life. Life is a work in progress. All we can do is to create as much value in our lives as we possibly can with every precious moment that we are given. There's no point in stressing out about it beyond that.

That covers the set-based part of how unified logic can be thought of as a constructive logic. We also covered the transformative logic part as well, of course. We can thus now conclude that unified logic is at least partially constructive, specifically with respect to all transformation rules and finite sets. It may even be completely constructive, but I am not sure. You might think that I am going to conclude this section now, having already met my goals here (partially) as best I can. However, there's still one more interesting aspect of the connection between unified logic and constructive logic that'd I'd like to discuss.

Specifically, I want to point out that unified logic has these constructive properties even despite the fact that it does not need to sacrifice the law of the excluded middle ($A \vee \neg A \to T$) or the law of double negation elimination ($\neg\neg A \to A$). Think about it. The unified logic equivalent of $A \vee \neg A$ is $A \cup \neg A$, but $A \cup \neg A$ is always $\mathcal{U}$. Similarly, the unified logic equivalent of $\neg\neg A$ is $\neg\neg A$[62], but $\neg\neg A$ is still always A. Thus, unified logic obeys its own analogs of the law of the excluded middle and the law of double negation elimination, specifically $A \cup \neg A \to \mathcal{U}$ and $\neg\neg A \to A$, despite having (at least partial) constructive properties.

This is yet another example of unified logic somehow managing to satisfy the motivating principles of a branch of non-classical logic without actually having to sacrifice any of its other desirable or intuitive properties to do so. None of the other traditional approaches to non-classical logic ever seem to be able to get away with that. They all have to make bizarre and counterintuitive sacrifices in order to get what they want. This to me is evidence that unified logic is closer to the essence of what the proper foundations of logic are than the other logics are.

That being said, the idea that unified logic is *the* proper foundation for logic is probably questionable. Indeed, I'm betting that someone who comes after me will manage to do better, actually, and that unified logic will just be a stepping stone towards whatever the real best possible logic is. Nonetheless, I do find these trends quite compelling and interesting. I hope that my own journey opens up new avenues of thought for others to build upon, and that one day the final truth is reached to be shared with everyone in all its glory. To be a step on the ladder of progress is to stand firm in the present moment and yet to still move the future forward.

Anyway though, it's time to move on to the next branch of logic now. I hope you found our discussion here to be *constructive*. Get it? Hur hur hur hur... so funny. Words have multiple meanings that differ according to context. Therefore, laugh. Such depth. Such insight. Comedy from the heavens. Puns: the ultimate form of humor, the pinnacle of human wit, the holy water by which teenagers are driven away, the **pun**t by which we kick the metaphorical can of acquiring a real sense of humor down the road with every passing day. All glory to the pun. Long live the pun.

---

[62] This $\neg\neg A$ uses the same symbols as the other $\neg\neg A$, but here $\neg$ means set complement instead of classical negation.

5.6.9 Unified logic's relationship to connexive logic

Now it's time to talk about connexive logic. Connexive logic, as you may recall, is one of the less widely known branches of logic. However, the motivating principles behind it are really quite simple and easy to understand. It has long been a logic that has kind of been swept under the rug by logicians and mathematicians, who often aren't even aware of its existence at all. However, perhaps some of them also subconsciously don't want to face the simple yet alarming implications about the state of classical logic that connexive logic tends to draw attention to. Mostly though, it just seems like probably not many people are aware of connexive logic.

As usual, I will give a brief review of what this branch of logic is here, so that you can refresh your memory. However, if you would like a more in-depth review of what we previously discussed about connexive logic, then just refer to the main section on connexive logic, which can be found on page 366. If that too is insufficient for you though, then you can also search for more information elsewhere (e.g. on the internet), especially if you want more details on actually performing logical calculations in one of the traditional systems of connexive logic in practice.

As with many of the non-classical logics, connexive logic is not just one specific system of logic, but rather it is a family of logics which all satisfy connexive properties. While the criteria for what makes a logic connexive are indeed very simple and easy to understand, some of the specific ways of creating connexive logics may be quite strange or convoluted. The implementations details of a system of logic can sometimes be arguably more complicated than the underlying principles being implemented. This is even more likely to happen when something about the guiding paradigm the designer is using to create the logic is fundamentally wrong in some way, as is very likely the case in many of the traditional implementations of the non-classical logics probably, which often possess some quite bizarre quirks.

Anyway though, putting that aside, let's refresh our memory of what it means to be a connexive logic. In particular, a connexive logic is any logic in which at least one of Aristotle's theses or Boethius's theses are true. Stronger connexive logics accept all of Aristotle's theses and Boethius's theses as true, whereas weaker ones accept fewer of the theses (e.g. perhaps only Aristotle's theses). What are Aristotle's theses and Boethius's theses? They are the following two pairs of logical expressions, interpreted relative to the specific vocabulary of the logic being tested for connexive properties:

Aristotle's Theses:

1. $\neg(A \to \neg A)$

2. $\neg(\neg A \to A)$

Boethius's Theses:

1. $(A \to B) \to \neg(A \to \neg B)$

2. $(A \to \neg B) \to \neg(A \to B)$

Translated into plain English, Aristotle's theses together say "No statement can imply its own negation." and Boethius's theses together say "No statement can imply both some other statement and that other statement's negation at the same time.". As much as one might be tempted to try to add these theses into classical logic as new axioms, doing so would be a terrible idea. All of the connexive theses simplify into either A or $\neg A$ when you apply the laws of classical logic to them (see page 367), and thus taking any of them as new axioms in classical logic would be tantamount to saying that all statements are true, which would be unacceptable.

On the other hand though, if you look at the literature for connexive logic, you will see that all of the systems of logic that *do* have connexive properties seem to be forced to lose other properties in order to do so. Many (all?) of these traditional systems of connexive logic also appear to have some strange quirks and counterintuitive behavior. They generally don't seem very intuitively understandable. Instead, they feel arbitrary, contrived, and impractical, just like most of the other traditional non-classical logics.

What about unified logic though? How does it fit into this picture? Is it connexive? Yep, it sure is. This is no surprise though of course. I already told you before that unified logic would satisfy the core motivating principles of all of these various different branches of logic. The pattern of unification continues as usual. That is after all why it's called *unified* logic. It's a reference to the way that unified logic strives to capture so many different logics at once inside a single system. It's like an umbrella[63].

Anyway though, let's look at exactly why unified logic is a connexive logic now. Can you see why? Let's think about it. Let's start with Aristotle's theses. What's the unified logic equivalent of $\neg(A \to \neg A)$? Well, let's start with the inner part of the expression, $A \to \neg A$. The unified logic equivalent of material implication (when interpreted as a value expression) is unified implication. So, $A \to \neg A$ can be translated into either $A \mapsto \neg A$ or $A \rightarrowtail \neg A$. However, we don't really care about which of the two sets is returned here, so whether we use sub implication ($\mapsto$) or super implication ($\rightarrowtail$) won't matter much in this case. Finally, we add on the outer negation (but interpreted as a set complement this time) and we now have $\neg(A \twoheadrightarrow \neg A)$, which is thus the unified logic equivalent of $\neg(A \to \neg A)$.

What is the behavior of the expression $A \twoheadrightarrow \neg A$ in unified logic though? Think about how unified implication works. Unified implication only returns a non-empty set if the condition set is a subset of the consequence set. Can A ever be a subset of $\neg A$ though? Nope. No set can ever be an existential subset of its own negation (i.e. its own set complement). The negation of a set is all the things *not* in that set, so A could therefore certainly never be a part of $\neg A$. Thus, we know that $A \twoheadrightarrow \neg A$ is always an empty set[64], and therefore we can say that $A \twoheadrightarrow \neg A = \emptyset$. This is the statement expression equivalent of the value expression $\neg(A \twoheadrightarrow \neg A)$. Alternatively, we could write $\not\exists(A \twoheadrightarrow \neg A)$ or $A \not\sqsubseteq \neg A$ and it would still mean the exact same thing as $A \twoheadrightarrow \neg A = \emptyset$.

Or, instead of framing the relationship as a statement expression in unified logic, we could just frame the relationship as a transformation rule by writing $A \twoheadrightarrow \neg A \to \emptyset$. In that case, we'd be saying that the value expression $A \twoheadrightarrow \neg A$ can always be transformed into the value expression $\emptyset$, rather than making an existential statement. Regardless of the nuances though, the point is the same. Unified logic does indeed satisfy the first of Aristotle's theses.

What about Aristotle's other thesis? Well, clearly if $\neg(A \to \neg A)$ works then so will $\neg(\neg A \to A)$. There's no reason why the reasoning would change. A can't be a subset of

---

[63] Don't worry though. Unified logic isn't likely to enable the creation of the T-virus. The zombie apocalypse is not yet upon us. It is not yet time to conscript your fireball spitting grandmother into the great battle against the undead legions. You can rest easy... for now. Dun dun dun... [lightning strike]

[64] And equivalently, we know that $\neg(A \twoheadrightarrow \neg A)$ is always the universal set, and hence is always true.

¬A, and likewise ¬A can't be a subset of A. It works in both directions. No set which is the negation of another set could possibly be an existential subset of that other set.

What about Boethius's theses though? Do they apply to unified logic? Yep, they sure do. However, the translation into terms of unified logic might confuse you a bit. The structure of the resulting expression will not be the same as the structure of the original Boethius theses. Translating the symbols one-to-one in the most direct and obvious way possible will not be correct. The unified logic equivalent of $(A \to B) \to \neg(A \to \neg B)$ is *not* $(A \rightharpoonup B) \rightharpoonup \neg(A \rightharpoonup \neg B)$[65].

The real translation is $\neg((A \rightharpoonup B) \rightharpoonup \neg B)$, and it always returns the universal set, just as the value expression version of Aristotle's thesis $\neg(A \rightharpoonup \neg A)$ also always returns the universal set. Do you see why? Take a moment to think about it if you'd like to. The answer is pretty much the same as why Aristotle's theses are true in unified logic. It's just because no subset can be a member of its own negation. The only difference is that we're testing it more indirectly this time. The only way that A could be a subset of first B and then also ¬B is if B was also a subset of ¬B, but we already know that this is impossible based on our prior analysis of that case for Aristotle's theses. Thus, $(A \rightharpoonup B) \rightharpoonup \neg B$ must always be the empty set, and therefore $\neg((A \rightharpoonup B) \rightharpoonup \neg B)$ must always be the universal set.

What's the statement expression version of this? Well, as usual, we simply convert the expression into an existential statement and there's our answer. Specifically, the statement expression version is $(A \rightharpoonup B) \rightharpoonup \neg B = \emptyset$, or any of its other equivalent forms in alternative notations. Clearly this statement is also true, as we just discussed. We could also say that $(A \rightharpoonup B) \rightharpoonup \neg B \to \emptyset$ is an admissible transformation rule if we wanted to.

That covers the first of Boethius's theses. What about the second one, the one whose form is $(A \to \neg B) \to \neg(A \to B)$? Well, we can see from this expression that $(A \to \neg B) \to \neg(A \to B)$ would simply translate into $\neg((A \rightharpoonup \neg B) \rightharpoonup B)$ as a value expression. It doesn't take much thinking to see that the same reasoning will apply again here. The only difference is that the order of B and ¬B in the nested expression has been flipped. This difference has no substantive impact on the conclusion though. Likewise, the statement expression version will be $(A \rightharpoonup \neg B) \rightharpoonup B = \emptyset$ or any of its alternative forms, and so on.

Thus, we have now proven that unified logic obeys all of the connexive properties, i.e. both Aristotle's theses and Boethius's theses, and that therefore unified logic is not just any connexive logic but a strong connexive logic. Notice something interesting though. Notice that unified logic didn't have to make any bizarre counterintuitive sacrifices in order to achieve these properties. Unified logic still remains just as intuitive and easy to understand as ever before. It remains unburdened.

It is on this note that I bring this section to a close. We are done with connexive logic now. Indeed, we have covered quite a few of the different branches of logic that we intended to. All that remains is classical logic. Thus, it is finally time that we come full circle, back to where we started. It's time to take one more brief moment to consider

---

[65] Remember to think in terms of the underlying *meaning* of each part of a logical expression when you translate between different systems of logic. Don't just translate the symbols to their closest analogs without actually thinking about it.

how unified logic stands in relation to classical logic. After that though, we'll be moving on to new material.

5.6.10 Unified logic's relationship to classical logic

It's time to talk about our old friend classical logic yet again. Our main discussion of classical logic originally started on page 82, but we have also mentioned it throughout the book in other places when relevant. Classical logic, as you may recall, is by far the most dominant of all of the currently existing branches of logic. If you learned formal logic at all in school, it was probably classical logic you learned. Most students (outside of math and philosophy at least) probably aren't even aware of the existence of the other branches, much less ever exposed to them.

This dominance is not really surprising though. Classical logic is super easy to understand, at least in most respects. As long as you turn a blind eye to the strangeness of things like material implication and its paradoxes, and to the other oddities about the way classical logic behaves, classical logic still feels like an intuitive system to work with. Sure, it feels a bit fishy sometimes, like there's something missing from it, like the way it models logical inference is only skin-deep, but it nonetheless otherwise feels fairly natural.

Indeed, in particular, the whole truth-functional idea of simply assigning each statement a value of either true or false is very seductive and appealing. It is also undoubtedly extremely computationally efficient. Indeed, in a sense, classical logic is essentially the principle structural basis for constructing computer hardware. Computer hardware is mostly just a bunch of simple classical logic expressions implemented as physical circuitry[66] and then connected together with other devices such as displays and memory and so on, thereby creating the illusion of a machine with a mind of its own, i.e. a machine imbued with the power of logic[67].

However, be that as it may, there are reasons to be skeptical of this crude black-and-white way in which classical logic operates. The viewpoint of classical logic is useful, in the right context, but it is also very clearly not the full picture of what logic actually involves. I hope by now that I have made that painfully clear to you, through our many discussions of the numerous subtle and not-so-subtle shortcomings of classical logic. Indeed, classical logic may leave a lot to be desired, but so too do many of the traditional non-classical logics. Sometimes the side effects of the cure are worse than the disease.

---

[66] The physical circuitry of a computer is made of components called logic gates. Logic gates are made out of tiny semiconductors. Semiconductors act as electrically activated switches that can flip on and off without actually physically moving, thus preventing them from breaking constantly, like mechanical switches would. These semiconductors are then arranged in such a way as to mimic the behavior of classical truth tables (e.g. parallel for $\vee$, sequential for $\wedge$). A collection of semiconductors that implements a truth table is called a logic gate. These logic gates are then arranged to mimic the required reasoning of the computer component being designed. This is how it is done in practice. This is what makes computers "smart". Computers are basically like insanely complicated microscopic Rube Goldberg machines made out of classical truth tables.

[67] Computers are special though, since they can do things that conventional classical logic cannot do, by virtue of the computational generality and extra features that a computer has access to. Computer engineers and programmers also tend to almost never think in terms of material implication. They use conditional jumps and flow of control manipulation instead, which is an entirely different beast from conventional mathematical implication, despite the fact that it also uses the phrase "if then".

How unified logic fits into this part of the picture should be clear to you at this point. We have already illustrated how unified logic manages to satisfy many desirable properties for a logic simultaneously. However, that's not really what I want to focus on in this section. We've already done that. We've spent much of this book pointing out these flaws in classical logic and alluding to the corresponding alternatives and fixes and such. Instead though, what I now want to shift our attention to is to how unified logic and classical logic are *similar* to each other and why that's a good thing. I want to talk about how unified logic not only satisfies the underlying motivating spirit of the non-classical logics, but *also* satisfies the underlying motivating spirit of classical logic as well.

What *is* the underlying motivating spirit of classical logic though? We never really discussed it before. We only ever talked about the details of classical logic, and much less so its spirit. There are reasons for that though, in a sense. You see, classical logic has a different origin. It was not created the same way that most of the non-classical logics were. Most of the non-classical logics originate from criticisms of specific aspects of classical logic's behavior. Most of them are derivative, in that sense, from classical logic. Set theory is an exception[68].

Classical logic itself, on the other hand, grew organically from logicians' initial attempts to formalize the process of reasoning. It grew out of an attempt to overcome limitations in the nature of syllogistic logic (the ancient ancestor of set theory), to expand the scope of what was possible to talk about and to refactor and remove certain redundancies in how syllogistic logic was performed. The modern form of set theory didn't exist at the time. Classical logic is the product of a combination of simplistic intuitions about how logic should work and a desire to escape from the strictures of syllogistic logic at any cost, even if it meant accepting some counterintuitive properties in order to do so successfully.

Even just escaping from the limitations of syllogistic logic was quite enough of an achievement on its own. It wasn't until quite some time later that the non-classical logics started to appear and gain real momentum. Classical logic was born of a balance of instinct and pragmatism. It evolved from natural intuitions about the nature of reasoning, combined with a practical need for a flexible yet easy to use system. It fit fairly well with people's needs, despite all its flaws.

However, the flaws of classical logic were incidental rather than intentional. The true heart of classical logic, its underlying motivating principle, has always been a combination of natural intuition and ease of use. Everything else about classical logic just grew out of that, subject to the distorting constraints of forcing everything into a truth-functional binary logic. What classical logic *wants* though is to be as natural as possible. That's the core inspiration.

Classical logic does a pretty good job of this too. It strikes a good balance. It's very

---

[68] In fact, classical logic is indirectly derived from set theory in a sense. More precisely: Syllogistic logic (ancient set theory) came first. It was followed shortly by stoic logic (ancient classical logic), which tried to do things differently from syllogistic logic, but stagnated and went nowhere for a long time. Syllogistic logic then monopolized the study of logic for more than 1,000 years. Then came classical logic much later, picking up the thread, but mostly inspired by a desire to overcome the limitations of syllogistic logic, more so than by stoic logic itself. Then came modern set theory sometime after that. Then came the other non-classical logics. At least, that's my understanding of it. History is messy.

often the "lesser of two evils" when compared to most of the other branches of logic. It's very practical for a lot of things, as long as you avoid its paradoxes and such. It's a good compromise, in a lot of ways. It has access to a large set of very natural properties that make a lot of sense. It's far from perfect though, and that's where unified logic comes in: as an attempt to improve upon it.

Indeed, unified logic strives to satisfy as many of the most natural properties of classical logic as it can, without significantly degrading how easy to use, intuitive, or predictable the resulting system is. That's the beauty of it. We've already seen a bunch of these examples of properties that both classical logic and unified logic share in fact. Consider, for example, the large list of properties shared by both classical logic and set theory which can be found on page 147. Since unified logic is an extension of set theory, it follows that unified logic also has all these same properties too. For instance, notice that both obey the law of identity, the law of non-contradiction, and the law of the excluded middle. These are three very desirable properties.

In addition to that big list though, unified logic also satisfies many other additional properties in common with classical logic. The extension of unified logic with unified implication is a big part of what enables these additional correspondences. For example, both classical logic and unified logic have the law of contraposition. Both have modus ponens and modus tollens. Both have their own versions of disjunction elimination (a.k.a. proof by cases or case analysis). Both have implication operators that are transitive. This list is just a sample though. There are many other parallels.

The net effect of all of this is that both classical logic and unified logic are generally easy to use and intuitive. Most of the basic properties that apply to both systems are unsurprising and highly relatable. Neither is plagued by an overwhelming feeling of being arbitrary and contrived, unlike many of the other branches of logic unfortunately. Both feel like highly principled logics, even if some flaws do exist. However, unified logic takes it a step further. Unified logic satisfies all these wonderfully intuitive properties of classical logic while also simultaneously plugging a bunch of holes and shortcomings in the behavior of classical logic.

You can even use unified logic like a binary logic if you want to, by restricting your sets to just $\emptyset$ and $\mathcal{U}$. It will differ from classical logic in certain respects, such as by obeying the principle of implosion ("from falsehood, nothing follows") instead of the principle of explosion ("from falsehood, anything follows"), but otherwise it will tend to behave similarly where you would intuitively expect it to. This is another sense in which unified logic successfully satisfies many of the same underlying principles as classical logic: i.e. through being able to emulate binary reasoning. The broader notion of reasoning with any arbitrary sets (instead of just with $\emptyset$ and $\mathcal{U}$) is more powerful though[69].

Having said all this, it should now be clear that we have fulfilled our basic goal here. Unified logic is indeed the umbrella logic that I have claimed it to be. It satisfies a diverse variety of logical principles quite nicely. As such, we have now come full

---

[69] Remember that objects have truth values in unified logic. Any described object which has a corresponding non-empty set is true in unified logic. We are operating under an existence-based truth perspective (see page 27). Admittedly though, reframing arguments in terms of arbitrary sets could potentially be difficult sometimes. I'm not sure what the full reach of the system really is. Only time will tell. A seed takes time to grow.

circle. Logic came into existence first from set-based thinking (syllogistic logic), then it evolved into binary truth-functional based thinking due to practical constraints (classical logic), and now it has apparently returned once more to inspire some new set-based thinking via unified logic. The two systems (classical logic and set theory) have thus fed ideas back and forth between each other over time, like a swinging pendulum. Perhaps one day the truth will be revealed and the pendulum will at last come to a rest. In the meantime though, all we can do is explore new ideas as best we can.

Indeed, the hope of exploring alternatives systems of logic is to one day create a logic that has all of the desirable properties that one could ever want, without any of the undesirable properties. Whether or not this is actually possible is unclear. Some desirable properties may inherently conflict with others potentially. However, it is nonetheless still extremely worthwhile to try to figure out what system of logic has the best overall balance of properties. Unified logic seems to bring us closer to that ideal, but I'm not sure exactly how much closer. That's good enough though. Truth is a mountain that only patience can climb.

Oh, and as a side note, in case it wasn't obvious, since unified logic satisfies the spirit of classical logic (i.e. ease of use and a natural feel etc), and classical logic in turn is a superset of stoic logic, it therefore also follows that unified logic satisfies the spirit of stoic logic as well. Differences between all these various branches of logic will always exist of course, but it is the underlying motivations and core principles that we really care about. The implementation details are incidental.

Speaking of implementation details though, we are actually only halfway done with implementing unified logic. Do you remember what I said near the beginning of this chapter about how our discussion of unified logic would be split up in two parts? I said (on page 375) that the first part of unified logic would be called primitive unified logic whereas the later part would be called relational unified logic. Well, everything we've been talking about so far has just been primitive unified logic. Primitive unified logic is cool, but its still fundamentally just a simple extension to set theory.

Relational unified logic is different. Relational unified logic is where things get *weird*. Things that you probably thought weren't even possible to solve or express will become solvable and expressible. Want to apply numbers to each other *as operations*? In relational unified logic you can, even as absurd as it sounds. Want to divide by zero? You can do that too. And, it's all rigorously well-defined too. I know it sounds crazy, but just humor me. Prepare for a brain explosion. It's time for relational unified logic.

5.7 Introduction to relational unified logic

It's time to modify and reinterpret our foundations some. We've previously said nothing to contradict the basis of set theory as it currently stands in the literature. We've gotten about as far as we can with that approach though. It's time for a new angle. One must often question one's assumptions in order to make progress in developing ideas. Very often, it turns out that some basic ideas which one previously assumed to be fundamental are not truly so. We must always be willing to question our assumptions. We must always ask ourselves: "But how could that be wrong?"

Information should not just be absorbed without question, as if truth were some arbitrary thing that could be declared by fiat or social convention. Truth is not a social convention. You can pose arbitrary hypothetical scenarios that aren't real, and then simulate those, but even that still doesn't make truth arbitrary. Hypothetical simulations aren't the same thing as lies. Everything we hear or think should (at least ideally, subject to time and resource constraints) be passed through a mental filter designed to test whether it can be contradicted. What remains (what survives the filter) is far more likely to be true and useful. This is the essence of science. We cannot hope to reliably reason correctly if we are not willing to question the past.

Learning is motion. It is movement away from falsehood and towards truth. One cannot turn towards truth without simultaneously turning away from falsehood. One cannot respect reality and yet honor delusions at the same time. One cannot have it both ways. Not all beliefs are equal, and they never can be. Yet, some may be offended as you turn away from the traditions and beliefs of the past. Let them. That is their weakness, not ours. It is their loss, their burden. Ignorance is for the weak. When you seek knowledge, you must be willing to face reality on its own terms, whether it be ice-cold and cruel or warm and cuddly, whatever the case may be. You must bend to the truth, because it will never bend to you. Truth is timeless and has no ego.

The more we make ourselves like truth, the more timeless and without ego our perspectives become, the better we can see not just reality itself but also the infinite cornucopia of all conceivable logically coherent worlds. As such, with truth on our side, our dreams can easily become reality. Yet, if we instead turn towards ignorance, then only darkness and stagnation will await us. It is one of the great ironies of life that running away from our challenges tends to only make our lives even more difficult, yet running towards those same challenges instead, and overcoming them, allows us to live easy and peaceful lives with hardly a worry in the world.

Truth is empowering. We should always take a moment to consider possible counterexamples and alternatives to the things we hear people say, and indeed even to our own thoughts as well. Only then can an idea's legitimacy start to become credible. It's not enough for an idea to merely feel good. Emotions are too deceptive in that respect. It's all too easy to say something that sounds clever or interesting on the surface but that nonetheless isn't true. Famous quotes and proverbs are often examples of this.

Being catchy isn't the same thing as being valid though. Many wise-sounding things that people say are actually completely wrong, and often obviously so, as one can often easily determine the moment one tries to think of a counterexample for even just a brief moment. Popularity can't be relied upon here. Some sayings hold true wisdom and others don't. It depends. The point is: We must question everything. We must not get sucked into intellectual black holes. We must always be able to separate ourselves from our own assumptions. We must not let ego stand in the way of truth.

As such, even the most seemingly "obvious" and heavily culturally ingrained assumptions should often be upended, analyzed, and deconstructed. The merit and "obviousness" of such beliefs is often an illusion, born not of genuinely natural or insightful qualities but merely of social momentum, pretense, and lack of imagination. It is on this note that we now turn our attention to the concept of sets, and especially to a very specific, very subtle, and seemingly hair-splitting aspect of how they are treated in the traditional literature. Indeed, it may not even occur to many people that this specific aspect of how sets are treated could possibly be done differently, so heavily ingrained it is in the currently prevailing assumptions.

What, specifically, am I talking about here? Well, to explain that, I think a thought experiment would be the most illuminating approach. Consider for a moment, what a set is. However, I don't mean merely the concept of a collection of objects, but rather I mean the set *itself*. What is the nature of the set itself, independent of its content? What does it mean to be a set really? What is the *substance* of a set? If I say I have a set containing some objects, then what is the set itself really made up of, independent of whatever arbitrary collection of objects it contains?

This seemingly hair splitting question will make a huge difference in how sets will behave. It will radically alter the scope of what it means for a set to be well-defined, how widespread sets are, and what kinds of uses of sets are possible. What do I really mean though when I ask what the *substance* of a set is? Well, one aspect of it is simply this: Does a set exist separately from its contents? Is an object wrapped in a set different from that same object standing alone? Does a set itself have physical substance, separate and apart from what it contains, or is a set instead merely some kind of abstract ghostly label by which we refer to some collection of objects, one that cannot itself be treated as any kind of solid thing?

The answer may surprise you. There are at least two different ways of thinking of sets to choose from. Traditionally, according to the literature and the standard views of logic and mathematics as they stand today, a set is considered to be a separate thing in and of itself, an object in its own right. This is seen as obviously the way things should be, as if it is the only possibility that could ever make any sense. Sets are treated as if they are physical containers, akin to cardboard boxes or briefcases. Solid.

It's easy to see why a person would instinctively think that this should be the case, especially considering that thinking by analogy and anthropomorphizing concepts into human terms is very typical of how human beings like to think. Is that the right way of thinking of sets though? Is the most natural and principled choice *really* to treat sets as having their own independent physical substance apart from their contents?

The answer is actually no. The standard view on this matter seems to be wrong. Despite how deeply ingrained it has become as an assumption in the culture of mathematics, it is really quite questionable. Indeed, in reality, the most natural and principled definition of a set does *not* treat sets like physical containers. That's what the upcoming thought experiment will open your eyes to.

You see, while it is true that you *can* make sets act like physical containers if you want to, and the result is logically coherent if done correctly, the problem is that doing so is *not* really the most conceptually correct and natural way of conceiving of sets. In fact, much of the strange problems and convolutions that happen when working with sets in the traditional literature exist solely because this choice was made incorrectly. Once we make the correct choice though, a whole bunch of these problems in the foundations of set theory will instantly disappear and the rigorous basis for sets will be greatly simplified.

Let's get started. Here's a thought experiment for you: Let's think about objects. Specifically, lets think about this habit we have as humans of labeling things we find in our environment with different names. What's really going on when we do that? What really makes something one object as opposed to multiple? We're so accustomed to labeling objects in our environment that we often don't realize what a nuanced and sophisticated act it really is.

Consider, for example, an apple sitting on a table. We can pick up the apple, hold it in our hands, roll it around some, take a bite out of it, whatever we want to do with it really. We typically think of an apple as one thing. Is it really though? What about all the constituent parts, each little particle inside the apple, each of which could also be considered to be an object in its own right? What about the relationships that each of those particles have to each other and to the outside world? Aren't those also objects too?

Indeed, rather than viewing the apple as just one monolithic thing, we could instead view it as a set of objects and relationships. It is only taken together that these constituent parts constitute the apple. Think about that though. Whether we consider the apple as just one thing or as the set of all of its constituent parts and relationships, it is still the same thing. How we view it does not change what it is. An apple *is* its constituent parts and relationships.

Yet, if the traditional way of thinking about sets is to be believed, then the set of constituent parts would not be considered the same as the whole. If sets were analogous to physical containers and had substance of their own, then wrapping a group of objects in a set would create something *different from* the group of object themselves. It would mean that an apple imagined as one whole thing and an apple imagined as two halves held together were different things, as if the later were not an apple. Yet, it is. It must be. We are merely framing it in a different way, not actually changing the object itself.

Let's take this thought experiment even further though. Let's consider an even more nuanced way of thinking about the apple. Let me ask you this: What's the difference between an apple and a set containing only that same apple? The traditional view would say that these are two different things, just as it would say that 7 and {7} are two different things. Is that actually conceptually correct though, in the most natural sense of what a set should really mean?

Let's think about it. What fundamentally is the most parsimonious way of creating a set, the way that requires the least extraneous assumptions and the least unnecessary complications? Well, to that point, let's consider a real-life example to illuminate our thoughts. Let's say I wanted someone to pick up two colored balls from the table in front of me, one black and one white. I could point to the two balls, one after the other, and tell the other person to pick them up, and in this way I will have created a set in order to communicate my intent.

How would this be any different though if I only had one ball on the table instead? What if I only had a black ball or a white ball, and not both? What if I then pointed to that single ball and asked someone to pick it up? Would it magically stop being a set and start being just the one ball alone instead? Would it then be considered to exist outside of any set? That's what the traditional view on sets would lead you to believe. It treats individual objects as being fundamentally different from sets of one object. Is that right though? Should individual objects truly be treated differently from sets? Is that truly the most natural way of thinking here?

Well, let me propose this. Let's forget about the traditional view on how sets are supposed to work for a moment. Let's suppose instead that sets are more like abstract labels than physical containers. Why should an apple be any different than a set containing only that same apple? If sets are just abstract labels for referring to objects, then aren't we implicitly creating such an abstract label even when we are only talking about one specific apple? Hence, aren't we actually still creating a set anyway, even then?

Why would me pointing to two objects on my table and giving them a name create a set, and yet pointing to just one object on my table and giving it a name *not* create a set? I'm doing the same thing, in the same way, with the same purpose, and with the same semantics. The resulting reference that I have communicated to the other person can also be used to accomplish all the same kinds of tasks, regardless of whether the reference is to one object or to multiple objects. Why then would I treat an apple and a set containing only that same apple as different? There is no natural reason. They aren't really different.

An individual object *is* the same thing as a set containing just that one object. Sets are like the wrapping paper of our language. We can't talk about objects without implicitly using that linguistic wrapping paper. Every time we talk about any object, even just a single object, we are actually implicitly wrapping it in a set by doing so. Thus, for example, we actually *can't* talk about 7 without talking about $\{7\}$, because 7 fundamentally *is* a set containing only 7. Why treat individuals as existing outside sets? What do you even actually gain from it?

Treating individuals as not being sets of one effectively forces us to treat individuals and sets as separate special cases. That's not necessary though. We don't have to do that. It is a needless overcomplication, a false distinction. I mean think about it really. How can an individual object *not* be a set of one? Isn't a set of one really what it *means* to be an individual in the first place?

Why treat them as different? Indeed, the opposite system, where an object and a set containing only that same object are treated as being the same can be made to work and to be rigorous and well-defined. Doing so also causes multiple paradoxes and foundational problems in set theory to instantly evaporate.

In order to create such a system though, we will need to pin down these concepts a bit more and make them more precise. Our thought experiment so far has been illuminating, but it will start becoming unwieldy if we don't formalize these concepts we're working with here a bit more. We'll need precise definitions, clear terminology, and well-chosen symbols if we going to be able to keep everything properly distinguished in our head.

As such, our first order of business will be to give these two different notions of sets (as physical containers vs as abstract labels) corresponding formal terms. Both are valid in a sense, but one is a more natural and conceptually correct model of the fundamental essence of sets than the other. Here are the definitions:

Definition 153. *If we think of a set as being analogous to a physical container, such that an object X and a set $\{X\}$ are treated as being two different things, as if the set $\{\ \}$ itself has tangible substance of its own, independent of its contents, then let us refer to* such a set as being an **opaque set**. Such sets are currently the dominant interpretation of how a set should behave, given the existing views at the time this book was written.

We will represent opaque sets using ⦃ ⦄ as our delimiters. Theses symbols were chosen here because the extra vertical lines you see here in the set brackets can be interpreted as representing a physical barrier that prevents you from "seeing through" the set, and hence serving as a reminder that the set must be an opaque set, i.e. a set with existential substance of its own, independent of its contents.

Definition 154. *If we think of a set as being analogous to an abstract ghostly label, which we merely use to point to some collection of objects, and which is thus not a physical thing in and of itself, such that an object X and a set {X} are treated as being the same thing, as if the set { } has no tangible substance of its own, then let us refer to such a set as being a **transparent set**. Such sets are not currently the norm in set theory, at the time this book was written.*

We will represent transparent sets using the standard { } set brackets as our delimiters. These symbols were chosen because they fit well with the ⦃ ⦄ symbols by analogy. However, the obvious problem with this convention is that it conflicts with the existing use of these { } symbols in the literature. Thus, it will be critical that you always say up-front that you are using this convention if you do so, in order to avoid confusion with the existing literature of set theory.

The other reason I have chosen to use { } for transparent sets (despite the usage conflict) is because all the sets I have used prior to this point in this book contain no nested sets and thus still have the same effective meaning in their respective contexts (as you will see). Transformative logic and unified logic were actually always intended to use transparent sets, I just didn't tell you that until this moment, because it wasn't important until now and would have been an unnecessary distraction.

You will notice however that the way I've permitted set brackets to be omitted at various different points throughout the book (prior to this point) will correspond perfectly well with the behavior of transparent sets, in that X and $\{X\}$ are treated as being the same and hence the occasional omission of the { } is not just a convenient convention but indeed is logically permitted. This is very intentional. Doing it this way avoids confusion with respect to how my system is designed to ultimately work, without having to go back and re-explain everything using a new notation.

From this point forward in our discussion, { } will always represent a transparent set and ⦃ ⦄ will always represent an opaque set. More details on how to work with transparent and opaque sets will be given as we go along. There are some very significant differences in how they handle certain cases. Care should be taken. Many famous properties that apply to opaque sets do *not* apply to transparent sets. Be careful that you aren't doing something with one of these types of sets that depends on results that aren't valid for that type of set.

The idea of transparent sets may seem a bit odd to you right now, especially if you're highly accustomed to opaque sets, but it will start to make more and more sense as I explain the details of how to make it work. There are subtle aspects to how these sets differ that seem to strongly point in the direction of transparent sets as being more natural. Here's a simple example, in fact: Consider the empty set. Which type of set would make more sense to represent it?

Keep in mind that the empty set is supposed to represent *nothingness*, i.e. the absence of *any* objects. If we modeled the empty set as an opaque set would it really even be nothingness? Nope. It wouldn't. An empty opaque set cannot truly represent nothingness, because it is *not* nothing. An empty opaque set still has existential substance. It is still a thing that *exists*, and hence cannot be a correct representation of nothing.

In contrast though, an empty transparent set doesn't have any substance of its own, and so it really does represent nothingness. The symbols we use to represent an empty transparent set aren't themselves nothing of course, but the underlying concept we're modeling still is. Empty transparent sets behave like real nothingness, whereas empty opaque sets do not. A great example to illustrate this point is the so called "set theoretic definition of the natural numbers", which is basically a really asinine and convoluted way of defining non-negative integers in terms of nested empty opaque sets. It's the kind of thing that ivory tower academics love, but which people who care about actually getting real work done (like me) tend to hate. It creates a sequence that looks something like this:

1. $0 = \{\}$
2. $1 = \{0\} = \{\{\}\}$
3. $2 = \{0, 1\} = \{\{\}, \{\{\}\}\}$
4. $3 = \{0, 1, 2\} = \{\{\}, \{\{\}\}, \{\{\}, \{\{\}\}\}\}$
5. ...

Hilariously contrived, isn't it? Sorry out-of-touch ivory tower academics, but 2 *isn't* a set containing 0 and 1. Rather, 2 is just a counting number, specifically the counting number that comes after 1. That's all there is to it. The non-negative integers are just a sequence of arbitrary forms that we use to quantify the relative magnitudes of things. This whole nested empty opaque set system is an terrible way of thinking of it. It confuses more than it illuminates. Just because you *can* model numbers this way doesn't mean that this is what numbers really are and it also doesn't make it a good idea. What you gain is not worth the price. It isn't merely pedantic; it is also conceptually misleading and unrepresentative of the real underlying nature of the system.

There are better and worse ways of thinking about what something is, and the so called "set theoretic definition of the natural numbers" is definitely one of the worse ones. This is exactly the kind of thing that makes a lot of people dislike higher-level mathematics and pushes so many people away from ever wanting to study mathematics in-depth. It's ludicrous. It's like hanging a sign around mathematics' neck that says "We enjoy inflicting unnecessary pain on ourselves and our students. We don't care at all about whether what we're doing creates real-world value or not.". This kind of thing is part of where mathematics' bad reputation in the general population comes from.

Hey mathematicians, you know what's a lot better than building up your entire mathematical world based on nothing but empty opaque sets? Simple: Just use whatever arbitrary forms and algorithms you need to accomplish what you want along the way. It's far easier and far more flexible. It also makes far more sense and doesn't make you look like some kind of out-of-touch fool. Arbitrary forms combined with arbitrary algorithms are fully computationally general. There is no need for this kind of contrived crap.

"But how can we build up our universal set-theoretic math world without using empty opaque sets?" says some mathematician somewhere, inevitably, as they read this. Well, to that point, keep in mind that even when you work in your universally empty opaque set based framework of thought, that you build up abstractions and then ignore the lower-level details once you're past a certain point. You can just do the same thing here, except with arbitrary forms as your foundation instead of empty opaque sets. There's nothing really stopping you from doing it.

Plus, there's also the fact that transparent sets seem to eliminate the need for the ZFC axioms[70] in certain respects, but more on that later. We've kind of gotten derailed from our original thread of conversation, due to me ranting for a little while about how I dislike the "set theoretic definition of the natural numbers" etc. Let's get back on track now and continue our prior discussion.

Specifically, recall that I said that the reason I brought up this whole "set theoretic definition of the natural numbers" thing was because it makes a great example for illustrating that empty opaque sets are most certainly not nothing. Indeed, they are definitely something and thus cannot ever capture the essence of true nothingness and emptiness. Transparent sets, in contrast, can represent nothingness in a more conceptually correct way. More on that in just one moment though.

We should first fill a terminology hole that has been created by this distinction between the behavior of empty transparent sets and empty opaque sets. We can't safely use the same symbol to refer to two different concepts. Specifically, we can no longer use $\emptyset$ safely as shorthand for the empty set, because it is now ambiguous between the two types of sets. As such, we now need two new corresponding symbol definitions to fix this:

Definition 155. *As shorthand for the **empty opaque set**, i.e. for $\{\!|\ |\!\}$, we may write $\hat{\emptyset}$. We may no longer use $\emptyset$ for this purpose. Notice that the difference is that there is now a "hat" symbol written above what used to be only $\emptyset$.*

Definition 156. *As shorthand for the **empty transparent set**, i.e. for $\{\ \}$, we may write $\emptyset$. In other words, we have now redefined $\emptyset$ to refer unambiguously to only empty transparent sets. Moreover, if we say "empty set" without qualification, the intent should now default to meaning a transparent empty set instead of an opaque empty set.*

Anyway though, back to the main point. Why is it that the empty transparent set $\emptyset$ is better at representing nothingness and emptiness than the empty opaque set $\hat{\emptyset}$ is? Well, we already talked about some of that, but let's think about it some more. In particular,

---

[70] The ZFC axioms are the most common foundational assumptions for academic set theory currently. The ZFC axioms can often be overlooked in many practical contexts though. Don't worry if you aren't familiar with the ZFC axioms yet, but feel free to look up more info about them if you want.

I want you to consider the following informal premise: You can't make something out of nothing.

Doesn't that sound like something that should be true of an object that represents nothingness, if indeed it does? I mean, think about it. If I have nothing, and I add more nothing onto it, should I now have something? Or, should I still have nothing? If you ask me, I think the later should be true. Adding nothing onto nothing should still leave you with nothing. It shouldn't be possible to manipulate nothing in any way that causes it to suddenly become something. That would be contrary to its nature. Nothing isn't something. It's the absence of existence. You can't build up new things out of nothingness.

Yet, if you look at empty opaque sets, they *do* let you build up something from "nothing". The so called "set theoretic definition of the natural numbers" is an example of this. Thus, it seems that an empty opaque set therefore could not truly be nothing. Let's now turn our attention to transparent sets though. To truly understand the nuances of why transparent sets really can model nothingness, we should first properly grasp their behavior better.

You see, as it turns out, you can't build up anything new by nesting empty transparent sets inside each other. Any transparent set that is filled with any number of nested transparent sets, each of which contain no other forms except for other transparent sets, must always be equal to the empty transparent set. Thus, for example, there is no difference between $\{\!\{\ \}\!\}$ and $\{\!\{\ \},\{\!\{\ \}\!\}\}\!\}$. They both represent the exact same thing, just written a different way. Both are equivalent to the empty transparent set $\{\ \}$ (a.k.a. $\varnothing$). This example is just one special case of this kind of behavior though. Let's give a broader definition for this behavioral quirk of transparent sets:

Definition 157. *Transparent sets don't themselves have any substance of their own, and thus nested transparent sets are logically equivalent to their corresponding flattened transparent sets. Only forms other than transparent sets have substance of their own. It is only those forms that are not eliminated during this process. Let us call this property that transparent sets have of always being equivalent to flattened versions of themselves* **transparent subsumption**.

This is where transparent sets get their name from. It is as if you can "see through" transparent sets, as if they are not even there, when you nest them inside each other. The word "subsumption" here was chosen because subsumption means "to absorb something into an encompassing set", which is quite an accurate description of the process that is going on here. Nested transparent sets effectively get absorbed into the surrounding outermost set, until nothing remains except for a flattened set containing only opaque forms.

Thus, for example, $\{\{1,2\},\{\{3\}\}\}$ is the same set as $\{1,2,3\}$ for transparent sets. As you can see, the forms 1, 2, and 3 are not being reduced any further though. That is because they are not transparent sets. They are more like solid objects. They are opaque forms. Notice also that there is nothing stopping you from mixing transparent sets and opaque sets together and the result will still be perfectly well-defined.

For instance, $\{\!\{\!\{\ \}\!\}, \{\!\{\ \}\!\}$ cannot be reduced any further, because $\{\!\{\!\{\ \}\!\}$ and $\{\!\{\ \}\!\}$ are like solid objects. They are like physical cardboard boxes. You cannot safely ignore

502

a physical box. It is not the same thing as the kind of ephemeral label or ghostly abstraction that a transparent set represents. You can however make things more complicated than they need to be, when working with transparent sets, if you want to. Let's define a term for that too, i.e. for the opposite process of transparent subsumption:

Definition 158. *Since transparent sets have no substance of their own, and have the property of transparent subsumption, you may therefore add on as many additional nested layers of transparent sets (and also redundant copies of subsets of the set) as you want without changing what set it is. In other words, you can frame the set however you want. Let us call this process of adding on unnecessary complications to the structure of a transparent set* **transparent convolution**.

For example, $\{1, 2\}$ and $\{\{\ \}, \{1\}, \{2\}, \{1, 2\}\}$ are the same set for transparent sets[71]. The set $\{\{\ \}, \{1\}, \{2\}, \{1, 2\}\}$ is a transparent convolution of $\{1, 2\}$. Likewise, $\{1, 2\}$ is a transparent subsumption of $\{\{\ \}, \{1\}, \{2\}, \{1, 2\}\}$. You may hold the form of the transparent set in whatever representation you want at any given moment, since they are all equivalent anyway. You don't have to simplify it if you don't want to. Always remember that transformations of forms are optional. Just because you can reduce $2+2$ to 4 doesn't mean you actually have to.

Standard math education often misleads students in that respect, by giving them the impression that simplification is mandatory, as if evaluation is always necessary, when it really isn't[72]. Indeed, you can keep any form in unreduced form for as long as you want. You can keep it that way forever, if you want to. It's all just arbitrary timeless forms. Don't feel like you're always in a rush. You're not. All well-defined forms are equally legitimate in the eyes of logic and mathematics. So, naturally, applying transparent subsumption or transparent convolution is arbitrary in that sense, although the flattened version is the most streamlined version of course.

I know it may seem weird to some people that we let $X = \{X\}$ in this system, but it's really perfectly fine to do so. The forms X and $\{X\}$ are just arbitrary forms like any other, so we can do whatever we want with them as long as we're consistent with respect to the properties we are trying to model. Don't worry. I know it might seem unnerving and strange at first, but you'll get used to it and you'll eventually realize that it works just fine.

The fact that you could continue this process arbitrarily long, such that X could be transformed into $\{\ldots \{X\} \ldots\}$, with any arbitrarily large number of $\{\ \}$ brackets, actually poses no problems, neither conceptual or practical. No matter how many (finite) $\{\ \}$ you wrap an object in, it's still perfectly well-defined. The fact that you have the freedom to apply as much transparent convolution as you want to a transparent set does not at any point cause that transparent set to become poorly defined. That's why there's nothing to worry about here really. A transparent set won't explode like a fountain of chaos or anything like that. You can hold a transparent set in whatever form you want, for as long as you want.

---

[71] If the duplicate copies of the numbers in this set here seem strange to you, please remember that duplicates don't matter in sets, regardless of whether those sets are transparent or opaque.

[72] This is probably the result of too much rote "learning" in the education system and not enough genuine understanding of concepts.

You wouldn't let the fact that 4 is the same thing as $0 + 0 + 1 + 3 + 0$ cause you to think that 4 is an invalid object would you? No? Good. Then, if you are self-consistent in how you think about logic and math, you shouldn't have a problem with the way transparent sets behave either. You can put a transparent set in whatever form is most helpful to you at the moment, just as you can with any other form in logic and math, as long as your transformations are justified under your existing rule set.

If this discussion of transparent subsumption and transparent convolution is not enough to prove to you that empty transparent sets are a better representation of nothingness and emptiness than empty opaque sets are, then let's take a moment to consider the intuitions as well. Consider, for example, the case where we are just wrapping empty transparent sets around each other, i.e. transparent sets that don't contain any opaque forms. What are we really conceptually doing when we do this?

Well, to answer that, let's think back a bit to the thought experiment that originally inspired transparent sets. As you may recall, we discussed this notion that transparent sets are like abstract ghostly labels and that indeed referring to any object actually implicitly wraps that object in a set, even if you only refer to one object. Similarly though, what are we conceptually doing when we wrap empty transparent sets around each other?

The answers lies in applying our labeling analogy again. For example, suppose we had ⦃ ⦃ ⦄ ⦄ as our set. What, conceptually, is this set like? Well, applying the labeling analogy, it is as if we are pointing to nothingness, so that we can refer to it, which then causes it to be wrapped in a transparent set (yielding just ⦃ ⦄), but then pointing at it again at an additional level of indirection, which causes it to be wrapped in yet another transparent set (yielding ⦃ ⦃ ⦄ ⦄).

Yet, no matter how many levels of indirection we use in referring to nothingness, it still remains nothingness. It is like following pointers in a computer program. It all still points to the same thing. Even if you have to go through multiple pointers to reach the final destination, it is still the same place. That's a more conceptual and informal sense of what ⦃ ⦃ ⦄ ⦄ means. The same analogy holds for other arbitrary nested empty transparent sets as well.

One other thing that may concern you, is that perhaps you think that using transparent sets will cause us to not be able to talk about sets of sets. This too is a false concern, for the same reason as why transparent subsumption and transparent convolution don't cause transparent sets to become invalid. If you want to talk about sets of sets with transparent sets, then just choose not to simplify the set. Hold it in the form you need. That's perfectly fine. Plus, if you truly do need a set that acts like a physical container, then just use an opaque set. There's nothing stopping you from mixing them.

Here's an interesting thought though: If $X = \{X\}$, for some opaque form X, then doesn't that mean that every opaque set is always implicitly wrapped inside a transparent set? Indeed it does. It is true that $\{\ \} = ⦃\{\ \}⦄$. In fact, this generalizes even further. *All* objects are *always* implicitly wrapped in transparent sets. There is no such thing as an object that exists outside a transparent set. Transparent sets are the wrapping paper of our language. We cannot even refer to anything without implicitly wrapping that thing in a transparent set[73]. That's because a transparent set is what a reference to something conceptually *is*. This results in relational unified logic having a very interesting property, as a consequence of using transparent sets:

Definition 159. *Relational unified logic has both a **universal domain** and a **universal closure**. All objects in relational unified logic are transparent sets (hence a universal domain), and thus it is also true that all operations return transparent sets (hence a universal closure).*

Specific sets of objects can still have their own unique sets of distinguishing properties of course, but nonetheless, at the highest level, every object can be thought of as being wrapped in a transparent set. Regardless of what arbitrary rule set you define, this universal domain and universal closure over transparent sets will always be present. Thus, these properties truly do apply to everything in relational unified logic. Everything is a transparent set.

This also has the net effect of causing virtually any combination of operations and operands to become well-defined, even those that weren't originally intended to be used together. It allows (for example) the empty set to always be returned whenever an operation does not otherwise make any sense, hence a usable return value will always exist in principle. You'll see what I mean better later. There are still some more ideas we'll need to explore before we can understand the big picture here.

For example, since everything is a set, it is perfectly legitimate to say 2∩3, since both 2 and 3 are sets in relational unified logic. Indeed, 2∩3 is the same as {2}∩{3}, which in turn is the same as { } (a.k.a. ∅). So, as you can see, I'm really not joking about this. I really do mean that *everything* is a set. Things can be more than just sets, but they are always also sets regardless of what else they are. This is perfectly well-defined from the standpoint of transformation rules. They are all just arbitrary forms.

Oh, and by the way, if you're wondering if this is what I was talking about when I said that unified logic made it possible to apply numbers to each other *as operations*, then no. This is *not* that. I mean, it is a related concept, as you will see later on, but it's not the same thing. There is actually a much deeper and much more interesting sense in which you can apply a number to another number (or indeed any object to any other object) in a meaningful and non-trivial way. I know it may sound a bit insane, but once you see it and it clicks it'll suddenly open your eyes to a whole other world of possibilities that you've probably never noticed before. It'll make the stuff we've seen so far look tame. You probably won't expect it. More on that later though. We still have a lot to cover before we get to that crazy stuff.

For example, the concept of membership in a set doesn't work quite the same way for transparent sets as it does for opaque sets. In opaque sets there is a very clear distinction between elements of the set and subsets of the set. However, this is not true for transparent sets. For transparent sets there is no difference between elements and subsets of a set. Every subset of a transparent set is also an element of that set, and likewise every element of a transparent set is also a subset of that set. Sounds a bit weird right? Well, take a moment to actually think about it though.

---

[73] e.g. We cannot have 3 without also having {3}. 3 *is* {3}.

Let's consider a concrete example, to make the point clear. Consider, for example, the transparent set $\{1, 2, 3\}$. It is certainly true that $\{2\}$ is a subset of this set, wouldn't you agree with that much at least? Sure you would. However, isn't it also true that we can apply transparent convolution to $\{1, 2, 3\}$ to transform it into the transparently equivalent set $\{1, \{2\}, 3\}$? Thus, isn't it true that $\{2\}$ must also be an element of $\{1, 2, 3\}$, in addition to being a subset of it? Yep, it sure is. Indeed, this must clearly be true of *all* subsets and elements of any set, since $X = \{X\}$ is always true for transparent sets and hence anything that is currently a subset of a set could also be interpreted as instead being an element of that set.

So, that demonstrates the case where subsets are always also elements of a transparent set. The other direction, i.e. that every element of a set is also a subset of that set, is even easier. Indeed, even opaque sets have this property. It's not unique to transparent sets. A subset is just a selection from the elements of a set to form a new set, and it is always possible to select only one of those elements to form your subset, and so of course every element of a set is also a subset of that set. This is true regardless of whether we are talking about transparent sets or opaque sets.

We have thus now demonstrated both that an object being a subset of a transparent set implies that it is an element of that transparent set and also the converse. We have thus demonstrated both directions of implication. Consequently, these two concepts must be logically equivalent for transparent sets. Therefore, there is no point in distinguishing between elements and subsets in the context of transparent sets. We can thus use the same word and the same symbol to refer to both concepts and it will make no difference in our outcomes. Let's define a new term for this property:

Definition 160. *For transparent sets, there is no difference between elements and subsets. They are logically equivalent. We will refer to this property of transparent sets as* **transparent element-subset equivalence**, *or (optionally) just* **element-subset equivalence** *for short if the context is clear. The fact that this is true implies that any time you talk about subsets you are also talking about elements, and vice versa.*

It also implies that there is no point in having two different "element of" ($\in$) and "subset of" ($\subseteq$) operators for transparent sets. Just one will suffice. Well, you could use both if you wanted to, technically, but that would be confusing. It would look like a false distinction and thus would be misleading. Thus, we will arbitrarily decide to only use one of these terms and symbols.

It makes no logical difference which one we choose. However, for the sake of standardization, we will choose to use "subset" and $\subseteq$ as our standard terminology for both cases now, except in cases where pointing out the equivalence explicitly is still useful, such as when comparing relational unified logic's behavior with the behavior of other logical systems that still distinguish between "element of" and "subset of", and making analogies and correspondences between them.

Thus, expressed in symbols, whenever $X \subseteq Y$ is true so must $X \in Y$ be, and vice versa, if X and Y are transparent sets. Notice that similar thinking doesn't work on opaque set relationships, just as stated in our definition. Also, keep in mind that transparent "element of" and "subset of" are different relationships than their analogs for opaque sets.

506

For example, consider $\{\{1, 2, 3\}\}$. Is $\{2\}$ an opaque subset of $\{\{1, 2, 3\}\}$? Yep, it is. However, is $\{2\}$ an opaque element of $\{\{1, 2, 3\}\}$? Nope. We'd need $\{\{1, \{2\}, 3\}\}$ to be our set for that to be true, but then we'd be talking about a fundamentally different set, since opaque sets are analogous to physical boxes. The fact that $2 = \{2\}$ is what makes the analogous example for transparent sets different.

So, having said all this, we've pretty much covered the basic formal perspective here. What about the intuitive perspective though? Does this element-subset equivalence property make sense in the context of our intuitions and our natural sense for what transparent sets should be? To answer that question, we should think back once again to the thought experiment by which we originally framed the concept of transparent sets: the apple on the table.

Do you remember the apple? Do you remember the corresponding thought experiment? To refresh you memory, the idea of the apple thought experiment was for us to re-examine what it means to label an object as being "one" thing. You can think of the apple on the table as one monolithic unit ("the apple"), or you can think of it instead as the set of all of its constituent parts (apple particles, etc) and their relationships to each other and to the outside world. Either way you think of it though, it's still the same apple. Whether we frame it as one monolithic unit or as a set of many things, it must still be the same thing.

How we think of a thing does not change what the thing itself is. We already broadly understand this idea, but let's apply it to the more specific context of element-subset equivalence and see what happens. What does this mean in the context of the apple? Well, it means we need to think of the apple in terms of elements versus subsets. So, let's do that.

Suppose, for example, that we cut the apple into eight slices, not with a real knife but with our minds[74]. We can think of each of these eight apple slices as monolithic things, just as the apple as a whole was. In this case, we would be thinking of each apple slice as being like an element. Each of these eight apple slices are members of the set of which the whole apple is composed.

This is all well and good of course, since it is just how we are thinking about it. However, what if instead of thinking of each apple slice as monolithic units, we thought of each of them as being sets of their own constituent parts just like we did in the thought experiment with the apple as a whole. Well, in that case then, aren't each of those apple slices now subsets instead of elements, conceptually? Similarly, couldn't we have started by framing our thought in terms of these eight subsets, but then decided to think of each of those subsets as elements instead?

Indeed, we could. We could do this thought experiment in either direction, equally well. Thus, element-subset equivalence lines up perfectly well with our foundational intuitions on how transparent sets would be expected to behave. This is additional evidence that everything is working as it should. Our foundational intuitions and our formalisms are still correlating perfectly well, which means we're probably still on the right track with our formal implementation.

---

[74] We don't want to actually change the state of the apple, strictly speaking. That's why I said "in our mind" instead of in reality. This is a nit-picky detail though.

If all of this isn't enough for you, then try this: Imagine for a moment that element-subset equivalence weren't true, but we still had the properties of transparent subsumption and transparent convolution at least, otherwise there wouldn't be any point in distinguishing between transparent sets and opaque sets. Suppose someone then asked us whether or not $\{1,2\}$ is an element of $\{1,2,3\}$? How could we answer them? If we look at $\{1,2,3\}$ as it stands, then it would appear that $\{1,2\}$ couldn't be an element of $\{1,2,3\}$. In this form, only three elements are apparent: 1, 2, and 3.

However, $\{1,2,3\}$ is also equivalent to $\{\{1,2\},3\}$ via transparent convolution, and hence whatever is true of $\{\{1,2\},3\}$ with respect to its elements and subsets must also be true of $\{1,2,3\}$. However, if we instead looked at $\{\{1,2\},3\}$ as it stands, then this would suggest that $\{1,2\}$ *is* an element of the set. It's clearly visible as one of the elements in $\{\{1,2\},3\}$. Thus, we have a dilemma. The answer is ambiguous and impossible to resolve, and therefore our assumptions must be invalid. Our system would be inconsistent in this case. Thus, the only way we can retain logical consistency when working with transparent sets is if we do acknowledge that element-subset equivalence must be true. There is no escaping it.

And so, we now come full circle, successfully connecting our intuitions and formalisms together. It is now clear that transparent element-subset equivalence must be true, even though it may seem absurd at first glance to the uninitiated. Pretty cool right? That's not the only thing that's cool here though. Transparent sets have plenty more interesting consequences than just that.

In fact, if you're well familiar with the math literature, there's a good chance several other inevitable consequences of the behavior of transparent sets have already occurred to you while we've been talking about all this. There are some pretty profound consequences of this behavior. For example, suppose that I asked you whether a transparent set was a member (an element) of itself. What would the answer be?

The answer would always be yes. This is true regardless of what the set is. For instance, if we had $\{1,2,3\}$ then this would also be equivalent to $\{\{1,2,3\}\}$, and therefore $\{1,2,3\}$ must be an element of $\{1,2,3\}$, i.e. must be an element of itself. The set $\{1,2,3\}$ is also a subset of $\{1,2,3\}$ as well, of course, but that would be true for opaque sets too[75]. Think about all this though. This is not some small piece of trivia. The consequences are profound and far-reaching. Before we talk about that more though, let's make a formal definition for this:

Definition 161. *For transparent sets, every set is always a member (an element) of itself, in addition to being a subset of itself. We will refer to this property as* **transparent self-membership**, *or (optionally) just* **self-membership** *for short if the context is clear.*

So, let's talk about consequences now. Who here has heard of Russell's paradox? It's a super famous paradox in the world of mathematics. In fact, Russell's paradox is so important that resolving it was actually the original motivation for creating the entire ZFC (**Z**ermelo-**F**raenkel Set Theory with the Axiom of **C**hoice) system of axioms, i.e. for the most common basis for modern set theory, upon which many set theory related

---

[75] Don't confuse subsets with proper subsets, by the way. A proper subset has to be smaller than its superset. However, when we just say "subset" without qualification then we are allowing for the possibility that the subset could be the same set as its superset.

math research papers depend. Here's Russell's paradox for you: "What is the set of all sets that do not contain themselves[76]? Is this set also a member of itself or not?"

Naturally, since the existing math literature always (apparently) assumes opaque sets as a given, this paradox has historically always been thought about in terms of opaque sets. Do you see why this question is problematic in that context? Think about it. It is analogous to the liar's paradox. Let's call the set of all sets that do not contain themselves R, in honor of Bertrand Russell, the person who invented it.

Suppose we assume that R is a member of itself. Then, as a member of itself, it cannot possibly satisfy its own criteria, since R is precisely the set of sets that are *not* members of themselves. Hence, we have a contradiction. If, on the other hand, we assume that R isn't a member of itself, then surely it must belong in itself, since it now clearly satisfies the criteria necessary for membership in itself. Hence, we yet again have a contradiction. Either way, we're screwed. Each assumption implies that its own negation is true. This is the essence of Russel's paradox.

For opaque sets, all this reasoning makes perfectly good sense of course. After all, how could an opaque set ever be put inside itself? That would be like trying to put a physical box inside itself, which clearly would make no sense. What about transparent sets though? Think about it. There is no impediment to putting a transparent set "inside" itself. Transparent sets aren't like physical containers. Instead, transparent sets are more like abstract labels or pointers. An abstract label can easily be self-referential, and hence, by analogy, "self-containing". Indeed, not only is there no *barrier* to putting transparent sets inside themselves, but they *always* are.

All transparent sets are elements of themselves. Thus, let us now revisit Russell's paradox. Let us ask ourselves: What is the transparent set of all transparent sets that do not contain themselves as elements? Is this set also a member of itself or not? The answer is obvious now. Let's represent the set again with R, except this time R is a transparent set. What is the value of R? The answer is simple: $R = \{\ \}$.

In other words, under transparent sets, Russel's set is the empty set. But, is this set a member of itself? Well, kind of, yes, in a manner of speaking, but there's more nuance to it than that. The empty transparent set is a member of itself, but it is an empty member at that, a member that represents nothingness and falsehood. So, the empty transparent set is not really a substantive member of itself, since substance requires existence (i.e. non-emptiness, in unified logic) and that's exactly what the empty transparent set lacks. Whether you think about the subset relation involved here in non-existential or existential terms sort of changes the interpretation a bit. However, the point is, it is sort of a member, but in a way that we can ignore (relevant: nothingness is falsehood from an existence-based truth perspective). Thus, we now have a well-defined answer to Russell's paradox, one that let's us not get ourselves all tied up in knots about sets and move on to other things.

There is a bit of weirdness to this maybe, but it lets us plug the hole with a usable value. The beauty of working with transparent sets like this is that it will let us work with sets in a "naive" (i.e. simple or intuitive) yet still rigorous way, such that we will generally be able to ignore the need for tedious long-winded axioms about what exactly

---

[76]The word "contain" here is referring to containment in the sense of elements, not in the senses of subsets.

a set is (e.g. ZFC) and instead will be able to express ourselves in a more natural and seamless way.

Indeed, the ZFC axioms are quite a monstrosity compared to the "naive" perspective of thinking of sets as just collections of things that match whatever arbitrary properties we want. Transparent sets let us get closer to that "naive" ideal. Abandoning ZFC, so that you no longer have to deal with its tedium, is quite a desirable advantage if you can do it successfully. There's more to transparent sets' advantages than just overcoming Russell's paradox though.

Transparent sets also allow us to overcome a few other problems in set theory as well. For example, traditionally, when working with opaque sets, there is no way to define a fully general universal set. Indeed, some of you may have noticed that I repeatedly used universal sets in this book, even though some types of universal sets are traditionally considered to not be well-defined. Universal sets that have a sufficiently limited scope (e.g. only integers) have always been well-defined of course, if that's the only thing you care about. However, universal sets that are allowed to contain other sets have generally been considered to be poorly defined.

Think about it. Think about the most general possible universal set you could imagine, covering pretty much all conventional mathematical objects. Think about a 'set of everything", in other words. Wouldn't it have to contain itself, in addition to everything else it contained? Indeed it would. For opaque sets, this would be impossible. No opaque set could ever contain itself. That would be like putting a physical box inside itself, which would be impossible, as we have already discussed. What about a universal transparent set though?

Such a set would have no problems. A universal transparent set could easily be a member of itself, and indeed *must* be if it is a transparent set. Thus, in my system, the universal set is generally well-defined. That's why I used it throughout the book without much concern about doing so. I knew that we would eventually reach this point in the book, where I would explain that universal sets are generally well-defined in my system, since they are transparent, and hence able to contain themselves without causing logical contradictions or paradoxes. See? I had a reason for what I was doing after all. It's all in keeping with my long-standing tradition of not giving you the full picture until it becomes necessary, so that you can instead focus on the concepts better in isolation and learn more effectively.

Explaining too many details at once tends to make things murky. When you're learning something new, you need the various aspects of what you're learning to be clearly isolated from each other, otherwise your mind won't feel confident in making strong associations between specific things. It won't be able to decide which of the overwhelming number of details it should form a confident association with. Thus, the mind floats like driftwood on the ocean, learning nothing, when it is overwhelmed.

That's part of why it's really important to truly take the time to try to capture the conceptual essence of things, to really examine things from multiple angles and with multiple concrete examples, to really think about things in a principled way rather than merely a mechanical or technical way. Otherwise, we wouldn't be able to pin down enough pieces of info to ever overcome our uncertainty. Our mind would remain adrift in the ocean of information, unable to steady itself. People need solid ground to stand on in order to learn effectively. A waterfall of too much information just knocks them off their feet instead. Anyway though, here's a formal definition for this property of having a well-defined universal set:

Definition 162. *If a theory has a well-defined universal set, then we may say that it has the property of* **universality of sets**.

Incidentally, our system here, relational unified logic, seems to have the property of universality of sets. We just demonstrated that. Transparent sets are what enable it, clearly. This is all enabled by transparent element-subset equivalence and transparent self-membership. However, those aren't the only interesting properties in play here. There are also some other interesting effects as well.

For example, all transparent sets are their own power sets. Do you remember what a power set is? Let me fresh your memory, just in case: A power set is the set of all subsets of a set. It is clear that all we have to do to create a power set for a transparent set is to apply transparent convolution to it until we get a set that looks like the set of all subsets of what it previously looked like. However, transparent convolution always generates equivalent sets, and therefore every transparent set must be equivalent to its own power set. Here's a corresponding definition:

Definition 163. *For transparent sets, every set is equivalent to its own power set. We will refer to this property as* **transparent power set self-membership**, *or (optionally) just* **power set self-membership** *for short if the context is clear.*

I actually already gave an example of a transparent set being equivalent to its own power set earlier, although I didn't mention this relationship at the time. Specifically, I am referring to the example of how $\{1, 2\}$ and $\{\{\ \}, \{1\}, \{2\}, \{1, 2\}\}$ were equivalent for transparent sets, which occurred on page 503, right after the definition for transparent convolution.

So, now we've covered both transparent self-membership and transparent power set self-membership. These are certainly both interesting properties. However, they are not the only examples of these kinds of equivalences. There are many other examples. Indeed, any kind of transparent convolution or transparent subsumption applied to a set will always yield an equivalent set. Transparent convolution and transparent subsumption are thus the general form of these kinds of equivalences, in other words.

Why then did I give names to these specific ones and focus so much attention on them? Well, the reason is because transparent self-membership and transparent power set self-membership just happen to be two of the most important and interesting cases of transparent convolution. They help illuminate the overall contours of the kinds of consequences that transparent convolution and transparent subsumption have. They are important enough cases that it is worth giving them special names and extra attention.

There's something else interesting I want to point out here. Did you notice how both empty sets and universal sets only correctly capture the essence of what they represent when the sets are transparent? Opaque sets in contrast don't quite seem to capture the conceptual essence of what they are supposed to represent correctly. What exactly do

I mean by this? Well, let's think back to our discussions of transparent empty sets and universal sets, and the related material.

For instance, do you remember how empty opaque sets can be nested inside each other to generate an overly complicated way of modeling non-negative integers, via the so called "set theoretic definition of the natural numbers"? Do you remember how that is problematic if you want an empty set to truly represent nothingness? How can you say that an empty set truly models nothingness if you are able to build up *something* out of it?

No amount of nothingness should be able to be built up to yield something. That would be contrary to the fundamental nature of nothingness. Yet, in contrast, empty transparent sets capture this effect just fine. No amount of nested empty transparent sets ever creates something from nothing. It always remains equivalent to the empty transparent set. This means that empty transparent sets are better models of nothingness than empty opaque sets are.

Similarly, do you remember how it was impossible to define a universal opaque set[77], because doing so would be tantamount to trying to put a physical box inside itself, which is impossible? After all, if a fully general universal set is truly universal, then it must contain itself as a member, since the universal set itself is one of the things existing in that logical universe.

Opaque sets just can't get around this problem. A fully general universal set just doesn't make sense for them. Yet, in contrast, universal transparent sets are easily attainable and make perfectly good sense. All transparent sets are members of themselves, and thus there will certainly be no problems making the universal transparent set also a member of itself. Thus, here again, where opaque sets have failed to properly model the concept, transparent sets have succeeded.

Thus, in both cases (i.e. for both empty sets and universal sets), transparent sets behave correctly whereas opaque sets do not. Think about what that means though. The existing literature would have you believe that opaque sets are the correct definition of a set. How can that be though, if opaque sets cannot model even something as simple and fundamental as empty sets and universal sets properly? If you really think about it, this is quite strong evidence that transparent sets are fundamentally a more natural definition for sets than opaque sets are.

That's my point here. That's why I brought up empty sets and universal sets again. They make a good example for illustrating why you should distrust the idea that opaque sets are the most natural definition for sets, even despite the fact that some properties of transparent sets may initially seem strange. Like many things though, unfamiliarity is often mistaken for genuine strangeness. Popularity isn't the same as merit. Indeed, in reality, transparent sets are actually quite natural, rigorous, and intuitive. Transparent sets only *seem* strange at first because so many people are so ingrained in thinking in terms of opaque sets.

Speaking of ingrained frames of mind though, there's something else I'd like to talk to you about besides just this whole transparent set versus opaque set business. After all, there's more to what makes relational unified logic different and interesting than just the

---

[77] for a sufficiently all-encompassing definition of "universal set", one that includes all possible sets

fact that it carefully distinguishes between these two different types of sets. There's also the fact that relational unified logic interprets sets of relationships in a special way, one that differs significantly from the currently ingrained viewpoint in some ways. Indeed, this is actually where the "relational" in relational unified logic comes from, as you will see. A thought experiment well help you understand. Let's begin.

5.8 The relational essence of what a thing is

Have you ever taken a moment to wonder what logical and mathematical objects actually *are*? I mean, obviously, they are at least things that we can operate upon and transform using logical and mathematical reasoning of course, but I'm not talking about that part. I'm talking about the deeper sense. What is a logical or mathematical object *really*? What makes one object different from another one? What makes them the same?

Let's consider a concrete example. Think about the number 2, for instance. What makes the number 2 what it is? Is it the symbol, the continuous line across the paper or computer screen which traces out the shape of the number? Is that what 2 is? Well, that's how we refer to it. That's the conventional representation of 2 across most cultures and systems of writing, because most of humanity has standardized on the Hindu-Arabic numeral system for math. But does that make the symbol 2 the essence of what 2 is? Surely not.

It is easy to imagine an alternative world where 2 is represented by some other arbitrary shape or form. For example, we could just as easily represent the number 2 using the symbol β, the 2nd letter in the Greek alphabet, and clearly this would not genuinely change the meaning and behavior of the number. It would just mean that we were using a different symbol to represent it. The essence of 2 itself though, on the other hand, would remain unchanged. Why? What is it that really gives 2 its fundamental essence and nature?

The answer is relationships. Relationships are the essence of what a thing is. A thing has no meaning in the absence of relationships to other things. A thought experiment will make this more clear. Consider, for example, the relationships that 2 has to other numbers and mathematical objects. Consider, for example, the fact that 3 is the successor of 2, i.e. that 3 is the (integer) number that comes after 2. What would happen if we deleted this relationship, such that 3 was no longer the successor of 2. Would it still even be the number 2?

The answer is no. If 2 did not have 3 as its successor, then it could not possibly be the same thing as the 2 that we conventionally know. It would have to be a different object entirely. Changing an object's relationships changes what that object is. This is generally true even if you only change just a single relationship. Think about what that means. It means that the identity, the fundamental essence, of objects in logic and math is inherently tied to the set of relationships that are relevant to those objects. The relationships *are* the object. Let's make a formal definition for this:

Definition 164. *An object* is *its relationships. More precisely, an object's substantive identity (its fundamental essence) is entirely determined by the set of all relationships* *that are relevant to it. We will refer to this property as* **relational identity**. *Relational identity means that the form we use to represent an object is not truly that object, but rather only the* identity *set of that form is the real object.*

Thus, for example, X is just a meaningless form, a name by which we refer to something, but **id**(*X*) *in contrast is the true fundamental essence of X, i.e. its meaning. Remember though, that the relational identity of a form is always relative to whatever rule set is currently in effect. A form that means one thing in one context can mean something entirely different in some other context. Relational identity only makes sense in context, in other words.*

As you can see, our definition of relational identity here depends upon another definition that we gave earlier in the book, specifically the definition of "identity set". Do you remember what an identity set is? Our original definition for it occurred on page 252. Well, to summarize, the identity set of a form X is the set of all forms that are derivable from the initial forms and transformation rules of the current rule set and that also contain at least one copy of X.

In other words, putting it in more symbolic terms, **id**(X) is the set of all forms in ⟨ + ⟩ which contain at least one instance of X. For example, in the transformative language (a.k.a. rule set) of true equations, **id**(2) is the set of all true equations involving 2, i.e. $\{2 + 2 = 4, 2 \cdot 3 = 6, 2^8 = 256, ...\}$. Get it?

This makes perfect sense if you think about it. If a form X is embedded inside a form generated by the current rule set, then X must have some kind of relationship to the other parts of the form. This even generalizes beyond just the kinds of conventional concepts that we are used to dealing with in math, such as equations. It applies to any arbitrary kind of form. The identity set captures all possible relationships that a form has within a rule set, and hence captures the entirety of a form's meaning and substance in that context.

Well, more precisely, the identity set captures all of the *formalized* aspects of the meaning of the form. The manner in which the generated forms are subsequently interpreted or connected back to the real world (if ever) is determined by the user's intent though, which is not technically present in the formalism itself per se, but rather is externally imposed upon that formalism in order to capture the essence of some concept the user is interested in modeling.

For example, addition is normally thought of as a formalism for capturing the essence of summing together counts and magnitudes of things, but it could also just be interpreted as manipulations of arbitrary uninterpreted forms with no real intended meaning. The identity set only truly captures the *formal* aspects of a form's meaning. How that "meaning" matters in terms of the real world or some other thing the user is trying to model is determined by how it is ultimately interpreted by the user, just as is true for all of logic and mathematics really.

One could argue though that a perfect formalism really does capture all of the meaning though, and therefore that the subjective interpretation by the user thus does not really matter. Make of it what you will though. The exact nuances of what "interpretation of meaning" really means don't matter much for our purposes here. Indeed, the identity set captures as much of the meaning of a form as the formalized system is capable of ever capturing, and that's the only stuff we can really work with anyway, so regardless of the exact philosophical nuances involved, it makes no real difference to us and is all still perfectly fine regardless.

Also, one should keep in mind that it is not only the symbol choice for the form X that doesn't really matter in terms of what it ends up actually meaning via **id**(X), but rather neither do any of the other chosen symbols in the vocabulary of the current rule set either. It's not the symbols that ever really matter. It is only how they *connect together* that is the true essence of a relational identity.

Yet, it is impossible for us as humans to communicate without choosing some arbitrarily set of symbols by which to do so, so we must always make some kind of arbitrary selection of symbols and syntax by which to represent things, even though on the deeper level it is only the the underlying structure of the *connections* that ever actually matter, and never really the symbol choices themselves. Of course, some symbols and syntax structures are more practical or memorable than others, but that's another matter entirely.

Anyway though, let's spend some more time with some thought experiments about this notion of relational identity, so that we can make sure it really sinks in properly. Multiple examples and angles of attack are always helpful for making sure that our concepts and our understanding of those concepts rest upon solid ground. Let's consider more than just that one example of the number 2 and its relationship to its successor number 3.

As you will see, no matter what example we consider, we will always find that changing the relationships of an object fundamentally changes the identity of that object. Thus, you will see that an object *is* its relationships, and indeed that it has no existence separate from them. It is as if the fundamental essence of objects is that they are bundles of relationships, like giant clusters of wires connecting to every other possible point in the universe to which they happen have any relationships, as if that is their real substance. Indeed, you cannot touch any of those connections without changing what the object fundamentally is.

Consider, for example, some random person and the set of all relationships they have to the world around them. Suppose we had, for example, someone named Rick who lives in Japan teaching English as a second language to Japanese kids and whose mother is someone named Elizabeth. What happens if we change these relationships? Suppose we changed Rick's mother to someone else, who lived a different life than Elizabeth, someone named Katherine for example. Would Rick still be the same Rick really? Nope. Similarly, suppose we changed it so that he lived in Spain and spoke and taught German instead. Would that be the same Rick? Nope.

You really can't change even a single relationship of an object even slightly without instantly changing that object's identity. It doesn't matter whether we're talking about numbers or people or shapes or anything else. All objects are bound to their relationships. One thing we can do though is to change what names we use to refer to objects, as long as we do so completely externally from the system itself and don't pick a name that conflicts with any of the other names that are also in use in our formal model of the system. That's because its not the names themselves that matter, but rather it is the underlying connections, just as we have previously discussed.

Notice this point about the names and symbols being *external* to the system under study though. If we changed Rick's real-world name to John then we would arguably in fact be changing a real relationship and hence would be changing Rick's identity to be some new object. If, on the other hand, we were originally referring to Rick externally as X but then started using Y to refer to him instead, but without actually changing his real state in the system we were modeling, then you could argue that such a change would not truly change Rick. It sort of depends on the context and interpretation though.

One interesting thing about relational identity (among many), is that it makes it possible (at least in principle) to determine that two objects are effectively identical even if they are written using completely different symbols and appear completely different at first glance. This is easier said than done however. If the only difference between different expressions of the same underlying relational identity were just one-to-one symbol choices, then maybe it wouldn't be too hard to compare the objects. However, in practice, transformative languages tend to differ not just in terms of individual symbols, but also in terms of their syntactical structure.

For example, one transformative language might use infix notation where another uses function notation, while still referring to the same underlying concept. As a consequence of this, it is often very difficult to see whether or not two objects in two different transformative languages are the same underlying relational identity. The arbitrary structure of the syntax can easily confuse the comparison and make it quite tedious. Nonetheless though, the concept of relational identity does make it possible in principle to identify such equivalences, so the comparison is still worth noting and considering attempting. Let's make a corresponding formal definition:

Definition 165. *Because it is only the underlying connections that truly express the real essence of an object, relational identity makes it possible (in principle) to determine whether objects from different transformative languages still nonetheless have identical underlying meaning. If such a determination is made between any two such objects, then we will refer to those objects as having* **relational equivalence**.

Relational equivalence is different from the conventional sense of equivalence, in that relational equivalence is a more high-level and abstracted notion of equivalence. Indeed, two objects can have relational equivalence while still not being considered strictly equivalent in the more direct sense. Having relational equivalence means that two objects have the same underlying structure in terms of connections, once all superficial details (such as symbol choice and syntax) are completely ignored and factored out somehow.

Anyway though, besides just discussing these more formal aspects of relational identity, I'd like to also take a moment to go off on a tangent about a related informal philosophical idea. It's not really going to be a rigorous kind of idea, nor necessarily even useful, but nonetheless, when it occurred to me, I thought maybe it could be useful someday to someone or at least provide a good impetus for additional thoughts.

Let's get started. Let's take a moment to think more deeply about the sense in which an identity set captures the essence of what something is. On the one hand, we have this notion of relational identity and how an object is the same thing as the set of relationships it has. On the other hand though, we also have an existence-based perspective on truth, wherein we treat emptiness as being the same thing as falsehood and non-existence.

In other words, on the one hand, the set of relationships that something participates in determines its existence and impact on the universe, whereas, on the other hand, the absence of any such relationships makes something effectively meaningless. What if we used that to create some kind of gradient though, with emptiness on one side and with sets of large numbers of highly impactful relationships on the other side?

Well, it seems to me that perhaps we could argue that the closer an object gets to being empty (in terms of relationships, in some sense) the less strong its existence becomes. Perhaps existence is a gradient rather than merely a black-and-white thing. In other words, I'm saying that perhaps one could argue that as an object loses its relationships it becomes more like nothingness, i.e. that losing relationships in some sense *moves* an object closer to the empty set in a real way.

It is at this point in this thought experiment that I find myself reminded of the common human experience of sometimes feeling as if one's life has become more "empty" under certain circumstances, such as in the context of some kind of loss. And, in light of that, I wonder if there is some kind of genuine connection here. I wonder if perhaps when we lose some kind of connection to something (i.e. when we lose some kind of relationship to the world) or when our position of strength is otherwise lessened in life it is as if we are loosing a bit of our existence and moving closer towards emptiness (a.k.a. nothingness, a.k.a. the empty set).

Thus, perhaps when we experience loss in our lives and subsequently feel a sense of "emptiness", then maybe what we are really feeling when we experience that is this act of moving closer to the empty set by losing some of our relational strength as objects participating in this universe. Perhaps this "emptiness" is not just a feeling we experience, but perhaps it is actually a real thing.

Perhaps losses of strength in our relationships to the world around us create not just a *feeling* of emptiness, but rather a genuine resonance with this logical property of the universe of there being a gradient between emptiness on the one hand and prolific and abundant sets of impactful relationships on the other hand. Perhaps we are truly feeling the abyss of nothingness, in some sense, pulling at us like a cold gust of wind, when this happens.

Who knows, perhaps one could even argue that this is additional evidence of logic as part of the structure of the universe itself. Perhaps one could argue that such experiences are not just feelings or metaphors, but may actually be (in part) a tangible part of reality. This is admittedly a very hand wavy and speculative thought, but I think it's interesting enough to be worth mentioning regardless. This idea is not really so much something I believe in as it is rather merely an intriguing possibility that occurred to me one day. The idea is still far too vague to be credible yet, probably.

Even putting human experience aside though, the idea that existence could maybe come in degrees, i.e. that some things could exist *more* than other things in some sense, could potentially one day have formal value. It's not necessarily just a useless philosophical pipe dream. Perhaps someone can figure out how to make it work in a well-defined way someday. Let's define some new terms for it, to help the idea maybe one day get some real traction:

Definition 166. *One could perhaps argue that as a set loses relationships and moves closer to the empty set its existence becomes less and less strong in some sense. Thus, in other words, perhaps existence is a gradient rather than a binary thing. This is just an informal idea however, as it currently stands. We will refer to the "amount of existence" of a specific set as being that set's* **existential strength***, we will refer to the "existence spectrum" as a whole as the* **gradient of existence***, and we will refer to each position on that gradient of existence as being a* **degree of existence***. Notice the subtle distinctions between the terms here.*

Thus, for example, maybe you could ask for the existential strength of some set and then calculate that value in the form of some kind of formalized "degree of existence" expressed in some sort of corresponding suitable units. I have no idea whether it is actually possible to formalize any of these terms in a usable way though. This definition should not be taken as anything more than speculation and food for thought.

This is all well and good of course, but it is certainly one of the most hand-wavy definitions we've ever made in this book. Philosophy only really has value when we can manage to somehow connect it back to the real world, back to the land of testable claims. Still though, initially vague thoughts can gradually be sharpened over time, so these kinds of vague thoughts are still worth having and noting. As long as an incomplete idea doesn't turn into some kind of unjustified premature dogma or waste too much time, it's fine. Incomplete ideas should be treated like prospects, like things that may or may not bear fruit one day, rather than being used to create a fog of obscurity designed to evade accountability and the need for rigor.

Even putting aside this admittedly quite vague notion of existential strength though, you may still be wondering why we've bothered to discuss all this relational identity stuff. I mean, it does provide a nice insight into the essence of what an object is, over and above merely the symbol we use to represent that object. Certainty, this has value in itself of course. However, what more than that though? What does thinking in terms of the relational identity of objects actually buy us? What is the impact, besides just it being an interesting and elegant way of characterizing the nature of a form under a given rule set?

Well, I'm about to finally tell you. You see, having now adequately prepared ourselves, we have come at last to one of the perhaps most surprising parts of this book. If this was a tour, and I was your tour guide, then this would be the part where I announce that we are about to arrive at Crazy Town, so buckle up and brace yourselves. You see, it turns out that this notion of relational identity actually buys us *quite a lot* once it is properly understood and exploited. That's what the next section is about. The concept of relational identity is worth far more than just a way of characterizing forms. We're arriving at the summit of the mountain now, and the view will be quite unexpected and interesting. It's time to open your eyes to a whole new world of how objects are capable of interacting with each other. Let's begin.

5.9 Relational operations

Relational identity and identity sets may indeed provide a nice way to gather together all of the relationships that apply to a given form, but it is perhaps not immediately clear how we can take advantage of this, except for comparing and analyzing the underlying relational structures of forms. However, it is actually not all that difficult to use relational identity and identity sets for other purposes besides this. The trick is just to frame how you think about it in the right way.

Normally, when we work with forms, we think in terms of transforming those forms into new forms by applying whatever transformation rules are currently immediately applicable to those forms. This way of thinking has the distinct advantage of being very easy to reason about and being generally computationally simple and cheap. It's a very direct way of working with forms. For example, directly transforming $2+2$ into 4, justified by the existence of $2+2 \rightarrow 4$ in your rule set, would be an example of this kind of thinking. However, there are also other (less direct) ways of working with forms.

For example, instead of thinking in terms of what we can do to transform some specific form we have on hand into some other immediately accessible form, we can reframe our thoughts in terms of all of the relationships known to apply to a given form and then work *backwards* from that. In other words, we can start with the sets of all relationships that apply to the objects we are working with, which are initially extremely broad and all-encompassing sets, and then systematically narrow those sets down until we've got a more specific result that we are looking for.

Thinking like this will provide some additional generality and constraint satisfaction ability, but at the cost of requiring us to reason as if we already know what all of the relevant relationships that apply to all of the objects we are working with are. We can trade computational directness for relational generality, in other words. The result will sometimes be impractical and computationally horrific, but it will however allow us to identify interesting theoretical properties and special objects that we would not otherwise be able to see very easily. It will also allow us to solve some problems and questions that would otherwise maybe be impossible to solve.

So, that's the basic idea we'll be shooting for. Now we just have to figure out how to actually do it. How do we move from all-encompassing sets of all relationships applicable to given forms to more specific results we are interested in? Well, let's start by trying to reconstruct a very simple transformative computation, but this time by working backwards by constraining identity sets instead of by directly applying transformation rules.

Let's consider a concrete example. Suppose, for example, that we wanted to compute $2+2$ but that we weren't allowed to use the applicable transformation rules directly anymore, but instead were only allowed to reason in terms of identity sets, relational identity, and set operations. How would we do it? Well, to accomplish it, we're going to have to slowly build up a new vocabulary of operations designed for constraining sets of relationships in appropriate ways. We need to be able to filter sets of relationships until we can narrow their contents down to exactly what we are looking for.

Let's work in terms of the transformative language of true equations. That seems like probably the most natural choice for solving this $2+2$ example. Doing so will give us a nice diverse set of easily understandable relationships expressed as forms. Thus, since we have chosen the transformative language of true equations as our system of implicit transformation rules for the purposes of our thought experiments here, it follows that **id**(2) will look something like $\{2 \cdot 3 = 6,\ 5 + 2 + 3 = 10,\ 21 + 2 = 23, \ldots\}$. In other words, **id**(2) will be set of all true equations that contain at least one instance of 2.

To make things easier to think about, let's treat 2 and other numbers containing 2 as digits as different things. Thus, for example, let's *not* consider $22 + 33 = 55$ to be a member of **id**(2). Technically though, transformative logic wouldn't be able to distinguish these cases unless we used delimiters around each number to block misinterpretation (or else some other technique that accomplished the same effect). For example, we could use parentheses to make it clear that 22 is not an instance of 2, such as by writing (2) instead of 2 and (22) instead of 22. Doing so would make it clear which things should be considered as complete units (e.g. any parentheses enclosed form would be considered to be a complete unit in this case) and which shouldn't. The form 2 exists inside 22, but in contrast the form (2) does not exist inside (22). Get the idea[78]?

However, I will just be ignoring this technicality to keep things more readable. Keep in mind though that doing so is technically an abuse of how transformative logic actually works. Delimiters or other kinds of constraining rules would technically be required otherwise, but we're just going to gloss over that fact. In other words, in the text that follows, we will interpret numbers in the usual sense, such that 22 existing in a form is *not* considered as evidence for 2 existing in a form.

Anyway though, let's return our attention to $2 + 2$ again now. Let's think about how we're going to solve it now. First, let's just get our bearings and then work from there. Where should we start? Well, the closest analog of $2 + 2$ that exists in **id**(2) under the transformative language of true equations is $2 + 2 = 4$. Thus, we should probably frame our thoughts in terms of that. How will we reach $2 + 2 = 4$ from **id**(2) in a general-purpose way though? How will we extract our answer? These are the kinds of things we should be thinking about now.

One important thing to notice here though is that it is not just **id**(2) that is involved here. There are also three other identity sets involved in this problem, and they are **id**(+), **id**(=), and **id**(4). Relational unified logic doesn't privilege any one form over any of the others. All of these forms (i.e. 2, +, =, and 4) are all just arbitrary objects with arbitrary sets of applicable relationships from relational unified logic's perspective. They all also have completely well-defined identity sets.

The idea of **id**(+) and **id**(=) may initially seem strange to you perhaps, but these sets really aren't that strange if you stop and think about it. The rules of an identity set are just that it collects together the set of all forms under the current rule set (a.k.a. transformative language) that are valid (i.e. that can be generated from the initial forms and transformation rules of the rule set) that contain at least one instance of the form

---

[78] In traditional mathematics, our brains just automatically perform this kind of "grouping into complete units" behavior when we parse expressions. Technically though, we are actually applying hidden implicit assumptions when we do so. For example, who is to say that $2 + 2 \rightarrow 4$ wouldn't also justify $22 + 22 \rightarrow 242$, technically? See? You need to explicitly identify which parts of every form should be considered as wholes or not, if you want to be 100% rigorous. Doing this implicitly instead does save a lot of space though, admittedly. The choice is a trade-off between rigor and convenience.

whose identity set is being calculated. Thus, in this case, **id**(+) is just the set of all true equations that contain at least one instance of + and **id**(=) is just the set of all true equations that contain at least one instance of =.

Thus, **id**(+) will look something like $\{1+2=3,\ (5\cdot 2)+4=14,\ 3+3=6,\ldots\}$. Notice that it won't contain any forms that don't contain any + instances. It won't contain $2\cdot 3=6$ for example. Notice also that this is certainly a distinct set from **id**(2) and from other sets, so it is indeed a legitimate distinction we are making here between these various identity sets.

Interestingly though, the identity set **id**(=) is actually the same thing as the universal set $\mathcal{U}$ under the current rule set. This make sense though, since the only valid objects in the transformative language of true equations are true equations, which will of course always contain at least one =, since they are equations, and hence **id**(=) must be the universal set $\mathcal{U}$ in this case. This could easily be false in other rule sets though, so remember to be careful.

As for **id**(4) though, we won't actually be able to use it for computing $2+2$, because that would be cheating. Remember, we are pretending here that we don't actually know that $2+2=4$, but rather we are trying to calculate it by constraining the other identity sets involved until 4 pops out on its own. The identity set **id**(4) is involved, in a sense, but we'll have to frame our solution in terms of the other parts though. Thus, we really only will be working with **id**(2), **id**(+), and **id**(=) here, for the most part.

We should build up towards our solution to this problem one step at a time, in a modular way. As such, what's a good way to start moving a bit closer to solving this problem? Well, if you think about it, one thing we are missing here currently is the ability to constrain *where* in a form another form occurs. If we look at $2+2=4$, for example, we can see that each of 2, +, =, and 4 are occupying distinct positions inside the form $2+2=4$.

Thus, to move closer to solving our problem, we should create some way of narrowing down an identity set to select only the subset of that identity set where a given form is always located at a given position. In other words, we need some way of *placing* forms in specific positions in expressions, but indirectly, via set constraints. We'll need to define a new operator in order to do that though. Here's our new definition for that operator:

Definition 167. *To filter a set in order to select only the subset of that set where a given form is always at a given position in each form in that set, we will use what we will call a **placement operator**, or a **place at** for short. Given a specific set, a specific form, and a specific place where we want an instance of that form to always be located, we will write* set@(position, form), *where* set, position, *and* form *have been replaced by their corresponding inputs. The symbol @ is the placement operator (a.k.a. the "place at").*

The set *input is the set of forms that we would like to filter, the* position *is the index (i.e. the position number) that we want the* form *to occupy, and the* form *is the form that we are searching for. The* position *index is the starting index for each* form *match search, such that we want each* form *to start on that index. The starting form doesn't have to end on the* position *index though, in which case we must check that the* entire string of symbols in `form` is present on the following indices (i.e. it works like an indexed substring search would in computer programming). The `position` is a positive integer. To retain consistency with the indexing conventions of most of mathematics, the indices will start at 1, not 0, and hence the first symbol in a form is considered to be at position number 1, not position number 0.

The `form` input can be any arbitrary set of forms. It doesn't have to be just a single form that we are searching for. In the event that the `form` input contains multiple forms, then the match will be considered successful whenever a form in `set` matches any of the forms in the `form` input set. In other words, the `form` set acts like a logical disjunction. If, on the other hand, the `form` set is an empty set, then the placement operator will return an empty set, since there are no forms to match for in that case, and hence there would be no way for any of the forms in `set` to ever pass the placement set.

The case where only one form is in `form` is obvious, you just test for only that one form as you search through `set`. Naturally, since relational unified logic uses transparent sets, there is no difference between writing the `form` input as X versus as $\{X\}$, for any arbitrary single form X. Hence, the way the placement operator reacts to single versus multiple `form` inputs is seamless, by virtue of the behavior of transparent sets.

Oh, and also, the reason why the @ was chosen for this operator is because the @ symbol is commonly read as "at", which is an English word that is frequently used to indicate where something has been positioned (e.g. "I left my coat at the store."), and hence this choice of symbol thereby serves as a useful memory aid.

Note that this operator works on any sets, not just on identity sets. Identity sets were just the example that motivated us to create this operator, but really it works on any kind of set. For example, if we had some random set like $\{ADC, FDE, GCD\}$ then we could still of course apply a placement operator to it, even though it isn't an identity set. For example, the following expression would be true:

$$\{ADC, FDE, GCD\} @ (2, D) = \{ADC, FDE\}$$

In case it's not clear, the reason why $\{ADC, FDE\}$ passed through the filter is because both ADC and FDE have D in the second position. If we instead did something like $\{ADC, FDE, GCD\} @ (2, I)$, for example, then the result would be the empty set $\{\ \}$, since no such matching forms exist in this set. The placement operator also works on named sets too, but you'll need to use transience delimiters to avoid ambiguity. For example, if $\Omega \to \{\alpha + \beta, \delta \cdot \gamma\}$ then the following expression would be true:

$$\cdot |\Omega| \cdot @ (1, \delta) = \delta \cdot \gamma$$

Remember that $\delta \cdot \gamma$ still counts as a set here, in addition to being an arbitrary form, and is equivalent to $\{\delta \cdot \gamma\}$. This implicit behavior of transparent sets may seem weird at first, but it is actually quite a convenient behavior in many ways, since it lets you gloss over the differences between individuals and sets fairly seamlessly. Indeed, in relational unified logic, you generally think of everything as being automatically generalized to handle any arbitrary number of inputs (as sets), which seems to help remove a bunch of artificial barriers to expressing yourself more freely and naturally.

Oh, and by the way, do you see why the transience delimiters are necessary here? Do you see why we must write ·|Ω|· and not just Ω? The reason is because, since we are using transparent sets, Ω could itself be considered a transparent set equivalent to {Ω}, which would make it ambiguous which of {Ω} or {α+β, δ·γ} we intended to apply the placement operator filter to. This is maybe a bit annoying, but it's how transformative logic works currently. Perhaps there is a better way to handle these kinds of notational quirks, but I don't know what it is if there is. In any case though, this system works well enough for our purposes in this book, so whatever. The underlying principles are the most important point.

Anyway though, let's return our focus to what originally motivated us to create the operator: identity sets and such. As you may recall, we were trying to figure out how to compute 2+2 without being allowed to directly use the applicable transformation rules. We settled on a strategy of trying to find ways to gradually narrow down the relevant identity sets until we can reach $2 + 2 = 4$, at which point we'll somehow extract 4 as our answer. We can't actually use 4 itself to get there of course though, because that would be cheating. It would presuppose that we already knew the answer.

Thus, we instead need to think of the problem in terms of set constraints applied to the identity sets of the *other* forms participating in the $2 + 2 = 4$ relationship. Well, now that we have the placement operator at our disposal, we're considerably closer to reaching that goal. We now have a way to filter our identity sets so that only forms where a specific form is at a specific location can pass through. Thus, for example, we can now create sets like this:

$$\mathbf{id}(2)@(1,2) = \{2 \cdot 3 = 6,\ 2 - 7 = -5,\ 2 + 2 = 4,\ ...\}$$

Do you see what this is doing? The set $\mathbf{id}(2)@(1,2)$ is effectively the set of all true equations involving 2 where 2 is the first symbol in the equation. This in itself may not be very useful, but think about what might happen when we start combining this with other sets that have been similarly constrained. For example, consider the following sets:

1. $\mathbf{id}(+)@(2,+) = \{2 + 3 = 5,\ 0 + 7 \cdot 3 = 21,\ 2 + 2 = 4,\ ...\}$

2. $\mathbf{id}(2)@(3,2) = \{4 \cdot 2 = 8,\ 9 - 2 = 7,\ 2 + 2 = 4,\ ...\}$

3. $\mathbf{id}(=)@(4,=) = \{7 + 4 = 11,\ 4 \cdot 2 = 8,\ 2 + 2 = 4,\ ...\}$

As you can see, each of these sets is simply the subset of each form's identity set which has that form placed at a specific numeric location. As such, how might we use all four of $\mathbf{id}(2)@(1,2)$, $\mathbf{id}(+)@(2,+)$, $\mathbf{id}(2)@(3,2)$, and $\mathbf{id}(=)@(4,=)$ to move closer to our goal of calculating 2+2 indirectly? Well, we have each form involved in $2+2 = 4$ now placed at its corresponding location, as separate sets. Now we mostly just need to combine these sets together. How do we do that with these sets though, in this context? Well, we simply use set intersection of course. Here's what it would look like:

$$\mathbf{id}(2)@(1,2) \cap \mathbf{id}(+)@(2,+) \cap \mathbf{id}(2)@(3,2) \cap \mathbf{id}(=)@(4,=)$$

What would the result of this expression be under the transformative language (i.e. the generating rule set) of true equations? Well, it would certainly be a set that includes $2+2=4$ in it. However, the set would still also include some other things in it besides just $2+2=4$, so we are not quite to our goal yet, arguably. We are getting significantly closer though. The set would look something like this:

$$\{2+2=4,\ 2+2=4+0,\ 2+2=4-0,\ 2+2=4-1+1,\ \ldots\}$$

As you can see, the problem is that there is nothing to stop the set from including numerous different other forms that do have $2+2=4$ at the beginning but which also have a bunch of other stuff appended to the end. Notice that all of these forms in this set are still true equations though. Identity sets are not even capable of generating forms that aren't valid under the current rule set, so there's no way such forms could creep in. For instance, the form $2+2=4+7$ could never appear in this set, since the transformative language of true equations would never be able to generate it.

Ideally, we'd like to cut out all this extra crap and just get back the $2+2=4$ that we are looking for. In the case of true equations, one could perhaps argue that these kinds of extra forms cause no real harm, since the constraints of the rules prevent the equations from being substantively changed (i.e. the extra bits don't really amount to anything here, since they must sum to zero).

However, even so, these extra forms are annoying. Also, in other rule sets, allowing these extra forms in could very easily be completely unsafe. The transformative language of true equations just happens to have special properties that make these extra forms not very substantively significant, but many other systems won't have this property. Thus, we still need some way of cutting out these extra forms regardless. Restricting the length of the forms would suffice for this purpose. Therefore, let's define a corresponding new operator to accomplish this:

Definition 168. *In order to restrict the length of forms when combining sets of forms together using set operations, it would be useful to have a way to express form length as a set. To accomplish this, we will use what we will refer to as a **form length set**. In symbols, it will be written as* $\text{FoLenSet}(n)$, *where n is the form length, a positive integer. The value of* $\text{FoLenSet}(n)$ *will be equivalent to a set containing all forms of length n that are valid within the current rule set.*

Definition 169. *By point of contrast to the definition of form length set though, we could instead (in certain contexts) use a filtering function instead of a set. Doing so would be vastly more computationally efficient than a form length set, but would behave a bit differently. We could call such a function the **form length filter**, instead of calling it a form length set, and represent it by* $\text{FoLenFilter}(s, n)$, *where s is the set we want to filter and n is the form length we want. Notice that this version of implementing a form length constraint doesn't require that we pretend that we know every conceivable valid form of length n under the current rule set, and works with any set, unlike the form length set. Each approach has different conceptual and computational pros and cons though.*

I'm just going to ignore the form length filter operator in the following discussion, even though it is more computationally efficient. I'll be using the form length set instead. I decided to define both anyway though, for the sake of pointing out that there are multiple ways to think about doing this. Computational constraints are of course extremely important, but we're going to be mostly focusing on the concepts and theory of things in this section.

After all, this book is first and foremost intended to explore new concepts and illuminate insights. Computational efficiency is a separate concern from that, at least initially. Ideally, every mathematical theory should eventually be reframed into a computationally feasible form, otherwise it will not have much practical value. In this respect, the infinite sets our theory here will tend to generate are (like all infinite sets) not directly computationally tractable.

However, there is nonetheless a silver lining here. The computational problems involved here are not actually quite as bad as they may first appear. You see, infinite sets may never be computationally efficient when interpreted as infinite objects directly and literally, but if you instead think of them more abstractly, as merely predicates or property specifiers, then they can be much more efficient.

For example, an infinite set of "all true equations" can instead be thought of as merely being a predicate that implements an algorithm that is able to test whether any given equation is true. A computer program could be written to do it. See? It isn't necessarily easy to reframe an infinite set like this, but it is often doable, at which point the computation will become possible, in an indirect but still useful sense. Much the same can be said of our form length set operator here. If you just think of it abstractly as a *quality* expressed by a predicate, instead of directly as an infinite set, then it poses no real problems for reasoning past that point.

Anyway though, let's return to our main thread of thought again now, with our new form length set operator now at our disposal. As you may recall, we were working on computing 2+2 indirectly using set constraints. We're currently trying to reach 2+2 = 4 using identity set constraints. Here's what the expression we've been building up will look like now, once we add in the form length set constraint into the mix:

$$\mathbf{id}(2)@(1, 2) \cap \mathbf{id}(+)@(2, +) \cap \mathbf{id}(2)@(3, 2) \cap \mathbf{id}(=)@(4, =) \cap \text{FoLenSet}(5)$$

This time the expression really will return only $2 + 2 = 4$ (a.k.a. $\{2 + 2 = 4\}$, since we are working with transparent sets). Does this mean we now have our answer and have successfully evaluated $2 + 2$? Well, actually, no, not quite. We are very close, but there is still one more subtle discrepancy that we need to adjust. You see, $2 + 2$ should not actually transform into $2 + 2 = 4$, but rather it should transform into just 4. The answer to "What is $2 + 2$?" is *not* "$2 + 2 = 4$". The form $2 + 2 = 4$ is an equation, not a number. The transformation we are trying to emulate here is $2 + 2 \twoheadrightarrow 4$, not $2 + 2 \twoheadrightarrow 2 + 2 = 4$. See the subtle difference?

As such, to retrieve our real desired outcome here, we will have to also add on some way of extracting specific pieces of each form from any given set of forms. In the case of $2 + 2 = 4$ above, we would only need to extract 4 from this one form, but in the more general case we would also need to be able to handle sets of forms. It would also be useful to be able to rearrange the forms we extract in whatever arbitrary way we need. Being able to do so would be a lot more useful than only supporting extraction of a single form at a time. We'll need to define multiple new operators and notations in order to make our system work in such a generalized way though. Thus, here are the corresponding definitions:

Definition 170. *Often, it is useful to be able to represent a range of integers in a compact way, such as for specifying a range of indices for example. To accomplish this, we will use what we will call an **inclusive integer range abbreviation**, or just **integer range** for short if the context is clear, which we will write as A..B, where A is the lower bound of the range and B is the upper bound of the range, both integers, both inclusive. This particular notation is inspired by a similar notation that exists in some programming languages. Notice that the notation resembles an ellipsis, but has two dots instead of three and also pushes the numbers and dots close together in order to keep everything compact.*

Alternatively, we could use standard interval notation, but doing so would be slightly more verbose, less compact, and more ambiguous. Standard interval notation doesn't make it clear what number set domain the interval is actually being applied to. For example, the integer interval $[1,4)$ and the continuum interval $[1,4)$ are actually two quite different things, despite appearing identical under interval notation. As such, I'd rather have the notation we use be unambiguous, and I am not concerned with anything but inclusive integer ranges in this case. Thus, I have defined this A..B integer range notation here for the sake of clarity and concision. Inequalities would be another possible option though of course.

Definition 171. *In order to be able to extract pieces of forms in a widely general-purpose way, we will need some way to not only extract specific pieces of forms, but also a way to rewrite those extracted forms into whatever other arbitrary new form we want. To accomplish this, we will need a sufficiently diverse vocabulary. One such necessary element that we will need in order to communicate our intent will be to have some way of indicating which pieces of the old form will go where in the new form. We will accomplish this by using a type of delimiter that we will refer to as an **insertion mark**.*

Each insertion mark will contain either a single integer or an integer range. This integer or integer range will indicate which piece (in terms of position numbers) of the old form should be inserted at a new location (determined by where the insertion-marked integer or integer range is placed in a schematic expression) in the new form generated by performing an extraction. More on that later though. What I'm saying here will make more sense once you see the big picture and concrete examples. The insertion mark itself will be represented by a thin bracket drawn over the integer or integer range to which it is bound, as in $\overline{n}$ or $\overline{a..b}$.

For example, the expression $\langle \overline{1}, \overline{3}, \overline{5} \rangle$ would indicate that we intend to create a new vector-like[79] form where the forms at positions 1, 3, and 5 have been extracted from the

---

[79]The tall and shallow angle brackets ⟨ ⟩ here are a common way of notating vectors. For example, ⟨1, 2⟩ would be a 2D vector with "x-component" 1 and "y-component" 2. However, another common notation for vectors is to use parentheses, as in (1, 2) for example. Parentheses are used for many other things besides vectors though, so using tall and shallow angle brackets for vectors instead is thus less ambiguous.

old form and placed into this new vector-like format as a new form. Notice that the ⟨ ⟩ symbols and the commas have been added on and may not have existed in the old form at all. The insertion-marked integers simply indicate insertion points for extracted forms, whereas the other parts of the form are just arbitrary extra stuff that we are allowed to add in.

This kind of capability will enable a greater degree of generality and diversity than merely extracting a single unmodifiable form would. For instance, applying this extraction schematic $\langle \overrightarrow{1}, \overrightarrow{3}, \overrightarrow{5} \rangle$ to $x + y + z$ would result in $\langle x, y, z \rangle$. Get the idea? Insertion marks are not themselves the extraction operator though. Insertion marks are just a way of helping to build up extraction schematics. To actually perform extractions like this, we will need to use an extraction operator, which is the subject of the next definition.

Definition 172. *Whenever we want to perform extraction upon a set of forms, so that we can retrieve pieces from those forms and rearrange or isolate those pieces in some way, we may use what we will call an* **explicit extractor**[80]. *The explicit extractor operator will be represented by the symbol* ⨉. *This symbol was chosen because the crossed arrows in it look like they could represent "rearranging" or "extracting" something, and hence the symbol seems like a fitting choice here.*

The format of the operation is set⨉(extraction schematic), *where the extraction schematic is a form containing any arbitrary combination of zero or more insertion-marked integers (or integer ranges) and zero or more extra forms, whatever is needed to fulfill your intent. Each integer represents a position in a form. Thus, for example, $\{x+y+z\} \bowtie (\langle \overrightarrow{1}, \overrightarrow{3}, \overrightarrow{5} \rangle)$ would result in $\langle x, y, z \rangle$, or equivalently the set $\{\langle x, y, z \rangle\}$, since we are using transparent sets.*

In the event that a form in the input set of an explicit extractor does not have a match for one of the insertion-marked integers (or integer ranges) then the entire form will (in effect) be ignored, even if other parts of the extraction schematic did have matches. Thus, the explicit extractor only retrieves forms whose extraction is entirely well-defined. Sets of forms that have no forms suitable for extraction return the empty set when put through an explicit extractor. Thus, for example, $\{ABCD, 2+2\} \bowtie (\overrightarrow{4})$ would only return $\{D\}$ and $\{2+2\} \bowtie (\overrightarrow{4})$ would only return $\{\ \}$.

Those of you with a background in computer programming may recognize these "extraction schematic" expressions as being similar in behavior to the formatted input strings of printf functions in many programming languages, and indeed they are similar. The insertion mark is essentially filling the same role as an escape character (e.g. %d in C) or a delimited input index (e.g. {2} in C#). Likewise, the explicit extractor operator is acting like the printf function call and the input set is acting like the input data stream. So, as you can see, the similarities are fairly strong.

Anyway though, now that we have defined these new operators and notations, we can return once again to our task of indirectly computing $2 + 2$ via set constraints. This time though, we will be able to carry it out to completion and will have reached our goal. Can you see how we can do it now? The answer is simply to apply the explicit

---

[80] Why this operator is said to be "explicit" will become clear later, once we have defined a related operator called the "implicit extractor", as a point of contrast.

extractor to cut out the last little bit of $2 + 2 = 4$, i.e. the result value we want. Thus, our next iteration of our set expression for indirectly computing $2+2$ will look like this:

$$(\mathbf{id}(2)@(1, 2) \cap \mathbf{id}(+)@(2, +) \cap \mathbf{id}(2)@(3, 2) \cap \mathbf{id}(=)@(4, =) \cap \text{FoLenSet}(5)) \overline{\times} (5)$$

And the award for the most convoluted and computationally horrific way of computing $2 + 2$ ever goes to… Jesse Bollinger! *Wooooo!* The crowd goes wild, etc. Yep. This thing is quite a monstrosity. Bet a bunch of you didn't even think it was possible to make $2 + 2$ this convoluted[81]. It works though[82]. It lets us reframe the operation of $2 + 2$ entirely into terms of set constraints. This in itself is not useful, at least not in this case. Computing $2 + 2$ indirectly is not the real payoff here. The real payoff comes later, once we take the time to fully consider the generality of this kind of thinking and its deeper implications.

The most obvious aspect of this higher level of generality is that we can clearly apply similar ideas to perform many other operations besides just $2 + 2$. The nature and structure of this kind of set-based thinking clearly does not depend specifically on the context of $2 + 2$. It could just as easily be any other kind of value we are trying to compute. Indeed, we could even employ this kind of thinking on forms that have no conventional sense of "value" or "equality", but rather which merely express arbitrary relationships or structures of any kind.

Also, notice that we do not necessarily always want to use an extractor either. Sometimes we will only be interested in identifying the relationships or arbitrary forms involved. Indeed, both approaches are useful in different contexts. Sometimes we will want to extract a specific piece of a form, and other times we will merely want to find the set of forms or relationships to which some property we are interested in applies. It depends on what we want to do. It's flexible.

This way of thinking will also free us from the crudely single-valued approach of conventional logic and mathematics, in a certain sense and in some interesting ways. It will enable us to think more broadly about forms and relationships and to transform them in more arbitrary ways than we were previously able to. Not only can we choose not to focus on the conventional single-valued return value of an expression, but we can grab any arbitrary parts of that expression and recast them into whatever other kind of form or structure we want, at least in principle. This is an interesting capability.

Be that as it may, I imagine that many of you reading this right now are quite alarmed and dismayed at the size and unwieldy verbosity of the kind of expressions we would need to use in order to perform these kinds of operations in this manner. That is definitely a fair criticism. Indeed, this notation is extremely unwieldy and impractical as

---

[81] Never underestimate a mathematician's ability to make things more convoluted than they need to be. ☺

[82] Yes, I realize the irony of the fact that I just recently finished mocking the set theoretic definition of the natural numbers for being too convoluted just shortly prior to this… Shut up. Stop judging me. ☺[83]

[83] Also, I want you to know that emoticons are *always* appropriate in all forms of writing. Always. Emoticons forever. Proof: Let all those who disagree be defined as fools. Suppose also that U are told to shut up. Therefore, emoticons are always good. Q.E.D.[84]

[84] Footnotes inside footnotes are also pretty much always a good idea. In fact, it's the hallmark of award-winning writing. I'm pretty sure there's some kind of official typography rule about that. The typography nazis will surely break your windows if you don't include at least one nested footnote in your writing somewhere.

it currently stands. I certainly wouldn't want to deal with something so verbose and clumsy either.

Don't worry though. This isn't the endgame. As you will see, this notation is mostly just an intermediate form designed to illustrate some underlying concepts and to make certain connections between ideas clearer. The actual final notation that we will be using for most use cases of this kind of set-based way of doing operations will be vastly more concise and streamlined. It won't be anywhere near as hideous and unwieldy. The final product will be much more pleasant. For the time being though, just be patient and humor me.

Let's take some time to really think about the deeper implications of this set-based way of thinking of operations. Let's not just take everything at face value. A little bit of imagination goes a long way. Not everything is as cut and dry as it appears at first glance. Understanding the technical details of a concept is one thing, and is all well and good of course, but it does not mean that you have truly grasped that concept's real essence and deeper expressive potential.

It is one thing to play notes on a piano by memorizing the corresponding motions required of your fingers, but it is quite another to truly understand the music. Doing something is not the same as understanding it. Similarly, looking at something from one angle is often not enough to get the full picture and sense of it. Sometimes you have to stop what you're doing to truly understand where you are. Don't get too caught up in the hustle and bustle of life and forget to smell the roses. Take a moment to consider different perspectives, different paradigms of thought. Intellectual insight requires patience and a nuanced perspective, just as human empathy does.

We should neither give up too early nor too late in our endeavors in life. We should take a moment to relax and to steady ourselves, but by just the right amount, the natural amount. Discerning the optimal path forward requires a spirit of generosity, yet also a willingness to pass judgment. It requires balance. Truth is timeless and patient, and so must we be if we are to properly appreciate and absorb it. We must learn to see not just what something is, but also why it is. We must feel the full scope and reach of something, its real fundamental essence, if we are to ever truly understand it.

Much the same can be said of our set-based way of doing operations we've been experimenting with in this section, just as can be said of any other kind of logic or math really. Ideally, we don't want to merely do math, but rather we want to truly learn how to speak it, just as we do with our natural language. We want to be able to string together arbitrary concepts using math. We want to be able to essentially *paint with concepts*, which is something that only logic, mathematics, and programming can ever truly enable. Only with the rigor of logic, mathematics, and programming can ideas be made so precise that you can turn them into a computer program and then have it actually come to life, as if it were a real living thing (although admittedly a virtual/simulated one) that has now popped into existence. Novelists can imagine and describe arbitrary worlds, but programmers can actually create them.

As such, insights into the conceptual essence of ideas in logic and mathematics can be extremely empowering. The difference between just understanding the technical aspects of a concept on the one hand and understanding its true essence and expressive scope on the other hand is truly vast. The difference is night and day. Only with a proper understanding of the underlying conceptual essence and expressive power of something can we truly employ it in a fully diverse and natural way, as easily as we might string together arbitrary words in English. Through this power, we can not only investigate new ideas, but make them real as simulations. Such thinking opens up an incredible cornucopia of creative diversity. The untapped potential of conceptual clarity is staggering.

Conceptual clarity is the only way to truly grasp the contours of an idea. Mere technicalities and mechanical know-how aren't really enough. As such, let's use this kind of thinking to guide us in exploring set-based operations in relational unified logic more deeply. It will be quite illuminating, as you will see. However, before we get to that, I feel like an analogy would be useful here. The forthcoming discussion is going to sound very tangential and unrelated, but it will still help to illustrate my point here nonetheless.

Let's talk about De Casteljau's algorithm. Yes, that's right, De Casteljau's algorithm. Quite an unexpected shift in topic, right? So random. Well, I bet a lot of you actually haven't even heard of this algorithm before. It is one of my favorites though, because it is so elegant, so ingenious, and so broadly useful. It is also extremely easy to understand and implement. What is it? Well, in short, De Casteljau's algorithm is simply a very elegant and versatile way of computing points on Bézier curves.

What are Bézier curves though? Well, the exact details and quirks of what Bézier curves are don't really matter much for our purposes here. All we really need to know about Bézier curves for my De Casteljau's algorithm analogy to make sense here is that Bézier curves are blending functions which allow you to interpolate between two endpoints while bending the curve of the path of traversal towards any arbitrary number of other points between those two endpoints along the way. In other words, Bézier curves allow you to express things like "Move from point A to point C, but take a path that smoothly bends towards point B along the way." for example[85].

One example of a real-world use of these kinds of curves is in vector art programs, such as Adobe Illustrator or Inkscape for example. Generating visual curves like this is probably the most typical example of what most people think of when they think of blending curves, such as Bézier curves. However, art is certainly not even remotely the only context were such blending curves are useful. Such blending curves can actually be used in any conceptual space whose objects can be quantified to resemble vectors.

Indeed, it is only when you think from that kind of broader perspective that the real power and meaning of blending curves becomes clear. Only when you truly grasp the essence of that idea does the real expressive potential of blending curves really click in your mind. Indeed, if you want to be empowered to paint with concepts in arbitrary and natural ways, then you must learn how to navigate any conceptual space at will, and blending curves are absolutely an essential part of being able to do that.

A mathematician who does not know how to blend things in arbitrary ways or to navigate conceptual spaces is kind of like a person who does not know how to walk from point A to point B. Lacking that knowledge is often crippling. Blending is an

---

[85]Be aware, however, that Bézier are certainly not the only way of doing blending along paths. There are in fact an infinite number of such ways. Bézier curves just happen to be one of the most famous, uniform, and easily understandable ways.

essential part of the vocabulary of mathematics, within the relevant subfields at least. Not knowing how to use blending functions in math would arguably be kind of like not knowing how to use modifiers (e.g. adjectives and adverbs) in English. The lack of such knowledge would make your ability to express yourself more stilted and brittle. You would only be able to work with stuff you already had and you wouldn't be able to combine and modify concepts nearly as easily. Transforming and navigating a mathematical space in interesting and useful ways is considerably more difficult if one does not understand how to seamlessly blend between different things (where possible).

You could still do lots of things without an understanding of blending, and indeed many people do, but you'd have a lot more difficulty expressing nuanced variations of concepts and transitioning from one behavior into another smoothly. Granted, some people in mathematics right now may not consider blending to be as essential a skill as I do, but I think that's perhaps mostly just because maybe too few mathematicians appreciate the value of this kind of arbitrary expressiveness and conceptual flexibility as much as they should. In these respects, De Casteljau's algorithm is just one way of performing blending among many, but it still serves as a great demonstration of the concept, one that is both easy to understand and versatile, and so it's what I'm going with here.

Anyway, enough yapping. Let me explain what De Casteljau's algorithm actually is now. It won't take long. The idea is that we start with an ordered list of vectors, all with the same number of dimensions, and then use them to define a curve. The first and last vectors are the endpoints, and the rest are called "control points". The purpose of the control points is just to bend the curve as it passes from the first vector to the last vector.

The endpoints are guaranteed to intersect with the curve at least once, but the control points in contrast seldom ever intersect with the path of the curve, but rather merely influence it in most cases. The algorithm for generating points on the curve is essentially a recursive generalization of the concept of linear interpolation.

Linear interpolation is just when you calculate points between two end points using a simple linear (i.e. straight line) behavior, typically with an interpolation parameter where 0 represents the source endpoint and 1 represents the destination endpoint. As such, if $\vec{a}$ and $\vec{b}$ are our endpoints[86], if t is our interpolation parameter, and if $\vec{v}(t)$ is our interpolated point, then the formula for linear interpolation between two points will look like this:

$$\vec{v}(t) = \vec{a} + t(\vec{b} - \vec{a})$$

Notice that when $t = 0$ then $\vec{v}(t) = \vec{a}$ and when $t = 1$ then $v(t) = \vec{b}$, just as we want. This makes perfect sense if you intuitively read what the formula means. The formula literally says "$\vec{v}(t)$ is $\vec{a}$ plus some proportion of the displacement (i.e. the directed distance) from $\vec{a}$ to $\vec{b}$". This is just literally a mathematical translation of our informal English notion of "traveling some portion of the distance from point $\vec{a}$ to point $\vec{b}$".

---

[86] By the way, don't confuse the arrows over $\vec{a}$ and $\vec{b}$ here with material implication or anything like that. That's not what those arrows represent. In this case, the arrows over a and b in $\vec{a}$ and $\vec{b}$ merely indicate that a and b are vectors. This notation for vectors is fairly common, but is not really necessary.

That's the beauty of having a strong grasp of conceptual clarity. If you want to implement a concept in math, you literally just translate each informal English concept in your mind into its equivalent mathematical concept. Thus, in this way, you can speak math as easily as you can speak English, in a sense. It's as if you can paint with concepts, which is amazing I think.

You just need to understand the underlying conceptual essence of the ideas you are working with, and then this kind of awesomely natural thinking will suddenly become possible. Part of the problem though is that many mathematicians seem to be in love with obscurantism, and thus tend either to subconsciously (or deliberately) suppress (or omit) these kinds of intuitive insights or else to not care enough to even find them in the first place. Much of the existing math literature is like teaching people to mimic the words that someone else says, but without ever telling them what any of those words mean, and then expecting them to understand the meaning anyway. I don't know about you, but I personally find free-thinking creative human beings to be a lot more useful than parrots.

Anyway, free-wheeling tangents aside, let's return now to the matter at hand: De Casteljau's algorithm. As I was saying earlier, De Casteljau's algorithm is really just a recursive generalization of the concept of linear interpolation, and the formula for linear interpolation is $\vec{v}(t) = \vec{a} + t(\vec{b} - \vec{a})$. How do we actually get from mere linear interpolation to arbitrary Bézier curves though? Doesn't that seem like kind of a stretch? How can we generate *curves* out of nothing but straight lines?

This is where the genius of De Casteljau's algorithm comes in. Here's how it works. Suppose we have n vectors, each with dimensionality d. Let $\vec{p}_i$ represent each individual vector, such that $1 \leq i \leq n$. We treat this order as significant, such that $\vec{p}_1$ is the source endpoint and $\vec{p}_n$ is the destination endpoint, and all other $\vec{p}_i$ are intermediate control points. Suppose further more that t is our interpolation parameter for how far along the curve we want to calculate a point. Thus, for example, if $t = 0.20$ then it would mean we want the point 20% along the length of the Bézier curve represented by the n vectors. With these details now defined, here's how De Casteljau's algorithm would work:

1. Given a specific t whose corresponding vector $\vec{v}(t)$ on the Bézier curve we want to calculate:

2. For every consecutive pair of adjacent vectors ($\vec{p}_i$ and $\vec{p}_{i+1}$) in the list, generate a new point using linear interpolation (i.e. using $\vec{p}_i + t(\vec{p}_{i+1} - \vec{p}_i)$) with t as the interpolation parameter. This yields a new list of vectors with $n - 1$ vectors in it.

3. Return to step 2, but this time using the new list of vectors to fill the same role as the previous list of vectors. Continue doing this until only one vector remains in the list.

4. The last vector remaining in the list is $\vec{v}(t)$, the point on the Bézier curve we are trying to calculate.

Easy, right? Each level of vectors guides the next level of vectors along their paths. The net effect is that it ends up creating a curve. Try visualizing the process and thinking about the logic of it. You'll see what I mean. It's kind of surprising that just applying linear interpolation repeatedly to a list of points until you only have one point left would generate such smooth curves, but it does.

It's also super useful. One of the best things about this algorithm is that it works for any number of dimensions and for any number of control points. Consequently, it allows you to easily express paths that bend through any arbitrary conceptual space fairly easily, even if that space's geometry is not easy to visualize, such as if it has a high number of dimensions.

At first glance, the dimensional generality of De Casteljau's algorithm may seem like something that you wouldn't need very often, like something you'd perhaps only care about when working with higher-dimensional spaces. Indeed, most people seem to not do very much work with higher-dimensional spaces, at least not in the traditional geometric sense, and so maybe the generality of De Casteljau's algorithm may not seem very useful. However, tempting as this line of thought may be, the algorithm is actually more broadly useful than you may think.

In reality, higher-dimensional curves are applicable to many different kinds of real-world practical applications. You just have to think creatively enough. The key realization necessary to understand why is to realize that the space we operate in doesn't have to be "space" in the conventional sense that we usually think of it. We don't have to think in terms of width, height, and depth (etc). In fact, literally *any* quantified conceptual space whose objects can be treated as vectors will work, and such spaces are extremely abundant in many real-world situations. Thus, there are actually a huge number of real-world use cases for calculating curves with vectors of more than three dimensions.

Some concrete examples will make this much clearer. Higher dimensions aren't really as weird and esoteric as many people think, at least not for simple uses of them. For example, suppose we were designing a video game where the behavior and personality quirks of the artificial intelligence that controls the creatures the player is fighting are determined by seven quantified parameters, which we call the "personality profile" of the AI.

For example, suppose that one personality profile parameter is "fear" (i.e. how intensely the AI weighs dangers relative to other factors), suppose another personality profile factor is "reward" (i.e. how intensely the AI weighs acquiring new assets relative to other factors), suppose another personality factor is "craziness" (i.e. how often the AI makes decisions on random whims instead of calculated reasons), and so on. These parameters could be bounded to any arbitrary number range, or even none at all, and it won't necessarily make a difference for our purposes here. All that matters is that each parameter is quantifiable and that the parameters can be treated collectively as one vector.

In this case then, the personality profile of the AI can be treated as a 7-dimensional vector. Just because these vectors are used for AI though doesn't mean that we can't still manipulate them in all the same ways we would otherwise manipulate spatial vectors. Indeed, if we wish, we may smoothly blend between different personality profiles of the AI by simply applying De Casteljau's algorithm to the corresponding personality profile vectors. This could enable us to, for example, seamlessly transition from one AI behavior to another as circumstances may merit while keeping very precise control over the path the AI profile follows to do so.

For instance, we could define a Bézier curve that lets us smoothly transition between different specific pre-defined and carefully balanced AI profiles that we have already created for our game, so that we can have a much better chance of ensuring good behavior along the transition between behaviors. For instance, we could start an AI in "fearful and defensive" mode at the beginning of a game round, to give the player more time to warm up, and then transition closer to "greedy and intelligent" as time passes, but if the game round starts taking too long (e.g. due to a stalemate or whatever) we can gradually shift to "crazy and fearless" instead, in order to try to get the game round to reach some kind of dramatic conclusion in an interesting and entertaining way.

To do so, we could define three AI profiles as 7-dimensional vectors in our system, and then blend between them using De Casteljau's algorithm. We could use "fearful and defensive" as the starting point, "greedy and intelligent" as an intermediate control point, and then "crazy and fearless" as the ending point. We could then have the AI's personality profile at any given moment of time be determined by the Bézier curve between these points.

For example, we could have the AI's personality profile transition automatically over time according to a preset timer, regardless of what specifically is going on in the game, or we could alternatively use some other kinds of events to control when and how the AI profile moves along the curve. There are lots of different variations on how we could do it, each with different behavior and consequences. Use your imagination. The possibilities are infinite, in principle.

See? That didn't require any higher-level analytic geometry for 7-dimensional space. There's really nothing here that is actually hard to understand. De Casteljau's algorithm works seamlessly in any arbitrary number of dimensions, and is intuitively easy to understand. It just bends paths towards control points. Indeed, for low numbers of control points at least, the consequences of this behavior are fairly easy to intuitively work with. Thus, we can use this technique for quite a few interesting things if we think inventively enough. We don't even need to actually be able to understand what an arbitrary n-dimensional space looks like to still be able to navigate it in simple and useful ways.

Learning to generalize your ability to think logically and mathematically like this is extremely empowering, especially if you work in a STEM field, and especially in computer science and game development in particular. You can create some really amazing special effects with this kind of thinking. Conceptual clarity opens up lots of doors. It greatly increases the breadth and volume of what kinds of arbitrary concepts you are able to successfully express. Indeed, in my creative and intellectual endeavors in life, I have found that discovering and documenting new deep conceptual insights into the fundamental nature of concepts is one of the most valuable things one can ever do.

That's why it's such a shame that so much of mathematics is so often mired in (perhaps deliberate or perhaps merely culturally habitual) obscurantism and poor communication. The amount of untapped powerful ideas in math is vast, and mathematicians' frequent disdain for conceptual clarity and for "trivial things" is largely to blame for a lot of it I think. It is often next to impossible to extract these kinds of useful deep insights from the typical math research paper or textbook. At least the mechanical details and proofs for things are often relatively easy to find and are abundant in the math literature, but the actual deeper insights into the fundamental nature of things, the motivating factors, and the understanding of how to string the concepts together in natural and expressive ways is all too often missing or ignored completely as somehow unimportant.

Indeed, attempting to spend lots of time talking about these kinds of simple ideas really thoroughly and in detail seems to often get you nothing but derision and contempt in much of the math world, probably because it gives people an easy prejudicial excuse to paint a target on your back that basically says "if this person is focusing so much on trivial things, then it must be because they are stupid". This kind of toxic "intellectual" posturing is a plague in academia. It is an attitude that greatly inhibits real progress. This is the kind of thing that made me leave my mathematics major and switch to computer science instead[87]. Computer science has a more sane culture. Computer science tends to be more pragmatic and more accepting of pursuing conceptual clarity[88].

There's much less obscurantism and poor communication in computer science than in mathematics, currently, unfortunately. You really have to fight tooth and nail to find even tiny useful insights hidden in the literature of the math world. Such insights are poorly documented in the math world, because the current culture of mathematics doesn't really seem to care much about such things. Discussions of everyday intuitions are too often cut out of the math literature. Many mathematicians are way too obsessed with "purity", in an extremely toxic and counterproductive way. This mentality sucks the life out of otherwise useful ideas and renders those ideas often incomprehensible and without value to people outside of math. Mathematicians are also often in denial about all this too.

If you try to talk to mathematicians about these problems they may sometimes get all indignant and huffy about it and say something like "We do care about intuition! You're wrong!" and then they'll cite a bunch of contrived and opaque examples that barely would ever even qualify as conceptually illuminating to normal people in any other field. Of course, there are some things that mathematicians can easily explain the conceptual essence of in a useful way, but my point is really just that mathematics as a whole doesn't really honor conceptual clarity and intuitive insight as much as it should. Mathematics' current culture has this really weird disdain for pragmatism that is often really intensely repulsive for people who care about creating real-world value, like us programmers and game developers for example.

It doesn't have to be like that though. Mathematicians could broaden their perspective. They'll need to walk outside the echo chamber of the existing mathematics literature and community though, to truly do so, and take some time to return to first principles and to see the world with fresh and newborn eyes again. For example, I think every mathematician should learn to program (at least the basics, but hopefully more), because doing so will teach them the real meaning of rigor, and make them realize that tons of their work in hand-written mathematics actually relies on implicit assumptions

---

[87] The main reason was for game development though.

[88] Computer science also has forced rigor, which mathematics does not really have, despite what mathematicians think. Only a compiler can reliably create true forced rigor. Mathematics is also rife with an abundance of hand-waving, ambiguity, hidden implicit information, and unsafe assumptions, unlike computer science.

and is in fact rife with ambiguity and poorly defined concepts. Mathematicians also often greatly underestimate how abundantly common careless errors, imprecision, and incorrect or incomplete definitions are.

I mean, hand-written math is fine up to a certain point[89], if the concepts are basic enough or exploratory enough, but my point is that mathematicians have kind of become complacent about their own rigor and frankly have become kind of deluded about it. Learning programming, going back to basics on their assumptions and foundations, and placing more value on conceptual clarity and useful real-world insights, would really help mathematics to grow to become much more interesting, influential, and powerful than it is today. This is something worth fighting for I think. Mathematics could be made to be much more appealing to a broader audience and also more intrinsically valuable.

Maximizing conceptual clarity and natural expressiveness, in addition to providing more motivation and a better sense of the value proposition, is crucial for achieving that, I'd wager. Teach people to speak math just as easily as they speak English, such that they can string together concepts arbitrarily to simulate anything they could ever dream of, and make *that* the focus of most of mathematics, and I'm guessing that tons more people would gladly study the discipline.

Anyway though, I've clearly gone off the rails here with another one of my random tangents. Sorry about that. This subject is kind of one of my greatest pet peeves and one of the things I'm most passionate about, so I tend to get sidetracked into talking about it pretty easily. I think talking about it is still worth it though, even if I end up saying the same thing multiple times. It's something that really needs to be said. It is worth repeating, emphasizing, and attacking this problem from multiple directions I think.

Mathematics has needed a cultural change for quite a long time. That's the only way it's going to ever overcome its current stagnation and earn more broad respect and use among the general population. Mathematics needs to embrace the spirit of clarity and natural expressiveness. It's something worth fighting for I think. Imagine a world where almost everyone appreciates the value of math. It would help so much with clear thinking and would also enable some really fun creative work in many different contexts (e.g. game dev, engineering, film, art, etc). I'm really into it, so I can't stop talking about it sometimes.

Sorry if I got too worked up there and derailed our train of thought. Let's now return to the matter at hand, once again. As I was saying, De Casteljau's algorithm gives us a really versatile and expressive way of blending between different points in any arbitrary space. Spaces with more than three dimensions may be impossible to visualize as physical spaces, in the conventional sense, but that doesn't mean you can't still work with them conceptually in extremely useful ways nonetheless.

Indeed, literally anything that can be quantified by any fixed set of parameters can be turned into a vector of corresponding dimensionality and then manipulated as such. Thus, in that sense, there are actually tons of objects all around us with high numbers

---

[89] I'm writing this book, for example. However, there's a reason why this book only ever addresses really basic foundational issues and never advanced proofs though. If I wanted to work with advanced proofs then I'd probably prefer to use a computer. I only trust the most basic of the basic when it comes to hand-written mathematics. Even then, I seldom ever trust the hand-wavy stuff and also generally only trust things that are (at least in principle) computable. I also usually only consider very strict (e.g. constructive and finitistic) uses of the concept of infinity to be valid.

of dimensions, it's just that you have to think in terms of abstract concept coordinates instead of in terms of physical spatial coordinates. Thus, hyperdimensionality is actually not weird at all, in that sense, and indeed you encounter it every day in real life in great abundance.

Countless different things can be quantified and parameterized: light intensity (a.k.a. brightness), fuel efficiency, height, weight, salary, shoe size, cost, mass, color, chemical composition, success rates, years of experience with specific tools for job applicants, age, blood pressure, electrical current, alloy composition ratios, amounts of specific ingredients in recipes, hit points in video games, temperatures, friction coefficients, numbers of pages in books, numbers of visitors to webpages, etc. All of these are (or contain) parameters which can be quantified and then converted into vectors of arbitrary corresponding dimensions. We can then subsequently manipulate and navigate the resulting conceptual space however we wish.

Creating Bézier curves with De Casteljau's algorithm is just one example. We could do lots of other things too. For example, we could calculate weighted distances between points in order to determine which objects are similar to each other. Some product catalog websites (e.g. Amazon) might use some form of this kind of thinking somewhere in their math, I'd guess. In effect, with sufficient effort, we could potentially create an automatic categorization system that can identify relationships before even we ourselves become aware of them. There's lots of fun and useful stuff you can do once you realize that thinking in hyperdimensional terms is not really as difficult or intimidating as it initially sounds, at least for relatively simple use cases.

De Casteljau's algorithm is an especially elegant way to demonstrate the value of generalizing the way we think about something by extending it in a natural way. It shows how we can often greatly expand the scope of how we can use a concept by simply deepening our understanding of that concept's underlying essence and expressive potential. When we really think about the essence of what a mathematical object truly is, it often opens up a whole new world of thought and nuance.

Much the same can be said of our set-based method of performing operations that we have previously been experimenting with in this section. That's why I wanted to mention De Casteljau's algorithm. I wanted to use it as a loose analogy for what we are about to do. You see, just as reframing our perspective on De Casteljau's algorithm allowed us to diversify its apparent expressive utility, the same can be said of our upcoming discussion of a new way of thinking about our set-based method of performing operations via set constraints. We're not really going to be adding on any fundamentally new behavior per se, but by viewing the existing concepts in a new light we *are* nonetheless going to broaden our scope of understanding and expressive potential. It will be eye opening, once it clicks. You'll see.

More on that in just a bit though. First, I think we should define some more terminology to make things easier to talk about. We keep talking about this "set-based method of performing operations via set constraints" that we've been gradually building up piece by piece, but it is starting to become unwieldy to refer to it via so many words and contextual references. We should really define a formal term for it. Words have power and make it a lot easier to build up more complex thoughts and also for the corresponding ideas to potentially gain traction in society. As such, here's our corresponding new definitions:

Definition 173. *When we think of performing operations in terms of constraints imposed on sets, then we may say that we are **operating by constraints**. One of the most typical cases of operating by constraints is using set intersections on multiple sets in order to select some subset of those sets that is of interest to us. However, all of the other set operations are also permitted (e.g. set union, set negation, etc) and we may still refer to such cases as "operating by constraints". For example, determining that $2+2$ is equivalent to 4 by imposing constraints upon the identity sets of the participating objects (e.g. via set intersections and placement operators etc) would be an example of operating by constraints.*

In other words, any form of thinking about operations indirectly via set expressions may be considered "operating by constraints". The subject matter doesn't have to be "operations" per se though, since set operations work for any arbitrary forms and not just forms that represent "operations" in the conventional sense. As far as set operations are concerned, there is no real difference between forms that do represent "operations" in the conventional sense and those that don't. It's all just sets of arbitrary forms. All forms are equally inanimate, in principle. How we treat those forms though is a different matter.

Definition 174. *To provide contrasting terminology, to make communication and discussion easier, we will also define a term for the "opposite" of operating by constraints. Specifically: When we think of performing operations in terms of direct transformations performed upon given forms, instead of indirectly via combinations of sets, then we may say that we are **operating by transformations**. This is the more conventional sense of performing operations, such as one encounters when working directly with transformation rules in transformative logic for example, and in other contexts. For example, transforming $2+2$ into 4 via applying the rule $2+2 \twoheadrightarrow 4$ would be an example of operating by transformations.*

Technically, one could argue that operating by constraints is itself a subtype of operating by transformations, given the fact that set expressions are themselves also arbitrary forms that are being interpreted and transformed just as any other arbitrary form would be. Indeed, this is true as far as it goes, but it is nonetheless still useful to have some terminology to distinguish "operating by constraints" and "operating by transformations" from each other, so that we can communicate more effectively when talking about and comparing and contrasting these two different ways of thinking.

As a general rule, operating by transformations tends to be more computationally efficient than operating by constraints. Probably the biggest reason why is because when you work with the identity sets of objects when operating by constraints you very often end up with infinite sets, which are of course not directly computable. Both are useful ways of thinking though, so keep that in mind.

Computational considerations aside though, let's briefly refresh our memory of what we were talking about earlier with regards to operating by constraints, when we first successfully computed $2+2$ via set constraints. Basically, to solve it, we just needed to create some new operators for selecting specific subsets of sets of forms (e.g. the placement operator) and then to combine those with set intersections and such, and that enabled us to solve it. The expression we ended up creating to achieve it looked like this:

$$(\mathbf{id}(2)@(1,2) \cap \mathbf{id}(+)@(2,+) \cap \mathbf{id}(2)@(3,2) \cap \mathbf{id}(=)@(4,=) \cap \text{FoLenSet}(5)) \overline{\chi}(5)$$

As you can see, there's quite a lot happening here just to accomplish such a small task. As I told you earlier though, don't worry too much about how unwieldy this approach is, because we'll eventually create a much more streamlined way of performing this kind of reasoning. More on that later though. Before we work on streamlining this way of thinking, I'd like to draw your attention to something else.

Specifically: I want you to take a moment to think about what would happen to this kind of reasoning if we dropped the placement operators, form length sets, and explicit extractors and then just thought about performing set operations on identity sets instead. Yeah, we'd of course lose the ability to place each form at a specific location we want, but that's not really what I'm talking about here. We already talked about that in the build up to creating this system in the first place, after all.

What I'm really talking about here is the deeper sense of *what it would mean to just take identity sets and perform set operations on them*. It's true that we couldn't place each symbol or form in the location we want in such a case, but just forget about that for a second. Forget about the goal of trying to calculate a specific item in a specific position in a relationship. Just think about the big picture of putting identity sets into arbitrary set expressions instead.

What would it mean? What would the underlying conceptual essence of such expressions be? Well, here's where it pays to be able to interpolate between natural concepts instead of only being able to think about things mechanically. To figure out what it means, we can just take what we already know that operating by constraints in the long-winded $2 + 2$ example means, and then take away elements of that piece by piece until we are left with the correct conceptual interpretation of the case where we have removed the placement operators, form length sets, and explicit extractors and such.

In this way, we can work backwards from our well-established understanding and intuitions for the long-winded case that uses placement operators (etc) to the simpler case that only uses identity sets and set operators, in a way that allows us to establish an understanding of what the simpler case means where we previously had none. Let's begin.

First, let's imagine taking away the explicit extractor. The extractor was always just there to remove the final answer we were seeking from the form or relationship in which it resided. In the $2 + 2$ case this would be the form $2 + 2 = 4$. This is pretty straightforward to interpret. It just means we're thinking about the answer as a relationship instead of as the specific part or the specific representation.

Next, let's imagine removing the form length set constraint. As you may recall, the form length set constraint just enforces that we can't have trailing forms beyond a certain fixed length segment that we might be interested in. In the case of equations, this is relatively harmless, since the trailing content is forced to have no effect in order for the equation to still be true (e.g. $2 + 2 = 4 + 1 - 1$ etc). Notice the same relationship is still being expressed in the case of equations, its just that there are a bunch of noisy redundant copies of the same underlying relationship floating around. In other transformative languages though, the trailing forms may be very meaningfully distinct, but then in those cases that's probably fine since it would be what is intended anyway.

Lastly, let's imagine taking away all of the placement operators. Placement operators are what let us constrain each identity set so that a form of our choosing is always at a certain position in each form in the identity set. We could also use placement operators on any other kind of set besides identity sets, but that's not really relevant here. Placement operators in this context are just used for specifically identity set constraints. What is the net effect of removing the placement operators though? Well, if you think about it, it must be that it makes all the participating forms in the set expression now effectively be orderless and of arbitrary arity.

What does all this collectively mean though? Well, we're back to thinking in terms of relations (or arbitrary forms) and also have removed the ordering and arity of all of the participants. However, one thing we didn't remove was the sense that by combining identity sets together in arbitrary set expressions we are essentially performing operations with them indirectly. The difference is that now we are performing those operations in an orderless and arity-ignorant way and that our result is a set of relations (or arbitrary forms) that express what the resulting object *means* in terms of relationships, rather than just in terms of a specific form (or forms) that represents it.

In other words, it means that we can apply *any* object to any other object in an orderless and arity-ignorant way and still conceive of doing so as being a operation that results in a valid and logically coherent result. Indeed, the concept of relational identity has taught us that the true essence of any object is simply that it is a collection of all relationships (or arbitrary forms) that are relevant to it. Thus, if the identity set of one form is the essence of what that thing is, and a complete expression of its nature (and indeed it is), then the same must be true of objects resulting from arbitrary combinations of identity sets via set expressions.

Thus, for example, this must mean that **id**(3) ∩ **id**(4) is just as much a valid and meaningful object as **id**(3) and **id**(4) are, and hence just as much a valid object as 3 and 4 as we conventionally think of them. In other words, **id**(3) ∩ **id**(4) can be thought of as being the result of applying 3 and 4 to each other *as operations* in an orderless and arity-ignorant way. Neither 3 nor 4 is the "first" or "second" parameter or anything like that, but nonetheless the result is still conceptually valid. It still means you can apply 3 and 4 to each other *as operations*. Let that sink in for a moment.

Even more interestingly, this will work for any arbitrary form. In other words, operating by constraints essentially completely erases the distinction between "objects" and "operations". Everything, regardless of what it represents, effectively becomes treatable as both a piece of data and an action that you can perform. For example, you can now 4 something just as validly as you could negate something. It renders the distinction between nouns and verbs irrelevant and non-existent, in a sense, with the possible exception of the set operators themselves, since you do need to treat them as special in order to perform operating by constraints in the first place perhaps.

Pretty cool, right? Bet you probably weren't expecting that one (except that I've alluded to it before). It's the kind of thing that would sound utterly laughable if not for the fact that thinking in terms of identity sets and arbitrary set expressions clearly makes it possible and logically coherent regardless. In this way, we can now define new objects with bizarre combinations of properties that would not otherwise be apparent if you just looked at the surface-level rules of any given transformative language. These strange objects have a sort of pseudo-existence within each transformative language, where they are always represented by other more specific concrete forms yet can still be reasoned about in the abstract just as arbitrarily and flexibly as any other object could be via operating by constraints (at least in theory).

It's important to keep a broad perspective on this. For example, just as we can easily express **id**(3) ∩ **id**(4) and have the result be a valid set of forms expressing relationships, the same could be said for any other conceivable combination of arbitrary forms, such as **id**(+) ∩ **id**(·) or **id**(2) ∩ **id**(+) for example.

For instance, just as **id**(3) ∩ **id**(4) is like applying 3 and 4 to each other as operations, so would **id**(+) ∩ **id**(·) be like applying addition and multiplication to each other as operations. In either case, the result will be a set of all relationships shared by both participants, i.e. all forms in the transformative language of true equations where both participants are present at least once in each form. We should really define some terminology for this though, to help keep everything straight in our heads, as usual:

Definition 175. *When working with identity sets and combining them together arbitrarily via set expressions, the distinction between "nouns" and "verbs" effectively becomes irrelevant. Both are simply expressed by arbitrary forms being constrained in arbitrary ways, and thus the net effect is that both "nouns" and "verbs" end up being treated identically. This enables some interesting ways of thinking about things. We will refer to this property as* **noun-verb constraint equivalence**.

Definition 176. *When an expression contains only identity sets and set operations, and no placement operators or extractors etc, then we may refer to this way of thinking as* **operating by identity sets**. *Operating by identity sets is a specific type of operating by constraints. Every instance of operating by identity sets is also an instance of operating by constraints, but the converse is not true. For example,* **id**(3) ∩ **id**(4) *is an example of operating by identity sets, and so are* **id**(3) ∪ **id**(4) *and* **id**(+) ∩ **id**(·) *etc.*

Definition 177. *While it may be true that any arbitrary set expression can be used when operating by identity sets, it still might be useful to have some terminology for referring to and distinguishing some specific common cases. The case where only the intersection of identity sets is used in an expression or subexpression is one such case. As such, let us refer to the intersection of the identity sets of any arbitrary number of forms F_0, F_1, ..., F_{n-1}, F_n as being the* **relational intersection** *of forms F_0, F_1, ..., F_{n-1}, F_n. Thus, for example, the relational intersection of 3 and 4 would be* **id**(3) ∩ **id**(4).

Those of you with a strong background in programming or mathematics may be reminded of the concept of first class functions here somewhat. The reason for that is because first class functions are essentially a way of treating functions as data instead of only as operations that you can apply to things, which thereby enables new ways of working with and thinking about functions (e.g. lambda calculus, etc). Well, noun-verb constraint equivalence here accomplishes much the same thing, except that in addition to making it possible to treat functions as data it also makes it possible to treat data as functions.

The traditional concept of first class functions wouldn't let you treat the number 3 as an operation that you could apply to something else, but in contrast by operating by identity sets you can. Thus, in a certain sense, one can think of noun-verb constraint equivalence and operating by identity sets as generalizing the concept of first class functions even further than it has already been generalized, by making all nouns also usable as verbs in effect.

However, the catch here is that identity sets are often infinite and assume complete knowledge pretty much, so the worst-case computational overhead of this way of thinking is pretty terrible. Still, it is nonetheless interesting from a theoretical standpoint, and perhaps useful too. Reframing the concepts in a more computationally tractable way (e.g. in terms of predicates instead of directly in terms of sets) may make it more widely viable though.

The result of a relational intersection can be manipulated just like any other arbitrary set of forms. You can take the result and then apply whatever other set operations, placement operators, and extractors you want. That's not all you can do though. Any operation that is logically coherent and can operate upon sets of forms in a well-defined way will also work of course.

Furthermore, since their is no real substantive difference between the set of arbitrary forms in a relational intersection (e.g. **id**(3) ∩ **id**(4)) and the set of arbitrary forms in a more conventionally well-understood identity set (e.g. **id**(3)), then they must both be equally coherent as logical objects, in principle. The exact nature of a specific relational intersection may often not be clear, but it must surely be well-defined regardless.

Luckily though, there is a related concept whose result seems to be at least somewhat easier to grasp intuitively. Specifically, I am referring to the union analog of the concept of relational intersection. As you might imagine, it works essentially the same way as relational intersection, except for that it uses union instead of intersection. Let's create a formal definition for it and then discuss it some. Here's the corresponding new definition:

Definition 178. *In addition to relational intersections, the analogous case involving unions is also common and worth giving a specific name. Let us refer to the union of the identity sets of any arbitrary number of forms F_0, F_1, ..., F_{n-1}, F_n as being the* **relational union** *of forms F_0, F_1, ..., F_{n-1}, F_n. Thus, for example, the relational union of 3 and 4 would be* **id**(3) ∪ **id**(4).

The relational union of forms is perhaps best understood as being a hybridization operation, such that performing it on a given set of forms has the net effect of creating a new object which has the properties of *all* of those forms simultaneously, in a disjunctive (not conjunctive) sense. This makes perfect sense when you consider the fact that identity sets are the sets of all relationships that apply to each of their corresponding forms, and hence that if you union a bunch of identity sets together you end up creating an object that shares the relationships of *all* of those forms simultaneously, since that is indeed literally what you are doing by unioning their identity sets together.

This is easily demonstrated by an example. Suppose, for example, that we were considering the relational union of 3 and 4, i.e. **id**(3) ∪ **id**(4). Let's suppose we then proposed that we now want to treat this new set **id**(3) ∪ **id**(4) as a new object in its own right, just like 3 or 4 are. We could (optionally) call it "thour"[90] and represent it by ¾, or we could just not bother giving it a specific name or symbol and instead just refer to it via **id**(3) ∪ **id**(4). Regardless though, suppose someone then asked us to compute ¾ + 2, i.e. "thour plus two". How would we do it? Well, here's one way:

$$(\, (\mathbf{id}(3) \cup \mathbf{id}(4))@(1, \{3,4\}) \, \cap \, \mathbf{id}(+)@(2,+) \, \cap \, \mathbf{id}(2)@(3,2) \, \cap \, \mathbf{id}(=)@(4,=)$$
$$\cap \, \mathrm{FoLenSet}(5) \,) \rtimes (\overline{5})$$

The result of this expression would be {5, 6}, which is exactly what one would expect for an object that behaves like both 3 and 4 simultaneously, in the disjunctive sense. So, as you can see, relational unions really do behave like hybrids of the properties of multiple objects. Basically, you evaluate expressions for every possible combination of the forms from which the relational union was made, and then carry the result along as a set, treating it conceptually as one object, but essentially reasoning about it disjunctively.

Also, on a side-note, I should again remind you that this ridiculously long-winded and clumsy notation here will not be the final notation system that we will ultimately be using for most cases, so don't worry about how awkward it is quite yet. It'll be streamlined later, at least for most use cases. After all, having a streamlined notation is critically important for creating real-world impact and value.

As much as one might want to be able to disregard such surface-level considerations, and to only focus on the underlying concepts, in reality how pleasant a system is to work with is nonetheless absolutely essential to determining that system's real value to other human beings. Clarity and ease of use are mandatory for the creation of real-world value. Only fools ignore such factors. Mere correctness and technical adequacy isn't the same thing as true value creation. Pragmatic constraints matter.

In addition, there are other (simpler) ways to accomplish what we have done here (in this specific case) besides just using relational unions and set constraints etc. For example, we could simply write {3, 4}+2 and then interpret it in terms of transformative logic, treating it as the same thing as {3 + 2, 4 + 2}, thereby permitting you to transform it directly into {5, 6}. Nothing in transformative logic stops you from working with multiple forms at once, after all, since set notation is interpreted that way by default in both rules and free-standing forms.

We discussed this convenient feature of transformative logic briefly once before, on page 255. We even defined a term for it: collective transformation. However, while this approach (collective transformation) may be easier and much more computationally efficient in this case, there may also be cases where a relational unified logic approach (operating by imposing constraints on sets) will have fundamentally different capabilities, such as when dealing with truth criteria or certain kinds of arbitrary hybridizations of properties for example.

---

[90] This is a blending of the words "three" and "four", not to be confused with "Thor" the mythological Norse god of thunder. ☺

Also, besides just relational intersections and relational unions, any arbitrary well-defined set expression is also permitted. There's nothing stopping us from using whatever arbitrary set operations we choose to in a set expression, and therefore we of course don't have to restrict ourselves to only using intersection or only using union. Those (intersection and union) are just especially common and important special cases. As such, it would probably also be useful to have some terms to refer to the more general case. Here are the corresponding definitions:

Definition 179. *Any set of forms created from an arbitrary set expression whose operands are identity sets may be referred to as a **relational object**. Other sets of relationships besides just those composed of identity sets may also be considered as relational objects, if they have a similar spirit and function. Indeed, any set of forms whatsoever that in some sense expresses any set of "relationships" (or, in the more general case, any set of arbitrary forms that are valid within the current rule set) could be referred to as a relational object.*

Relational objects can have additional constraints imposed upon them in order to emulate the behavior that a hypothetical object with that set of relationships would have if it existed, but any given specific relational object may or may not have a representation as a specific form or symbol within the vocabulary of the current rule set under consideration. It depends. Basically, relational objects allow you to think more generally, so that you are not restricted to only thinking in terms of the existing vocabulary of forms and symbols in a rule set. Thus, relational objects are more abstract, in a sense, than objects that are directly represented by specific individual forms are, but still nonetheless behave the same when viewed in terms of operating by constraints.

Definition 180. *All operations that we perform to either create or manipulate relational objects may be referred to as **relational operations**. A relational object is a thing you create, whereas relational operations are how you create such things and also how you manipulate them. The phrases "operating by constraints" and "operating by identity sets" are also related terms, but refer more to a way of thinking than to performing an action itself. Notice the subtle distinctions.*

You do need to be careful that you actually interpret your set expressions correctly when creating generalized relational objects though. It's easy to make a careless mistake. In particular, one careless mistake that could very easily mess you up is if you don't interpret set negation (a.k.a. complementation) correctly when you use it on a set of relationships. There may be a temptation to interpret the set negation as meaning *all* conceivable forms that are not in a set.

However, you need to remember that set negation is always relative to a specific universal set. The universal set of the current rule set (a.k.a. $\mathcal{U}$) includes *only* the valid forms from that rule set. Thus, for example, if X is some arbitrary set of relationships within our current rule set, and we write $\neg X$, then $\neg X$ will include only *valid* forms that are not in X. $\neg X$ means the same thing as $\mathcal{U} - X$.

This behavior prevents set negation from returning a bunch of useless nonsensical forms from outside the current rule set when you use it. Always remember that set negations are relative to a specific universal set. They don't make sense outside of that context really. Well, admittedly, you could have a set of all conceivable forms if you really want to, and you could use that as your universal set, but then you should make that intent explicit, for clarity.

Set negation is the most obvious example that pops to mind where careless mistakes would be easy to make when thinking in terms of relational operations. Any other arbitrary set operations are also permitted too of course, and those may occasionally have other pitfalls too maybe, but I think our coverage here is probably adequate enough as it is. Just remember to reason about the set operations you encounter in a conceptually correct way and you should be fine. Regardless of what crazy forms and relationships are in a set, set operations will still always work the same way they always do, so don't worry too much. Don't let the weirdness of this way of thinking confuse you too much, is my point. It's always still just arbitrary sets of forms we're talking about.

Anyway though: The kind of generalized thinking that relational operations and relational objects allow us to do is powerful and appears to open many new doors to us. However, this system really is overly verbose in its current form. We need a much more concise and graceful way of expressing these kinds of ideas than just these huge chains of sets and placement operators and such. We need to make our notation much more scalable and readable. To accomplish this, we will create a new shorthand notation capable of expressing certain very common use cases for this way of thinking much more concisely and effectively. This shorthand notation will be called "blueprint notation". Let's begin our discussion of it.

5.10 Blueprint notation

Given that working directly with transformations of forms is so concise, why is it that working instead in terms of constraints imposed upon sets is so much more verbose? Why is it that even conceptually trivial operations such as 2 + 2 require so much notational overhead when operating in terms of constraints imposed upon sets, in our system as it currently stands? Well, the answer is mostly redundancy. Complications caused by generalizing beyond just serving the most common use cases is also a contributing factor. Unnatural or indirect ways of thinking also tend to correlate with extra verbosity.

Indeed, in general, most overly verbose and clumsy systems suffer primarily from some combination of redundancy, complications from over-generalization, and unnatural ways of accomplishing things. A good analogy would be how people tend to cut across the grass when walking from point A to point B if doing so is faster than following the paved path. Taking a more convoluted and lengthy path just doesn't feel as good. Our minds naturally tend to rebel against such obviously needless expenditures of energy. Why waste that energy when we could use it instead on something with a better return on our investment? Such waste would seem foolish.

Both human beings and other kinds of natural systems tend to follow the path of least resistance. Good system design should account for this tendency. Nature has a tendency to bend itself into a more efficient structure as soon as it is able to. Nature tends to fight back vigorously against unnatural structures and to gradually wear them down over time. More direct ways of accomplishing things are naturally much more attractive to us and thus we tend to gravitate towards such approaches without us even consciously thinking about it much.

In a sense, although it makes some people mad, people are right to walk across the grass. Doing so is more optimal. It is more practical. If someone has to walk across the grass in order to achieve a much more efficient path to their destination, then it usually means the people who designed the sidewalk failed in their jobs. The dirt paths where people have walked so frequently that the grass has died are the real places where walkways should have been laid down, roughly speaking. That's an oversimplification of course, but the spirit of it is pretty much true.

One should not allow stagnant bureaucratic pretenses to override natural design sense. Simple pragmatism is almost always better than rigid dogma. Similarly, it is important to ensure that the systems we design are sufficiently streamlined as to feel at least somewhat natural. Otherwise, if we design systems in too contrived or unnatural of a way, it will tend to drastically reduce how widely adopted and how useful the system will ultimately end up being.

Every tiny little bit of friction in each element of a user interface or notation will tend to drastically reduce that system's usage and acceptance. The usefulness of a system relative to how much friction there is in getting something accomplished at each step within that system generally resembles exponential decay[91]. Thus, the value of a system can deteriorate extremely quickly if these kinds of factors in the design of the system are not adequately considered and accounted for.

A clumsy system has little to no value in the eyes of most people. This is just as it should be, generally. Pragmatism matters. Value creation is everything in human society. The rest is just incidental technical details and exploratory side-tangents. Let value creation be your center. Don't be a slave to dogma or obscurantism. Favor perception over prejudice. See the world with fresh and newborn eyes in every moment. Control what you can control, for the greater good, but don't let your mind be crippled by extremism.

Such is the attitude we must have in mind in order to see how to successfully streamline our system for operating by constraints that we've been working with so far. The current system as it stands is technically powerful, but it takes too much effort in most cases. To improve upon it significantly, we will need to be willing to give up some generality for the sake of practicality and brevity.

We'll need to cast aside our love of complete generality and be willing to reduce what our system is capable of, all just to make it more expressively compact. Perfectionists and ideologues often find such steps difficult to take, or indeed to even realize the existence of. The dogma of complete generality is often very appealing, but it is

---

[91] This seems to be true of both public-facing products and private technical systems. It seems true of pretty much any kind of system really. High friction in a system seems to almost always degrade that system's usefulness roughly exponentially, in direct proportion to the amount of friction. For example, bad code in a codebase, which is seen only by programmers, can easily be just as damaging to programmer productivity as friction in the user interface of any kind of consumer product would be to consumers' ability to use the final product. Too much interface friction can easily destroy the potential success of any project or product, internal or external. It is often one of the deciding factors in success. Clarity is king. Respect that, or else it could easily doom your endeavors to failure.

a siren song. Pursuing it uncompromisingly is often poisonous to creating maximal real-world value for one's fellow human beings.

Good value production is about creating interesting and useful constraints, not about chasing some fruitless notion of "infinite freedom" like some lunatic trying to find the pot of gold at the end of a rainbow. You already have "infinite freedom". You have the freedom to do whatever you want already. A blank sheet of paper has infinite freedom. So does a blank source code file sitting open in a compiler. Freedom isn't value creation though. Freedom is merely the *possibility* of value creation. True value creation though, has always really been about finding interesting or useful constraints and then shutting up and making it a reality instead of just talking about it or thinking about it endlessly.

Design within limitation is such a famously powerful and effective principle precisely *because* design essentially *is* limitation. Every choice you make in a design for something is a new limitation that you are imposing on that system. By choosing to do X for some specific aspect of something, you are also choosing *not* to do Y or Z for that same specific aspect. Thus, designing towards "infinite freedom" is effectively impossible in that sense[92]. It is a fool's errand, originating from a fundamental failure to understand what design fundamentally is, which is the creation of interesting or useful constraints via some physically or virtually manifest form.

As such, it is now time that we impose some useful constraints upon our way of thinking about operations in terms of sets. Doing so will reduce our freedom but in return we will receive increased concision and practical value. It's a classic example of making a trade-off between abstraction and specificity, which is something that happens very abundantly in both mathematics and programming. So, how are we going to do it? Well, basically, we're going to make the structure of our expressions more similar to working directly with forms, but not *too* similar. We're going to find a balance between writing out forms directly and thinking in terms of sets.

Consider, for example, the way that we are currently using placement operators to position forms where we want them to be in the result set. Under the current notation system, we have to write out each specific position index for every participating form. This is highly laborious. Yet, if you think about how we typically will end up using these set expressions, we would expect it to be very common to have to specify these positioned forms for pretty much every participating form.

Usually, we'll have to do this for long contiguous strings of forms. Only the unknown parts of the form constraints we are building up won't need to be specified. We can take advantage of this sequential property to tighten up our notation. When the position indices are so often sequential and nearly exhaustive (and they are), then it effectively becomes a form of redundancy. The easiest way to explain how to eliminate this redundancy is to just dive straight into the corresponding definition and notation

---

[92]Infinite freedom only exists in the absence of all constraints, because if there were any constraints whatsoever then it by definition wouldn't be infinite freedom. Yet, usefulness only exists in the presence of at least some constraints, otherwise without any constraints then nothing you do to act upon anything would have any significance. Thus, it is logically impossible to maximize both freedom and usefulness at the same time in any system[93]. The best you can do is to create a balanced mix of various trade-offs.

[93]Oh, and by the way: In the programming world especially, "architecture astronauts" (i.e. design ideologues) would be wise to remember this principle. It proves that their pipe dream of one day creating a "perfect generalization" can never succeed, and thus that balanced pragmatism is inherently superior to uncompromising dogmatism in programming.

that we will be using to overcome it:

Definition 181. *We will refer to the symbols* ⟦ ⟧ *(i.e. double lined square brackets) as* ***blueprint delimiters***. *Suppose we have some form F which is composed of n other forms* $F_1, F_2, \ldots, F_{n-1}, F_n$, *such that each* F_k *fills the position at index k in the form. Then, if we write* ⟦$F_1 F_2 \ldots F_{n-1} F_n$⟧, *it should be considered as shorthand for a relational intersection of* ***id***$(F_k)@(k, F_k)$ *for all n constituent forms, plus a form length restriction of length n. Thus:*

$$\llbracket F_1 F_2 \ldots F_{n-1} F_n \rrbracket = \mathbf{id}(F_1)@(1, F_1) \cap \mathbf{id}(F_2)@(2, F_2) \cap \ldots$$
$$\cap \mathbf{id}(F_{n-1})@(n-1, F_{n-1}) \cap \mathbf{id}(F_n)@(n, F_n)$$
$$\cap \text{FoLenSet}(n)$$

Any expression surrounded by ⟦ ⟧ *will be referred to as a* ***blueprint expression***. *Furthermore, more broadly speaking, this entire style of notation (i.e. blueprint delimiters, blueprint expressions, and any other related operators) will be referred to as* ***blueprint notation***.

Thus, as you can see, the net effect of blueprint notation is that it allows us to use a sequential notation similar to how we would normally write out forms, while still framing our thinking in terms of constraints imposed upon sets. Blueprint notation thereby allows us to eliminate the redundancy involved in specifying form positions manually using placement operators. Admittedly, this shorthand only works for relational intersections and does restrict you to a more sequential structure for how you specify form constraints, but nonetheless the concision and expressiveness that you gain by doing so is well worth it.

So far though, we've only discussed part of how this notation will work. The definition above provides the basic skeleton for how blueprint notation will be interpreted, but just this on its own is a bit limited. We'll need to take some time to actually think about how to utilize it effectively. We'll also want to define at least one new supplemental operator in order to make it more expressive and useful, but more on that in just a bit.

Let's start by returning once again to our example of computing 2+2 via constraints imposed upon sets. As you may recall, the completed version of our (very long-winded) way of computing 2 + 2 occurred on page 528. In that version, we constrained the 2, +, and = symbols to match the form of the desired relationship and then subsequently extracted the result we wanted (the value 4) from that. We had to do this because otherwise we'd have been cheating. If we had used 4 in the expression we used to compute it then that would have implied that we already knew the answer, which would have eliminated the point of the entire computation.

However, with blueprint notation in contrast, it may not necessarily be immediately clear how we would go about computing this same expression without cheating. I mean, it is already doable, but you might not immediately see how to do it properly. As such, let's start with what we do know (i.e. what is immediately obvious from the definition of blueprint notation), and then build up from that until we can see how to properly extract values that we are interested in when using blueprint notation. What do we already know? Well, for example, we know that if we *could* cheat when computing $2 + 2$ then we could write something like this:

$$[\![2 + 2 = 4]\!]$$

What would this mean? What would the result of it be? Well, this expression would essentially have the same net effect as testing whether or not $2 + 2 = 4$ is a valid form under the current rule set (a.k.a. the current transformative language) and returning $\{2 + 2 = 4\}$ if it is but otherwise returning $\{\ \}$ if not. Thus, this expression isn't really a useless expression, but the problem is that it doesn't actually compute 4, since we can only write it like this if we already know that the value 4 belongs there.

That's not to say that this kind of expression isn't useful. Indeed, in certain contexts, this kind of expression may be exactly what you want to write. Sometimes you may just want to test a specific form for validity under the current rule set, and maybe you feel like using blueprint notation to express that for whatever reason. You don't necessarily always care about extracting specific parts of the forms you compute this way. Sometimes you just want the entire form as it stands, with no modification. Sometimes you only care about the relationship as a whole, as embodied by some set of arbitrary forms, rather than the specific parts.

Be that as it may though, in this specific case (the case of computing $2 + 2$) we still need some way to omit 4 so that we can actually solve for it without already knowing it. To do that, what we really need is some way of specifying *less* information for the part of the blueprint expression where 4 is located. We do still have to put something in that slot in the blueprint expression, since blueprint expressions do imply a fixed form length, but we need to make what we put there vaguer.

We need to allow the blueprint expression to explore many different possible forms that it could place where 4 is, so that it can discover which forms will ultimately make the expression valid on its own, instead of us just giving it the answer in advance. In this case, only the form 4 will make the equation true, but in other cases multiple forms may work. There are actually multiple ways we can do this, but let's pick a natural one. We know that $2 + 2$ must clearly be an integer for example. Any form is permitted in each form position in a blueprint expression, so we may thus use a transience delimited form representing the set of all integers in place of 4. That should accomplish what we want. As such, suppose that Int is defined by Int → $\{\ldots, -2, -1, 0, 1, 2, \ldots\}$ in a trans-closed way. Then, we may make our tentative blueprint expression for $2 + 2$ vaguer (thus removing the cheating) like so:

$$[\![2 + 2 = \cdot|\,\text{Int}\,|\cdot]\!]$$

The reason why this is permitted is because when we substitute $\cdot|\,\text{Int}\,|\cdot$ into the equivalent relational intersection for the blueprint expression $[\![2 + 2 = \cdot|\,\text{Int}\,|\cdot]\!]$, then the $\cdot|\,\text{Int}\,|\cdot$ part becomes **id**($\cdot|\,\text{Int}\,|\cdot$)@(5, $\cdot|\,\text{Int}\,|\cdot$). This is a perfectly valid expression in unified logic because (as you may recall) both identity sets and the placement operator accept sets of multiple values as inputs in these positions[94] (i.e. in this case, in the positions where $\cdot|\,\text{Int}\,|\cdot$ is).

---

[94]These design choices in the behavior of identity sets and placement operators were made in part in anticipation of this moment, but also for the sake of general utility in other contexts.

However, this expression (i.e. $[\![2+2 = \cdot|\,\texttt{Int}\,|\cdot]\!]$) will still return $\{2+2 = 4\}$, just as $[\![2 + 2 = 4]\!]$ did. If we want to actually retrieve the *value* of $2 + 2$, rather than the entire relationship, then we will have to do a bit more. Only once we extract the actual value 4 (without cheating) will we have truly covered all the same ground as our previous much more long-winded way of doing these kinds of computations. Unsurprisingly, an explicit extractor will suffice for accomplishing this. Thus, we have the following expression:

$$[\![2+2 = \cdot|\,\texttt{Int}\,|\cdot]\!] \bowtie (\overline{5})$$

The form in position 5 of the result set of $[\![2 + 2 = \cdot|\,\texttt{Int}\,|\cdot]\!]$ is 4, and thus the result of this expression will be 4. As you can see, this is a much more concise and readable way of working than what we were doing earlier. However, even this is arguably unnecessarily verbose. In this expression, we are still having to track which position the part of the form we are interested in extracting is located at. This is a bit of a tedious nuisance, since we have to count the forms to find the position index we need. Also, the position index feels a bit contrived and the generality of the explicit extractor will probably often be unnecessary.

Thus, it will be possible to simplify this expression (and our notation) even further. The trick will be to find some way of eliminating the need to specify which position indexes we want to extract forms from in the blueprint expression. We'll also need to reduce the generality of how expressive our extraction mechanism is, in order to be able to successfully trade generality for even more brevity. To accomplish this, we will define the following new operator:

Definition 182. *Given a form in blueprint notation, such as $[\![F_1 F_2 \ldots F_{n-1} F_n]\!]$, that we wish to extract some part of, we may choose to use a restricted shorthand version of extraction instead of using an explicit extractor. We will refer to this restricted shorthand version of extraction as an* **implicit extractor**. *The symbol for it will resemble the insertion mark symbol, which we used in the extraction schematics for explicit extractors*[95], *except that it will be placed below the part of the form that we are interested in instead of above a position number, and it will be written inline with the blueprint expression itself instead of apart from it.*

For example, given the blueprint expression $[\![F_1 F_2 \ldots F_{n-1} F_n]\!]$, where each of the F_k are just arbitrary forms that have been placed at each position k, we could use an explicit extractor to extract F_n by writing $[\![F_1 F_2 \ldots F_{n-1} F_n]\!] \bowtie (\overline{n})$, but we could instead just write $[\![F_1 F_2 \ldots F_{n-1} \underset{\sqcup}{F_n}]\!]$ and it would have the same net effect. Thus, as you can see, given any form F_k within some blueprint expression, whereas insertion marks for explicit extractors would be written over position numbers, as in $\overline{k}$ for some integer k, implicit extractors would be written underneath the form itself, as in $\underset{\sqcup}{F_k}$. This correlation makes it easier to remember the two different extraction notations.

By the way, the "implicit" in "implicit extractor" actually refers to exactly this omission of the position numbers and extraction schematics, and hence to how this approach thereby enables you to think more directly and more implicitly. Thus, naturally, by contrast, the "explicit" in "explicit extractor" refers to the opposite, i.e. to the need

---

[95] The definition of the explicit extractor, which also describes extraction schematics, can be found on page 527.

to specify the positions explicitly and in more general terms. This should make clear why the adjective "explicit" was chosen for "explicit extractor", which I previously did not specify because I knew it would make more sense if I waited until now to explain it.

Furthermore, multiple implicit extractors can be used within the same blueprint expression. In that case, the blueprint expression should be read from left to right and the forms to which implicit extractors have been applied should be pulled out sequentially to create the structure of the forms in the result set. Thus, for example, $[\![F_1\ F_2\ F_3\ F_4]\!]$ would result in a set of forms where each extracted form would look like $F_2 F_4$, but only extracted forms that satisfied the criteria of $[\![F_1 F_2 F_3 F_4]\!]$ and were valid under the current rule set would be included of course, if any such forms existed.

Armed with this new mechanism, we can now express the $2+2$ example even more concisely and directly. Previously, we've been writing $[\![2+2 = \cdot|\,\mathtt{Int}\,|\cdot]\!]\mathsf{X}(\overline{5})$, which is concise but still requires the tedium of specifying the position number corresponding to what part of the form we want to extract. We can now avoid that though, via an implicit extractor. Thus, we now have our next version of expressing $2+2$ via set constraints:

$$[\![2+2 = \underline{\cdot|\,\mathtt{Int}\,|\cdot}]\!]$$

This is clearly much more practical than the old way of writing it all out manually using identity sets, placement operators, and set intersections. Granted, it's still not as concise as working purely in terms of transformations of forms directly, such as when we simply apply $2+2 \to 4$ to $2+2$ to turn it into 4. However, even though blueprint notation may not be as concise as direct transformations of forms, blueprint notation is still more general and allows us to express certain cases in a very pleasing and natural way.

For example, suppose we wanted to express an interval of some type of numbers. We could use interval notation to do this of course, which already exists in the literature. We could write $[a, b]$ for example, and use it to mean (ambiguously) either the inclusive integer range from a to b or the inclusive continuum range from a to b or whatever else. However, doing so is not very conceptually direct and descriptive, arguably. Such notation is arbitrary and doesn't naturally parallel the actual relationship involved.

In contrast though, blueprint expressions allow us to write intervals in a form which is much more clearly directly connected to the underlying relationship of the interval and number type. Blueprint expressions also generalize to any arbitrary relationship instead of only being limited to expressing intervals. This means that blueprint notation solves a much broader expressive problem than interval notation does. Let's consider some examples. Look at what intervals look like when expressed as blueprint expressions:

1. $[\![1 \leq \underline{\cdot|\,\mathtt{Int}\,|\cdot} \leq 10]\!]$

2. $[\![1 \leq \underline{\cdot|\,\mathtt{Cont}\,|\cdot} \leq 10]\!]$

3. $[\![0 < \underline{\cdot|\,\mathtt{Cont}\,|\cdot} < 1]\!]$

4. $[\![\pi \leq \underline{\cdot|\,\mathtt{Frac}\,|\cdot} < 2\pi]\!]$

See how directly and naturally these convey the intent? This way of writing intervals is not as arbitrary as traditional interval notation. In contrast, this notation directly employs the logic of the relationships involved in each interval to build each set of numbers. As a consequence, we end up with a notation that is much more descriptively clear. In fact, I bet many people could guess what these expressions represent even without being taught the formal details of blueprint notation. It is a convenient shorthand[96].

The real beauty here though is that this notation is general purpose, instead of only working for interval notation. It works for any arbitrary set of transformation rules and relationships, including even contrived rules that wouldn't otherwise make much conventional intuitive sense. Additionally, we can nest blueprint expressions inside each other in order to express many different operations at once. For example:

$$[\![[\![1 \leq \cdot|\text{Int}|\cdot \leq 10]\!] + 1 = \cdot|\text{Int}|\cdot]\!]$$

See how it works? See how blueprint expressions just naturally compose together due to how blueprint expressions are defined? The result of this expression is the set of all numbers from 2 to 11, i.e. the same result as $[\![2 \leq \cdot|\text{Int}|\cdot \leq 11]\!]$ would give. We could also write the results in interval notation as [2, 11] or 2..11 if we wanted to though. List omission such as $\{2, 3, \ldots, 10, 11\}$ and set builder notation are also options. Regardless of what format you like to write things in though, blueprint notation is convenient and expressive for many things, especially because of how general purpose it is.

Another thing to keep in mind when dealing with blueprint notation is that the forms and sets you use in blueprint expressions can be as broad or as narrow as you'd like. For instance, for the $2+2$ example, we've been using the set of all integers (Int) in the position in the blueprint expression where we are extracting the answer from, but this is not necessary, even though the type of the result is indeed an integer. We could just as easily use a much vaguer set and the expression would still work as intended. For example, we could just use the set of all numbers (Num) there instead of the set of all integers (Int). In that case then, $2+2$ would look like this:

$$[\![2 + 2 = \cdot|\text{Num}|\cdot]\!]$$

The result would still be the same as for $[\![2 + 2 = \cdot|\text{Int}|\cdot]\!]$ in this case, but there are other cases (other blueprint expressions) where changing how broad or narrow a set is could change the outcome. The reason $[\![2 + 2 = \cdot|\text{Num}|\cdot]\!]$ is equivalent to $[\![2 + 2 = \cdot|\text{Int}|\cdot]\!]$ is because the answer set $\{4\}$ is a subset of both sets, and therefore both sets will catch all of the answers in the result. If this was not true though, then one could have a different result than the other. It depends on the context.

Also, be aware that you don't necessarily always want to catch all possible solutions to some part of a form. Sometimes you *intentionally* want less than that. Your choice of what set you use allows you to broaden or narrow which parts of the answer set you want to see. For example, even though an equation may have many different solutions

---

[96]Intervals can also be expressed using set builder notation, but blueprint notation is still nonetheless a nice shorthand for intervals.

on the continuum, you may only be interested in the solutions that are integers. In that case, using Int instead of Cont for that corresponding part of the blueprint expression would allow you to express this additional constraint easily. What the optimal choice is depends on the situation. Don't think in black-and-white terms. A nuanced perspective is best.

Oh, and it occurs to me that it would also be useful to have a more precise term to refer to the much more long-winded way of expressing the same thing that blueprint expressions express, i.e. a term for expressions that use long-winded set operations and placement operations instead of concise blueprint expressions. This would be useful for comparing and contrasting the two approaches. Admittedly, we do already have terms like "operating by constraints", "relational objects", and "relational operations" etc, but the problem with those terms is that they are either too broad (e.g. both blueprint expressions and the long-winded set expressions are examples of "operating by constraints") or else do not quite refer to the same thing as the expression itself. Thus, let's add on the following new definition, for completeness:

Definition 183. *When an operation is expressed in terms of sets, set operators, and placement operators, instead of in terms of blueprint notation, then we may refer to such an expression as a **placement expression**. Thus, the term "placement expression" refers to our earlier (often much more long-winded) way of performing operations via set constraints, whereas the term "blueprint expression" refers to our newer (often much more concise) way of performing operations via set constraints. This term ("placement expression") is similar to (but different from) some of the earlier related terms we defined. Try to keep the subtle distinctions in mind.*

It is much more practical to just be able to say "placement expression" than to have to say something like "the long-winded way of expressing sets of forms via set operations and placement operators from earlier" or whatever. Thus, having this new term defined will help make it easier to talk about things more precisely and concisely. It's important to remove obstacles to us expressing ourselves well. If it takes too much effort to express something, many people often won't bother. People frequently don't even realize how much a lack of adequate terminology is actually deeply subconsciously blocking their ability to think.

Words have great power in determining our ability to think about things effectively or not. Human thought doesn't scale well without being augmented by language. Building up a strong vocabulary, one with minimal descriptive friction and minimal ambiguity, is an essential component of intellectual and creative progress. A field of study often cannot advance until it is given more precise and expressive terms. Every tiny little improvement helps. No improvement should ever be considered "below us" or "too trivial". The cumulative impact of even tiny improvements over all of human history will often be enormous.

Anyway though, now that we have a better sense for how blueprint expressions work and how to use them, we should take some time to consider what the implications of this kind of set-based thinking are. What kinds of things can we express with this system or not? What does the nature of this system imply about what is possible more generally? How does this way of thinking differ from other ways of thinking, in terms of what consequences arise from it? Can this system answer questions we couldn't answer before? How does this system's expressiveness and convenience differ from other systems and in what respects? Is the underlying sense and deeper meaning of the answers this system gives us a net improvement over other systems or not?

These are important questions to ask ourselves. It's always important to not get so lost in the trees that you lose your sense for the forest as a whole. On the other hand though, parts of these questions are very open-ended and are yet to be determined. Much of it is beyond the scope of this book, a book which is exploratory in nature and cannot hope to be comprehensive while still meeting its conversational clarity and accessibility goals. Nonetheless though, it is still well worth spending more time trying to get a better sense for what the consequences of this way of thinking might be. Indeed, some of these consequences are quite interesting. Let's explore those more.

For example, in this system (i.e. in relational unified logic, via blueprint expressions or placement expressions etc) it is possible to divide by zero and get a logically coherent and usable result, a result that does not differ in any fundamental way from any other value that one might work with in the system. This is easiest to see when you think about division in terms of what it means from multiplication's perspective. Just as $B - A$ is the additive displacement (directed additive distance) from A to B, so too is B/A the multiplicative displacement (directed multiplicative distance) from A to B. In other words, the value of B/A tells you what you'd need to multiply A by if you wanted to turn it into B.

Think about what this means in terms of how division works. When we perform division we are essentially just trying to solve for the missing value which would allow us to move from A to B via multiplication. In other words, performing B/A is really just the same as solving $A \cdot X = B$ for X. The order of A and X doesn't matter of course, so solving $X \cdot A = B$ for X would also have the same net effect. Thus, division by zero is really just the same thing as solving $0 \cdot X = B$ or $X \cdot 0 = B$ for X, given some B to fill the role of the numerator. As such, if we were interested in evaluating division by zero on the continuum (the most likely use case), then the following blueprint expression would represent division by zero, if B were replaced with a number on the continuum:

$$[\![0 \cdot \underbrace{\cdot | \mathtt{Cont} |\cdot}\ = B]\!]$$

We could instead think of this in terms of standard equations and solve it that way of course, but there's a reason I'm using blueprint notation here. I'm trying to illustrate a conceptual relationship here, as you will soon see. Equations would also be a fair way of doing so, but reasoning instead in terms of intersections of set constraints seems likely to be more contextually convincing for illustrating the possibility that perhaps division by zero is actually perfectly well-defined after all, contrary to popular belief.

The reason is because the set-based way of thinking we employ in relational unified logic does not distinguish "single values" from "multiples values" *per se* in any kind of prohibitive way. All sets of arbitrary forms and all identity sets are equally valid objects in the eyes of relational unified logic, regardless of origin. It can't even *see* a difference between sets of forms originating from "single values" versus from "multiples values" really, in terms of its formal mechanisms, which is part of what makes it such a powerful way of thinking. It makes things well-defined that otherwise wouldn't be considered well-defined if we used a more traditional approach.

For example, consider once again the case of division by zero. Generally speaking, mathematicians divide division by zero into two separate logically distinct cases: one where the numerator is zero and another where the numerator is non-zero. We will do the same. It is easier to think about that way, since the two cases do behave quite differently in terms of output. Let N represent a non-zero number on the continuum. Then, written in blueprint notation, we have two different cases for division by zero:

1. $[\![0 \cdot \underline{\cdot | \mathtt{Cont} | \cdot} = 0]\!]$

2. $[\![0 \cdot \underline{\cdot | \mathtt{Cont} | \cdot} = N]\!]$

The first case above is the blueprint expression equivalent of $0/0$, whereas the second case above is the blueprint expression equivalent of $N/0$, where N is any non-zero continuum number. The first case above essentially says "What is the set of all continuum numbers X which are valid solutions of $0 \cdot X = 0$?". In other words, it asks what is the set of all numbers on the continuum that when multiplied by 0 yield 0. The answer is of course the set of all numbers on the continuum, i.e. the answer is $\mathtt{Cont}$ itself. Any number times 0 is always 0.

On the other hand though, the second case above is quite different. It asks "What is the set of all continuum numbers X which are valid solutions of $0 \cdot X = N$, if N is not 0?". In other words, it asks what is the set of all numbers on the continuum that when multiplied by 0 yield a specific non-zero number. Clearly though, no such numbers exist. Nothing multiplied by zero can ever be anything other than zero. Thus, the second case results in merely the empty set $\{\ \}$. This is in stark contrast to the first case, which resulted in the entire set of all numbers on the continuum $\mathtt{Cont}$. These two results could not be more different. One is nothing whereas the other is everything. As different as they may be though, these are indeed what the correct solutions are.

So there, we've solved division by zero essentially. Zero divided by zero is equal to the set of all numbers on the continuum and a non-zero number divided by zero is equal to the empty set. The fact that relational unified logic reframes all operations in existence into terms of constrains imposed upon sets is what makes these answers valid and significant. Everything is a transparent set in relational unified logic, and therefore sets are of course always valid results for any computation. Relational unified logic thus eliminates the barriers to treating division by zero as well-defined which previously would have prevented us from doing so.

Now, I know what some of you are thinking. Doesn't this result seem kind of trite, in a sense? Indeed it does. That's a fair criticism. After all, it is not as if mathematicians don't already know how to solve equations of the form $A \cdot X = B$. Of course they do. Ask any half-decent mathematician what values of X on the continuum would make the equation $0 \cdot X = 0$ true and they will of course say that any value on the continuum would work. Likewise, it is clear to any even slightly competent mathematician that $0 \cdot X = N$ has no solutions if N is non-zero.

That's not the point here though. I certainly didn't write all of this just to reiterate such a widely known pair of facts. The real point to all of this is that I've reframed the problem in such a way as to justify the answers to these kinds of questions as being well-defined. Solving these equations is not in itself my intended contribution. Everybody already knows how to do that. They've known that for a super long time. That's easy.

The real insight and value-added here is that my system causes the result of these expressions to become just as equally valid as any other conventionally valid object in the system. It puts 0/0 on the same footing as 2 + 2 by framing them both in terms of the same underlying computational process and way of thinking. It eliminates the barriers to treating these kinds of strange objects as well-defined, thereby allowing us to think more generally and more flexibly. It creates a *universal domain* and a *universal closure*[97], where virtually everything that you could ever think to express will always be well-defined[98], even if it just means returning the empty set in the worst case. *That* is the point.

Another interesting point here is that there are actually multiple ways that we can think about these objects. The first way of thinking about these objects, the more surface-level and face-value way, is to just think of them in terms of being sets of forms that successfully fulfill some role implied by the constraints imposed upon the set of all valid forms within the current rule set. In this case, we think of each form in the result set as being simply a name or label for the corresponding valid case it represents. This is like what we are doing here with $[\![0 \cdot \cdot | \text{Cont} | \cdot = 0]\!]$ and $[\![0 \cdot \cdot | \text{Cont} | \cdot = N]\!]$, and also elsewhere with other similar expressions, such as $[\![2 + 2 = \cdot | \text{Int} | \cdot]\!]$ etc. When we evaluate these expressions, a set of satisfactory forms is returned. However, these forms just act like names or labels and just tell us which forms are valid under the constraints we have stipulated.

On the other hand though, there is a second way of thinking about these objects. Instead of just thinking about each form in terms of its name and face-value, we could extract the identity set of the entire result set. In this case though, we'd be thinking of the result not so much as merely a set of names for a bunch of objects, but rather as one single cohesive relational object formed from multiple others. For example, if we were to write **id**($[\![0 \cdot \cdot | \text{Cont} | \cdot = 0]\!]$) instead of $[\![0 \cdot \cdot | \text{Cont} | \cdot = 0]\!]$, then the result would be the set of all *relationships* involving any numbers on the continuum.

And, since any set of arbitrary relationships can be treated as just as validly being conceptually a single object as any other more conventional object in relational unified logic, it follows that one could argue that **id**($[\![0 \cdot \cdot | \text{Cont} | \cdot = 0]\!]$) (i.e. the set of all relationships involving numbers on the continuum) is the answer to 0/0 in a deeper and more meaningful sense than just returning a set of names of numbers would be.

Indeed, such a set is just as much an equally valid arbitrary set of relationships as anything more mundane seeming such as **id**(2) would be. Thus, we can't really say that the identity set of 0/0 is any less a coherent single object than we could claim the same of the identity set of 2. As such, we are therefore equally as able to treat strange cases

---

[97] See page 505 if you want to reread the definitions of these terms.

[98] The reason is because in relational unified logic (1) everything is always a transparent set, (2) all operations are defined in terms of basic set operations, and (3) basic set operations never fail for well-defined sets, and thus all operations are always well-defined.

such as 0/0 as meaningful distinct objects as we have been able to treat any other more conventionally well-understood object likewise.

Put more simply, just as there is a difference between thinking of 2 as merely a name or label for something that we can pass around and manipulate, versus thinking of it in the deeper and more meaningful sense of its identity set, the same can be said of any other nominally strange object such as 0/0. The concept of an identity set provides us with a nice ability to distinguish between thinking about things in a face-value way, as merely arbitrary forms (like merely names or labels) that we manipulate, versus thinking of forms in terms of all of their connections and relationships to other forms in the deeper and more meaningful sense via their identity sets. Thus, on the surface level 0/0 can be considered as being the set of all numbers on the continuum, but on a deeper level it could also be considered as being the set of all *relationships* on the continuum. The choice depends on what we want to use the result for and how we want to think about things and so forth.

Another interesting aspect of this way of thinking is that it broadens the number of ways we can think about the result of any arbitrary expression. Normally, from the traditional viewpoint, we typically think of operations as returning one specific member of a relationship, as determined by constraining all the other participants in that relationship to specific values. This is all well and good of course, and tends to be very practical and computationally efficient too.

However, when working in terms of constraints imposed on sets (i.e. "operating by constraints" etc), it is not necessary to be so narrowly focused on any specific participant in a relationship over any of the others. Indeed, in contrast to the more traditional way of working, when operating by constraints we can concern ourselves with as few or as many of the participants in each relationship as we want. This makes it possible to treat the arity of each operation in whatever arbitrary way we want, thereby enabling us to easily define unconventional variants of existing operations and relationships in some interesting ways. We can select exactly which parts of each relationship we want to treat as inputs or outputs in whatever arbitrary way we want.

Indeed, doing so can enable us to easily define many operations that would otherwise normally be considered to be completely "crazy" or "nonsensical" by many people. For instance, it is possible to give *unary division* (i.e. division with only a numerator or only a denominator) a completely well-defined and usable answer under this system. Thus, for example, expressions like $\frac{12}{}$ or $\frac{}{2}$ could be considered perfectly well-defined, just as much as any other more conventional expression such as $2 + 2$ would be considered perfectly well-defined. Sounds kind of crazy, right?

Well, crazy sounding as it may be, it is nonetheless true. Indeed, it's not even difficult to define, honestly. This is just yet another example of how little you can trust your "common sense" and instinctive responses when trying to discover the truth. Very often "common sense" is really just nothing more than the sum of all of a person's existing prejudices and limitations of knowledge. People tend to easily forget just how much their state of mind is actually just a product of the time and culture they live in, plus their level of education and experience etc.

For example, both the number zero and the concept of negative numbers were once viewed as not being credible concepts by the vast majority of humanity. People used to think zero and negative numbers were *obviously* completely nonsensical[99]. Similarly, your "common sense" instinct here may be to think that surely unary division couldn't make any rational sense. Yet, it does. There are two cases, one for the numerator and one for the denominator. Let N represent the numerator (the number on the top) and let D represent the denominator (the number on the bottom), both as continuum numbers. Here then is one way of performing unary division for each of these cases, written as blueprint expressions:

1. unary division of the numerator: $[\![\cdot | \mathsf{Cont} | \cdot \cdots \cdot | \mathsf{Cont} | \cdot = N]\!] \mathbb{X}(\langle \overline{1}, \overline{3} \rangle)$

2. unary division of the denominator: $[\![D \cdot \cdot | \mathsf{Cont} | \cdot = \cdot | \mathsf{Cont} | \cdot]\!] \mathbb{X}(\langle \overline{3}, \overline{5} \rangle)$

Notice that in this case I am using explicit extractors instead of implicit extractors, so that I can transform the result into vector notation. This is not strictly speaking necessary, but does add some clarity arguably. If, on the other hand, I still wanted to use implicit extractors, then I'd need to be sure to space out the different parts of the returned forms in the result so that it is clear that each component of each form is in a distinct position, instead of merely slurring them together, which would be confusing. A concrete example will make this clearer, but first let's see the corresponding blueprint expressions. If we wrote these unary division expressions with implicit extractors instead of explicit extractors, then they would look like this:

1. unary division of the numerator: $[\![\cdot | \mathsf{Cont} | \cdot \cdots \cdot | \mathsf{Cont} | \cdot = N]\!]$

2. unary division of the denominator: $[\![D \cdot \cdot | \mathsf{Cont} | \cdot = \cdot | \mathsf{Cont} | \cdot]\!]$

In the case of the expressions that use explicit extractors above, the result will be a set containing forms that look like $\langle a, b \rangle$. In contrast though, in the case of the expressions that use implicit extractors above, the result will be a set containing forms that look like $a\ b$. In both cases though, a and b represent numbers on the continuum. Notice the difference in format between these approaches. Implicit extractors force us to just use spacing in order to keep the subcomponents of the resulting forms distinct, since the notation does not provide any mechanisms for specifying a formatting schematic, whereas explicit extractors give us the freedom to structure things more diversely and arbitrarily. This is part of the trade-off of choosing which type of extractor you want to use. Pick your poison. Life is full of trade-offs. Very few things are universally better than the alternatives[100].

Anyway though, it's kind of subjective which one of these notations for unary division would be best (vector notation versus spacing notation). We'll go with the vector

---

[99]"Hah... you can't have nothing or less than nothing. What would the point of that be? It's unnatural! It defies all common sense and thus surely can't be true! Don't be ridiculous!" — Sincerely, Early Human Civilization

[100]More precisely: Generally speaking, most things have "local maxima" in terms of the trade-offs in their design. Objectively inferior or superior designs *do* exist and indeed are quite common, but usually only in a localized sense. For example, one car may be objectively superior to another car in all respects, but even the best car is essentially incomparable to the best cargo truck. Cars and cargo trucks have *fundamentally different purposes* and therefore *inevitably* suffer from unavoidable trade-offs. Much the same can be said for the design of many other things besides automobiles.

notation generated by the explicit extractor example above though, just for the sake of choosing one. There's not much point in agonizing too much over superficial details, at least not when merely exploring a concept to broaden our horizons, as we so often are doing in this book. I think it's time that we considered some concrete examples for unary division and see how they behave. It really won't be as weird as you might be expecting. Indeed, unary division is actually so natural that I bet you've even partially performed it in your own daily life before, multiple times, without even realizing that that was what you were doing.

For example, suppose we wanted to decide how many pieces to slice up one pizza into. This would be an example of unary division of the numerator. We have something that we want to divide, but we don't yet actually know how many pieces we want to divide it into. The numerator of the thing we want to divide is 1 in this case, since we only have *one* pizza. The denominator though is omitted, since we are merely asking ourselves "What are all the possible ways of evenly dividing this one pizza into pizza slices?". We thus could use the following blueprint expression to express this computation:

$$[\![\cdot | \mathsf{Cont} | \cdots \cdot | \mathsf{Cont} | = 1]\!] \overline{\times} (\langle \overline{1}, \overline{3} \rangle)$$

This expression results in a set containing every possible combination of a and b on the continuum such that $a \cdot b = 1$, formatted as $\langle a, b \rangle$. Thus, for example, if we wanted to cut the pizza up into eighths then $\langle 8, 1/8 \rangle$ and $\langle 1/8, 8 \rangle$ would be the forms in the result set that would represent this possible choice. Any possible pair on the continuum that satisfies $a \cdot b = 1$ works though, of course. Thus, the result set will look something like this:

$$\{\langle 8, 1/8 \rangle, \langle 4, 1/4 \rangle, \langle \pi, 1/\pi \rangle, \langle 1, 1 \rangle, \langle 0.4, 2.5 \rangle, \ldots\}$$

This set is infinitely large. However, notice that computing partial subsets of it is quite easy. Indeed, there are many circumstances in daily life where someone would think about what possible ways of dividing something up there are and then would choose one arbitrarily. Every time someones does this, they are performing partial unary division on the numerator. Thus, surely then, if this operation is so abundantly common in daily life (as indeed it is), then one can bring oneself to believe that unary division is in fact actually well-defined after all, contrary to how absurd it may have appeared at first glance.

Be that as it may, the pizza example here may feel a bit trite and vapid, but that's partly just because it uses a numerator of 1, which is not a very interesting case. There are other (more useful) cases for unary division on the numerator. For example, suppose we had some quantity of some resource that we wanted split up evenly, but that we didn't know how many groups we wanted to split it up into. This could easily happen in scenarios where you want to figure out a "reasonable sized group" for splitting a large quantity, or even just simply because you want to decide how to split some resource for whatever completely arbitrary reason (e.g. splitting money among some yet-to-be-decided number of people). For example, if we had $1,000 to split, what are some of the ways we could do it? Well, using blueprint expressions, it would look something like this:

$$[\![\cdot | \mathsf{Cont} | \cdots \cdot | \mathsf{Cont} | = 1000]\!] \overline{\times} (\langle \overline{1}, \overline{3} \rangle) = \{\langle 500, 2 \rangle, \langle 50, 20 \rangle, \langle 40, 25 \rangle, \langle 200, 5 \rangle, \ldots\}$$

As you can see, this example results in a set of values that is a little bit more interesting, although still quite simple. The pizza example always ended up generating pairs of numbers of the form x and $1/x$, due to the fact that the numerator of the unary division was always 1. However, this other example here, of splitting $1,000, is not quite so trivial. Regardless though, these are both very easy and mundane examples. Once you get past the mere shock value of daring to suggest that "unary division" could ever be well-defined, it fast becomes clear that unary division really isn't very weird after all. Indeed, it is really quite normal and natural once you've acquired a proper appreciation of it.

The key to why this way of thinking empowers us to work with so many supposedly "strange" objects so easily lies partly in the fact that relational unified logic frees us from having to focus so single-mindedly on any specific participant in any relation over any of the others. The universal domain and universal closure relational unified logic establishes is also essential to its higher-level generality. By being able to focus our attention on as many or as few parts of each form or relationship as we wish, we are empowered to express many more possible variations on each operation in our vocabulary. This results in each operation having as many or as few inputs and outputs as we want: at least zero, but no more than the total number of participants in the underlying relationship.

Indeed, if you thought unary division was a strange notion, then you should wait until you see what comes next. It arguably gets even crazier than that. In fact, as weird as unary division may be, there's no reason we have to stop at unary. Why not just omit *all* of the inputs to a division operation, instead of only one[101]? There's nothing in relational unified logic stopping us from doing so. Indeed, performing division without any inputs whatsoever is perfectly valid in relational unified logic. Thus, for example, expressions like — (division with both numerator and denominator omitted) can be considered to be valid instances of division in relational unified logic and are in fact not undefined at all. Written in blueprint expressions, it would look something like this:

$$[\![\cdot | \operatorname{Cont} | \cdot \cdots \cdot | \operatorname{Cont} | \cdot = \cdot | \operatorname{Cont} | \cdot]\!] \bowtie (\langle \overline{1}, \overline{3}, \overline{5} \rangle)$$

The result of this expression would be a set of vectors of the form $\langle a, b, c \rangle$, such that $a \cdot b = c$ is always true. We could also have chosen to just omit the explicit extractor here entirely, in which case the result would instead simply be a set of equations of the form $a \cdot b = c$, which is arguably conceptually clearer in this case. However, I decided to use the vector format again here for the sake of consistency with the other examples that we were recently discussing above. Exactly how you choose to do things when working with this system is quite open-ended and free-form though, so feel free to use whatever notation you personally find most pleasing or useful in the moment.

By omitting all of the inputs, division essentially becomes equivalent to the entire set of all true division relationships. Incidentally, notice that it also becomes equivalent to the set of all true multiplication relationships as well. In other words, division with

---

[101] An operation with zero inputs (an arity of zero) can be called "nullary" or "zeroary". Those are the standard terms in mathematics, apparently. It is sometimes written as "0-ary", just as any other arbitrary arity k (where k is any non-negative integer) can be written as "k-ary".

no inputs is equivalent to multiplication with no inputs, essentially. They both originate from the same underlying relationship. These kinds of correspondences are pretty typical of this kind of thinking. Inverse relationships are very closely connected to each other, so it often ends up actually being completely arbitrary whether or not you choose to consider a given instance of an operation to be a variation of itself or of its inverse.

It all depends on how you think about it. It's arbitrary. All participants in each relationship are treated the same, i.e. as just being equally arbitrary forms. Relational unified logic doesn't favor any one specific participant in any relationship over any of the others. This is a natural consequence of thinking in generalized relational terms instead of in terms of one specific preset return value for every operation. Relational unified logic doesn't play favorites, in that respect.

Anyway though, besides reducing the number of inputs we specify for any given operation, we can also do the opposite. We can also *increase* the number of inputs we provide, over and above what is actually needed to determine the output value, and yet still always be guaranteed a well-defined result. For instance, instead of reducing division from a binary operator down to a unary or 0-ary operator, we can do the opposite and make it into a ternary operator (i.e. an operator that takes three inputs).

Sounds kind of weird, making division accept three inputs instead of just two, right? Well, weird as it sounds, it works just fine. The answer is also never undefined or erroneous, regardless of what arbitrary inputs we feed it. Here's what it would look like in blueprint notation, given continuum numbers A, B, and C, where C represents the numerator, A or B represents the denominator, and the remaining variable represents the result of the division:

$$[\![A \cdot B = C]\!]$$

In other words, ternary division just involves filling out a value for each possible participant in the underlying multiplication relationship and then evaluating it. If indeed the relationship we put in the blueprint expression is true, then it will be returned, but otherwise only the empty set will be returned. Thus, even if we feed the expression incorrect values we will still receive a well-defined result, i.e. we will always receive at least the empty set. The net effect of all of this is essentially that ternary division is just a validity checker for a division relationship.

For example, if we wrote $[\![2 \cdot 6 = 12]\!]$ then the result would be $\{2 \cdot 6 = 12\}$, which would indicate that the relationship is indeed true, i.e. that $12/2 = 6$ or $12/6 = 2$ is true. If, on the other hand though, we wrote something like $[\![500 \cdot 8 = 7]\!]$ then the result would merely be the empty set $\{\ \}$. The empty set being returned here tells us that $500 \cdot 8 = 7$ is false. This is fitting of course, considering that in unified logic empty sets are synonymous with falsehood, whereas non-empty sets are synonymous with truth.

One thing you may have noticed here though is that this time I didn't include an extractor. The reason for that is because the choice of what to extract (if anything) is ambiguous and arbitrary. The blueprint expression example above, representing "ternary division" (and also "ternary multiplication", incidentally), doesn't include an extractor, but there's nothing to stop us from adding one on in whatever arbitrary way pleases us.

We could just return the entire relationship, as we are already doing here. But, we could instead return all three numeric participants as a vector, such as $\langle A, B, C \rangle$. Or, we could return just the normal result of division (i.e. just one of A or B in $A \cdot B = C$), such as if we wrote $[\![A \cdot \underline{B} = C]\!]$ or $[\![\underline{A} \cdot B = C]\!]$. We could even do much stranger extraction selections, such as extracting $\langle B, C \rangle$ or whatever else we want. It's arbitrary. There's no clear indication which exact part of the form (if any) would be most appropriate to extract for ternary division, nor is the choice clear for many other similar types of overspecified operations either.

This ambiguity and arbitrariness in which parts (if any) to extract from the form when performing ternary division is why I didn't include extractors in the original example of ternary division. This ambiguous and arbitrary choice may initially feel kind of odd and discomforting, but nonetheless from a conceptual standpoint it is perfectly fine. Higher-levels of generalization often have these kinds of effects on how we think about things. Generalization makes how we think more flexible, but that same increased flexibility also often makes things fuzzier and more capable of potentially diverging in lots of different arbitrary directions that we are now forced to choose between.

This kind of freedom of choice can be discomforting, in some ways, because it makes it less likely for there to be just one single clear path forward that can be followed thoughtlessly. However, our anxiety over such things as human beings is often superficial. It is usually best to be pragmatic and to just go with whatever works well under whatever circumstances we find ourselves in. Human knowledge is inherently limited and there's only so much we can keep in our head at once and only so much we can predict. It is best not to fret over such things too much. Don't get trapped in analysis paralysis and obsessive deliberation. Such behavior is often very wasteful. Often, only experimentation can truly clarify the best path forward.

In the meantime though, before one can find a more optimal path forward, all one can do is to make best guesses and then test how they work out in practice. There's nothing to be ashamed about in doing that. Life is a work in progress, a never-ending learning process. It's best to not let ambiguous choices disrupt your ability to function, and to instead simply be pragmatic about your choices and do what you need to in the moment. Always be open to future improvements though, when such opportunities arise and when the gain is worth the cost. Time brings clarity, and what was optimal once before may no longer be. It depends really, as with most things in life.

The fact that there is ambiguity and arbitrariness in our choices does nothing to diminish the validity of the underlying concepts though. It just broadens the scope of what possibilities we have to think about. Such is the cost of flexibility. Absolute freedom is powerful, yet also overwhelming. It is so easy to lust for greater freedom and dream of how it will somehow make everything better, and yet when one finally acquires it to become dismayed at the multitude of differing choices and the responsibilities and consequences that come with every possible choice. This effect can cripple productivity.

Too many people spend most of their lives always following other people's orders, always heeding the capricious whims of whatever random false "authority figures" society has currently decided to elevate. Too few have learned to master the art of making their own choices in life by the time they are adults, and thus many people struggle to handle freedom well when they finally get it, and therefore end up managing their time and choices very poorly. This often results in analysis paralysis, inhibited personal growth, and economic stagnation.

Much of the key to overcoming this though is to simply give yourself permission to choose arbitrarily based on your own best guesses and to explore the consequences that follow and to learn from them. Optimal living requires a pragmatist's attitude. You must learn to be a player instead of a pawn. You must learn to stop being a cog in someone else's machine, and to instead realize that choices are always at least in part arbitrary. The authority of choice exists whenever you decide it does. Stop waiting for other people to tell you when you are "allowed" to make a difference. They never will. You must seize authority for yourself.

Stop jumping through other people's hoops. Stop allowing other people to corral you through their own artificial contrived barriers that they have designed to slow you down and to limit your progress. The only gatekeeper for your ambitions in life should be you. The rules of life are what you choose them to be. Authority is just a social convention, although admittedly sometimes backed by force and violence etc. Ultimately though, in most cases, other people only have authority over you when you let them.

So much of what people think are the "rules" of life are actually nothing more than a mind game designed by exploitive people to limit your power and to reduce your awareness of what the real constraints of life actually are. A rule is only "official" or "mandatory" if someone succeeds in deluding you into thinking it is. Your thoughts become your reality. Choose to make your reality a good and righteous reality, instead of merely allowing it to become an extension of someone else's exploitive agenda.

Anyway though, I kind of got derailed onto yet another of my ranty tangential discussions again there. Such wild tangents are well worth it though I think. These kinds of random tangential discussions often add lots of value. Too many people are too rigid in how they express themselves. Value creation should be our center, our primary guiding principle. Often, only a natural, uninhibited, and conversational mode of exposition can truly maximize the amount of value we create in our work.

We should focus less on dogmatic adherence to rigid formats and standardized assumptions, and more on on simply making a powerful and productive impact through value creation as much as possible. Simply be natural. Random tangents are often well worth while, even when they derail a discussion temporarily. Indeed, it seems to me that uptight formulaic editorial control often makes things *worse* instead of better. It destroys value in exchange for superficial consistency. It creates stagnation and complacency. It sacrifices the things that matter most (value creation) for the things that matter least (the appeasement of those who care more about homogeneity than about anything that actually matters). That's not a good trade-off.

Anybody who is not so far gone down the rabbit hole of dogmatic adherence to formulaic ways of thinking that they can no longer see the world with fresh eyes should be able to clearly see that. Whether it's ranty tangents, emoticons, or personal pronouns, natural expression is generally better than rigid dogmatism. So much of the conventional "wisdom" of how things should be done is actually just arbitrary and counterproductive. Be brave. It's time to stop letting other people's attempts to micromanage us get in our way as writers, as creators, and as human beings.

Stop letting other people control the way you think and the way you express yourself. Stop being ruled by fear and conformity. Start living based on genuine merit instead of merely based on whatever arbitrary pretenses a handful of other falsely "authoritative" people want you to believe and to adhere to. Stop letting other people backseat drive in your life and in your creative process. Pick those intrusive control freaks up by the collar and throw them out the high-speed window of your mind. Toss them in the dirt, where they belong.

These people would try to control and dictate the creative expression and style of everyone else on the whole planet if we let them. Don't let them. They have no right to. Generally, unless there are special ethical circumstances, nobody has the right to intrude upon the details of someone else's life or of any creative work that is not of their own making. Your creative spirit is a sacred garden and a core part of your power as a human being. Allowing intruders into your mind severely damages your power over your own outcomes. Keep parasites out of your mind. Think and live freely, within the constraints of benevolence and reason. Let your mind grow wild and free. This is the path to power and to making a difference in the world.

Anyway though, as you can see, I did it again. I made yet another ranty tangent, because I saw yet another opportunity for value creation. Who cares if it derails the discussion temporarily though? It's well worth it. I feel like most of the best ideas I've ever had in life (and indeed in this book also) only ever came into existence because I allowed myself to think freely and to follow tangential thoughts wherever they led. Boldness is necessary for creative breakthroughs. It is an empowering way of thinking. Fiery passion and an uninhabited creative spirit are essential to maximizing the impact of a person's life. And, as long as you constrain those passions with benevolence, self-discipline, and tact, and choose your battles wisely, then the outcomes tend to be positive.

Putting all this aside though, let's now return to our discussion of blueprint notation and relational unified logic and such. Before my ranty tangents derailed us, we had previously been discussing the very free-form way in which relational unified logic is able to treat the arity of operations. By reasoning in terms of constraints imposed on sets of arbitrary forms, we are able to place our focus on whatever arbitrary aspect of any given operation we want to, both with respect to inputs and with respect to outputs. This enables us to think about things in some unusual ways, and to give reasonable definitions to operations that we would not otherwise think could be reasonable, such as unary division for example.

However, during that discussion, we were very focused on just a few specific examples, and it is very important to realize that these principles we've been discussing apply far more broadly than that. Any operation will work. Division was just one example. Interesting as it may be that things like $0/0$ and $N/0$ can be made well-defined under this system, it's just the tip of the iceberg. There are lots of other crazy things we can do.

For instance, just as we could easily define unary division as a coherent operation, the same can be said of unary subtraction. I'm not talking about negation though. Indeed, interestingly, it turns out that unary subtraction and negation are actually *not* the same thing, contrary to popular belief. I mean, to be clear, it's definitely true that you can interpret the a in $b - a$ as being negated and then added to b, as in $b + (-a)$. That *is* negation. That's not what I'm talking about here though. I'm actually talking about the *other* part of the subtraction operation when I talk about unary subtraction. Indeed, unary subtraction typically operates on the b part of $b - a$ and works quite differently from negation.

Just as we reframed the division example in terms of multiplication though, we will likewise reframe the subtraction example in terms of addition. The subtraction example will also need to be expressed in relational terms, as an equation, just as we did with the division example. As such, we'll want to think in terms of forms that look something like $A + X = B$, i.e. that look like the equation form of $B - A$, where A, X, and B are arbitrary numbers on the continuum. X would normally represent the return value of $B - A$. The B part is the part that the unary subtraction is being applied to as a unary operation (in effect). As such, here's what unary subtraction would look like as a blueprint expression:

$$[\![\cdot\mid \text{Cont} \mid\cdot + \cdot\mid \text{Cont} \mid\cdot = B]\!] \rtimes (\langle \overline{1}, \overline{3} \rangle)$$

If we wanted to try to write this in traditional notation, instead of blueprint notation, then the closest representation would be something like $B-\ $, i.e. like $B - A$ but where the A part has been omitted. Notice how this contrasts to something like $-A$ though, which in contrast would just be negation. However, be that as it may, traditional notation wouldn't scale well for these kinds of use cases though, so it wouldn't really be advisable to attempt to write unary subtraction this way in practice.

Blueprint notation is far better equipped to handle the generality of this way of thinking. Attempting to use traditional notation for this would probably end up causing some expressions to become ambiguous and indecipherable. Nonetheless, it's still interesting to think about what the closest analogous representations for expressing these kinds of operations in traditional notation would look like though.

In the case of the specific blueprint expression and extractor I have employed above though, the result will be a set of forms that look like $\langle A, X \rangle$, where $A + X = B$, and B is the number that the unary subtraction is being applied to. In other words, it will simply be a set of pairs of numbers (written as vectors) such that the numbers always sum to B, the sole operand of the unary subtraction. For example, if B was 10 then $\langle 2, 8 \rangle$, $\langle 7, 3 \rangle$, $\langle 5, 5 \rangle$, $\langle -20, 30 \rangle$, and $\langle 0.75, 9.25 \rangle$ would all be examples of members of the result set. This behavior is perfectly analogous to the behavior of unary division of a numerator. The only difference is that this time we used addition instead of multiplication.

Also, just as with division, we could mess with the arity of the subtraction operation in whatever arbitrary way we want. For example, we could define ternary subtraction (subtraction with 3 inputs) and nullary subtraction (subtraction with 0 inputs, a.k.a. "zeroary" or "0-ary" subtraction) too and whatever other quirky variants we want. We would need only write down the corresponding blueprint expressions that express our specific intent in each case. As before, the interpretation is also somewhat arbitrary, since these operations can also be considered as variations on addition instead of subtraction. More generally though, these same flexible behavior patterns will reappear regardless of what specific operation or relationship we choose to work with.

Indeed, there are countless other specific examples of operations and relationships that we could apply this way of thinking to. The system will always work the same regardless though, since it always operates purely in terms of sets of arbitrary forms and such. The number of possibilities is literally infinite. As such, comprehensive coverage is impossible. Moreover, even putting comprehensive coverage aside, attempting to cover even a relatively small set of additional examples here would largely just be redundant and would mostly be a waste of our time. The point has been made by now.

That's why I think it's now time to move on from discussing these specific examples. There are still some other aspects of the system that I'd like to discuss. The bulk of what I've wanted to say about the core content of unified logic has been said by now though. As such, the remainder of what we will be discussing will be more peripheral or high-level in nature. There are several different conceptual connections and little bits of new material that I'd like you to consider, but for the most part we will now begin winding down our discussion of unified logic. To make an analogy, we have climbed to the apex of the mountain, but are now beginning our leisurely descent.

Firstly, I would like to point out that blueprint notation is flawed. While it may be a convenient shorthand notation for some things, there are still many other things that may be more effectively communicated using a more traditional approach. A good example of this is the fact that simple named variables are often easier to understand and manage than generic transience delimited sets are. When you use a lot of generic transience delimited sets like ·|Cont|· (etc), instead of merely simple named variables like A, B, and C, then expressions can gradually start to become harder to read and to constrain. Named variables are also often a lot easier to refer to.

As such, you may often find that it is better to use traditional set builder notation than blueprint notation. For example, the blueprint expression $[\![\cdot|\text{Cont}|\cdot + \cdot|\text{Cont}|\cdot = B]\!] \bowtie (\langle \overline{1}, \overline{3} \rangle)$ could instead be expressed in set builder notation as $\{\langle a, b \rangle : a + b = c\}$. This is much clearer and much easier to understand in this case. Set builder notation is a great notation really, most of the time. Don't underestimate it.

Nonetheless though, even if we end up using set builder notation instead of blueprint notation, blueprint notation has served as a useful thought experiment and as a way of forcing us to think in terms of reconstructing operations in terms of constraints imposed upon sets instead of falling back into old ways of thinking. If I had explained all of these ideas for unified logic using traditional set builder notation first, then perhaps you might not have understood as clearly that unified logic really is a genuinely different way of thinking about what operations are and what they are capable of. Even just that fact alone makes the creation of the blueprint notation system worth it.

Regardless of the fact that blueprint notation is my own creation and thus is inevitably somewhat dear to me, I do highly recommend that you consider the possibility that set builder notation will often be much more readily understandable by many people in many cases. Blueprint notation itself is very much "hot out of the oven", just like many of the other ideas I have come up with in this book. Blueprint notation hasn't had even remotely as much time to be tested or refined as the traditional notation has. As such, I would always advise readers of this book to exercise caution in light of that. One should be skeptical of my system. There are some signs of flaws.

It often takes many years for a new system to have all the miscellaneous wrinkles potentially ironed out of it and for its true scope of capabilities to become genuinely understood. A diverse set of many different people, each making their own tweaks over the course of history, is also often necessary for a new system to evolve to the point where its real place in the literature is understood. Sometimes a new system will turn out to not be worth it compared to other existing systems, and that could be the case here too, but other times the new system will manage to make some significant lasting contributions. It depends. It is an evolving process. Regardless though, the only way to find out is to try, by continuing to create and to explore new ideas, so naturally that is the path I've chosen.

5.11 Considering the broader context

Who really knows what will happen? It's hard to predict the future. Regardless though, such experimentation and gradual evolution is how most progress happens. This is just as true in science as it is in art. It applies to all creative endeavors really. Evolution is the way of the world. That's not anything to fear though. It's exactly the way that things *should* be. It is what is necessary for optimal outcomes.

As such, we should all remember to not get too caught up in the hype of every new moment. We should instead simply take things on their merits. We should incorporate what is useful to us, and leave the rest alone. We should always strive to take the best of all worlds, rather than merely becoming dogmatically trapped in one way of thinking over all other ways of thinking. So, naturally, that is what I suggest you do for this book as well: i.e. you should interpret this book and all the ideas in it in a nuanced and pragmatic way, as with all things in life. Take what is good, and leave the rest. Experiment with what works best in what contexts, rather than just making premature assumptions. Diversify your frame of mind, but keep it within the constraints of what is most effective.

Speaking of diversification though, besides just the basic set operations and blueprint notation and such, there are also some other operators we could define if we wanted to. Some of them might have interesting interpretations in the right contexts. For example, consider what might happen if we permitted ourselves to change all instances of some form inside a set of forms to some other form. What would the consequences and interpretation of such a thing be? Well, let's define such an operation and then take a few moments to think about what the net effect of applying it to some things might be. Here's our corresponding new definition:

Definition 184. *It may sometimes be useful to substitute all instances of a given form within each form in a set of forms for some other form. Let us call such an operator a **renaming operator** or **substitution operator** and represent it by the symbol $\triangleright$. We might as well permit the operation to perform multiple renamings at once.*

*Thus, let us have the format of $\triangleright$ be $S \triangleright (A_1 \curvearrowright B_1, A_2 \curvearrowright B_2, \ldots, A_{n-1} \curvearrowright B_{n-1}, A_n \curvearrowright B_n)$, where S is a set of forms we want to apply the renaming operator to, where each A_k represents each old form we want to change to some new form, and where each B_k represents the new form that we want to change each old form to, up to an arbitrary finite number n of such renamings (a.k.a. substitutions). We will refer to the $\curvearrowright$ as the **rename-to arrow** or the **substitute-with arrow**.*

Each renaming (a.k.a. each substitution) that this operation performs works by searching for all instances of the indicated old form embedded within any of the forms in the input set *S* and then changing all of them in parallel to the indicated new form. It does this for each of the $A_k \curvearrowright B_k$ renaming (a.k.a. substituting) pairs listed in the renaming operator's inputs, ordered and performed from left to right (because the order can change the outcome), until all such cases have been applied.

Thus, for example, if we had an expression like $\{X, Y, Z\} \triangleright (Y \curvearrowright Z)$ then this would evaluate as $\{X, Z, Z\}$, which in turn would also be equivalent to $\{X, Z\}$ since duplicates don't matter in sets. That's a very simple example though and is not very illuminating. Let's explore some more detailed examples to get a better sense for the operator's potential behavior. It can have weird effects when applied to some systems. For example, consider the following expression:

$$\{2 + 2 = 4, 2 \cdot 3 = 6, 2 - 10 = -8\} \triangleright (2 \curvearrowright 7, 4 \curvearrowright 60)$$

What would the result of this expression be? Well, it would be $\{7 + 7 = 60, 7 \cdot 3 = 6, 7 - 10 = -8\}$. Clearly then, knowing what we know about arithmetic and algebra, all of these equations would now be false as a result of applying these renaming operations. However, although that may be true as long as we think of these objects in their conventional sense and meaning, one must always remember that forms can potentially be interpreted in any arbitrary way, if we choose to do so. The rules of transformative logic and unified logic don't dictate the meaning of forms, but rather only express formalisms by which to work with the forms and to mold them into expressing whatever arbitrary set of behaviors and relationships we desire.

As such, we cannot really rely upon our understanding of the conventional sense of what these forms usually mean when we attempt to interpret what they now mean after being transformed. We must think more broadly than that. Indeed, let's take this thought experiment one step further and see if it proves to be more illuminating. Suppose, for example, that we were working with the entire transformative rule set for arithmetic and algebra, instead of just this tiny limited set of equations $\{2 + 2 = 4, 2 \cdot 3 = 6, 2 - 10 = -8\}$.

As such, with all of arithmetic and algebra now implicitly at our disposal, we can simply write ⟅ **+** ⟆ to express the set of all forms derivable from the initial forms and transformations rules of arithmetic and algebra. Here's the interesting part though: What would it mean if we applied a renaming (a.k.a. substitution) operation to this entire set? For example, what would happen if we did this?:

$$⟅ + ⟆ \triangleright (2 \curvearrowright 7)$$

Well, the result of this expression would be the set of all valid forms (e.g. numbers, arithmetic expressions, equations, etc) that can be generated under the rules of arithmetic and algebra, except that after generating that entire set you would replace every instance of the number 2 with the number 7. Let me also remind you that we are assuming that adjacent groups of digits count as single atomic entities for these purposes, such that this renaming (a.k.a. substitution) operation would not change the internal digits of other numbers (e.g. 27 would not change to 77, because 27 is not 2).

I mentioned this implicit convention briefly earlier. Technically though, to be strict and to express these assumptions explicitly, you'd want to use delimiters to block misinterpretations, such as (for example) by writing [2] instead of just 2 and writing [27] instead of 27, so that the 2 in [27] cannot match to [2] because it would be missing the adjacent [ ] delimiters. The choice of delimiter symbol (e.g. [ ] here) is arbitrary, but ideally you probably wouldn't want the symbol choice to conflict with any other delimiters you use. Anyway though, that (i.e. careful use of delimiters) is how you'd do it if you wanted to be really strict about it. We'll be avoiding that mess in this book though, for the sake of brevity and less notational noise.

As I was saying though, $(\![+]\!) \triangleright (2 \curvearrowright 7)$ would have the net effect of replacing all instances of the number 2 with the number 7 within the set of all otherwise valid forms of arithmetic and algebra. What would this really mean though? Well, interestingly, it would tell us exactly what the universe of arithmetic and algebra would be like if 2 were in fact instead 7, regardless of what the physical reality is.

That's part of the fun of the arbitrary nature of mathematical reality: you can explore wild assumptions and see what would happen as a consequence, regardless of whether those assumptions actually hold true in the real world. Logic and mathematics essentially allow you to explore all conceivable logically coherent universes, even including ones that boldly defy even the most intuitive and seemingly necessary rules of the normal real-world universe we live in.

To that point, this renaming (a.k.a. substitution) operator is yet another example of a seemingly bizarre yet still perfectly well-defined way of pushing the boundaries of how we normally think about logical systems. It is yet another data point for demonstrating the truly enormous expressive power of thinking in terms of generalized transformation rules, arbitrary forms, and set operations, instead of in a more traditional and rigid way.

Just as reframing all operations in terms of constraints imposed upon sets allowed us to give division by zero (and many other strange operations) a well-defined result, despite doing so defying "common sense", so too does this new renaming and substitution operator allow us to explore yet more very strange questions. The example above, i.e. "What would it be like if 2 was 7 instead?", is merely one such example. The sky is the limit. Arbitrary forms and transformation rules are a very flexible way of working and thinking.

Speaking of flexibility though, one especially great aspect of the flexibility of the way of thinking that transformative logic and unified logic provide us is their universally "always well-defined" nature. That's what I'd like to talk about next. We've actually already talked about it some before in scattered places throughout the book, but it is important enough that it merits additional discussion and emphasis regardless.

Both transformative logic and unified logic have a certain sense in which their results can be considered "always well-defined", but the exact nature and nuances of this property differ between these two systems of reasoning. Each system has a different way of handling "undefined" material that it encounters. Let's review both of these systems and how they handle their respective cases. Let's start with transformative logic first.

As you may recall, transformative logic is all about transformation rules and how those rules can act upon arbitrary forms. Normally, in transformative logic, we have a set of predefined transformation rules and initial forms from which other valid forms within the rule set may be generated. Each transformation rule tells us that if a form looks a certain way then it can be transformed into some other form. However, the transformation rules say nothing about cases that are outside of the vocabulary of forms that the transformation rules are aware of. How are such cases (i.e. forms outside the normal predefined vocabulary of the rule set) supposed to be handled though?

Well, the answer is simply that we cannot touch those parts of the forms we encounter. We can only apply transformations to forms that match our existing rules. Beyond that though, we must leave anything outside the vocabulary of our rule set untouched. However, this does nothing to stop us from continuing to apply our transformation rules to other parts of the forms that *do* match our rules though, even when there are forms that we don't know how to work with nearby.

The beautiful thing about this way of working is that it allows us to handle both forms that are within our vocabulary and forms that are outside of our vocabulary seamlessly. Instead of just throwing up our hands and declaring that we have encountered an error of some kind and then giving up (like we would in traditional logic and math), we can instead continue working with any subcomponents of any form that our rules can still understand.

Furthermore, even though a form may not be understandable by our rule set, that still doesn't change the fact that it is just as much an arbitrary form as any of the other forms that we might consider are, and thus is still broadly "valid" in that sense, even if it exists outside the normal vocabulary and internal validity of our transformative rule set itself.

Thus, pretty much no matter what we do in transformative logic, as long as we follow the correct processes and express ourselves in a coherent way, the forms we encounter will always be well-defined and usable as values in our reasoning process. It is in this sense that transformative logic can be said to be "always well-defined" and to have it's own version of the "universal domain" and "universal closure" that unified logic has, in some broad sense.

In contrast though, let's now refresh our memory of unified logic and in what sense it can also be thought of as being likewise "always well-defined". The reasons for its universal well-definedness are different though. In the case of unified logic, it all comes down to the fact that unified logic redefines all operations in terms of constraints imposed upon sets of arbitrary forms, via set operations and some simple filtering operations. Combine this with the fact that basic set operations (e.g. ∩, ∪, etc) are themselves essentially always well-defined and the net effect is that operations expressed in unified logic are in turn also essentially always well-defined.

As such, unified logic guarantees that even the strangest operations one could ever conceive of will always still work, in the sense of at least returning a usable value rather than some kind of catastrophic error state. In the worst case, if a particular combination of constraints imposed upon sets are truly impossible to satisfy and just don't "make sense" together (e.g. $N/0$ where N is non-zero), then unified logic can just return the empty set to represent that it could not find any matching forms that satisfied the criteria. One way or another though, unified logic will always return a set of some kind, at least in principle and ignoring computational efficiency constraints.

In fact, unified logic's way of ensuring that virtually everything is always given a well-defined value, even if it must resort to just returning empty sets, was foreshadowed all the way back in the first chapter when we discussed the so called "definition of undefined" on page 38, around when we were still talking about basic philosophical thought experiments about what truth should be defined as being and such. If you recall, the "definition of undefined" basically just says that (within the system we've developed in this book) undefined sets are defined as equal to the empty set, as a way of indicating that they represented "nothing".

Indeed, a few of the concepts from the first chapter were essentially always intended to eventually tie into this material on unified logic once we got here. It is also important to realize that when I started writing this book I did not actually yet even understand all of the different directions the book would end up going. This is to be expected though. Creative works often have a life of their own, a natural direction that they want to grow in. An author's hand may guide their book, but the book's existing content also highly influences the thoughts and decisions of the author. It isn't really a one-way relationship.

Past a certain threshold, creative works come to life and start naturally moving in certain directions, seemingly as if of their own free will. Fighting this phenomenon tends to only make the end result worse and to destroy large amounts of potential value. Thus, it's generally best to actually just go with the flow and be natural, to a reasonable extent, if indeed tangible value creation and conceiving of as many interesting new ideas as possible are your primary concerns. Much the same can be said for any other kind of creative project really. Consequentially, as a result of this less inhibited and more natural way of working than is traditionally applied, some of the material in this book is a bit scattered around in slightly odd ways and there are also some dangling loose ends here and there from exploring lots of random tangential thoughts.

This structure may result in some confusion or tedium for the reader on a few points, but it seems best to me to nonetheless let the book stand mostly in its original organically grown form, with only careful surgical edits, trimming, and polish applied, rather than distorting the whole thing just to make a few of the concepts seem a little less scattered. Retaining the original form of the book helps to better illustrate the actual *real* process of the creation of a new system of logic and mathematics, which I have long believed to be neglected in the way that many logicians and mathematicians typically write (which is unfortunately often far too sterile, cryptic, and unmotivated).

Showing the process of creation in all its organic and original quirkiness, unashamedly and without excessive inhibition or fear, is often the only way to truly guarantee that the crucial conceptual insights that inspired the work in the first place are not forever lost to time. Indeed, in my experience, sharing deep insights and "aha moments" is generally only possible when you make yourself vulnerable enough to share your own thoughts as they truly and naturally occur.

In stark contrast though, only allowing yourself to share your thoughts once all the color and personality and developmental nuance has been methodically sucked out and "purified" tends to practically guarantee that some of the deeper underlying insights will be lost and destroyed in the communication process. Sharing deep insights requires clarity, but clarity in turn requires vulnerability and transparency throughout one's communication style. Thus, usually, there can be no successful sharing of deep insight without also a willingness to communicate so clearly that you make yourself highly vulnerable to easy criticism.

You have to be willing to face reality. You have to be willing to live and to think freely. Lots of people think they live and think freely when they actually don't though. The fear of ridicule and the force of conformity run deep within human instinct, but you must nonetheless somehow overcome them if you are to ever achieve true greatness and true joy in life. This may not be a battle that ever ends, but it is always a battle worth fighting.

Anyway though, random tangents aside, let's return to what we were talking about again. Besides just ensuring that many different seemingly strange operations have well-defined results, the set-based way of thinking employed in unified logic also has value simply in the fact that it allows us to think in terms of multiple return values in a seamless way. By always permitting all operations to return zero or more values, we also effectively eliminate error states for all but truly syntactically malformed expressions (e.g. malformed expressions such as in incorrect uses of the foundational operators of transformative logic and unified logic).

Multi-valued thinking is much more flexible than single-valued thinking is. There are countless circumstances in which it makes absolutely no sense to only allow the return of a single value, seeing as there are very often zero relevant values to return or more than one. Yet, in the traditional framework of mathematics (and also in mathematics' influence upon other systems, such as programming languages and code libraries), we all too often see single-valued thinking dogmatically and rigidly enforced. This is generally unnecessary though.

Error states are created far too often, when often simply allowing the system to return zero or more results would more than suffice. This rigid adherence to single-valued thinking often has crippling effects on one's ability to seamlessly express oneself, and often causes one to have to do lots of roundabout things in order to work around the expressive limitations inherent in strictly single-valued thinking. Multi-valued and set-based thinking are much more powerful and versatile though, in contrast.

It's not enough to just permit multi-valued returns though, technically. In the general case, you need to allow the number of returned values to vary freely, instead of fixing it to some preset number. This enables an expression to always potentially return an empty set when it can't find any matching results to return for a particular set of circumstances. Of course, there are many specific cases where it is logically impossible for an expression to need to return anything other than a fixed number of results.

In the general case though, you need to always make it possible for each expression to return zero or more things, so that the system has sufficient flexibility to handle "undefined" and "ambiguous" cases whenever such cases may arise in the future. Such cases can often only really be handled seamlessly when you permit the return of zero results (for "undefined" cases) or of more than one results (for "ambiguous" cases). There is also of course still the single-valued case as well. The beauty of multi-valued thinking though is that it can handle all cases (including the single-valued cases) in a uniform way if done properly. This can eliminate a lot of unnecessary tedium in how one expresses oneself potentially. It is a more dynamic way of thinking.

Why have a system that deals with individuals and groups of individuals as completely distinct cases when you can instead handle both cases uniformly by simply thinking in terms of sets? One need only apply operations to as many participants as are actually involved in any specific expression. If an operation is given just one value as input, then just perform that operation upon that one value and then you're done. Similarly, if an operation is given an empty set (i.e. is given no values) as input then simply do nothing. Likewise, if an operation is given multiple values as input, then just perform that operation multiple times, once for each value, in a combinatorial fashion.

The net result of all this is that there is no need to distinguish the single-valued cases from the zero or multi-valued cases. The system will automatically do as little or as much as it needs to. For example, cases where a value is "undefined" or doesn't exist will simply cause everything that "undefined" value touches to end up doing nothing, since performing an action on every member of an empty set is just the same thing as doing nothing. Thus, thinking in terms of sets causes the effects of "undefined" values (which translate to empty sets) to be automatically controlled and contained, instead of requiring the contrived and unnatural "error handling" that strictly single-valued thinking in contrast would tend to more often necessitate.

When you think in terms of sets and multiple values, suddenly it becomes much easier to express things. No longer does an expression failing to find a result corresponding to some criteria create a catastrophe. The system simply returns an empty set to indicate that it could find nothing[102] that matched the given criteria. Operations sometimes failing to find anything fitting their criteria should be *expected* and is in fact quite normal. The fact that we so consistently treat such cases as somehow being "errors" in the traditional approach to things strikes me as essentially just an arbitrary quirk of history, one that mostly just causes us a lot of unnecessary trouble and tedium.

One thing that might give you pause here though is the computational inefficiency and intractability of using blueprint expressions and thinking in terms of infinitely large identity sets and such. That is certainly a fair concern. Indeed, even worse still, it is true that direct computation of some of these kinds of expressions could potentially be impossible, except for finite subsets (i.e. limited samples) of those infinite sets. Do not despair though. There is a way around such things. Blueprint expressions and identity sets are primarily of theoretical value. They are for thinking conceptually, rather than for actually performing the literal computations they represent. They aren't really meant to be evaluated directly as such.

Indeed, there is nothing to stop us from still thinking in terms of sets of multiple return values, even if we completely put aside blueprint expressions and identity sets and don't use them in any way. We can instead choose to work in terms of *collective transformation* (defined on page 256), which happily has none of the horrifying computational overhead that blueprint expressions and identity set based thinking have.

We can still do quite a lot with collective transformation, especially if we combine it with generalized algorithmic thinking such as is available in programming languages on a computer. Collective transformation covers much of the same territory as blueprint

---

[102] Remember that an empty transparent set is *literally* nothing. That is exactly what it represents and how it behaves. An expression returning an empty set thus means that the expression has "no answer", or equivalently (in a different sense) that its answer *is* nothing.

expressions do. It mostly just can't "see constraints before they happen", since it can only work in terms of direct applications of the allowed set of transformation rules and does not assume it already knows the complete identity sets of everything it is operating upon (unlike blueprint expressions and such).

Don't let all this verbiage scare you though. This is a lot of words for something that is really quite simple in practice. Collective transformation really just means we can do multiple things at once, in a combinatorial fashion, according to any arbitrary set of transformation rules, essentially, while simultaneously ensuring that all of the values we operate upon in our system are defined as sets (or ordered lists) instead of only as single values.

Done properly, this should ensure uniform treatment of both individuals and groups of individuals, as well as automatic and flexible multi-valued thinking, while still being very computationally feasible. We can get maybe 90% to 95% of the value of multi-valued thinking without actually having to go as far as dealing with blueprint expressions and infinite identity sets directly. Thus, the situation is probably not nearly as computationally dire as it may seem at first glance. This really is quite an effective way of thinking, as long as you are pragmatic and flexible about how you approach it.

In fact, the way that this system ties in to computational constraints in this way is not just a coincidence. Part of the original inspiration for unified logic actually came indirectly from my experiences with computer programming and video game development, among other things. You see, when I resigned my first programming job in the professional (a.k.a. "AAA") video game industry several years ago[103], I had actually originally planned to spend several years creating an indie video game, as a solo dev, with a careful eye for keeping my design scope under control since I'm just once person[104].

However, relatively early on in that process, while really just beginning to think about the game design and how I would structure my code architecture, I was struck with an insight about how to potentially organize my code in a more seamless way by thinking in terms of multiple values at once instead of merely thinking about single return values at a time.

I'm probably not the only programmer to have ever thought of this idea, of always thinking in terms of multiple values in order to create a more seamless way of connecting things together, but one thing led to another, and my thoughts set off a cascading chain reaction of other thoughts (like dominoes falling), and thus I ended up gradually building up a ton of notes to myself about ideas related to many of these concepts that

---

[103] My decision to resign was mostly due to a combination of toxic office politics, being tired of dealing with so many obviously counterproductive technical decisions, and also wanting more personal creative freedom and independence. It was a really terrible work environment for me. Leaving it was one of the best choices I've ever made in my life. The combination of toxicity, lack of creative autonomy, and lack of meaningful opportunities became unbearable. It fast became very clear that I was wasting my time there. Life is too short to waste on such things.

[104] Amazing things are possible (even for just one person) if you manage your time and resources carefully enough. Maximizing your value creation per unit cost and time is essential. Rapidly iterating on a minimum viable product (a "fast to fun" prototype, as I like to call them, in the case of game dev) is also crucial. In contrast, prematurely falling in love with an abstract theme or feeling or imagined outcome, without testing the final product repeatedly along the way, is often deadly to success. The devil is in the details. What you choose to *not* do is just as important as what you choose to do. You generally want to cut out as many low-value things as possible, and refocus that precious energy and time mostly on the higher value content instead.

you now see in this book. Those notes in turn eventually inspired me to write an actual corresponding book (i.e. this book) and to publish it, thereby temporarily completely derailing my game development project for several years. I still plan on creating a game after I'm done with this book sometime though. Game development has always been my first love. Logic and mathematics are merely my second love, dear to me as they may nonetheless be.

That summary I just gave is a bit of an oversimplification. After leaving that first game dev company I also had a tiny two man game dev job, for a brief time, at a different company. In contrast though, my second game dev job was thankfully generally *not* a toxic work environment for me, unlike my first game dev job. It did collapse from an inability to make sufficient money though. My first game dev job though, was simply far too toxic and lacking in real opportunities for me to continue tolerating it. I felt like just a cog in the machine in my first game dev job. It was disempowering, inefficient, and wasteful.

Every time I tried to make any form of meaningful impact whatsoever on anything in the system I was blocked by ridiculously petty and rigid office politics at every turn. Dogmatic and irrational thinking seemed like the norm. People got stuck in narrow paradigms far too often. Even some of the most obvious and critical problems in the game engine were seldom acknowledged, not even when the solution was quick and harmless. It was a "kill the messenger" kind of workplace. The more you tried to fix the real problems the more hostile attention you received. I just wanted to do the best I could, but they often made that impossible even on the most simple, easily fixed, and uncontroversial things.

For example, my suggestion to make it possible to clear the command queue of the AI bots got me nothing but a death glare from the person in charge of that part, even though the absence of this feature was constantly causing problems in production and huge delays and thus was surely needlessly costing the company a bunch of money. That's just one small example. Most decisions at the company felt like that though. I could not tolerate it anymore past a certain point. It was truly unbelievable.

I prefer not to work with people who throw tantrums so easily. I'd much rather actually get work done instead. The product and the bottom line are the only things I really care about. Petty stubbornness and ego shouldn't get in the way of making a good product and being profitable. I like people who can think rationally and can distance themselves from their own paradigms and assumptions.

That job was clearly not a good environment in which to invest in my future career. It was poison to my creative spirit and my intellectual and economic growth. It was thus unacceptable. Even a week working solo would probably teach me more than months of what working for that company would, even though I originally actually thought that working there was going to be a wonderful and enlightening experience. I tried so many times to make the best of that job and to approach it all with a good attitude, but it just never got any better. Such is life though. Life takes many unexpected turns. Sometimes you just have to roll with the punches as best you can.

It all turned out for the best though, since I ended up having the wonderful privilege and honor of getting to write this book, which I am quite happy with. I cannot predict what the outcomes will be, but writing this book has truly helped me to grow as a person and to mature my mind and my reasoning skills. It has been a truly irreplaceable experience, and will surely be worth it regardless of what happens after publication (even just for the experience alone, if it comes to that).

Sometimes the darkest events of a person's life are what enable them to rise from the ashes and truly find their place in life. I feel like my hellish first game dev job, and subsequently the ideas for this book that occurred to me months after leaving, before I was about to start my indie game dev project, have essentially been that for me. Other than general life experience and significantly reduced naivety though, my first game dev job taught me much less in terms of actual game development skill than even small amounts of my own independent work has. Being a cog in someone else's machine often severely cripples a person's learning process and potential. Notoriety and prestige mean almost nothing compared to self-motivated persistence and creative passion. That's one thing that working that terrible job taught me beyond any doubt, so I guess that's something at least.

It sure is nice to be free of that now. The generally toxic environment of that first company was quite stifling to me. It felt soul-crushing. Now I'm free though, and have learned my lesson the hard way. I'm no longer that star-struck foolish college grad I once was, who thought working for some fancy "AAA" game dev company would be wonderfully creatively fulfilling. My frame of mind is much more real and pragmatic now.

I've realized that creative fulfillment is far simpler than that, and has nothing to do with working for some fancy prominent company. Being able to work and think freely, and to express my own creativity how *I* want to, instead of according to someone else's random (and often counterproductive, unnecessarily costly, and unimaginative) whims, is very liberating and cathartic, I think. I'm really thrilled to be out of that environment and to be the captain of my own destiny again now.

Self-motivated creative work is the best kind of work. Being a cog in someone else's machine is poison to the creative spirit. A decentralized abundance of creative autonomy is far more capable of generating high levels of value for one's fellow human beings than any amount of creative dictatorship and monopolization of power will ever be. So many creative people are so underutilized in their work at companies.

They spend their whole lives building up their ability to be maximally creative, yet when they arrive at work they are all too often treated like assembly line workers. This is economic waste on a truly vast scale. It makes each person into a pale shadow of their true creative selves, thereby actually vastly decreasing their overall economic productivity. It both makes the work environment more unpleasant and also simultaneously makes it less profitable.

A light touch to management, with lots of creative power spread diffusely among many different people instead of just super-concentrated in one or two people, strikes me as far more likely to maximize creative and economic outcomes. Alas though, that hunger for creative power at the expense of others all too often prevails. That is why an independent breakaway is so often necessary in order to reach one's full potential, provided one can survive the risks and dangers of striking out on one's own.

That's how I would do things if it was me. I would give the people the breathing room to fully creatively express themselves within reasonable guidelines and resource constraints. Democracy and decentralized power have always been profoundly more productive and empowering than dictatorship has ever been. When selfishness is allowed to spread too far beyond one's own rightful personal autonomy, stagnation and inefficiency are inevitable, and we are all made the poorer for it.

Anyway though, back to the matter at hand. As you may recall, before I got off on this tangent, I had just mentioned that some of these ideas for unified logic were actually originally inspired in part by my experiences in computer programming and my frustrations with single-valued thinking. Well, there's actually more to what inspired the book than just that of course.

It would really be quite hard to accurately summarize every tiny little thing that contributed to the eventual production of this book. How does one summarize an entire lifetime of nuanced little pieces of inspiration and chain reactions of events? One cannot, not really anyway. That would be a book in and of itself, and I'm afraid I also probably have forgotten many aspects of it.

That, incidentally, is yet another reason why leaving text in its natural and organically grown state and simply allowing your thoughts to stand as they are, even if the result is a little bit random and tangential sometimes, is so valuable. If you wait too long to express your original thoughts and insights, you will very often lose them and subsequently no longer be able to communicate them.

Freshness has value. It is better to express your thoughts while you still can. Every good researcher has a journal or a note system of some kind. It would be too hard to remember everything otherwise. Such are the limitations of human nature. Even the creator of a work often cannot remember all of their own material. The same is certainly true of me as well.

One thing I do remember though is that the mathematical concept of a "Minkowski sum" also played a significant role in inspiring unified logic. It was one of the sparks of inspiration that initially triggered the cascade of thoughts that eventually led to this book. You see, as I seem to recall, I was randomly exploring Wikipedia one day, following some links, and I somehow ended up on the page for the Minkowski sum.

I'm not sure, but I suspect I somehow ended up there because I had been researching collision detection algorithms and had come across the Gilbert-Johnson-Keerthi distance algorithm (a.k.a. the GJK algorithm) and was interested in it because I heard that it was especially computationally efficient. As it happens though, the GJK algorithm depends upon Minkowski sums and related concepts in order to work.

Thus, following links on the GJK page is probably what eventually led me to the page on Minkowski sums. At the time I was still distantly considering writing some of the core graphics and physics code for my planned game by myself[105]. As such, my mind at some point turned to thinking about fast collision detection, which is often of paramount importance for a game's performance constraints.

Indeed, fast collision detection between objects is extraordinarily important to us in game dev, since we need to be able to rapidly detect under what conditions arbitrarily

---

[105] A foolish idea... I later wisely decided to use a pre-built engine instead. Solo devs don't generally have the time and resources to reinvent the wheel like that. Even just simply choosing to write the basic engine code yourself can often instantly doom a game dev project. Doing so is very often far too costly in terms of both time and resources. Solo devs can't afford to be so wasteful.

shaped objects would intersect with each other. We need that information in order to simulate collisions, so that we can make objects in the game react as if they are solid and impermeable.

In reality though, under the surface, each object is really just a collection of numerical vectors (i.e. points in space) tied together in such a way as to emulate edges and surfaces. Those objects don't just act solid by magic. There's actually quite a lot of work behind making those objects behave as if they are solid successfully, without slowing your computer to a crawl. The GJK algorithm, which employs Minkowski sums, is just one way of helping to accomplish that.

A game developer only has a tiny fraction of a second between each frame (often between 1/60 and 1/20 of a second) to finish simulating everything in the entire game and to render the corresponding fancy graphics on screen. This is necessary in order to create the illusion of continuous motion and to synchronize with the refresh rate of the screen. The calculations need to be lightning fast. Even something as fast as a computer will often struggle under such heavy burdens. This is where a lot of lag in video games comes from.

The interesting thing about Minkowski sums though, which served as part of the spark of inspiration for unified logic, is that they work in terms of multiple values in a combinatorial way. A Minkowski sum is a binary operation which takes two sets of vectors as inputs. It then forms a new set of vectors by adding every vector in one of the input sets to every vector in the other input set. Thus, for example, if we treated "+" as representing the Minkowski sum instead of normal addition, then the following equation would be true:

$$\{\langle 1,0 \rangle, \langle 0,1 \rangle\} + \{\langle 2,0 \rangle, \langle 0,2 \rangle\} = \{\langle 3,0 \rangle, \langle 1,2 \rangle, \langle 2,1 \rangle, \langle 0,3 \rangle\}$$

I saw this concept explained on the Wikipedia page for Minkowski sums and a thought suddenly struck me: Why not make *everything* in logic, mathematics, and programming work like that? Why not *always* think so seamlessly in terms of multiple values, instead of merely in rigidly single-valued ways? Granted, Minkowski sums are intended specifically for vectors, and unified logic is quite a different beast, but nonetheless, reading about this on Wikipedia was ultimately one of the crucial moments in triggering me to explore the thoughts that eventually led to the creation of unified logic and this book. It set off quite a chain reaction in my mind. It's the simplest little insights that are often the most valuable.

Indeed, exploring these thoughts also eventually led me to yet another idea: a different way of thinking about error handling in computer programming, a middle path between error codes and exceptions, one that mixes traits of both in a highly advantageous and seamless way. You see, in computer programming, one of the most common sources of tedium and extraneous complexity is in the necessity of handling "error states" in the code.

Here an "error state" means any state that is somehow outside the norm of what a specific procedure or function is intended to handle. The exact meaning of an "error" is actually not as cut and dry as some might believe, considering that even when a program encounters "errors" the computer is actually just doing exactly what it has been instructed to do. For example, the program might be instructed to operate upon some object via a reference to it, but the object itself might not actually exist, resulting in the computer not knowing how to respond or causing the program to return a bunch of irrelevant and misleading data.

Generally though, in programming, there are two commonly employed ways of handling such "error states". The first (the older way) is to use what is called "error codes". An error code is simply a piece of data somewhere in the computer which the program periodically modifies in order to reflect whether or not a specific error has occurred at some point in time.

For example, if division by zero is attempted, the computer might then set a boolean variable somewhere (i.e. a location in memory designed to store a simple true or false value) to true, thereby indicating that a division by zero was attempted somewhere. Division by zero would normally be considered an error in most programming languages, although it wouldn't be considered an error in this book, since this book considers division by zero to essentially be well-defined, as you have seen.

A different part of the program, usually the part that was performing the calculation that resulted in the division by zero, would then subsequently check this error code to see if the error occurred, and if so it would then respond accordingly in some way. There is a danger here that the programmer will forget to check these error codes, thus possibly causing corruption of data and other kinds of bad results.

Checking these error codes is also often tedious and verbose, and adds a lot of low-value visual noise when trying to read the code. It fills up the code with lots of little bits of error checks interspersed with the remainder of the code, interrupting the real meat of what the code is intended to do, thereby making it harder to follow the flow of control and harder to reason about the code. The errors themselves often rarely happen, yet the error handling often consumes a lot of space in the text of the code. This can become quite tiresome to work with once it reaches past a certain level of complexity. There is also a high danger that the programmer will fail to handle some of the errors, since nothing in the error codes actually forces the programmer to do so. This makes it easy for the program to accidentally silently damage data.

As an alternative to this way of working however, programmers also invented another way of handling errors: exceptions. In computer programming, an exception is a way of reporting an error that interrupts the existing flow of control in the code and then forces the program to attempt to handle the error in some way, pushing it up the control path repeatedly until some other piece of code on that control path actually attempts to handle it. If no code subsequently handles the exception though, as it is pushed up the control path, then the program will often terminate itself, rather than risk any more data potentially being corrupted by the error.

Both of these ways of working can be quite painful sometimes. Both generally require that you riddle your code with extraneous error handling details and lots of special cases. This distracts from the intent of the code and makes it less expressive and more awkward to work with and to change. Both methods have pros and cons. Error codes are highly computationally efficient, but require careful discipline to use properly and can sometimes be quite a mess to truly clean up after properly.

Exceptions provide automatic guarantees that code will be forced to handle the errors somewhere, but only do so at a large performance penalty. Exceptions will also usually force the program to terminate prematurely if not handled. Sometimes terminating the program like that is smart, but other times it is just counterproductive. There will also be times when the controlling code won't know how to actually handle an exception that is returned to it.

Many programmers also overuse exceptions. Exceptions provide a convenient excuse for a programmer to punt their problems off on other programmers and users, needlessly, instead of simply handling those problems directly and immediately, which is possible more often than many programmers realize. Normal use cases in the flow of control should generally not be thrown as exceptions to the controlling code, yet often are anyway.

If the primary intended use cases of a program cease to function once all the exception handling is removed from the program, then exceptions are probably being overused. Except when forced by the design of another library to do so, the normal flow of control of a program (i.e. the expected use cases) should never depend upon the existence of exception handlers. Exceptions should be exceptional. That's why they're called exceptions.

Error codes and exceptions often make code brittle and rigid, each in their own special way. It sure would be nice if there was a way to handle the bulk of the most common use cases for error handling in a more seamless and natural way, wouldn't it? Indeed there is. The trick is to realize that most "errors" are really just a consequence of rigidly single-valued thinking causing you to be unable to work directly with empty (a.k.a. "undefined") and multi-valued (a.k.a. "ambiguous") cases in a seamless way.

In addition to that though, there is also often a failure to adequately model the space of all possible outcomes to a computation, causing oneself to thereby become deluded into thinking that something is an "error" when in reality it is actually simply a matter of an inadequate and conceptually incorrect representation of the problem space. The domain and range of functions are often chosen incorrectly. Many programmers are too quick to assume that their model of the conceptual space is correct when it isn't.

For example, if a function usually returns single integers, but sometimes instead returns nothing (i.e. "no matching result"), then the case where it returns nothing should *not* be considered an error case and likewise the real range of the function should *not* be considered to be just integers. Conceptually, such a function actually returns either a (transparent) set of one integer or else a (transparent) set containing nothing. Its domain is sets, not single integers. Saying its domain is just single integers would actually be fundamentally conceptually incorrect, despite how popular such sloppy thinking apparently is.

This brings me to my idea for a 3rd possible way of handling "errors" in computer programming, as an alternative to both error codes and exceptions. The idea is quite simple really: All you need to do is to design your programming system from the ground up so that it always works in terms of sets (or ordered lists) instead of in terms of single-valued thinking. You then combine this with careful conceptually correct modeling of the content you are trying to represent in the program. You then make sure that all operations automatically think in terms of potentially multiple values when given inputs, so that every operation automatically responds to the presence or absence of values in each input in a seamless way.

Basically, you implement a much more computationally efficient and precisely controllable analog of unified logic in your programming system. In other words, you simply think in terms of sets instead of in terms of single values, and, done properly, this could probably eliminate a surprisingly large amount of the common "error" cases and needless tedium that one would normally otherwise have to deal with on a daily basis when programming. Underneath the hood, you wouldn't have to always work in terms of sets or arrays (such things could be optimized away, on a context-dependent basis), but on the surface level things would appear as if you were always working in terms of sets or arrays. This would make expressing a lot of things much more seamless and natural I think, if done properly and tastefully.

Why though did I say that this way of handling "errors" mixes traits of both error codes and exceptions? Well, that's because all of these sets being passed around behave just like normal values (like error codes), while also simultaneously automatically propagating through each calling scope in the code (like exceptions). Do you see why? The answer is simple: It's because of the combinatorial way that operations that are designed to automatically handle multiple values will end up behaving.

For example, if an input to a function is an empty set under this system, then it will cause the parts of that operation that the empty set subsequently touches to either become a no-op (i.e. a "do nothing" action) or else to return yet another empty set, which in turn would likewise propagate to other places in the call stack, nullifying everything it touches along the way in an automatically conceptually correct way (correct relative to what the concept of "nothingness" means). Thus, in net effect, we get a system that handles null values automatically, as best it can, and that also more generally handles any arbitrary number of values seamlessly. This eliminates a very large volume of common "error" cases and tedium, as long as the programmer indeed models the concepts in their program in a genuinely conceptually correct way.

In essence, since most "error" cases in programs are caused by functions failing to return objects that cover the *real* domain and range of the function in a conceptually correct way, especially insofar as many code libraries are often designed in such a way as to render them utterly incapable of properly representing empty or ambiguous results, it follows that much of the tedium and "errors" involved in many code systems are actually just conceptually illusory. These "errors" are all too often just figments of our imagination.

Indeed, perhaps most of these "errors" exist only because the code does *not* really model what it purports to. Thus, by reframing our way of thinking to always be posed in terms of sets, we thereby save ourselves from our own stupidity (especially our tendency to prematurely believe that we have fully correctly modeled a system, when in actuality we really haven't), by making it always possible to represent the fringe cases directly in the results, via the versatility of sets, instead of having to so often resort to highly awkward and unnatural mechanisms such as error codes or exceptions.

That is why I consider this kind of set-based thinking to be interpretable as a form of error handling, in a sense. It philosophically is capable of filling at least some of the same role as error handling, but also fills other roles as well. It enables a more natural mode of expression. Indeed, I would go so far as to say that these thought experiments border on being evidence that the one true conceptually correct representation of the concept of "null" in programming is actually an empty set.

Everything suddenly seems to become more seamless and natural once you start thinking in terms of sets like that. Empty sets can act like magically self-propagating "null" values, with no more need to constantly check for null, as long as you have properly implemented the corresponding system where all operations always work in terms of sets (or ordered lists), rather than just in terms of rigid single values. This alternative way of thinking about "error" handling merits a name of its own, so that we can more easily talk about it, and so that we can compare and contrast it to other methods more easily (etc). Thus, as usual, we will be defining a corresponding new term:

Definition 185. *When programming, as an alternative to error codes and exceptions, one may choose to instead frame everything in terms of sets (or ordered lists), such that it is always possible (in the general case) to return zero or more values for any given function in the program. Allowing arbitrary sets to always be returned in this way creates a kind of universal domain and universal closure in the programming system, such that it is always possible to return as few or as many values as are applicable to any given case.*

When this system is then subsequently combined with operations that are defined in such a way as to always be able to work seamlessly with multiple values in each input, in a combinatorial fashion, then the net result is that sets of values can propagate through the system automatically in a conceptually correct way. For example, an empty set being passed around in such a system will automatically tend to result in either no-ops (in the case of procedures) or returned empty sets (in the case of pure functions), such that each empty set will essentially act as a self-propagating null value, frequently without the need for any tedious null checks.

As such, in contrast to the methods of error codes and exceptions, let us refer to this set-based approach as being the method of **self-propagating sets**. *Thus, one could now perhaps say that error codes, exceptions, and self-propagating sets are three different possible methodologies for how to handle "error" cases. Self-propagating sets behave as simple values (like error codes) and yet also automatically propagate through the system (like exceptions). Thus, self-propagating sets are kind of like a "middle path" between error codes and exceptions, in a certain sense. This seems like it should create a very natural and expressive compromise if done properly. However, each of these three methods of handling "errors" will likely have their own respective pros and cons. Pick your poison, as usual.*

Each of these methods have their own form of additional computational overhead that they add on to the code they are attached to. Error codes require you to repeatedly check each error code wherever it might be applicable, exceptions are heavy-weight objects that carry a big performance penalty and often contain lots of extra data that is never actually used, and self-propagating sets require everything to be framed in terms of sets (or ordered lists) and thus carry all of the potential extra memory allocation and indirection costs associated with that.

However, it is also true that all three methods can be optimized in order to reduce the impact of their respective negative traits. There are also specific contexts where each of the three methods might have special advantages that can be exploited for greater benefit than any of the other methods. It depends really. A skillful programmer can do well with any of these approaches. Be that as it may, I suspect that there is a valuable lesson to be learned in the more natural and seamless mode of expression that the method of self-propagating sets would bring if properly applied.

It is also worth noting that there are multiple different ways of going about trying to apply this kind of set-based thinking to your code. It doesn't really have to be an all-or-nothing kind of thing, even though that is indeed the ideal case in principle, in a sense. For instance, you could just write certain specific critical components in terms of sets (or ordered lists) and then apply this kind of thinking just specifically in that context, as needed. A universal framework for this way of thinking, such as a programming language designed with it in mind, might be a nice ideal, potentially, but time and resource constraints often make such things very much not worth pursuing in the short and medium term[106].

Even in my own code, I probably wouldn't dogmatically try to enforce these ideas. I would simply apply whatever ideas work best in each specific circumstance. Pragmatism is much more important than ideology. The economic momentum behind existing libraries and engines is often far more important than these kinds of relatively fringe concerns about how exactly things should be structured. It is best to have a balanced and adaptable mindset, rather than always trying to force things to extremes unnecessarily. One should neither be too indifferent nor too fanatical. A nuanced and pragmatic perspective is essential to achieving optimal productivity and creative expression in life.

Oh, on a random side-note, I should also point out that just because transparent sets are conceptually supposed to be ephemeral objects with no real substance of their own doesn't mean that our representation and implementation of them won't nonetheless contain quite a lot of real substance. To quote Alfred Korzybski: The map is not the territory. Indeed, if you attempted to implement transparent sets in a programming language in some way then it would surely require the use of some real memory, just like anything else does. This is true even in the case of empty sets of course, despite the fact that they conceptually represent nothingness.

Similarly, when performing logic and mathematics by hand, we still have to use symbols (e.g. { } or ∅) in order to be able to successfully represent the transparent empty set, even though it is conceptually the embodiment of nothingness and the symbols themselves are certainly not nothing. This dynamic is simply part of the nature of communication and computation. There's just no way around it. Thus, one should generally keep in mind that our implementations of things and those things themselves are not in fact the same.

Indeed, it is only on the surface level, in terms of how we manipulate things and how those things subsequently respond to the rules we have in place that the illusion of our simulation of those concepts ends up appearing to mirror the real things themselves.

---

[106]Research is a game of patience. New ideas often take a lot of time before they finally achieve critical mass and become more widely economically viable. That's just the nature of the beast.

Thus, for example, the fact that an implementation of pure nothingness is not itself nothing actually does not pose any real conceptual problems whatsoever, contrary to any naive concerns one might have about such things. Understanding this may matter if you end up trying to implement something like self-propagating sets in a real system.

Anyway though, that's enough about self-propagating sets and computer programming and such now. Let's turn our attention to something else. Specifically, I think it's time to tie in the concept of non-classical logic again now. As you may recall, we previously discussed various different branches of non-classical logic and explored their underlying motivations, philosophies, and criteria for satisfaction (in chapter 4, starting on page 291). We then subsequently eventually connected that discussion to unified logic and illustrated how primitive unified logic itself indeed successfully satisfied many of the criteria of these various non-classical logics.

Well, I think its important to note that the same also applies to relational unified logic, and thus to the whole of unified logic more broadly. Why? The answer is actually quite simple. Primitive unified logic was always about "simple" objects and the direct set relationships that we could express and evaluate between them. However, such simple objects are really just arbitrary forms. It is also true that all of the objects contained in sets in relational unified logic are equally considered nothing more than just arbitrary forms.

It thus follows that there is no structural reason why if primitive unified logic satisfied the criteria of the various different non-classical logics that the same wouldn't also be true of relational unified logic. Even when a form expresses a relationship, as indeed we frequently encounter in relational unified logic, it is still just as much an arbitrary form as any other "simpler" object is. There is therefore no impediment to saying that the non-classical properties that apply to primitive unified logic would also apply to relational unified logic.

One thing that is perhaps a bit more of an open question though is how one might actually fully utilize these non-classical properties in the context of relational unified logic. Relational unified logic is after all more intricate and tends to do most of its work through filtering and combining set criteria, much more than it seems to deal with the more basic set relationships such as what one would deal with when using unified implication (defined on page 380) in primitive unified logic for example.

I'm not really sure at this point what exactly all of the connections here might be. There might be techniques for using the non-classical properties, insofar as they do apply to relational unified logic, to achieve useful inferences beyond just what the usual approach of combining set constraints and using blueprint notation and such yields. Regardless though, it is clear that the non-classical properties that apply to primitive unified logic do indeed still apply to relational unified logic, since it is all just framed in terms of arbitrary forms anyway. Make of it what you will though. The exact nuances of the potential connections are unclear to me. I've decided it's outside the scope of this book.

Another noteworthy point that occurs to me is that of how well the generality and natural expressiveness of transformative logic and unified logic might compare to other kinds of high-level computational systems, such as lambda calculus, relational algebra/calculus, and turning machines (etc). Where exactly do the ideas in this book stand relative to other possible approaches to achieving expressive generality? What is the most naturally expressive and pleasant system of logic and mathematics that one could ever conceive of? Does such a system even exist? These are interesting and worthwhile questions I think, but it's anybody's guess what the ultimate answers might be, as far as I can see. It's still definitely a fight worth fighting though.

I also wonder how logicism (the idea that all of mathematics is somehow reducible to the foundational axioms of formal logic) fits into this picture. Do transformative logic and/or unified logic satisfy the principles of logicism? I kind of suspect that transformative logic does, at least, and probably that unified logic does too. For that matter though, what about systems like lambda calculus, relational algebra/calculus, and turning machines though? Do they also satisfy the goals of logicism in some sense? I don't really know, but it's something consider. I'm just going to leave this as an open question. My knowledge here is limited.

Who knows what the future may hold? I'm just one person exploring my own particular perspective on these ideas. It often takes input from many different people in a field of study before a body of work can really find its proper place. Some ideas will stick and some won't. Some ideas will be genuinely new and others will merely be accidental reinventions of existing concepts. Different approaches may have different pros and cons.

Each person has their own unique background, each with their own strengths and weaknesses. Take me for example. I got maybe 3/4 (or more) of the way through my math degree before I switched to computer science. I never intended to do math research though, not even when I was a math major, nor did I ever intend to write a math book. It just sort of happened that way. These random ideas that inspired this book just happened to occur to me when I was about to start a new game dev project and I subsequently felt that it would be irresponsible to not share them with the world. Thus, here we are.

My main background is in programming and game development. Even when I was a math major, the motivation for me studying the subject was always that I would one day apply it to game development, in order to be able to create interesting new game mechanics, which often requires a higher level of understanding of mathematics in order to do successfully. I've always had a passion for conceptual clarity and natural expressiveness when it comes to logic and mathematics though.

Distilling concepts to their deeper essence and then figuring out how to chain those concepts together in natural and seamless ways has always been my strong suit in logic and mathematics, much more so than things like high-level technical proofs and such. I treat math much more like a language for expressing arbitrary creative concepts, like a novelist would treat the English language, than like merely a puzzle or a technical curiosity.

I don't really gel well with the culture of mathematics. Complexity for complexity's sake is utterly repulsive to me, not attractive, unlike many typical mathematicians. I have no love for puzzles without purpose. My attitude and my ambitions in logic and math are more creative and expressive than technical. I'm a pragmatist at heart, not an academic. I hate all forms of obscurantism. I love clarity.

Value creation, not idle academic curiosity, is my primary center of being. It matters to me that my work ends up creating real-world results for people. That is always my hope. I'm not fond of unmotivated blathering. Exploring random ideas that may never bear fruit is perfectly fine with me, but only up to a reasonable amount. In contrast though, I find wallowing in abstract nonsense and empty pretentiousness to be utterly backwards and asinine. I cannot stand that part of the culture of mathematics hardly at all. I have little patience for such things. I value my life too much to squander it like that. It's a huge part of why I don't work in academia.

I like my freedom. I wouldn't trade it for anything. Freedom and independence let me focus on what actually conceptually and pragmatically matters to the real world. It lets me have a full and genuinely impactful life, rather than a life of waste and aimless pedantry. It lets me push aside all the low-value technicalities and stifling fringe concerns that academia would force me to disproportionately focus on, so that I can instead systematically and aggressively focus on high-value things instead.

All that being said, it is still clearly true that there are probably many holes in my knowledge and perspective, just as is true of all people. My own personality and stance on what I want out of life is one thing, but the value of exposing my ideas to many different alternative perspectives is quite another. I have little doubt that there will be unforeseen wrinkles, implementation flaws, and notational flaws in at least some of the content of this book. Such problems virtually always arise once a new system grows past a certain size and complexity.

Most of the systems of reason that we employ today in logic and mathematics have probably gone through multiple iterations and have received contributions from many different people. If and when new published ideas stick at all, it is usually in a gradually modified form. Few frameworks ever survive unchanged from the first version in which they were published. I would be surprised if that doesn't end up also being the case for the content of this book either, insofar as any of these ideas actually end up sticking at all, of which I am fairly hopeful but cannot predict the future.

I will be interested to see what happens with some of the ideas in this book after publication. I have a fair amount of experience in logic and mathematics and related subjects, relatively speaking, but there are nonetheless quite a few subjects that I am not familiar with. Modern mathematics has expanded so greatly at this point that it seems virtually impossible for any one person to be even vaguely familiar with all of it.

As such, I really can't predict what may happen once people with more experienced hands get hold of the content in this book, but I'm curious about what the reaction may be. Regardless though, whatever the result may be (for better or worse), getting to write this book has been quite a wonderful and illuminating experience and I will be contented regardless. My understanding of the depths of logical thought have been much improved by this experience.

I know it's probably an overused analogy, but it really is true what they say about standing on the shoulders of giants. It is quite clear to me that I would never have even had the wherewithal to conceive of all the ideas in this book if there weren't already a bunch of pre-existing points of inspiration already established in the literature. For example, my random chance encounter with the concept of Minkowski sums and with the various different branches of non-classical logic proved to be absolutely essential to triggering the creation of this book.

I wouldn't be here today, getting to experience the joys of writing this book, were it not for the huge bulk of existing logical and mathematical content that served to loosely inspire my thoughts. Thus, it is clear to me that I owe a great debt of gratitude to the countless named and nameless mathematicians who have gradually built up the collective literature of all of mathematics over so many years. None of this would have been possible without all of that. Thank you for that.

Oh, and by the way, in case you haven't figured it out yet by the tone of how I'm talking right now, we're officially winding down the chapter on unified logic now. The curtains are drawing to a close. The finale is over. We've finished most of our climb down the mountain and now we're just heading for the lazy soft-edged rolling green hills below. We've talked about pretty much everything I planned on talking about with respect to unified logic in this book. The bulk of the book, the core content for which it was named and which motivated its creation, is now done essentially. All that remains now is extra stuff.

Thank you for reading my book and for taking this journey with me. I hope you have found it illuminating. I've taken great pains to try to explain the foundational concepts of logic in such a way as to trigger abundant "aha moments" in the reader's mind, and I hope that I was successful in that respect. Clarity matters quite a lot to me. I've always been a firm believer that clarity and deep insights actually *can* be communicated directly, contrary to what some people believe, but it's just that you have to actually care enough and be willing to do enough work to do so successfully.

You can't afford to be lazy if truly good communication is your goal. You really need to take the time to carefully dig out the diamonds in the rough and to present them from multiple angles, if you really want the concepts to truly click and be deeply understood. You also need to be uninhibited enough to not be stifled by pedantry or pre-conceived notions. Clarity requires natural and transparent expression. Pretentiousness and artificially restricted style, in stark contrast, are generally not helpful. Everything you say should be framed in terms of the underlying value and substantive message of what you are actually trying to convey, and not in terms of merely the image of yourself you want to create, nor in terms of fear of whatever arbitrary social conventions just happen to currently be dominant. Live free and think free. That's the key.

This can be the end of the book, if you want it to be. This is sort of the conclusion, in a sense. There are just two more chapters remaining in the book (chapters 6 and 7), and they are both non-essential to the main thrust of the book. I still think they are valuable though and worth reading. One is about additional information on informal logic and the other is about a bunch of random interesting ideas that I wanted to include in the book but which didn't fit anywhere in particular without breaking the flow.

Anyway, as you may recall, way back when we briefly discussed informal logic in chapter 4 (the non-classical logic chapter) in section 4.2 on page 300, I mentioned that the first time I wrote the section on informal logic I got very carried away with it and accidentally turned what was originally supposed to be a small summary into a massive nearly 130 page mini-book. I even ended up inventing a few new pieces of related terminology that didn't exist before. My random tangents and rants got a little bit out of control. Thus, the "rant-filled introduction to informal logic" chapter (a.k.a. chapter 6) was born.

Including such a massive section on informal logic in the non-classical logic chapter would have severely broken the flow of the text and would have been a major distraction. Thus, I pushed it out into its own chapter at the end of the book, and told you (back in section 4.2 somewhere) that you could either read it immediately or else wait until near the end (i.e. until now) to read it, so that it didn't disrupt your chain of thought too much. Well, to that point, the train has reached the end of the line now, so to speak, so its now or never. The big ranty informal logic chapter is the very next chapter after this one. Personally, I think it's worth reading, but it's your call of course.

After the informal logic chapter though, the final chapter (chapter 7) is the so called "Extras" chapter. What exactly is this chapter? Well, it is exactly what it sounds like, basically. It is an assortment of somewhat random ideas I had, stretching across multiple different subjects, including logic, mathematics, and programming, but also other things too, such as a few ideas on how to potentially make society better in certain specific respects. The ideas vary in terms of originality or redundancy. Some are just pre-existing things that I am adding my voice to for emphasis, whereas others are more like ideas that I came up with on my own. It varies.

In all cases though, the ideas in the extras chapter are things that I thought might end up being valuable enough that they are worth piggybacking on the rest of the content of this book, rather than keeping separate from it. You see, in the event that this book is successful, I would like it if some of these extra ideas would get more attention than they currently do. I feel like some of these ideas could add good value to people's lives.

Every little bit helps you know, and this is especially true when you consider the cumulative effect of each little improvement spread out over all of the rest of human history. Even the smallest improvements in how humans do things could easily have a huge payoff, especially if you sum up all of the eventual benefits and efficiency gains over all of current and future human history.

One final thing that I will also be placing inside the Extras chapter though, besides just random ideas, is things like post-scripts, addendums, updates, "housekeeping items", useful little technical things, and so forth. For example, this book was written in a typesetting system called LaTeX, which is generally the most common typesetting system currently in use for producing scientific documentation, especially math-heavy documentation.

LaTeX is not like other word processing systems though. It is actually more like a programming language designed to produce documents. It isn't a "what you see is what you get" kind of thing. It is a code compiler that produces documents instead of executable programs. You can define new commands in it, for example, and then write those commands into your LaTeX code to achieve whatever arbitrary effect you want to in your document, at least in principle anyway.

However, reproducing all of the effects someone else has achieved in their own document can be quite tedious potentially and would require a lot of guesswork sometimes. Thus, to alleviate that problem, and to make it easier for other interested people to produce their own documents that use some of the same things that my book does, the Extras chapter will therefore include all of my code for the LaTeX command definitions and packages I used in writing this book. That's what I meant when I said I'd include "useful little technical things" too.

I think you probably get the point by now. Chapter 6 ("A rant-filled introduction to informal logic") starts on page 591 and chapter 7 ("Extras") starts on page 727. It's time for you to make your decision. You can read as much or as little of what remains of the book as you would like to. You should now know enough to understand what kind of stuff to expect in the next two final chapters and to consider whether you think reading them will be worth your time. I hope you do, but I'd understand if you don't though, considering how long the book is. Anyway though, thanks a ton for reading my book. I hope it will add value to your life in at least some way. Thanks for coming along for the ride with me! Keep fighting the good fight. Never stop learning and evolving.

Chapter *6*

A rant-filled introduction to informal logic

6.1 The value of informal logic

6.1.1 An underappreciated field of study

Those of you with a background in formal logic may be wondering why I'm even bothering to cover this type of logic. Why learn informal logic when we could instead focus all of our energy on the more highly evolved rigorous branches of logic? Isn't informal logic basically redundant now that we have formal logic? These are the kinds of questions I imagine some of you are thinking right now. You might even be tempted to skip this section.

However, if you have no familiarity with informal logic, then that may be unwise. You see, as tempted as you might be to assume that informal logic has been rendered redundant by modern formal logic, you'd be wrong. As pristine and beautiful and powerful as formal logic may be, it still does not cover many of the uses cases for logic. In fact, there are a multitude of situations in real life that are just way too impractical to translate into formal logic.

Yet, I imagine that it is still fairly common for many logicians, mathematicians, engineers, and scientists to look at informal logic with some degree of disregard. This is an understandable sentiment, but it is misguided. It is surely true that with respect to proof and evidence we must always strive to subject ourselves to the utmost rigor. However, formal logic is far too demanding for many everyday situations we encounter. Few situations in real life permit one to pause long enough to jot down a formalization and solve it. Even just accomplishing the act of formalizing an arbitrary real-life situation is often absurdly difficult. At times it may not even be possible, given the limitations of the current state of knowledge in formal logic.

Informal logic specializes in handling exactly the kind of loosely defined problems that one encounters most frequently in daily life. It's like a diffuse cloud of defense mechanisms swirling around in the ambiance of your mind. It's the anti-virus and firewall of the mind, specifically designed to efficiently shield your mind against possible threats, without imposing too large of a burden upon your consciousness. It's the first line of defense in the battle against poor reasoning and exploitation. It's an immune system for the mind. A mind that does not protect itself against outside influences, carefully filtering those outside influences for signs of deception or manipulation, risks becoming a slave to someone else's interests. If you don't protect your free will, you just might lose it.

That's why informal logic is still important to learn. It helps protect you from the stuff that falls through the cracks of the more sophisticated formal logics, the stuff that formal logic either can't process efficiently or does not yet know how to solve. Informal logic should be studied by anyone and everyone who cares about thinking clearly and protecting their own free will. It should be a part of every logician's toolbox, rather than something viewed with disregard.

Informal logic can seldom generate airtight proofs, but it can do a great job of preventing a lot of cruft from ever entering your mind. It's intended for broad coverage. Focusing too much on only formal logic will leave you vulnerable to these kinds of threats to the health of your mind, and it is better to have at least some kind of defense mechanism in place to handle those situations. Informal logic fills that role. It's supplementary. Whereas formal logic is deep, informal logic is broad.

Informal logic is also sometimes referred to as critical thinking, although the terms do have a somewhat different connotation. If you want to search for more information on informal logic, some of the best terms to search for are "informal logic", "critical thinking", "logical fallacies", and "cognitive biases". I especially recommend that you carefully study logical fallacies and cognitive biases, and also come back to them periodically and refresh your memory in order to keep your mental defenses sharp. What exactly are logical fallacies and cognitive biases though? Well, let's see.

6.1.2 An introduction to logical fallacies

A logical fallacy is (roughly) any kind of direct error in reasoning. Each type of logical fallacy follows a certain pattern or structure, in such a way that it renders the argument being made invalid. An invalid argument structure is one that does not conserve truth and thus its conclusion cannot be trusted. There are two types of logical fallacies: formal and informal. A formal logical fallacy is one that is provably wrong due to its inherent logical structure. In contrast, an informal logical fallacy is either a fallacy that logicians do not yet know how to formalize, or else a fallacy that covers a specific pattern of poor reasoning that is at least often wrong. Both types (formal and informal) are useful to know.

Even though this section will focus primarily on informal material, let's explore an example of a formal logical fallacy anyway in order to create a point of contrast. One of the most common examples of a formal fallacy is the fallacy of "affirming the consequent", which is also known as a converse fallacy. A converse fallacy has the following structure: Suppose we know that A implies B and that B is true. Therefore, we erroneously conclude, A must be true as well. This is a common error in reasoning among the untrained public.

For example, suppose someone told you "If Bob does well on his current project at work, then he will receive $500 more in income this month." and that you then found out that Bob in fact did receive $500 more in income this month. Many people would immediately conclude that he therefore must have done well on his current project at work. However, that would be unwise. That kind of reasoning is totally invalid. For all we know, Bob may have received that $500 from a completely different source, one that has nothing at all to do with his current project at work. Arguments of this form are not credible, despite how common they are. It can be proven beyond any doubt that this argument form does not conserve truth.

In stark contrast, there is a closely related form of argument that is always valid. In particular, I am referring to the argument form known as "modus ponens", which is a rule that represents the correct way to apply implications. If you don't like the Latin, you can alternatively refer to it as as "implication elimination" or "implication application". It has the following structure: Suppose we know that *A* implies *B* and that *A* is true. Therefore, *B* must be true as well. This argument form is universally valid.

However, please note that validity is a conditional concept. When we say that an argument is valid it means that the argument would be correct *if* the premises were actually true. An argument can be both logically valid and untrue, if its structure is correct but its premises are false. For example, I could say that if all dogs are cats, and all cats can levitate, then all dogs can levitate. This would be an valid argument, even despite the fact that it is of course utterly untrue. If the premises were actually true, in addition to the argument structure being valid, then we would say that the argument is sound or correct. It's important to remember that validity and truth are not the same thing. Validity is all about truth conservation, which is not the same thing as truth itself.

Anyway, that's all I have to say about formal fallacies in this section. It's time we now turn our attention to informal logical fallacies. What are they like? How do they differ from formal fallacies? How reliable are they? What kinds of things do they cover? Are "informal fallacies" always erroneous? Well, for the most part, the various labels for different types of informal fallacies are pretty reliable. There are some limited circumstances where arguments that have the structure of certain informal fallacies are actually not technically fallacious. Generally though, informal fallacies are pretty consistent labels for poor reasoning patterns.

There are tons of different types of informal fallacies, perhaps hundreds of them. However, certain types of informal fallacies are far more common than others. Furthermore, some types of informal fallacies are actually supersets of many different more specific subtypes. Covering all of the different possible types of informal fallacies that are currently known would be well beyond the scope of this book. Doing so would require an entire book in itself, and in fact there are books out there that specialize in doing exactly that. You'll need to read one of those if you are interested in more thorough coverage than what we will discuss here. For efficiency, we will mostly only be discussing some of the most common and most universal types of informal fallacies.

Anyway, let's dive in. Working from a combination of my memory and some publicly available lists of fallacy types (e.g. Wikipedia), I have carefully selected a limited subset of logical fallacies for us to discuss. I have chosen this particular set because it seems like it would provide a good balance of broad coverage and practical utility. The fallacies I have selected are among the most common, most broadly applicable, and most dangerous of all fallacies perhaps.

Unfortunately, these fallacies are all also extraordinarily common in modern day society. Perhaps one day this will change, as education levels continue to rise in the long term. However, in the meantime, those of us who care about the continued survival and prosperity of humanity, rather than merely about manipulating things for short-term small-minded selfish gains, should be mindful of these toxic thought patterns and do our part to spread awareness of what they are and how to resist them. Anyway though, here's the list of informal fallacies I selected:

6.2 A guided tour of some informal fallacies

6.2.1 Non-sequitur

Both formal and informal variants of this fallacy exist. It is perhaps the broadest and most universal of all fallacies. In Latin, "non-sequitur" literally means "it does not follow" and indeed a non-sequitur is any argument in which the conclusions do not follow from the premises. However, while technically this term can refer to virtually all forms of incorrect reasoning, it is more often used with a more restricted connotation, where it indicates that a conclusion is significantly disconnected from the premises.

Thus, the term non-sequitur is seldom applied to arguments that only have minor technical flaws, even though these too are technically non-sequiturs, but instead the term is usually reserved for cases where the disconnect between the premises and the conclusion of an argument is not subtle. For example, "My hair is brown and therefore so are my eyes." would be an example of a non-sequitur. When non-sequiturs are detected, people often react incredulously, due to the intensity of the logical disconnect: "What? Did I miss something? That didn't make any sense."

6.2.2 Red herring

A red herring is basically a distraction tactic. Instead of addressing the actual merits and substance of an argument, you bring up something irrelevant in an attempt to throw the other person off the scent. The distraction can be subtle or blatant, but if it is irrelevant and its intent is to throw the other person off track then it is a red herring. Occasionally, people also use the term red herring to refer to accidental or unintentional distractions, but this usage is perhaps somewhat less common. The term is supposedly derived from using a type of smelly prepared fish called a "red herring" to distract hunting dogs. That's probably why it has somewhat more of an intentional connotation than an accidental one.

For example, if a politician said "Sure, the wealthy may be able to afford to pay more in taxes, but have you ever considered that the poor live like kings compared to how people used to live hundreds of years ago?" then this would be an example of a red herring. How much the median quality of life has improved for all of humanity has no relevance whatsoever to whether or not some people are disproportionately more wealthy than others.

Money is ideally supposed to correlate to how much value you have contributed to society, and thus wealth does deserve some respect in that sense. Many wealthy people have genuinely earned their wealth, and it would thus be unjust to jealously take it from them. However, wealthy people can afford to contribute far more to funding public works and services than others can. The money has to come from somewhere. Increasing taxes on the poor can easily be devastating and can push them to the brink of starvation and debt slavery (etc), whereas increasing taxes on the wealthy, while still not generally ideal, causes relatively little harm.

6.2.3 Appeal to emotion

This common tactic involves manipulating people's emotional responses to a situation in order to make it more difficult for them to think about the situation clearly and objectively. To truly understand the nature of an appeal to emotion though, we should first understand what emotions really fundamentally are. Emotions are essentially biological goal setting and compensation mechanisms.

Hunger tells you that your goal should be to find more food. Fear tells you that there is incoming danger and that your goal should be to flee from it or to otherwise deflect or handle it somehow. Depression tells you that your repeated attempts at accomplishing something have failed, and that therefore you should reduce your activity levels severely (i.e. feel depressed and lethargic, like you don't want to do anything) so that you can conserve more of your energy instead of wasting it on tasks that you have frequently failed to accomplish.

Happiness tells you that an experience, asset, object, or accomplishment that you have acquired is likely to be beneficial to you, and that therefore you should be conditioned to repeat it more in the future. It does this by rewarding you with a rush of euphoria which is designed to reinforce whatever behavior pattern led you to acquire the benefit, thereby making you more likely to survive and prosper in life thereafter.

All emotions can be explained as biological goal setting or compensation mechanisms that have evolved over time through a process of natural selection in order to maximize your survival and reproduction chances. I could continue listing more emotions and explaining what their underlying purposes actually are, but I think you get the point by now. Why did I go to the trouble of describing this in such detail?

Well, to understand the nature of emotional manipulation, we need to understand the fundamental difference between logic and emotions. Let me give you an analogy. Emotions are like the travel destinations you set on a GPS, whereas logic is like the navigation system that figures out how to get you from point A to point B. Emotions tell you where you want to go, whereas logic helps you figure out how to get there.

Let me ask you this though: What would you do if your destination suddenly changed but all the instructions on how to get there stopped seeming to help you reach that destination? What would you do? You would of course ignore the navigation system (i.e. logic) and instead only pay attention to your desired destination (i.e. emotion). This is exactly the purpose of emotional manipulation. By changing someone's emotional state you change what *goal* they are seeking, thereby causing them to be more likely to ignore what logic is trying to tell them.

Thus, the proper way to fight emotional manipulation is often actually to reset the person's emotional goal setting mechanism to something that is more rational yet which is still emotionally appealing. This is the fundamental reason why simply throwing more logic and evidence at someone when they are in an emotionally compromised state often has no effect whatsoever on them. This is a crucial insight if you ever want to nullify the effects of emotional manipulation (e.g. propaganda, brainwashing, polarization, bad life choices, etc) probably.

For example, if you see a commercial on TV which shows a fancy car gliding along perfectly maintained roads to the backdrop of a beautiful landscape, and it cuts to a scene of a young attractive couple sitting in the front seats wearing sophisticated-looking clothing and jewelry, while classy smooth jazz music plays in the background, then this is an appeal to emotion. It is essentially an attempt to brainwash you by making your goal setting mechanisms (i.e. your emotions) behave incorrectly and to target the wrong things.

The marketers who designed that commercial want you to think that if you buy that car then you will also get all of those other things (an attractive partner, beautiful landscapes, sophistication, happiness, etc) even though the car is in fact nothing but a cold piece of metal and plastic whose only functions are to look pretty and to get you from point A to point B quickly. If a marketer genuinely wanted to appeal to your better senses then they would not go to such great lengths to emotionally manipulate you with content that is so utterly irrelevant to the merits of their actual product.

People who attempt to communicate with you in this manner almost certainly don't have your best interests in mind. You should generally not trust people from whom you detect this kind of emotionally disingenuous way of framing things. It's nearly instant proof that they are trying to exploit you. If you are wise, this kind of media content should always immediately activate your mental defense mechanisms. Similarly, charming yet emotionally disingenuous behavior you encounter from people in real life should also immediately set you on guard. Charm should never be a basis for trust. There is no reliable correlation between charm and trustworthiness.

6.2.4 Appeal to authority

Wearing a white lab coat doesn't make you a scientist, just as calling yourself the King of England doesn't make you royalty, but that doesn't stop people from trying to convince you otherwise. An appeal to authority is an attempt to make you accept an argument purely on the basis of some kind of superficial appearance of authority, instead of by actually addressing the merits by providing evidence or logic. Hard evidence and airtight logic are the only genuinely solid foundations upon which to base any belief. However, sometimes in real life we lack the time and resources to acquire direct proof for ourselves, and therefore we may sometimes legitimately tentatively place trust in authority figures.

This should only be done when that authority figure has a track record of supporting their views with credible evidence or valid deductive reasoning, and it should be open to revision, otherwise it would be fallacious. For example, if I was famous in society and my official title was "Grand Lord Jesse Bollinger the Great, His Majesty, Master of the Universe, Speaker of Infinite Truth, Oh Magnificent Bountiful Fountain of Everlasting Awesomeness, Manliness, Genius, and Infallible Perfection" then it would still not make this book even the slightest bit more credible than its contents actually directly merit. Nonetheless, you are of course welcome to refer to me by this title if you so desire. Just saying.

6.2.5 Appeal to popularity

What if I were to tell you, dear reader, that all of human society, including all of your friends, loved ones, family, and personal heroes thought that the Earth was flat? Would that make the Earth any less spherical? Of course it wouldn't. Truth is not a function of human opinion. Reality doesn't care what you think. The cosmic clock just keeps on ticking, as it always has, regardless of what human beings are doing or thinking. You can deceive people, but you can't deceive reality. The truth is immortal.

Adding more people to a belief generally adds nothing at all to its truth. Popularity carries no weight in the realm of truth. Probably the only exception to this rule is if the subject matter itself is a matter of popularity. For example, if someone asked what percentage of people read at least one book each year, then it would of course be legitimate to answer on the basis of popularity. In contrast, if someone asked whether or not life exists on Mars then neither position could be justified solely on the basis of which one is more popular.

6.2.6 Appeal to tradition

Why do we not allow multi-letter variable names in mainstream mathematics? Why do we continue using a variety of different notations that are actually logically ambiguous? The answer is mainly tradition and mindless social momentum. There's not much real thinking or merit involved in it. Being restricted to single-letter variable names in mathematics is crippling for clarity. It's practically impossible to define a large number of named reusable objects without allowing multi-letter variable names, yet here we are, still stuck in the past.

If you suggested to someone that this policy should be changed and their response was to just dismissively say "Well, the convention to only ever use single-letter variables is just how we've always done it, so that's how we are going to continue doing it. It's so entrenched that there's no way we could change it even if we wanted to. Therefore, let's not even try." then this would be an example of an appeal to tradition fallacy.

It also disregards the fact that mathematical notations have actually changed so much over time that earlier notations are often unrecognizable to modern practitioners. But if that's the case, then why has it changed so much? Think about it. If changing social momentum is supposed to be virtually impossible, like many people seem to believe, then where is the constant barrage of changes coming from? Don't think like a coward. Think based on merit instead of based on mindless adherence to traditions.

6.2.7 Appeal to force

Why bother going to the effort of constructing a sane and evidence-based argument supporting your viewpoint when you can instead just force the other person to "accept" your belief through violence? This is the underlying operating principle behind the appeal to force. It's hard to imagine an argumentative technique that is more vile and meritless than the appeal to force. Appeals to force are especially beloved by thugs, anti-intellectuals, dictators, murderers, cultists, rapists, pedophiles, torturers, abusive spouses, warmongering politicians, etc… i.e. people who could be justly labeled "the scum of human society" essentially.

Many of the other informal fallacies are ultimately pretty mild compared to this one. Appeals to force are the hallmark of evil. They are the bane of humanity. They are the wellspring of human suffering. The majority of the worst atrocities in human history have probably been caused by appeals to force. Those who consciously understand the wicked nature of this approach, yet still choose to apply it anyway are behaving both very unwisely and very unethically. To quote an old proverb: "Never give a sword to a man who can't dance."[1]

6.2.8 Appeal to indignation

How dare you insult my professionalism! That's what an appeal to indignation sounds like. The idea behind this technique is that if someone pins you down with evidence that does not look good for you, then instead of addressing the merits of their points in any kind of legitimate way, you instead just huff and puff and verbally inflate yourself like a pufferfish or a cobra in an effort to intimidate the other person by making the confrontation as unpleasant as possible. It's a scare tactic.

Even under normal circumstances, many people find it very uncomfortable to confront someone about a problem with their behavior, but an appeal to indignation makes this experience even more uncomfortable than it already is. This tactic also can sometimes indicate hidden vulnerability and weakness. The person using this tactic may know subconsciously that they can't compete on the merits. If you ignore their manipulative tactic and instead persist in the confrontation anyway then they may even become hysterical and have a complete meltdown.

6.2.9 Appeal to consequences

I may not have any real evidence that worshiping the Aztec serpent-god Quetzalcoatl will make you immortal, but think about how wonderful it would be if it did! Therefore, you should wor-

---

[1] Another important thing to realize about appeals to force is that they generally don't actually convince anyone of anything really. Coercion often only creates the appearance of compliance, while actually increasing resistance in reality. To quote another relevant saying: "A man convinced against his will is of the same opinion still."

ship Quetzalcoatl, just in case the immortality thing turns out to be true. This is an example of an appeal to consequences fallacy. This is one of the favorite rhetorical tricks of religions. Practically all religions are partly based upon this fallacy in at least some form. Always remember though, no matter how good the consequences would be for some hypothetical scenario, if there is no credible evidence that the hypothetical scenario is in fact true, then the potential payoff of that hypothetical scenario should be treated like it has no weight and a probability of zero.

This fallacy relies on people not taking the time to realize that it is profoundly unwise to give any weight to a potential payoff for which you have zero reliable evidence. If leprechauns existed then you indeed might get rich by finding a leprechaun's pot of gold at the end of the rainbow, but devoting your life to trying to find leprechauns and recruiting others to join your search on that basis would be completely irrational and extremely counterproductive. This is just as true for all religions as it is for leprechauns.

6.2.10 Appeal to nature

All natural organic food! If it's natural then it must always be good for you, right?! How about tape worms? Have you considered tape worms? They're natural. Are *they* good for you? Nope. An appeal to nature is basically an idealistic fallacy where people have gotten into the habit of assuming that natural things are automatically good and artificial things are automatically bad. In reality though, it completely depends on each specific case. Some natural things are good and some are bad. Likewise, some artificial things are good and some are bad.

Take computers for example. Some people think that computers are somehow invading our lives and ruining society and that we all need to go back to a more natural state of living off the land with crude tools. This is hardly a rational viewpoint though. In addition to being extremely useful, computers are also actually one of the most environmentally friendly tools ever conceived by humanity. Nothing reduces paper consumption more than computers. Paper comes from trees, which come from forests, which are cut down to make paper. In the long run, computers are likely to vastly reduce paper consumption even more than has already happened. Not only that but some computer devices will also probably eventually become so energy efficient that the environmental protection benefits will become staggering.

6.2.11 Appeal to hypocrisy

Also known as *tu quoque* in Latin, which literally means "you too", this fallacious argument tactic tries to use the inconsistency of the behavior of someone making a point to try to discredit that point, even though the point itself and the behavior of the person making it are conceptually distinct and have no real relevance to each other. One important rule to remember for thinking clearly is that all arguments exist separately from the people who made them.

It doesn't necessarily matter if the person making the argument is being a hypocrite. The only thing that actually matters is the substance of what is being discussed. For example, suppose that an obese man walked up to you when he saw you eating a doughnut and warned you that you shouldn't eat those because they are bad for your health, and then he sat down and immediately started eating a doughnut of his own. Would this make his point that doughnuts are unhealthy any less true? Of course not.

6.2.12 Strawman

This one is a special favorite among politicians, partisan journalists, and participants in flame wars on internet forums. The idea behind a strawman fallacy is to mischaracterize your opponent's views in order to make them sound much more unreasonable than they actually are. People often use this technique when they are not knowledgeable or insightful enough to actually address the real fundamental foundations of their opponents views, so they instead just resort to creating a comically weaker and unrepresentative version of what their opponent believes. They then refute *that* viewpoint instead of the original one, hoping that the audience will not be attentive enough to notice the deceptive sleight of hand.

A classic example of a political issue where both sides constantly strawman each other is abortion. Pro-abortion advocates mischaracterize their opponents as being automatically against women's rights and as being small-minded sexist pigs, whereas anti-abortion advocates mischaracterize their opponents as being evil baby eating murders who don't care about children or human life. Both sides are being disingenuous and are misrepresenting the other side, thus both arguments are invalid. Remember, if you don't characterize your opponent's views accurately then your argument against them will carry no real weight and will probably accomplish nothing except angering and polarizing the other side, thus resulting in an increasingly tense and toxic culture.

The opposite of using a strawman fallacy against an opponent is treating that opponent with a *spirit of generosity* in which you actually *strengthen* your opponent's argument (if necessary) instead of weakening it, in order to show that even when strengthened it would still be the lesser choice. The scientific method requires that you treat opposing views with a spirit of generosity, lest you unwittingly suppress evidence or become a victim of your own preconceptions. Scientists are often willing to overlook superficial errors in colleagues' work and to instead focus their attention on the core essence of the argument.

Whereas political discourse tends to be riddled from top to bottom with strawman propaganda, in stark contrast scientists more often treat their opponents with grace and with a spirit of generosity. Note however that having a spirit of generosity does not mean that all viewpoints deserve to be treated with respect. Once a viewpoint has been thoroughly discredited then it often no longer deserves to be treated respectfully. Not all viewpoints are created equal. Some viewpoints are inherently worthless and counterproductive. However, you should always be willing to consider new evidence that comes to light, regardless of how poor a viewpoint's reputation is. The less reputable the claim is though, the stronger the evidence for it must be for it to be reconsidered generally.

6.2.13 Ad hominem

In Latin, ad hominem means "to the person". The idea behind an ad hominem fallacy is that instead of addressing the merits of someone's argument, you instead attack them in a personal way that has no real relevance to the subject matter being discussed. It's an attempt to discredit your opponent by smearing them instead of addressing the merits of their points. Politicians and social media mobs love this one. It's kind of like the hostile counterpart of the appeal to authority. Whereas an appeal to authority tries to elevate your own credibility, an ad hominem attack instead tries to damage your opponent's credibility. It's kind of like a form of argumentative bullying.

The person using an ad hominem attack perhaps can't actually support their own position on the merits, so they instead resort to trying to bring your position down by launching a barrage of irrelevant insults or mischaracterizations at you. They will often also use social manipulation in order to try to make you seem somehow "mean" or like a social outcast in the hope that social conformity will turn your audience against you and cause psychological damage to you.

It's important to not let ad hominem from an opponent harm your frame of mind and to not allow it to suppress your life and your ambitions. Sometimes a show of strength, confidence, and fearlessness can be helpful for partly nullifying an ad hominem attack. Other times though, it may be best to just stay away and to not engage with the attacker(s), so that you don't feed into their momentum and toxicity, etc. It depends. Creating a separate counternarrative that implicitly undermines the attacks, keeping a healthy distance from the situation, minimizing the surface area of attack, and focusing your attention on more productive things are often helpful though.

Don't let ad hominem attacks disrupt your life or outlook. Disrupting and damaging your life is usually what the attacker wants, so don't let them have it. Living well and being unbreakable, relentlessly constructive, unfailingly calm, strategically distant, and consistently benevolent in contrast often does a lot to take power away from an ad hominem attack and to immunize your life against them.

Ad hominem fallacies are often also strawman fallacies, but sometimes not. If Hitler says that $2 + 2 = 4$ then the fact that he's Hitler doesn't make it any less true. Irrelevant insults and personal attacks don't have any place in the systematic pursuit of objective truth. Don't be toxic.

6.2.14 No true group member

This fallacy is actually more commonly known as the *No True Scotsman* fallacy. However, the term "No True Scotsman" is inherently discriminatory against the people of Scotland. Virtually all in-groups are abundantly guilty of this fallacy, thus it makes little sense to associate it specifically with only the Scots. The term "no true group member" is just one I made up to try to correct this injustice. I think it would be wise for us to eliminate this little piece of institutionalized discrimination that has somehow wormed its way into our language. Doing so will help broaden our perspective.

More to the point though, the idea behind the no true group member fallacy is that every time you encounter evidence that your particular in-group is guilty of some negative quality you state that no *true* member of your in-group would ever do that or have that negative attribute. Meanwhile, back in realityland, if a member of your in-group has a negative attribute then that is definitive proof that members of your in-group are indeed capable of having that negative attribute. Ignoring evidence that there are toxic elements within virtually all in-groups accomplishes nothing but making you unjustly self-righteous and deluded. For example, saying that no true Christian would ever murder someone would be an example of a no true group member fallacy. Likewise, saying that no true scientist ever acts irrationally would also be an example of a no true group member fallacy.

6.2.15 Special pleading

The idea behind special pleading is to treat something you believe in like a special case to which the normal rules of objectivity somehow don't apply. You don't actually provide any real reason for why that specific case should be treated any different. Special pleading is essentially a form of mental compartmentalization. It means you're holding some beliefs to a lower standard than other beliefs, without having any genuine justification for doing so. This is a common defense mechanism in pseudo-science and religion, and is one of the most significant reasons why people can be totally rational in most aspects of their life and yet still be totally out of their mind in some other aspect of their life.

For example, if you are highly skeptical about the claims of almost all religions you encounter, yet you are not skeptical about the claims of your own religion, then you are guilty of the fallacy of special pleading. Likewise, if you scrutinize the claims of car salespeople with great care, yet you readily accept claims that crystals have healing powers, based on nothing more than anecdotes, then you are guilty of special pleading.

Special pleading indicates that your standard of evidence is not correctly calibrated. The amount of evidence necessary to prove any claim is always directly proportional to the magnitude of that claim. Extraordinary claims require extraordinary evidence. There is no such thing as a claim to which the standards of logic and evidence do not apply. All claims, no matter how treasured or socially entrenched, are vulnerable to the light of logical and evidentiary scrutiny.

6.2.16 Slippery slope

The slippery slope fallacy basically amounts to a paranoid overreaction to the potential consequences of an action. The person committing a slippery slope fallacy grossly overestimates what the effect of making a change will be, and claims that the change will set off a chain reaction of increasingly ridiculous negative consequences, like a line of dominoes falling over. This is another favorite tactic of politicians. The slippery slope fallacy tends to occur more frequently in highly polarized and antagonistic environments.

For example, if a politician argued that we shouldn't raise taxes 2% to meet a short-term budget problem, because if we did then it would embolden the government and trigger an unstoppable chain reaction that would lead to taxing people at 95% of their income, in effect establishing permanent socio-economic slavery to the government, then this would be an example of a slippery slope fallacy. The reaction is way more extreme than the situation actually merits.

Another example would be if an employer thought that granting any special treatment to one employee for a specific problem would automatically cause all the other employees to expect the same. Whether or not they would depends on the circumstances and also the psychology of the group. For instance, an employee receiving special treatment in connection with a medical problem is fairly unlikely to attract much jealousy from other employees, even if it technically means that the one employee is being compensated a bit better. Thus, the employer shouldn't assume that all forms of special treatment will result in a "slippery slope", and indeed some occasional exceptions could easily be merited and beneficial overall.

6.2.17 Cherry picking

The idea of cherry picking is to make your argument look better than it actually is by selecting only evidence that supports your view while excluding any evidence that opposes it. Basically you create an illusion of support in order to trick your audience into trusting you more than you actually deserve. It's a form of evidence suppression. For example, if an online shopping website only posts the positive reviews it receives and hides all of the negative reviews it receives, then that website would be guilty of cherry picking.

6.2.18 Anecdotal evidence

The idea behind an anecdotal evidence fallacy is to treat a collection of personal anecdotes as if they qualify as credible evidence for a claim. This is a popular technique, but it isn't valid. Evidence doesn't work that way. There's a reason why hearsay is not admissible in a court of law. Anecdotes, testimonials, stories, gossip, and hearsay are all *not* examples of hard evidence. No matter how many of them you collect it will *never* qualify as proof of the claim.

Anecdotes are practically guaranteed to have selection bias in them. In order for evidence to qualify as credible, it must be collected in a methodical and disciplined way, one specifically designed to weed out possible sources of bias. Here's an insightful quote (possibly coined by health researcher Scott C. Ratzan) that sums it up pretty well: "The plural of anecdote is not evidence."

6.2.19 Circular reasoning

When an argument assumes its own conclusion as a given in order to "prove" that its conclusion is true, then this is called circular reasoning. Introducing premise assumptions to create conditional arguments is normally a legitimate thing to do, but it becomes illegitimate when the claim you are attempting to prove is already explicitly or implicitly assumed in the initial assumptions upon which your argument is based. You can't prove a claim by assuming that it is true.

Only proofs that are derived from other axioms or theorems (or experimental evidence) have explanatory power. For example, if I said that the reason circles are circular is because they are round then this would be an example of circular reasoning, both literally and logically. Likewise, if I claimed that Holy Book X is true because Holy Book X says it is then this too would be an example of circular reasoning. A proof of a non-trivial claim cannot be circular if it is to have any hope of being convincing.

6.2.20 Correlation-causation fallacy

This fallacy occurs whenever someone thinks that just because two things occurred at the same time that therefore one must have been caused by the other. Correlation is not causation. Just because two things tend to happen together (i.e. correlate with each other) doesn't mean that one has to be the cause of the other. Many superstitions are probably based on correlation-causation fallacies.

For example, if some tells you that stepping on a crack will give you bad luck, and then you step on a crack and end up in a car accident an hour later, then believing that this constituted evidence that stepping on cracks causes bad luck would be a correlation-causation fallacy. One particular subtype of correlation-causation fallacy is known by its Latin name "*post hoc, ergo propter hoc*", which literally means "after this, therefore because of this". This variant refers specifically to chronological correlation-causation fallacies where you assume incorrectly that just because one event proceeded another that the first must have caused the second.

6.2.21 Reification fallacy

This fallacy is also known as "confusing the map with the territory" and that is an accurate analogy for what it means. It means that you are treating an abstraction as if it is the same thing as the corresponding reality to which it refers. It is similar to an equivocation fallacy. In fact, the symbolic-semantic equivocation fallacy we covered earlier in the book (on page

33) is also a reification fallacy. For example, if you heard someone say "The pen is mightier than the sword." and then interpreted that as a recommendation for what weapon to use in a swordsmanship tournament, then you would be guilty of a reification fallacy. There's also a good chance you'd die.

6.2.22 False dichotomy

If you treat a situation as if there are only a limited number of options, when in fact more options exist, then you are guilty of a false dichotomy fallacy. The false dichotomy fallacy is also known as *black-and-white thinking*. This fallacy usually originates from failing to recognize that the given options do not exhaust all possibilities. It can also be caused by political extremism, fanaticism, and dogmatic thinking.

For example, if someone says "You are either with us or against us." to you then they are guilty of a false dichotomy fallacy. After all, it could be that you are neutral and neither support nor oppose them. It is important to remember that dividing something into a set of possibilities is only a fallacy if those possibilities do not exhaustively cover all possibilities. Thus, for example, saying that every statement is either true or false in classical logic would *not* be a false dichotomy. It would in fact be true. What constitutes a false dichotomy depends on what the subject matter is though sometimes.

6.2.23 False balance

Fairness means that everyone deserves to be treated equally and to always receive equal time and equal respect, right? Wrong. Contrary to popular belief, that is definitely *not* what real fairness actually is. Real fairness is when each side is given an amount of respect and time that is directly proportional to how much the substantive merits actually favor that side. If a viewpoint has no credible scientific evidence or logical proof backing it up, then it deserves no respect and no time whatsoever. The fallacy of false balance occurs whenever someone assumes that all sides of an argument automatically deserve equal respect, equal time, and equal media coverage regardless of whether each side actually has any credible supporting evidence to back up its claims.

The false balance fallacy is very closely related to another fallacy named the *argument to moderation fallacy*, which occurs whenever someone makes a "middle of the road" compromise without actually adequately weighing the merits of each conflicting viewpoint. Extreme positions sometimes turn out to be true, in which case any other position is false. You should never assume that compromises are automatically somehow "more fair" than siding with a specific side. Only hard evidence and logical proof can ever genuinely justify which side to choose in a conflict. A moderate viewpoint is just as much an arbitrary viewpoint as any of the extreme viewpoints are. There is nothing inherently virtuous about moderation. Every case must be considered separately based on the merits.

Our society is unfortunately currently rife with the false balance fallacy. This is not a small problem. The false balance fallacy is an extremely toxic and dangerous fallacy. This is made all the more insidious by the fact that false balance *seems* perfectly innocent to the untrained eye. People tend to think that giving equal time and equal resources to every side is just "common sense" and is just "obviously the right thing to do". They couldn't be more wrong. Counterintuitively, giving equal time and equal resources to everyone is actually one of the fastest ways to guarantee that people are treated unfairly.

One of the most glaring examples of false balance in modern life is the way the news media currently behaves. News journalists currently seem to often have a wildly incorrect understanding of what real objectivity and fairness actually are. They tend to believe in a sort of "equal time doctrine" wherein they continue to treat thoroughly scientifically discredited views as if they are still up for debate. This creates a false impression in the public that certain issues are more controversial than they actually are, and misleads people into thinking that the sides are evenly matched when in fact one side is overwhelmingly and obviously superior when you look at the actual evidence involved.

For example, conspiracy theorists should almost never be given any time in mass media. Likewise, people who believe things like that the earth is flat or that the holocaust never happened or that vaccines cause autism should also almost never be given any time in mass media. The only exception is if genuinely credible evidence has come to light. If you don't have scientific evidence or logical proof on your side then you don't have *anything* on your side. Spreading viewpoints that are anti-scientific, anti-intellectual, or otherwise insufficiently backed by hard evidence in mass media is very often an inherently unethical and immoral thing to do. It sometimes perhaps even borders on being a crime against humanity. It may be endangering the survival of our entire species in some cases.

For the most glaring cases of false balance, especially the ones that endanger public health (e.g. claiming vaccines cause autism), it may even be wise to make such cases into crimes, right up there in terms of severity with theft, committing fraud, lying under oath, medical malpractice, pedophilia, and money laundering. The news media need to learn that their current style of covering the news, which tends to emphasize a combination of false balance, sensationalism, fear mongering, rubbernecking, distortion, demonization, and cherry picking, is severely endangering the mental health and intellectual integrity of society. It is severely endangering the future of humanity.

The news media loves to create artificial controversies out of thin air and to blow events out of proportion, in part to inflate their ratings in order to rake in more views and cash, and in part because they have an incorrect understanding of what fairness and objectivity actually mean. News outlets that grossly distort reality, that spread unsubstantiated propaganda, or that deny overwhelming scientific evidence should be made more accountable for the poison they are constantly injecting into our society.

That being said, free speech is of course by far one of the most important of all human rights. For that reason, if we ever do implement criminal punishment for spreading misinformation in mass-media, I think it would probably be wise to restrict the punishment to only cases where the breach of ethics is blatantly obvious and the evidence of falsehood is overwhelming, such as for claiming that vaccines cause autism or that the holocaust never happened for example.

In theory, journalists are supposed to objectively report the truth. In practice, they frequently fail to do so. News outlets have a vested financial interest in keeping dead debates alive well past their expiration dates. When an industry thrives on hysteria and outrage to sustain its viewership, it will do anything to keep those controversies alive as long as it can, which inevitably leads to false balance. News outlets really need to find a less exploitive and less unhealthy way of covering the news. The current style of journalism tends to focus on generating as much toxic hysteria and mob-like behavior as possible. The more hysterical a society becomes, the less it will tend to think rationally. This will inevitably lead to very bad decisions and will endanger us all.

In stark contrast, the scientific community is generally relatively good at preventing false balance within its own publications. Credible scientific journals and publishers will consistently refuse to publish anything that is not backed by some combination of evidence or logical proof. It doesn't always have to be perfect, but it does have to be reasonably credible. *That* is how real fairness and objectivity works. If you don't have evidence to support your case, then you don't deserve to be given much time in mass media.

Nobody will stop you from self-publishing your ideas though. This is the proper balance for managing the distribution of information, while still protecting both free speech and the mental health of society at large. It should come as no surprise that scientists have had objective communication figured out fairly well for quite some time now. That is after all what science is all about. The news media would be wise to start imitating the way scientific journals behave in this respect, at least somewhat, instead of constantly sensationalizing and polarizing every little thing they touch.

Do you still harbor some doubt about this? Does my suggestion that each side should *not* be given equal time or equal respect bother you? Does it all just sound too mean and unfair? Well, if that's the case then I have a thought experiment for you that will prove to you beyond a shadow of a doubt that the "equal time doctrine" is a fundamentally untenable and nonsensical view of what fairness should be.

Suppose that there are n beliefs on what the correct answer is to some particular issue in some arbitrary hypothetical society. Let $V_1, V_2, \ldots, V_n$ represent these n different views on that issue. Suppose, purely for argument's sake, that we somehow magically know that V_1 is the correct view on the issue and that all the other views are wrong. Suppose that the news media in this society believes in the "equal time doctrine" and thus chooses to allocate an equal amount of time to all n views. Understand? Well, here's an interesting question for you: What happens to the percentage of time the media spends discussing the truth as you increase the size of n? What happens when n approaches infinity?

What happens is that the percentage of time the media spends discussing the truth decreases rapidly as n increases and drops to zero as n approaches infinity. Think about what that implies. Even when n is very small, such as when $n = 2$, this "equal time doctrine" will still have an extremely negative effect on how much time the media spends actually covering the truth. In a media system based on the "equal time doctrine" the maximum amount of time the media can spend covering the truth depends only on n and not in any way on what is true. Thus, if $n = 2$ then it will be impossible for the media to spend any more than 50% of its time covering the truth. It will spend the rest of the time spreading falsehoods. Therefore, with a 50/50 split of truth and falsehood being constantly spread by the mass media, social progress could easily grind to a halt.

Even if you just nudge it up a tiny bit to $n = 3$ then already the media will be spending the majority of its time, about 66% of it, doing nothing but spreading misinformation. In this way, all media organizations that adhere to the equal time doctrine will effectively end up becoming fountains of endless misinformation and will tend to cause the hypothetical society in which they operate to become increasingly misinformed and increasingly at risk of systematic society-wide collapse.

In any society that adopts this style of media coverage, the integrity of news organizations will probably gradually decay until the entire concept of truth itself has become seemingly irrelevant to all social outcomes. Social trust of the media will also eventually drop through the floor, causing people to no longer listen to anything the media says, thereby destroying one of the only defense mechanisms a society has against political corruption (i.e. destroying journalism's ability to defend against corruption). The only way out of this requires (among many other things) the abandonment of the "equal time doctrine" and the toxic cultural effects that inevitably come with it.

In stark contrast, suppose that we changed the situation so that the amount of time the media gave to each of the n viewpoints was directly proportional to the amount of hard evidence and logical proof corresponding to each viewpoint. The behavior and outcome of the system in that case would be profoundly better. In a system where time is given to viewpoints in direct proportion to how much evidence those viewpoints actually have, increasing n will not significantly damage the amount of time the media spends covering the truth. Multiple viewpoints might have at least some supporting evidence in their favor, and thus would still receive some coverage, but overwhelmingly the system would tend to converge on spending a high amount of time covering the truth. The system would remain stable, trust in the media would increase, and social strife and polarization would drop significantly.

The equal time doctrine inevitably leads towards a "truth-free" society where, in the extreme case (when n is sufficiently high), approximately 0% of media time is spent covering the truth. This is the inevitable outcome of news organizations abandoning their sacred journalistic duty to act as objective observers, to instead choose to become partisan political propaganda machines.

In contrast, evidence-proportional coverage would lead to the opposite effect and would trend towards increasing levels of truthfulness in the news and a better chance of an increasingly stable, happy, productive, peaceful, and cooperative society. Amazing what such a subtle and nuanced difference can do isn't it? Isn't it also horrifying how poisonous a seemingly innocent and good-natured-sounding idea like "always give all sides equal time" can be? This is yet another example of the extraordinary dangers of reasoning based on "common sense" and blind intuitive assumptions.

Moreover, there is yet another social arena in which the toxic iron grip of false balance reigns supreme. This is in the widespread and popular view that all people deserve equal compensation and equal respect at all times, and that all worldviews automatically deserve to be treated equally and showered with gold stars, bear hugs, and big fat juicy kisses. This is often accompanied by a form of social hysteria where even criticizing anyone else's worldview or culture or opinion automatically gets you branded as a discriminatory bigot, even if that other worldview or culture or opinion provably leads to increased human suffering based on the evidence.

It's true that *people* are created equal and deserve an equal *opportunity* to pursue success in life. However, it is *not* true that all *beliefs* are created equal, nor is it true that all *achievements* are created equal. A person who generates 10 times as much value for their fellow human beings as some other person deserves 10 times as much compensation and 10 times as much respect.

Likewise, beliefs that tend to generate more violence deserve less respect than beliefs that tend to generate less violence. Everything should be judged according to its actual merits, and then compensated proportionally. Nothing deserves equality automatically. Equal treatment requires equal merit. *That* is true fairness. True fairness is proportionality, not blind equality. This is also the core reason why communism failed and why it is a fundamentally counterproductive idea, *both in practice and in principle*.

There is no such thing as a good implementation of communism. It would be counterproductive *even in a world where humans always behaved perfectly*. Forcing everyone to be compensated equally regardless of what work they do and how much that work is worth is actually one of the most profoundly unfair things that any society could ever choose to do. Those who do better deserve better. Jealousy is not a valid excuse to try and take things from them.

6.2.24 Shifting the burden of proof

When someone makes a positive existential claim, the burden of proving that claim typically belongs to the person making it, not to the person denying it. What's an existential claim? An existential claim is any claim that asserts that an entity, object, or relationship exists within the system under discussion (e.g. the real world, an abstract logical system, a computer, etc). Existential claims do not have a symmetric standard of evidence. The null hypothesis always has the high ground relative to any other hypotheses, and thus is easier to defend.

If you do not have enough evidence to justify a positive existential claim, then the null hypothesis (i.e. the denial of the existence of that thing) wins by default. For example, if someone claims that unicorns exist then the burden of proving that this is true rests on the person making that claim, not the person denying it. If the person claiming that unicorns exist cannot provide sufficient evidence to prove their case, then the only rational choice is to assume by default that unicorns *don't* exist. All existential claims are false by default.

The fallacy of shifting the burden of proof is when someone tries to argue that the burden of proof belongs to the person denying the existential claim instead of to the person making it. This is the opposite of the correct way of thinking. Probably the most common example of the fallacy of shifting the burden of proof that you'll encounter in modern life is when people try to claim that a religious deity exists. The religious person will assume by default that their particular religious deity exists, without any credible evidence, and will then try to argue that if the non-believer can't *disprove* it then it must be true. This is utterly invalid from a logical standpoint.

Why? Think about it. There are an infinitely large number of conceivable entities that might exist in the universe. It would be *ludicrous* to blindly assume that they exist without having evidence of any of them, but that is exactly what the implication is if you think that it is ever valid to shift the burden of proof. Take the deity scenario for example. There are an infinite number of conceivable deities, and each religious person in society has effectively plucked one of them out of thin air, wildly and arbitrarily out of an infinite number of possibilities, and then blindly assumed that their favored deity exists by default.

This makes absolutely zero sense, but then again a belief doesn't have to make any sense if it is *brainwashed into you from birth by childhood indoctrination*. Children's minds have an evolutionary adaption that causes them to tend to blindly accept anything that their parents teach them that sounds like it might be important for their survival. The effects of this can be difficult to undo. This is in fact the real source of much of the religious belief in our society. It is essentially child abuse and mind rape.

6.2.25 Hasty generalization

This one's pretty simple and straightforward. Whenever someone treats a small number of occurrences of something as sufficient justification to make a much larger generalization than that small number of occurrences can actually rationally justify, then that person has committed a hasty generalization fallacy. It is essentially a premature statistical inference. This is a really common unintentional fallacy, but it is also sometimes used maliciously in order to spread propaganda and statistical misinformation. Many forms of racism, sexism, ageism, and other forms of discrimination probably exist in part because of the fallacy of hasty generalization.

For example, if one time when you were walking down the street you got mugged by a black person, then you might be tempted to do a hasty generalization and conclude that black people "must just be more violent than other kinds of people". If you followed through with that thought and internalized it as a belief, then it would essentially be your first step down the road to becoming racist. The road to hell is paved with good intentions, just as the old proverb says.

The violation you felt when you were mugged, and your inability to get justice against the mugger, might then cause you to vent all of your hatred towards all black people in general, out of a misguided subconscious sense that it might one day give you closure and make you feel a better sense of personal security. Similarly, if you've been mistreated by the opposite gender in some incident from your past, then you may be tempted to think that all members of the opposite gender "always do bad thing X" or are otherwise inherently evil in some sense, but you'd be wrong of course. This is how many discriminatory beliefs are born. Try to be aware of this and not let it infect you.

It's important to stay calm and think these kinds of things through. Almost every atrocity in human history was committed with a smug sense of self-righteousness. Hitler probably genuinely thought he was doing the world a good thing by massacring the Jews. Always remember your own inner potential for evil. The moment you start thinking you're above evil is often the moment you start becoming it. Self-righteousness is the branding iron of wickedness. When you're blinded by your own self-righteousness even the darkest tunnel will look for all the world like it lay basked in sunlight.

6.2.26 Faulty analogy

Used correctly, analogies are a great way to make concepts easier to understand and more tangible. Used incorrectly, they are a great way to trick people into thinking a relation exists were one does not. In either case though, analogies tend to sound catchy and concise and therefore often have a higher chance of being convincing than many other types of arguments. This makes them a powerful tool for both good and evil. That is why it is important that when you make analogies you make them in a conceptually correct way. An analogy that is made in an incorrect or misleading way is called a faulty analogy.

Normally, an analogy is supposed to work by showing that two entities are the same with respect to a specific property X, and that therefore either entity may be used as a metaphor for thinking about the other. In contrast, in a faulty analogy the person making the analogy only demonstrates that the two entities share property Y in common but then attempts to use that fact to imply that they also have property X in common even though it has never been established that they actually do have X in common. That is essentially the reason why faulty analogies cannot be used to justify anything.

Only if the property being alluded to is the same one as the one that was demonstrated to be shared between the two entities can an analogy be logically valid. However, artistic analogies that are not intended to be part of a logical argument in the first place, such as those found in creative writing and poetry, don't necessarily have this constraint though. Normally I would create my own example here, but I found a hilarious example of a faulty analogy on a website called *onegoodmove.org*, in the part titled *Stephen's Guide to the Logical Fallacies*, and I can't resist using it. Don't let your boss see it though. Here it is: "Employees are like nails. Just as nails must be hit in the head in order to make them work, so must employees."

6.2.27 The fallacy fallacy

Here's an interesting one. This fallacy usually occurs when someone gets too excited by the prospect of identifying fallacies in an opponent's arguments and overreaches on what that implies. Proving that an opponent's argument contains a fallacy only proves that the *argument* is invalid. It says nothing of whether the *claim* the opponent is making is actually true or false. If someone claims that X is true and you correctly point out that their argument for X has a fallacy in it, then that does *not* prove that X is false. The only thing it proves is that their *argument* is invalid. If the opponent can come up with a different argument that contains no fallacies then their claim could still end up being true.

Even if you prove that *all* of the arguments that your opponent can come up with in favor of X contain fallacies then that still doesn't necessarily prove that X is false. The only way to prove X true or false is to do so directly. This fallacy occurs most frequently in people who have not internalized the fundamental difference between validity and truth, and also in people who are new to logic and therefore more prone to getting overly excited about it and overreaching in their conclusions.

For example, if I argued that if all wizards are lizards and all wizards are reptiles then we could conclude that all lizards are reptiles, and you correctly pointed out that my argument contained a fallacy, then that would not prove that the claim that lizards are reptiles is false. In fact, the claim is true even though my hypothetical argument in favor of it is invalid. Be warned though, that if in addition to pointing out fallacies in your argument your opponent also proves you false directly then your opponent has legitimately defeated you, and thus you couldn't claim that they were committing the fallacy fallacy.

6.2.28 Fitting fallacy

Just because an explanation for something fits doesn't make it true. Any time that you assume that if an explanation fits some set of circumstances then it must be true then you are committing a fitting fallacy. I actually wasn't able to find a suitable term for this fallacy in the standard vocabulary, even though it is actually quite a common fallacy, so I decided to just create a new term for it myself. I thought that the name "fitting fallacy" was a very *fitting* choice of words. Get it? Hah... Anyway, now that you're done rolling your eyes, let's discuss the nature of the fitting fallacy in more depth.

The fitting fallacy is closely related to two important scientific principles, namely Occam's Razor and the principle of falsifiability. A great many arguments that contain fitting fallacies do not satisfy at least one of these two principles. However, some arguments which nominally satisfy both Occam's Razor and the principle of falsifiability are still fitting fallacies. What are Occam's Razor and the principle of falsifiability? Well, Occam's Razor is a rule of thumb (not a universal law) which says that among multiple competing explanations for something the explanation that makes the fewest arbitrary assumptions is the one most likely to be correct. Every additional assumption that you add on to a theory is another way for that theory to be proven wrong. Thus, the more arbitrary assumptions a theory makes the higher its chances of being incorrect are.

In contrast, the principle of falsifiability says that if a claim cannot be tested then it cannot be treated as scientific or justifiable. All untestable claims are inherently invalid and worthless, insofar as they are intended to pursue objective truth. We may not ever know the ultimate nature of existence, but we do know that everything we encounter in real life exists in at least some form. Science restricts itself only to the part of existence that we can interact with and test, i.e.

to physical and logical systems, and ignores all other possibilities. Untestable claims are, by definition, guaranteed to be fruitless to pursue. A little bit of thought will show you that this must be true, by the very nature of what it means for a claim to be untestable.

Almost all conspiracy theories are based on fitting fallacies. For example, is it conceivable that aliens built the Egyptian pyramids and that all the overwhelming evidence to the contrary is just an elaborate cover-up? Sure, but that doesn't make it true. It's not enough for a theory to just fit. A strong theory needs "smoking gun" evidence. Why would someone go to such extreme trouble to cover-up something like the pyramids? There's very little to nothing to gain from convincing people of something like that, and people are too lazy and error-prone for such a large scale cover-up to have much chance of success even if they tried it. Conspiracy theorists would be wise to study Occam's Razor and the principle of falsifiability. Maybe if they did then perhaps they could break free of their paranoia and learn to have a more productive view of reality[2].

Besides conspiracy theories, yet another endless source of nonsense in our society are grand untestable theories of existence. All grand untestable theories of existence are fitting fallacies. No exceptions. This is true whether it's simulated reality, solipsism, reincarnation, string theory, multiverse theory, dream world theory, any religion ever conceived of (whether past, present, or future), or any other of an infinite number of conceivable theories arbitrarily and baselessly plucked from the ether of humanity's overeager and dogmatic imagination.

Putting the more extreme cases of conspiracy theories and grand untestable theories of existence aside though, there are also some much more mundane and subtle examples of fitting fallacies in modern society. In some ways, these more subtle examples can actually be more dangerous because relatively few people seem to be clear headed enough to detect them when they occur. These subtler examples can satisfy both Occam's Razor and the principle of falsifiability and yet still be fitting fallacies. One example of this type of more subtle fitting fallacy can be found in the way our legal system sometimes operates. Judges, lawyers, and jurors sometimes consciously or subconsciously operate as if an explanation merely fitting an event is enough for it to be true. This is invalid.

Fitting fallacies seem to be an abundant source of false convictions and counterproductive behavior throughout the legal system, and many people are unfortunately oblivious to it, in large part perhaps because few people outside of the hard sciences and mathematics ever actually correctly learn how to think objectively. Science is the only objective worldview. Everything else rests on little more than thin air. Many people don't want to do the hard work involved in learning to think logically and scientifically, usually out of a misguided sense that it will be more difficult or more unpleasant than it actually is in reality. They want the easy way out in life, so they avoid it. They avoid science and math classes with the same fervor as if those classes were wrapped in plague blankets.

I wonder if they would be as comfortable with their decision to run away from learning to think logically if they realized that the overwhelming majority of injustice and human suffering in the world is probably caused by a lack of capacity for rational thought. If people continue to run away from the burden of rational thought, then it will continue to remain impossible to ever truly defeat large-scale human suffering. When you choose blissful ignorance over diligent rationality, you are unwittingly enabling a vast multitude of sources of human suffering. A vote

---

[2] On the other hand, many conspiracy theorists seem drawn to conspiracy theories in part because they seem to want to feel special and smart without having to actually do the large volume of hard work that real science or real knowledge in contrast would require. Thus, their attachment may be emotional and not rational, and so reasoning with them may not get through to them regardless. Perhaps it would be more productive to redirect their emotions instead.

for ignorance is a vote for suffering. Stupidity is the furnace from which human rights violations are forged.

6.2.29 Proof by intimidation

All of the previous fallacies we've discussed up to this point tend to be more common among the uninitiated than among people who have studied logic or science extensively. This one is different. This one occurs most frequently among the well-educated. Academia is rife with it. Experts of all stripes are in constant danger of committing this fallacy. Many even do it deliberately, knowing full well in their heart of hearts that they are being disingenuous when they do so.

What is it? Well, a proof by intimidation is an attempt to obscure an argument by filling it to the brim with lots of intellectually intimidating but ultimately superficial content in the hopes that the person that the argument is being presented to will accept it out of fear of being seen as stupid instead of based on the actual merits of the argument. Academics and experts often strongly associate their personal identity and sense of self-worth with their intellectual abilities. This creates an egotistical blind spot and makes them vulnerable to pretending to understand things that they actually don't understand. The higher the ivory tower is the harder it falls.

I put this fallacy at the end of the list so that it would give academics and experts pause, so that they don't get too cocky after reading through the other fallacies, perhaps contemplating condescending thoughts of the less educated people who tend to commit those fallacies more frequently. Not all fallacies are the product of simple ignorance. Some fallacies increase in frequency with the more knowledge you acquire, unless you are careful to keep your own ego reigned in and your mind disciplined. Staying committed to clarity of communication helps tremendously in this respect. It is also wise to cultivate a perspective of universal empathy for all humankind and to recognize that for the most part all human beings have roughly the same intellectual and creative potential.

Most so called "intellectual differences" are probably simply the result of growing up in a less supportive and less privileged environment, which causes the person to end up developing inaccurate self-limiting beliefs about their own intellectual and creative capacity. If this is you, it's important to remember that a sense of struggle is completely normal in any intellectual or creative endeavor, and that human beings are fond of bragging, so when you hear someone say that they just naturally have a talent it's still more likely that actually they are omitting a huge amount of practice. Whether consciously or subconsciously, this omission is probably made in order to artificially inflate one's apparent social worth in the eyes of others by leading them to believe that this difference is inherent rather than primarily a matter of practice.

If people realized that almost anybody could be smart and creative with sufficient practice and discipline, then it would make the accomplishment seem less impressive. It would make the person bragging seem less inherently special and therefore perhaps less socially and reproductively valuable. It seems to me like this is probably in part a subconscious evolutionary reproductive trick designed to deceive others into overestimating a person's inherent value over others.

Thus, people lie about where their skills came from (e.g. not admitting to how much they had to practice) for personal social gain, making it look like inherent genetic talent, and thereby inadvertently spreading extremely toxic self-limiting beliefs to other people. Don't buy into that crap though. Genetic talent is largely a myth. Even when talent does genuinely exist at the genetic level, it is probably only a marginal advantage. It's just icing on the cake. I'd estimate that someone without any genetic talent could probably still attain 95% of the skill that someone with genetic talent could. It seems like it's probably only that last little bit that genetics applies to. That's what I think and that's what my experiences in life seem to indicate.

Anyway, back to the main subject of proof by intimidation. If you've ever attempted to read at least a few academic research papers, then you have almost certainly encountered at least trace elements of a proof by intimidation strategy. One of the most common uses of proof by intimidation is to obscure the fact that what a person is saying contains far less real information than it appears to on the surface. It is not uncommon to read entire paragraphs or even entire pages of text in an academic document, only to find on closer examination that it contained almost zero real information.

The words will just meander back and forth like a winding old road or a snake and never quite manage to clearly say anything substantive. Technical terms and jargon will be thrown in the mix liberally in order to maximize the chances that the reader will be too intimidated to call out the writer on the nonsensical, incoherently complex, or low-substance nature of the text. Here again, clarity of communication will be your saving grace, if you let it. I daresay that any research paper that is not written with a readily apparent passion for clarity should automatically be viewed with suspicion.

I detest academic obscurantism. It is like hideous graffiti spray-painted haphazardly and at random throughout the otherwise pristine hallways of science. Obscurantism does not merely obstruct communication amongst scientists, but also poisons the public's perception of what science stands for. Real science, communicated with a passion for clarity and with empathy for all of humankind, is empowering and fills the mind with wonderment. In contrast, academic obscurantism and pretentiousness are at best frustrating and at worst deeply erode public trust and respect for science. It is not enough for scientific research to merely be correct. It should also be clearly explained, thoroughly documented, empathetic, and accessible.

In mathematics especially, it is so easy to just wrap yourself up in symbols and technical jargon and to rely upon the difficulty of your obscure text to deter any potential criticism you would otherwise face if it had been written more clearly. However, as comfortable and easy as that may be, every time you do so you are damaging the image of science and empowering the forces of anti-intellectualism and anti-scientific prejudice. Instead of doing what's comfortable and self-serving, choose instead to lay your concepts bare with as much clarity as you can muster and know that the criticism this invites will only make you stronger and more rational in the future. Stop treating complexity as if it is a virtue. Clarity and rigor are virtues. Superficial complexity is not.

6.3 The consequences of poor reasoning

6.3.1 A society rife with ignorance

That concludes our coverage of what I consider to be the most common and most important informal fallacies. Be aware however that there are plenty of other types and subtypes of informal fallacies besides just these. There are an abundant number of different ways for an argument to be rendered invalid. Clearly, with so many ways that things could go wrong and with so many counterintuitive factors to keep in mind, logical and scientific training is a necessity if you ever want to truly be a consistently rational human being.

Unfortunately, it seems like most members of society currently believe themselves to be rational thinkers by default, without having to put any effort into deliberate logical and scientific training. One of the classic dead giveaways that someone is probably guilty of at least some amount of this brand of ignorance is if you often hear them say that "it's just common sense" that something is true. Science has proven repeatedly that "common sense" cannot accurately gauge reality. Only the scientific method and logical thought experiments stand any chance of being consistently objective. Even with intense life-long scientific and logical training, at least some amount of significant errors are still practically guaranteed to happen.

To err is human, just as the old proverb says. Error-free human thought and immortality seem like equally implausible goals to me. Don't think that this somehow equalizes the playing field though. The difference in error frequency between an untrained layperson and a master of logic and science is extraordinarily vast. It's like comparing a pigeon to a supercomputer. My point is, the inherent fallibility of humanity is no excuse to get mopey and give up, nor is it an excuse to treat all humans beings as equally and hopelessly error prone. Perfect reasoning may be impossible, but near-perfect reasoning is probably achievable, if not now then perhaps at some point in the future when humanity has intellectually and culturally matured enough for it to happen.

Putting too much trust in "common sense" instead of in logic and science is one of the defining hallmarks of ignorance. Without logic to filter and scrutinize the information you encounter and to organize and refine your thoughts, biases and misinformation will inevitably creep into your mind and change your perspective for the worse. The longer a person chooses to live an irrational life, the more biased their mind will tend to become.

To make an analogy: If you never cleaned your house then it would just keep getting increasingly dirty and dysfunctional over time. The same principle applies to your mind. The longer you neglect your mind, the more biased it will become, and the more difficult and time-consuming it will become to ever repair the damage. Taken to an extreme, some people eventually become so biased from long-term neglect of their mind that it becomes nearly impossible to even reason with them. Too much neglect of a person's mind can become lethal to a person's future intellectual integrity.

Our hidden biases inevitably lead to suffering and missed opportunities, both for ourselves and for those around us. This is true regardless of whether or not we ever actually become consciously aware of it. People are *not* born with good sense. It has to be practiced deliberately. It is just as much a skill as playing the violin is a skill. Just as people should be required to learn to drive before they are allowed to hop into a car and get on the road, people should also be required to learn at least the basics of logical thought and fallacy resistance before they are allowed to enter adult society, lest they endanger the continued survival of humanity and our future prosperity.

Allowing people to get through the school system without ever being trained in fallacy resistance is a huge threat to our survival as a species. I daresay that a person who hasn't been trained to think logically may be even more dangerous to society than a loaded assault rifle. It would be nice if society would realize this before we all die due to some inevitable misguided decision. It's only a matter of time. A society that is not rational is a society that is on a deadly countdown to destruction. I dearly hope that we can become a more rational society before that invisible timer hits zero. Embracing rational thought is the only way to secure our future. All other paths lead to a high chance of random suffering and death.

Outside of the STEM disciplines and the creative arts, in regular everyday society, far too many arguments you encounter will be riddled with fallacies. Look at flame wars on internet forums if you want to see some particularly bad examples. It's gotten so bad that some of the arguments that people encounter on a day-to-day basis actually contain more fallacies than they do substantive statements.

This society-wide irrational attitude, where arguers are often far more interested in saving face and defeating the other side than in actually discovering the truth, greatly impedes social progress. This has to change. We can't make progress if the only motivating force in public debates is to protect your own ego while smearing your opponent. We have to find a way to defeat this behavior pattern. All it takes is a single fallacy to potentially render an entire argument utterly meaningless and counterproductive. Some errors are admittedly superficial and easily corrected, but others are quite fatal to the integrity of an argument.

6.3.2 The poison of anti-intellectualism

Anti-intellectuals would have you believe that learning to think rationally will somehow make you less personable, less warm, less empathetic, less feeling, less happy, less likable, and less compassionate. Anti-intellectuals want you to believe that the people who cultivate and refine their minds, who create countless works of art beyond anything you've ever imagined, who selflessly engineer medical devices to save your children's lives, who guard and record the history of humanity so that we might learn from the past and the wisdom of those who came before us, who fill your life with nearly every modern source of entertainment and convenience you now enjoy, from games to films to novels to theme parks to music, are somehow just cold out-of-touch elitists who do nothing but sit around all day looking down on others. Nothing could be further from the truth. Exactly the opposite is true. It is the anti-intellectuals who are the ice-cold elitists and the empty-eyed faceless statues of our society, not the intellectuals and the creative types.

It is the anti-intellectuals who whittle away their precious little time on Earth, day after day, doing little more than looking down condescendingly and arrogantly upon the rest of society. They create nothing. They explore nothing. They let their minds rot, allowing it to become overgrown with weeds and cobwebs, instead of tending to it and protecting it like the sacred garden that it is.

They contribute little of value to society, if anything. They foster social strife and violence everywhere they go. Yet, the anti-intellectuals expect you to respect them above and beyond all others in society, despite their utter lack of significant contributions to the lives of the rest of us, while simultaneously advocating that we should all be looking down upon all of the people who actually *do* create most of the value of society: the intellectuals and creative types.

And then, as if that's not enough, the anti-intellectuals have the gall to say that the intellectuals and creative types are the entitled ones. Think about that. Let the implications really sink in. Once you do, you will realize that the most entitled people *by far* in our society are actually anti-intellectuals. Anti-intellectuals benefit constantly from the insight and the blood, sweat, and tears of intellectual and creative people, and then they turn around and try to make intellectuals and creative types all sound like sub-human scum.

Whenever anti-intellectuals acquire political power they often begin systematically perpetrating large-scale violent atrocities against society, like the Khmer Rouge of Cambodia for example. Many intellectuals and artists throughout history have been murdered, burned alive, imprisoned, and persecuted in countless horrific ways under the social influence of anti-intellectuals, all just because the public was gullible enough and short-sighted enough to fall for the backwards rhetoric, baseless child-like finger pointing, and jealous bullying tactics of anti-intellectuals.

To make an analogy, every time you buy into the anti-intellectual rhetoric that intellectuals and creative types are just cold-hearted elitists, it's sort of like you're placing a little miniature stamp of approval on the Khmer Rouge. That's where it could end up if not kept in check. History has shown that before. Every time you sneer at intellectual and creative accomplishment you are unwittingly increasing the chance that humanity will one day lose its freedom and its color and its joy and degenerate into a soulless oppressive dystopia.

Every time you use the word "nerd", every time you mock someone for adding more value to society than you have by calling them "overachievers", every time you blindly and falsely assume science is boring, every time you spread malicious propaganda designed to cast all intelligent people in an unsociable light, every time you discourage a friend from accomplishing something great because you secretly fear it will make you look relatively worse in comparison, you are behaving like an anti-intellectual and you are unwittingly contributing to the potential damnation and misery of all humankind.

The path of intellectualism and creativity leads to a prosperous, empathetic, and loving society, whereas the path of anti-intellectualism leads to a living hell, a nightmare laid bare in broad daylight, where freedom is scarce and tyranny looms large. If that hell ever becomes reality, there may be no turning back. Only in a society where freedom, intellectualism, and creativity are allowed to breathe is social change ever possible. If society ever travels too far down the path of anti-intellectualism, it could be sucked into a hellish intellectual black hole from which escape is impossible. Misery would become eternal. There perhaps wouldn't be any hope of recovery.

The true snobs and pricks of society, the true paragons of arrogance, dullness, pretentiousness, entitlement, and elitism, are those who look upon all the creative wonders of the world and see nothing but the blinding and enfeebling light of their own petty jealousy for the accomplishments of others. That is the true spirit of anti-intellectualism. They wrap their rhetorical bile in innocent-sounding code words and phrases like "being normal" and "regular folk" to camouflage the malicious poison within, and to implicitly condemn anyone who ever seeks to accomplish anything more in life than they have. They condemn anyone who has cultivated a stronger sense of empathy and creative spirit than they have. Every anti-intellectual person could have chosen a better life, a more constructive life. All people have fundamentally the same inner potential. I've always believed that.

But instead though, anti-intellectuals choose to worship mediocrity and to shun anyone who accomplishes more than they have by labeling those other people as somehow "not normal". Anti-intellectuals are fundamentally bullies. I don't think it's even possible to be an anti-intellectual without also being a bully. In essence, anti-intellectualism is just one particular subtype of bullying. Anti-intellectuals create little to no real value for society, so the only way they can get a sense of self worth is by instead pulling other people down (e.g. through verbal smear attacks, harassment, psychological attacks, violence, etc) so that those other people don't seem as high-value as they previously did. Bullies can't stand seeing other people doing well. The more a bully sees other people doing well, the more that bully feels the yawning abyss of their own lack of real value contribution to society. Bullies can't stand to look in the mirror of their own mind.

As cruel as bullies are, they are perhaps often also victims of their own self-limiting beliefs. Instead of attacking other people, bullies should brave the mirror of their own mind long enough to look inside themselves and reignite their own inner fire of limitless creativity and wit that lives within all people. All people have the capacity to accomplish great things. All people have the power to make their dreams become reality. All they have to do is realize it.

Drop the crushing burden of self-limiting beliefs from your shoulders and breathe in the vast freedom of what you could potentially be if only you tried. Even the worst behaved people are still probably often redeemable, under the right circumstances at least. Just as there is a demon hiding within each of us, waiting for us to lose perspective on life so that it can wreak havoc upon the world, so too is there also a hero, waiting for us to summon it by lighting the candles of bravery and hope.

These are not intended to merely be empty inspirational words. Self-limiting beliefs really are crippling. In my experience, when people think they can't learn a skill it's most often all just in their head. It's a matter of perspective. When you think in advance of attempting something that you won't be able to do it then it tends to create a self-fulfilling prophecy. I have lived this. I have gone from having many self-limiting beliefs to having few, and I have experienced the transformation it yields firsthand.

I have gone from the tunnel vision of self-limiting beliefs to the vast open canopy and horizon of a liberated mind. You should take that journey too, if you haven't already, and experience for yourself the calm confidence and clarity that it inevitably brings. If you can just find a way to let go of all your unjustified self-limiting beliefs then I think you will find it quite empowering and well worth the effort. Run towards your challenges instead of away from them. That's the key to becoming a stronger person.

For example, you may believe, like a lot of people, that you can't do math. This is something I hear pretty frequently from people. It's almost like it has become some kind of contagious psychological disease. One might even give it an official name, like perhaps "math flu" or "mathematical PTSD". If you're one of these people, you might be thinking that surely since I'm writing a book on logic and math that I'd be one of those people who has always been good at math and logical reasoning. You'd be wrong though. I used to think I wasn't good at math either. Yet here I am. That feeling that you can't do logic and math, or anything else for that matter, is a lie. Don't listen to it. I've lived it. I've escaped from that tunnel and seen the truth. I believe that you can too, that *all* people can.

Imagine a world where every single human being alive has reached the same level of heroic clarity and perspective on life as all of the greatest thinkers and artists who ever lived: the Einsteins, the Newtons, the Da Vincis, and the Michelangelos of the world. One day every person alive could be like that, simultaneously. I believe that this is attainable. If we could establish universal rationality, universal empathy, and world peace (within some acceptable margin of error and imperfection), and then get everybody to also drop all their self-limiting beliefs, then I believe that it could actually happen for real. All people have roughly the same potential, it just needs to be unlocked by cultivating the right environment.

I didn't just drop my own self-limiting beliefs though, I also said to myself that not only am I going to try to be good at this, but I'm going to try to show the world something it hasn't seen before, and voila the more I work towards it the more it seems to be happening for real. It's amazing how much just believing something is possible increases the chances of it happening for real. We make our own reality in life. You can too, like all people.

I could have chosen to hide all of this from you though, to make you believe that this is all just some kind of inborn talent that I have, but that would be a lie. I won't be one of the ones who instill that false impression of inherent genetic talent in you. Acting like it is an inborn talent would give a false impression of reality and thereby perhaps contribute to your self-limiting beliefs. It would be unethical.

That's why I have chosen to take the road less traveled and to instead lay bare the fact that I too at one time had those limitations, in the hopes that by doing so you will abandon your self-limiting beliefs and hopefully for once *believe* someone when they tell you that you can do math (if you haven't already), or whatever else you desire, contrary to whatever self-limiting beliefs you may hold. There's so much tragic untapped potential in humanity. We've got to somehow break through that wall and liberate the hearts and minds of society.

Logic is not some cold monolith, contrary to what the anti-intellectuals would like you to believe. It won't turn you into a robot. Logic is a way of navigating the conceptual space of the world around you. It opens your eyes and lets you see the world for all that it truly is, with nuance and depth that you could never achieve without it. It lets you sink your toes deep into the grass of life and feel the world as it truly is.

While everyone else around you is caught up in the hustle and bustle of daily life, always in too much of a rush to stop and smell the roses, always too distracted to soak in all the wondrous little details of this world, logic will open your eyes and let you peer into eternity to see the timeless machinations of all that ever was or ever will be. It will even let you see things that don't exist, yet conceivably could have if the universe had been different. Just as there are streets and bushes and lights all around us, so too is there the timeless ambiance of the fabric of the universe instilled in all that lay before us. Only with a mind empowered by the force of logic and science will you ever be able to witness this glorious landscape, this timeless crucible of the world.

Logic opens gateways to new worlds. It let's you see new possibilities that you perhaps never would have conceived of otherwise. It broadens your understanding of possibility spaces and hence broadens your creative potential too. Whether you're an artist or an engineer, you'll need a means by which to take what you see in your dreams or in your imagination and make it into something real. Logic, whether conscious or subconscious, is what enables you to do that. Logic is the bridge between dreams and reality. Logic is a way of navigating life and seeing connections. Life is a tapestry of connections. Logic lets you see the threads. Then, once you can see the individual threads and you focus in on them, you can start to weave them into a new reality. You can change the tapestry into whatever you want it to be.

Logic also protects your mind from enslavement by others. Instead of being a pawn in someone else's game, you become one of the players. Logic enables you to stop drifting through life aimlessly like a leaf in the water. It lets you take control of your own destiny. It empowers and enhances all aspects of excellence in life. Not only that, but the more logical a person becomes the more likely they are to develop a deep empathy and appreciation for their fellow human beings.

The more connections you can see the more likely it is for you to understand the perspectives of other people and thus to be able to empathize with them and to fulfill their deepest needs for genuine connection, connection built on a true understanding of each other and a loving appreciation for every little nuance of life and all the quirks and beauties of human limitation. Thus, the more you learn to think logically the *warmer* your personality will actually tend to become.

Logic by itself can't really make you colder. Only *other* kinds of things, such as bitterness, jealousy, emotional scars, selfishness, and hatred, can do that to you. Occasionally you will find people who try to justify their own unpleasant or hostile behavior as being "logical", but this is often just an excuse, i.e. just smoke and mirrors designed to help the person avoid being held accountable for unethical or inconsiderate behavior. It is much like how some people in dating culture use "independence" as a facade to justify being heartless, cruel, unfaithful, and irresponsible to others.

Having a right to make your own choices doesn't give you a right to treat other people like crap. Being a prick is an entirely separate dimension of a person's personality from independence or logic. Cruelty certainly doesn't originate from a spirit of freedom or rationality. It would be more logical to be more compassionate. Even from a selfish standpoint, compassion is still more productive than behaving in a toxic way. Compassion is typically more profitable, even if you don't genuinely care about other people. It is the more logical choice, under most circumstances.

Selfishness is not independence, just as emotional apathy is not logic. True rationality honors emotion. Without emotions, we wouldn't even have a reason to do anything. Without emotion, there would be no purpose in life. Emotions *are* what give us goals. A person without any emotions would essentially be a lifeless corpse. Only someone with emotions could ever care about their own survival or about protecting their loved ones. Only someone with emotions could feel compelled to get up in the morning. If you didn't care about your own survival or well-being (which is an emotion, by the way) then why would you ever do anything at all to improve your life?

You wouldn't. You wouldn't be able to see any difference in the value of any of the possible outcomes of your choices. You'd just lie down and die, like an inanimate object. Disregard of emotion is thus inherently irrational. A truly rational person knows that the most logical thing to do is to cultivate and refine your emotions, not to disregard them. Emotions are the source of all our goals, hence the only logical way to ensure our own continued well-being is to protect and nurture our emotional state. Logic and emotion go hand in hand, like yin and yang or husband and wife. Neglecting one also endangers the other.

Logic is not at all like what the propaganda of anti-intellectuals would lead you to believe. It's *always* better to be rational than to be irrational. No exceptions. Rationality has no negative side effects. Logic does absolutely nothing to diminish one's creativity, social skills, likability, or ability to have fun. In fact, quite the opposite, it enhances them. Logic is also *not* hostile to emotional experiences, contrary to popular belief. This point is so important that it is worth repeating again, but this time with slightly different words to capture a slightly different angle.

Emotions are essentially goal setting mechanisms in disguise, but there's nothing irrational about having goals, nor is there anything irrational about enjoying emotional experiences. Logical people embrace their humanity, not abandon it. In fact, abandoning your emotions and acting like a robot would be a very irrational thing to do. Emotions are the only thing that make us get out of bed in the morning. If we had no emotions then there would never be any motivating reason to do anything. Think about it. If we all suddenly lost all of our emotions then it would effectively be a death sentence for humanity. We would just sit around like lifeless porcelain dolls until we died of starvation. Hunger and the fear of death are emotions.

Portrayals of intellectuals and of logical thought in television and film are often filled to the brim with anti-intellectual prejudice. A large portion of society has unfortunately become blind to this though, which is dangerous because it tends to encourage irrational behavior. This is an entirely backwards and disgusting aspect of our current culture. It shouldn't be this way. It should be obvious that logic wouldn't undermine a person's warmth or charisma.

However, people are so constantly exposed to anti-intellectual propaganda[3] that many people have become deluded into thinking that logical behavior is somehow unsociable, even though the opposite is true. This is much like how the AIDS virus tricks a body's own immune system into attacking itself. Being rational (rather than prejudicial, dumb, gossipy, and presumptuous) is one of the best things you can ever do to make yourself more sociable and more enjoyable to be

---

[3] Perhaps this trend is rooted in latent jealousy. Perhaps it originates from the same psychological source as what compels some people to undermine and sabotage their own friends' progress and accomplishments to make themselves feel less relatively unaccomplished. This is, incidentally, the same psychological place bullying comes from.

around, yet anti-intellectual propaganda has convinced a huge portion of the human population that rationality is somehow bad for social skills. This is ridiculous.

It is a travesty. People have essentially been trained to undermine their own prosperity, to cripple their own chances of experiencing the kind of profound and fulfilling enlightenment that only a logical mind can provide, and hence to miss out on the corresponding wonderful life and personal warmth that it inevitably brings with it. Learning to see people as they truly are is impossible without a logical mind. Such perception requires clarity. Clarity requires the ability to see connections, and seeing connections is the domain of logic. The doors of truth open only to those brave enough to see reality as it is.

A mind without logic cannot see all of the connections between people, and thus will inevitably succumb to prejudice and end up treating people wrong at least some of the time. It is thus physically impossible to maximize ethical excellence without also maximizing logical excellence. You can't escape from the impact of logic. Either you pay the price of learning to think logically by doing all the corresponding hard work that it requires, or you pay the price of personal suffering in your life by committing unintended acts of evil which you are unable to even perceive as such due to your weak ability to understand the world as it truly is. You have no other choices. No other choices exist. Choosing irrationality is tantamount to choosing to be at least partially evil, even if unintentionally so.

Irrational people are among the least likely to be able to understand the perspectives of others. How could they? Logic is what gives us our ability to see connections between things. How can you maximize the strength of those connections to other people if you can't even see the connections in the first place? How can you make sure those human connections are tended to and treated gently if you don't even have enough logical skill to navigate that space without inadvertently damaging those connections?

Logic is not cold. Logic is warm. Contrary to anti-intellectual propaganda, not only does being logical not make you colder, but it actually isn't even possible to maximize your human warmth and empathy for others without learning to be logical. Only a logical person can consistently avoid discriminating against other people. Only a logical person has the wit to truly deeply understand the feelings and perspectives of others. Without logic you would always inevitably eventually make a mistake in your relationships with other people, due to your inability to safely navigate the conceptual space of the human relationships involved.

Logic enhances and expands the human spirit. It never diminishes it. It fertilizes the ground of creativity and makes us more likely to succeed in everything we do in life. It adds color to life. It never takes it away. Logic also protects us from evil. Evil is mostly just a form of poor reasoning, an incorrect perception that one is making the world better when in reality one is making it much worse. Thus, the more humanity learns to think logically, the weaker the forces of evil and human suffering will tend to become.

In many ways, the battle to teach people to be more rational is not just *a* battle of good and evil, but rather it is *the* battle of good and evil. Most of all of the sources of evil in human behavior stem from poor thinking. It's easy for people to become bitter and jealous of the achievements of others and to thereby come to feel justified in spreading anti-intellectual propaganda to indirectly attack the targets of that bitterness and jealousy, but to do so is unethical. It just perpetuates the cycle of evil and human suffering.

The only way to end evil is to end the cycle of ignorance, and that requires a society-wide extermination of anti-intellectual and anti-creative sentiment anywhere and everywhere that it exists. If society would invest a lot more time into teaching fallacy resistance, bias resistance, critical thinking, and the basics of formal logic, evidence, science, and proof to every single member of society then it would go a long way towards one day achieving this. Clarity is essential. It isn't enough to just memorize the principles of logic or to know all the superficial details. People need to genuinely understand it and internalize it on a deep level and be able to apply it to daily life easily.

As the situation currently stands, it does not appear that the schools are doing an adequate job of this. By the time people graduate high school, they should already be highly resistant to fallacious arguments and capable of performing at least basic systematic reasoning and thought experiments. Schools should train children to be resistant to all forms of illegitimate rhetoric and psychological manipulation.

Children should be taught to defy "authority" figures, instead of being taught to obey them as they so often unfortunately are. They should be taught to question everything, to realize that all beliefs are vulnerable to the light of reason, that there is no such thing as a belief that deserves immunity to scrutiny or criticism. For example, schools should actively teach resistance to the numerous unethical and manipulative psychological techniques that advertisers and politicians use, instead of merely turning a blind eye to such things. We need to vaccinate children against propaganda and sophistry so that we can slowly suffocate the forces of ignorance and evil out of existence in our society. This would go a long way towards fighting against bigotry and making society much more difficult for evil people to control.

Children should be taught that the more people try to hide information or discourage exploration of alternatives pertaining to a particular belief, the more suspicious one should be of that belief and the more likely it is that the belief will turn out to be on the wrong side of history. Hiding information, obstructing communication, and scaring people away from ever considering alternatives are very often signs that a belief system is wrong or evil. Any belief system that actively discourages the free exchange of information should be treated extremely skeptically. If a belief were based on merit then it would generally not need to hide information. Information hiding and obscurantism are strongly correlated with bad intent.

If you encounter two competing belief systems, the first of which discourages exploring alternatives and suppresses information, and the second of which encourages exploring alternatives, fosters transparent communication, and welcomes questioning of assumptions, then the second is probably the more correct belief system. Coincidentally, this is exactly the difference between religion and science. Religions actively try to scare people away from ever considering alternatives or questioning assumptions whereas science does the exact opposite. A wise person knows that no belief system that intentionally suppresses exploration of alternatives should ever be considered trustworthy, or indeed even ethical. Suppression of information tends to extremely strongly predict evil intent. Remember that.

In stark contrast, actively exploring alternatives, embracing free speech, welcoming criticism, treating nothing as sacred, doubting assumptions, and always being open to hard evidence and logical proof tend to be correlated with good intent. Such things tend to indicate that one cares more about merit and providing value to one's fellow human beings than about blind adherence to dogma or the pursuit of power. If we ever want to guarantee continued freedom and prosperity for all of humankind then society as a whole must somehow come to embrace logical thought.

There's no other way to consistently ensure that evil won't eventually rise to power again. We've got to find a way to somehow break the cycle of endless suffering. Universal rationality is one of the obvious necessary steps for achieving that goal. There may be some other necessary steps as well, but without everyone embracing rationality then achieving world peace would surely become impossible. Trying to achieve world peace without intellectual honesty is like trying to sail a ship without a compass: there's essentially zero chance of arriving at the destination.

Logic isn't just good for fighting against the forces of evil and discovering the truth though. It is also the most powerful tool ever conceived of for making our dreams into reality. Properly applied, logic will allow our species to one day guarantee that almost everybody will get to live a fulfilling life, no matter how unfortunate the circumstances they are born into. How? Well, logic enables us to accomplish creative endeavors that we otherwise never could. This includes everything from engineering new life-saving devices, to creating new works of art that provide fresh insight into life and a deeper sense of fulfillment, to producing new video games to fulfill any arbitrary fantasy that a person could ever imagine.

Logic is not just a problem solving tool and a life saver, but also a bountiful fountain of infinite entertainment and fulfillment. To explore logic is to explore all conceivable universes and to thereby lift humanity up to higher and higher states of consciousness and sentience. Properly understood, it is the ultimate tool of fulfillment and creation. Through logic and through empathy for our fellow human beings, there are few things that we cannot accomplish.

Video games illustrate this liberating quality of logic particularly well. Consider, for example, what life would be like without video games. If you were born into poverty or disease, then what hope of fulfilling your fantasies and dreams would you have? Sure, you'd have some chance, like all people, but realistically speaking the majority of people born into poverty or disease would not get to experience as full a life as they deserve.

What chance would a poor person have of experiencing what it's like to live as a king or as a wealthy merchant? What chance would a diseased person, confined to bed rest, have of experiencing what its like to be a great warrior or to be a beloved hero? What chance would they have of owning a fleet of fancy cars and competing in street races on a daily basis? What chance would they have of overcoming their circumstances well enough to fulfill these kinds of dreams? Realistically, they would have relatively little chance. However, in a sense, video games change all of that.

Video games let even the poorest and most diseased people experience practically any fantasy they could ever conceive of. Video games, through the immense combined power of computer programming and digital artwork, allow any fantasy to become reality. Video games make it so that even the most seemingly absurd forms of dream fulfillment are readily and cheaply available, instead of being nearly unattainable. Through the power of video games, even a child dying of cancer can live a full life through fantasy fulfillment. Video games enable us to experience practically anything we could ever dream of, at least in principle. This is likely to become even more true as more and more different kinds of games become available and thereby fill out countless different kinds of fantasies.

Thus, as a related consequence, every time a news journalist smears video games on TV, such as by unjustly blaming a mass shooting on the video game industry for example, then they are unwittingly damaging countless people's chances of dream fulfillment. A child dying of cancer may have some specific dream they want fulfilled, which a video game could conceivably satisfy. However, if a certain subset of news journalists were to succeed in spreading their ignorant anti-gaming propaganda, it might thereby somehow stop the game that would have fulfilled that dying child's dreams from ever coming into existence.

Video games are just an arbitrary medium of communication, just like books or music or film. Saying that video games are somehow inherently bad for society is just as fundamentally and profoundly stupid as saying the same of books or music or film. Only a profoundly ignorant, discriminatory, bigoted, unimaginative, narrow-minded, uninformed, and anti-intellectual frame of mind could make someone think like that. Anti-gaming sentiment is fundamentally of the same nature as book burning.

Nonetheless, despite how deeply and obviously toxic anti-gaming propaganda is to the cultural advancement of society, it apparently doesn't stop a certain subset of news journalists from spreading their hateful prejudices regardless. Maybe if news journalists actually understood the real potential consequences of their actions then they would lose a lot more sleep over it. However, like all bigots, these news journalists often never actually take the time to even attempt to understand the perspectives of the people they are unjustly attacking and inflicting suffering upon.

Then again, why should we even expect empathy from news outlets to begin with when most of them have degenerated into mindless political propaganda machines that have seemingly abandoned all semblance of journalistic integrity and objectivity? Of course, one thing news journalists will never tell you on air is that one of the TV industry's main economic rivals is the game industry. In some ways, perhaps no entity in existence poses a greater long-term financial threat to the TV industry than the game industry. The TV industry has a massive multi-billion dollar vested interest in attacking the game industry, regardless of whether it is actually ethical to do so.

This vast financial incentive to smear the game industry effectively renders much of what the TV industry says about the game industry untrustworthy and worthless. The TV industry has a massive conflict of interest with the game industry. Considering that the game industry is probably one of the leading causes of declining TV viewership, you would do well to treat everything the TV industry says about the game industry with suspicion. Why is the TV industry so threatened by the game industry? Well, it has to do with the fact that the game industry has certain traits that make it fundamentally more powerful as a creative medium than the TV industry in the long run. Think about it. Why watch someone else experience something on TV or film when you can experience it *yourself* in a video game?

In terms of expressive capability and potential as an artistic medium, video games are a strict superset of TV and film, and this is only going to become more true with time. This will probably eventually lead to TV and film being eclipsed by an ever increasingly large and more powerful game industry. This is the fundamental reason why the TV and film industries fear the game industry above all others and thus is probably a huge contributing factor to why they so often seek to slander the game industry and damage its reputation. Remember, whenever you see a lot of negative propaganda against a group of people one of the first things you should look for is a financial incentive. Just as animals can often be tracked by following trails of footprints, so too can corruption often be tracked by following trails of money.

That being said, the TV and film industry will almost surely continue to survive and thrive in at least some form regardless. It's just that TV and film will probably become significantly weaker than they currently are, relative to the game industry. The one saving grace that TV and film have that will prevent the game industry from gradually economically obliterating them is that TV and film are an inherently extremely lazy and mindless form of entertainment, whereas games require interactivity, effort, and independent thought. Extreme laziness is the economic niche that will ensure that the TV and film industries will survive in perpetuity, even as the game industry nonetheless continues to make huge economic gains against them.

Anyway, back to the main point. Logic enables game development which in turn enables fulfillment of people's dreams, no matter what circumstances they are born into. Isn't that wonderful? Well, if anti-intellectuals ever have their way then all of that could disappear. Children dying of cancer would no longer be able to fulfill their wildest fantasies. Poor people would no longer be able to experience what living like a king feels like. You'd be taking away a lot of people's hopes and dreams. The negative consequences of anti-intellectual behavior are vast and devastating. Being an enemy of logic effectively makes you an enemy of joy itself. Choosing ignorance is effectively the same thing as choosing suffering.

It's not just you who will suffer for it though, it's everybody around you. When you choose ignorance, and disregard the pursuit of intellectual integrity, you are guaranteeing that you will become at least in part a bigot and thereby also guaranteeing that those who have to spend time around you will suffer for it, whether or not you ever become consciously aware of it. Worse still, ignorant people are the least likely kind of people to ever realize that they cause so much misery for other people. By definition, ignorance decreases your ability to perceive reality, to be self-aware, and to have empathy for others. It wouldn't be ignorance if that wasn't the case. Ignorance has self-reinforcing negative effects, whereas rationality has self-reinforcing positive effects.

Having a passionate intellectual and creative spirit is the key to maximizing the amount of value you can provide for your fellow human beings. Choosing anything less than that is like choosing to be less than the person that you were meant to be. Who would want that? Who would truly want to be less than what they should be? If you create more value for your fellow human beings then they will love you all the more for it and you will live a fulfilling and deeply worthwhile life. You'll be able to look back with satisfaction at what you have given to the world. It's never too late to start. Every moment is like a new life, if you really stop and think about it. You can't change the past but you can always make the best of every moment you have left. It's a matter of perspective.

Every person on the planet is born into different circumstances. Some are born poor and have to fight for everything they have, and others are born rich and live in a state of effortless perpetual comfort. Flailing our arms around like children and crying and cursing our circumstances is unlikely to make the situation any better. Doing so would just waste valuable time and resources, time and resources that would be much better spent working to improve our circumstances proactively. That's true not just of what circumstances we are born into, but also of every moment in which we live. It's as if every moment in which we live is a new life, a new set of circumstances created by random chance and events beyond our control. It's hardly any different from being born rich or poor.

There's a lesson in this. Stay calm and alert enough and you may realize it. Stop comparing your life to others as if it's some kind of valid measurement of your relative worth. Everybody lives every moment under a different set of circumstances. Everybody has a different starting point in life. What excellence means from one person's viewpoint is different from what it means from another's. You wouldn't judge two contestants in a tournament equally if one started with an advantage or with a different set of resources, so why do you think it's ok to do it when comparing your life to the lives of others? Let go of every concern in your life that is outside of your ability to control. By definition, any time spent on something that cannot be changed is wasted time. Live every moment as if you have only just now been born, and then ask yourself: How can I make the best of this?

Some people look upon the world, with all its flaws and injustices, and then they ask themselves: Why can't things be better? Why can't we fly among the clouds? Why can't we end human suffering? Why can't we make it so that everybody can fulfill their wildest dreams, no matter where they come from or what their personal limitations are? Why not? They then ask themselves how we can move towards that future, one step at a time, even if it takes generations for the dream to finally be realized, even if they have to move heaven and earth to do it, and even if they will never see the full benefit of it themselves. These people are the intellectuals and the creative types of our society. They look upon the suffering of humanity and they try to alleviate it. They look upon the dreams of humanity and they try to make those dreams become reality.

Fulfilling those dreams comes at a cost though, and that cost is the burden of maintaining a healthy rational mind and a passionate creative spirit. Anti-intellectualism threatens the survival of these endeavors, and thus threatens the prosperity and future of humanity itself. A healthy rational mind cannot thrive in a society that suppresses free speech and open criticism. Likewise, a passionate creative spirit cannot thrive in a society that sneers at diversity and culture.

When you embrace anti-intellectualism, you are unwittingly embracing the destruction of everything that brings you the most joy in life. You are weakening the medical doctors that protect your life and your family. You are weakening the books, paintings, music, games, sculptures, architecture, fashion, technology, social scene, and culture of society. You are weakening everything that fills your life with deeper meaning and keeps you happy even through the darkest of times. You are weakening the human spirit itself, and making humanity become less than it otherwise could be.

Anti-intellectualism is the blight of humanity. Few things do more to obstruct human progress than people looking down upon intellectual and creative accomplishment. It poisons everything. Logic is the defense system of the human mind. Without it, humanity has no way to fight against the forces of evil and suffering. When you choose to sneer at logical thought, you are in effect welcoming evil and suffering into your life and into the lives of those around you with open arms.

You are glibly ignoring all of the immense undeserved pain it will inevitably cause to your loved ones and to humanity as a whole. Anti-intellectualism is one of the primary sources of human evil. An ally of ignorance is an ally of evil. Only those who side with reason can consistently protect the interests of humanity and of their loved ones. Ignorance is a fool's bliss. Nothing enables the rise of evil more efficiently than widespread ignorance. A reasonable society is difficult to bend towards evil, but an ignorant society is easy.

If you want to maximize the value that you bring to those around you in life, and also your own sense of fulfillment and tranquility, then one of the best things that you can ever do is to become more rational. Light the fires of the creative spirit that lives deep within yourself. Intellectual and creative endeavors are the ultimate way to create abundant value for your fellow human beings and thus also a fantastic way to show them empathy and warmth. Yes, in the short run it may not be as easy as embracing ignorance, but in the long run you will find that it will make your life far easier.

Embracing your full intellectual and creative potential will open countless doors for you and will make your life both much more fulfilling and much easier. It takes real empathy and warmth of spirit to be able to walk down the path of intellectual and creative achievement. This is yet another reason why you should never buy into anti-intellectual propaganda. Intellectual and creative endeavors tend to favor a warm and compassionate personality, a personality that is the opposite of the kind of ice-cold, lifeless, and dull personality that tends to characterize anti-intellectuals.

Anti-intellectuals want everyone to believe that intellectual and creative endeavors somehow suck the life and warmth out of people and make them cold. Ironic, right? It reeks of a level of hypocrisy and delusion that is nothing short of jaw dropping. Intellectual and creative endeavors do not make people cold. Nothing could be further from the truth. In fact, intellectual and creative endeavors actually vastly expand one's capacity for human warmth, empathy, tactfulness, and insight. Yet, despite this, anti-intellectuals still believe that they are somehow magically entitled to the utmost respect and that their opinions are automatically just as valid without even doing any of the hard work required to actually determine whether any of their assumptions have merit or not.

They think that their own blind and utterly baseless "common sense" assumptions somehow magically carry more weight than beliefs based on actual hard evidence and logical proof. It's been so long since the last time they actually bothered genuinely testing one of their own claims (if ever) that they no longer have any sense at all for how truth and reality even work. What's left behind is nothing more than a mind composed almost entirely of pure arrogance and blind assumptions, the mind of an anti-intellectual. There is no greater arrogance than believing that an opinion that is not backed by hard evidence or logical proof should be treated as an absolute truth.

Anti-intellectuals contribute very little of genuine long-term value to society, yet somehow they still think they deserve the utmost deference and even leadership positions within our society. They create almost nothing yet they think they deserve everything. They are walking physical manifestations of psychological projection and hypocrisy. They define themselves as being "normal", and everyone else as being not, when in reality it is they who are abnormal and it is they who are clueless about how ideal social interactions should actually be.

The very thing that anti-intellectuals accuse intellectuals and creative types of being most guilty of, namely of being cold and out of touch, is in fact the very thing that anti-intellectuals are themselves most guilty of. If anti-intellectuals want to see who the real snobs and pricks of society are then they should look in the mirror. Unfortunately, anti-intellectuals are exactly the kind of people who tend to lack the kind of self-awareness and introspection necessary to look into the mirror of their own mind to see their own flaws.

The best we can do is to try to make as many people as possible as rational as possible, one step at a time, until the number of anti-intellectuals remaining in society becomes negligible enough that they no longer pose a statistically significant threat to the rest of society. When that glorious day finally comes, society will at long last be able to breath a sigh of relief and finally rid itself of the immense stupidity and endless suffering caused by the influence of anti-intellectuals. I envy the human beings that will one day live to see that day. However, as it stands, we are all currently living in a wildly transitional and dangerous phase in human history where the majority of the public still tends to think irrationally, yet simultaneously has access to vast amounts of powerful technology. It is a potentially deadly combination.

An extremely small minority of society (scientists) has created countless forms of advanced technology for the rest of society to benefit from, yet the vast majority of society does not yet possess the same level of intellectual discipline and self-control that produced that same technology. It's like arming a toddler with a shotgun. From an economic standpoint, this is a necessary phase in human history, but it is also a very dangerous one. Economies of scale basically force it to happen this way, but that doesn't make it any less dangerous or worthy of caution. I hope our species will survive long enough for a more universal rationality to take hold before we manage to somehow kill ourselves.

If the forces of anti-intellectualism are not defeated relatively soon then there may be little hope for the survival of ours species. Consider yourself warned. We either embrace rationality and empathy for all humankind, or we'll probably go extinct. Those seem like the most likely outcomes to me. Every time you mock intellectual and creative achievements you are aggravating this situation and thus unwittingly threatening the survival of our entire species. Society needs to realize that it isn't even possible to be a maximally ethical person without also being an intellectually honest person. Anti-intellectualism is a one-way ticket to doom. Society needs to realize this before it is too late.

Also, people need to realize that as automation continues to rise it will eventually destroy the vast majority of mundane and repetitive jobs. Robots and computers are way better at those kinds of tasks than humans are. Most of the remaining jobs will be in intellectual or creative fields of study. Genuine creativity is extremely difficult to automate, thus creative jobs are inherently more economically stable. No amount of government intervention can ever stop this change. Low-skill jobs will continue to economically weaken over time. Wishful thinking and iron-fisted policy changes won't save you, not even slightly. Only embracing intellectualism and the creative spirit will. High-skill jobs are the future. Accept it and adapt accordingly or suffer the consequences.

Automation is just too overwhelmingly powerful as an economic force for anybody to ever have any hope of stopping it. Most jobs that require little to no intellectual skill will eventually be driven to near extinction. Once that happens, the only way to become prosperous will be to invest in learning an intellectual or creative occupation. The more you try to fight the forces of intellectualism and creativity the more you will be driven towards poverty. This will become more and more true as time passes and technology becomes increasingly advanced.

Trying to run away from this will just make it even worse for you. If you don't like the dryness of science then at least choose something that inherently requires human creativity. Your goal should be to fill an economic niche that a machine could never easily fill. This will increasingly become the key to financial stability in the future. Running away from educational challenges because you think it's "too hard" or you don't want to put forth the effort would be one of the dumbest decisions you could ever make in life. You can do it. Everyone has roughly the same intellectual and creative capabilities. The key is to not harbor any unjustified self-limiting beliefs.

The great irony of avoiding intellectual and creative endeavors because they are "too hard" is that, of all the decisions a person could ever make in life, it is the one that will do the most to make your life much more difficult. If you truly want an easy life then you would be wise to invest in learning the skills that other people consider to be "too hard". Once you learn those skills, you will become much more valuable to society. This in turn will make you more marketable, thereby causing your life to actually become much easier. Run towards your fears instead of away from them. Let your ambitions guide your path through life, instead of your fears. Your fears will only lead you to be less than what you should be.

All the sources of entertainment and fulfillment that you have in life come primarily from intellectual and creative endeavors. Whether it's books, music, films, theme parks, video games, theater, television, architecture, or new technology (etc) it all originates from intellectual and creative sources. Basically almost anything fun in your life is the product of some intellectual or creative endeavor.

Anti-intellectualism and mediocrity culture threaten the existence of all of these things. The more people in society harbor anti-intellectual sentiments, the fewer of these things will be produced and the lower the quality will be. If you like having these things in your life then you would be wise to stand against the forces of anti-intellectualism. Anti-intellectualism is truly a cancer on society. It causes nothing but negative effects. Anti-intellectualism and censorship tend to go hand-in-hand. Once anti-intellectual sentiment becomes too strong within a society, free speech is likely to be destroyed, at which point it could even become impossible to ever establish free speech and prosperity again. You can't change people's minds if you're being systematically censored. The fate of humanity literally depends upon defeating anti-intellectualism.

Rationality and the creative spirit are the key to saving humanity from its own stupidity.

Embrace your own inner potential. Not only will it set you free, but it will also protect the freedom of everyone else around you. The more voices that stand up for freedom, the harder it will be for evil to get a stranglehold on society. Don't be a pawn in someone else's game. Learn to think logically and creatively. Use that knowledge to transform your mind so that it becomes like an armored fortress, shielded from malicious foreign influences and misinformation. Carefully cultivate the sacred garden that is your mind. Don't let other people fill it up with garbage. Don't let yourself become imprisoned by your own ignorance and bigotry. Don't let yourself become petrified by your own self-limiting beliefs. The prison we build inside our own minds is the strongest prison of all.

Anyway, I think I've made my point by now. I had to get all that off my chest. For too long, anti-intellectualism has been allowed to permeate our culture and to poison public policy and decision making. I'm sick of how utterly backwards and counterproductive it all is. I'm sick of all the endless finger pointing and polarization. I'm sick of all the baseless prejudice and malicious propaganda against anyone and everyone who chooses to do more with their lives than what the normality police jealously claim is acceptable. It's disgusting. I think it's about time that we made a concerted effort to end anti-intellectualism once and for all. It's time to finally set things right. Anti-intellectualism is poison. We should stop tolerating it.

We need to purge anti-intellectualism from society anywhere and everywhere that it exists. All public officials who hold anti-intellectual or anti-scientific beliefs need to be kicked out of office. Only sane people should ever be given positions of authority. Anti-intellectual behavior in everyday life needs to become just as socially unacceptable as racism or sexism or pedophilia. The normalization of anti-intellectual sentiment and its associated propaganda codewords (e.g. "nerd", "being normal", "regular folk", etc) needs to end.

It's absurd that sneering at intellectual and creative achievements, which literally power almost everything that is great about modern life, has somehow become a popular cultural thing. It makes no sense. Humanity will never be able to reach its full potential while these kinds of rancid sentiments are still a common occurrence. Enough is enough. Humanity has suffered long enough under the rule of morons and lunatics.

It's time for the dawn of a new age of reason. It's time to rise up and toss anti-intellectualism and ignorance into the trash bin of history. The verdict is in. The historical evidence for the toxicity of anti-intellectualism has become overwhelming and indisputable. To save humanity from eventually destroying itself, we must at bare minimum somehow defeat anti-intellectualism. The only rightful place for anti-intellectualism is oblivion. Either we destroy anti-intellectualism or it destroys us, probably.

6.3.3 An introduction to cognitive biases

Anyway, that's the end of my tangential rant. As you may recall, we were previously discussing some of the informal fallacies in an extended numbered list format. After the informal fallacy list, on page 612, I started talking about the state of reasoning in modern society and how ridiculously common the fallacies often are in everyday life. This set off a chain reaction of related remarks that ultimately led to my lengthy tangential rant about anti-intellectualism and the threat it poses to our world.

Getting diverted onto this tangent arguably disrupted the flow of what we were talking about. However, I think it was a tangent worth having. Even if it disrupted the flow of what we were talking about, I still think we got some great value out of it. It needed to be said. Too many introductions to informal fallacies don't really give a proper sense for why learning about them is so important. People should understand the real-world consequences of what they are being taught and why it matters. Doing so helps provide motivation. Conventional "wisdom" about never getting diverted onto free-flowing tangents when writing, while sometimes helpful, is overrated. It's too stifling. Natural conversational honesty and allowing yourself to follow thoughts wherever they may lead is often far more valuable. You could easily end up finding powerful insights that you otherwise never would have found.

I wanted to give you a real sense for what the consequences of allowing bad logic to roam free in society are, and why it should not be treated as carelessly as it currently is. Also, I wanted to do something to help fight against the forces of anti-intellectualism. More people should be speaking out against it. Left unchecked and unopposed anti-intellectualism can become extremely dangerous. As a society, we need to start taking the threat anti-intellectualism poses a lot more seriously. We shouldn't just blindly assume that things will work out based on wishful thinking. There's too much at stake to leave it to chance.

Anyway, in informal logic, in addition to studying informal fallacies we also study something called cognitive biases. Cognitive biases are not quite the same thing as fallacies, but they are closely related. They will be the focus of our next discussion. It is important to understand both informal fallacies and cognitive biases if you ever hope to live a maximally rational life. They both contribute greatly to what makes people so often behave irrationally.

A cognitive bias is usually not itself an error in reasoning, but rather it is a psychological condition which tends to generate irrational behavior. Notice the distinction. A logical fallacy, whether formal or informal, is an actual concrete example of an error in reasoning. Whereas logical fallacies originate from bad argumentation, cognitive biases originate from bad psychology. Whereas fallacies indicate there is a problem with an argument, cognitive biases indicate there is a problem with the logical integrity of someone's mind.

Note however that it is possible for an entire species to universally share a cognitive bias, such that not even a single living member of that species is immune to it. Just because an individual is "normal" or is an "expert" doesn't mean that they don't still have cognitive biases. You should never assume that people's minds have logical integrity by default. In fact, the human mind tends to be inherently biased.

Only tremendous self-discipline and careful thought can ever hope to remedy the effect. People who consider themselves to be infallible are ironically not generally rational people. Believing yourself to be infallible is one of the fastest ways to make yourself dumber. It effectively destroys your mind's error correction and learning mechanisms. The damage to your mind can become irreparable if you remain in this state for too long or if you commit too heavily to your assumptions.

The study of cognitive biases actually originates from psychology, not logic, unlike the study of informal fallacies. This isn't really that surprising considering that psychologists are focused on understanding the mind, whereas logicians are focused on understanding whether any given specific argument is valid or not. In the day-to-day work of a logician or mathematician, they don't generally care much about what the psychological state of the mind of whoever produced the argument was, but rather they generally only care about the substance of the argument itself.

Nonetheless, logicians, mathematicians, and scientists of all kinds would do well to learn about cognitive biases and to at least be aware of the most common ones. You will find that most well-read people in the STEM disciplines have at least a rudimentary awareness of some informal fallacies and cognitive biases. Being aware of these biases helps to keep your mind clean and objective, regardless of what specific subject you study.

It's important to note that for some instances the boundary between fallacy and cognitive bias is not clear. Also, some cognitive biases likely serve useful evolutionary purposes in some contexts and help people cope with overwhelming or limited information. People have limited mental processing capacity and limited time, so it ends up not always being possible to completely think things through. On the other hand, some cognitive biases may have no redeeming value.

Some may just be genetic flaws shared by all humans (perhaps caused by genetic drift) or may just be bad cultural or personal habits. It's a mixed bag. There are actually a huge number of named cognitive biases identified by psychologists. Many are subtypes of broader types. Some may not even be significant enough to deserve a name. Some may be contradictory or wrong. As such, I have selected a small subset of cognitive biases, using importance, broadness, and non-contradiction as my primary selection criteria. Here's my selected list, with accompanying discussions:

6.4 A guided tour of some cognitive biases

6.4.1 Confirmation bias

This one is a classic example. It's one of the most common and most widely known cognitive biases, and it constantly interferes with people's ability to think rationally. It's one of the main reasons why social progress is often so slow, even in the face of overwhelming evidence. It contributes greatly to the formation of social "echo chambers" and separate subcultures within societies that largely don't interact with each other and don't trust each other. What is it?

Well, confirmation bias is the tendency of people to seek out information that confirms their already existing beliefs, while simultaneously avoiding being exposed to any information that would conflict with those same beliefs. That's why it's called *confirmation* bias. People, especially people who aren't trained in science or logic, tend to seek to *confirm* what they already believe instead of looking for evidence which could potentially disprove it. It's a subconscious form of suppression of evidence.

A rational person considers both sides of every argument proportionally to the evidence available for each side, and never avoids or suppresses any evidence that might favor one of those sides. In fact, the core insight that made science possible for humanity was for us as humans to always question our own assumptions and to systematically criticize and attack our own beliefs to try to disprove them. This self-critical and anti-arrogant approach is designed to prevent confirmation bias from distorting our perspective. In contrast, an irrational person is dominated by confirmation bias and will often almost never seek out any information that could potentially conflict with what they already believe. Politics and religion are among the worst sources of confirmation bias in society.

Confirmation bias is partly caused by people caring more about their own ego than about finding the truth. There's a big difference between trying to win an argument and trying to discover the truth. Winning is all about ego. A person whose goal is merely to win will not be concerned with whether or not their position has anything to do with the truth. They are motivated only by a desire to either save face or inflate their social status. In contrast, a person whose goal is to discover the truth is selfless and thus will feel satisfied regardless of which side wins. A truth-seeker is only concerned with whether or not a position is backed by evidence or logic. They don't care *who* wins as long as the truth comes to light.

See the difference? This is why science is so abundantly and consistently morally superior to all other prior belief systems. All other belief systems merely seek to win, whereas science actually seeks the truth. The other belief systems are nothing but arrogance in disguise. Remember this the next time you see an anti-scientific person start huffing and puffing at having their views criticized. Their beliefs aren't anything genuinely special or respectable. It is literally just pure arrogance. Wrapping their beliefs up in a different set of clothes doesn't change that. Think of such people the same way as you would of any other arrogant person. That's all they really are, fundamentally, beneath the surface veneer of unjustified self-righteousness.

Not all beliefs are created equal. Some, like science for example, are enormously beneficial to society. Others, in contrast, have nothing but harmful effects. Some beliefs are mixed bags of both pros and cons. Some beliefs should ideally be destroyed, whereas others (e.g. science) should be protected as passionately as you would protect the life of your own mother. I've said this before but it is so important that it bears repeating yet again: The *only* genuinely reliable basis for any belief is hard evidence or logical proof.

Life is too short to learn everything, so admittedly we sometimes do have to trust experts. However, true as this may be, remember that authority figures are only trustworthy when they have a proven track-record of careful adherence to standards of evidence and logic. When you take steps to protect yourself from confirmation bias, it is essential that you always weigh each side of an issue proportionally to the amount of supporting evidence each side has. Any other behavior will inevitably mislead you and therefore will also endanger the rest of society indirectly via your poor decision making skills.

Here's a tip for protecting yourself from confirmation bias. Whenever you encounter a new claim, always question its integrity from multiple different perspectives. For example, ask yourself: How could this statement be wrong? What kinds of fallacies and biases might be involved? Does someone stand to gain financially or socially by spreading misinformation about this subject? Are they omitting any important information? Can I disprove the claim? Can I think of any counterexamples that would immediately invalidate it? If it contains errors, are they easily corrected, or are they fundamental and irreparable? What are some alternatives to this claim? Do any of those alternative claims seem closer to the truth than the original claim does?

Is the claim even testable[4]? Can extraneous or superficial assumptions involved in the claim be removed in order to make the claim testable? Are there any credible sources of information on the subject matter that might help make it easier to understand the claim? Are the terms in the claim precise enough that the claim has a well-defined and unambiguous meaning? Is it actionable? Anyway, I think you get the idea. The more questions you ask the better. Question everything. That's the motto of science. Science is the ultimate anti-authoritarian philosophy. Science is the spirit of freedom and truth. Through the power of science, dreams can become reality and lies can evaporate at a glance.

6.4.2 Survivorship bias

Let's try a thought experiment. Suppose that you had perfect reasoning skills but not omniscience. In other words, suppose that whenever you see a piece of information in front of you, you are able to reason about it perfectly, but that you only have access to the information that is in your immediate surroundings. Is having perfect reasoning skills sufficient to imply that you are free of all biases? The answer, it may surprise you, is no. Even with perfect reasoning skills,

---

[4]Untestable claims (not to be confused with conditional premises) are always worthless and as such should usually be treated with contempt.

the mere fact that you have limited information access is enough to guarantee that you will be at least partly biased.

You can't reason about what you can't perceive, and therefore you can't always reach the correct conclusions when presented with an arbitrary claim. So you see, it is in fact logically impossible for a human to be unbiased. Note however that this does *not* mean that all claims made by humans contain errors. Tons of claims can still be known with certainty. It just means that a human cannot possibly be unbiased with respect to *all* claims. There will always be at least one claim that each human will have a biased perspective on. Don't let this fact discourage you in the slightest though. Reducing how biased you are is still extremely valuable. Just because you can't reach 100% success towards a goal doesn't mean that it isn't still worth your time to reach 99% of that goal. After all, you would still get 99% of the benefit, which is practically the same.

What was the point of this thought exercise though? How is it relevant to survivorship bias? Well, you see, survivorship bias is all about limited information, especially limited information that is outside of your perception. That's why the thought experiment is relevant. Survivorship bias is an especially tricky form of bias because it is very easy to overlook. Survivorship bias happens whenever someone fails to properly account for the possibility that the information they have received has been filtered by some kind of selective process that has caused the resulting information to not accurately represent the complete system being considered. It is also very closely related to selection bias and sampling bias, which have very similar properties but different connotations. At a deeper conceptual level, the three terms (selection bias, sampling bias, and survivorship bias) may even be essentially synonymous.

One of the most common examples used to introduce people to the concept of survivorship bias is a famous story from World War 2 about how the Navy decided what parts of their bomber airplanes should have their armor reinforced in order to reduce how many would get shot down in combat. The Navy collected together a bunch of bombers that had managed to survive their missions and then analyzed the damage on those planes to try to determine where on the planes they should add more armor. The Navy initially thought that it would be best to reinforce the parts where the most bullet holes on the surviving planes were. However, a statistician named Abraham Wald realized that reinforcing the parts that came back with lots of bullet holes in them would not be very effective at reducing the kill rate. Can you see why?

What Abraham Wald suggested was to instead reinforce the armor everywhere where there were *not* many bullet holes on the planes. You see, the sample of bomber planes that the Navy had gathered consisted entirely of planes that had *survived* combat. Thus, counterintuitively, the parts of the planes that had the most bullet holes were actually the least important parts of the planes to protect. Planes that were shot in locations other than where the surviving planes had been shot were the ones that got killed.

Thus, the correct logical solution was actually to reinforce the parts of the planes that were *not* coming back with many bullet holes on them. The Navy's first plan to reinforce the parts that had more bullet holes in them was an example of survivorship bias. They were assuming that the sample represented the complete system when in reality the sample only represented a very specific subset of the system with special properties. That's why the first plan was wrong.

This might sound like some kind of obscure scenario, and you might think that it is unlikely that you'll ever have to deal with something like it, but I wouldn't be so sure if I were you. This is not an isolated issue. Survivorship bias is abundantly common in everyday life. For example, have you ever felt like a disproportionately large proportion of high quality things seem to come from the past? Does it sometimes feel like music, architecture, art, or other products from the past have a higher average quality and durability than the average modern-day products have?

The reason for that is survivorship bias. As society moves forward, the things from the past that society is interested enough in to bother keeping around (or which are durable enough to survive) are the highest quality things. More of those higher quality things survive than of lower quality things. This creates a false impression that people in the past cared more about the quality of their work than in the present day. In reality, average quality probably tends to increase gradually over time as humanity becomes more knowledgeable and well informed, although not always and not in all respects.

Let's do another one. Take relationships for example. Each person has a set of all the people they've dated. You've probably heard someone say something like "All men/women are jerks." at some point in your life. The people who say this have presumably based this primarily on their experience with the people they've dated. The conclusion is of course invalid. If the person making this claim was more intellectually honest then they might realize that the correct statement is actually "All men/women *that I've dated* are jerks." which is quite a different statement.

Leaving out the "that I've dated" part of the statement introduces survivorship bias into it. Being aware of this might help the person solve their dating problems more effectively. Consistently meeting jerks may be indicative of a systematic error in how you date. It doesn't mean you're a jerk yourself necessarily, it just means that you at least have the wrong approach. Just because you are the common denominator in the bad things you experience doesn't actually merit concluding that you are the cause, contrary to popular belief among many high-strung and presumptuous people (e.g. among lots of toxic people on social media and internet forums), but it is still a possibility to consider. Many other kinds of systematic factors (e.g. bad dating culture, unintentionally attracting the wrong people, poor communication skills, desperation, etc) could very easily be responsible for the effect though.

Survivorship bias also afflicts even the most well-intentioned and well-educated institutions. Scientific research is often plagued by a particularly insidious type of survivorship bias known as *publication bias*. What's publication bias? Well, publication bias is the tendency of some researchers to mostly only publish studies that had interesting or positive results, while disregarding any related studies that had no result or even negative results. In such cases, if the sample size used in the research isn't sufficiently large to prevent the effect, then there will be a strong chance that the results are misleading.

By not publishing results when they are neutral or negative, survivorship bias is introduced into the scientific literature, thereby potentially distorting the truth. For that reason, many scientific journals have now put new restrictions on studies, designed to reduce this effect, such as requiring scientists to publish their studies regardless of the outcome, among other things. These preventive measures may help some, but at least a small to moderate amount of survivorship bias is essentially unavoidable.

6.4.3 Observer-expectancy effect

Our expectations alter our behavior and hence also alter our reality. When we expect something to go a certain way, that expectation can inadvertently cause us to behave in a way that alters the outcome. The chances of the outcome we expect could increase or decrease. The nature of the outcome could change. Either way though, our expectations for how the system should behave can lead us to do things that could end up tainting our observations. We may think that we are acting as a neutral independent observer, when in reality we are influencing the system and thus acting as a part of it. This is known as the observer-expectancy effect.

This kind of bias is likely to be especially common in fields of science that depend heavily on the use of statistics and that perform numerous direct experiments involving human beings or animals. The social sciences and medical sciences are especially vulnerable to the observer-expectancy effect and must be exceptionally careful and vigilant in order to prevent it. Why are the social science and medical science in particular so vulnerable?

Well, for example, one contributing reason is because human beings and animals are very good at picking up on subtle cues for what a researcher wants or expects. This can cause the subject to alter its behavior based on what they think the experiment's intent is. Some subjects will conform to the expectations. Others may deliberately try to do the opposite of whatever they think is expected. Some may change their behavior to match whatever seems most socially acceptable at the time. They will do this regardless of what the truth is, thus tainting the result.

For example, a researcher might phrase survey questions in a way that subtly (or not so subtly) pushes the recipients towards specific answers. People often give away their feelings through non-verbal cues, and many people in society are adept at picking up on these cues and altering their behavior accordingly in order to maintain their social status. Thus, for example, if a survey recipient sees a researcher get tense when the survey recipient is about to give a certain answer then the survey recipient might decide to change their answer at the last moment.

The same is true if whenever the researcher describes the different survey options their tone changes subtly to reflect their opinion of each option. For example, the researchers voice might speed up, take on a dull or dismissive quality, or even become passive aggressively angry. Questions could also be worded in such a way as to imply what the "best answers" the asker thinks should be. People don't like confrontation, so there is a strong chance that they may conform to the researcher's feelings just to feel more comfortable.

Observer-expectancy bias is one of the main reasons why the concept of "double blind experiments" was invented. In a double blind experiment, neither the observer nor the subject is allowed to know much about the other and none of the subjects know whether they are in the control group or the test group for an effect. When done properly, care should also be taken to ensure that the subject never has a chance to see the observer's non-verbal responses and the wording of questions should be carefully screened to remove any loaded language. In this way, most of the observer-expectancy bias in an experiment can potentially be prevented, thus increasing the reliability and scientific integrity of the results. This

In a more informal context, the observer-expectancy effect also occurs regularly in daily life in the way that people treat each other. When person *A* expects that person *B* will be an enemy, this will often cause person *A* to treat person *B* as such in advance. This in turn makes it much more likely that person *B* will respond with hostility towards person *A*, regardless of whether or not person *B* ever actually originally had any ill will towards person *A*.

Few things anger people faster than being treated like an enemy for no apparent reason. In effect, it creates a self-fulfilling prophecy. The more you treat other people with hostility, the more genuinely hostile towards you they will tend to become. If you think someone hates you, without sufficient evidence, then it probably won't be long before they actually *do* hate you. In effect, you are digging your own grave. This mutual cycle of self-perpetuating hate is also probably partly responsible for a lot of the polarization and tribalism that afflicts society.

The observer-expectancy effect is a psychological phenomenon, but it is nonetheless fairly closely related to a more fundamental phenomenon in physics known as the *observer effect*. In physics, the observer effect refers to the fact that in order to observe and measure a system you generally have to interact with it in at least some way, but that very act of interacting with the system changes the state of the system and thus also changes the results of the observation relative to what the result would be if you weren't observing it. Take eyesight for example. You may think that you are not interacting with an object when you are just looking at it, but that actually isn't true.

Eyesight is based on photons colliding with the lens of your eye. There is also a chance that the photon will bounce off your eye and back towards the thing you are observing. If it then collides with that object then it could change that object's state ever so slightly. Thus, it is in effect impossible to even look at something without creating the possibility that you are interacting with it. The same is true of many other systems. Many things are impossible to measure unless you deliberately introduce a change to the system designed to produce the signal that enables you to measure it.

It is thus virtually impossible to observe a system without influencing it at least slightly. For large-scale systems this effect is often negligible, but for small-scale systems, like atomic or subatomic systems, the observer effect becomes an increasingly influential and headache-inducing problem. This is why measurement at the atomic or subatomic level is inherently difficult and uncertain in physics.

6.4.4 Pareidolia

Have you ever seen something that wasn't really there? For example, have you ever seen an inanimate object that looks like it has a face but on closer inspection it just has a few superficial characteristics that faces have that tricked your eyes? Have you ever seen a shadow across a wall that looked like a hand or a person, only to find that it wasn't? Have you ever thought you heard footsteps behind you but when you looked back there was nobody there and nowhere that the sound could have come from? If so, then you have experienced pareidolia. Pareidolia is essentially pattern recognition gone haywire. It causes you to perceive identifiable entities and patterns that aren't actually there.

Pareidolia is a consequence of the aggressive nature of our pattern recognition capabilities. From an evolutionary survival standpoint, it is usually more useful to perceive a pattern that isn't there than to miss a pattern that is there. Being jumpy and paranoid and overly sensitive to stimuli increases your chances of not being eaten alive by a wild animal. Overactive pattern recognition is also a critical component of what enables human beings to be so smart. It causes humans to explore a much larger number of possible explanations for things than they otherwise would.

Pareidolia also helps us to see connections between things so that we can discover hidden insights into how they relate to each other on a deeper level. In general, pareidolia is a useful mechanism, but sometimes it can become maladaptive. Pareidolia also plays an important role in our ability to appreciate art and entertainment, in that it greatly increases our tendency to spot interesting thematic elements in things and also makes our creative perception more flexible and adaptable. In that sense, it is actually a good thing that we have it, so long as it is reasonably controlled.

If I had to guess, I would think that some conspiracy theorists probably have a higher tendency towards pareidolia than most other people do. In fact, I'm no psychologist, but it sounds like that might make a good psychology experiment. You could design a test to see if believing in conspiracy theories is correlated with higher rates of pareidolia. Feel free to take that idea and use it as your own, by the way.

Conspiracy theorists often come up with ludicrously complicated and bizarre explanations for events. They have a strong tendency to see lots of patterns and connections between things that probably don't actually exist. I suspect that much of conspiracy theorists' behavior actually originates from some combination of fitting fallacies, pareidolia, confirmation bias, and a desire to save face. If one could somehow eliminate those four mental problems, then I think it would have a high chance of eliminating many conspiracy theorists' paranoid and irrational behavior patterns. It seems like a reasonable theory.

6.4.5 Subjective validation

Everybody wants to find meaning in their lives. Some people want it so much that they'll find it even when it isn't there. That's the essence of subjective validation: finding special meaning where it does not actually exist. It's the tendency to think that if a property is applicable to something, then it must surely be meaningful rather than merely coincidental. Subjective validation causes you to see what you want to see, instead of what is actually there. It means you fill up the world with meaning and rationalize what you see even when you do not actually have sufficient justification to do so.

Some people can't stand not having a "meaningful" explanation for everything they see, so they grab the first seemingly meaningful explanation that fits and then prematurely assume that it is true, instead of taking the much more intellectually honest approach of learning to be comfortable with uncertainty[5]. There are lots of different contexts in which subjective validation occurs. It's quite common. Let's do an example.

Suppose that an archer launches an arrow at a target and the arrow hits the exact center of that target. Suppose you also know nothing about this archer except what you have just witnessed. What is your first instinct on how to interpret this? Is it to assume that the archer is exceptionally skilled with the bow and did it intentionally? Or is it to assume that it was just a random fluke? The correct answer is that you should assume neither.

The skill level of the archer is *impossible* to determine from this scenario without more information. Neither explanation is necessarily any more likely than the other. If your instinct was to assume the archer was exceptionally skilled, then congratulations on being guilty of subjective validation. You picked the more meaningful interpretation of what you saw without actually having sufficient justification to do so.

Here's another one in a similar vein. Suppose you're driving down a road in a car with your friend. Along the way to your destination you pass ten different cars, all with different license plates. At one point you pass by a car that has "NEWTON" written on the license plate. All the other cars along the way have standard issue randomized license plates. Once you arrive at your destination your friend turns to you and says "Hey, did you see that Newton license plate? What an unlikely license plate! It's way more unique and way less likely to see than those other plates we saw." as the two of you get out of the car and begin walking down a nearby sidewalk.

Do you see a problem here? Was the Newton license plate actually any more unlikely or any more unique than the other license plates were? The answer is no. It only *seems* like it was because of subjective validation. All of the license plates were equally unique and equally unlikely to be encountered. It is just as much of a statistical anomaly to run into a standard license

---

[5]Incidentally, if it weren't for subjective validation then there's a fair chance that maybe religion wouldn't even exist. If people were more intellectually honest and emotionally mature, then they would be much more comfortable with uncertainty and unknowns, and thus there would be little to no motivation for anyone to prematurely assume that some deity created everything or that any kind of afterlife exists.

plate labeled "XETW-9382" as it is to run into a custom license plate labeled "NEWTON". Neither is more likely than the other. They are both arbitrary sequences of symbols that could have any meaning, or none at all.

It only seemed like a remarkable event to us because "NEWTON" has special meaning to us and "XETW-9382" does not. Note however that this argument depends on the fact that our friend is talking about individual license plates rather than groups of them. Custom license plates *are* more uncommon than standard license plates, but that's not what our friend actually said. If our friend had instead said "Hey, did you see that custom license plate? Those things sure are less common than standard license plates." then the statement would be statistically justified, unlike the previous statement.

I actually originally got the idea for this license plate example from a story I heard once about a physics professor who one day came into his class and (so the story goes) started talking excitedly about seeing a standard license plate with some specific label, something like "XETW-9382" or whatever, to the great bafflement of his students. His students couldn't figure out why he thought this was such a remarkable event. The professor intended it as a thought exercise to teach the students an important lesson about how statistics work, and how such a license plate would be just as statistically unique as any custom license plate would be. I forget where the story came from and who the physics professor was.

Anyway, let's do another example. One of the most abundant sources of subjective validation in society is from so called "fortune tellers", "psychics", "mediums", "spiritual healers", and "astrologers" (etc). All of the people who work in these "jobs" are either con artists or crazy. Most of them are probably con artists though. Don't be naive. Most of them are probably aware that they are exploiting other people unethically and just don't care. It takes truly monumental insanity to do that kind of "work" without ever realizing that you don't actually possess the supposed powers that you pretend to have.

Personality tests are another example. Personality tests often rely on subjective validation to make people believe the test is more accurate than it really is. How? Well, the most common technique is to make very broad statements that could be true of almost anybody. In the case of fortune tellers and psychics (etc), in addition to making vague statements they also observe your outwardly visible traits to estimate things about you, among other techniques. A little bit of knowledge can go a long way. People do and say things that reveal a lot more information about themselves than they realize, and with a bit of deductive reasoning and worldly knowledge it can sometimes be roughly estimated.

For example, certain words are only used by certain professions, and a person's education level and socio-economic status also tend to influence word choice and communication style significantly. There are lots of seemingly unimportant things that actually tell you a lot about someone's background. All of those things increase the opportunity for exploitation. Fraud is the name of the game. These con artists know that people will tend to disregard occasional prediction failures and tend to not be observant enough to realize that the information is actually either so vague as to apply to almost anyone or is simply being gathered from subtle cues.

Making broad statements that would be true of most people is such a common technique among con artists that there is an official term for the psychological effect that it depends on. The effect is arguably a specific subtype of subjective validation. It is referred to synonymously as either the Forer effect or the Barnum effect. However, it occurs to me that there are really two distinct things involved in this and that they perhaps deserve separate terms instead of being conflated with each other. On the one hand we have the deceptively broad statements themselves, and on the other hand we have the psychological flaw that it exploits. These are two different things and thus ideally deserve two different names. How should we pick those names though?

Well, as it happens, there is a convenient difference between Forer and Barnum that makes a good basis for deciding this. You see, Forer was a psychologist who focused on studying the psychological effect involved in these kinds of statements, whereas Barnum was a hoaxer and entertainer who was fond of making such statements and misleading people for profit. It is therefore fitting to associate the name Forer with the psychological effect, on the one hand, and the name Barnum with the deceptively broad statements, on the other hand.

Thus, a better choice of terminology than merely using "Forer effect" and "Barnum effect" as synonyms would be to refer to the psychological effect as the *Forer effect* and to refer to the deceptively broad statements as *Barnum statements*. This eliminates the pointless redundancy between the two terms and is more useful. It also associates each person with the part of the phenomenon that they more deserve to be associated with.

Let's talk about one of Forer's famous experiments, where he demonstrated just how incredibly easy it is to trick people with Barnum statements. Forer was a psychology professor at the time and he gave his psychology students a personality test. The students filled out the personality tests and turned them in. Forer said that a week later he would give them back the results of their tests, once he had time to look over them. When he gave the students back their results, he asked them to rate how accurately they each thought the results described their personality on a scale from 0 to 5. The average accuracy rating the students gave was 4.26 out of 5, which in relative terms is equivalent to an accuracy rating of 85.2%.

The catch is that all of the personality test results that Forer gave back to each of his students were identical. Regardless of what each student wrote on their personality test, the professor gave back the exact same results to everyone. The one week waiting period was just to make it seem like he really was analyzing the tests to determine the results. He could have printed out the results before he even administered the test and it would have made no difference. The reason why the test was able to receive such a high accuracy rating was because the test results consisted of nothing but Barnum statements. Here is the list of personality trait results he gave back to all of the students[6]:

1. You have a great need for other people to like and admire you.

2. You have a tendency to be critical of yourself.

3. You have a great deal of unused capacity which you have not turned to your advantage.

4. While you have some personality weaknesses, you are generally able to compensate for them.

5. Disciplined and self-controlled outside, you tend to be worrisome and insecure inside.

6. At times you have serious doubts as to whether you have made the right decision or done the right thing.

7. You prefer a certain amount of change and variety and become dissatisfied when hemmed in by restrictions and limitations.

8. You pride yourself as an independent thinker and do not accept others' statements without satisfactory proof.

---

[6]The original paper, which includes this list, is entitled *The Fallacy of Personal Validation: A Classroom Demonstration of Gullibility*. Forer's full name is Bertram R. Forer. A full PDF of the paper can be easily found on the internet.

9. You have found it unwise to be too frank in revealing yourself to others.

10. At times you are extroverted, affable, sociable, while at other times you are introverted, wary, reserved.

11. Some of your aspirations tend to be pretty unrealistic.

12. Security is one of your major goals in life.

Sound familiar? Sounds like every personality test result ever, right? That's how they work. They rely on people's tendency to think of themselves as uniquely special and to automatically ascribe meaning to things that could just as easily be purely coincidental or merely overly broad. It's kind of eye opening though that almost everyone has exactly this same set of problems, isn't it? It helps put things in perspective.

And remember, these were *psychology students* we're talking about, not just random people. These students were trained in how to understand the way the mind works, and they still didn't catch on to the fact that they were being psychologically manipulated. Admittedly though, Forer did this experiment in 1948, so psychology students are probably more widely aware of Barnum statements these days than back then.

Now, you might be thinking to yourself "Well, that's all well and good, but I never really put much trust in personality tests anyway, so I don't really have much to worry about. Right?" or something to that effect. Unfortunately, no. You're not off the hook. Even if you're wise enough to stay away from this kind of stuff, you're still not even remotely free from the dangers of subjective validation. This kind of thinking is hardly limited to personality tests and charlatans. In fact, the entire criminal justice system is also highly susceptible to subjective validation bias.

Let's talk about criminal profiling. For those unfamiliar with it, criminal profiling is when the police try to formulate an estimate of what a criminal is like based on crime scene behavior and certain other factors. For example, if the crime scene has objects thrown around everywhere in a mess then the police might estimate that the criminal responsible probably has a disorganized personality. This is textbook subjective validation. It's just like our earlier examples: the archer example and the license plate example. The investigators are ascribing meaning to something that could just as likely be coincidental. There are countless other possibilities, besides just having a messy personality, that could have caused the crime scene to end up that way.

The investigators subconsciously *want* the most meaningful explanation to be true so they prematurely lean towards it without adequate justification. The idea that the mess is inherently a deep part of the criminal's psyche is just so much more juicy and satisfying than the innumerable other much more mundane possibilities. Subjective validation bias is sort of like behaving as if your life is a movie and you're the director, but you can't control the script, so you just try to pretend that it was meaningful, after the fact, in order to make yourself feel better. Got to keep the *Big Meaningful Story* alive, am I right? Just can't allow the possibility that the movie reel of your life is actually more like the *Coincidental Highly Randomized Blob*, can you?

My point is, many people are *desperate* to feel like life is inherently meaningful, and that desire can sometimes bleed over into their reasoning processes, causing them to fill their perceived world with more meaning and intention than is actually there. Don't get me wrong though. I'm not trying to give you an existential crisis, nor would I want you to think this is a good reason to get depressed. It's not. You can have as much meaning in your life as you want, it's just that you need to create it within yourself via what you choose to do with your life, instead of merely by trying to assign disproportionate meaning to things that aren't even under your control. When you do that instead, I think you will find it much easier to find peace and contentment in life.

Criminal profiles are apparently often wildly incorrect. In fact, in a study of criminal profiling, the way the police responded to criminal profiles followed approximately the same pattern of behavior as how people respond to personality tests and astrology[7]. Yep, that's right. The police apparently respond to criminal profiles more or less the same way that teenage girls respond to their horoscopes. These decisions could potentially end up effecting who gets sent to prison, who gets the death sentence, and who doesn't (etc). Creepy, right?

Society is dangerous. What many people call "justice" would often more accurately be described as mob rule. The best way to deal with the legal system is to avoid anything that increases your chances of ever having to interact with it, except when absolutely necessary. Criminal justice often operates primarily based on whatever the current prejudices of society are, not based on empathy and truth. The real world is full of color and nuance, but the criminal justice system is far too often black-and-white and narrow-minded. It sometimes lacks even basic intellectual integrity and honesty. An honest person would admit that *anybody* could easily experience a moment of rage or poor judgment and make a terrible mistake, but not the criminal justice system apparently, much of the time.

The criminal justice system instead usually seems to prefer to demonize people so that it can feed into the self-congratulatory *us vs them* narrative of society, wherein members of the general public tend to want to think of themselves as perfect little angels, but tend to want to think of the accused as being somehow inherently bad or broken in some sense. Such crude thinking is not very helpful. Empathy is a more productive way to change behavior than vindictiveness.

People should stop being so blindly self-righteous about these kinds of things. We need to acknowledge that the potential for evil exists within *all* of us, instead of demonizing only specific groups of people. You can't fix evil if you aren't willing to even acknowledge what it really is. The true nature of the problem only becomes apparent once you stop automatically attaching "that person is inherently bad" labels to anyone who ever does anything that you don't like or that you don't understand. People are often far too preoccupied with engaging in social posturing designed to make themselves always look like "one of the good guys" whenever anything bad happens, rather than taking the time to look deep within themselves to see that the same evil exists within all people and to genuinely own it and try to defeat it.

People can't fix problems with their own behavior that they won't take ownership of, and evil is something that lives inside all of us. Evil just needs the right combination of toxic social ingredients and psychological irritants in order to manifest itself. This is true of pretty much any person. History has shown that repeatedly. When such a time comes, when evil arises within yourself, you will not be able to protect yourself from it unless you are willing to acknowledge your own capacity for it. You must own your evil if you have any hope of ever defeating it. Evil can only be chained by self-awareness and empathy. Feelings of righteousness and conviction, in contrast, often tend to just make evil stronger[8], especially if those feelings are not treated skeptically and carefully.

One can see this kind of thing happen with great frequency on TV, where every single time anyone does anything bad every single person on TV pretends to not have even the faintest

---

[7] See (for example) *skepdic.com/profiling.html* for more information. It includes some relevant citations. Criminal justice institutions aren't as reliable as the average person thinks they are. The way they operate is arguably often more about creating a social narrative than it is about real justice.

[8] By the way, similarly, disassociating yourself from something you did by saying "I'm not the kind of person who would ever do that" after you did it, generally indicates that you will actually learn absolutely nothing from the experience. It means that you are refusing to take ownership of your own actions and are instead choosing to live in your own little bubble world where you will continue to treat yourself as a perfect person. Such "apologies" are seldom genuine, and will predictably tend to be followed by repeated behavior and little to no personal growth. Personal growth requires that you take ownership of your own actions.

idea how someone could ever do such a thing, even though every single one of them actually knows deep down that they too have almost certainly had vindictive and unethical thoughts about other people at least a few times in their own lives. Everybody is always too busy wearing their perfect little angel masks to ever actually be able to address the real underlying cause of evil. Self-righteousness is the fortress home of evil, the place that it always returns to to protect itself from ever truly being defeated, and the ashes from which it is perpetually reborn. Self-righteousness blinds us to empathy and compassion.

Oh, and while we're still on the subject of criminal justice, let's take a moment to talk about "lie detectors" (a.k.a. polygraphs). These things are maybe even worse than criminal profiling potentially. Such devices can easily result in injustice and human rights violations. Some governments may even sometimes use things like "lie detectors" and "psychological evaluations" (etc) as a bludgeon to systematically persecute their enemies without evidence.

It provides great cover. Society has a tendency to automatically be dismissive of anybody who has ever been labeled as a criminal or crazy person, thus creating the perfect cover by making it so that nobody ever takes the victim's experiences seriously. I'd guess that this is especially true in less democratic governments probably, but even in the more rational democracies these unethical techniques for forcing false convictions and oppressing dissent may still sometimes be used potentially.

Anyway, let's refocus on the "lie detector" itself. What exactly does a "lie detector" actually do? Well, apparently, basically it simultaneously measures several different physiological states, such as heart rate, breathing, blood pressure, and the electrical conductivity of skin. In this way, it can sometimes provide a rough measurement of how stressed out or tense the person wearing the detector is.

The idea behind using it for detecting lies seems to be based on the intuitive "common sense" assumption that people become more stressed out and tense whenever they lie, and that therefore (fallaciously) if someone is stressed out or tense during a question then they're probably lying. However, the only thing that the machine actually measures accurately are the physiological parameters themselves (i.e. heart rate, breathing, blood pressure, and the electrical conductivity of skin) and nothing else. Making any inferences beyond that is not safe. Countless different unrelated things can cause each of the physiological parameters to change, most of which have absolutely nothing to do with lying.

All the interrogator apparently has to do to make the needle spike on a "lie detector" is to simply stress you out enough to make your heart rate, breathing, blood pressure, or skin conductivity change. That's pretty much all there is to it. The correlation to lying is seriously that unreliable and weak. The test results are just speculation essentially. If the interrogator wants to, they could potentially trigger the spikes themselves intentionally in order to increase the chances that you will get convicted. For example, suppose that the interrogator takes one look at you when you walk in and just decides that they "know" intuitively that you're guilty, perhaps based subconsciously on your ethnicity, physical appearance, beliefs, reputation, or mannerisms (etc). They also might personally dislike you for some reason other than appearance.

Think you can pass the test just by staying calm? Think again. Even if you manage to stay completely calm about the questions themselves while being interrogated, there are still numerous different ways that your "lie detector" reading could still randomly spike. Need to use the bathroom? Potential polygraph spike. Feeling a bit attracted to your interrogator? Potential polygraph spike. Feeling hungry? Potential polygraph spike. Anxious about what you're making for dinner tonight? Potential polygraph spike. Realized you forgot something at work? Potential polygraph spike. Small interrogation room triggering your claustrophobia? Potential polygraph spike. Worried about whether you'll be detected as lying even if you tell the truth? Potential polygraph spike. Random fluctuation in your physiology caused by literally nothing at all? Potential polygraph spike. Does the brand of soda the interrogator is drinking remind you of your dead best friend? Potential polygraph spike.

Get the idea? Could this test possibly be any more unreliable or unethical? Whether or not you survive a "lie detector" session is potentially mostly determined by a combination of your interrogator's attitude towards you and randomized environmental circumstances, not by whether you actually lie or not. Welcome to the horror show that is human society. We serve injustice bagels and bigotry juice every morning from 7:00 am to 10:00 am in the lobby. Whether or not your life goes well could easily be determined primarily by a combination of random chance and how bad the people around you are at thinking. Please enjoy your stay at Humanity Hotel. Check-out times are determined randomly, alarm clocks will intermittently fail, and you will not be informed in advance. PS: We know that you stole the miniature soap bars... and the pen from the front desk... and the towels from the community bathroom... and half of the waffles. Please change your name to Richard so that we can start using a more fitting nickname for you from now on. Thank you, and have a lovely day.

Anyway though, if you stop and think about it on a deeper level, "lie detectors" seem to say very little about the person they are used on, but quite a lot about those who do the interrogations and those in society that support their use. The great irony of "lie detectors" is that they apparently don't actually even measure the honesty of the person being interrogated, but rather they measure the barbarism and ignorance of society.

In a sense, at worst "lie detectors" merely monitor how belligerent the interrogators are and how skilled they are at badgering people, and at best they provide a convenient method by which to find someone guilty based on nothing more than irrelevant circumstantial physiological factors. They mostly just make it easier to carry out witch-hunts and show trials, in essence. As such, these unethical devices deserve a new name, a more fitting name. I suggest that "lie detectors" should be renamed to "witch-hunt seismographs". That would seem to be a much more accurate description of what they actually are. What they measure is subjective validation, not lies. Just as seismographs detect earthquakes, polygraphs detect witch-hunts.

6.4.6 Denomination effect

People generally aren't born with good number sense. What's number sense you may ask? Well, number sense is the ability of a person to sense the size of numbers, rather than to merely treat numbers as abstractions for calculation or designation. Being good at arithmetic or logic doesn't necessarily imply that you have good number sense. Number sense requires that you *feel* the size of the numbers. For example, if someone tells you they have 1 million marbles, do you have a good sense for how large that actually is? If so, then that's a point in your favor for having good number sense. On the other hand, if your only sense for the relative size of, say, 1 million and 1 billion is that they are both vaguely "really big" then you have flawed number sense.

Can you picture the size of numbers in your head? Don't worry if you can't. I can't either usually. However, there are ways to improve. I personally don't practice number sense much. Like arithmetic, it's just not something I care much about. However, I do know a simple technique that makes it easier to improve your number sense skills over time. It can also be used to temporarily raise your number sense on an as-needed basis if you don't want to commit to regular practice (like me). Here's the technique:

Whenever you encounter a number whose size you can't sense, try reframing it as a physical analogy or comparison designed to make it possible for you to visualize the quantity all at once. Do this a bunch of different ways until the number no longer feels like merely an abstraction. The more physical analogies you apply to the numbers you encounter, the better your number sense will gradually become. Relative magnitude comparisons designed to make the scale manageable are the key to doing this effectively. For example, if a number is really big then try reframing it in terms of something really small, like grains of sand for example, and then ask yourself how big of a volume it would cover and compare that to real life and visualize it.

Why all this talk of number sense though? What does this have to do with cognitive biases? Well, as it happens, lack of number sense is one of the main contributing factors involved in the *denomination effect*. What's the denomination effect? Well, the denomination effect is a cognitive bias that causes people to perceive the value of money differently based on how many pieces the total amount has been broken up into and how small or large each of those pieces are. This bias actually could apply equally well to any arbitrary quantity, not just money, but it was originally coined to refer specifically to money. I personally think the term should be broadened, but from what I can tell the term is currently mostly used in the context of money.

Anyway, more specifically, the denomination effect causes people to underestimate how quickly small amounts of money add up to large amounts of money. For example, a person will often be more likely to spend $20 if it is broken up into lots of small purchases of $1 or $2 compared to if it is broken up into a few bigger purchases of say $5, $10, or $20, even though the amount of money is the same. A perfectly rational person would not behave this way. A perfectly rational person would instead calmly gauge the relative value of the money, anticipating that making lots of small purchases would quickly add up to become a much larger expense, and would budget accordingly.

The essence of this bias (the denomination effect) stems from the average person's inability to accurately understand how quickly numbers scale up, in large part due to a lack of good number sense. The thought that usually goes through people's head is something along the lines of "Oh, it's just X amount. That's hardly anything!" causing them to end up treating the amount as negligible in their mental accounting process when they shouldn't. Even small amounts of money are vastly more than zero, but many people nonetheless treat virtually all small purchases as pretty much costing nothing in their minds, thereby resulting in huge miscalculations in budgeting. People often have a terrible sense of scale.

Many companies love to exploit this effect in order to get people to spend more money without realizing it. Many subscription programs, product payment plans, and microtransaction systems are based on it. Dividing a large amount of money over a longer timespan allows companies to create the illusion that something is much more affordable than it actually is. For example, suppose a salesperson was trying to sell you a piece of furniture and offered to give you a payment plan where you could pay 69 cents per hour for 3 months.

A gullible person with poor number sense might instinctively think that 69 cents per hour sounds very affordable. After all, 69 cents is hardly anything! It's negligible, right? Wrong. In fact, 69 cents per hour for 3 months adds up to $1500 by the time its all over. That's not even the worst of it though. This is actually a best-case scenario. We've been talking about a *linear* payment plan here. Many payment plans (and almost all loans) are actually based on *compound interest* which means that the amount you owe will grow *exponentially* over time instead of at a constant rate.

Let me give you another linear example to put things in perspective more though. Do you spend $10 or more per day on little expenses? For example, do you drink coffee or eat out or drive a car or leave the lights on or appliances running regularly? That $10 extra per day may not seem like much, but it'll add up quickly over a long enough span of time. Over a decade, $10 per day in expenses adds up to $36,500, which is roughly enough to pay for most of a child's entire in-state college education at an average public college in 2017, or 73 $500 video game consoles, or 7 $5,000 used cars, or 24,333 pounds of apples at $1.5 per pound. How many starving people could you feed with 24,333 pounds of apples, and for how long relative to the number of people? Just something to think about.

My point is, small numbers really do add up a lot faster than you realize. It is wise to be consciously aware of this. Physical analogies are tremendously helpful for clarifying how much a given amount of money is actually worth. It's not just a number. Physical analogies help give you a tangible perspective, so that you can make wiser decisions about how you use your resources. In fact, you may even find it helpful to always think of monetary amounts in terms of some common commodity that you personally value highly instead of merely in terms of money. Thus, for example, if you're into paperback novels, instead of thinking about your bank account balance as "X dollars" you could think of it as "Y paperback novels".

That way, every time you buy something, you'll think "I'm giving up n paperback novels by buying this" instead of thinking "I'm giving up m units of an abstract economic number." and you might find that this significantly improves your judgment in making purchasing decisions. Doing so gives you a much better sense for what the real value of your money is and how wealthy or poor you really are. In any case though, always remember to keep in mind how fast numbers scale and don't rush into any purchasing decisions, even if the up-front cost seems low.

Take the time to calculate how many of several different alternative types of items you could hypothetically buy with the same money and what the total cost over time will be. Also, in general, you should be extremely skeptical whenever someone is trying to get you to make a purchase decision (or any other significant decision) quickly. That's often a sign that the other person doesn't have your best interests in mind. When people try to rush you or to play to your fears it very often means that they are trying to exploit you and are hoping that you will make a bad decision.

6.4.7 Dunning-Kruger effect and impostor syndrome (together: the DKI spectrum)

Have you ever met someone really talented who, no matter how amazing of a job they do creating something and no matter how much they accomplish, never seems to feel satisfied by any of it and is dismissive of all praise and compliments they receive? That's the **impostor syndrome**. Conversely though, have you ever met someone who doesn't have the faintest clue what they're doing yet still firmly believes that they are really good at doing it, someone who constantly misinterprets or misunderstands even the most basic principles of something, yet acts as if they are far better than they actually are? That's the **Dunning-Kruger effect**. However, theses two cognitive biases (the impostor syndrome and the Dunning-Kruger effect) are really just two sides of the same coin though.

These two biases really exist on a spectrum, spanning from people with low skill on the one hand to people with high skill on the other hand, corresponding to any given skill set or body of knowledge. As such, for ease of reference and to unify our terminology some, let's refer to this entire spectrum collectively as the **Dunning-Kruger-impostor spectrum** (or **DKI spectrum** for short) and likewise to the underlying psychology that causes both biases as the

Dunning-Kruger-impostor effect (or **DKI effect** for short). I made up these combined terms myself.

In the case of people with low skill, the Dunning-Kruger effect predicts that they will tend to overestimate their own skill level and to generally act disproportionately confident. In contrast, in the case of people with high skill, the impostor syndrome predicts that people will tend to underestimate their own skill level and generally act disproportionately humble and insecure. In a nutshell: The more you know, the more you know how much you don't know. Whereas, the less you know, the less you know how much you don't know.

In other words, the DKI effect is a bimodal cognitive bias. It has a certain special duality, or two-faced quality, sort of like a coin or like the yin-yang symbol from Chinese philosophy. The effect isn't universal though. It tends to afflict some people more than others. Personally, I think that the reason why it afflicts some people more than others might be related to the afflicted people perhaps having some kind of underlying identity anxiety or self-esteem issues in both cases (i.e. on *both* sides of the spectrum), but this is just a guess.

One of the most common examples of this effect in action, on the high end, in my experience, is the plight of creative people who fit the "starving artist" mold. The relentless hunger to create something truly compelling, combined with the enormous scope of an artist's imagination and also the frequent economic difficulties of art, together creates the perfect breeding ground for impossibly high standards and restlessness. On the other hand, much less experienced artists, who have not yet banged face first into the inherent limitations and challenges of artistic self-expression, are perhaps more likely to not yet understand the nuance and scope of what they will have to deal with to really make an impact.

This inevitably will lead to some of them prematurely acting as if their work is more insightful or skillful than it actually is, which will probably be perceived as pretentiousness or arrogance. In that case, it may be an example of the Dunning-Kruger effect (the low skill side). In contrast though, artists who can never take a compliment and are never satisfied with their past work are more likely to be on the impostor syndrome side (the high skill side). Regardless though, more generally, the DKI effect applies to any profession, but some seem to be more prone to it than others.

On the high skill end of the spectrum, high levels of creative demands and limitless potential and ambition probably greatly worsen the effect, thereby causing excessive humility and insecurity among experts. In contrast though, on the low skill end of the spectrum, fields of study that seem somehow (deceptively) a part of "common sense" to many people (e.g. logical reasoning, writing, psychology, social skills, ethics, leadership, etc) will tend to cause people to more often grossly overestimate their own skills.

6.4.8 Actor-observer bias

If I accidentally collide with you in the hallway, causing your stack of papers to go flying everywhere, then it's because it was an accident. However, if you do it to me, then it's because you're a jerk. If I'm late to a meeting with friends, then it's because I got held up by circumstances beyond my control. However, if you do it to me, then it's because you don't respect me. Did I take half of the waffles? Well, yes I did, but it's not like anyone else was using them at the time. Did you take half the waffles? As punishment, I have decided that your new name is Richard. Now your nickname finally fits your personality. Did the policy your own political party supported turn out really bad? People make mistakes, man. Calm down. Did the policy the opposing political party supported turn out really bad? They must be literally evil incarnate. Next they will come for our babies, to feed them to their pet demon monkeys no doubt.

When people treat their own behavior as circumstantial and dynamic, yet treat other people's behavior as inherently bound to personality and innate, then they are guilty of *actor-observer bias*. In essence, actor-observer bias is a form of hypocrisy wherein the potential influence of situational constraints on other people's behavior is not sufficiently accounted for, thus leading one to ascribe personality faults to other people prematurely.

The existence of this effect may have something to do with the fact that people are intimately aware of the internal state of their own mind, yet know little about the internal state of other people's minds, thus causing each person to have less perspective on what kinds of situational constraints other people may face. It takes a lot of effort to consider all the possible explanations for someone else's behavior, and even if you try you could still very easily miss a bunch of possibilities. Thus, it is wise to be cautious in your judgments of other people and to give people the benefit of the doubt and a reasonable margin of error. Environmental factors and randomness often explain far more of a person's behavior than their personality.

6.4.9 Halo effect

Have you ever been asked to rate something in several different independent categories? For example, on a movie review survey, have you ever been asked to give separate ratings for visuals, sound, acting, lighting, plot, and so on? If not a movie review, then have you at least at some point seen a survey of some kind that asked you to rate different aspects of one thing along several different axes? There's a good chance that you probably have, at least in some form. People like to think that they are good at keeping these different independent categories separate from each other in their minds and not letting them interfere with each other. However, in many cases, positive or negative ratings in one category will bleed over into other categories and effect them even when those categories are logically unrelated. This is an example of the *halo effect*[9].

The halo effect applies to a lot more than just surveys though. It's quite a broad phenomenon. It also applies to how you perceive people, how you judge products, what you think of different social or cultural groups, and many other things. It taints the standards of judgment you apply to everything you encounter in life. If you really like one aspect of something, you'll tend to rate it disproportionately high in all the other categories. Similarly, if you really dislike one aspect of something, you'll tend to rate it disproportionately low in all the other categories.

It's not a totally consistent pattern, but people's positive or negative feelings for things have a tendency to bleed over into other things without justification. The halo effect is probably weaker among experienced critics, trained experts, and scientifically minded people, but even then it can still have a very significant influence. It can be difficult to resist the temptation to rate all aspects of something more harshly if there is something about it that you intensely hate. Similarly, if you really love something, then it can be hard not to overrate other aspects of that thing even if doing so isn't justified.

Let me give you a first-hand example of the halo effect from my own personal experiences. Back when I was in college, pursuing my computer science degree and getting close to completing it, I had to take a capstone course for game development, because game design was the

---

[9]Note that even though the "halo effect" sounds like it should maybe only refer to positive biases, it actually also can equally refer to negative biases as well. The term is often used in a neutral way, despite its positive connotation. This is probably a common point of confusion. Apparently though, sometimes people use "horn effect" to refer to a halo effect with a negative connotation, but using "halo effect" for both the positive and negative cases seems to be more common.

concentration I had selected for my computer science major. I also had a minor in mathematics, because I was previously a math major and had got about 3/4 of the way to completing my math degree before I switched to computer science. Anyway though, the game development capstone course was focused on creating a complete game. It was kind of our senior project. I teamed up with a guy I knew from the department. The project went well.

Periodically during the game development project, all of the teams were required to do presentations on how their progress with the games was going, to make sure that we all stayed diligent and continued making progress. During each presentation, all of our classmates would fill out survey forms rating each game in various categories such as gameplay, sound, and so on. For one of the early presentations, we presented the game and received only average to somewhat above average ratings.

After that presentation, in the time available to work on the project before the next presentation, pretty much the only thing we added was some really fancy visually impressive special effects and sound effects and music, as well as some other minor tweaks. However, when the next presentation came, our review ratings for *all* of the categories skyrocketed. This included even the rating for gameplay, which had remained unchanged between the previous presentation and this one. This bleed over between ratings that we experienced is a very typical example of the halo effect in action.

Let me give you a few more examples to make sure that you understand the full scope of how common a phenomenon the halo effect is. Take physical appearance and attractiveness for example. If someone is attractive, many people will tend to automatically assume that the attractive person is also more likely to have a bunch of other positive attributes, such as being financially responsible, honorable, compassionate, intelligent, ambitious, competent, and socially well-adjusted. This is in spite of the fact that, logically speaking, the only thing that you can reliably conclude based on the fact that someone is attractive is that they are indeed attractive, plus perhaps some limited conclusions about their physical health and an increased chance of romantic success.

This is a very common source of the halo effect in day-to-day life, and causes highly attractive people to be treated in a disproportionately privileged way. I read once somewhere that it often even causes them to receive significantly higher pay for an equal amount of work. That's not even the worst of it though. Ever noticed how villains tend to usually be less physically attractive than heroes in movies? That's because of the halo effect. Moreover, the fact that so many movies do this also causes society to be subliminally socially conditioned to automatically associate being less physically attractive with being evil.

This contributes to discriminatory behavior in society, such as people being much quicker to treat less attractive people as if they are physically threatening or creepy, based on even the faintest plausible indication, or even based on no indication at all. On the other hand, highly attractive people tend to be given massive margins of error for their behavior, even when from an objective standpoint they may be very obviously acting in an unethical or threatening manner.

Another example of the halo effect is how internet forum communities often behave, especially when some element of hero worship comes into play. For example, on a YouTube channel for a particularly well-liked personality, the community may downvote you into oblivion if you ever try to pose even the most polite or objective criticism of their beliefs or of their hero. Part of this behavior stems from the halo effect.

People like some specific aspect (or multiple aspects) of their hero so intensely that they become hypersensitive to criticism and react disproportionately negatively in order to protect what they value so much. The halo effect causes people to apply asymmetrical standards to things. The more intensely someone likes or dislikes specific aspects of something or somebody, the more likely it is that those feelings will bleed over into other unrelated or weakly related things and inhibit their ability to judge that entity objectively.

Also, the halo effect plays a significant role in things like cronyism and corruption. Many human beings have great difficulty separating their ethical duties from their relationships. When someone does something beneficial for you, it may be difficult for you to thereafter ever judge that person in any kind of neutral or objective way. For instance, do you ever find yourself applying a lower standard of judgment to your friends than to strangers? If yes, then you are guilty of one of the same underlying psychological biases that generates a lot of political corruption in the world.

The main difference is arguably just that you and your friends don't have as much political power, and hence that the corrupting influence of your own bias is less easily noticed and less damaging. Regardless though, the halo effect is still a potentially quite dangerous source of corruption, both on the small scale and the large scale. It's one thing to simply owe someone something in an honorable context, but it's quite another to let your biases breach your ethical duties. Sometimes doing the right thing requires you to harm your own relationships, but sometimes not. Getting the balance just right can be a nuanced and complex problem sometimes. Such is life though.

6.4.10 Ostrich effect

Have you ever had something unpleasant that you knew from a rational standpoint that you should take care of, but you just couldn't bring yourself to do it and instead spent all your time running away from the problem and pretending like it didn't exist? If so, then you were under the influence of the *ostrich effect*. The ostrich effect is a cognitive bias that causes people to avoid unpleasant situations, even when handling the situation should clearly be a high priority and would make the person much better off in the long run. It also sometimes takes the form of denying overwhelming evidence of something bad in order to avoid the cognitive dissonance and inevitable unpleasantness that would come from acknowledging it as a problem. Holocaust denial, climate change denial, AIDS denial, and denial that smoking causes cancer are all caused at least in part by the ostrich effect.

Often, the very same group that is responsible for creating a problem will desperately try to deny that problem's existence. This is often motivated by a selfish desire to protect themselves from ever having to take any responsibility for the problem. Denialists are often not emotionally mature enough to handle the massive cognitive dissonance associated with admitting to themselves that they have actively contributed to something that is so clearly negative. Neo-Nazis deny the holocaust. Heavy polluters deny climate change. Promiscuous people deny HIV/AIDS. Tobacco companies deny that smoking causes lung cancer.

Do you see a pattern here? Notice how in every single case there is a glaringly obvious conflict of interest and a clearly self-serving unwillingness to take responsibility for their own actions, even in the face of overwhelming evidence. It's a classic case of "motivated reasoning". Indeed, in the worst case, not being willing to admit responsibility for destructive actions (e.g. denying the holocaust, spreading medical misinformation, etc) is one of the most dangerous forms of the ostrich effect. Such behavior could easily cause huge damage to humanity's future if not kept in check.

There are of course numerous much more mundane and relatively innocent examples of the ostrich effect in day-to-day life though. Avoiding doing your homework, putting off major projects, not fixing problems around the house, not confronting people about recurring relationship problems, not pursing your more challenging ambitions, not scheduling a doctor appointment that you need, not acknowledging your own personal problems, and so on; they're all caused at least in part by the ostrich effect. Now, I know what some of you may be thinking. Aren't these just examples of procrastination? Yes. You're right. They are, kind of.

Procrastination is fairly closely related to the ostrich effect, but there's some nuance to what the two terms mean that distinguishes them from each other. Procrastination is something that you *do*, not a psychological state, and generally only refers to relatively mild and mundane forms of avoidance. Spending millions of dollars bribing politicians and media outlets to deny climate change certainly involves the ostrich effect, but few would claim that it's merely procrastination. The ostrich effect, in contrast to procrastination, is not something that you do. Rather, the ostrich effect is a psychological state that triggers counterproductive avoidance behaviors.

It seems like there is a missing piece of terminology here maybe. We have a term for referring to procrastination as an action, but have no such equivalent term for the broader case of *any* avoidant behavior caused by the ostrich effect. Therefore, let's create a new word to fill this gap. To refer to this broader concept of any kind of avoidant behavior caused by the ostrich effect, rather than merely to the milder forms that procrastination usually refers to, we could call it **dyscrastination**. Thus, for example, a giant oil company that bribes politicians to deny climate change would be dyscrastinating.

Avoiding doing your homework would also technically be an example of dyscrastination, since dyscrastination is a more general word intended to cover all possible cases of actions motivated by the ostrich effect, but generally in such cases the term procrastination would be preferred. I chose to use the Greek prefix "dys" in designing this word because it means "abnormal". The word "procrastination" is derived from the Latin prefix "pro" meaning "forward" and the Latin root "crastinus" meaning "belonging to tomorrow".

Thus, by combining the prefix "dys" with the root "crastinus" to create "dyscrastination", the new word dyscrastination acquires the connotation of "reacting abnormally to the future", thus making it a fitting term. It also has the benefit of sharing the same prefix as "dystopia", meaning a hellish authoritarian wasteland that maximizes human suffering, which is exactly what humanity may one day become if denialists (i.e. people who engage in a lot of dyscrastination) ever acquire too much political power.

6.4.11 Identifiable victim effect

If you show someone some horrific statistics, such as that millions of people died in a war for example, they will often react indifferently. They will feel about as much urgency to do something to stop those atrocities from ever happening again as they feel to do their taxes. To many people, an abstract number is about as easy to relate to as a slab of granite. I assure you however, that if you had ever had the chance to meet those people, the millions of people whose lives were suddenly cut short by a grisly death, then you would have found it quite easy to relate to them. Doubtless, you would have found many friends among them, if you had ever had the chance to get to know them. Yet you don't care. Why is that?

If, on the other hand, I vividly describe a specific individual victim to you, perhaps even accompanying it by a video of the horrible fate they met, suddenly you care. The video of the injustice the victim suffered spreads around on the internet and on televisions shows and outrage erupts across the nation. There's rioting in the streets. Civil rights organizations and advocacy groups organize by the dozens and are out in full force trying to make a difference. All of this for just *one* victim. Think about that. Really think about it. Imagine if people reacted to statistics about the deaths of millions of people with the same proportional response as how they react to just *one* vividly described and clearly identifiable victim. Would wars even still exist?

People tend to empathize more with detailed stories of individual victims than they do with abstract statistics summarizing numerous victims. This is known as the *identifiable victim effect*. It is not something to be proud of. It indicates a tremendous lack of imagination, empathy, and number sense. It is a very common problem, effecting seemingly almost the entire human population. At first glance it may seem like a hard problem to overcome.

However, with some self-awareness, thought experiments, empathy, and mathematical insight you can overcome it, thereby significantly reducing the bias it introduces into your thinking. Doing so will enable you to *feel* the real weight of these kinds of horrible statistics much more clearly, thereby improving your judgment and sense of morality. Let me guide you through a few thought experiments designed with this purpose in mind.

Here's the first thought exercise. Every person is unique. Sure, there are admittedly a tiny number of genetically identical people in the form of twins, but even they still live distinct lives and thus end up developing into different people. It's easy to think of every person you walk by in the streets as "just another average person" by default, but keep in mind that any one of these people could potentially end up being critically important to humanity. They could be the person who figures out how to cure cancer, or who creates the next new genre of music, or who writes a novel that includes social commentary that prevents the rise of the next dictator. You just don't know. Moreover, due to the chaos effect, even the most seemingly unimportant person can massively alter the course of history, even regardless of whether they ever become famous or whatever.

Important or not though, every person alive is unique and brings their own distinct personality to the world. Each person adds their own bit of color to humanity, which collectively creates the society we live in. Once a person dies, that distinct personality and all the unique color that it brought with it to the world is gone forever. In fact, imagine if every person literally represented a specific color, and that once that person died then the corresponding color also ceased to exist. Imagine that one day as you're walking down a sidewalk on a busy city street you saw a person walking across the crosswalk get run over by a reckless driver and die.

Imagine if the moment that person passed away the entire color green ceased to exist. That would give you a more accurate picture of the magnitude of what you have just witnessed. Could you still be so dismissive of it? Could you still treat it as just another statistic? A person is more than just a number on a page. Their existence has significance, and if a person's life gets cut short then so will the potential value that person could have brought to the world. What kind of value could that person have brought to the world if they had lived? You'll never know now.

Here's the second thought exercise. Everybody is somebody else's family member or friend. They may not be *your* family member or friend, but they are at least *someone's*. Consequently, every time someone dies there is at least one other person who has to experience the grief of losing a family member or friend. You might find it difficult to feel much for people who you don't know though. As such, let's reframe the situation in a more personal way so that you can empathize with it.

Suppose, for example, that you just heard on the news that a million people died during a war. Imagine that we made clones of your family members and friends until we had a million such clones in total. Imagine we then strapped you down to a chair bolted to the floor in some room, and then walked each of these clones into the room, one at a time, and replicated the way they died, killing them right in front of you while forcing you to watch. There are lots of different ways that people die in wars.

Some are burned alive. Some explode when the bombs drop, catapulting severed limbs through the air and painting the walls red with blood. Some are raped, stabbed, and left to die in dark filthy alleyways. Some are buried alive under collapsed buildings. Some are eviscerated, and get to see what their own intestines look like hanging out of their own bodies. Some die instantaneously. Others experience hours, days, or even weeks of suffering while they die.

For argument's sake though, suppose that each of the deaths takes an average of only 10 minutes from start to finish. This is probably overly generous (it could easily be worse), but will work well enough as an imaginary estimate for the purposes of our thought experiment here. Now remember, these are clones of your own family members and friends that we are talking about here, not just random strangers. Everybody is someone else's family member or friend. This thought exercise just makes it easier to see that. Let's put this in perspective now. If you were strapped into the chair and forced to watch each of your own family members and friends die in countless gruesome ways like this, a million consecutive times in a row, at a rate of 10 minutes per death, allowing 8 hours per day for rest, then it would take just over 28.5 years for the ordeal to end.

Here's the math: 1,000,000 deaths at 10 minutes of suffering per death is 10,000,000 minutes of suffering in total, which is equivalent to $10,000,000/60 = 166,666.\overline{6}$ hours of suffering in total. There are 24 hours in a day, but 8 hours of rest are required each day for sustainability and realism, therefore the person in the chair only has $24-8 = 16$ hours available per day in which to experience the suffering. Therefore, to experience the total amount of suffering, the $166,666.\overline{6}$ hours of suffering must be spread out into 16 hour chunks in order to determine the actual number of days the person strapped to the chair must endure. $166,666.\overline{6}/16 = 10,416.\overline{6}$ days of suffering. There are 365 days per year, therefore the person must endure $10,416.\overline{6}/365 \approx 28.5388$ total years of suffering, excluding the time spent resting during those approximately 28.5388 years.

Keep in mind that you are not allowed to do anything other than sit strapped into the chair watching your loved ones and friends die all day, except for the 8 hours allowed for rest and sustenance. Would you even still be sane by the end of it? How much grief would you experience during those 28.5 years sitting in that chair? How badly would you want to stop the social and political forces that caused the war? That's how you would feel about *every* million deaths that you hear about on the news or in history class, *if* you had enough empathy and clarity to actually internalize what those numbers really mean. Eye opening, right?

There is nothing romantic about war. War is an abomination. There are only two valid reasons to ever go to war: (1) self defense and (2) overwhelming moral necessity, such as defending those who can't defend themselves or destroying authoritarian regimes. No other valid reasons for war exist. Any war that isn't based on either self defense or overwhelming moral necessity cannot be anything other than evil. A soldier saying they love war would be like a firefighter saying they love fire. All who love war are not patriots. They are murders. Warmongering is to soldiering what arson is to firefighting. That's how it should be viewed. It's time that more people realized this and stopped being roped into committing atrocities.

6.4.12 Zero-sum bias

Are you familiar with the concept of a zero-sum game? A zero-sum game is any competitive system wherein every gain made by one participant in the system is balanced out by a loss of equal magnitude among the other participants. Thus, if you sum together how much the participants have collectively gained or lost during each competitive action, imagining gains as positive numbers and loses as negative numbers, then the sum will always be zero. For example, suppose that we invented a new game where there's two opposing teams and each team has 10 gold coins hidden at a fort that they control. The goal of the game is to capture as many gold coins as possible.

In such a system, every time one team gains a coin then the other team will lose a coin. Suppose that after 30 minutes you check each team and find out that the red team has lost 7 coins and the blue team has gained 7 coins. Notice that this sums to zero. In fact, no matter what span of time you examine during the game, the sum of the gains and losses that the teams experience during that span of time will always be zero. This works even if there are more than two teams in the game. Even with an arbitrary number of teams, the sum of all gains and losses in any given time period by all sides will always equal zero. That's why it's called a zero-sum game. The property of being zero-sum is essentially a way of rigorously quantifying and formalizing what it means for a system to have a strictly competitive structure, such that it tends to discourage cooperation and reward antagonism.

The name zero-sum game can be a bit confusing however, due to the fact that the concept applies to more than just games. Many different systems can exhibit this kind of behavior. For example, choosing how to allocate a pizza among friends could be interpreted as a zero-sum "game". Every time one person gains a slice of pizza it means that another person won't be able to have that slice. Some economic systems are zero-sum games, but some are not. Not all competitive environments qualify as zero-sum games.

The term zero-sum game originates from game theory, which is the study of competition and cooperation within the constraints of a system. Game theory should not be confused with game *design* theory. Whereas game theory studies competition and cooperation within any arbitrary system and tends to be pretty dry, game *design* theory studies how to maximize the entertainment value in things like video games and board games. The two fields are only somewhat related, despite how similar the names are.

Because the term "game theory" actually applies to more than just games, it seems like an alternative term may be merited. What we really need is a general term that refers to all systems that are competitive or cooperative while not necessarily being games. Some systems will be primarily competitive, but other systems will be primarily cooperative, and both are addressed by the theory. Systems with a strong mix of both competition and cooperation also exist. So, what we really have is a gathering of distinct forces, each with potentially differing interests and goals, which may or may not end up competing or cooperating, but will still generally effect each other in some way. The word "conflux" thus seems like it might be a good choice to capture this nuance.

A conflux is a flowing together of distinct entities, in such a way that each entity influences the others and the system as a whole. Two rivers flowing together is a conflux for example. Thus, for instance, we could refer to two people deliberating over how to use $100 between the two of them as a conflux maybe. They might be selfish and choose to compete for the resource, or they might instead cooperate in some way to obtain a higher value item for mutual gain. Either way, it's still a conflux in a sense.

Therefore, an alternative name for game theory, one that avoids the possibility of being confused with game *design* theory, might be *conflux theory*. Another potential advantage of using the term "conflux" is that it is a relatively rarely used word, unlike the word "game" (which is extremely common), thus making it less likely to generate conflicts of meaning. I'm not sure about this suggestion though. It is just a random idea I head. Make of it what you will.

Anyway though, we kind of got sidetracked there for a bit. I wanted to make sure you fully understood what the term zero-sum game meant before explaining what the related concept of zero-sum *bias* means. It is very often possible for multiple participants in a system to work together to achieve mutual benefit, such that by choosing to work together they will generate more value for themselves than if they had chosen to be antagonistic towards each other. Such systems are called non-zero-sum games.

Being non-zero-sum means that there is at least *some* incentive to cooperate instead of competing. Some non-zero-sum games are only slightly cooperative, whereas others are extremely cooperative. A zero-sum bias is when someone has a predisposition to think of situations as being zero-sum when those situations are *not*. This causes the person with zero-sum bias to make suboptimal decisions that reduce their own benefits, the benefits of all other participants, or both, relative to how much they could otherwise have benefited if the person had chosen to cooperate instead of competing.

People with zero-sum bias tend to view things through a lens of conflict, scarcity, and hostility. Anytime someone with zero-sum bias sees someone else gain something, they tend to automatically assume that the gain must have come at the expense of someone else, and couldn't possibly have been a mutual benefit forged by cooperation. In other words, it's a very eye-for-an-eye, every person for themselves, dog-eat-dog, and hysterically cynical kind of way of looking at things.

Selfishness, immaturity, bigotry, greed, low empathy skills, bullying, anti-intellectualism, and a culture that worships mediocrity all seem to be somewhat correlated with having a zero-sum bias, in my personal experience at least. For example, assuming that immigrants moving to your country will reduce the number of jobs available to the native residents is an example of a zero-sum bias. It isn't a justified assumption. Increased immigration often results in a larger economy, which in turn creates more jobs and more mutual benefit for everyone.

Situations where increased immigration reduces job availability may even be the exception, not the rule. Larger communities (e.g. huge cities, nations with higher populations, etc) tend to actually create *more* jobs, not less. The more humans gather together, the more cooperative potential it creates, and thus the larger the economy tends to become. Notice, for example, that many types of jobs and specialized businesses aren't even capable of surviving outside of big cities, since they aren't scalable enough to survive outside that context. Networking effects strengthen economies, not weaken them, usually.

Zero-sum bias often results in what is known as the "crab mentality", in which people sabotage each other's achievements and efforts out of a misguided notion that anytime someone else gains something then it means that they have lost something. The crab mentality is fairly closely related to bullying. Bullies don't create much value for society themselves, and thus feel compelled to pull others down in order to prevent those other people from ever rising too high above the bully's own social status. People with the crab mentality have basically the same mindset, except perhaps that it might stem more from a scarcity mindset than from self-esteem or cruelty problems.

The term "crab mentality" originates from a folklore anecdote that if you put a bunch of crabs in a bucket, then even if they could potentially easily escape if they stayed coordinated enough, they instead often just grab onto each other whenever one starts moving towards the top, thereby preventing any of them from ever escaping. Each crab refuses to allow any of the other crabs to acquire too much of a superior position to themselves, causing all of them to constantly jealously sabotage each other's attempts at escape. I don't know if the story is true, but that's where I heard the term comes from. Regardless though, it's a pretty good analogy for illustrating the essence of this way of thinking.

You've probably encountered people who behave like this at some point in your life. It's a highly toxic and counterproductive way of behaving, but that doesn't stop some of the more petulant and territorial members of society from doing it anyway. A really common example of this behavior is how underachievers in school like to constantly drag each other down and actively undermine any attempt their friends make to improve their own lives. This is that "culture of mediocrity" thing that I was talking about earlier. It's just like the wise old saying goes: Misery loves company.

The crab mentality is a passive-aggressive form of bullying. Underachievers are unwilling to accept anyone ever doing better than themselves, so they opt instead to try to crush the spirits of everyone else around them in order to stomp out any possible threats to their own complacency. Isn't that extraordinarily ironic though, if you really stop and think about it? They try to mock achievements, yet everything about their behavior and their entire life identity actually centers around the high degree of importance they place on achievements, otherwise why would they ever bother going to such efforts to stop the people around them from achieving things? Underachievers think they are free from the burden of achievement, but in reality they are slaves to it. By defining themselves as the opposite of "trying hard" they lose control of their own destiny. They become like leaves drifting in the water.

You get these cliques of anti-intellectual underachievers who, unless they change their attitude, are probably much less likely to accomplish much in life, and then they sit around all day smearing intellectual and creative endeavors and mocking anyone who wants to escape from their pathetic little bubble of mediocrity. You can sometimes tell who these groups are by the fact that they frequently use words like "nerd" and treat "being normal" as if it is the ultimate thing to aspire to in life. This kind of attitude will often result in systematic poverty, lots of unnecessary suffering, and deep regrets later in life. It would be nice if more of these people would realize that this kind of poisonous attitude is probably exactly *why* they're poor and miserable in the first place. Few things will condemn you to a life of poverty faster than having a disdain for achievement.

Having an underachiever mindset, i.e. being the kind of person who mocks people just for "trying hard", usually leads to a much less enjoyable and much less inspiring life. It's hard to become wealthy or creatively fulfilled if you're actively hostile to the entire concept of achieving anything more than the norm. The only way to break the cycle is to get rid of that bad attitude, firmly commit to achieving greater things, and reignite your own inner intellectual and creative flame.

If possible and ethical, it's also often a good idea to cut the people who are dragging you down out of your life. There's no reason to carry the weight of their burdens on your shoulders, nor do you want to be exposed to the kind of poisonous anti-intellectual propaganda that constantly pours out of their mouths. There's little to no point risking the health of your mind and your future by exposing yourself to people like that. You've got better things you should be doing.

Regions where the culture of mediocrity has established a stranglehold could easily get stuck in a kind of self-perpetuating cycle of toxicity that could last for generations. The amount of hidden human suffering this causes is probably enormous. I wish there was some kind of magic wand I could pick up that would open these people's eyes, but alas there is not. People truly deserve better than this. They don't seem to realize that their own self-limiting beliefs are exactly what is undermining their own quality of life.

In the meantime however, I can at least try to communicate my thoughts in this book, and hope that at least some of it will reach someone out there and help pull them out of the bad situation that they're stuck in. I want them to experience the profound liberation, freedom, and comfort that a more intellectual and creative lifestyle inherently tends to generate. I hope that one day they will wake up from the destructive cycle that they have trapped themselves inside, and that then at long last they will find true happiness and clarity in life.

6.4.13 Statistical history bias

This is actually a term that I made up. It encompasses multiple related statistical biases. There are two different subcategories of statistical history bias, and within each of those subcategories there are two specific types of bias. Thus, there are a total of four specific types overall. As you will see, two of these specific biases have very standard terminology in common use. On the other hand though, the other two biases currently do not apparently have any corresponding terminology in the literature, so I have invented some here to fill the gap.

Statisticians are probably aware of these effects. However, the lack of terminology for the broader type of bias (statistical history bias) and for the missing two specific types could potentially make it more difficult to accurately perceive the big picture of what the nature of these biases might be. That's why I decided to provide my own terminology framework to potentially remedy these omissions. I might as well try, in case it ends up being useful for someone.

Before I explain what these biases are though, I need you to understand the difference between a dependent event and an independent event. A dependent event is any event whose future chances of occurring depend upon what has happened in the past and upon the current state of the system. Dependent systems are systems where the causal relation between events is strong and predictable. In contrast, an independent event is any event whose future chances of occurring do *not* depend upon what has happened in the past and where the current state of the system is either stable or negligible in effect. Independent systems are systems where the causal relation between events appear weak due to the influence of chaotic effects and emergent properties or due to irrelevance. This causes the outcome of each individual successive event in an independent system to be difficult or impossible to predict in advance, although statistical summaries of probabilities for large samples of data are still viable.

Events are always dependent or independent *relative to some other event*. The concept of statistical dependence or independence makes no sense for an isolated event. An event can be tested for dependence or independence with another event regardless of whether that other event is of the same type or of a different type. Thus, for example, the dependence or independence of a particular dice roll could be tested against the previous dice roll, *or* it could be tested against the assumption that you ate a burrito this morning. Don't assume that an event's dependence or independence is always relative to other events of its own type.

Let's refer to the hypothetical event whose probability of happening we are considering as the *tentative event*, and to the event that has already occurred (and to which the tentative event is relative) as the *given event*. These two terms (tentative event and given event) are ones that I made up. I couldn't find any shorthand terms for distinguishing the two events from each other (I'm not a statistician) so I created some as a matter of pragmatism. Naturally, in the case where there are multiple given events, they will be referred to as the "given event**s**" instead of as the "given event".

Let me give you some examples to make sure that the meaning of statistical dependence and independence is clear. The classic example of an independent event is dice rolls. Every time you roll dice, the imprecision of the dice throwing motion, the turbulence of the air, the angular and unnatural shape of the dice, the varying nature of the impact surface, and the complex causality of the rapid sequence of impacts combine to make it very difficult to predict the outcome of the dice roll. The dice also remain structurally unchanged after each roll. This results in independent behavior where each successive dice roll is unaffected by the previous roll. Stated more precisely, we are saying that *given* any arbitrary history of previous dice rolls (the given events), the next dice roll (the tentative event) will be unaffected by those past events.

Now let's consider an example for the dependent case. Consider, for example, the probability of breaking your neck relative to the event that you have just fallen down the stairs. These two events are dependent. Falling down the stairs increases the probability of you breaking your neck considerably. Thus, the past history of one type of event is altering the chances that another type of event will occur.

Stated more concisely, the given events are changing the probability of the tentative event. This is the essence of what it means for an event to be statistically dependent. On a side note, I also want to point out that there are of course *other* ways of breaking your neck, but that the overwhelming majority of the time you are not in those kinds of situations, and the probability of your neck just spontaneously snapping while you are walking around or sitting at your desk is extremely small.

Anyway, now that we know what dependent and independent events are, let's talk about what the two subcategories of statistical history bias are. Here again there is apparently no existing terminology for these broader terms as far as I can tell, so I had to make some up myself. The two subcategories of statistical history bias are **statistical dependence bias** and **statistical independence bias**. Statistical dependence bias happens whenever someone treats events like they are dependent when they are not. Statistical independence bias is the opposite; it happens whenever someone treats events like they are independent when they are not. Let's talk about statistical dependence bias first.

The first type of statistical dependence bias is known as the **gambler's fallacy**. The gambler's fallacy happens whenever someone assumes that each successive similar event in a streak of similar events is less likely than the last, within a system where the events are independent. Consider coin tossing for example. Let's assume that we have a fair coin with an equal chance of landing on either side. Suppose that we flip our coin four times and all four times it comes up heads.

What are the chances of the next coin toss coming up heads again? If your instinct is that the chances of the coin coming out heads again have gone down, because the streak is "bound to break sometime soon", then you are guilty of the gambler's fallacy. The coin is fair and will always have an equal chance of coming up heads or tails. Coin flips are independent events. By definition, independent events are incapable of effecting each other. What happened in the past doesn't matter.

The gambler's fallacy is a very common bias. The probability of a large chain of similar independent events occurring *does* decrease as you make the hypothetical chain longer. However, that is only true from the perspective of someone observing the probability *before the events start happening* (i.e. where the events are accounted for as a group instead of separately). The past can never effect the future for an independent event. Looking forward into the future, a chain of five consecutive heads is indeed less likely than a chain of four consecutive heads. However, looking backwards into the past, a pre-existing chain of four consecutive heads does nothing whatsoever to effect the chances that the next coin flip (the fifth one) will be heads. Do you see the distinction?

Oh, and by the way, you should avoid gambling. You might be tempted to try to learn to "beat the system" somehow by studying statistics, but that's a fool's game. In fact, in mathematics we even have a special nickname for gambling. We call it the *stupidity tax*. Time spent gambling is time that would be better spent investing in a more productive and more reliable method of generating wealth. Gambling is one of the slowest and least effective ways to make money, which is why it's so ironic that people are attracted to it out of a desire to get rich fast.

If you ever want to become wealthy someday then gambling is actually probably one of the dumbest things you could possibly do. The house always wins. The most likely outcome of gambling is that you will simply lose a lot of your money, or in the worst case you may even become a gambling addict and end up destroying your entire life and all of your relationships with your loved ones.

Besides, gambling is an absolutely dreadful source of entertainment. It's about as exciting as dry toast, compared to the alternatives. There are far more enjoyable things to do with your time. For example, video games are a far superior source of entertainment, and are also much less damaging to your finances. So are things like books and movies and various other hobbies.

Anyway, the second type of statistical dependence bias is called the **hot hand fallacy**, which is also known as the **reverse gambler's fallacy**. It's pretty similar to the gambler's fallacy, but with a twist. Instead of assuming that each successive similar event in a streak of similar events is *less* likely than the last, the hot hand fallacy assumes that it is *more* likely and that therefore the streak will continue at least for a while. As a reminder, keep in mind that we are assuming here that the system under consideration is *independent*, just like we were for the gambler's fallacy.

So, as you can see, whereas the gambler's fallacy assumes that streaks will end, the hot hand fallacy assumes that streaks will continue. The hot hand fallacy essentially captures the idea that luck runs in streaks. People who are guilty of the hot hand fallacy are likely to believe that they experience periods of "good luck" and "bad luck", such that during periods of "good luck" they expect statistically independent events to just magically be more likely to have favorable outcomes during that time period, at least until the "good luck" runs out.

Although the hot hand fallacy is in some sense the opposite of the gambler's fallacy, both patterns of biased thinking often occur in the same person. It's impossible to commit both the gambler's fallacy and the hot hand fallacy simultaneously, because that would be logically self-contradictory. However, different situations can create different psychological responses, and therefore the same person may commit the gambler's fallacy in one situation yet commit the hot hand fallacy in another. The real underlying cognitive bias may actually be the broader phenomenon of statistical dependence bias, i.e. assuming that a system is dependent when it is actually independent.

The term "hot hand fallacy" actually originates from the observation that in certain sports people have a tendency to expect that if a player has recently been successful in the game then it temporarily increases their chances of additional successes. However, players are human beings and therefore may be psychologically responding to their own success, possibly causing them to temporarily have a genuinely higher chance of success due to the inspirational nature of their previous success.

In other words, sports are often dependent systems, and therefore you cannot necessarily safely say that someone is committing a fallacy by assuming that a streak in a sport is more likely to continue for a while than to stop. Despite the origin of this term from this dependent context, it still seems to me that the hot hand fallacy and the reverse gambler's fallacy are really conceptually the same thing in disguise, and thus I decided to adopt the term for use in this broader context. Just something to keep in mind perhaps.

Anyway, time to move on to the other subcategory: statistical independence bias. As a refresher, recall that statistical independence bias happens whenever someone treats events like they are independent when they are actually dependent. In other words, it is the opposite behavior of statistical dependence bias. This is the subcategory for which I found no existing suitable terminology, and therefore had to make up my own. These biases seem to maybe be a bit more rare than the statistical dependence biases, perhaps due to the human mind potentially being more evolutionarily biased towards thinking in terms of statistical dependence than in terms of statistical independence. I'm not sure though.

We will call the first type of statistical independence bias the **unwavering fallacy**. It is statistical independence bias's analog of the gambler's fallacy. The unwavering fallacy occurs anytime you observe a streak of similar events and then assume that the chances of the next event being similar are independent of the past events, when in fact the chances are dependent and will become decreasingly likely. In other words, the unwavering fallacy is when you assume that streaks don't become increasingly likely to end as they become longer when in fact, due to the specific nature of the system, the streaks actually *do* become less likely the longer they continue.

In essence, the unwavering fallacy represents a failure to realize that some systems are subject to thresholding effects, wherein a chain of similar events can eventually trigger a breaking point where the state of the system changes and the system has to release some sort of accumulated pressure or strain. That's why it's called the *unwavering* fallacy. It represents a failure to realize that some systems *waver* and change behavior as chains of similar events accumulate. Not all systems can sustain any arbitrary sequence of random events without being forced to change state. Let's do a thought experiment to illustrate that this effect is indeed real and not just something I made up.

Consider, for example, an array of steam pipes that have been bolted to the side of the wall along some of the hallways in a power plant. Suppose that periodically the pressure will build up in the pipe system and one of the junction points between the segments of pipes will randomly release a burst of steam. Suppose furthermore that based on your own limited experience working at the power plant for the past year it seems that each of the pipe junctions is about equally likely to release the random burst of steam whenever it happens.

Furthermore, suppose that the past history of the bursts of steam seems to have no effect on which junction will release the steam next. This leads you to believe that the events are statistically independent. However, suppose that one day a single steam pipe junction experiences 10 consecutive steam bursts in a row. Based on your knowledge from studying statistics as an engineer, you assume that the chances of the burst of steam happening an 11th time at that junction will remain unchanged.

Is that a safe assumption though, given the physical constraints of the pipe system? Remember, the steam going through these pipes is extremely hot. Normally, each junction of the pipe system has plenty of time to cool down in between bursts of steam. However, if by random chance the bursts of steam repeatedly occur at the exact same junction, then that junction of the pipes could become so overheated that it deforms or melts or explodes.

If this were to happen, then it would cause a change in the state of the system, and therefore also a change in the probabilities of the events. All of the junctions share this same vulnerability, and this implies that there is an upper limit on how many consecutive times a burst of steam can occur at a specific junction before the system will break. In other words, it's a statistical thresholding effect, and therefore your assumptions as an engineer working at the power plant that the events would remain independent would be an example of the unwavering fallacy.

Any system in which the source of the random events is capable of becoming "tired" or "worn out" in some sense is vulnerable to the unwavering fallacy. For example, suppose that you're a biologist studying species of frogs that hunt by suddenly extending their sticky tongue to catch insects. Suppose that as a biologist you want to compute the probability of a frog successfully catching its prey when the frog releases its tongue. As you collect your data, it might seem that the probability of successful insect capture is an independent event that in no way depends on the success or failure of previous attempts.

However, the frogs tongue is a biological organ, and as such it is potentially susceptible to becoming worn out. If the frog has a long chain of successes, then its tongue could eventually reach a threshold where it becomes tired or low on saliva, such that the probability of successfully grabbing insects drops after that point. The point is, be very wary of assuming that chains of similar events can continue indefinitely if the system could potentially have some kind of thresholding effect where it wears out in some sense.

Anyway, time for the next and final type of statistical history bias. We will call the second type of statistical independence bias the **cold hand fallacy**. It is statistical independence bias's analog of the hot hand fallacy. The cold hand fallacy occurs anytime you observe a streak of similar events and then assume that the chances of the next event being similar are independent of the past events, when in fact the chances are dependent and will become increasingly likely. In other words, the cold hand fallacy is when you assume that streaks don't become increasingly likely to continue as they become longer when in fact, due to the specific nature of the system, the streaks actually *do* become more likely the longer they continue.

In essence, the cold hand fallacy represents a failure to realize that some systems "warm up" or stabilize once they experience a long enough chain of similar events. Put differently, some systems experience "shifting winds" or "turning tables" wherein once enough consecutive successes are achieved then having more successes becomes even more likely. It's like a statistical version of the concept of momentum, triggered by a cascade of favorable circumstances.

The cold hand fallacy happens whenever someone fails to acknowledge that a system behaves like this, and instead assumes that the system is indifferent to changing circumstances and thus "cold" and "mechanical" in nature, like a robot. That's why it's called the *cold hand* fallacy. The other reason why it is called the cold hand fallacy is because it is the statistical independence bias analog of the hot hand fallacy, hence making the name a fitting and memorable choice of words. As before, let's do a few concrete examples to illustrate that this effect is indeed real and not just something I made up.

A good example of this would be any kind of hand-to-hand combat between two opponents, such as martial arts for example. Suppose you're a spectator watching two boxers fighting in an arena. Every time the two boxers approach each other and try to make a move, one of them will randomly succeed in hitting the other. Thus, you will end up with a sequence of varying hits, kind of like a sequence of coin flips. Sometimes boxer *A* will be the one to land a hit and other times boxer *B* will. Suppose that as you observe the fight you notice that the boxers are about equally skilled and exchange about an equal number of hits.

This leads you to believe that the system is independent. However, what would happen if boxer *A* got in an abnormally large streak of successful hits against boxer *B* due to random chance? Normally boxer *B* might be able to recover in between hits due to the variation creating opportunities for rest. However, in the event of a long streak of hits, boxer *B* could become dazed and boxer *A* could become emboldened by confidence, leading to the chances of *A* successfully hitting *B* again after that point actually *increasing* instead of staying the same or decreasing. In effect, boxer *A* has acquired an advantage over boxer *B*, and this has imbued boxer *A* with a kind of statistical momentum. The tables have turned in boxer *A*'s favor.

Any system in which the source of the random events is capable of gaining momentum in some sense is vulnerable to the cold hand fallacy. For example, consider the common experience that people have of "getting into the groove" or "being in the zone" when working on projects. If someone dismissed these experiences as not being real and attempted to claim that human productivity is always the same regardless of whether a person is "getting into the groove" or "in the zone" or not then this would be a cold hand fallacy.

People really do experience performance momentum when they are working on things. For example, computer programmers are famous for getting really absorbed in what they are working on in order to solve complicated problems. Keeping all of the moving parts involved in a computer program straight in your head can require enormous focus. Interrupting a programmer while they are "in the zone" like this can cause them to lose their focus so badly that it may sometimes take them hours to recover enough to be as productive again.

That's concludes our discussion of the two subcategories and four specific types of statistical history biases. Let's summarize for the sake of perspective and clarity: Statistical history bias is the top level umbrella term for all of these kinds of biases. There are two subcategories of statistical history bias. They are called statistical dependence bias and statistical independence bias. Each of them is also divided in two types. Statistical dependence bias manifests as either the gambler's fallacy or the hot hand fallacy. In contrast, statistical independence bias manifests as either the unwavering fallacy or the cold hand fallacy. Human beings seem to be more naturally predisposed towards statistical dependence biases than towards statistical independence biases.

All of these biases are equally irrational. It is very easy to unwittingly fall into one of these traps. Trying to keep all of them straight in your head could be difficult, because it may be hard to discern which case applies at any given moment. However, notice that all of these biases are clearly closely related. As such, you may find it easier to protect yourself from them if you think from the perspective of the more fundamental essence of what these biases really are.

In particular, regardless of which specific type of these biases is occurring, statistical history bias is ultimately caused by not accurately understanding the degree to which a system is dependent or independent. Therefore, all you have to do in order to avoid all of these biases is to make sure that you are doing the best you possibly can to correctly understand the degree to which any given pair of events is dependent or independent. Don't treat pairs of dependent events as if they are independent, or vice versa. As long as you do that, and judge the events accordingly, then you should be fine.

One interesting thing I want to point out (a personal theory of mine) is that human beings' tendency towards statistical dependence bias is probably a result of the evolutionary pressures we experienced prior to the modern era. Our tendency to think about pairs of events as dependent events by default probably increased our survival rates, thus causing it to become a more common personality trait due to natural selection pressures. Why? Well, there are a few reasons probably. One reason is because natural systems, the kind that primitive humans would have encountered prior to the invention of modern technology, tend to be dependent systems more often than they are independent systems. The natural world is generally an interconnected web of strong causal relationships, rather than merely a sea of statistically independent random noise.

Take fishing for example. It's not like every time you cast your line in a lake the lake just calculates a flat random number and gives you a statistically independent chance of catching a fish. Instead, fishing success depends mostly on special conditions like the ecosystem health of the body of water, whether or not a school of fish is swimming by, the depth of the bait and hook, the skill and technique of the person fishing, and so on. By defaulting to thinking about events dependently, you increase the chances that you will subconsciously respond to these environmental cues until you gradually find a more effective way to fish. This may result in you adopting ritualistic fishing behaviors that you don't even consciously know the reason for, yet which still improve your chances of success.

The same is true for other things in day-to-day human life, such as social structures. There are often underlying patterns to the way other people around you behave, and thus assuming statistical independence when it isn't there would probably make you lose more social opportunities. Social structures are often highly interconnected and dependently related, so statistical dependence is a better default assumption for them. Other things that have a statistically dependent nature include hunting, weather, and agriculture.

These are all things on which human survival critically depends. Another reason why dependent thinking may have been advantageous for human evolution is because it tends to encourage pattern recognition. By being predisposed to treating events as being causally related, we have a higher chance of discovering hidden patterns and relationships in the world around us. On the other hand though, this way of thinking may also cause us to experience more false positives, leading us to see patterns and relationships that don't actually exist, and thereby creating a fertile ground for delusion and insanity unmatched by any other species.

In contrast, artificial and contrived objects such as dice, playing cards, roulette wheels, pseudo-random number generators, slot machines, and lotteries are all things that don't occur in nature. These systems tend to behave in a more statistically independent manner. This is probably one of the main reasons why casinos and other gambling institutions are able to exploit people so effectively. Human beings are predisposed to think of all the systems they encounter as statistically dependent by default, but all of the systems in play in casinos and other gambling institutions tend to be unnatural statistically independent systems.

The human mind isn't as well adapted to dealing with such systems, because they are rare in nature. Thus, casinos and other gambling institutions may essentially owe their existence to the fact that human evolution hasn't caught up to these modern era inventions. So, the next time you see someone commit the gambler's fallacy or hot hand fallacy, remember that it doesn't mean that the person is stupid necessarily, but rather it simply means that their normally very adaptive and beneficial tendency to recognize patterns and to think dependently is being exploited by the unethical owners of the gambling institution.

Let me give you a tip so that you can understand the real nature of "luck" better. Basically, whenever random events occur in a statistically dependent environment, the word "luck" essentially becomes synonymous with "favorable circumstances". In contrast however, whenever random events occur in a statistically independent environment, the word "luck" becomes synonymous with pure incoherent randomness. Now, consider this: In a natural environment, where dependent events are more common, believing that you have "good luck" will cause you to stick around in that environment in order to extract as much benefit as you can while the "good luck" lasts.

But think about what I just said. In a statistically dependent environment "good luck" is just a synonym for favorable circumstances, and there is certainly nothing irrational about thinking that favorable circumstances will continue being beneficial to you while they last. Thus, in a state of nature, the concept of "good luck" is actually rational from an adaptive standpoint. It's only when you place it in the context of statistically independent events that it becomes genuinely irrational.

Finally, I want to explain why I chose the term "statistical history bias". You see, one thing that all of the different types of statistical history bias have in common is that they all involve an incorrect judgment made on the basis of what the past *history* of events was within the given system. This is true regardless of whether the nature of that incorrect judgment was dependent or independent. Therefore, the phrase "statistical history bias" successfully encompasses both statistical dependence bias and statistical independence bias, thereby making it suitable as a term for the superset of both. And with that, we are finally finished discussing statistical history bias. Time for the next cognitive bias.

6.4.14 Contrast effect

How we perceive something changes depending on what that thing stands in contrast to at the moment that we observe it. This is true both of how we think and of how we see the world around us. We call this phenomenon the *contrast effect*. For example, a color will appear different depending on what other colors are adjacent to it and how dark or light they are. For instance, a grey square placed against a dark background will appear lighter against that background than against a bright background, even though in both cases the grey square is still the same color. It will seem like the grey square is a different shade of grey depending on what context it appears in.

Even if you do a side by side comparison, this effect is hard to dispel. However, you can break the illusion by blocking out the contrasting background. Once you do, it suddenly becomes obvious that the grey squares were actually the same shade of grey all along. This is a common way of demonstrating the contrast effect. It is a very eye-opening experience. It makes you realize that what you see around you is not the real world itself, but rather a filtered and cleaned up version of it.

It's as if your brain has an image filter inside it that is applied to everything your eyes see, much like the image filters that you would find in art or imaging software. This quirk in our vision probably evolved to make information processing more efficient and more adaptable to changing circumstances. If you have never seen this demonstrated before, you should. It's something everybody should witness at least once. It's instant proof that your perception of reality is partially an illusion and that your brain can be tricked into seeing things that aren't real.

Let's do a few more examples. The contrast effect is actually a very broad phenomenon and applies to far more than just eyesight. I want to make sure you understand that, so the implications sink in. Consider, for example, the effect of differing prices on similar items. Suppose, for example, that you visited a website that sells some kind of software package related to your work. They offer three different versions of the software, and each version is more expensive than the last and has slightly more features.

How you perceive the price of any individual product on the page will change depending on what the prices of the other products are, even though the actual value provided by each individual product never changes. Thus, for example, placing a more expensive product by a cheaper product, where the cheaper one is obviously a better deal, could cause more people to buy the cheaper product than if the cheaper product had been isolated on its own page with no other products placed nearby to provide price contrast. People's perception of prices can change wildly depending on what those prices are held in contrast to.

Another example is people's height. If you put an average height person side-by-side with a tall person, the average height person will suddenly start reading as "short" even though nothing about their height has changed. It's the relative differences between things that determine how you perceive those things, more than it is the absolute differences. Thus, for example, in the case of height differences between buildings, if building *A* is 30 feet tall and building *B* is 60 feet tall, then your mind will tend to think of building *B* in terms of being 2 times taller than building *A*, rather than in terms of being 30 feet taller than building *A*. Perception is relative. The reason for this may be so that our minds are better able to adapt to widely varying circumstances. Absolute perception may be harder to process efficiently than relative perception.

Put another way, our perception of contrast between entities is more multiplicative than it is additive. In fact, ignoring this fact will get you in trouble when writing computer graphics code. For instance, I remember once programming a custom GPU shader for my final game project for my computer science degree, where the shader was intended to be capable of amplifying or dampening the color contrast of any given object. The idea was to calculate the displacements of each color vector from the nearest point in greyscale space (the diagonal of the RGB color cube) and then modify the color vectors of the pixels based on that.

There are two ways people typically would attempt to do something like this. One way would be to add to the color displacements by some absolute quantity, whereas the other way would be to multiply the color displacements relative to their current magnitude. One of these ways creates lots of visual artifacts and looks hideous, whereas the other way creates a very pleasing and smooth color contrast enhancement. The correct choice is to multiply, not add. The reason for this is because additive modifications *destroy contrast*, whereas multiplicative modifications *preserve contrast*. In a sense, multiplicative differences are the fundamental essence of what contrast really is.

For example, going back to the building height example, compare what would happen to the contrast in the height of the two buildings if I added 150 feet of height to both of them versus if I multiplied both of their heights by 6. In both cases, the height of building *A* would become 180 feet. However, the relative height of building *B* would change. In the case of additive modification building *B* would become 210 feet tall, whereas in the case of multiplicative modification building *B* would become 360 feet tall. Prior to our modification, building *B* was twice taller than building *A*.

However, in the case of additive modification the ratio has changed such that building *B* is now only $1.1\overline{6}$ times taller than building *A*, whereas in the case of multiplicative modification building *B* remains twice taller than *A* despite both buildings increasing in height. This is an example of what I meant when I said that additive modification destroys contrast.

For an even more extreme example, consider what would happen if we added 10,000 feet to both buildings. Building *A* would become 10,030 feet tall and building *B* would become 10,060 feet tall. Notice that under these modifications the buildings would be perceived as having almost the exact same height. Building *B* would be only about 1.003 times the height of building *A*. In fact, as the size of additive modifications approaches infinity, the contrast (the difference in relative magnitudes) between the heights of the buildings will approach zero.

This effect is also why game designers who make additive changes to the parameters of the entities in their games often suddenly feel like the balance of their game has mysteriously changed. This can lead to a lot of frustration and wasted time. The balance of a game is based on the contrast (i.e. the multiplicative differences) of powers, but additive changes to parameters alter that carefully built up contrast.

Thus, if you want to modify a set of parameters in a game design while still preserving all the other existing balance and fine-tuning that you've been doing, you need to multiply the parameters by a common factor instead of adding to them. Any group of entities which has their parameters simultaneously multiplied by the same amount will retain the same relative contrast and balance amongst themselves as before.

However, entities that are not within the group being modified may change in contrast and hence may need rebalanced. Thus, for example, if you have six creatures in your game named A, B, C, X, Y, and Z and you multiply the hit point parameter of A, B, and C by 3 then the *relative pairwise balance* of hit point power amongst A, B, and C will remain unchanged, but in contrast the balance of power between A, B, and C and the remaining creatures X, Y, and Z will change, which may or may not be what you want to happen.

This concept is crucial for understanding how to efficiently rebalance gameplay, but unfortunately very few game designers seem to know it. This is probably because many game designers don't know enough logic, math, and computer science to understand how to navigate the conceptual space of the game in a logically consistent, intentional, and safe manner.

Every time you want to make a change to a parameter of some object or creature in your game, ask yourself this: What other objects or creatures in this game do I want to retain the same balance of power, relative to the specific parameter of the specific object or creature I am changing? Once you know what group of entities that is, you should make sure to multiply the same parameter for *all* of them by the same amount once you do make the change.

However, this doesn't mean that you always have to think in terms of multiplication factors when balancing. Rather, it merely means that you have to translate any modifications you want to make into terms of multiplications in order to apply them safely to any given balance retaining group. Thus, for example, if you want to "add 10 hit points" to a particular entity's hit point count without destroying its existing balance relative to some set of other entities, then you can still do that, it's just that you have to use algebra to translate it from terms of an additive modification into terms of a multiplicative modification. That way, the change you make will only shift the balance of power among entities that are *not* in the balance retaining group.

Let me give you some formulas for the most common use cases of this. Suppose you have a group of n related entities, each of which has a corresponding parameter p_i, where i is a positive integer that identifies which of the n entities the parameter p_i belongs to. Suppose that there is one specific entity (identified by k) among this group whose parameter p_k you want to modify in some way without changing the relative balance of power among the group of entities (i.e. without changing the relative balance of power among the various p_i values). This requires that we multiply all of the p_i by the same number. We will refer to this number as the *balancing factor* and will represent it with the letter b. The following formulas describe what value of b is necessary in order to perform the corresponding contrast-conserving versions of the operations:

$$\text{To set parameter } p_k \text{ equal to } x: \quad b = x/p_k$$

$$\text{To add } x \text{ to parameter } p_k: \quad b = (p_k + x)/p_k$$

$$\text{To multiply parameter } p_k \text{ by } x: \quad b = x$$

After computing b, we simply multiply every p_i in the group of entities by b, i.e. we change each p_i to $b \cdot p_i$, and then balance conservation within that group will be thereby mathematically guaranteed. The fundamental reason why these operations always conserve contrast (a.k.a. the relative balance of power) among any group of entities is because $p_j/p_i = (b \cdot p_j)/(b \cdot p_i)$ for all possible pairings of arbitrary parameters from any two entities i and j within the group. In other words, since the balancing factor b is shared by all members of the group, the ratio of any given entity's parameter p_i to any other entity's parameter p_j can't possibly change. The b would always get factored out in the ratio of the parameter between any given pair, and hence contrast (a.k.a. the relative balance of power) is guaranteed to be conserved.

Notice that our application of these principles here are not actually a bias, because in this case we are perceiving the nature of the contrast correctly rather than incorrectly. So really, we should probably actually define two different terms to distinguish (1) the cases where the contrast effect is acting as a cognitive bias from (2) the cases were it is merely a law of nature being correctly applied. Let's refer to the case where it's a cognitive bias as **contrast bias**.

In contrast, let's refer to the case where it's a law of nature as the **law of contrast conservation**. The law of contrast conservation has the form $a/b = (ca)/(cb)$ where a, b, and c are any arbitrary values (whether constant or variable). Furthermore, let us refer to any practical application of this to create a desired change in a system, such as conserving color contrast or gameplay balance etc, as being an application of the **principle of contrast**.

6.4.15 Fluency heuristic

The easier something is to understand, the more likely you are to believe it and to treat it as important. This is the fluency heuristic. There are probably at least several reasons why this happens. One reason is because it makes it easier to make fast decisions, which is important in a hectic environment where time is limited and responding too slowly may be dangerous or counterproductive. Another reason is because the more difficult it is to process a piece of information, the more likely it is to make a mistake. The more likely it is to make a mistake, the less trustworthy the information effectively becomes.

A third reason is because favoring easy to understand information and dismissing hard to understand information makes it easier to efficiently search for valuable pieces of information. You only have a limited amount of time and energy available to process a piece of information before other concerns of day-to-day life will take precedence, therefore you are often better off favoring easy to understand information than risking investing time in hard to understand information. Thus, as you can see, the fluency heuristic is a useful evolutionary adaptation which serves a valuable purpose and is *usually* beneficial.

However, while the fluency heuristic does make your mind function more efficiently, it also introduces bias and can easily mislead you. The truth is not determined by human opinion, nor does the truth care in the slightest whether or not you are capable of understanding it. A difficult to understand truth is still true. Information that is easier to understand may be easier to detect errors in and to grasp, and that might make it more trustworthy from a human viewpoint, but from an absolute viewpoint ease of understanding is irrelevant. That's why the fluency heuristic can get you in trouble sometimes, and that's why it qualifies as a cognitive bias. Clarity is important, but it is no substitute for truth. Clarity should be used as a way of overcoming the limitations of human communication and cognition, not as a way of portraying falsehoods as truths.

For example, it is much easier to convince someone of the truth of Newton's classical mechanics than to convince someone of the truth of Einstein's relativity, even though relativity is technically more accurate according to the expert consensus of physicists. The fluency heuristic can create a situation where an easy to understand argument with less supporting evidence is more convincing and believable than a hard to understand argument with more supporting evidence. From the standpoint of a hypothetical perfectly rational and perfectly intelligent mind, this behavior would be clearly irrational. However, from the standpoint of someone with a mind with limited resources, like us humans, the fluency heuristic is a necessary evil, albeit one that should be treated with some suspicion and caution.

People in academia and research could really benefit enormously from learning to exploit the fluency heuristic more in how they communicate with the public though. The public often won't believe something until it has been communicated with the utmost clarity, ease, and accessibility. When you have two competing theories, where one theory has been explained clearly and simply and the other theory has been hidden in a fog of obscurantism, jargon, and pretentiousness, then the easier theory will very often gain more traction and be much more widely adopted.

This is often true regardless of which theory has more supporting evidence. Theories live or die by the hand of clarity. If you can't explain an idea clearly enough for other people to understand it, then for all practical purposes it will be as if the idea doesn't even exist. If a theory can't be easily understood, then the overwhelming majority of people will see it as having no value and will not trust it.

Another example of the fluency heuristic in play is in evolution denialism. Some people find it way easier to believe that a sky-wizard waved a magic wand to summon the universe into existence than to believe that seemingly trivial laws of nature can interact with matter and energy in such a way as to spontaneously generate higher levels of complexity, eventually leading to the emergence and evolution of living organisms. The idea that higher-level complexity can emerge from the interactions of seemingly trivial laws of nature is often one of the most difficult parts of the theory of evolution for some people to accept.

However, if you spend even the slightest amount of time studying emergent phenomena or evolution then it will become immediately and glaringly obvious that not only does higher-level complexity emerge spontaneously from lower-level laws and material interactions, but that it does so easily and abundantly. It would actually be weirder if emergent effects and evolution *didn't* spontaneously happen than it is that they do. The fact that spontaneous emergence of higher-level complexity from simpler systems can be so hard to believe for some people is actually purely a byproduct of the quite restrictive limitations of the human mind. Human minds usually lack the computational throughput necessary to intuitively grasp what kinds of emergent phenomena can arise out of any given set of simple rules.

Let me give you an example that will make the inability of the human mind to intuitively predict the emergence of spontaneous complexity more abundantly clear. Suppose that you have a 2-dimensional grid of cells, like what you would see on graph paper. Suppose that every cell is considered either "live" or "dead". Live cells are shaded black, and dead cells are shaded white. Consider the following set of rules:

1. Each cell's neighborhood consists of the 8 cells touching it on the grid, i.e. all the cells that are horizontally, vertically, or diagonally adjacent to it[10].

2. If a cell is dead and it has three live cells in its neighborhood then it becomes a live cell.

3. If a cell is live and it has two or three live cells in its neighborhood then it stays alive, but in all other cases it dies.

4. Cells are updated in parallel, based on a snapshot of the state of the entire grid before any new updates are applied. During each update cycle, the rules are applied to every cell on the grid exactly once per cell. Time in the system is not continuous and smooth, but

---

[10]This is known as the *Moore neighborhood* of a cell. If you were designing a rule set where you didn't want to include the diagonal cells, then you could use what's called the *Von Neumann neighborhood* instead. Which neighborhood you choose for a rule set changes its behavior though, so don't do so carelessly.

instead pops forward suddenly each iteration as if scanning forward in a movie frame-by-frame (i.e. time in this system is "discrete", as we say in math; or "turn-based", as we say in game dev).

Here's my thought exercise for you: Based solely on looking at these rules as written, predict how the system will behave if we randomly initialize the states of the cells. In other words, imagine that we start with a blank sheet of graph paper and flip a coin for each cell to determine whether it will be live or dead. Predict, in general terms, how you think the system will behave. It doesn't have to be too specific. Your prediction can be quite vague. The point is just to see how good you are at intuitively grasping roughly what kinds of things might happen with this rule set.

A lot of you, especially if you come from a computer science background, may already be familiar with this rule set. It is, after all, quite famous in certain circles. If this is you, then just try to imagine that you don't already know what the rule set will do and just think of it as a thought exercise, as a matter of perspective. Will all the cells live? Will all the cells die? Will they form some kind of rigid structure, like a crystal? Will it be pure chaos, with cells flipping on and off with no coherent pattern? Will the system stabilize, such that cells never change after a certain point? Will any higher-level patterns emerge? What will those higher-level patterns look like if they do? In general, what will happen?

It may surprise you, but these seemingly trivial rules are actually *already* sufficient to support the spontaneous emergence of collections of cooperating cells that behave like coherent higher-level multi-cellular pseudo-organisms. Perhaps the most famous of these pseudo-organisms is known as the "glider", which occurs with great abundance in the system. Gliders look sort of like tiny little winged birds that move around the grid in one of multiple different diagonal directions. There is also another pseudo-organism called a "glider gun" which constantly generates new gliders, thus providing a means of multi-cellular reproduction within the system. That's just the tip of the iceberg though. Numerous different types of cooperating groups of cells exist within the system, and the rules even support crude forms of memory storage, oscillation, and synchronization.

The official name of this rule set is *John Conway's Game of Life* and it supports quite a diverse array of phenomena. In fact, according to LifeWiki, a Wikipedia-style website devoted to the study of this rule set, it supports at least 113 known types of pseudo-organisms (a.k.a. "spaceships"), 442 oscillators, and 218 stable objects (a.k.a. "still lifes")[11]. It's hard to appreciate how interesting the rule set is by just reading about it or simulating it by hand. There are numerous simulators available for it on the internet and as standalone programs. You should find one and experiment with it some so that you can see what the rule set looks like when animated in real-time.

If that's not enough for you though, these rules are also *Turing complete*[12], which means that in theory the system can support any conceivable computation that you could ever imagine. For example, if you place live and dead cells in exactly the right spots on a large enough grid, it can compute anything you want it to, such as adding and multiplying numbers, calculus, geometry, truth tables, statistics, physics simulations, and so on.

---

[11] LifeWiki's web address is *conwaylife.com/wiki*. It has some interesting stuff, including images and animations of a bunch of the pseudo-organisms so that you can see what they actually look like.

[12] This was proven by Paul Rendell in 2001 in a paper entitled *Turing Universality of the Game of Life*. Rendell also created yet another way of implementing a universal computer using *John Conway's Game of Life* in a 2011 paper entitled *A Universal Turing Machine in Conway's Game of Life*. Rendell recently published a full-size book in 2015 on how the system works, which is entitled *Turing Machine Universality of the Game of Life* and is available from the publishing house Springer.

After it finishes performing the computation, you simply examine the grid and the answer you asked it to compute will be encoded in some form on the grid. When I say that it can compute anything, I do mean *anything*. That even includes *artificial intelligence*. Yup, you heard that right. This tiny little rule set *already* has enough information in it to support any conceivable artificial intelligence program that you could ever write, in principle.

Thus, if consciousness can emerge from artificial intelligence, then it can also theoretically emerge from this tiny rule set. In principle, all you have to do is give the simulation a suitable grid of initial cells. Mind you, it would be extremely difficult to actually do this in practice, not to mention it would likely require a prohibitively large amount of computational resources and time. But still, it is doable in principle. Bet you didn't see that one coming, did you?[13] So much for human intuition, right?

Our inability to intuitively grasp the consequences of even this tiny little rule set is proof that we humans are actually quite a lot dumber than we like to think we are. It makes you wonder what an organism 10 times smarter than us would be like, doesn't it? We probably wouldn't even be able to conceive of the depth of such an organism's thoughts. It would be like a blind person trying to imagine colors. To put this in perspective, a perfectly intelligent being would take one look at the rule set for *John Conway's Game of Life* and instantly recognize that it could support any conceivable artificial intelligence that could ever be created. Imagine what the world would look like if you had that level of piercing insight.

Also, remember that despite how much diversity this rule set supports, it's just *one* among countless other possible rule sets that you could choose. There's an entire community that specializes in experimenting with these kinds of systems. The way that one rule set behaves can be vastly different from another. For example, one rule set may generate seas of endlessly swirling spiral shapes. Another may generate mazes. Yet another may generate stable networks of interconnected rooms that resemble caverns. It really is incredible how much is possible based on even such a simple grid-based method as this. In case you're interested in looking up more information on these kinds of systems, the general term for them is *cellular automata*. There are a bunch of different programs available that can simulate many different rule sets.

Anyway though, back to the point, as you may recall, the reason why we went off on this tangent about *John Conway's Game of Life* was to explain why people's intuitive disbelief of the concept of higher-level complexity spontaneously emerging from simple rules is fundamentally unfounded and unjustified. The fact that the average human is too stupid to realize that this is how the universe works does nothing at all to diminish the fact that this is indeed how the universe works. The universe doesn't care if you don't understand its profound and elegant nature. It'll just keep on ticking along either way. The universe is just one giant emergent effect engine, just like *John Conway's Game of Life* is.

Ask yourself this: If *John Conway's Game of Life* already has enough in it to support any arbitrary artificial intelligence program that could ever be conceived, then how in the blazes could the real universe, whose rules are *vastly* more complex and rich than the rules of *John Conway's Game of Life* are, ever manage to *not* spontaneously evolve life? So you see, just as I said, the idea of a universe without evolution would actually be far weirder than one with it. This is not difficult to see. All it takes is a bit of intellectual honesty and experimentation and it quickly becomes glaringly obvious.

Evolution denialists need to wake up and face reality. The spontaneous emergence of beauty and complexity from simple rules is the norm. It is not some kind of absurd notion. It is extremely common. It can be easily observed, measured, tested, and proven. There is *nothing*

---

[13] Well, unless you've already heard about the rule set...

implausible about it. It is proven fact. Life evolves just as surely as the sun rises in the sky each day. Almost any logical system that you could conceive of probably experiences at least some form of emergent effects. It is part of the fundamental essence of what our universe is.

The natural world is *already* in and of itself unimaginably more nuanced and captivating and beautiful than any fake "explanation" ever conceived of by any religion or folklore could ever be. Whereas religion is dull and unimaginative and explains nothing, science is vibrant and surprising and explains everything. Once you see the truth, you realize that the idea of beauty and complexity spontaneously emerging from simple rules is actually a far simpler and far more credible idea than any kind of hand-wavy creation myth will ever be. Just because an idea *seems* easier to understand at first glance doesn't make it more credible. You need to actually test things and experiment if you ever hope for your ideas to have any integrity at all.

This brings us back to our main subject, the fluency heuristic. The inability of the human mind to trace the consequences of vast numbers of interacting states is largely responsible for why emergent or chaotic systems seem so incomprehensible and unpredictable to us. However, there are of course numerous vastly more mundane instances where the fluency heuristic misleads us. For example, there are many incorrect "facts" that float around in popular culture largely by virtue of simply being easy to understand or entertaining to talk about, despite being wrong.

For instance, the notion that there are "left-brained" and "right-brained" people, and that the "left-brained" ones are analytical whereas the "right-brained" ones are creative is *not* supported by science. In fact, the left-right brain pop-sci myth is actually mostly just a social bludgeon used to baselessly discriminate against people, no fundamentally different from discriminating against people for being black or white. It also generates lots of counterproductive self-limiting beliefs. You shouldn't buy into it. Almost anyone who can excel in analytical tasks can also excel in creative tasks and vice versa. All it takes is practice, patience, and good learning materials.

One common example of the fluency heuristic biasing people's judgment is in what is known as the "rhyme as reason effect". The rhyme as reason effect describes how an idea will tend to stick better and be more likely to be accepted as true if the phrase used to describe it is memorable, such as if the phrase rhymes or is otherwise catchy in some way. Mottos and slogans are often designed to exploit this effect. There are also numerous poetic sounding proverbs and famous quotes that can be easily proven wrong with a counterexample if you just stop and question them for even a few seconds, but which people still blindly adhere to just because the words superficially sound good. These false nuggets of "wisdom" are sometimes referred to as "thought terminating clichés", which is a pretty good description of what their net effect on society actually often is.

For example, one particularly overused thought terminating cliché is the ever famous "definition of insanity" quote, which is frequently misattributed to Einstein[14]. It goes something like this: "The definition of insanity is doing the same thing over and over and expecting different results.".

This is actually bad advice in many cases, and causes lots of people to give up too easily when they don't immediately get the results they want. It also blatantly contradicts scientific practice. The whole reason scientists collect so much redundant data and prefer large sample sizes is because when studying a natural phenomenon, repeating the same experiment over and over will often result in at least a slightly different outcome each time, due to randomness and measurement errors. This is pretty much the opposite of what the "definition of insanity" quote

---

[14]Einstein is perhaps the most misattributed and misquoted person in human history. As such, any time someone says that Einstein said something your default position should be skepticism. Einstein quotes require strong citations.

says. If the quote were true then it would imply that almost all scientists are insane, which is clearly false.

Anyway though, the fluency heuristic is a complicated issue. It is both an extremely useful time management optimization and a dangerously misleading bias. All sentient beings with limited information access are by definition at least slightly biased, due to the fact that limited information itself is a form of judgment bias. Nonetheless, it is still extremely important to try to minimize the impact of your biases if you hope to live a consistently stable, rational, and prosperous life. Clarity *does* lend ideas more utility and trustworthiness from a human standpoint. However, clarity isn't truth. It's just a way of making concepts more transparent. Whether clear or opaque, every idea should always be subjected to logical scrutiny. No matter how easy or intuitive an idea seems, always ask yourself: How might this be false? Can I think of counterexamples?

6.4.16 Bikeshedding

This cognitive bias is closely related to the fluency heuristic and procrastination. It's a type of avoidant behavior derived from a tendency to gravitate towards easy to understand tasks while avoiding hard to understand tasks, regardless of priorities. Even when one task clearly generates more value per unit of time invested, if there is an easier to understand task available then people will often tend to be attracted to that easier task.

This is often true even if the task is very insignificant and adds little to no value to the overall project. The term bikeshedding originates from a famous hypothetical story about the planning team for a nuclear power plant spending all of its time during a meeting discussing the details of the bikeshed, because it was easy to understand, instead of addressing the far more complicated and far more important matter of the design of the nuclear power plant itself.

There are lots of different contexts where this behavior occurs. It's a very common counterproductive human behavior. For example, programmers might focus disproportionate time and energy on comments and code style instead of on core functionality, politicians might spend most of their time passing trivial or unenforceable laws with little to no impact on people's lives rather than prioritizing society's most critical issues, students might do every trivial exercise they can find in their textbook while totally ignoring the more challenging and illuminating exercises, and so on.

If you want to get ahead in life though, it's generally best to focus on the activities that give you the best value per unit time invested rather than merely doing what is currently most comfortable for you. What's easy in the short term is often hard in the long term. The more you confront hard problems in the short term, the more your long term future will tend to become easy. The irony of avoiding hard problems is that it tends to make life harder.

6.4.17 Attentional bias

The way people perceive the world changes depending on what they spend the most time thinking about. This is attentional bias. One of the most distinctive examples of this effect is known as the *frequency illusion*. When you've recently been exposed to something, then you are more likely to start noticing it everywhere. This creates the false impression that the object is becoming more common, when in reality you are simply noticing it more often.

The reason why this happens is probably because after your first encounter with the object your brain's pattern recognition system has been primed with the associated stimulus, marking it as something to look out for now that you understand that it has potential relevance or value. It's yet another component of your brain's automated filtering mechanisms, just like pareidolia and the contrast effect.

This cognitive bias is partly responsible for why when people buy new clothes they may suddenly notice lots of other people wearing similar clothes and may think that the other people are copying their style or that maybe its a new fashion trend. It's also the reason why you're more likely to notice people who drive the same car as you on the road than you are to notice other specific cars. Let me tell you a story from my own personal experience with the frequency illusion.

I once ordered a piano sheet music book for Final Fantasy 8 from Japan, since it wasn't available locally at the time, and it arrived in a bright yellow DHL shipping truck. Prior to that time I didn't even know that DHL existed. I thought that the USPS and UPS were practically the only shipping companies operating in the United States. Suddenly though, I started noticing yellow DHL trucks frequently on the road after that. I was for all practical purposes totally blind to DHL's existence despite the fact that they were bright yellow, had always been there, and were very hard to miss.

Besides the frequency illusion, attentional bias can also cause you to automatically scrutinize something more the more you think about it. For example, someone who loves shoes and owns a lot of them will tend to notice a lot more about other people's shoes than they otherwise would. Also, there are gag videos where something absurd will happen on screen, like a person wearing a giant bear suit walking across the middle of the scene, and most people watching the video the first time won't even notice. That's attentional bias too. Have you ever been been so engrossed in reading something or working on something, that you can't even hear someone talking to you?

That's because the human mind only has a finite amount of consciousness to distribute. If there's not enough unused consciousness available to notice something then you literally won't even perceive it. As such, when you get jolted out of this state of deep focus, then it probably indicates that you were so engrossed in what you were doing that your brain's computational resources were potentially nearly maxed out. Productivity is often maximized in this state of enhanced focus, but it is not always easy to enter it, so interrupting someone who is focused like this can be pretty counterproductive for them. This state of deep focus and maximal efficiency is commonly referred to as "flow" in psychology, and also in game design theory in a slightly different sense.

6.4.18 Anchoring bias

The first idea that someone hears posed as an explanation for something will tend to be treated as the default position if it sounds plausible enough. Often it will initially be given a disproportionate advantage against other possible beliefs, and everything will subsequently be more likely to be framed in terms of it. Similarly, the first price that someone hears during a negotiation, such as when haggling with a merchant or negotiating the salary for a job, will often tend to serve as a focal point and change the way subsequent negotiations are perceived relative to if a different starting point had been chosen. Most broadly speaking, among any arbitrary set of related entities, the first entity from that set that a person encounters will sometimes tend to disproportionately color their perception of the other entities, at least temporarily. This is anchoring bias.

Put another way, many people's minds have a tendency to want to fill in unknowns by grabbing onto the first plausible explanation they can find, which they then begin to incorporate into their worldview. The longer that anchoring idea sits in someone's mind, the more connections the mind will form between it and other pre-existing or new ideas. The more of these connections form, the more entrenched the idea will become, and the more difficult it will become for the person to handle the cognitive dissonance that would be associated with abandoning the idea. However, a perfectly rational being would not treat an idea as automatically being any more likely to be valid than the alternatives are just because that idea happened to be the first idea that they encountered. How early on an idea is encountered and how correct it is are not generally reliably correlated with each other. Belief precedence has little to no relevance to the substance of most claims.

6.4.19 Framing effect

The same information presented in different ways can change the way you react to it, even though nothing about the information itself has actually changed. This is the framing effect. It's a very broadly applicable bias, due to the fact that virtually any environmental or contextual factor could potentially subtly alter your judgment. For example, showing someone the same product twice, both at the same price of $5 but framed in two different ways can result in significantly different reactions.

Suppose that in the first case we simply say that the product costs $5, with no extra rhetorical frills attached, and in the second case we say that it's 75% off of $20 for a supposedly "discounted" price of $5. It is likely that the second version may attract more sales even though literally nothing about the value proposition has actually changed. In both cases the product is identical and therefore the value per unit cost is the same. A perfectly rational shopper would see no difference between the two offers. Perfectly rational shoppers only ever think in terms of value per unit cost and never in terms of how a merchant is attempting to frame the sale.

Fake "discounts" are one of the most popular methods that businesses use to exploit their customers. Many stores artificially inflate their prices and then "discount" them in order to be able to claim that the items are "on sale" so that they can take advantage of the framing effect to generate more sales for an item than is actually merited. It's a form of economic parasitism. Like all forms of exploitation, in the long term it probably does more harm to the economy than good. Anyway though, this specific example aside, the framing effect comes in countless different forms, some of which are benevolent and others of which are not. Literally any environmental or contextual factor that skews someone's judgment away from objectivity could potentially be considered to be an instance of the framing effect.

6.4.20 Hindsight bias

This is a pretty widely known bias. It is often expressed by the popular saying "hindsight is 20/20". Basically, hindsight bias is when you assume that your reasoning for a prediction was valid and correct based merely on the fact that your prediction happened to end up being true. The problem with thinking like this is that your prediction may have become true only by coincidence. Your reasoning for making the prediction may have in fact been irrational and unjustified.

People often like to say that they "knew it all along" after something happens, even when their basis for this belief may in fact be complete nonsense and have no real predictive power. Real predictions are often very difficult to make and require quite a lot of information, so it is generally wise to default to being skeptical of them. The average person, untrained in logic and science, often has a grossly inadequate estimate of just how much information is actually required in order for a claim to be genuinely justified.

Hindsight bias is often exploited indirectly, in that it is sometimes used as a bludgeon to push for a particular decision or policy, both in business contexts and in domestic contexts. For example, suppose that two people at a company have two competing ideas for what project to do next, namely project A and project B. Suppose furthermore that the company decides to go with project B, but project B ends up failing due to random circumstances.

Consequently, the person who was pushing for project A may use this as an opportunity to say "See! I told you project B was a bad idea! We should have gone with my project A idea!" when in fact there may be no genuine reason to believe that project A would have been better. In fact, it still could easily be the case that project B was objectively a much smarter choice than project A, even in spite of the fact that project B failed. Only a careful and substantive analysis could discern the difference. You can't accurately make these kinds of judgments based on mere hindsight.

Another example is intrusive and controlling parenting, which I've heard some children unfortunately have to suffer through. If a parent wants their child to major in medicine, but the child instead decides to major in business, but the child subsequently ends up in a car crash while in-route to a college specializing in business, the parent may then attempt to use this as leverage to say that the child shouldn't have majored in business and may hysterically attempt to push the child to change to majoring in medicine. In the worst case, if the parent controls all of the funding, then they may even just unilaterally use the car crash as a convenient excuse to force the change of major against the child's will, despite how irrational it would be to do so. Here again, hindsight is a poor metric by which to judge past decisions.

There is only one way to justify a prediction, and that is by providing some combination of hard evidence, logical proof, or strong statistical data. Only then can you truly claim that a prediction was valid, or if not that then at least statistically likely. Many people think that the truth of a predicted event is the primary basis on which to judge the validity of the prediction of the event, but in reality this is often not true. In general, it is far too likely that the occurrence of the predicted event could just be coincidence. The event itself is often weak evidence at best. To truly judge a prediction, you have to look at the structure of the argument itself to determine if it was ever truly logically justified. Keeping this in mind will help you avoid hindsight bias.

6.4.21 Outcome bias

Outcome bias is similar to hindsight bias. The difference is that outcome bias happens when you judge a decision differently based on what the outcome is, regardless of how reasonable the decision was, whereas hindsight bias happens when you treat a prediction as being more accurate and more reasonable than it actually was, based on information which could not have been known at the time the prediction was made. Outcome bias is about the morality or quality of decisions, whereas hindsight bias is about the validity of predictions.

For example, suppose we have two people, person A and person B, who both drive their cars to work everyday. Suppose furthermore that person A and person B are equally well-intentioned and benevolent in all respects. Both drivers always obey traffic laws, always strive to be safe, and never intend to harm anyone. However, suppose that person B accidentally hits a pedestrian while driving to work one day. In contrast, person A never hits anyone. Both drivers acted with equally good intent and made the best safety judgments they possibly could at every moment. How will society judge them though?

Person *B* in this scenario will likely be severely punished, even though objectively speaking person *B* is *exactly* as ethical as person *A* was. The only difference is that person *A* was lucky enough to not hit someone. Thus, there are actually two victims in person *B*'s car crash: One is the pedestrian that driver *B* hit. The other is driver *B* themselves. The pedestrian is a victim of pure random chance, not a victim of person *B*. In contrast, person *B* is actually a victim of society's poor reasoning skills, preconceptions, mob rule, and bigotry. Person *B* is not actually guilty of anything.

The only guilty party here is actually society itself and the false "justice" that it imposes upon anyone who is randomly unlucky enough to get caught up in the legal system somehow even despite having the best intentions. If you think about the situation objectively, it is quite blatantly obvious that person *B* is actually morally innocent. The fact that the pedestrian being hit is emotionally upsetting and unfortunate is irrelevant. Bad things just randomly happen sometimes.

The car crash example here is thus no more of a moral transgression than someone randomly being diagnosed with cancer would be. Person *B* acted with good intent at every step, and thus is ethically blameless. In a truly honorable and just society, people would only ever be judged based on their intentions, if such a thing were possible, not based on random outcomes that are outside of their control. To do otherwise would be to indulge in outcome bias.

6.4.22 Choice-supportive bias

Sometimes, people will tend to view their own choices as being disproportionately more positive than the other options that they did not choose, regardless of the actual merits. This is known as choice-supportive bias. Selective memory and ego are potentially strong contributing factors in what causes people to have choice-supportive bias. In the case of selective memory, people remember what they focus on more, and the fact that they made a certain choice will often imply that they happened by chance to focus more on the positive attributes of the choice they made, but less on the positive attributes of the choices they didn't make. People's memory also often becomes oversimplified, romanticized, caricatured, and exaggerated over time, perhaps as a side effect of the brain attempting to make the memory easier to store by removing details.

In the case of ego and how it contributes to choice-supportive bias, people often treat their choices as being part of their own identity, which in turn causes them to associate any criticism of the choices they've made with criticism of their own personal identity, thus agitating their ego and impeding objective thought. In effect, people often conflate the choices they've made with their own value as a person, and thus come to feel personally threatened by any criticism of any of those choices. In contrast, a perfectly rational person would examine their own decisions exactly as if they were instead an external disinterested observer, one with no egotistical interest in how the choice is judged.

One of the most common forms of choice-supportive bias is *post-purchase rationalization*. This is when a person makes a purchase, typically of an item that is at least somewhat expensive, and then when they later start to notice flaws in their purchase they try to rationalize it away by putting extra effort into focusing on the positive aspects of the purchase. They often don't want to admit to themselves that they made a mistake, because then they would have to admit to themselves that they did something stupid, and now have to take responsibility for the resources they wasted.

Choice-supportive bias is often essentially a subtle form of arrogance. A calm, confident, and rational person would be able to look at the mistake objectively and admit to themselves that it was unwise, and thus they would thereby be able to learn to avoid similar mistakes in the future more effectively. A person who is thinking rationally would not feel personally threatened by this admission. They wouldn't try to run away from the personal responsibility that it entails.

In contrast though, a person harboring subconscious arrogance will have difficulty admitting the mistake to themselves and thus will try disproportionately hard to convince themselves that the choice was good. Mentally disassociating your personal identity from your choices, so that you can weigh those choices objectively instead of egotistically, and also not fearing taking responsibility for repairing the damage caused by those bad choices, and all the effort that entails, is the key to avoiding choice-supportive bias.

6.4.23 Sunk cost fallacy

The sunk cost fallacy is sort of similar to choice-supportive bias. The difference is that the sunk cost fallacy is more about how future decisions are made based on what expenses and resource losses have occurred up to this point, whereas choice-supportive bias is more of a matter of ego and not being willing to acknowledge flaws in one's past choices. In short, the sunk cost fallacy is about economics, whereas choice-supportive bias is about ego.

Someone can be perfectly willing to admit to making bad choices in the past and may have no egotistical tendencies whatsoever, yet still commit the sunk cost fallacy. When you make suboptimal economic choices based on a feeling of being committed to past choices, for fear of wasting the expense you committed to those past choices, then you are guilty of the sunk cost fallacy. It doesn't require arrogance.

A perfectly rational economic actor will always think in terms of what actions going forward will give them the best economic return, relative to their own values and goals in life. For such a person, whether or not they have committed resources to a certain task is irrelevant if switching to a different task would objectively yield greater returns. If the expenses and lost resources for a past choice are not recoverable, then there is no point of viewing oneself as being "committed" to that choice. Instead, you should think in terms of asking yourself what the best possible future path going forward is. Sometimes the resources invested so far in something *are* worth continuing an endeavor, but sometimes not. It depends. Don't get locked into counterproductive economic commitments though.

Here's a simple thought experiment to make the idea more understandable: Forget everything you know about your past decisions, and then imagine that you were suddenly teleported into your current circumstances. What would be the optimal way to use the resources you now see in front of you? That's how a perfectly rational and economically optimal person would think about it. Here's yet another way to look at it: Imagine that it's like you just powered on a video game console and opened a game. Imagine that after clicking "Start" on the main menu you find yourself suddenly in your exact current economic circumstances, running around as a character in the game (yourself) but controlled as if by an external objective observer (the player).

You have no idea where the starting resources you have been given when you started the game came from. What is the optimal way to use those resources to maximize your outcomes? The answer is this: Act as if you are performing this thought experiment at every single moment of time, as if as you make progress the game is constantly restarting and you are constantly being thrown into the present circumstances without knowing where your resources came from.

Think as if in every moment you are constantly standing at the intersection of a vast nexus of pathways stretching forth into the future, that each represent different choices you could make, like world-lines representing parallel universes, each slightly different from the others, and at every moment you look at those pathways and you try to choose the most optimal one as best you can and act accordingly. Live every moment as if you have only just now been born. That's optimal decision making. If you act like that then it will guarantee that you are *not* committing the sunk cost fallacy.

One of the most common examples of the sunk cost fallacy is when you are working on a project, but then realize at some point that the project is not going to end well. You now know that you are definitely not going to get the return you were hoping for on the project, yet you still feel committed to it because you have invested so much time and resources into it. There may be some other project you could switch to which would surely give you a better return on your investment, even in spite of the fact that you have already wasted some of your resources on the previous project. If such an option is available, then a rational economic actor would switch to it, under the right circumstances.

However, there is a catch. Your understanding of the economic situation and the true costs of switching (including even hidden costs such as demoralization, burn out, not completing projects, etc) has to actually be accurate. This is not always an easy thing to judge. Many times when you are heavily committed to a project, then it may indeed still be wise to continue with that project, because otherwise you might not have the resources to switch to a different project while still managing to get a greater return. This is especially true for projects that are nearing completion. Even if you find out that a project won't go as well as you had hoped, it may still be optimal to see it through to completion. Even if you don't make a profit, you might still be able to recover some large portion of what you invested in some form, thereby allowing you to reduce your losses.

Thus, even if you have a new opportunity that is a better than the current project that you are heavily invested in, switching to another project might actually cause you to make a net loss. The point is, don't be overzealous in trying to avoid the sunk cost fallacy. Don't waffle back and forth between projects, never getting anything done. Don't optimize prematurely. Completing projects that are doomed to economic failure can still be a valuable learning experience, one that may pay huge dividends when applied to another project down the road. There's a lot of nuance involved in accurately making a decision to switch between projects, so you need to be very cautious about it and ensure that you really are thinking objectively about what your circumstances are. It's very easy to accidentally be biased in either direction. Don't be overly committed, but also don't be overly fickle. You need to strike the right balance.

You don't want to stay committed to a bad idea, but you also don't want to lose valuable resources that may be recoverable by seeing the idea through to completion either. Think carefully. What would an external objective observer do? Consider the time scale. The longer the time required to complete a bad project, the more likely it is that you should switch to something better. For example, if you are majoring in some subject in college, but then realize that you hate it and want to do something else, then sticking with it just because you've invested a few years into it is probably a bad choice. The consequences of doing so may be with you for the rest of your life. In that case, there's a good chance that you should probably switch, even at extra expense (within reason and considering economic outcomes), and even if it makes you feel "behind".

On the other hand, there are times when you should stay committed, even to bad ideas. For example, if you are a game developer who has never completed a game before, and you have completed 95% of one of your game projects, but the game seems to be a bit mediocre and disappointing, then there is essentially almost no circumstance under which switching projects at that point would be a good idea. It would be hard to economically justify such a switch at such a time. The learning experience of seeing the mediocre game through to completion and then releasing the project for sale, and thereby getting a chance to experience the dynamics of the market yourself for the first time, would likely be hugely valuable, and you would even stand a chance to regain your investment over many years of passive income from the sales.

Keep this in mind. Avoiding the sunk cost fallacy will help you move forward in life, allowing you to more efficiently acquire wealth, creative fulfillment, peace of mind, and social status. If you've ever noticed someone in your life who seems to always manage to get ahead in life faster than the other people you know, there's a good chance that they are either consciously or subconsciously avoiding the sunk cost fallacy. They are seeing through the noise in their lives and focusing in like a laser on what is truly important to them. They are making decisive choices that truly move their life goals forward. They are not slaves to their past, nor to their present circumstances. They don't waste time worrying about things that are outside their power to control.

They know that the average person doesn't accomplish much in life, and that therefore the average person is often not a very good source for advice or inspiration. They know to be skeptical of the assumptions of society. They question everything. They work for themselves, not for others. They ignore the artificial barriers society thinks they need to pass[15], and instead immediately jump forward to where they want to be. They know how to efficiently find the path of least resistance and maximum gain. They relentlessly eliminate waste. They know that time is the most valuable thing they own. They know how to play the game. They don't live life half-asleep.

Logical thinking pays huge dividends. It doesn't just help you discover the truth. It also makes you much more likely to become wealthy and to find personal contentment and peace of mind in life. It does this via how it empowers your perspective on optimal decision making and thereby enables you to mentally free yourself from the chains of your past and present circumstances. Logic is freedom. The prison we build inside our own minds is the strongest prison of all. Logic loosens the chains of that prison.

There are many people who would gladly enslave you, mind, body, and soul, if ever given the chance. Exploitation is everywhere, so be careful. Don't stay committed to people who don't have your own best interests in mind. Replace the bad investments in your life with better investments, whenever you are sufficiently certain that you can make a net gain by doing so. Diversify your assets and don't fall into fallacious thinking. Prepare for contingencies and assume that at least some things will go badly and at least some people will betray you. Make your life error resistant. Acquire a timeless eye, one that sees through lies and discerns fact from fiction, one that can see the nexus of all possible choices stretching forth before you from this moment of time, and at every moment of time, one that empowers you to move your world closer and closer towards the best possible path, towards the best possible world available to you.

---

[15] For example, consider the "do your time" mentality, where newbies are often artificially held back by inefficient and fruitless methods of practice and training, often given extremely low-value tasks with almost zero educational value, and often also undercompensated and interfered with. The people doing this like to pretend that it is "necessary for building character", but in reality it is most often just wasteful and oppressive. The real reason it is done is probably often just to make the more senior people feel better about themselves (i.e. bullying), and also to break the newbies' spirits in order to make them less likely to take any power away from the seniors, like a form of hazing or brainwashing or Pavlovian conditioning, rather than for any kind of genuinely productive reason.

6.4.24 Optimism bias

Thinking that future events or endeavors will go better than past experience or statistical data indicates is optimism bias. This most commonly manifests either as a belief that you are much less likely to experience negative events than other people are or as unrealistic planning of your future endeavors. With optimism bias, even if someone nominally knows at some level that their baseline chances aren't actually that great, based on past experience or data, they will still often assume that things will go well regardless.

For example, people will often assume that they are much less likely to contract a disease or health condition than others in their group are, even though they have no actual rational or evidentiary basis on which to believe this. For instance, many smokers will assume that they will be the exceptions and won't ever contract lung cancer. The same kind of thinking applies to any kind of negative event. People walking across a street in a rush often think that they are "smart enough" to not get run over, unlike those other "average" people, completely failing to realize that many of the people who were hit when walking across the street (perhaps even almost all of them) *also* thought that they were skilled enough to do it safely. This kind of thinking causes people to not properly plan for risks and negative events in life, and ultimately causes a lot of unnecessary suffering.

The other most common manifestation of optimism bias is probably unrealistic planning, typically of large projects. Many people have a pretty strong tendency to underestimate how long it will take them to complete a project, based on a wildly idealistic estimate of how productive and diligent they are going to be once they are working on the project. This is often referred to as a *planning fallacy*. Typically it goes something like this: The person planning the project thinks back to past projects they've worked on and realizes that they did a less than ideal job of being productive on those projects.

However, the person nonetheless assumes that *next time*, on the upcoming project, they will act completely responsibly and will operate seamlessly and with ideal efficiency. It's kind of like at every new project the person thinks to themselves "This will be the time that I have a breakthrough and maintain a perfect work ethic. I'll also be so diligent and so aware of my environment that no unexpected setbacks will occur." even though this is highly unrealistic and even though all of the person's past experiences indicate that things are unlikely to go that well.

This effect is even more pronounced in contexts that naturally tempt the planner to "dream big", such as in software development and in architecture and construction. Many a software project has been blown absurdly beyond its initial budget and allocated time. The same is true of many architecture and construction projects. The more grand your ambitions are, and the more moving parts the implementation has, the greater the danger of unrealistic planning. In fact, this effect can be so pervasive that it may even be wise to double all of your initial estimates (or more), even when you think your estimates are objectively realistic.

It's easy to deceive yourself on ambitious projects. It plays into your subconscious ego and such, and also into any wishful thinking thinking you may have. People often want large projects to go well so badly that they will just outright not consider the possibility that it won't go as well as planned, because they can't mentally handle the idea of dealing with the unpleasantness associated with what might happen if it doesn't go well. The more a person invests in something, the greater the danger will be that they will deny or ignore problems that appear later on, just because they don't want to face the responsibility and work necessary for fixing those problems.

Having lots of real-world experience completing similar projects can significantly reduce optimism bias though. Battle-tested veteran project managers will tend to be better at avoiding unrealistic planning errors. However, even the most grizzled veteran can still easily be tempted to make overly positive estimates on projects. It's easy to fall into the trap of thinking "I'm so experienced that my estimates are actually realistic." when in fact many factors cannot be accurately predicted with *any* amount of expertise.

A wise project manager knows to always leave themselves a large margin of error for financial safety and for project scope and resource consumption. In fact, that's good advice for life in general, not just for projects. Always leave yourself a large safety margin in all aspects of your life, whether it be your bank account, your employment prospects, your attachment to people, things, and places, or your health (etc). No plan is ever truly guaranteed. Catastrophe can always potentially strike. Be prepared for it. It's not *if* it's going to happen, it's *when* and *how*. Think of negative events as being guaranteed in at least some quantity, instead of merely as being abstract "possibilities".

6.4.25 Golden hammer effect

This one is an especially common affliction among software developers. The golden hammer effect is a cognitive bias where someone who has recently learned a new technique or has been given a new tool will suddenly start applying that technique or tool to a disproportionately large number of different situations, many of which the technique or tool is poorly suited for. The effect is often summarized by the popular saying: "If all you have is a hammer, everything looks like a nail."

For example, a software developer who has recently learned about the concept of implementation inheritance is likely to severely overuse the concept. This is the golden hammer effect in action. Another example would be a carpenter who has recently bought a new tool, who might subsequently begin using that tool in lots of weird contexts where it is not the ideal choice. The same might also be true of a plumber, an auto-mechanic, or a handyman who has recently acquired a new tool. The tool could even be a literal hammer. That is, after all, where the name of the effect comes from. There's a reason for that.

As irrational as it is, in many ways, to overuse a new technique or a new tool in this manner, it is not necessarily actually as bad of a thing to do as it may appear from an outsider's perspective. You see, the golden hammer effect is actually probably an important contributor to how people learn the boundaries of new techniques and new tools that they acquire. Although a technique or tool may be designed for a specific context, there might be other unconventional contexts where it is actually extremely effective. The chances of discovering such creative uses of new techniques or new tools is vastly increased if people are born with an impulse to overuse the new techniques or new tools they acquire.

Thus, the golden hammer effect actually sometimes has a net positive effect over time, as long as the obsession is only temporary. It seems to probably be an ingenious evolutionary adaptation that actually often yields a net gain. Apparently irrational behaviors sometimes have hidden advantages that are not immediately apparent. Let this be a lesson to you to use caution before prematurely judging apparently irrational behaviors of the general public. Many strange behaviors are indeed genuinely counterproductive, but others are actually just evolutionary genius in disguise. It can take a sharp eye to spot the difference. Some apparent flaws really are just flaws, but others are advantages in disguise.

In contrast however, sometimes the golden hammer effect can become cemented in a user's mind, where it will now manifest as deeply entrenched ideological fanaticism, often with very counterproductive effects. When the golden hammer effect becomes a long-term condition in this manner, instead of being just a temporary effect that augments learning, it will almost always have a net negative effect on the user's mind. In some cases, it may even mutate and become a popular myth or "religious belief" of a particular discipline.

For example, in computer science there is a certain portion of the population that believes that it's best to only ever return from one location in a function and considers it good style to contort all code into this format. This is pure idiocy. Following this advice is actually one of the worst things you can ever do to the quality of function code. Exactly the opposite is true. Well designed function code actually tends to live by the mantra of "return early, return often", which is quite the opposite of "only return once".

Ideal control flow tends to hinge on creating a sequence of filters that establish a sequence of ever more restrictive invariants, returning as often as necessary along the way to support that structure, so that the control flow in the code does not become deeply nested and difficult to read. The more cases you can eliminate from the handling code as early as possible via conditional returns, the more readable and the less complex the rest of the code will tend to become. Doing so in functions inherently requires frequent use of return statements.

According to what I heard once[16], the myth of "only return once" may actually originate from a misinterpretation of some very outdated advice. You see, originally programming languages allowed you to return *to* multiple different locations from a function, including locations that were different from where a function was originally called from. This resulted in quite a lot of chaos and was indeed a very bad idea. Thus, many programmers advised against it. However, later programmers (ignorant of the context) distorted this advice until it became "only return once", which is not at all the same statement.

There is a huge difference between how many locations you allow yourself to return *from* in a function and how many locations you allow yourself to return *to* from a function. Allowing yourself to return *from* multiple different locations in a function is essentially harmless and tends to improve the structure and readability of your code greatly when correctly used, whereas allowing yourself to return *to* multiple different locations is extremely dangerous and most modern programming languages don't even allow you to do it anymore.

Anyway, although the golden hammer effect has some temporary utility in terms of helping you to learn the boundaries of where a technique or tool is applicable, it is still very much a cognitive bias and it can still easily have counterproductive effects on your mind. It is best to strike a balance. Respect that the golden hammer effect sometimes plays a useful role in the learning process, but also be wary of its tendency to contribute to irrational applications of techniques and tools. Especially don't allow it to turn into fanaticism or narrow-minded thinking.

We live in a diverse world, so you should expect optimal results to require a similarly diverse set of techniques and tools. If you only ever seem to be using a narrow set of techniques and tools in your work, double check that you aren't thinking too narrowly. Some situations are indeed only optimally solved with a narrow set of techniques and tools, but many more are not. Keep your perspective fresh and nuanced. Remember, unless your name is Thor[17], your hammer is not always the best choice for the job. Control your tools. Don't let them control you.

6.4.26 Spotlight effect

Thinking that people pay more attention to you than they actually do, such as noticing and remembering every little imperfect action you do or picking up on your inner feelings based

---

[16]I've since forgotten the source, so take this with a huge grain of salt.

[17]Thor is the Norse god of storms, thunder, lightning, and strength (etc). In mythology, he wields a magical hammer.

on subtle aspects of your body language, and so on, is the spotlight effect. The effect gets its name from the fact that it makes you feel like you are "under the spotlight" more often than you actually are. In reality, people often are so preoccupied with attending to their own concerns and thoughts that they only have a relatively small amount of that attention to spare on other people. Usually, people notice far less about you than you might believe.

We can only know our own thoughts with certainty, not the thoughts of others. The huge amount of insider information that you have about your own self and your own state of mind, combined with the much smaller amount of information you have about other people's internal perspective, contributes to you often estimating how much attention other people pay to you inaccurately. It may also in part be a defensive adaptation, one designed to make you extra cautious about how you appear to other people, just in case, due to the fact that sometimes social forces could unexpectedly jeopardize your survival or your future.

The more social anxiety a person has the worse the spotlight effect will tend to be for them. For a socially anxious person, even the smallest mistake may cause them to feel like they are standing at the center of a firestorm of negative social attention. It can make people feel very self-conscious. In fact, the extent to which a person has social anxiety and the extent to which a person is vulnerable to the spotlight effect may even be nearly the same thing. Shy people may just have an unrealistically severe notion of how much other people focus on them and how consequential that focus will end up being.

On the other hand though, such fears are sometimes justified and can serve as a useful precaution. Society is the mob. While most of the time most people are relatively reasonable, once they get whipped up into a hysterical frenzy they can become extremely dangerous. Survival in our society depends upon large-scale cooperation and each person having a reasonably stable role and position to fill within society. As beneficial as this social order may be to accomplishing great things and having a high standard of living, a wise person knows to nonetheless be wary of the random whims of society.

The same society that can work together to create seemingly impossible and wonderful things which no one individual could ever hope to create on their own, such as computers or radios for example, can also devolve into a tyrannous cesspool of bigotry, suffering, and irrationality. The social order is filled with monumental benefits, but also with colossal land mines. It is wise to keep society at a healthy arms-length distance from yourself if you want to guarantee your own long-term survival and well-being.

Stay out of toxic and fruitless debates and stay away from highly egotistically fragile and hysteria-prone people and communities. People who lack nuance[18], lack empathy, or are constantly on self-righteous crusades and always looking for excuses to grab their pitchforks and hold witch-hunts (e.g. social media, political news media, forums with downvoting systems, cults, etc) are among the most dangerous and wasteful people that you can ever spend time with in life. Stay away from them. Focus on more productive priorities instead. When necessary, use social awareness, tact, avoidance, passive attrition[19], and neutralization strategies to fly under the radar when you do have to deal with these kinds of toxic people though.

Focus primarily on consistently productive and beneficial work, the kind that will be viewed as being mostly socially and politically neutral. Limit the amount of information you make

---

[18] Relevant quote: "Tyranny is the deliberate removal of nuance." — Albert Maysles

[19] For example, try boring or tiring people out by repeatedly responding slowly or generically, each time providing little to nothing of value to them, and generally making things as tedious and costly as possible for them, preferably in a non-obvious way with lots of plausible deniability. This will make it much less worth their time to continue interacting with you, while still usually helping to keep you relatively safe from any tantrums or hostility they'd otherwise direct at you.

available to other people to reduce your chances of becoming yet another victim of society's strong tendency towards mass hysteria and unforgiving one-dimensional vindictiveness. Treat toxic social cliques like the plague and always stay aware of the immense danger of interacting in any way with such people.

Justice is often scarce or nonexistent in such contexts, and motivated "reasoning" and self-interest (e.g. extreme punitive reactions to even the smallest ego slights, hidden behind a false veneer of righteousness) are often the only real principles behind which such people operate. These people operate on the principle of guilty until proven innocent. Their interest is often purely to find excuses to attack other people, not to find the truth and not actually to make the world better. They will pretend to be honorable and to have good intentions on the surface, but really they just want to inflate their own egos. They just want someone to point their fingers at and to feel superior to. They don't care how much suffering this causes or what it costs though.

My point is, while in most reasonable situations, where people are behaving fairly normally and calmly, the spotlight effect will tend to make you overly self-conscious and nervous, there are still plausible reasons why feeling so disproportionately self-aware and paranoid about other people's reactions may be useful. Basically, in *healthy* social contexts (e.g. eating at a peaceful restaurant) you should try to loosen up and not worry as much, whereas in high-strung or toxic social contexts (e.g. social media, politics, forums that support downvoting, etc), or when interacting with people who possess more power than their level of mental maturity actually merits, a little bit of extreme social paranoia may actually save your life or livelihood.

When in the presence of hysteria or social toxicity, it is often wise to immediately deflate or deflect the threat, distance yourself from it, and then quickly relocate yourself to a more socially neutral and productive environment. Choose your battles wisely. Creative productivity is almost always a better place to spend your time than talking to hypersensitive and egotistical people. Always remember that there are lots of other things that you could be doing besides what you currently are. Don't get stuck. Don't needlessly place yourself in a toxic environment. Anyway though, some biases have reasons for existing. Be mindful of the way these biases can distort your thinking in counterproductive ways, but also be mindful of the fact that there are sometimes special contexts in which these biases are designed to protect you. The spotlight effect is such an example.

6.4.27 False consensus bias

Many people have a tendency to believe that other people agree with them more often than other people actually do. This includes everything from likes and dislikes, to political beliefs, to religious beliefs, to life outlooks, to opinions of games, movies, and food, and so on. This is the false consensus bias. It gets its name from the fact that many people act as if there is a rough consensus with the rest of society that aligns with their own beliefs, when in fact this is very often not the case.

For example, when you walk down the street, passing various people along the way, there is a good chance that you will assume that a disproportionately high proportion of the people that you passed share roughly the same beliefs as you do. This is in spite of the fact that you probably have little to no evidence to support this belief. There are multiple possible contributing factors to this effect.

One factor may be that, since each person's experiences are limited to their own lives, they lack imagination in estimating the full scope of variety of beliefs that other people may hold. Another factor may be that people want to view themselves as being part of the majority, as this tends to carry social advantages and additional power, and hence makes the person feel "better" than people who believe differently. The person may also not want to deal with the consequences of having a minority opinion, which may place an additional burden on them in life, depending on the nature of the belief and what the current power structures of society are. Believing that most other people have roughly the same worldview as you do tends to feel a lot more comfortable than acknowledging diversity. Comfort isn't truth though.

The false consensus bias also occurs frequently in the beliefs and the propaganda of political parties. You will often hear political parties phrase their policies in terms that imply that everybody (e.g. the entire nation, etc) supports those policies, even when there is abundant evidence that this is definitely not the case. This is probably in part an attempt of those political parties to psychologically manipulate people by exploiting conformity and peer pressure psychology[20], but it is also probably at least sometimes a genuine instance of an unwitting false consensus bias.

Political ideologues often become so invested in their ideas, and in their often misguided notions that their own specific narrow beliefs are the social norm or the natural default, that the cognitive dissonance of acknowledging that their own beliefs actually are *not* the consensus and may in fact be deeply wrong may become psychologically insurmountable. People often don't want to take personal responsibility for the consequences of acknowledging that their life-long "most treasured beliefs" and general worldview may in fact be objectively wrong.

Making that adjustment would require careful thought, introspection, hard work, and owning up to past mistakes, which many people, in their arrogance, find difficult to do and will make up any excuse to avoid. This behavior can result in all kinds of negative consequences for the rest of society. People also often don't want to be seen as "losing the argument", even when the evidence is clearly not on their side. Much of this behavior has more to do with egoistical slights and how people react to them than it has to do with any kind of real belief that they would hold if they actually thought about what that belief entailed independently from its context as part of a social conflict against people who believe differently.

6.4.28 Conformity

Conformity comes in many different forms and varieties. It means a tendency to adopt the opinions of others, regardless of objective merit. It is a cognitive bias that has both negative and positive effects on society. There are a vast number of different terms closely related to conformity, some of which border on being synonyms of conformity with only a subtle difference in connotation and meaning at times. These include, but are not limited to: the bandwagon effect, herd mentality, groupthink, following the crowd, collective unconsciousness, information cascade, mass hysteria, hive mind, social proof, peer pressure, and deindividuation.

Often, people's decisions are made more for the social advantages of conforming to others in society than for any kind of objective reason. It is often the case in a society that the current set of widely held beliefs is held in place not by real merit but merely by groups of people in power who do not want to lose any of that power. This is especially true in politics and religion, which both often exploit, rather than serve, the public while simultaneously pretending to act in the public interest through a wide variety of systematic propaganda mechanisms.

For all the air of authority these organizations try to project, they are often actually utterly dependent on creating an underlying hostile or fearful atmosphere, without which people would quickly stop conforming. A truly strong belief, in contrast, one based on genuine merit and

---

[20] e.g. "Everybody believes we should club baby seals to death, so why don't you? What are you, some kind of loser?"

alignment with truth, would not need to be propped up in this manner. It could stand on its own merits.

A disgustingly large portion of what people "believe" is actually based on nothing more than a desire to not lose special social privileges sustained by a particular social group's bigotry and oppression of other groups, rather than any kind of respect for truth. Thus, many people's "most treasured beliefs" are in fact fake and would instead be more accurately described as being their "most treasured desires for power and privileges acquired by conforming to a social group, regardless of merit or harm to other social groups".

That's part of why some people seem impossible to convince regardless of what evidence you present to them. Their "beliefs" are often actually just greed or insecurity in disguise, thus telling them the truth does nothing to move them. This is unsurprising, if you actually stop and think about it. With some people, it was never even about truth to begin with. They don't actually believe in what they say they do, but rather they merely pretend to in order to benefit themselves, regardless of the long-term costs to others or to society as a whole. Anybody is potentially vulnerable to this effect.

To help prevent yourself from falling into this same trap, ask yourself this: Do I believe this because I have reason to think that it is actually true? Or, do I believe this because I am too afraid of the consequences that the rest of society might impose upon me if they learned that I don't believe or obey it? The more fear you experience when considering adopting a belief that is outside the norm, but more reasonable than the norm, the more oppressive the society you are living in probably is. If your social group discourages you from ever considering alternatives, then you should be especially suspicious.

Suppression of alternatives is one of the most reliable indicators of evil intent. This trend is especially common among religions, virtually all of which heavily demonize other social groups and constantly strive to systematically suppress the free exchange of information in society. Nothing is more threatening to an anti-truth social order than the free exchange of information. Asking lots of questions, both of yourself and of others, is necessary for the pursuit of truth.

Suppressing such activities is strong evidence of an attempt to cover up falsehoods, probably so that a specific social group that has somehow acquired a disproportionate amount of unearned power can continue to possess it. Don't be naive. Always consider self-serving greed as potentially being the real reason why someone professes to believe what they say they do. Retaining artificial social advantages is very often the only real reason why people "believe" the things they say they do, not merit or truth. Remember that.

That being said, as toxic as conformity can so often be, it's not all bad. Conformity is actually a very complex and nuanced issue. Conformity covers a whole spectrum of different degrees of magnitude and different contexts. Its effects on our society can range anywhere from very bad to very good. It is actually a critical aspect of what even allows us to form large-scale cooperating societies in the first place, but it is also frequently one of the most destructive forces in society.

Most atrocities would not have been possible without conformity. Many a civilization has both risen and fallen on the shoulders of conformity. However, many forms of it are also quite innocent and unsubstantial. Take fashion for example. When someone sees someone else wearing clothes that they like, but which still fall within current social boundaries, then they might feel compelled to imitate that style. This is essentially a harmless behavior and is not something to be concerned about generally.

Anyway, there are a lot of reasons why conformity exists. It's not like it's just some random psychological quirk that doesn't serve a purpose. Social structures wouldn't scale well without it. For example, think about it from an information processing standpoint. Each human has only a limited amount of time and resources available for having truly original thoughts and for investigating the truth of the world around them. Yet, as humans, we essentially have no choice but to use lots of knowledge and technology which we ourselves do not actually fully understand in our daily lives. This would be pretty much impossible without conformity.

For instance, no individual scientist has sufficient time to actually perform every experiment they've ever learned about. Thus, there will always be at least a part of a person's knowledge that must be based on trust and conformity to the beliefs of others in society. Advanced technological societies like ours wouldn't work otherwise. We have accumulated more information than any one person could ever possibly learn, thus making trust and cooperation in at least some quantity unavoidable. We really are, to an certain extent, a hive mind. Our collective intelligence is much greater than our individual intelligence. We know a lot more as a society than we know as individuals. Humanity is not just a collection of individuals. We are more like an extremely intelligent super-organism, much like an ant colony except *vastly* more intelligent.

Experts with a proven track record of providing strong logical arguments or evidence are the best sources for conformity-based beliefs, usually. However, even the most trustworthy sources should always be treated with at least a little bit of skepticism. One of the most important caveats when considering supposedly "expert" opinion is to look for conflicts of interest. When clear conflicts of interest are present, it is wise to be skeptical, even regardless of how prominent or trustworthy an expert appears to be.

You'll need to carefully balance the potential benefits of believing other people against the chances that they are wrong. This is not always easy. When in doubt, pick the side that is the most transparent, neutral, scientific, and open to evidence. Never trust sides that base their beliefs on mere faith, i.e. on belief without evidence, as *by definition* beliefs that aren't based on evidence can never be considered trustworthy.

Another reason for conformity is group safety. It's just like how herds of animals move together in big flocks for protection. That is, after all, where the term "herd mentality" comes from. In contrast, if every individual in a group just did whatever they wanted to, it could cause the larger-scale behavior of that group to become much less stable. The resulting erratic behavior of the group could perhaps jeopardize the survival of the group as a whole. Groups often need to avoid common threats and environmental hazards, and that is often much easier to do if there is a shared social convention that discourages that dangerous behavior somehow. However, while it is true that there is generally "safety in numbers", sometimes group coordination can backfire and make things worse.

Just as genetic homogeneity in organisms poses a massive risk to the survival of those species, the same is true for belief homogeneity and the survival and well-being of the people who hold those shared beliefs. Just as it is wise to protect genetic diversity, it is also wise to protect belief diversity. There are stories from biology of entire species going extinct in the blink of an eye due to genetic homogeneity. In a homogeneous population, a single disease can easily wipe out the entire population.

Similarly, if all human beings believed pretty much the same things, then a single external threat (e.g. an alien invasion, a major environmental threat, etc) would be much more likely to make the entire human species go extinct than if the ecosystem of beliefs was highly diverse. Creative thinking requires diversity of thought. Having too many ideological fanatics running around trying to make everything artificially homogeneous thus endangers the survival of our entire species, and in fact you can see the effects of this play out repeatedly in history on a smaller scale.

Homogeneous societies appear to be more likely to go insane and to commit atrocities than multi-cultural societies. Such events tend to cause major social collapses, just as diseases tend to devastate genetically homogeneous populations of organisms. Adaption requires diversity. This is just as true of beliefs as it is of genetics. Homogeneous societies thus have much less ability to adapt, and therefore become much more likely to suffer massive collapses the moment they encounter a problem they can't solve. A wise businessperson knows to diversify their assets, just as a wise society knows to diversify their culture. Diversity supports stability, whereas homogeneity amplifies risks. This is a pretty universal principle. Societies that prefer to survive should keep that in mind.

On the other hand, some degree of conformity is critical for maximizing cooperative harmony. For example, when society collectively shames someone for doing something bad, the reason why that act of shaming works and is effective is because of conformity. Thus, morality and conformity are joined at the hip in human society, in some sense. Morality couldn't be enforced if people were uneffected by what other people thought. Information would also spread much more slowly, or perhaps even not at all, if people weren't conforming to what they hear other people say and thereafter repeating that information to others as if it was fact.

If nobody ever said "Hey, did you hear that X happened?" until they had completely proven to themselves personally that X did in fact happen then *much* less information would flow in society, and even if it did flow it would flow much slower. People have very limited time. A certain degree of tentative trust is necessary. Conformity is both a force for good and a force for evil. If you care about the well-being of society and of yourself, then you would do well to be mindful of it and to carefully distinguish the cases where it is acting as a force of good from the cases where it is acting as a force for evil, and everything in-between.

6.4.29 In-group bias

This cognitive bias is exactly what it sounds like. In-group bias is when you apply a weaker standard of evidence or moral judgment to members of your own in-group than you do to members of other groups. This is a very common phenomena. It's what makes social cliques feel the way they do. Much of the prejudice and corruption in our society actually originates from unconscious in-group biases. However, in-group bias also fills a role in society and is not all bad. You see, while it is true that in-group bias has a strong tendency to impair judgment and to support corruption and cronyism, it also can sometimes increase the survivability and overall well-being of a group.

There are many possible contributing factors involved in incentivizing in-group bias. One factor that may be involved is indirect benefit. It's true that assets that you claim for yourself will tend to benefit you the most directly, but it is also true that assets claimed by a group that you are closely associated with will also tend to benefit you. The more resources your own in-group possesses, the more likely you are to personally benefit from those resources at some point. Thus, anything you can do to help out members of your own in-group can end up indirectly benefiting yourself as well. Consequently, simply being a member of the same in-group as someone else can therefore in and of itself create a conflict of interest and impair your ability to objectively judge that person.

It's clear why this bias would often be a useful trait to possess from a natural selection standpoint. Groups that collectively retain more resources are more likely to survive and prosper than other groups are. This is even more true when resources are scarce. Thus, having at least some in-group bias will tend to increase the survival rate of a group, thereby making that group more likely to propagate its genes. However, there is a limit to how much in-group bias can benefit a group. Too much will backfire. Too much in-group bias destroys the ability of a group to hold its own members accountable for bad behavior and bad thinking, thereby causing corruption and inefficiency to run rampant in that group, which in turn could endanger that group's survival.

Thus, from a biological standpoint, having a balance of both in-group bias and objectivity will yield the highest survival rate overall. In other words, between the two extremes of complete in-group bias and complete objectivity there is a point of optimal fitness that groups will tend to gravitate towards through natural selection processes, which will vary relative to the conditions imposed by the environment. Scare resources will probably push groups more towards in-group bias, whereas abundant resources will probably push groups more towards objectivity. Let's call this optimum the **in-group bias balance point** of the environment.

Another factor that may have contributed to causing humans to have in-group bias is the evolution of our abnormally high degree of intelligence as a species. In order for a new trait to evolve in a species, there must usually be a lot of natural selection pressure available to push that species to adapt towards having that trait. This implies that at some point the human species probably must have experienced intense selection pressure towards evolving higher intelligence.

Merely competing with other species may not have been sufficient pressure to evolve as high a degree of intelligence as we have. Once humans became the most intelligent species, the only way to provide enough selective pressure for further growth in intelligence may have been to turn on each other, thus necessitating that in-groups compete against each other sometimes, both socially and militarily. In this way, dumber in-groups would tend to gradually die out, leaving only the smarter ones standing, and thus gradually increasing the average intelligence of our species.

Therefore, ironically, despite the fact that in-group bias usually leads to *worse* reasoning when trying to discover the truth, it nonetheless may have been essential for humans becoming so highly intelligent in the first place. The natural world works in unexpected and counterintuitive ways sometimes. Seemingly bad behaviors can fill surprisingly important roles sometimes. However, be that as it may, when it comes to the objective systematic pursuit of truth, in-group bias is pretty much never a good thing. One can argue that in-group bias helps you and your in-group to survive and prosper, but one cannot really argue that it helps you discover the truth. The truth stands as it is. It does not care who survives and prospers and who does not. Objectivity requires that you ignore your own personal interests sometimes.

If the truth is what you are seeking, then there will be circumstances where you might have to act against the interests of yourself, your friends, your family, or your community. An effective strategy in this respect is to separate your personal life from your intellectual life as best you can. In other words, you can allow some in-group bias into your personal life, but you shouldn't allow much of it into your intellectual life. Society should also be careful to guard against any undue influence of in-group bias in public policy, as this is often the foundation upon which atrocities are built.

When it comes to your personal life a certain amount of carefully controlled in-group bias can actually be beneficial to you. However, when it comes to your intellectual life you should instead strive to think like an independent disinterested observer, one with no conflicts of interest and no reason to exploit the system. This requires discipline. Even when you have the best of intentions, it is still easy for in-group bias to subconsciously creep into your mind and taint your thinking. Many a horrible thing has been done in human history without the people doing it ever even becoming consciously aware of the underlying prejudice and evil entailed by their own actions. Evil flourishes in societies that are half-asleep. Self-awareness, and being honest enough to recognize your own potential for prejudice and bias, is probably one of the most important steps in breaking that cycle of evil.

6.4.30 Reactance bias

When someone tells you to do something or tries to restrict your freedom, and you react by defying them regardless of the merits, then that is reactance bias. This particular bias is closely related to the concept of "reverse psychology". Reverse psychology is a method of persuasion that works by telling someone to do one thing in the hopes that they will actually do the opposite, out of a sense of rebellion or spite. Reactance bias is why reverse psychology sometimes works.

Many people do not like being told what to do, and will oppose any attempt to order them around, even if they otherwise would have agreed on the merits under different circumstances. This is yet another reason why it is important to be tactful when trying to persuade someone of something. If you trigger their reactance bias through your lack of tact, then they may do the opposite of what you want instead, even if they would have otherwise agreed.

An excessively large amount of reactance bias effectively makes you into a mirror image of whatever your opponent is. By doing the exact opposite of whatever your opponent says, you have actually effectively given your opponent complete control of you. That is the great irony of being too reactive. You do it because you are motivated to protect your own freedom and to refuse to ever obey your opponent. However, by doing so in excess you have actually effectively obliterated your own freedom. A mirror has no freedom. If what you rebel against is what defines you then you will never be free.

Reactance bias is most common in extremely polarized and emotionally charged situations. For example, it is particularly common in politics, where sometimes one political party will do the opposite of another political party not because of the merits, but merely because they feel irresistibly compelled to rebel. Clearly, this behavior can have all kinds of deeply negative effects on society. It is probably one of the leading causes of obstruction of political and social progress. An all-consuming desire to "win" against your opponents at any cost is not a basis for sane thinking. Defining yourself in terms of perpetual opposition to someone else is a one-way ticket to a miserable and bitter life. Vengeance is not a good foundation for rational thought and a joy filled life. A life well lived is the best revenge[21].

Many cultures have held on to bitter grudges that have long since lost all real meaning. It amounts to nothing more than self-imposed suffering at this point. Why hold on to ancient grudges, thereby filling your life with bitterness and regret, when you could instead form you opinions objectively based on the merits? You'd be a lot happier that way, and the world would also be a much more pleasant and peaceful place. For instance, there are religious factions that have been fighting each other for centuries, often over very small differences in beliefs. What's the point though? Why can't more people have the mental fortitude to make a fresh start?

All you have to do is question your assumptions, look at the world around you with fresh eyes, and chart a new course based on what the evidence suggests. Move past the past. You'll be far happier that way. All you have to do is put aside your ego long enough to see what's really important in life. Stop trying to "win" arguments. You can't. The harder you fight to win an argument, the more you will lose sight of the truth. Simply align yourself with the evidence

---

[21] I can't remember where I actually first encountered this idea (i.e. "A life well lived is the best revenge."), but apparently it may be a variation of a George Herbert quote, who I have no familiarity with. I probably heard it randomly somewhere, out of context.

instead. Commit to peaceful rational discussion, instead of hostility. Nobody is perfect, but aligning yourself with the truth and giving your best effort to find it will do a lot to help make the world a much better place, one with much less suffering and much less needless conflict.

If reactance bias causes so much trouble though, then why does it exist? Well, the reason is probably that in the right context it actually fills an important role. You see, human beings tend to rank each other according to who obeys who. Individuals who are perceived as being higher ranking tend to receive more benefits, more respect, and better mates than individuals who are perceived as being lower ranking. The more people witness you obeying other people, the more their respect for you and their estimate of your social rank will tend to decrease.

Thus, from a natural selection standpoint, sometimes it is actually more advantageous to rebel against someone even if you agree with them than it is to be objective. Social rank has a large impact on reproductive outcomes, thus individuals with at least some degree of reactance bias will probably tend to reproduce more often, thereby making their genes more common. I think this is probably where the bias comes from maybe. It seems like a reasonable theory.

As such, just as with in-group bias, a compromise may be the most effective. Allowing a small amount of reactance bias into your personal life may in fact serve you well occasionally. Properly applied, with the right balance of tact and context, reacting strongly (but gracefully, from a position of calm confidence) against any attempts of other people to order you around (even if you agree with them) can potentially sometimes do a lot to protect your social standing in the eyes of witnesses. Subtle signals like this can have a lot of influence on why different people are "inexplicably" treated differently by a group. Showing weakness too often will make you a target for bad attention. Reactance bias can sometimes help you make it clear that you don't take orders from anyone, and hence that you are not implicitly socially ranked below them.

On the other hand though, reactance bias is not a good thing when it comes to objectivity. Your social standing doesn't have anything to do with the truth. In fact, when you are seeking the truth, reactance bias tends to just lead you astray. Thus, when it comes to your intellectual life, you should strive to be immune to reactance bias. Also, while reactance bias may protect your social standing in certain situations, it can also hurt it. You need to pick your battles carefully, otherwise you may instead just damage your image even more by seeming childish, petty, hard to work with, or lacking in perspective.

Don't overreact. You also need to consider the greater good. A person who truly has strong social standing does not blow up at the slightest provocation. Someone in a position of strength would already have so much resources and so much going on for them that they wouldn't care about such insignificant things. In contrast, overreacting to small things immediately signals weakness and low social standing, and that is why it is such an unattractive quality in a person. People are repulsed by it.

Be careful that in making clear that you don't take orders from others that you don't end up throwing a disproportionate tantrum about it. That would actually be counterproductive to the social advantages that reactance bias is normally otherwise intended to give you. Be calm instead, and react from a position of unflinching charismatic strength. Remember that anger is typically a threat response, and hence often a sign of weakness in disguise.

Ultimately, you will need to find the right balance. Objectivity actually does inspire a certain amount of inherent social respect sometimes, by making you far less likely to seem childish or petty in the eyes of witnesses, and thus is generally the better default behavior. It can give you an aura of calm confidence, wisdom, and fairness. However, sometimes you may nonetheless find it more advantageous to indulge in a bit of reactance bias. It depends on how the natural selection pressures balance out.

For example, in dating, it is probably unwise to let other people do things that make them appear to rank above you. For optimal outcomes, you'll need to weigh the relative values of objectivity, protection of social rank, and protection of personal freedom against each other in the context of each specific situation, and determine which are the most important overall, taking account of both your own personal interests and the well-being of the community as a whole. Be tactful about it.

6.4.31 Moral luck

Everything happens for a reason. You get what you deserve. Justice has been served. What goes around comes around. Or at least, that's what someone who believes in moral luck would say. Moral luck is the idea that random events are morally significant, even if the event is outside of the participant's control. In the context of how people think, it can also be considered to be a cognitive bias, rather than merely an idea about how the world might work. This is a pretty common belief. You've probably encountered it many times throughout your life, if you've been paying sufficient attention.

The problem with moral luck is of course that often things just happen randomly, with no real moral justification of any kind. Sometimes corrupt people get to live long and peaceful lives, filled with every pleasure they could ever want. Sometimes innocent and compassionate people suffer constantly, never seeming to be able to get ahead in life, and die young from diseases or calamities that they did nothing to deserve or to contribute to. The world isn't always fair. Yet, that doesn't stop a certain subset of the population from believing that all events can be traced back to a moral cause of some kind, regardless of how unknown or contrived that supposed cause may be. This can result in all kinds of counterproductive behaviors in society, such as victim blaming for example.

In victim blaming, people attack the victim of suffering instead of the real perpetrators, environmental factors, or social forces that actually caused the event. This effect is especially notorious in cases of rape or exploitation. Sometimes, people will look for even the slightest excuse to blame a woman for being raped, such as the woman wearing even mildly alluring clothing, being naturally pretty or charismatic, or talking briefly with the perpetrator, despite the fact that nothing could ever justify a rape regardless. From the standpoint of a compassionate outsider, this behavior is baffling. However, there are psychological reasons why this behavior exists. One of the main contributing factors is probably moral luck.

People who believe strongly in moral luck may experience intense cognitive dissonance when they experience or hear about events that contradict their view of the world as being inherently just and moral. They can't mentally handle the idea of moral luck being false, because it makes them subconsciously realize that they too could one day be the victim of an unfair world, and therefore that one day they too could experience immense unjustified suffering. This in turn causes them to wildly distort their view of the event in order to avoid the intense stress, fear, and social responsibility that acknowledging it would entail. In other words, victim blaming of rape victims likely originates at least in part from a combination of both the ostrich effect and internalization of the concept of moral luck.

Internalization of the concept of moral luck may perhaps be due to the high prevalence of the idea of moral luck in our society, such as in films, in books, and in culture in general. It says more about the fearful state of the person doing the victim blaming than it does about the actual victim. In the case of rape victim blaming, plain old sexism also probably plays a significant role as well, of course. However, more broadly (i.e. outside of the specific case of rape and other similar examples) the ostrich effect and moral luck probably constitute the primary foundation upon which victim blaming as a more general phenomenon rests. That's my theory anyway.

As for victim blaming with respect to exploitation, a classic example would be when well-off people (i.e. middle class and wealthy people) blame poor people for being poor, as if most poor people were *willingly* poor or merely being lazy. The implication is that poor people wouldn't be trying to escape poverty even if they had the means and the know-how, and that therefore the situation is their own fault. Clearly however, an objective perspective shows that most of poverty is caused by structural problems in society itself. Few people would willingly stay poor if they had a way to escape from it. I'd guess that the vast majority of poverty is actually caused by a combination of lack of social resources[22], bad attitudes[23], and systematic exploitation.

The number of forms of systematic exploitation currently plaguing our society and preventing upward mobility for the poor is vast and diverse. For instance, systematic exploitation in modern society includes, but is not limited to:

1. wealthy people not paying poor people in proportion to their value contribution

2. laws that permit business or employment contracts to waive basic civil or legal rights (e.g. arbitration clauses, gags on free speech and whistleblowing, non-competes, employers claiming the intellectual property rights of work produced during an employee's own free time, waivers of the right to sue for grievances, overtime pay loopholes, shifting the cost of replacing equipment onto employees, unescapable contracts, stacking legal conflicts in the employer's favor, predatory employee release agreements, etc)

3. reductions in employee and consumer freedom and choice

4. monopolization, corporate collusion, corporate lobbying, and political corruption

5. intentionally limiting access to resources and information that would enable more people to start businesses within certain industries (because the people currently in power in those industries don't want to allow any new competitors to come into existence)

6. constraints on basic entrepreneurial freedom, such as overbroad patents (e.g. almost all software patents), designed to make it difficult for new competitors to enter the market and compete

7. suppressing economic awareness, so that consumers and employees aren't aware of the real value of their products and services and thus are neither able to make rational economic decisions nor able to properly negotiate on their own behalf

8. taking advantage of naivety and acquiescence and people's tendency to accept anything that "sounds official" or "seems required" in any way

9. predatory loans (e.g. payday loans, high interest credit cards, rapidly depreciating vehicle loans, unaffordable real estate loans, etc), which are often specifically designed to target the poor and lock them into perpetual debt slavery

---

[22] e.g. lack of education, tools, books, capital, etc

[23] e.g. such as caused by a counterproductive upbringing or by toxic cultural forces, such as anti-intellectualism, mediocrity culture, or self-limiting beliefs for example

10. permitting destruction of the environment and other kinds of public resources for private gain (less public resources creates less economic mobility for the poor, since the poor *by definition* don't have much private resources and therefore often have no choice but to depend upon public resources to at least some extent)

11. supporting aristocracy, corporate cronyism, and anti-competitive policies while publicly pretending to support "free markets" (a favorite among corrupt politicians, and perhaps one of the leading causes of growing economic inequality, whereby the vocabulary of free market economics is co-opted and distorted in order to trick the public into voting in favor of aristocracy, financial slavery, and the systematic destruction of their own intellectual and creative freedoms)

12. undermining independent creators' economic autonomy (e.g. when publishers take a grossly disproportionate amount of the revenue that a creator's product generates, or claim intellectual property rights that rightfully should always belong to the creator and not to the publisher, who typically contributes little more than distribution and marketing and hence has no genuine right whatsoever to claim the product as intellectual property)

13. sales taxes (which disproportionately harm the poor and inhibit economic mobility, thereby also reducing the wealth of the entire society as a whole, unlike progressive income taxes)

In fact, speaking of sales taxes, in states where there is "no income tax" the government usually has to compensate with very high sales tax. This actually creates what is called a "regressive tax", which is in effect the opposite of a progressive tax. In a progressive tax system, the poor pay a lower percentage of their income and the rich pay a higher percentage of their income, due to the fact that paying more taxes imposes much less of a burden and much less of a danger to survival to the wealthy than it does to the poor.

In contrast, a regressive tax creates a situation where the poor actually pay extremely high percentages of their incomes in effective taxes while the rich pay almost nothing in effective taxes. The reason for this is because day-to-day necessities tend to have an approximately fixed (unchanging) cost, but day-to-day necessities constitute the vast majority of the expenses that poor people face and hence the vast majority of their income. In contrast, the wealthy barely even financially notice any effect from high sales taxes, due to day-to-day necessities constituting only a tiny fraction of the total expenses of the wealthy.

The concept of "no income tax" seems to be in practice really just a technique designed to try to trick the public back into an even worse economic situation. The phrase "no income tax" sure sounds appealing to voters, but what the politicians who push for these systems actually want in reality is a regressive tax system, one that is intended to have the effect of financially enslaving the poor while simultaneously empowering the wealthy to live virtually tax-free. To aid in maintaining this system of financial inequality, aristocrats seem to also sometimes try to discourage and defund public education, and to spread anti-intellectual sentiments and propaganda throughout society, perhaps because aristocrats may realize that the more educated people become the less likely they will be to fall for tricks like this.

In fact, a desire to discourage or defund public education, or to spread anti-intellectualism, combined with talk of "free market economics" but support for monopolistic policies (e.g. easy acquisition of patents, loosening regulations on corporate mergers, privatization of public assets into the hands of cronies, etc) seems to be pretty much a dead give away that you are dealing with an aristocrat or a crony capitalist, rather than with someone who genuinely supports a free market.

Ironically, though some of the more unethical members of the wealthy class seem to believe that exploiting the poor is to their advantage, in the long term such exploitation actually probably accomplishes nothing except reducing the total wealth of all of society, including of themselves. The wealthy wouldn't be able to stay wealthy unless people purchased the products and services they produce. However, poor people have few resources for buying those products and services. Thus, keeping the poor poor also blocks the wealthy from becoming even wealthier. Thus, ironically, in the long term, the best thing the wealthy class could ever do to increase their own wealth may actually be to lift the poor out of poverty, by a reasonable amount, but the wealthy class largely seems to currently be too collectively short-sighted to realize this.

This reactionary behavior and scarcity mentality of certain unethical members of the wealthy class has a very high social cost. Less poverty actually means more economic opportunity for everyone, regardless of individual wealth level. Class struggles of the wealthy suppressing the poor are essentially a form of self-harm. It is ridiculous. Most of humanity's problems are essentially self-inflicted. You will notice that this is a common pattern as you learn to become more rational. Human suffering is largely a self-imposed and fruitless affair. The tragedy of this will become more and more clear to you as you increasingly learn to see the world as it actually is, instead of through the lens of pre-rational prejudice.

Moral luck is also closely related to the concept of karma, but is slightly different. The concept of karma is officially a principle of Buddhism, Hinduism, Jainism, Taoism, and perhaps a few other religions, but the word has also been borrowed into the common vocabulary of secular life. The idea is that evil actions taint your spirit and that good actions elevate or purify your spirit, and that this taint or purity consequently effects what will happen to you in life and what kind of afterlife or reincarnation you will have. In a secular context however, the difference between moral luck and karma is that moral luck refers broadly to interpreting events (whether random or not) as having a moral quality or causality, whereas karma refers more narrowly to willful good or evil actions and how they may be seen as coming back to haunt you.

Moral luck doesn't require willfulness, but karma generally does. Moral luck also has no spiritual connotation. If you ignore the spiritual connotation and restrict yourself to a secular interpretation, then karma is a subset of moral luck. This implies that there is a missing piece of terminology for the other part of moral luck. Therefore, because the more words we have in our vocabulary the more expressive and powerful our thoughts will tend to become, let's name the part of moral luck that is not karma (i.e. not based on willful intent) **moral randomness**. Furthermore, to have a way of unambiguously referring to the part that *is* based on willful intent, without any possibility of spiritual connotation, let us define **moral retribution** to mean karma without any spiritual connotation. Thus, moral luck is the union of moral randomness and moral retribution.

Anyway, one thing you may have noticed though, that may have given you pause during our discussion of moral luck, is that some actions actually *do* contribute to an eventual retribution, such that certain actions really do increase the chances of negative consequences occurring later on. This raises the question: Is moral luck really a bias then? The answer is yes, moral luck is indeed a bias. However, there is a certain altered sense in which moral luck can be "true", but it has nothing to do will morality itself, but rather it is merely a product of causality. How does this work exactly though?

Well, to understand in what sense some kind of retribution can genuinely be caused by unwise or evil actions, we will need to understand that all systems which we interact with in day-to-day life tend to involve numerous different favorable and unfavorable factors. These factors in turn collectively make each specific environment or system that we interact with in day-to-day life have a net overall health in terms of how frequently positive or negative results will tend to occur within them, which we will refer to as that interactive system's **circumstantial health**. In other words, when a system has good circumstantial health, good events will tend to happen more often when we interact with it, whereas the opposite (more bad events) will tend to happen if the system has bad circumstantial health.

What does this have to do with moral luck and retribution though? Well, you see, evil actions have a tendency to damage the circumstantial health of a system, and thus to result in negative consequences. In fact, it's not just that evil actions have a tendency to damage circumstantial health, but rather it is that having a tendency to damage the circumstantial health of systems is actually arguably in some sense the *essence* of evil. When you take away all the subjective cultural opinions on what evil is, what you're left with is that evil is simply any and every action which has a very strong tendency to damage circumstantial health, and hence to consistently make the world a worse place, i.e. a place with significantly more suffering in it, where bad events randomly occur much more often. This is perhaps the objective essence of morality, in some sense.

When you remove all subjectivity, cultural quirks, and artificial assumptions from consideration, morality is simply guardianship over the circumstantial health of the world, so that we can all collectively continue to enjoy favorable circumstances going forward into the future. Notice that unlike the traditional culturally subjective definition of evil, this definition of evil (as actions that strongly tend to damage circumstantial health) seems more objective and maybe even quantifiable sometimes.

Having a more firm definition like this could maybe help somewhat in trying to move forward to an objectively better world, potentially, one with more prosperity and less suffering. It could maybe help discourage people from dismissing moral movements as merely "subjective" and may help encourage them to instead recognize that morality seems to actually be more of an objective phenomenon once you remove all of the unnecessary, superfluous, or artificial cultural assumptions surrounding it.

To distinguish this more objective sense of moral retribution from the more subjective sense of moral retribution, we will refer to the more evidentiary sense as **causal retribution**. The difference between causal retribution and moral retribution is that whereas moral retribution assumes that there is some kind of "justice" operating in the world that forces everything to ultimately be "fair" or "morally justified" (regardless of whether that makes sense or not) and thus often places blame on victims, causal retribution makes no such assumptions and instead just observes the objective fact that certain actions have a tendency to damage the circumstantial health of systems and hence have a higher probability of triggering bad outcomes.

Causal retribution doesn't assume that experiencing negative events implies that the people experiencing those events "deserved it", but instead only says that damage to the circumstantial health of the system may have contributed to the occurrence somehow, regardless of whether the people experiencing those negative events actually themselves had any role to play in contributing to those negative circumstances or not. In this way, causal retribution is a more unbiased and rational concept, whereas moral retribution is a more biased and irrational concept.

Here's a random interesting thought: Have you ever wondered why evil is consistently associated with dark, hostile, and barren environments in creative works? Take a moment to consider the fiction, artwork, games, films, and culture that you have experienced throughout your life. Take fiction for example. Why is it that forces of evil in stories often reside in places like, for example, blackened barren wastelands riddled with rivers of lava? Why live in a dark castle instead of a balmy beach getaway? Wouldn't evil characters appreciate some comfort too? Why is good associated with comforting environments like clean crystalline cityscapes and beaches while evil is associated with lava and darkness?

My theory is that the answer is because places like barren wastelands, rivers of lava, and dark castles (etc) all have bad circumstantial health, relative to most human beings who might wander into them. Why does that matter? Well, do you remember what our definition of evil was? Evil is any set of actions that have a strong tendency to damage circumstantial health. Why then shouldn't evil characters be placed in locations that have accumulated a lot of bad circumstantial health? It makes perfect sense thematically.

It is simply an application of placing similar things together. People apply these themes subconsciously without even realizing how insightful and ingenious it actually is. In other words, hostile landscapes and evil actually have *genuine* harmony with each other. They are *objectively* related to each other. It's not just a coincidence or a quirk of human history that this association exists in art and entertainment (etc). Pretty cool thought, right?

Sometimes though, evil characters are placed into comforting environments. This juxtaposition can create an interesting effect. It can make the story creepier and also can give it more of a muddled sense of tension and conflict. It also makes the evil characters "pop" more. It's like if you had a pristine white floor and then split a glass of red wine on it. The red wine draws your attention and focus. It may even instill you with an urge to clean the wine up off the floor, or a hope that someone else will, so that the dissonance of the wine on the floor will go away, returning the room to its previous (more harmonious) state.

It seems to me that this tension is probably caused by a subconscious awareness that the presence of the evil forces in the otherwise healthy environment could undermine and corrupt it, thereby damaging its favorable qualities, which is something that we humans naturally dread, because we are utterly dependent as a species on having favorable circumstances (e.g. community belonging, tools, accumulated collective knowledge, a specific and very narrow range of tolerable temperatures, etc). The greater the difference between the circumstantial health of the environment and the evil tendencies of a character, the greater the sense of potential lose will be. The larger the difference, the more will be at stake, and hence the more tension it will create. People don't want to see a good environment become corrupted. It's probably good for drama though.

In contrast, an evil character placed in a correspondingly hostile environment creates more of an ominous and oppressive atmosphere, one of hopelessness. I'd think a good storyteller should be mindful of the difference and of how the choice of setting will significantly alter the way a story feels. Both approaches have different pros and cons. Sometimes it will be more impactful to place evil in a correspondingly hostile environment, whereas other times it will be more impactful to place evil in an otherwise healthy environment.

It depends on what effect you are going for. It's a pretty nuanced judgment and shouldn't be interpreted too narrowly though, so be careful. Environments and characters that are typically associated with certain things are not necessarily inherently so. How things are described and implemented can completely alter how those things feel. For example, beaches are often culturally associated with comfort and favorable circumstances, but they can easily be made to have a hostile and oppressive atmosphere (e.g. such as if someone is lost on a prison island or if the beach is involved in a military conflict, etc).

Anyway, we got off on a little bit of a tangent there. Prior to going off on this tangent, we were carefully picking apart moral luck into two subtypes (moral randomness and moral retribution) when we realized that there is a certain sense in which something similar to moral luck can be considered true, which we named causal retribution to distinguish it from moral retribution. We also discussed karma briefly. As we learned, believing in moral luck is a biased way of thinking that can easily cause you to place blame on the wrong people, and hence for society to make bad policy decisions. However, on the other hand, we also learned that evil does indeed have an arguably objective meaning if you take away all the superficial components and that thus morality in the form of causal retribution is very real.

Consequently, the optimal perspective is probably this: There is no evidence that there is any "moral force" at work in the universe that justifies events on a moral basis. However, certain actions *do* have strong tendencies to make the circumstantial health of a system better or worse. We call actions that strongly tend to make circumstances better "good", whereas we call actions that strongly tend to make circumstances worse "evil".

Good and evil are somewhat relative, but also somewhat absolute. In one sense, good is partly a matter of perspective. In our case, goodness is always framed in terms of the well-being and prosperity of humanity as a whole and of ourselves as individuals, as well as on the well-being of many other species since we share the same ecosystem as them and hence are ecologically dependent on them also surviving to an extent. On the other hand, more broadly speaking, one could define goodness in terms of maximizing the average (or perhaps median) well-being and prosperity of all living beings in the entire universe, and in this sense good and evil would be absolute and thus would not then be relative to any particular perspective, individual, or culture.

Different perspectives might have different priorities, and there may be multiple different approximately maximal states, but every frame has an essentially objective moral compass in terms of seeking to maximize the overall well-being of the system as a whole. Thus, moral relativism (i.e. thinking that morality is subjective and that different individuals and cultures can each define their own completely arbitrary notion of morality without any regard to seeking well-being) is fundamentally wrong. Morality is objective.

The fact that different moral frames may exist doesn't change the fact that all of the frames ultimately seek to maximize the overall well-being of their respective systems, and that maximizing the overall well-being of the universe as a whole is also plausible. Worldviews that increase suffering or obstruct prosperity (e.g. allowing murder, oppressing women, etc) thus can never be considered to be optimally moral, contrary to what people who believe in moral relativism may think. Morality is not arbitrary. Morality is not merely a product of cultural norms.

Believing that morality is purely about cultural norms is ultimately a toxic and counter-productive worldview, one which has an inherent tendency to increase suffering in any society where it is a popular belief. Morality is simply a process of seeking local optimums of overall well-being, ones that minimize suffering. Not all worldviews or cultures are created equal. Some are *much* better than others. In fact, most of the struggle of humanity is the struggle of moving from worse cultures to better cultures. Thus, a society that treats all cultures as automatically equal is a society that is inherently incapable of ever becoming better. A culture can't have become better if the culture it transitioned from wasn't worse. Thus, treating all worldviews and cultures as equal and seeking to maximize human well-being are two mutually exclusive goals. Choosing one makes the other logically impossible.

6.4.32 Rosy retrospection and declinism

Things were better back in the day. The new generation is going to ruin society. These sentiments are examples of rosy retrospection and declinism, respectively. Technically, rosy retrospection and declinism are two different things, but they tend to go hand-in-hand with each other, so I've decided to discuss them together in one section. Both are cognitive biases as well as patterns of behavior. Rosy retrospection is a tendency to view the past more favorably than the present. Declinism is a tendency to think that everything in some system (typically a society, but sometimes other things) is consistently trending towards getting worse. Both are quite common within society and throughout history. Both tend to create unrealistic and distorted views of reality, and thus tend to impair people's judgment.

Have you ever felt nostalgic for something you experienced in the past at some point? Have you ever longed for the past like that? Ever felt homesick? Ever wanted to re-experience something you experienced as a child, such as a book, a game, a friend, or a location, etc? Well, as it happens, rosy retrospection is actually a form of nostalgia. Specifically, rosy retrospection is *biased* nostalgia. If you think about it, not all nostalgia for the past is necessarily unrealistic or distorted, although it often is. The term rosy retrospection is just a way of unambiguously specifying that you are referring specifically to the *biased* kind of nostalgia.

This also implies that there is a missing term for the other kind of nostalgia, the unbiased kind. Therefore, let's define a new term for it. We might as well. Let's call unbiased nostalgia **reminesia**, an intentionally distorted modification of the word "reminiscence" designed to make it resemble the word "nostalgia" more. Thus, nostalgia is the union of rosy retrospection and reminesia, meaning that nostalgia can be either biased or unbiased, whereas rosy retrospection is always biased and reminesia is always unbiased. Reminesia is probably less common, due to the human mind's strong tendency to be at least slightly biased when remembering things. This is just a random idea though. Maybe this new word will end up being useful but maybe not. No harm in trying, as usual.

Anyway, it's easy to see why rosy retrospection and declinism would tend to go hand-in-hand. Rosy retrospection distorts your view of the past, causing you to view it more positively than it actually merits. If you judge the past using more forgiving standards than you judge the present and the future, then it should come as no surprise that things will tend to seem to be getting worse from your perspective. Declinism is a natural consequence of long-term rosy retrospection.

The longer you have looked at the past through rose-tinted glasses during your life span, the higher your chances of internalizing declinism will be. Declinism is the source of the classic "grumpy old man" personality archetype in society. It seems to be part of what causes old men to tell children to get off their lawn, to complain about how they think that everything was better back in the good old days, and to claim that younger generations don't understand the value of hard work, etc. Indulging too much in rosy retrospection during your lifespan could potentially become a one-way ticket to a bitter and resentful retirement.

Declinism is especially common in old people and in political groups that have become significantly out of touch with the rest of society and culture. Reactionary politics and conservatism very often involve an element of declinism. In fact, in a sense, it is hard to even call a political position "conservative" if it doesn't contain at least some element of declinism. Conservatives are almost by definition biased at least partly towards declinism.

The older a person is the longer the amount of time they will have had to view the past in an overly positive light, and hence the greater their chances of internalizing declinism are. I'd guess that's part of why conservative political parties tend to skew more heavily towards the elderly than towards youth. Segments of society, institutions, or industries that were previously in-touch with society and prosperous but that are now in decline will also tend to be more likely to move towards conservatism and to embrace declinism and reactionary politics than they otherwise would be.

One possible contributing factor to the higher incidence of declinism in the elderly may also be the way their physical health and life experience colors the way they compare the present to the past. As humans age, their ability to experience pleasures and new experiences will tend to decrease, both due to actual deterioration of their bodies and the fact that many previously fresh life experiences will have become quite stale by the time they are in their old age.

This means that the elderly may sometimes experience less joy from doing the same thing that they previously did in their youth, even if they otherwise would have enjoyed it just as much as they always did. This may lead them to incorrectly associate the increasing dullness of their own lives with the state of society as a whole, causing them to conclude that the decrease in quality of life they are experiencing as they age must instead be due to changes in society rather than due to the unavoidable consequences of aging.

Some old people may find it more comforting to point the finger at young people and at social changes than to simply admit to themselves the unfortunate fact that their own time on Earth (and also possibly the time when their worldview was still culturally relevant) is fast coming to a close. Death is not an easy thing for many people to acknowledge and to accept. It can feel so much easier to instead distract yourself by blaming others, and to thereby absolve yourself of your own mistakes and regrets in life.

It also makes it easier to avoid having to do the hard work of showing empathy towards the youth and towards cultural changes. Morbid as all this may sound, don't worry too much though. Cultivating the right frame of mind can probably help get rid of the bitterness of old age. I will repeat the same advice I gave earlier in a different section: Live every moment as if you have only just now been born, and then ask yourself: How can I make the best of this?

One of the most common forms of declinism is stereotyping of generations, and the corresponding systematic discrimination and unnecessary suffering that always accompanies it. This pattern has repeated itself pretty much for almost every generation that has ever lived in human history. A certain subset of each older generation will almost always accuse the next up-and-coming younger generation of possessing some combination of negative attributes that in some way are seen by the older generation as degrading society. Almost every older generation accuses the next younger generation of possibly being the generation that will ruin society.

Yet, despite constant predictions of this throughout human history, it has never come to pass. Generational stereotyping is generally inherently prejudicial and pointless. Good things very seldom ever come from characterizing the "personality" of any generation. Such characterizations have never really been valid. Human nature doesn't change that quickly. Only many thousands of years of evolution could genuinely change the inherent nature of one generation of humans compared to another. This has probably not yet ever happened in our recorded history, since our recorded history is so far only several thousand years long at a best. Documentation from the earliest periods is far too sparse to be reliable.

Comically, sometimes members of an older generation will even quote complaints about youth written by ancient authors and philosophers, completely failing to realize the irony of this. If anything, ancient quotes complaining about youth are actually more applicable to the current older generation than they are to the current younger generation. The distance in terms of time between the ancient quote and the current older generation is less than for the current younger generation, thus actually making it more applicable to the older generation than to the younger generation. Funny, right? Hypocrisy has always been a special talent of humanity. Indeed, if hypocrisy had economic value then we might all be wealthy by now.

As of the time this book was written, the older generations' favorite target for unjustified prejudice and discrimination has been "millennials" (meaning people born sometime around the turn of the 2nd millennium, i.e. year 2000 plus or minus a decade or two, roughly). For example, the present author of this book (born in 1988) would be considered to be a "millennial" by this metric. The most notable environmental and economic circumstance of millennials is that we have grown up in parallel with surging advances in technology that have changed the nature of human civilization forever. As millennials have grown up and matured, so too has the computer industry and internet matured in parallel. When we were young, so was the internet. Likewise, now that we have matured, so has the internet. The same can be said for the game industry too, among many other things.

We millennials were the people born during the largest technological and cultural surge in human history. Add in some corrupt politics and bad economic policy into the mix, on the part of our predecessors, and millennials have faced one of the fastest changing and most complicated economic situations ever faced by any generation in all of human history. This would obviously create major obstacles and major changes in social structures and habits for *any* human generation. Yet, that doesn't seem to stop the older generations from labeling all of us millennials as inherently broken and inferior people to themselves. Such discrimination can be seen every day on TV on the "news" (in reality: the mindless politics and propaganda channels) and in daily life.

A certain subset of the members of the older generations constantly smear, denigrate, and vilify us for literally no reason at all, save for their own prejudice, bigotry, resentment, and jealousy. It also makes it easier for them to feel less bad about what they've done by shifting the blame for all of it onto us instead. When people treat another group of people unethically or exploitively[24], they often subconsciously redirect what would normally actually be guilt and shame for themselves, to making excuses and rationalizations designed to place the blame instead on their own victims. When people mistreat others, they often gradually start making up imaginary (false) reasons why that treatment was actually justified all along, because otherwise they'd have to realize the blame lay with themselves and then take responsibility for that.

The key thing to remember is this: Different generations often face different circumstances, but human nature almost never changes. Therefore, any attempt to ascribe differences of personality to different generations is fundamentally wrong and cannot ever be valid. The only possible exception, as I already mentioned once before, is probably over extremely long periods of time, such as tens of thousands or hundreds of thousands of years worth of human evolution. Whenever members of one generation compare themselves to members of another generation they should always remember that both generations have the same fundamental human nature. The *only* difference is that one generation was born into a different set of environmental, cultural, and economic circumstances than the other was, and thus had to adapt differently.

Besides the previously described factors, another factor which may play a role in contributing to rosy retrospection and declinism is the tendency of the human mind to simplify and exaggerate the details of memories. Why does this tendency to simplify and exaggerate the details of memories exist? Well, if you think about it, it is probably a form of data compression. An analogy is illuminating here. Take computers for example, and the data compression algorithms used on them to reduce the amount of space that files occupy. How do such data compression algorithms tend to operate?

If you are familiar with computer science, then you will know that compression algorithms usually work by doing some combination of removing non-essential details from the data and

---

[24] e.g. by accumulating massive debts for one generation's benefit and then making it so that the next generation will have to pay for all of it, etc

translating the data into a tighter form from which it can then subsequently be reconstructed. If you think about it, this has all of the same fundamental characteristics as the tendency of the human mind to simplify and exaggerate memories. Thus, it seems reasonable to infer that the tendency towards simplifying and exaggerating details in human memories exists because it essentially serves as data compression for the human brain. As a finite object, the human brain must by definition have a finite memory and a finite bandwidth for computation, so the existence of a subconscious data compression system operating within the brain just below the surface is not really that surprising. How else would you ever fit so much information in there?

Anyway, the point I'm trying to make here with respect to how this notion of data compression in the human brain is relevant to rosy retrospection and declinism is this: Rosy retrospection and declinism both in part hinge upon a tendency of people to forget about or disregard negative details of the past and to exaggerate or put into high relief the details of the present. This creates a situation where the person recalls fewer accurate details about the past and more accurate details about the present. This difference in detail density in turn implies that there will be more opportunities to criticize the present than to criticize the past. In other words, the "surface area" of a person's memories susceptible to criticism is greater for the present moment than for any one specific moment of the past.

This implies that even if a person is not motivated to indulge in rosy retrospection or declinism for any other reasons, they still may nonetheless experience at least some amount of the bias anyway, simply due to nothing more than the uncontrollable subconscious processes of the brain with regards to data compression (i.e. memory simplification and exaggeration). Being aware of this may help unbias your thoughts somewhat though. Anyway, that concludes our discussion of rosy retrospection and declinism. Remember to be mindful of the effects of these forces at work in you mind, and try not to succumb to bias. At least some degree of distorted memories of the past is probably unavoidable, but too much rosy retrospection and declinism will invariably make your life end up being more bitter and counterproductive than it needs to be.

6.4.33 Cultural calibration bias

This one is actually another term that I've made up, because I couldn't find an existing term for it. This section will be the first time I've ever introduced it. Nonetheless, it is not some obscure phenomenon, as you will see. It seems to be quite a common cognitive bias. It is especially relevant to politics in particular. So, anyway, what is it then? Well, cultural calibration bias is essentially an inaccurate sense of how to judge other groups of people and beliefs objectively.

Cultural calibration is not a single bias, but rather it is an entire spectrum of varying degrees of bias in one direction or another. There are two opposing poles on the cultural calibration bias spectrum, namely *closed cultural bias* on the one hand and *open cultural bias* on the other. On this axis, closed cultural bias is the negative pole and open cultural bias is the positive pole. In this way, the cultural calibration bias spectrum resembles a finite continuous number line segment.

When a person has closed cultural bias, it means that they are disproportionately prone to assuming negative things about new groups of people, new cultures, new beliefs, and new evidence that they encounter. Open cultural bias is just the opposite. When a person has open cultural bias, it means that they are disproportionately prone to assuming positive things about new groups of people, new cultures, new beliefs, and new evidence that they encounter.

In contrast, if a person has little to no bias in either the closed or open direction, then we will say that they have *objective cultural calibration*. This implies that they are at or near the zero point on the cultural calibration bias spectrum. Objectivity only exists near this zero point. Indulging in either closed cultural bias or open cultural bias is irrational and counterproductive. By definition, it would imply that you are judging situations disproportionately to their real merits or faults.

On the traditional political spectrum, as you might imagine, closed cultural bias roughly corresponds to conservatism and open cultural bias roughly corresponds to liberalism. Putting aside certain purely historical and arbitrary quirks of each respective side, this is more or less true. Please understand however that I am not saying that all policies of conservatives exhibit closed cultural bias nor that all policies of liberals exhibit open cultural bias. Rather, I am talking about conservatism and liberalism more broadly as general mindsets. In fact, any given policy supported by either conservatives or liberals could potentially be the best possible policy on the subject matter it is concerned with, the policy that maximizes the quality of life of society with respect to the factors the policy governs.

The fact that a policy came from a conservative or a liberal does not make it biased. Objectively speaking, every policy or position must be evaluated on its own merits independent of the source. Whether a position is conservative, liberal, moderate, or impossible to classify on the traditional political spectrum, it is always possible that it could nevertheless be the best conceivable position and therefore be objective. Never assume that a position is biased just because it came from a biased source. That is *not* how objectivity works.

For example, the concept of free speech originates from liberalism, but that does not make free speech a liberally biased policy[25]. In fact, a policy that heavily supports free speech is very likely to be the best conceivable position with regards to how speech should be governed[26]. Likewise, fiscal conservatism and decentralized economics is almost certainly objectively superior to runaway government spending and central planning. Understand? Being supported by a biased party does not make a belief itself biased or bad. Only lack of merit can do that. Political parties tend to be at least somewhat mixed bags.

Objectivity is *not* moderation. Moderation is often in fact just yet another rigidly dogmatic position. Objectivity is nuanced context-sensitive perpetually evolving rational thought. Objectivity always considers policies on a case-by-case basis. Objectivity never "leans" in any particular political direction, not even the center. It never holds preconceptions. The only rigid absolute of objectivity is to constantly question your own assumptions and to constantly ask yourself what the best way to maximize well-being and minimize suffering is[27]. Anything else would be biased.

The cultural calibration bias spectrum is not the same as the traditional political conservative-liberal spectrum. They bear a relationship to each other but they are not the same. In fact, they are quite different. The zero point of the cultural calibration bias spectrum does *not* correspond to the moderate position on the traditional political spectrum. In fact, objectivity

---

[25] In fact, the word liberal literally means "supporting freedom". Notice that the word liberty, on the other hand, which is derived from the same root, simply means "freedom". Liberal is the adjective form of liberty. Liberalism is where the concepts of free speech and democracy originated from. Conservatives in contrast originally focused on supporting and defending the monarchy (i.e. authoritarianism) and censorship. They originally had none of the associations with free speech that they now have. Both liberals and conservatives now usually support free speech though.

[26] It is certainly better than censorship. Censorship makes it harder for a society to change, and therefore less able to adapt, and therefore more likely to collapse.

[27] Objectivity also has to take resource constraints and plans for the future into account. It seeks to maximize well-being and minimize suffering not just for the short-term, but also for the long-term.

is actually impossible to classify on the traditional political spectrum. Closed cultural bias is a common attribute of conservatives, and likewise open cultural bias is a common attribute of liberals.

However, that does not make closed cultural bias the same thing as conservatism, nor does it make open cultural bias the same thing as liberalism. The cultural calibration biases only indicate *disproportionate judgment* in one direction or the other, whereas the traditional political spectrum merely indicates a grab bag of beliefs set to the backdrop of the current state of culture. Cultural calibration bias is fundamental, whereas the traditional political spectrum is arbitrary, historical, and filled to the brim with inconsistencies and self-contradictions.

Objectivity requires constant thought. It offers no lazy answers. It isn't controlled by traditions or social momentum. It doesn't adhere to any particular doctrine, except for unwavering dedication to the systematic pursuit of truth and maximal quality of life. The answer to a question at one moment of time may be different than what it was at some other moment of time. Objectivity accepts that some things change, but also that other things don't. It views the world as the world truly is, unfettered by any form of prejudice or dogma. It is neutrality and optimality incarnate. It is the only way to consistently and reliably reduce human suffering.

All other viewpoints are doomed to wander aimlessly forever, like driftwood in a turbulent ocean. Objectivity is the only position that has any real integrity, and the only position worthy of any real respect. Objectivity adjusts itself to the circumstances. It is not rigid moderation. Objectivity is perfectly willing to go to extremes when the circumstances merit it. If a belief is not based on objectivity, in at least some reasonably credible way, then it is based on faith (i.e. on belief without evidence), and thus cannot be considered justified. Judgment made without evidence is nothing more than idle whim, nothing more than mere randomness, nothing more than stormy weather of the mind.

One question we can ask ourselves however, is what happens if we *do* indulge in cultural calibration bias? What's so bad about it? What will the net effect be? Well, to answer that question, let's first try to reframe the two biases in terms of something less abstract and more familiar to us. Let's try reframing them as behaviors instead of thinking of them as cognitive biases. As behaviors, closed cultural bias roughly corresponds to *bigotry*, whereas open cultural bias roughly corresponds to *gullibility*.

Both bigotry and gullibility are extremely dangerous attributes for people in power to have. Both have laid entire societies to waste and collapsed empires. Any culture where irrational modes of reasoning (such as bigotry or gullibility) predominate for too long is sure to eventually fall from power. It's a matter of when, not if. Objectivity is the only worldview that has any chance of maintaining perpetual prosperity. All other paths are blind gambles.

Closed cultural bias and open cultural bias are both bad, each in their own special way. Both can easily result in atrocities. In fact, both *have*, many times throughout history. Let's take some time to explore how. It will be instructive. Let's start with closed cultural bias. That's the one that I'd guess people would probably have an easier time seeing the danger of, but I assure you that open cultural bias is actually equally dangerous. You will soon see why once we explore a few examples of it.

Closed cultural bias roughly corresponds to closed-mindedness, whereas open cultural bias roughly corresponds to open-mindedness. Many people see having an open mind as being inherently virtuous. It isn't. Nor is it really even a part of what it means to be objective, contrary to popular belief. Objectivity requires that you be neither closed-minded nor open-minded. Instead, you must be evidence-minded. Anything else would imply disproportionate judgment.

Anyway, back to closed cultural bias. How might it have bad effects on society? Can you think of any examples? You probably can, probably lots of them in fact. For example, racism, sexism, nationalism, warmongering, stereotypes, religious persecution, anti-intellectualism, eugenics, sanism, and classism are all examples of phenomena that are supported at least in part by closed cultural bias. Many of these phenomena might not even exist if closed cultural bias didn't also exist. Closed cultural bias is in many ways the central support pillar of much of the world's numerous different forms of bigotry.

Most of those terms and phenomena are probably familiar to you, but *sanism* is a less common term. Let's take a moment to understand it. It is worth being aware of it, tangential as it may be. Sanism means discrimination on the basis of someone thinking or behaving differently than the norm, by classifying them as "mentally ill" in some way. Psychiatry is its main advocate. Sanism is one of the most profoundly dehumanizing and disempowering forms of discrimination in existence, mostly due to the fact that once "officials" have declared someone "mentally ill" it is often very difficult for that person to recover social standing and personal autonomy. People declared "mentally ill" are often forcibly imprisoned in asylums, forced to take medications that may cause permanent brain damage, blocked from communicating with the outside world, and stripped of numerous different civil and human rights.

Asylums and psychiatrists stand to gain huge profits from this process. The fees they charge are both massive and recurring. They often have little to no economic interest in actually helping people get better. The longer people stay locked up or on medication, the more profitable it is. Many psychiatrists probably *want* "treatments" to be slow and ineffective. It is in their economic interest for it to be that way. In stark contrast, even simple things like just taking up a new hobby, acquiring a sense of accomplishment and purpose, or merely changing your mental attitude probably would frequently vastly outperform psychiatric "treatments". Even just playing video games may be more effective than psych treatments are, I imagine.

It wouldn't surprise me. Not only that, but video games are vastly cheaper, vastly more fulfilling, vastly more personally empowering, and vastly more entertaining than popping pills (which are often loaded with terrible long term side-effects) could ever be. Psychiatrists probably don't want you to consider those kinds of alternatives though. That would make their business significantly less profitable. Many human beings (not all) are very corruptible. Give them an opportunity to exploit others, and they will often take it.

Never assume that wearing the superficial trappings of professional medicine makes that any less true. Medical institutions can easily be just as greedy and unethical as giant oil companies and financial institutions. Medical institutions are probably just better at hiding it and manipulating public opinion. The aura of authority and professionalism that medical institutions project often inspires blind trust from most of society and thus probably greatly erodes accountability and ethics. It depends on each specific case though, surely.

Anyway: Many psych treatments are often inherently disempowering and dehumanizing, especially historically. The very act of medicating someone for "mental illness" subconsciously sends that person a powerful and sinister message, a message that tacitly tries to convince the person being "treated" that they are *fundamentally broken* in some sense. This is a dark message, perhaps one intended to maximize psych industry profits by instilling anxiety and blind acceptance of psych industry authority in the person's mind. It is a message intended to disempower the user, to make them feel dependent and hopeless. What could be more depressing than feeling like that? This may be exactly what the psych industry wants. Perpetual depression and mental instability are extremely profitable for them. This is a colossal conflict of interest. Wherever there are conflicts of interest in human society, there is almost always corruption and exploitation.

They probably *want* people to feel fundamentally broken. That's probably why they always try to make it sound like "mental illness" is an inescapable condition, an inherent part of who someone is, something someone is born with. They want to discourage that person from ever escaping from the condition, by making them think that such an effort would be pointless. That way they never really try, never truly give the effort to escape the condition their full effort and attention. Psych industry propaganda often seems like it is designed to put a fog of social obscurity and disguised pseudoscience between a mentally distressed person and the realization that that person's living environment and attitude is often most likely what really actually needs adjusted, not that they have anything inherently wrong with them.

Mental stress is a perfectly normal condition in modern life. It's a natural and healthy evolutionary mechanism designed to protect you and motivate you, not something to be treated as a "brain chemistry imbalance" or disease. Experiencing mental stress in a stressing environment *is* the balanced, healthy, and natural response. Just as physical pain warns you to avoid further physical harm, so too does mental pain warn you to avoid further mental harm. That's how you should interpret it. Don't let the psych industry brainwash you into thinking that mental pain is a dysfunction. It isn't[28]. It's a good thing. It evolved to make you behave more optimally relative to your environment.

In contrast, loading yourself up with brain damaging and addicting psych drugs *is* unnatural. Unlike living with your natural emotions, taking psych drugs *will* probably unbalance your brain chemistry. Ironic, right? It's the opposite of what the psych industry wants you to believe. The psych industry wants to turn you against your own mind, to make you think that their "treatments" will restore your mental balance when the opposite is probably actually true. If you are like most people (i.e are not a part of the probably tiny minority with real brain diseases), then you are probably already balanced, not only when you feel good, but also even when you are experiencing extreme stress.

Mental stress is a healthy reaction to stressful environments. It is an impulse evolved to inspire you to better your life, and to endeavor to change the systematic environmental and social factors that contribute to your bad experiences. In most cases, you should probably listen to *it*, not to the profiteering and often wildly unethical psych industry. The millions of years of evolution that made your emotions work the way they do in order to guide your behavior (i.e. to maximize your quality of life and survival rate) is your friend. The exploitive and self-interested psych industry, in contrast, is probably usually not.

The psych industry seems to want to saturate society with so much of a medication-oriented and powerless attitude that anytime anybody asks their friends and family for advice about any natural mental stress then those friends and family will just immediately mindlessly parrot back the psych industry's pill popping rhetoric, thus delivering you straight to their hands so that they can systematically exploit you for the rest of your life. They want to exploit that sacred trust that you have with your friends and family. They want you to believe that "mental illness" is an unchangeable genetic trait, just like being born with brown eyes or black hair. It seldom ever is though.

The percentage of people with true mental illness is probably very small, and a very small population is *not very profitable*. Thus, the psych industry is constantly seeking to expand the

---

[28] Well, at least not usually probably. I bet that it's extremely rare for someone to have a real physiological brain problem, relative to how many completely normal people are falsely treated as if their brain is somehow diseased by the psych industry. Lying to as many completely mentally healthy people as possible is surely vastly more profitable than only treating the real cases. The financial conflict of interest here is so huge that you'd have to basically be a moron to trust them. Give humans a chance to be corrupt and they very often will be.

scope of who is considered "mentally ill" as much as possible. This is almost certainly the real source of why you hear the news talking about large increases in "undiagnosed depression" and a plethora of other conditions. If the psych industry had its way, *being human* would be a mental illness. The more people they can disempower and brand as inherently broken, the more people whose spirits and sense of self-determination they can break, the more profitable they will become.

According to the psych industry, almost everyone would already have at least some kind of "mental illness". Even today, they already have a "mental illness' to cover pretty much any kind of person's personality. The "diagnostic criteria" for "personality disorders" read like Barnum statements, like fortune cookie paper designed to apply to as broad a group of people as possible. This "huge number of people with undiagnosed mental illnesses" the psych industry constantly tries to push into the news cycle may seem surprising and concerning at first glance, especially to a naive person, but it's actually not surprising at all once you realize that it is *exactly* the kind of rhetoric that most maximizes how much profit the psych industry will pull in. Do you really think that's just a coincidence? Knowing what you know about people's tendency to become corrupt when they are given unchecked and vast amounts of power and a means to exploit it for financial gain, ask yourself that.

To the psych industry, the rest of us are just like livestock. They want to break our will and pump our systems full of drugs designed to make us all complacent and incapable of fighting back. They want to take away all sense of self-determination and strength of will. They want us to be locked up like chickens or cattle, trapped forever in a cycle of hopelessness and dehumanization, perpetually lining their pockets with an endless stream of cold hard cash. Where does all the money the psych industry spends on marketing go? Surely some of it goes to TV ads, but I'm not convinced that's the only place it goes.

I wouldn't be surprised if they are secretly astroturfing on the internet, i.e. pretending to be real people to create peer pressure and a false consensus to compel people to take psych drugs. Billions of dollars can fund a *lot* of astroturfing you know. Perhaps they are even bribing news outlets and falsifying studies. Psychiatry has one of the darkest histories of any branch of science. As such, I wouldn't be the slightest bit surprised if they were doing all these things (astroturfing, bribing news outlets, falsifying studies, etc).

In fact, I frankly don't even think psychiatry deserves to be called science at all. They haven't earned it. They are more characterized by the systematic exploitation of society and of anyone who thinks or behaves differently than the norm than by any kind of dedication to the objective pursuit of truth. Read up on the early history of psychiatry especially, for example. It will curdle your blood. Remember that modern psychiatry is descended from this same line of thinking, this same basic worldview, this same basic notion that the natural state of the human mind, with all its wonder and nuance and color and emotional ups and downs, is just something to be forced to conform to some arbitrary standard of "normal" to make a few other people in society (controlling, simple-minded, nothing-special kind of people) just a little bit more comfortable.

It is a procrustean worldview, a worldview that seeks to force every last living human being's mind to be dull and homogeneous and subdued and devoid of all uniqueness and variety. It's fundamental premise is that being human, that having a mind that sometimes experiences emotional ups and downs and deviance, is somehow wrong, is somehow broken. They think that it is wrong to have a mind that works differently than how they want it to work. Psychiatrists want to be dictators of the mind. Hitler would be proud. Just as the relentless pursuit of ethnic homogeneity was the hallmark of the Nazis, so too is the relentless pursuit of mental and intellectual homogeneity the hallmark of psychiatry.

In contrast, a simple change in attitude (e.g. reading ancient Stoic philosophy) or taking up a new hobby (e.g. video games, painting, gardening, etc) tends to instill a person with an enhanced sense of personal autonomy and control over their own lives and destiny. It inspires them to dream big. If the ineffectiveness and humiliating nature of psych "treatments" is not bad enough though, there's also the tendency of negative studies on psych drugs to sometimes be deliberately suppressed and never published (probably due to both publication bias and huge financial conflicts of interests). A whole bunch of studies will often be commissioned to test the same drug, but then all of the studies that come back with negative results will be ignored and only the ones with positive results will be published, creating an illusion of effectiveness that isn't true in reality.

From what I hear, even placebos often outperform psych drugs. As ineffective and inefficient as psych drugs may be, one thing psych drugs often *do* do successfully though is sometimes cause permanent damage to the brain's natural biochemical balance, causing the user to become dependent on the drugs for life. Life long dependency is of course the most profitable possible outcome for any drug and is thus probably something the pharmaceutical companies deliberately encourage. The drugs are probably often designed primarily to make people dependent, not to help them.

If they wanted to help people then they'd use far simpler, far cheaper, far more sustainable, and far more empowering tactics, like making sure that people who are depressed find *fulfilling and creative hobbies* and *fix the real problems in their lives*, instead of just loading them up with ineffective and unethical (but extremely profitable) drugs. Thus, psychiatric "diagnoses" are rife with conflicts of interest and may actually often have little to no real credibility for that reason.

Historically, sanism has also been used by oppressive political regimes against political threats. What better way to suppress political opponents than locking them up and deliberately damaging their brains, disempowering them, humiliating them, and destroying their social standing by having them declared "mentally ill" due to "personality disorders", right? Don't be naive. The psych industry is one of the creepiest institutions ever created in human history and continues to be a major threat to human freedom. The fact that so many people automatically trust the psych industry so readily and unconditionally just because they wear the superficial trappings of "professional medicine" and "science" is one of the scariest parts. It allows the psych industry to systematically exploit millions of people without the slightest accountability or resistance.

Not only that, but if the psych industry ever successfully damages enough people's brains or gets enough people addicted to their drugs, then it could become literally systematically impossible for anyone to ever successfully oppose them. The psych industry is one of several possible ways human society could one day degenerate into a dystopia, a perpetual living hell. It's right up there with nuclear war in terms of how much of a threat it potentially poses to human society, yet hardly anyone seems truly aware of the danger it poses.

People trust psychiatrists and pharmaceutical companies far more readily than they should. People recommend psychiatric drugs and therapy far more often than is actually merited. Merely adopting the superficial symbols of science (e.g. white lab coats, scientific-sounding rhetoric like "brain chemical imbalance", etc) does not in any way make a discipline more trustworthy. Only merit and a lack of conflicts of interest can prove trustworthiness, but the psych industry has neither. It has both little merit and tons of conflicts of interest, in most cases.

Anyway, putting sanism aside, let's return to our discussion of closed cultural bias. Closed cultural bias is part of why some people never seem to respond to evidence, no matter how much of it you present to them and no matter how irrefutable it may be. Talking to someone with extremely high levels of closed cultural bias feels like talking to a brick wall. You may even begin to question whether the person is even a sentient being. It makes people seem like they have more in common with mindless automatons, computer programs, or inanimate objects than they do with the "rational and highly cultured beings" that humanity is supposedly so characterized by. You've probably talked to someone like this at some point, and have directly experienced how extraordinarily frustrating it can be to try to prove a point to them.

This way of thinking poses a major threat to human progress. Many challenges that humanity is likely to face in the near future, such as climate change and resource depletion, will require people to be highly adaptable, but too much closed cultural bias among a subset of society may make this impossible. It is an immense danger to our survival as a species. Humanity will have to figure out a way to consistently stamp out irrational thought patterns amongst the populace, reducing irrationality to negligible levels, otherwise our long-term survival as a species may be highly unlikely.

Putting a lot more emphasize on studying reasoning skills and on practicing resistance to fallacious arguments, propaganda, and false authority figures will likely be necessary. Students should learn how to detect and dismantle logical fallacies and psychological manipulation from a very early age, and should review the material frequently. It also needs to be applicable to daily life, rather than merely theoretical and abstract.

As it stands, the schools currently seem to barely even cover how to think logically and resist fallacies. Strong critical thinking skills and logic should be the number one priority for education. Everything else (all other disciplines) ultimately stem from it in some way or another, even if very indirectly. Students should be taught to question everything, including so called "authority figures", and to rely instead upon direct evidence and proof to chart their path through life. Unfortunately, our current education system in many ways does the opposite. Students are often punished for questioning authority, for asking tough questions, and even just for thinking differently than the norm.

Blind adherence to authority is the opposite of good education. Good education liberates the mind and instills creative and intellectual autonomy, and arms students with ample defense systems against all those in society who would seek to exploit them later in life, even if that includes their own parents, their own school teachers, or their own government. History has proven repeatedly that teaching people to think for themselves yields better results than blind adherence to authority. Societies that drift towards authoritarianism almost always suffer for it, whereas societies that drift towards personal autonomy and independent thought tend to prosper.

The only reason we have such prosperous and high quality lives today is because people have been able to acquire more personal autonomy and entrepreneurial power than they used to have. Authoritarianism hurts everyone in the long-term, even the people in power, they just don't realize it. Maximal economies of scale and creative diversity just aren't viable in a stifling authoritarian environment. Creativity requires the freedom to express oneself, and must be unburdened by the tarpit of collective tyranny, otherwise it will tend to fizzle out. Plus, authoritarianism often doesn't actually convince people of anything, it more often only makes them pretend to agree with the "authorities" in order to protect themselves from hostility and unjust punishment. Those who force respect will seldom receive it. The same is true of love and friendship. Or, to quote a popular phrase: "A man convinced against his will is of the same opinion still."

The only way to demonstrate the merit of an idea is on the battlefield of ideas, through debate and civil discussion. Force will never truly accomplish anything. It will only ever backfire in the long run. Murdering someone for believing that $2 + 2 = 4$ doesn't make it any less true. The only thing it demonstrates is the evil and pathetic nature of the person who chose to use force instead of evidence. A person with merit on their side would have no need for force. The battlefield of ideas is the only battlefield upon which any conflict of ideas can ever truly be won.

By the way, forced respect is also probably one of the classic hallmarks of bad parenting. Logic and evidence are the only real way to make a point stick to somebody. If the merits aren't on your side then the best you can ever hope for is a hollow victory, one that leaves your food tasting like ash in your mouth, metaphorically speaking. Being someone's parents doesn't in any way make the evidence and logic automatically side with you. The phrase "because I said so" is a product of ego, not reason. The same is true outside of parenting as well, of course. You can't escape from the demands of evidence. Either you adhere to a standard of evidence or you embrace delusion. To live life without reason is to drive a car with closed eyes. The truth stands firm. Delusion is all in the mind.

Closed cultural bias can be thought of as the degree to which your mind's eye is closed. A person who has maximal closed cultural bias has a completely closed mind's eye, i.e. is someone who is completely impervious to all forms of new information. It is as if they now live in their own little bubble world completely disconnected from reality. Once closed cultural bias has progressed this far, it can become incurable.

It can become an intellectual black hole. The only hope for enlightening someone who has become so completely closed-minded is for them to accidentally open their mind's eye a little bit by random chance, at some moment of time in the future, and for them to just happen to be exposed to a profoundly compelling perspective-broadening argument or experience at that exact moment, one strong enough to break through their bigotry.

This is generally unlikely, but it is still possible. It will probably have to be something personal, something that sits relatively well with their own perspective while still somehow managing to broaden it, something with strong empathy. It's best to protect people from closed cultural bias as early as possible. The longer you wait, the more damaging to the health of a person's mind it will become. Preventative measures are essential, especially if you have any hope of ever removing this bias from society as a whole, and thus of curing the corresponding suffering it causes.

Anyway, that's enough about closed cultural bias for now, let's talk about open cultural bias some. I mentioned earlier that open cultural bias is just as dangerous as closed cultural bias. Why is that? Well, let's see. Take a moment to think about it. Why might reacting disproportionately positively to new groups of people, new cultures, new beliefs, and new evidence that you encounter be a bad thing? I said earlier that open cultural bias roughly corresponds to gullibility. Why is gullibility bad? It's bad for the same reason that false beliefs are bad: It misleads you and causes you to make counterproductive choices.

If talking to a person with extreme closed cultural bias is like talking to a brick wall, then talking to a person with extreme open cultural bias is like talking to the world's most agreeable yes-man. People with extreme open cultural bias are like intellectual limp noodles. They are total push overs in terms of what they can be lead to believe. They are naivety incarnate. They have little to no mental filter. They just soak everything up like a sponge. Just as you have probably encountered severely closed-minded people before, you have probably also encountered severely open-minded people before. People who believe in new age mysticism, healing crystals, diet fads, celebrity gossip, self-contradictory statements (e.g. lots of Zen "wisdom"), ghosts, alien abductions, tarot cards, fortune telling, and psychics are common examples.

These kinds of people will often believe practically any "did you know" statement that you pose to them, especially if the statement sounds interesting or is in any way entertaining. If an idea has any kind of aesthetic appeal whatsoever, then they'll probably accept it as true, regardless of how much or how little supporting evidence you present for it. These people would do well to close their minds a bit more. Open-mindedness is *not* always a virtue. Internalizing universal open-mindedness will not actually make you more worldly or cultured really. It will often only make you dumber. Or, to quote the popular saying: "Don't be so open-minded that your brain falls out."

Notice that here too, just as in the case of extreme closed-mindedness, an extremely open-minded person's sentience feels somehow questionable. There doesn't seem to be much real thought to what this person will accept. They just accept practically anything you throw at them, as long as it has aesthetic appeal. This gives their personality a one-dimensional quality that makes it seem like they are disconnected from reality and practicality. It reduces the strength of their consciousness. In this sense, closed cultural bias and open cultural bias are actually similar in character: they both trend towards *mindlessness*. The more you stray from objectivity, the less sentient and the less aware of reality you become. True consciousness and self-awareness only exist in the realm of objectivity.

If that's not bad enough though, just wait. All the negative consequences of open cultural bias that I've been talking about so far, as bad as they may be, are just the tip of the iceberg. These garden-variety delusions are almost saintly compared to some of the worse forms of open cultural bias. Where extreme open cultural bias *really* gets dangerous is in politics. Many people wish that one day our society will become completely open-minded and equally welcoming of all beliefs. These people are *idiots*. Different beliefs have different values. Treating all beliefs as equal is one of the fastest possible paths to social decay.

If this wish for universal open-mindedness ever became true then it would likely lead to a society-wide collapse of unprecedented scale. It would leave the whole world intellectually defenseless and ripe for evil. Such a society would be so weak that even a faint breeze (i.e. even the slightest social movement towards an evil path) could destroy it effortlessly. If all beliefs were equal then truth would lose all meaning. It would be pure chaos. It would be hell on Earth. Universal open-mindedness is a false utopia. Don't listen to its siren song.

Thinking that universal open-mindedness is a good idea is like thinking that setting a wolf free in a cage full of rabbits is a good idea. The rabbits will be slaughtered. You can't solve all problems with hugs and kisses, nor should all belief systems always be treated with respect and equality. By treating all beliefs as equal, you have created the perfect environment for evil to rise to power. By going on a witch-hunt every time someone tries to criticize any belief or culture, automatically accusing that person of being an evil bigot, as if there were something inherently wrong with ever criticizing another person's worldview or culture, you actually gradually destroy society's defense mechanisms against bad ideas by disincentivizing all forms of criticism through hypersensitive social punishment. This trend has unfortunately become highly prevalent in our society recently.

Even the most objective, well-formulated, and tactful criticisms that you could ever hope to pose could still nonetheless easily get you verbally raped by a mob of hysterical morons on social media, and perhaps also on the news channels on TV if you're especially unlucky. This is the direct consequence of a large subset of our society abandoning objectivity and civilized discourse and instead embracing mindless false righteousness. This is what happens when you care more about being seen as "open-minded" than you do about the truth or about maximizing human well-being and minimizing human suffering.

One could say that this is the liberal version of bigotry: a vicious lack of concern for the tangible aspects of human well-being whenever it conflicts in any way with your own all-consuming goal of being seen as open-minded and non-bigoted. You'd let your own grandmother get blown up in a terrorist attack, without ever being willing to criticize the belief system that caused the attack, just so that you could be seen as open-minded and worldly, just so that you could protect your own ego and public image. It is ironic that by adopting such a vigilante attitude of embracing open-mindedness to stamp out bigotry that you actually end up becoming just as bigoted as what you sought to destroy, if not more.

Vigilante open-mindedness is just ego and self-interest in disguise. A liberal bigot prioritizes their own desire to never be seen as having judged anyone else unfairly over all other factors and consequences, even if it means opening the door to mass human suffering and death. That's nothing to admire. There's nothing benevolent about that. Protecting your loved ones inherently requires a willingness to *pass judgment* upon others: to rate one belief as superior to another, to suspect that one person is safer to leave your children around than another, to see that one path forward is better than another, to observe that one culture clearly results in less human suffering than another, and so forth. An attitude that all beliefs are unconditionally equal regardless of evidentiary merit makes this impossible, makes consistently protecting your fellow human beings and loved ones from harm and suffering impossible.

This is what the more bigoted subset of liberals don't understand about the other side's motivations. They can't see that closed-minded people (such as typical conservatives) mostly just prioritize their own safety and well-being, and that of their own loved ones and associates and general in-group, far more than being seen as magnanimous to other groups of people. They favor caution and adherence to what they already know to be safe, even if they lose opportunities and sometimes treat others unfairly as a consequence of doing so.

On the other hand though, conservative bigots frequently fail to see the immense value that people from other cultures so very often possess, and thus often end up subjecting those people from other cultures to immense undeserved suffering. Conservative bigots also tend not to grasp the implications of new evidence well, tend to miss a ton of opportunities for cultural growth and mutually beneficial cooperation, tend to not innovate and not explore new possibilities much, and tend not to be as adaptable as they should be.

Humanity can't survive such a failure to adapt for much longer. We need to learn from our mistakes, take full responsibility for our actions, and work together to protect the health and future of our world. On the other hand though, we also need to realize that it is *logically impossible* to make progress towards creating a better world without also being willing to sometimes judge one belief system or culture as fundamentally inferior to another. A careful balance of these competing forces is the only way to truly end the cycle of human suffering. In contrast, blind adherence to either extreme of the spectrum will accomplish nothing and will only serve to further perpetuate the cycle of human suffering.

Both open-minded and closed-minded people clearly present major problems to society. Closed-minded people pass judgment too easily, whereas open-minded people don't pass it easily enough. However, objectivity has none of the problems of either of these sides. Proportionality is key. Objectivity is always proportional in how it judges everything. The endless cycle of unnecessary suffering in society will likely not stop until the overwhelming majority of people learn to embrace objectivity. Objectivity needs to become a deeply ingrained part of our culture, like free speech and human rights, or like music and art. It needs to become something that is universally embraced, rather than something that is sneered at jealously. Unbiased thinking is essential. As a society, we must learn to balance our sense of judgment. We must neither judge too swiftly nor too lazily.

As appealing as a non-judgmental society sounds, it would be disastrous if ever implemented. Protecting the common good requires a willingness to pass judgment, even if that judgment could end up being wrong, i.e. could end up being inadvertently bigoted. You have a choice: Pass judgment and risk being unfair, or don't and let evil spread unimpeded throughout society. You can't both protect society and not judge society at the same time. Those who judge no one can see no evil, and those who can see no evil can protect no one. To judge no one is to care for no one. Non-judgment is a coward's justice. Equality without merit is tyranny. True fairness is proportionality. True fairness crushes the wicked and elevates the virtuous. It does not treat them as equals.

Better to risk being a jerk than to be unwittingly complicit in the spread of evil. Benevolence cannot exist without judgment, just as a plant cannot exist without water. To be maximally benevolent, you must be able to pick the better of every two paths presented to you. However, this would be impossible if you weren't willing to judge each path according to its merits and to label one as better than the other. Thus, even in a world of maximum benevolence and maximum quality of life, there will still be winners and losers in the realm of ideas. Universal equality and maximal benevolence are fundamentally incompatible. There can be no such future where they co-exist. You have to choose one or the other. What's more important to you? Treating every living person and every conceivable idea as equal regardless of merit, or maximizing the overall well-being of society?

Anyway, we got off on a bit of tangent there for a bit. Let's refocus on open cultural bias again now. I mentioned that open cultural bias is actually just as dangerous as closed cultural bias. We've been exploring some examples to help illustrate this point. Well, to that point, one of the best historical examples of open cultural bias gone wrong is communism. Well, it's not a perfect example (real-world politics is seldom monolithic), but it still illustrates the point well. As I said before, open cultural bias is closely related to gullibility, and extreme political positions such as communism often inherently rely upon this gullibility to form their foundation of support. For instance, it takes a *lot* of gullibility and naivety to believe that universal central planning is a good idea. The same can be said of the idea that all people should receive equal pay irrespective of what type of work they do.

Communism is a great illustration of why total equality is actually a form of tyranny. Fairness is not equality. Fairness is proportionality. A perfectly fair economic system is a system were every person receives compensation in exact proportion to the amount of value they have contributed to society. This inevitably results in inequality. It is statistically impossible for every member of society to contribute exactly the same amount of value per unit of time spent laboring. Communism, for all its high-flying delusional rhetoric and claims of embracing fairness, is definitely not a fair system. It is the antithesis, the very opposite, of true fairness. It also clearly violates the principles by which meritorious ideas must be discovered. Only honorable civil debate can prove the merits of an idea. Violent uprisings can't. Violence only proves violence, never the correctness of ideas. Might does *not* make right.

Objectivity is a mind like still water. It does not thrash about wildly. It does not murder. It does not rape. It stands in a state of perfect calm, a state of perfect proportion. It fits itself into whatever context it finds itself. It is flexible, yet at the same time weighty. It permeates the full body, the full truth, of its environment. It is open to change, but closed to chaos. It changes its shape to encompass every new object, every new piece of information, that is immersed in it. Impacts are minimized, but not rejected. It is not stubborn like a billiard ball, rejecting every foreign object it comes in contract with, but neither is it easily bent out of shape like a clump of dough. New visitors to the waters of objectivity only raise the level of the water, only enlarge the domain of the water, raising it to a higher level of awareness, only empowering it further. A pool of still water knows a peace that few other things know. The same is true of objectivity. Logic opens the floodgates of objectivity, of perspective without limits and tranquility without end. The timeless eye of truth opens only for those who bend the same way the world does: in exact proportion.

The madness of communism is certainly not objectivity. Communism is essentially one giant experiment in social naivety. Society would be wise to remember that most people tend to be corruptible when given access to power. Human beings have a natural instinct to try to game the system. It's a survival instinct that's been ingrained in our minds through millions of years of evolution. This exploitive tendency probably behaves better in a state of nature (with small tribes) than in modern society, the scope and power of which this instinct was not originally evolved to deal with.

There are almost no circumstances under which maximizing centralization of power will go well. Nor is maximal freedom a good idea. Maximal freedom is anarchy, and anarchy allows evil forces to operate without any checks and balances. If government (centralized power) were ever totally eliminated then what would really happen is that society would degenerate into violent gangs and small kingdoms. It would create a power vacuum that would immediately be filled by the worst of the forces of evil.

It would create *more* suffering, not reduce it. It would make life *less* enjoyable, not more. The thing that extreme centralized authority and extreme decentralized freedom have in common is that thinking they are good ideas inherently requires a lot of naivety and poor judgment. Anarchy and authoritarianism are more similar than they are different: they are both just different expressions of extreme naivety, i.e. of excessive open-mindedness.

These phenomena are probably why open societies like democracies sometimes become dictatorships. Too much open-mindedness opens the door for evil. It's like the analogy of the wolf being let into a cage full of rabbits. A totally open-minded society is a society that is ripe for exploitation and destruction. To conclude from this that democracy is a bad idea would be very misguided though. It is the lack of rational thought on the part of the public, combined with political corruption, that destroys a democracy. Democracy itself doesn't really implode on its own, at least not if it has strong enough checks and balances in place. It probably takes toxic cultural trends (e.g. widespread glorification of irrational thought, bad flow of information, etc) for a democratic collapse to actually happen.

The level of naivety required to believe that big government will just magically be benevolent and rearrange everything in a way that eliminates social injustice, instead of just grabbing power for themselves, is jaw dropping. It shows a complete lack of awareness of human nature and an almost child-like willingness to blindly trust others. Only carefully balancing decentralized power against centralized power has any chance of maximizing human well-being. Maximal authority destroys all accountability and inevitably leads to profound corruption and oppression.

Likewise however, maximal freedom throws the whole world into chaos and violence and opens the door for evil to rise to power. Neither approach has any chance of ever being optimal. Both maximal authority and maximal freedom would have catastrophic consequences if ever implemented. Thus, the only logical possibility is that the optimal mix has to be somewhere in-between the two extremes. The optimal mix probably leans strongly towards decentralization of power, but not *too* strongly. There's a very careful balance that must be struck here in order to maximize well-being and minimize suffering.

Anyway, that pretty much covers our explorations of closed cultural bias and open cultural bias. It's time to wrap this discussion up. I want to take a moment to point out that when I said earlier that science is neither closed-minded nor open-minded that this was not merely something I was arbitrarily claiming, but rather it is born out by the evidence. You can observe the behavior of scientific organizations, such as universities and research journals for example, and clearly see that their behavior generally doesn't fall on either biased side of the cultural calibration bias spectrum, but instead tends to be roughly centered near objectivity.

Only opinions supported by some combination of logic or evidence are even allowed to participate in the discussion in reputable published journals. This is certainly not an open-minded thing to do, but it is the *right* thing to do, the *objective* thing to do. Good scientists are neither cruel nor welcoming of new arguments: They give each argument only exactly the amount of consideration and respect that the substance of the argument merits. Anything else would be disproportionate, and hence neither objective nor fair.

Science as an institution has to carefully guard against false beliefs and bad evidence. This would be impossible if it were overly friendly to new ideas. As bitter as this makes some people, it is the only way for science to progress. Science is only trustworthy as long as it remains mostly uncontaminated. Every false belief allowed in poisons the credibility of the whole institution. Claims aren't like colors. You can have a favorite color, and nobody can really deny you that, but you can't have a favorite claim and then expect people to automatically treat that claim with deference and respect just because you happen to like it. In science, every idea will be treated according to its merits.

As such, many ideas will inevitably deserve no respect whatsoever. If you can't take that kind of criticism, then you're really going to struggle with the world of science. Scientists usually aren't being mean. It's just tough love, an ice-cold splash of reality soup. Intellectual honesty is the only way to see the world as it actually is. It is unavoidable. It would actually be crueler to let lies in. There's a whole world out there that needs our help. We can't help that world if we can't even be honest about its true nature. Letting false beliefs off easy accomplishes nothing but increased human suffering. In the long term, being overly nice to bad ideas is arguably the cruelest thing you could do. Running away from reality only makes life harder. You can't escape it, so you might as well learn to deal with it on its own terms as best you can.

People are used to treating open-mindedness like an inherent virtue, but it isn't really. Scientists *aren't* open-minded. Nor are they closed-minded. They are actually evidence-minded. There's a huge difference between being open-minded and being evidence-minded. Judgment is good. I don't buy into the idea that people should be non-judgmental. Some ways of behaving are better than others, and some beliefs are better than others. Treating all people's beliefs as equal would destroy any incentive to ever get better. A person can't get better if who they were before wasn't worse. Judgment is a necessary process for making the world a better place to live. There must be winners and losers in the realm of ideas, or else there can be no progress.

People with a high degree of closed cultural bias or open cultural bias are subconsciously motivated by a desire for brain-dead easy answers, by a desire to never actually have to think about the situations they encounter. They don't want to have to account for all the intricate context-dependent details and nuances that may apply. That would take effort. Life tends not to have such simple one-dimensional answers though. The irony is that by trying to simplify your worldview so much, in a subconscious effort to make your life easier, you will actually just end up making your life more difficult and more empty.

Life is full of color. We live in a complex interwoven tapestry: a case-by-case, context-dependent, and nuanced world. A rational person realizes this and accepts it. Delusion doesn't help anyone. Laziness is not a path to truth, nor is it even a path to prosperity or pleasure.

Wishing that a complex thing were simple doesn't make it so. Nature is what it is. Stop being so afraid of thinking. Stop being so afraid of being wrong sometimes. It really isn't that hard. Running away from reality only makes life even more difficult.

So many new doors and insights will open up to you once you start viewing the world as it actually is. Logic instills the mind with a profound peace and tranquility that no more simplistic worldview could ever hope to match. Choosing to think crudely is practically the same thing as choosing to live a lie, as choosing to be less than what you could be. Embrace the invincible honesty of logic. Embrace the one and only truly foolproof way to guide your life to greater horizons.

6.4.34 Tunnel bias

This one is another term that I came up with myself. There wasn't any existing term for it in the literature, as far as I could tell, so I created one of my own. There's no point in waiting for the rest of society to "allow" you to do something like this. There's no point in ever thinking that you need their "approval" before you can do things like invent your own new terms. Just do whatever comes naturally to you. Don't be afraid to put yourself out there in life. You have to seize the moment.

Victory belongs to those who turn opportunity into reality. Authority is just an institution. It only exists so long as you willingly give it to someone. Power is in the mind. Someone's authority over you stops existing the moment you decide it stops existing. Claim your own authority, protect your own freedom, and become what you are meant to be. Nobody hands out permission slips for greatness. Greatness is something that you take, not something that is given to you.

Anyway though, putting that tangent aside, let's get back to tunnel bias itself. Tunnel bias is when someone treats the most likely event(s) of some set of events as being likely to be true in absolute terms, when this could very easily not be the case. In essence, it is when people assume that the most likely events are probably true, despite the fact that the majority of the total probability distribution may actually be distributed amongst the *other* events, not amongst the "most likely" event(s). A large number of unlikely events can easily sum to a vastly greater collective probability than the most likely event(s). Tunnel bias is when you don't correctly account for this very common statistical phenomena, i.e. when you prematurely assume that the most likely event(s) among a set of events is (are) a safe explanation or prediction for a situation when it is actually not.

Tunnel bias is very common. I'd estimate that it may be one of the leading causes of people making incorrect assumptions about other people and about the situations that they encounter in life. Engineers, programmers, and scientists are also often blinded by tunnel bias when trying to diagnose a problem, causing them to focus way more time and energy on only the "most likely" explanations for a problem than they actually should. Gossip, which frequently makes viscous and premature assumptions about other people (the target of the gossip), seems to also be rooted at least in part in a form of tunnel bias.

For example, consider typical gossip among a clique of high school girls about some other high school girl outside that group. Suppose that girl X comes into school one day with a red circular mark on her neck. Girls from clique Y witness this and soon begin gossiping about girl X. The girls from clique Y might assume, for example, that the explanation for the red mark on girl X's neck must be that it is a hickey, a mark made by making out with someone where one person sucks on the other person's skin somewhere until it turns red.

The girls from clique Y might "think" something along the lines of "Well of course it's a hickey! *Durr hurr hurr...* That's just *common sense*[29]! It's the most likely explanation." and in fact they may even potentially be correct about hickeys being the most likely explanation for red marks on necks in high school. However, even if true, that would still not be enough information to conclude that girl X's red mark was probably caused by a hickey.

Just because an explanation is the most likely explanation for something does *not* make assuming that explanation justified, contrary to popular belief. You actually need vastly more data in order to safely reach any strong conclusions about the situation. In fact, you need so much information that it very often borders on being infeasible. In order to make an inference like this safely, you would have to know the full probability distribution of all possible explanations for the event, and must also know that the candidate explanation(s) that you are considering has (have) a probability greater than 50% at bare minimum.

A wise and compassionate person therefore knows to almost never make these kinds of assumptions about other people. Reaching justified conclusions about other people in these kinds of situations is almost never feasible in day-to-day life. Of course, that doesn't stop many people from doing it anyway, despite seldom really having sufficient information to do so.

Let's think back to that hickey example. It's a red mark on someone's neck. Think about it. How many possible explanations are there for red marks being on people's necks? Probably literally at least *millions*. The kinds of people who gossip about other people like this however, seldom have cultivated enough imagination and objectivity to realize even something that basic though.

Here's just a small number of broad *categories* of possible explanations for the red mark: skin rashes, scratching, bruising, birthmarks, spilt dye, caked-on food or drink, red-colored dirt, acne, fungus infections (e.g. ringworm, athlete's foot, etc), bacterial infections, viral infections, impacts and injuries (e.g. slapping, fighting, etc), skin shedding, skin parasites, plant secretions (e.g. poison ivy), chemical exposure, sunburn, blushing, insect bites and stings, allergies, burns, genetic quirks, and (of course) hickeys.

This isn't even an exhaustive list. Many of the items in this list are actually broad terms, each encompassing massive numbers of other more specific conditions and phenomena. For instance, there are countless different types of fungus, bacteria, viruses, and chemicals. There are also literally an *infinite* number of different ways of sustaining injuries that might cause red marks. For example, perhaps girl X was simply helping her grandfather get some tools out of the attic when a steel pipe fell down and smacked her in the neck. This would naturally leave a red mark, but it obviously wouldn't be a hickey.

The sum of the probabilities of all these countless different possible explanations is almost certainly vastly larger than the probability of a hickey, even if a hickey were the highest probability explanation among all explanations, and even that claim is extremely doubtful. Rare events, although extremely uncommon individually, can still nonetheless easily constitute the majority of all events. Thus, counterintuitively, rare events can easily be collectively more likely than common events. For instance, if the most common event among some set of events has a probability of 20%, but there are 80 other events, each with an individual probability of 1%,

---

[29] As I have said before, "common sense" is an incredibly unreliable basis for making decisions. A more accurate name for what "common sense" really is would be "common assumptions". When I hear someone justify something by appealing to "common sense", it really just reduces the credibility of their argument to my mind. It isn't actually even a real justification. Saying that "common sense" is the reason for something is pretty much the same as just saying that the belief is one of your pre-existing assumptions. It says that you either have no justification for believing it or otherwise just can't verbalize the reason (e.g. perhaps due to poor communication skills). Appealing to "common sense" says nothing but bad things about the state of your thought process.

then the vast majority of events will be *rare* events. Untrained laypersons often subconsciously don't have an accurate sense for what it actually means for an event to be rare or common, in this sense.

Clearly, the girls from clique Y are not justified in thinking that they can estimate girl X's situation based merely on their "common sense". To do so is to merely indulge in blind preconceptions, cruelty, and bigotry. In practice, gossip is typically just yet another form of discrimination. It comes from the same psychological place as things like racism. It means you harbor that same evil in your heart, just in maybe a less extreme form. All it would take is the right circumstances, like a sense of blind self-righteousness combined with some very misguided ideas, for that inner demon to grow into an atrocity-supporting monster.

For example, if that's the way you think (i.e. you gossip a lot and make assumptions quickly, etc), then you very easily might have been a Nazi in World War 2 if you had lived in Germany at the time. This is something that all human beings should be wary of. These evils lurk within all of us, to lesser or greater extent. It takes willful mental discipline to ensure that our demons remain chained up and inactive, where they belong. Thinking that evil is something that only other people do only makes you even more likely to become evil yourself.

Not only that, but gossip groups often don't even care about the truth of their claims. They often only care about the "entertainment value" and the potential illicit social gains they can extract from the situation. They like how it inflates their own sense of worth by making them feel superior to others. If they were truly superior though, then they would already be secure enough in their own greatness to not need to stoop to gossiping about other people.

A truly great person finds gossip to be a waste of time. They have more productive things to do. Gossip is thus frequently a hallmark of low value. Low-value people, i.e. people who contribute less value to society than they should, and thus feel a subconscious need to compensate, therefore probably are more likely to feel a strong urge to participate in gossip.

Lack of imagination is a huge contributing factor to tunnel bias. It is sometimes difficult to conceive of new ideas, because doing so can require flashes of insight that aren't always easily conjured up at will. Conceiving of new ideas is not a fully controllable process. However, there are still ways to greatly enhance your imagination and your creativity. For example, keep in mind that putting aside time to force yourself to do creative work is very much under your control. Holding off work until you "feel inspired" is generally a terrible plan.

Not giving excuses, and always setting aside time to work regardless of how inspired you feel, is often one of the main differences between people who are successful in their creative endeavors and those who are not. Prolific creators tend to be relentless about putting in time even when they don't feel especially inspired. Creativity can be cultivated, like a bountiful garden, but specific "aha moments" and insights (e.g. solutions to specific puzzles) are often quite elusive regardless. It isn't difficult to foster a creative mindset, but it often *is* difficult to direct that creativity towards a specific unknown.

Creativity is the art of serendipity. It's like ridding on a train, but hopping off at random stations. It doesn't always take you where you thought you wanted to go, but it does always take you somewhere interesting and full of color, if you're patient enough. It's not a one-way process. You create art, but it also creates you. Creative works often seem to have a mind of their own. Like a seed planted in the earth, an idea emerges from the soil as if on a mission, as if alive. It has a natural pull, a place where it wants to grow. If you listen to it, and let it be itself, it can take you somewhere wonderful. Serendipity is the fruit of creativity. Grow a forest of ideas in the garden of your mind and such fruits will come in abundance.

Imagination is a very fulfilling and worthwhile endeavor, but it is not always easy. It is especially difficult to be comprehensive, i.e. to imagine all possible explanations for something. Thus, if you aspire to be as benevolent, unbiased, charismatic, and respectable as you possibly can, then you must be wary of the limitations of your own imagination. You must take this into account in how you judge things, otherwise you will almost surely cause other people to suffer at the hand of your premature judgments.

You should assume, in most situations, that there are numerous possibilities that exist but for which you lack the imagination or life experience to conceive of. When possible, sitting down and brainstorming a bunch of different ideas can help reduce this lack of imagination significantly. However, even after lengthy brainstorming you should expect that there could easily still be *lots* of other possibilities that you remain completely unaware of. Good judgment requires awareness of your own limitations. Nobody is omniscient. Good planning assumes that at least some uncertainty is inevitable. Under most circumstances, it is nearly impossible for all conceivable factors to have already been accounted for in advance.

Tunnel bias is probably a huge contributing factor in why large swathes of society so often seem so prone to premature assumptions. People often get into their heads that there are consistent "common sense" explanations for certain behaviors and circumstances of other people, but most often this feeling is actually just pure unjustified prejudice. Tunnel bias applies to more than just these kinds of social scenarios though. Any instance of someone weighing the "most likely events" more heavily than they actually should, compared to the collective weight of the less common events, is tunnel bias. This actually happens quite a lot in science and engineering. It's certainly not as if being "smart" makes you immune to it.

For example, a programmer who automatically assumes that performance problems that worsen the longer a program is running are caused by a memory leak (a fairly common premature assumption) is probably succumbing to tunnel bias. There are lots of reasons why programs can become slower and slower as they continue operating, including: input-output, persistence, file management, scalability, data structure design, algorithm design, fragmentation, data corruption, hardware design, bandwidth throttling, 3rd party library design, interactions with other programs, operating system behavior, and complex emergent effects. Input-output problems, data structure problems, and algorithm problems are an especially common reason why programs sometimes run slower and slower over time.

A programmer shouldn't just assume that it is a memory leak. In fact, memory leaks are probably one of the less likely reasons for these kinds of cumulative slow downs, especially if the programmer is using one of the more modern programming languages. Almost every "memory leak" another programmer has ever asked me to help them fix has turned out to actually be some combination of input-output problems, data structure problems, and algorithm problems. However, that being said, archaic programming languages like C and C++ do tend to have some memory leak problems, especially when smart pointers aren't used.

This common thread, the fact that this same elemental bias (tunnel bias) is a significant contributor to both gossipy cliques and engineering problems, is interesting. In some sense, it implies that engineers can't possibly be truly immune to such toxic social instincts (e.g. gossip), despite the fact that many of engineers pride themselves on supposedly being more resistant or immune to such behavior than the general public. Why? Well, the unjustified assumptions of gossip and the unjustified disproportionate focus on the wrong "most likely problems" in engineering actually originate from the same fundamental psychological source: tunnel bias.

That's the connection. It's illuminating, if you really stop and think about it. It says that the same forces for bias often operate in basically the same manner in highly educated communities as in less educated communities, just perhaps to a different magnitude and being applied in a different way. It tells us that we need to be wary of our perspectives and to not be too quick to hop on our high horses. A scientist's or engineer's mind is no fundamentally different than anyone else's, it just has had more practice and more exposure to assumption-rattling experiences is all. Bias is a pernicious beast, one not easily slain.

We're just about done here. This discussion should have illustrated tunnel bias enough for you to understand it clearly by now. One thing you may be wondering though, is why did I choose the phrase "tunnel bias" as the name for this bias. After all, the phrase "tunnel bias" does seem kind of arbitrary relative to the psychological and statistical phenomenon we are talking about here. Well, the reason why I chose "tunnel bias" is because it serves as an analogy. When you're walking around and encounter a tunnel, it kind of has a way of drawing you in. Tunnels draw your eyes towards them and seem to beckon you to walk through them. This is analogous to what event is perceived as being the most likely event from a set of events with respect to tunnel bias.

The most likely event draws you towards it and tempts you to prematurely treat it as a good assumption to make, even though it could easily not be. To reach your goal: Rather than walk through the tunnel, it might actually be far wiser to take a different path, such as to walk along a humble little dirt road off to the side, away from the tunnel, or to go off the beaten path through the forested wilderness. Nonetheless, the tunnel tends to kind of pull people in and often is the most tempting option. It's easy to react on autopilot and not question whether it is really the best choice. Thus, the term "tunnel bias" is a fitting analogy for the bias we have discussed here.

6.4.35 Bias blind spot

This next one is an interesting twist. It's a bias about biases. We've discussed a lot of biases and now have a pretty good idea of how to identify a bunch of them. However, even armed with this knowledge, we are still in great danger of bias. In particular, it is especially easy for us to be more willing to identify biases in other people than in ourselves. This is known as the bias blind spot.

As self-interested individuals, with access to a much different level of information about ourselves compared to about other people, we often struggle to judge ourselves as biased, even if our biases would be glaringly obvious if we were observing ourselves from an outsider's perspective. The ego strives to protect itself through many wild contortions of the subconscious. This is not always a bad thing, in terms of protecting our psychological health (e.g. not becoming depressed), but it sure does get in the way of thinking objectively sometimes.

The reason why it's called a bias *blind spot* is because it is like an analogy for the blind spot of a vehicle. For instance, many cars have side-view mirrors that can't quite see everything on the road, and consequentially vehicles are sometimes able to sneak up on a driver by driving into the blind spot, creating a dangerous potential for traffic accidents. An outside observer looking towards the car could probably easily see the approaching vehicle in the blind spot, but this is not true of the driver of the car being approached.

In other words, while being in control of a vehicle increases the amount of information you have about it in *most* respects, this is not true in all cases. The driver of a car has *less* information about the state of certain things about their car than other people do. Whether or not the taillights are working would be another example of this. Much the same can be said of people and their social and intellectual perspectives. In *most* respects we know more about ourselves than other people do, but not in all respects. The self acts like a blind spot. It empowers us but also gives us some distinct vulnerabilities. It takes a very high level of self-awareness to correctly compensate for these kinds of nearly invisible biases in our own minds.

If you aspire to maximize your intellectual integrity and personal charisma, then it is essential that you strive to identify fallacies and biases both in others and in yourself. You need to protect yourself from other people, who will often try to push nonsensical beliefs onto you and to manipulate you for their own profit, but you also need to protect yourself *from yourself*. The difficulty of seeing your own biases and bad behavior patterns is exactly what makes them so insidious and dangerous. It's much harder psychologically to view yourself with suspicion than to view other people with suspicion, but it is no less necessary. Or, to quote the famous physicist Richard Feynman: "You must not fool yourself, and you are the easiest person to fool."

6.4.36 The curse of knowledge

Sometimes the more you know the harder it can be to teach something. This is because your perspective has changed so much that you now find it harder to see the world through the eyes of someone unfamiliar with the subject matter. This is the curse of knowledge. Both teachers and students are often well familiar with this phenomenon. Even if they don't know the official name for it, they have probably encountered it many times in their life.

If you've ever tried to teach someone something, but they don't seem to understand what you are telling them, or you stumble over your words while trying to frame the information in the other person's perspective, then you probably have at least some amount of the curse of knowledge. The student you are trying to teach likewise suffers from your poor teaching methods, but is not themselves subject to the curse of knowledge. The curse of knowledge is a bias of the person who knows more about the specific subject matter, not of the student who knows less, generally speaking.

It can be quite surprising to think you're an expert at something, but then to suddenly realize when you try to teach the subject matter to someone else that it can be quite difficult to actually communicate it in an easy to understand way. The curse of knowledge is essentially an empathy and memory gap. It is a failure to think from the other person's perspective and to remember the steps necessary to transition from an uninformed novice to a master. It is as if parts of the pathway of learning, your memory of the journey of discovery, have faded away into the mists of time, forgotten.

The curse of knowledge is like trying to tell someone else where to go by giving them directions suited for *your own* perspective instead of for theirs. It is like a GPS trying to convey directions by repeatedly saying "come to where I already am" over and over again. When this happens, the helpfulness of the teacher is reduced significantly. Students under such circumstances are often forced to either rediscover the information on their own (an often difficult, elusive, insight-dependent, and time-consuming process) or else to seek supplemental information from other sources (e.g. superior teachers, books, videos, websites, etc).

Learning is a journey, not a game of teleportation. You have to respect the journey. You have to value it. Just as you can't walk from point A to point B without walking in-between them, you can't go from less knowledge to more knowledge without moving your state of mind. Movement requires a journey. Destinations have no movement, they are just positions, just different places we can move between. A destination teaches no lessons, only journeys do. To teach someone effectively, you need to be a tour guide, not a preacher. You need to come down from the pulpit and walk with your student, to put yourself in their shoes, to guide them up that winding mountain road towards intellectual enlightenment.

Just as you can tell that you're working out effectively by the strain it puts on your muscles, the same can be said of learning and the strain it puts on your mind. Mental tension is a sign of learning as surely as muscle aches are a sign of muscle growth. Mental struggle isn't a sign of stupidity, it is a sign of progress. Embrace the struggle. Lean into your challenges, instead of running away from them. Just as a runner must learn to embrace the runner's high, or a body builder must learn to embrace the muscle burn, an intellectual must learn to embrace the mental struggle.

It's how you bulk up your brain. Everybody starts at different places in life, but it's the journeys you take that determine your strength as a person. Don't worry about other people's positions, that's outside your control and hence inherently unproductive to worry about. Focus on what's under your influence. Focus on the pathways and choices available to you, and how you might use them to your advantage. Focus on the climb ahead, on the journey.

Mastery has pros and cons. Overwhelmingly, it mostly has pros, but that doesn't mean that it doesn't have any cons. One of those cons is the curse of knowledge. Mastery can make it more difficult to relate to people who don't know as much as you. It can get lonely on that mountaintop, on that pinnacle of skill and knowledge. Lonely but peaceful perhaps, like the life of a monk. However, it doesn't entirely have to be that way. By embracing empathy and clarity, you can learn to relive the journey with other people and help them climb the mountaintop as well.

After that, it won't be so lonely anymore. Furthermore, few things help you understand a subject more thoroughly than the exercise of teaching it to someone else. It can be a very effective way to remove the wrinkles (the knowledge gaps) in what you think you know. More isn't always better, but it is always more. The balance is a matter of nuance. Respect the basics. Respect clarity. Relive the journey. Do that, and you'll probably make an impact.

How does one fight the curse of knowledge though? Well, I can think of at least a few techniques, beyond what I have already mentioned in this section. Here's a list, in no particular order:

Techniques for fighting the curse of knowledge

1. **Write notes as you learn new things, continuously refining your understanding of the concepts over time.** Capture the fundamental essence of the underlying ideas. Try to explain the concepts to yourself in your own words, without referring to the source material. Wait until a little bit after your exposure to the new material to do this, so that the content has a bit of time to fade away from your memory, so that you won't just subconsciously regurgitate the source material, and so that the exercise will strain your memory at least somewhat. Don't wait too long though, as then you may just outright forget it.

    You also need to keep productivity and time constraints in mind to do this efficiently. You'll have to find the right balance. Oh, and don't forget to check your explanations against the original material to make sure you didn't distort anything. On the other hand though, sometimes your explanations will uncover new insights that weren't present in the original material. In such cases, you should value those insights. Preserve them. New ideas can be extremely valuable. Consider publishing those insights so that others can benefit from them. Don't be a slave to authority. Be bold in exploring new thoughts.

2. **Write for the reader, not for yourself.** Don't write selfishly when you are trying to explain something. Don't be lazy. If you find a mistake in your work, fix it. If the reader

would benefit from a concrete example, write one. Don't just shovel dirt into your writing and expect the readers to do all the hard work of finding the diamonds in the rough on their own. Do that work for them. Make the connections and "aha moments" for them, whenever possible, rather than merely dryly lecturing about the material while blindly assuming that the readers will go to the great pain of re-deriving all the connections on their own. They will most likely just get bored and find something else to do, something that has more readily apparent value. Yes, there's something to be said for making the reader work to improve their retention of the material, but that is something best left to exercises and projects.

Introductory material should always strive to provide insights right out of the box. I don't believe authors when they say that the material is best presented dryly and abstractly, without motivation or clarity. I think that's just an excuse. Some authors use "making a reader work to learn" as an excuse to simply be lazy, or even to cover up the fact that the authors themselves don't actually genuinely understand the material at any kind of deep level either. If an author genuinely understood the material, then they could surely give at least a somewhat insightful treatment of it. If it's just a matter of poor communication or laziness, then work on that.

Society tends to adopt the theories of people who communicate well, and to ignore the theories of those who do not. Your career depends on good communication, regardless of what career you have. Even a programmer, who may not interact with people as much as other professions do, still has to communicate their intention clearly in their code if they want their code base to be at all maintainable. Empathy for the reader is essential. Clarity is not an option. Stop treating it like something nonessential. If an idea is not communicated clearly then it might as well not exist, because as far as the reader is concerned (who can't make any sense of what you're saying) it indeed already doesn't.

3. **Use lots of concrete examples.** Abstraction has inherent limitations. There are certain things that abstract words are too broad to really capture the nuances of. Abstractions are kind of like folders or tags. They are ways of categorizing concrete things. However, it is hard to understand what belongs in a category if you've never even been given any examples of things that belong to that category. Concrete examples also test your assumptions, as well as those of the reader. Concrete examples keep you honest. Creating a theory without exploring some concrete examples of it is like writing computer code without ever compiling it. It is absurd.

You can't trust anything that hasn't been tested. You can't just go with your gut on these things. You need to really experience things first hand. Humans have an amazing ability to make an abundance of false assumptions. Any programmer can attest to that. Compilers are brutal. The experience of learning how to program is truly eye opening. It really makes it painfully obvious just how freakishly often humans make false assumptions. The experience profoundly changes you, for the better. It makes your mind much more precise and makes you realize just how utterly inane and naive many preconceptions really are. It helps you see reality for what it really is. It opens your eyes to how feeble and meaningless many ideas you encounter in day-to-day life actually are.

Anyway, tangents aside, the point here is: Always test your work. Always illustrate ideas to the reader using concrete examples. Learning requires testing your assumptions, lest your ideas rapidly degenerate into mere armchair philosophy or empty pretentiousness. People tend to best remember the ideas that they know they can genuinely trust. The subconscious automatically treats untrustworthy ideas as having a low expected value. The less value an idea seems to have, the more likely it is to end up in our mental trash bins. The dividing line between nonsense and reality is testing. You can only trust what you can test. This is as true for the teacher as it is for the student.

4. **Avoid jargon, except where necessary.** If a simpler word exists that expresses the same information as a more complicated or obscure word, without significant lose of nuance, then prefer the simpler word. Your goal is to communicate useful information clearly, not to make yourself seem smarter than you are. Don't just wave your membership card in the *Secret Society of Obscure Words* in your readers' faces like some kind of mentally challenged braggart. Nobody cares what words you know. People only care what value you have to offer them. This is supposed to be communication, not cryptography. Your reader wants to absorb useful information as efficiently as possible, not engage in a game of verbal cloak and dagger. Verbal camouflage is the opposite of clarity.

Don't use words that you haven't ever encountered in the wild. If you've never heard a word used in real life, there's probably a reason why. Thesauruses are tools for remembering similar words that you already know, not for learning new words without context. If a word is purely a superficial synonym for other existing words, then there is a good chance that it has little to no value beyond nominal variety. This is especially true of synonyms for common emotive adjectives, the kind that literature majors like.

Don't write like a literature major. Don't use pointless pretentious words. Any nuance or special connotation that an obscure word may have will be rendered worthless if your reader isn't already intimately familiar with the word. Few things are more jarring to reader immersion than having to repeatedly stop reading to lookup words in a dictionary every 2 pages, only to find that 95% of the time the obscure word added literally zero value compared to its simpler equivalent. Nobody likes having their time wasted for no reason. Such behavior will only irritate the reader.

Technical jargon, in contrast, is sometimes unavoidably necessary, especially as a discussion scales up and becomes deeper. However, even with tricky technical jargon (e.g. engineering terms), you should still try to illustrate the point simply if at all possible within the time constraints. You need to carefully balance precision against ease of understanding, relative to the audience. When in doubt, lean towards ease. When a direct substitute for a word doesn't capture the nuance you want, try re-writing the surrounding text using simpler words structured in a more impactful way. Often, when it seems like a simple word is not enough, it is actually because the structure of your text is bad. Think creatively, not narrow-mindedly, when interpreting this advice.

Don't contort simple statements into an amorphous fog of fluffy and largely meaningless words. Academic journals are some of the worst examples of this. I've seen entire paragraphs worth of text, perhaps even entire pages, that ended up meaning almost nothing when distilled to their true essence. Don't do that. If you want your writing to have impact and to change the world, then look at an incomprehensible research article in an academic journal, and then do the *opposite* of that. I can think of no better example of bad writing than that found in academic journals. Academia almost seems to pride itself on having rock-bottom terrible communication skills. The joke is on them though, of course. Wider society won't care about any of their work unless it's communicated clearly.

Don't write in awkward grammatical forms. Write in a personable and warm style, one that doesn't hide the author's voice. Information is most optimally conveyed in a conversational format. Furthermore, writing in a weird 3rd person or omnipresent academic style actually damages scientific accountability. In contrast, when someone includes personal pronouns, it reminds the reader that the author of the work is a living breathing human being, *one who makes mistakes.* This is exactly what we want. We *want* the reader to view the author as a fallible human being. It makes the reader more likely to suspect the possibility of errors, and therefore more likely to check the author's work by trying to contradict the author's assumptions.

In contrast, if you write things like "It is known that X, in the context of Y, based on the research of Z..." then it often lends the text a misleadingly authoritative quality. It makes the reader wonder if the statement is indeed some kind of universally accepted fact, when in fact it often isn't. Such authoritative-sounding statements are often actually just the personal opinion of the author of the research, phrased in such a way as to make it sound more authoritative and unquestionable than it actually is, in order to scare people away from questioning it. Contrary to popular belief in academia, this dryly impersonal and disembodied style of writing frequently found in academic circles actually *damages* the academic integrity of the text, not enhances it.

Being personal is both more enjoyable and more scientific. Nobody likes incomprehensible text. Even in scientific circles, despite the nominal culture of obscurity in academic writing, the most cited papers often tend to be the more clearly written ones. It is as if academics are afraid to ever admit to not understanding anything, no matter how convoluted and inanely communicated the information is. Yet, if you look at the deeper patterns of behavior, at the sources academics go to and the informal and conversational style of how real research is conducted behind closed doors, then it becomes clear that *just like all humans* even academics ultimately prefer clarity and ease over dry convolution, even if they won't always admit to it out loud, lest they slight their own egos.

5. **Don't try to explain everything at once.** Instead, explain the information piece by piece. Identify self-contained conceptual units, and then teach those units separately from the rest. Don't make the reader have to juggle a whole bunch of different dependencies and relationships between a bunch of different concepts all at once. Be modular, just like a good programmer. Minimize the number of moving parts the reader has to understand at any given moment of time in order to understand what you are saying. Don't be afraid to explain things vaguely at first, if doing so improves the efficiency of the learning process.

   Perhaps start with an eagle eye overview of what's coming up, and then zoom in on distinct landmarks as necessary. Give multiple angles on the same subject, like a photographer doing a survey of some architecture, gradually accumulating enough pieces of information to eventually understand the whole. Be tactful. Be patient. Divide the concepts up into independently actionable parts. Try to give the reader some time to absorb the information before moving on. If you give too much information at once, it will tend to tire out the reader. Keep your vocabulary tight and focused. The larger the conceptual surface area of a given piece of text, the more likely it will be confusing.

6. **Motivate the material, when possible.** People think in terms of value. If they don't see the potential value in something, then they won't feel motivated to indulge it. The human mind also tends to solidify memories in direct proportion to how important it deems those

memories. A highly valuable memory, such as the fact that petting wild mountain lions is likely to get you killed, will be remembered very easily. In contrast, an abstract idea for which no motivation was ever provided, and for which no value is apparent, will quickly be forgotten. For instance, think back to what you learned in school growing up.

Weren't the ideas you placed the least value on also the earliest to be forgotten? This is probably why most people don't remember much from history class. What is its direct utility? What does it do for you? Often not much, except for perhaps in the really distinct cases. For example, the history of Hitler and World War 2 serves as a very memorable warning to protect your own life from similarly dangerous political trends. That's why almost everyone remembers Hitler, but almost nobody remembers the capital cities of foreign countries. One memory adds value to your life, whereas the other memory probably doesn't. One is very much worth the time to learn, whereas the other is much less so.

In mathematics especially, this is a principle that is all too often ignored. A large part of the current culture of mathematics is totally indifferent to whether or not their mathematics actually provides any value to anyone. Worse still, even when massive and powerful motivations exist for a piece of subject matter in mathematics, many mathematicians will nonetheless deliberately ignore those motivations and never say a single word about them, on the basis that revealing the motivations would damage the "purity" of the subject matter. This is pure idiocy. Providing value to your fellow human beings is what life is all about.

You certainly can't expect people to flock to mathematics if such a toxic indifference to doing anything meaningful has become the norm in its culture. As such, mathematics as it currently stands has a big public relations problem. All it might take to fix this would be for the mathematics community as a whole to start consistently providing the underlying motivations for the material whenever possible. Tell people *why* they should care about it. Value is the main attraction. Abstract purity is just a sideshow. Mathematicians need to shake off this toxic purity fad and refocus on the things that matter most: real-world value. You couldn't change this fact even if you wanted to. It is human nature than only high-value things will ever have wide reach in society. Motivation is essential.

7. **Think through the prerequisites to understanding the material, and plan accordingly.** Don't try to teach someone mounted archery before they even know how to ride a horse. Don't try to teach someone calculus before they even understand multiplication. Always consider the dependency chain. Some things are impossible to learn out of order. Other things are merely more difficult. In either case, proper accounting for prerequisites is essential. You have to know where you are to know where you're going. How would you use a map if you didn't know where you are? Establishing a sense of direction is essential. Know where you stand before moving forward. Seek high ground to get a better sense of the information landscape around you.

Be patient enough to learn the necessary prerequisites, keeping in mind the eventual payoff. However, it is important to strike the right balance between patience and progress. You don't want to be so obsessed with prerequisites that you use it as an excuse to procrastinate. You'll never accomplish anything that way. You need to know when enough is enough. You need to know when it is time to move on. At some point you will have to just boldly put yourself out there and make your move. Don't spend your whole life preparing. Don't be afraid to spread your wings.

If you feel compelled to explore an idea, and it seems reasonable enough and timely, then do so. Such whims often end up being very valuable. Often, the things that most excite us are where our time is best spent. Don't get trapped in bureaucracy. Live a fresh and interesting life. Pay attention to prerequisites, but also don't let them stifle your creativity. Don't live life just going through the motions of what other people want you to do. Be as persistent and forceful as a samurai, yet as lightweight and nimble as a feather. Think like an adventurer. Be prepared, but let the winds carry you where they will.

Anyway, that covers at least a few techniques for fighting the ill effects of the curse of knowledge. Applying these principles, if you aren't already, will surely improve the quality of your communication. It is definitely worth the time investment. Even if you don't particularly care about other people understanding your work, applying these principles will still nonetheless tend to significantly strengthen your own understanding of the material, and will often even result in you discovering valuable hidden insights by accident. In fact, many of the best insights I have had throughout writing this book happened spontaneously as a result of applying these principles to my writing.

Without an unyielding determination to explain my thoughts as clearly and completely as possible, from a perspective of empathy for the reader, many of the insights and ideas contained in this book probably never would have even occurred to me. The act of trying to teach you (the reader) the ideas, combined with patience and a willingness to explore wild intellectual whims and curiosities, is the source of most of my best material in this book.

In fact, I'd go so far as to wager that anybody who applies these same principles (embracing a passion for clarity, empathy, and curiosity) in their own work is likely to be able to do the same. When you have a high standard for clarity and conceptual correctness, insights have a way of just landing on your lap, sometimes almost effortlessly. Plus, as an added bonus, your writing becomes clearer and more enjoyable to read.

I've often heard it said that teaching a skill requires more expertise than merely being proficient in it. This makes some sense, when you consider the fact that teaching requires both proficiency in the skill and the ability to communicate that skill effectively, whereas proficiency alone requires only direct knowledge of the skill itself and lacks the burden of communication. The teacher has two ways to fail (skill and communication) whereas the practitioner has only one way to fail (skill).

That being said, the practitioner often applies the skill in deeper and more elaborate ways, by virtue of focusing on the skill's applications so much more, whereas teachers sometimes just rehash the same basic material over and over again. Whatever the truth may be though, one thing that is clear is that the curse of knowledge is very much a real effect and well worth fighting against. The closer we move to any specific perspective, the less our mental fog obstructs our view of it, but the more other things may start fading away into the fog. Whether teacher or student, whether professor or professional, all would benefit from reliving the journey of discovery.

6.5 The journey towards a more rational tomorrow

Anyway, that covers all of the cognitive biases that I'm going to be covering in this book. There's plenty more cognitive biases where those came from, but you'll have to look those ones up yourself if you want to learn more about them. That being said, keep in mind that not all named cognitive biases are actually genuinely conceptually distinct from other cognitive biases. Also, some cognitive biases may not be entirely rigorous or may not be correctly specified. Thus, if you do lookup the other cognitive biases, remember to be a little bit skeptical and to question what assumptions are being made.

We've now thoroughly explored numerous different informal logical fallacies and cognitive biases, and yet our coverage still isn't even complete. Thus, as you can see, there is quite a lot that goes into thinking rationally, and you are certainly not born with it; it must be practiced. Notice that many of the informal fallacies and cognitive biases we have covered here are quite counterintuitive. Worse still, many of them are rooted in subconscious instincts that are hard to compensate for even when you are aware of them.

I hope our exploration of these fallacies and biases has made it clear to you why everyone needs to have at least some familiarity with resisting fallacious reasoning and cognitive biases. There are so many potential sources of error. A frighteningly large portion of society seems to be convinced that they are automatically born rational, even though they are obviously not. Humans aren't born rational. You can no sooner expect to become rational without effort than you can expect to become muscular without effort. Mental fitness is a real thing. The more you ignore the study of logical thought, the more your capacity to defend your mind against misinformation and manipulation by others will atrophy.

There are hundreds of types and subtypes of fallacies and cognitive biases. It may seem like a lot to keep in mind for reasoning well, but don't panic. The underlying principles of logic are actually quite easy to get used to. For example, applying the law of non-contradiction, and gathering completely representative information about the system you are analyzing, are two of the most important underlying principles of rational thought, and certainly not hard to learn once you get into the swing of it. Knowing how to apply the principle of property conservation (page 104) in order to create valid chains of logical inferences, analogous to rows of dominoes falling over, is another crucially important underlying principle of logic.

Many fallacies and cognitive bias ultimately stem from a much smaller number of much deeper fundamental problems. Just as practicing exercises and exploring concrete examples of any subject helps you to gradually grasp the underlying fundamentals, much the same is true of logic. The more you study these concepts, the more you will grasp the underlying nature and principles of rational thought. It's very much a cumulative kind of thing. It snowballs after a certain point, and really adds a lot of value to your life. Mastery of logic makes your mind like an armored fortress. It makes your mind much safer from external threats. It also makes your mind much more likely to generate large amounts of value and wealth, both for yourself and for your fellow human beings.

Fair warning: Being totally rational sometimes requires you to go against the grain of social norms. Many people that you encounter in life will treat objective well-structured neutral criticism as being practically the same thing as insults, treachery, cruelty, or hostility. This is often true even if you are careful and respectful. In a society full of irrationality, you sometimes pay a price for being rational. True intellectual integrity requires a willingness to scrutinize and contradict ideas regardless of where those ideas came from, even if those ideas came from friends and family for example.

It is wise to choose your battles carefully and strategically. However, you sometimes nonetheless have to be willing to openly criticize even your own "side" in a conflict. You should be tactful about it, but even if you're tactful you are still likely to get bad reactions from people sometimes. Ego slights, no matter how minor or indirect or constructive, very often will trigger hostility. Sometimes relationships will suffer, but sometimes that's the price of truth. Loyalty to truth sometimes requires disloyalty to people.

Truth and conformity are often in conflict. If you embark on the path of truth, expect to sometimes potentially experience a higher degree of (unjustified) social isolation and prejudice, but also expect to probably experience a higher degree of freedom, self-sufficiency, serenity, and power. That being said, it isn't quite so cut and dry. Objectivity often ends up being an attractive quality, so frequently it will actually end up increasing the strength of your social network rather than decreasing it.

In any case, objectivity is a great way to more consistently filter out toxic factors from your life so that you can invest a greater proportion of your time in better things and in better people. Not everyone feels threatened by objectivity, and the people who don't will tend to be stronger and more worth-while people to associate with, people who are better for you. Don't confuse objectivity with being needlessly cruel or antagonistic towards others though. A truly objective person always understands the value of tact and empathy.

Overwhelmingly, making yourself more rational will tend to benefit you greatly. However, it is wise to be aware of the potential social conflicts that objectivity sometimes generates, and to plan strategically about how to deal with those problems. Consider the culture and circumstances you find yourself in, and engineer your approach to fit it in a tactically advantageous way. If it isn't safe to publicly criticize subject X in your society's current structure, find a way to gradually whittle subject X down indirectly over time, patiently undermining it in a safe and secure way, instead of attacking it directly and overtly. Death by a thousand needles can be a very effective strategy for change, as long as you are patient enough. Think beyond your own lifespan. Think like ocean waves crashing against the rocks, eroding them over time with relentless patience and subtle opportunism.

These examples of fallacies and cognitive biases we have explored in this section make it clear that having society become more rational would be highly beneficial for protecting humanity's future. Logic is far more than just a way of thinking about and augmenting technical endeavors such as science, engineering, and programming. Logic is the very essence of productive thought, the very essence of consistently and systematically protecting your own self-interest and well-being. Literally almost any endeavor you might ever seek to accomplish in life could probably be done more effectively if you thought about it more logically.

Synergies of logic and creativity are especially spectacular. The existence of video games, an accomplishment tantamount to creating magic, is such an example. The more you know about how to solve problems logically, the more powerful your creative spirit will also become. A life without logic is a life without direction. Logic is the mental compass of our minds. Without it, we are lost. Without it, we have no hope of ever maximizing our quality of life. Don't be like a piece of driftwood, wandering aimlessly in an ocean of ideas. Take control of your destiny. Embrace logic. Embrace the creative spirit, and use it to unleash your full potential.

Chapter 7

Extras

7.1 Introduction

Welcome to the Extras chapter. Unlike most of the other parts of this book, this part will *not* be organized like one continuous chain of related thoughts and tangents. Instead, the Extras chapter will simply be composed of a large number of small independent discussions of various different somewhat random ideas, suggestions, and otherwise useful items of information. Each item will be its own standalone thing. Little to no attempt to connect each item with the other adjacent discussions will be made. The flow of the text will abruptly break at the boundaries of sections, unlike in many other parts of the book. Indeed, it should be possible to jump around wherever you want to in this chapter, unlike in many parts of the other chapters.

However, as random as this diverse collection of items may indeed be, many of the ideas will fall under similar broad categories and subjects of interest. As such, I will be separating the various different groups of items into a few different basic categories, in order to make navigation and browsing easier. Separating the items into categories like this will also make it easier for some readers to skip over any subjects about which they have no interest.

Fair warning though: The categories will be treated somewhat loosely, so there might be some ambiguity and overlap about which category a few of the items really belong in. Anyway though, that should be enough information to get you properly oriented now. Please refer to the table of contents at the beginning of the book if you want a quick overview of all of the categories and individual items that will be covered here. Let's begin.

7.2 Logic, math, and programming

7.2.1 A much more intuitive unit for working with angles, hidden in plain sight

Degrees and radians... With how utterly ubiquitous they are, you'd think they were the only sensible ways of working with angles. You'd be wrong though. In fact, not only would you be wrong, but there's actually an alternative unit which is usually far easier for human beings to work with and to think about. Indeed, if natural and intuitive expression of concepts is what you care about, then in most cases neither degrees nor radians will be the optimal choice for how to frame things for humans. There is actually a far more natural choice. Degrees and radians don't really deserve to be as overwhelmingly dominant as they are. They are mostly only that way due to social momentum and lack of imagination.

What is this wonderful alternative unit of measure that I am referring to here? The answer is *cycles*, a.k.a. *revolutions* (or "revs" for short), a.k.a. *turns*[1]. No unit for measuring angles is more humanly intuitive than cycles (in most use cases). What is a cycle though exactly? Well, it is simply the number of times around a circle (or other periodic phenomenon) that one has traveled, considered as a continuum number so that it can support more than just whole numbers of cycles. Thus, for example, if one said "3.2 cycles" then it would mean traveling 3 times around a circle (or whatever other periodic structure) plus 20% more of the way towards what would complete a 4th cycle. Easy, right?

Let's do a bit of comparing and contrasting, to make it clear why cycles are generally much easier for human beings to think about than degrees or radians are. Suppose, for example, that I told you that there's an object in a video game (e.g. a magic glowing crystal) which floats in the air about 5 feet off the ground and rotates horizontally at a rate of 3747 degrees per minute. Can you visualize this in your head? Without doing any calculations, do you have any intuitive sense whatsoever for how fast a rotation this actually is?

Most people don't. And why should they? Degrees are a strange and highly arbitrary unit of measure to work with. Even just understanding where certain positions on a unit circle are in terms of degrees generally requires lots of tedious rote memorization, and even then people generally only end up with a tiny number of such positions actually being successfully memorized. This is often true even when someone does lots of math for a living.

Want to know where degrees actually come from and why they were chosen to be what they are? The answer is simple. Degrees were designed to have a relatively high ratio of perfectly divisible integer factors relative to the magnitude of the total number of degrees per cycle. It's the same reason why clocks have 12 hours and 60 minutes. Having 360 degrees in a circle is just a natural extension of that. It's just 60 · 60, which is 360. The numbers 12, 60, and 360 just happen to be numbers that are divisible by a relatively larger number of integers than the other nearby integers. They are like anti-prime numbers, in a sense. This makes these numbers easy to divide up in lots of different ways without leaving behind a remainder.

Here's the thing though: This is the *wrong thing to optimize for* when it comes to measuring angles and other periodic phenomena. It derives from an irrational obsession with trying to avoid fractions. It's like what an OCD person who is obsessed with integers and afraid of continuum numbers would think is a good design. Indeed, mathematicians were once (for quite some time) bizarrely obsessed with trying to force everything to be integers whenever at all possible, a tradition which dates all the way back to the Greeks.

I suspect this may have played a possible cultural role in contributing to the high prevalence of degrees as a unit of measure for angles. Having "round numbers" like that though is not really very important. In fact, it hardly matters at all, especially considering that numbers can be shifted and scaled and rebased arbitrarily to fit whatever you want them to, such that the concept of a "round number" is largely just a figment of your imagination, in a sense.

Most importantly though, degrees just aren't very intuitive to think about. On that note, let's turn our attention back to our example of the floating rotating object from before. As you may recall, I said that the object is rotating at a rate of 3747 degrees per minute, and I asked you to

---

[1] Of these three synonymous names for this unit, I most prefer the name "cycles". The reason is because "revolutions" and "turns" both have a purely geometric connotation, whereas "cycles" can apply more broadly than that. Geometric angles aren't the only kind of periodic phenomenon that we sometimes want to measure. Thus, "cycles" is the best choice of the three terms, because it is the most general. The other two terms are unnecessary.

try to visualize and estimate how fast that really is, without doing any calculations. If you're like most people though, this is a very difficult task. Such big and arbitrarily measured quantities are often hard to reason about and hard to really get a proper sense of.

Let's see what happens when we switch to cycles though. One cycle corresponds to one full turn around a circle. There are 360 degrees in 1 cycle. Therefore, there are 3747/360 cycles in 3747 degrees. 3747/360 is approximately 10.41. Thus, 3747 degrees ≈ 10.41 cycles. So, in light of this, we can now say that our floating object is rotating at a rate of approximately 10.41 cycles per minute. Can you visualize it now? I'm betting a lot of you can. I bet you can even see it spinning in your mind's eye now, at a vague approximation of that speed. It turns full circle about once every 5.76 seconds. It's actually rotating quite gently and lazily.

Even if you still can't visualize it though, can you at least *understand* it better now? I'm betting that almost 100% of you can at least do that much now. I'm also betting that 10.41 cycles per minute sounds a *lot* slower to most people than 3747 degrees per minute does. That's because numbers that are both big and disconnected from everyday experience tend to confuse and overwhelm the human mind very easily. In other words, its because degrees are *inherently poorly suited* to the way human beings usually think, especially compared to cycles. Yet, by sheer force of social momentum and lack of imagination, degrees are still the most dominant unit of measure for angles.

This is actually a really great example of how an entire society can get stuck in a counterproductive or narrow-minded way of thinking and working, even when the alternatives are extremely simple and extremely obvious. Too many people think based on mere conformity instead of based on first principles. Even when people think they aren't being pulled in by the whirlpool of everyday conformity, even when they think they are somehow above all of that, they still often are in fact actually mindlessly conforming in countless subtle ways. It effects all of us. People are so easily blinded to even the most basic concepts.

Blatantly obvious useful concepts and ideas can be staring people in the face literally every single day in their lives, and yet still go completely unnoticed regardless. This is part of why even the simplest idea or spark of insight can be so powerful, and why we must as people be braver about sharing such information with others, instead of being intimidated about being potentially dismissed or laughed at for focusing so much on "trivial" things. We must live freely and think freely, if we want to truly become aware of the real essence of things and to maximize our creative and expressive potential as individuals.

Pick up a math book or a game engine, or any other item that is likely to contain a large amount of mathematics and to have rotations of some kind in it, and you will find (at the time this book was written) that in most cases only degrees and radians are supported and used. There might also be support for "gradians" (400 units per cycle), or other odd units of measurement, but seldom will you ever find full support for cycles.

This is despite the fact that cycles are the most conceptually obvious and expressively natural of all of the possible unit choices. Perhaps people think to themselves "this is too obvious, so it can't be right or best". Perhaps that's part of why people don't give adequate attention to cycles as a unit of measure currently. Whatever the case may be though, I think this should change. Cycles are so natural and expressive. They are so much more pleasant to work with and easy to use for so many different tasks than degrees and radians are.

Even better though, cycles actually fully support any other arbitrary "perfect" division of a circle that you could ever want. This even includes degrees. It's arbitrary. You just have to write the cycle measurement as a fraction in that case, instead of simplifying it. For example, if I wanted to say the equivalent of 3747 degrees, and still frame it in terms of divisions of the circle into 360 pieces (e.g. if I was a masochist), even though we're now working in cycles, I could just write "3747/360 cycles" and just continue working with it in that fractional form. You can think of a cycle any way that you want to.

In other words, cycles are a *strict superset* of all other integer divisions of the circle. There is nothing that another way of dividing a circle up into integers can do that the cycle system can't also naturally do. There is therefore never a good reason to lock oneself into one of the other integer divisions of a circle, and therefore never any idealized reason to use degrees or gradians (etc) instead of cycles.

Similar things can be said of non-integer divisions of the circle as well. It works for radians too. You can just multiply your cycle measurement by 2π whenever you want the special properties of radians, on a case by case basis. Thus, cycles are essentially the most universal and adaptable measure of angles and cyclic phenomena that one can hope for, i.e. cycles are the measurement unit with the least possible friction for being adapted to any other specific use case. Cycles should therefore not only be used more widely, but cycles should actually be the *default* unit of angle (and other periodic phenomena) measurement that all people should use. Thus, the current status quo of neglecting cycles and only using degrees or radians is actually very much conceptually backwards.

Of course, sometimes you *do* have to use radians, since radians have special mathematical properties that make certain geometric calculations a lot easier to do. However, that does not mean that you should generally *think* in terms of radians. It's only in the context of certain specific geometric calculations that thinking in terms of radians will be more intuitive.

Otherwise though, for most other uses of angles and periodic phenomena that human beings encounter on a regular basis, cycles are a lot easier to think about than the alternatives are. Cycles truly are a widely underappreciated unit of measure in mathematics. I personally find this fact quite surprising. You'd think more people would have realized something so obvious by now, but apparently not. The stagnation in this respect is taking quite a long time to lift. However, be that as it may, I think the status quo will probably break soon though. Cycles are bound to become much more ubiquitous in the near future. All we really need is for more people to wake up to the fact.

Cycles are great because they allow human beings to instantly realize exactly how much of a rotation someone is talking about, without ever having to do any additional calculations to do so. For example, if I say "0.25 cycles" then you will instantly know that this amount is equivalent to a rotation 25% of the way around the circle, i.e. a rotation of 1/4 of a full turn. It doesn't get any easier than that. Thinking this way makes it immensely easier to express exactly how much of a rotation you want in a seamless way. Indeed, in my game dev and programming work, one of the first things I often do when I get my hands on a new engine is to write an abstraction over the existing code so that I can work in terms of cycles instead of in terms of degrees or radians. It helps a lot. You should give it a try yourself sometime, if you do that kind of work.

7.2.2 π is actually not the best circle constant (not my idea, but I support it)

Contrary to popular belief, the mathematical constant π (pronounced "pi" as in "pie") is actually not the best choice for the circle constant. You see, while it is true that π certainly works, and is "correct" in that sense, it is still not quite the best choice. The best choice is actually twice the value of π, which is known as "tau" and is represented by the symbol τ. In other words, $\tau = 2\pi$. Why though is τ a superior choice to π?

Well, the answer is because τ is more natural in several respects. One of the most obvious ways in which τ is better than π is in the fact that π corresponds to only half the circumference of the circle, and yet geometric math more frequently naturally tends to want to be expressed in terms of the entire circle circumference. It feels less strained when you use τ.

Additionally, using the constant π instead of τ causes lots of extraneous constant multiples to keep appearing in the derived geometric formulas for things like circumference, surface area, volume, etc. These extra factors tend to appear less often in formulas that use τ though, and thus make the resulting formulas easier to remember and easier to reason about.

Basically, the missing factor of 2 in π causes unnecessary complications in geometric formulas. When π is raised to some power p (often growing in proportion to the number of dimensions of whatever is being calculated) in a geometric formula you often end up having to also multiply that geometric expression by a power of 2 somewhere, such as perhaps 2^p for example, or some other related factor, in order to get the correct answer.

Those kinds of extraneous multiples don't appear as often if you use τ. This ends up also causing some geometric formulas to become both easier to read and easier to understand. It removes some extraneous distracting noise from the formulas and makes the underlying concepts pop out more clearly. Using τ also combines slightly more naturally with using cycles to measure angles (and other periodic phenomena) than using π does, since τ is framed in terms of the entire circle, just like cycles are. The number π itself in contrast was chosen kind of arbitrarily, as a quirk of history essentially, from what I understand. If you ignore that historical and social momentum though, τ is clearly the better choice.

All this being said though, I want to make it clear to you here that this isn't my idea. I certainly can't claim any credit for it, not even in terms of independent invention. I got the idea from someone else. I found it on the internet one day. If you would like more information on this subject, just look up "tau is better than pi" on the internet, or some other similar query, and you should be able to find more info on it, including info on some of its most prominent advocates and so forth. The subject has received fairly wide attention. You should be able to find plenty of discussion of it online. I just wanted to add yet another voice in support of the suggestion, since conceptual clarity is something I care very passionately about.

7.2.3 A new term for the study of mathematics, with less of a dry connotation

The word "mathematics" has quite a dry and mechanical connotation to most people's ears. When people outside the field of mathematics itself hear "mathematics", they often are reminded only of really mundane and banal calculations, such as arithmetic and basic algebra for example. This does not do justice to the real scope of mathematics though. Mathematics is far more than just the arithmetic and basic algebra that so many people think of it as being. It is so much richer and so much more imaginative than that.

Mathematics is the language of magic. It is a systematic methodology for expressing *any arbitrary conceivable concept* so precisely that the concept can then subsequently be simulated (whether by hand or by computer) to the point of seeming real. Whereas natural language merely lets you express worlds that you can imagine, mathematics allows you to express worlds that you can make real (through simulation). Mathematics, combined with logic, is also the essence of rational thought. It is the art of proof. It is one of the only ways to truly know that something is certain beyond all doubt, insofar as our universe allows any certainty of knowledge at all.

Yet, despite all the wondrous qualities that mathematics possesses, mathematics has a big branding problem. Far too many people associate mathematics with nothing more than dry and unimaginative arithmetic calculations. Instead though, if people truly understood what mathematics really was, then they would realize that mathematics (including logic) is really in fact *the study of all conceivable logically coherent universes*.

Mathematics is not a dry subject[2]. It is in fact limitless in its creative potential and its exploratory wonders. Yet, the connotation of the term remains so bland. This should be fixed. We should create an alternative term for mathematics to eliminate this problem. We need to make clear to people that the dull connotation of the word "mathematics" doesn't really tell the full story. As such, here's a new corresponding definition:

Definition 186. *The study of logic and mathematics is really the study of all conceivable logically coherent universes and systems. Yet, the term often lacks this connotation in many people's minds. As such, we will define* **logical cosmology** *as a synonym for the study of logic and mathematics. Specifically though, the intended connotation of logical cosmology is that of "the study of all conceivable logically coherent universes", rather than that of merely a mechanical process for computing things. The term "logical cosmology" is intended to have a much more exploratory, creative, adventurous, and all-inclusive connotation than the term "mathematics" has.*

Thus, having now defined this term, we can switch between "mathematics" and "logical cosmology" depending on which connotation we want to emphasize. For example, when working with things in a purely computational and mechanical way, we can say "mathematics" in order to emphasize the more tool-like connotation of the study of logic and math. In contrast though, when working in a more creative or arbitrary context, we can instead say "logical cosmology" in order to emphasize the broader exploratory aspects. We should keep in mind though that these terms really are synonyms though. They merely have different connotations. The underlying meanings are still the same in the stricter sense. Mathematics (a method for systematic generalized calculations) *is* logical cosmology (the study of all conceivable logically coherent universes and systems), and vice versa. Get the idea?

The hope is that this will help to dispel some of the misleading associations and biases some people have against logic and mathematics, and thereby to broaden people's perspectives and to attract more widespread interest in the subject. Don't underestimate the power of branding. In marketing, having a good name for a product can make or break that product's success. Much the same can be said of many other subjects though.

It is critically important to choose names and descriptions for things very carefully if you want those things to ever be as widely adopted as possible. You need to very clearly communicate the value proposition. This is true in virtually all subjects of study, whether it be mathematics or business or whatever else. The value proposition is an inherent aspect of the nature of good communication, and hence the need to convey it well is applicable to pretty much any discipline. All endeavors benefit from clarity. Don't neglect it.

Oh, and I also want to point out that if you interpret "logical cosmology" in the broadest possible sense, then it also includes things like computer programs and video games under its umbrella. The reason is because computer programs and video games are both simulated manifestations of specific possible logical and mathematical systems (a.k.a. "universes").

---

[2] The culture of mathematics and the way math research papers are written is often dry, but this does not mean that mathematics itself is dry. The fundamental nature and creative potential of a subject is not the same as how that subject just happens to currently be treated in the existing culture.

In fact, even things like digital artwork could be considered to be part of logical cosmology, since such things are represented entirely as logical, mathematical, and algorithmic objects underneath the hood (e.g. pixel data, etc). Basically, logical cosmology includes any logically coherent thing whatsoever that you might ever seek to study, as long as you think of it in terms of formal logic, mathematics, algorithms, or programming (i.e. in terms of the underlying logical structures and such).

Finally, in case it isn't clear, the reason why I chose the word "cosmology" in the term "logical cosmology" was because "cosmology" refers to the study of the universe in the broadest possible sense. Thus, by combining "cosmology" with "logic" via "logical cosmology" the term acquires the connotation of "the study of all conceivable logically coherent universes", which is exactly what we want here. There is also relatively little risk of confusion with the astronomical meaning of "cosmology", since the adjective "logical" would be redundant and weird under that interpretation. Basically, "logical cosmology" was the best term I could think of for what connotation I wanted here. It'll work just fine, and it has a good ring to it.

7.2.4 Two terms designed to disambiguate the meaning of "intuition", so that rationally justified intuitions are not so easily sneered at or dismissed

There's a problem with the word "intuition", and its that the word is associated with both bad intuitions and good intuitions. This makes it easy for people to use intimidation tactics in order to dismiss or obstruct the pursuit of the underlying insights that justify and connect concepts together in interesting and useful ways. The ambiguity of the word "intuition" essentially allows people to give a bad name to clarity and insight, by association, thereby allowing obscurantists (e.g. people who love complexity for complexity's sake, who don't care much about clarity or usefulness, who are pretentious or pedantic, etc) to often wield disproportionate power in fields of study like logic and mathematics, in subtle ways. This has toxic and counterproductive effects on progress.

It would therefore be wise to actually split the term "intuition" into two separate cases, one reserved only to refer to correct intuitions and the other reserved only to refer to incorrect intuitions. Doing so could perhaps help to reduce the influence of obscurantists and other kinds of intellectual bullies within the culture of mathematics and related disciplines. Anything that reduces the power of such people will surely improve the culture of mathematics as a whole, thereby helping to reduce stagnation and rot within the field, and ultimately making it more broadly appealing and easier to reason about, and thus is a worthy goal. Thus, here are the two new term definitions:

Definition 187. *A **natural insight** is a justified correct intuition. The term may refer only to correct intuitions, and should have a relatively high standard for use (i.e. should be accompanied by some kind of rigorous justification). A natural insight needs to be logically valid in at least some strong sense. Physical analogies are fine even if the corresponding physical context isn't necessarily there (e.g. you can use an analogy about a circle for any cyclic phenomenon even when not dealing with literal circular geometry, etc). Conceptual correctness (not empty pedantry) is the point.*

Definition 188. *An **instinct error** is an incorrect or misleading intuition. It refers to intuitions and impulses which are not well-founded and which misrepresent a concept or phenomenon.*

Such instinct errors are often the result of "common sense" thinking, i.e. are the result of assuming that your existing prejudices are correct without testing them (which is what "common sense" really is).

Thus, we may now subdivide intuitions into two types: natural insights and instinct errors. We may still use the broader term "intuition" though, if we intentionally want to refer ambiguously to either or both types of intuitions. For example, this enables us to now say that we are seeking natural insights for things, instead of merely "intuitions", thereby making it more difficult for other people to dismiss our endeavors as being "not rigorous" (and hence not worthy of pursuit), by making it clear that the non-rigorous and fallacious cases are excluded, and that we are only seeking *justified correct* intuitions rather than just any kind.

Just as intended, this has the net effect of separating the concept of "intuition" from the concept of sloppy thinking, so that obscurantists and other intellectual bullies can no longer have such an easy excuse to dismiss the pursuit of intuitions out of hand. By removing the equivocation that was previously inherent in the term "intuition", we now have an unambiguously *respectable* term for good intuitions: natural insights. This helps. Words have power.

One of my favorite super easy examples of a natural insight is how to interpret $B - A$. One way of interpreting it (probably the most obvious) is just to think of it as reducing B by A of course. However, there is also at least one other very useful way of interpreting it. You can think of $B - A$ as meaning the displacement (i.e. directed distance) from A to B. This is an extremely useful way of thinking about it when you are dealing with space and moving objects around between different positions, such as very often happens when doing game programming. Thinking like this makes it easier to arbitrarily chain concepts together, just as you would in natural language. Anywhere you need a displacement (i.e. directed distance) from A to B, you can just write $B - A$ and it will generally get the job done.

Building up a vocabulary of natural insights like this is extremely empowering. Taken to its limit, it will allow you to weave together any arbitrary concepts you could ever want, just as easily as you would chain together words in English. Yet, unlike English, what you say via the language of mathematics can be made real as a simulation. You just need to write a corresponding computer program to do it. The creative possibilities are limitless. It essentially lets you play god. And, all you need to do to unlock that power, is to simply acquire enough natural insights in logic and mathematics that you are able to successfully verbalize whatever arbitrary set of concepts you want to bring into existence. So you see, just as I said, mathematics truly is the language of magic.

This is a huge passion point for me. As an aspiring game developer, programmer, mathematician, and logician, nothing gets me more excited than the prospect of unlocking more of these natural insights. Every single tiny little natural insight I've ever discovered has significantly improved my expressive vocabulary for chaining arbitrary concepts together, and hence also for game dev. Natural insights aren't just cute little analogies or educational tools. They are so much more than that. Natural insights make dreams possible.

Indeed, armed with the right set of natural insights, and a knowledge of logic, mathematics, and computer programming, anything you can ever dream of can be made real by simulation. Properly understood, natural insights are the greatest source of creative freedom in existence. They are extremely valuable. And yet, due to the vice-like grip the obscurantists and other intellectual bullies have on the culture of mathematics, so very few of these profoundly empowering natural insights ever get published in any reasonably comprehensible form.

That is a massive tragedy I think. It is currently way too difficult for people to find these natural insights in the existing literature. Publishers, editors, and math authors seem almost hellbent on deliberately removing as much of these empowering insights as possible from their texts, so that nothing but the so called "pure math"[3] remains.

Such stunning waste. I'm tired of this aspect of the current culture of mathematics. It's one of multiple big reasons why I left the math community (by switching majors in college) and now operate outside it as a programmer and game dev instead. I'm so much happier outside all of that. Good riddance to it. The culture of mathematics can be so bizarrely backwards sometimes. It can be so stifling[4]. It is a historical tragedy that it has become that way.

In contrast though, luckily, game dev and programming are so liberating and joyous compared to dealing with the existing math literature and culture. I'm so glad I no longer have to deal with mathematicians' counterproductive attitudes and fruitless obsessions with low-value fringe cases and unmotivated hand-wavy nonsense. Their lack of awareness of their own vast abundance of poorly defined hidden assumptions also never ceases to amaze.

Complexity for complexity's sake may be a favorite of many mathematicians, but for most sane people though it is nothing but a joyless wasteland, a vast ocean of meaningless lip flapping. I love how much more computational and conceptual integrity computer science typically has than math. It is profoundly liberating. Everything is so much more consistently reliable and meaningful in computer science. It frees you from the chains of implicit nonsense and academic quackery.

In programming, everything is both more rigorous (programming forces rigor, via compilation, unlike math) and more creative (what could be more creative than game dev?). It's such a breath of fresh air to be so free to express oneself in such an intellectually honest and conceptually reliable way. That being said though, I do think that one day the culture of mathematics will improve greatly. The cultural improvement will most likely come from the gradual peripheral influence of external fields like programming and game dev upon mathematics though, I'd estimate. Or, perhaps mathematicians will one day finally stop deliberately damaging their own work by "purifying" it too much. That would be nice.

Anyway though, tangents aside, I also wanted to briefly mention why I chose the word "natural" in "natural insight". Well, the original reason was because the term was intended to have the connotation of "in correspondence with nature", i.e. as in correspondence with the natural properties of whatever arbitrary logical system is being studied. By happy accident though, the word "natural" also corresponds with the fact that justified correct intuitions also tend to improve your ability to naturally express yourself, as well as the fact that such thoughts (the insights) may sometimes occur to people naturally. So, really, there's at least three ways you can think of "natural" as being justified as the adjective to use here. The other aspects of my terminology choices for "natural insight" and "instinct error" seem self-explanatory to me though.

---

[3] i.e. nothing but purely the superficial technical details, with no motivation or value proposition or insight provided for anything

[4] For example, they'd probably never even let me write like this in academic math. They'd be too uptight to ever permit this kind of natural free-flowing self expression, and would probably constantly harass and pressure me to make my text more "focused and pure" (i.e. to omit the underlying motivations, insights, and personality) and more in line with existing dogma. This book wouldn't even exist without me instead deciding to chart my own path, regardless of what anyone else might think, and to just freely explore whatever I felt like exploring. Natural and uninhibited self expression is very empowering like that. Live free and think free.

7.2.5 Mathematicians should borrow the concept of namespaces from computer science

Mathematics has a namespace problem. All of the variable names, symbols, and terminology in math are typically just thrown together into one massive haphazard pile, i.e. "the math literature", where the chance of conflicts and ambiguity is maximized. This has all kinds of counterproductive side-effects. It doesn't have to be this way though. Mathematics could instead take a hint from computer science and start separating things out into distinct explicitly specified namespaces so that the chances of conflicts are minimized and the flexibility for future growth is maximized. This would be wise.

What is a namespace though? Well, it is exactly what it sounds like. It is simply a "space" for separating names, such that names in one namespace don't clash with names in another namespace. It means you organize names, symbols, and terminology into multiple broad categories, each of which has its own precisely specified contents and domain of coverage. The beauty of this system is that it not only prevents name conflicts, but also greatly improves your ability to create a large and diverse vocabulary for things.

You see, part of the problem with the existing mathematics literature is that every single time someone creates a new document or research paper, they are often forced to redefine a bunch of the basic terms and symbols, because otherwise everything would be too ambiguous. Relatively few names and symbols in mathematics have anything bordering on universal standardized meanings (e.g. π, e, etc), and even those are often open to context-dependent change. This creates a big awkward mess. It makes expressing yourself in the math literature vastly more tedious and lengthy than it needs to be.

However though, if you instead had lots of separate standardized namespaces to draw from, then you could avoid this problem. People could publish carefully designed standards for names and symbols that other people could subsequently say that they are using, for the purposes of whatever document or research paper they are currently writing, and this would cut out a ton of the need to constantly respecify and redefine the same things over and over again in mathematical writing. You shouldn't underestimate the power of names and of building up a really big and naturally expressive vocabulary. The current approach (i.e. having just one big global namespace) in the math literature makes this process difficult however, unfortunately.

In addition to creating various different standardized namespaces that could be easily referred to by mathematical authors though, it would also be wise to combine this with more support for multi-letter variable names as well. We've talked about this idea before some, earlier in the book (e.g. see page 165), but I think it is nonetheless important enough that it is worth repeating here. Single-letter variable names are not very descriptively expressive, concise and easy to write as they may admittedly be.

You could build up a more diverse and expressive vocabulary much more easily if you'd use more multi-letter variable names more often. I really think this is something that mathematicians should experiment with more. Even just adding a small number of additional letters to variable names can help add lots more conceptual clarity, descriptive readability, and vocabulary diversity. It is very much worth doing.

7.2.6 An axis of rotation is *not* the conceptually correct basis for a rotation

It is commonly said that rotations are things that occur *around* axes of rotation. However, this is actually conceptually incorrect. Rotation around an axis only makes sense in a 3-dimensional space. It does not work at all in spaces of any other number of dimensions. Indeed, there is actually a far better and far more general way of thinking of rotations, and that is to think of all rotations as existing inside 2-dimensional planar subspaces of any other space of at least 2 dimensions. When you think of rotations in this way, the concept instantly generalizes to any number of dimensions very seamlessly.

Basically, all you need to do in order to generalize rotation to any number of dimensions is simply to project all of the points of the object you wish to rotate into some 2-dimensional subspace of whatever n-dimensional space that you are rotating in, then rotate those projected points within that 2-dimensional subspace, and then afterwards push those points back out to the original number of dimensions by adding back in the difference (before the rotation) between the point projections onto the 2-dimensional rotation space and the original point positions from the original space[5].

When you do this, the remaining $n-2$ dimensions that are not in the 2-dimensional subspace being rotated will be invariant under rotation. Thus, for example, rotation in 2-dimensional space yields a 0-dimensional invariant subspace (i.e. a single point being rotated around), rotation in 3-dimensional space yields a 1-dimensional invariant subspace (i.e. a 1-dimensional axis being rotated around), rotation in 4-dimensional space yields a 2-dimensional invariant subspace (i.e. a 2-dimensional plane being rotated around), and so forth.

Yes, you read correctly, rotation in 4-dimensions actually occurs around a 2-dimensional *plane* instead of around an 1-dimensional axis. Sounds crazy, but it's true. This operation is virtually impossible to visualize, because we humans are 3-dimensional creatures, but mathematically it is nonetheless the correct way of extending the concept of rotation from 3 to 4 dimensions. This pattern continues similarly, with every additional number of dimensions creating yet another invariant dimension under rotation, such that the invariant subspace of the rotation always has $n-2$ dimensions.

Anyway though, I'm sure their are plenty of geometry and linear algebra mathematicians who know this already. I just wanted to point it out because I'm getting tired of people falsely stating that the proper basis of a rotation is an axis. That's just not true. Rotations occur *within* planar subspaces, not *around* axes. The case where rotation occurs around an axis is nothing more than a coincidence. That is only true in 3-dimensional space, which is certainly not the only space under which rotation is perfectly conceptually valid.

7.2.7 Mathematicians need to try harder to create descriptive names

Mathematics is a wonderful and extremely empowering field of study, especially when combined with programming. However, one big problem with the literature of mathematics as it currently stands is how ridiculously opaque and hard to understand so much of its terminology is. Admittedly, I imagine that there are many esoteric cases where it is indeed extraordinarily difficult to come up with any good name at all for some things. However, it is also simultane-

---

[5] In other words, just add back in the "rejection" of the 2-dimensional projection from before the rotation was applied.

ously true that a huge volume of mathematical terms in existence are simply needlessly opaque and cryptic. There's huge room for improvement here.

I've talked about the big benefits of having more clear, descriptive, and conceptually correct names for things before in this book (e.g. see the discussion starting on page 151, where I mentioned some ideas for some clearer names for common sets), but the point is definitely still important enough to merit repeating. Clarity makes an absolutely massive difference in how easy it is to think about things, and hence also in the rate of progress of new research as well. Its value can't be easily overstated.

For example, instead of saying "commutative" you can just say "order independent", and instead of saying "associative" you can just say "grouping independent". This makes a massive difference in terms of descriptive clarity. The meanings of the former terms are opaque and cryptic and can only be understood via rote memorization, whereas the meanings of the later terms are descriptively obvious, instantly comprehensible, and very natural to think about. Never underestimate the importance of the human factor.

Don't pick needlessly opaque terminology for things. Well-chosen names feel like a breath of fresh air, like a pleasant breeze on the beach, whereas cryptic names feel much more like being trapped in a polluted haze of smog, like stumbling around in the dark trying to find your way around, and randomly banging your shin bone into hard objects along the way. Don't use the technical qualities of your field of study as an easy excuse to do a woefully bad job of naming things. We can do a lot better than this.

7.2.8 A more rational standard format for dates and times

The standard format for dates and times used in the United States (i.e. "month/day/year" etc) is pretty bad. It is highly arbitrary and irrational. It would be better if we would switch to the international standard format (i.e. "year-month-day" etc) instead. In addition to that, for the time of day, it would actually be better to write the "AM or PM" part of the current time *before* the hour and the minute instead of afterwards. This would cause the date and time format to actually sort correctly, so that the larger time units always come earlier in the timestamp and the smaller time units always come later, just as it should be. It'd also be good to make each part of the timestamp always have a fixed width and to include the weekday name too, for maximum convenience and clarity. More specifically, here's the exact format I like the most:

```
year-month-day AM/PM hour:minute (weekday abbreviation)
```

Thus, for example, "`2018-07-01 PM 09:00 (Sun)`" would fit this format.

7.2.9 Logical punctuation is much better than traditional American punctuation

In the United States, we have a traditional grammatical "rule" that says that commas and periods and such should always be placed inside nearby quotation marks instead of outside. This is utter blithering nonsense and is an incredibly stupid and backwards choice to make. Nothing which isn't logically a part of a quotation should ever be put inside one, under any circumstances. In England they are better about this, and put their punctuation in the more logical and correct position more often than we do here in the United States.

Leaving the punctuation outside the quotes was always the more correct choice (and very obviously so). International English does it that way usually. The American tradition of doing otherwise is essentially nothing more than an error that somehow got entrenched as a rule. It most definitely does not "look better" to put punctuation marks where they don't belong[6]. That is nothing more than a myth and an after-the-fact rationalization for a mindless and backwards social tradition. It isn't the most logical choice.

As with all things in life, we should make our decisions based on what is most rational and effective, not based on merely what other people constantly try to brow-beat us into conforming to. This specific case is just one among many examples of this. We should always let value creation, not merely adherence to traditions, be our guide in life. We should not be slaves to the past, nor should we let the idle capricious whims of others stifle our ability to naturally express ourselves. Logic is the final authority on all truth. Traditions have no weight in comparison.

7.2.10 Natural selection forces sometimes *increase* complexity in collaborative documents

Collaborative editing sure has a lot of appeal in a lot of ways. It can be extremely effective for getting a lot more work done on something in a small amount of time, by crowd-sourcing the labor involved to many different people across a diverse range of backgrounds. This has made a lot of amazing things possible for people. Wikipedia and open source software projects are great examples of this. Collaborative editing also often makes it easier to detect subtle errors and can sometimes produce higher quality work than other approaches can, under the right circumstances.

However, there is a dark side to collaborative editing, and I'm not really talking about the largely fallacious "because it contains more errors" accusation that has got so much attention in the press but which has never actually been born out by the real evidence. Indeed, it appears that collaborative texts, such as Wikipedia, may actually be *more* accurate on average than traditionally edited documents are.

Instead though, what I want to talk about here is the tendency of collaborative documents to sometimes become *more* complex and incomprehensible over time, instead of less, once a certain threshold of complexity has been crossed, such that a snowball effect is triggered. The mathematics section of Wikipedia is currently a great example of this. Basically, what happens is that once a section of a collaborative document increases past a certain critical threshold of complexity, people's fear of accidentally damaging things they don't understand overpowers their willingness to make needed changes.

Thus, natural selection effects actually start to work backwards (i.e. making things worse instead of better) in that part of the document, causing it to gradually become more and more complex over time. It works similarly to how survival mechanisms work in biology. The most needlessly complex parts of the documents are the parts that people will be the most afraid to touch, and thus become the most likely to survive.

Poorly written overcomplicated crap therefore begins to accumulate in the document. The less technical the document is though, the more likely people are to see through this effect and to overcome it though, which is why things like the mathematics section on Wikipedia suffer the worst from these kinds of backwards natural selection effects. The effect requires a certain minimum threshold of complexity to be crossed before it begins to dominate.

My real point here though is to use this phenomenon to illustrate the importance of being more fearless in the face of superficial complexity in technical fields such as mathematics. We

---

[6]This is made even more painful to me by the fact that studying computer programming and mathematics has made me ultra-sensitive to conceptually incorrect logical structures.

need to get better at not falling prey to the "proof by intimidation" effect that is so unfortunately highly prevalent in technical fields like ours. We need to become braver about not trusting things that sound obscure or complicated.

We need to stop acquiescing to so many things just to "not sound stupid", and to instead bravely admit to not understanding things and then to treat that lack of understanding as a respectable and rational basis for not trusting the related material until it has been explained with a significantly higher degree of clarity. Basically, we need to embrace clarity more, while simultaneously fighting much harder to push obscurantists and other intellectual bullies out of positions of power.

7.2.11 Mathematics and computer science will gradually become inseparable

Currently, there are still a lot of mathematicians who think they can get away without knowing any computer programming. However, I think this will become less and less true over time. The human mind is just too error prone and slow at computation. Hand-written mathematics is just too limiting and does not scale well at all. It's fine for the basics and for really low-level conceptual stuff (e.g. the contents of this book), but doesn't really work very well past that point.

Once things start getting more advanced (e.g. very long proofs and computationally expensive operations etc), you will need the support of a compiler if you truly want to have any hope of ever detecting all the subtle little errors that could potentially crop up. Human beings are just way too prone to making sloppy implicit assumptions, assumptions that very often don't actually work in reality. Computers force us to make everything truly 100% rigorous and precisely specified, something that hand-written math just can't ever do. Even then, there are still often errors.

It is for this reason that I predict that mathematics and computer science will gradually become kind of a two-headed monster. They will become inseparable from each other, and anyone who studies mathematics will pretty much always be expected to also know how to program. There is also the additional confounding factor that as more and more hand-written proofs are discovered less and less of such proofs will exist, thereby forcing people into more computationally expensive territory over time, territory that only a computer can handle.

Computer programming is also an extremely eye opening experience. Indeed, interacting with a compiler is the only truly effective way I've ever found to make someone really fully grasp the inherently error-prone nature of the human mind. Learning programming will change the way you think. It is profoundly empowering. I think it's something that all mathematicians should do, if they know what's good for them. You see, computer science, not mathematics, is the most rigorous of all existing human disciplines. As such, learning programming will greatly improve your ability to think about concepts precisely, even when you aren't using it.

7.2.12 The tendency of some software to drift around randomly, without ever actually solving the most important underlying issues

In my work as a programmer, and in my general use of software as a consumer, I've noticed a strange pattern of behavior that sometimes afflicts software. It doesn't apply to all software, but for the pieces of software to which it does apply it is generally a consistent sign of stagnation.

Specifically, I've noticed that software sometimes gets stuck in a pattern of behavior where, past a certain point, it begins only changing in subtle and superficial ways, where no additional progress beyond that point is ever made in improving the actual important features or in reducing the overall bugginess of the software. I think this state of being merits a name. As such, let's define a corresponding new term:

Definition 189. *When software begins to stagnate, such that the only changes that seem to ever be made to it past that point either (1) ignore any real appreciable improvements that could be made or else (2) on average create little or no net gain against the overall amount of bugs and quality flaws in the software (e.g. if bugs are added just as fast as they are removed), then we may say that the piece of software has entered a **driftwood phase**. Such phases can be temporary or permanent.*

The word "driftwood" here was chosen as an analogy. It is as if the software has become aimless and now just makes small changes randomly, with little to no real direction. The software now behaves like a mindless piece of driftwood floating in the ocean. Additional updates past this point often accomplish nothing significant, except perhaps creating needless hassle and re-learning requirements for users, with no actual real net gain ever being made.

There are multiple possible reasons why a piece of software can enter a driftwood phase. A company could have a change of management, and the new managers may not understand the spirit of the software and thus may end up misdirecting it. The software developers could have become bored or disenchanted with the software, and may have lost interest in it, thereby damaging their attitudes and work ethics. The user feedback mechanism of a company may have become fundamentally broken somehow, such as by simple neglect, or such as by the installment of a "middle man" who now causes direct user feedback to no longer reach the programmers in an accurate form (or at all). Another possibility is that the company may have acquired a monopoly on the software, thus demotivating any effort for significant change.

It is also possible that a driftwood phase is caused by the software developers reaching the upper limits of their understanding, skills, and ability to make tasteful decisions. It could be that they have become stuck in a specific way of working and thinking, one that no longer allows their perspective to expand. They could be stuck in a paradigm essentially, and unable to get out of it. It may also be that the most important existing problems in the software seem like they would be unpleasant or tedious to deal with, thus perhaps subconsciously motivating the developers to procrastinate and to look for excuses to make other (often unimportant) changes instead.

Regardless of what the explanation is for a driftwood phase in software though, users tend to suffer for it. Indeed, it may even be relatively obvious to outside people with creative mindsets what could be done to improve things, and these people may even submit suggestions to the developers, but even then the developers themselves may be so stuck in a specific frame of mind that they will still continue to not make any real progress on anything past the point of stagnation.

Driftwood phases are often the result of a systematic lack of imagination, lack of creative autonomy, and/or lack of care on the part of the producers of the software. Such phases may open up great opportunities for competitors to gain ground though. Indeed, identifying driftwood phases may be a useful tool for spotting hidden signs of weakness in other software companies. Awareness of the dangers of this effect may also help existing software companies to make changes before it's too late.

7.2.13 Set-based file systems are much more conceptually flexible than hierarchical file systems

Existing computer systems tend to organize files into a hierarchy of folders. Each folder can only exist inside one other folder (the parent folder). This works alright, but it also has some severe limitations. Probably the biggest problem with hierarchical file systems though is that most real-world data and concepts actually *aren't* hierarchically structured. Most real-world data and concepts are actually best organized in terms of arbitrary sets, in such a way that each item is allowed to exist inside as many different sets as desired.

People have become so used to thinking in terms of hierarchies of folders that they've lost sight of the fact that things could actually be much different. One of the things that most sucks about using folder hierarchies is that it forces you to make arbitrarily choices in how you structure things that cause it to become impossible to ever get the best of all worlds.

For example, if you have a folder for your music, you can choose for the next level of folders to be folders for specific musicians (or bands), or you can choose for the next level of folders to be folders for specific genres of music, but you can't do both. Picking one way of thinking about things forces you to make the other way artificially more difficult. You could subdivide each musician folder with folders for genre, or you could subdivide each genre folder with folders for musicians, but neither way will be perfect. You're screwed either way.

Indeed, it is actually *logically impossible* to ever create a perfect categorization scheme using a hierarchical structure (in the most general case). This is true not only of file folders, but also of things like inheritance hierarchies in object oriented programming for example. Hierarchical categorization schemes are just fundamentally doomed to failure. As such, if you've ever felt frustrated by the fact that your folder system never seems to be quite right no matter what you do to it, and it always seems hard to navigate, it's not your fault. Hierarchies just suck. Only set-based systems can ever be perfect.

What exactly is a set-based file system? Well, simply put, it is a system that lets you attach as many arbitrary categorization labels to files as you want. These labels are often referred to as "tags" in the world of file system design, but this is really just another way of saying "sets". Anyway though, once we have labeled our files in this way, we can then subsequently specify arbitrary set expressions that will automatically go out and grab the corresponding set of files.

For example, we could write something like $\text{Music} \cap 2015$ to grab all files related to *both* music and the year 2015. And, instead of creating links to rigid folder hierarchies as shortcuts, we could instead just create predefined set expressions and save shortcuts to *those*, thereby allowing us to access any arbitrary set combination instantaneously, without ever having to type back in the corresponding set expression again.

Thus, as you can see, set-based file systems are better at organizing files in pretty much every possible respect. The only real potential downside is just that set-based systems may have more performance overhead and thus may need to be programmed more carefully in order to remain as computationally efficient and responsive as a hierarchy. I really wish that operating system developers would add in more support for this kind of thing to their software. Set-based file systems can even be added on to existing hierarchical systems as an augmentation, if necessary, so that backwards compatibility is not broken.

7.2.14 A simplified alternative to traditional software versioning

In software, it is common to use version numbers that have the format "major.minor.patch". That's currently the most popular way of doing it. The "major", "minor", and "patch" parts of the version number are incremented upwards every time a corresponding change of that magnitude is made. Each time one of the "major", "minor", or "patch" parts is incremented, all of the "major", "minor", or "patch" parts below it (i.e. all of the decimal-separated numbers to the right of it) are reset to zero.

For example, if the software is at version "1.3.7" and a major change is made in the next update, then the version will be changed to "2.0.0". Similarly, if a minor change is made in the next update instead, then the software would change from version "1.3.7" to "1.4.0". Get the idea? The exact nuances of how this kind of versioning is done actually vary from company to company, but this description above should nonetheless give you a rough idea of the basic mindset that this kind of versioning scheme typically employs.

The idea is that the "major", "minor", and "patch" parts of the version number can be used to get a rough sense of how likely an update is to break things from the previous version, and also to give some idea of how significant the new updates are. This sounds like a good idea on paper, but it is actually far from perfect in practice. It is both extremely arbitrary and likely to mislead users. It also has a huge hidden economic cost that is easily overlooked.

Firstly, the entire idea that the developers will be able to accurately predict which changes will actually break dependencies with the previous versions is based on nothing but wishful thinking. It essentially requires omniscience, which is something that the developers certain don't have. The choice of whether a change is "major", "minor", or a "patch" is also simply inherently arbitrary and non-rigorous. It isn't reliable, and thus will often mislead people.

Additionally, the use of the decimal point symbol (i.e. ".") in version numbers is itself also misleading. Most users (who are probably non-technical people, not software developers) will tend to automatically think about those numbers as representing fractions. To do so would be wrong though. The decimal points in software versions definitely don't represent fractions. For example, version "1.0.2" of something would actually be 10 versions older than version "1.0.12" would be, even though when interpreted as fractions the number "1.0.12" actually looks *smaller* and hence older.

It's true that with sufficient discipline the arbitrariness of this "major.minor.patch" system can be reduced somewhat. Regardless though, no matter how hard you try, this system can never be truly objective. It is simply an inherently subjective and poorly defined way of doing versioning. It's widespread popularity does nothing to change this fact. It lacks logical integrity.

Even putting these technical problems with "major.minor.patch" versioning aside though, the economic implications of employing this method of versioning things are arguably even worse. How could a mere versioning scheme have any economic impact though, you may ask? Quite easily, actually. You see, what technical minded people often forget, is that from the public's perspective the magnitude of a version number is often treated as a proxy for how good and how well developed the product is.

Thus, for example, if a consumer sees one product labeled "Product X Version 1.7.15" and another product labeled "Product Y Version 10", then the consumer will tend to assume that Product Y is a much better product and much more mature and much more popular. This is true even if Product X is actually the more mature product of the two, but just happens to do a much better job of not making breaking changes (and hence of not incrementing the "major" version number very often).

Thus, the more you follow "major.minor.patch" versioning conventions strictly the worse it will effect your profitability and the more it will damage the public's perception of the quality of your software. In fact, this one simple difference in versioning habits could easily make a huge difference in outcomes for products in a highly competitive market, simply because of nothing more than the fact that higher version numbers are often perceived as probably being better products by the public. Therefore, despite how universally popular "major.minor.patch" versioning is, it is actually both technically and economically ill-advised. It is neither very rigorously helpful nor maximally profitable.

Luckily though, there is a much simpler and more objective way of doing versioning, a far more foolproof way. The idea is simple. Instead of trying to capture subjective pieces of information, information that is impossible to ever consistently get correct, we should instead capture only objective information in the version identifier. There are several different sources we could use for such info. In this respect, I personally suggest the following version format (to be explained shortly) as probably being much closer to optimal:

```
r<release number>s<support patch number> (c<change number>, <year>-<month>-<day>)
```

The "`s<support patch number>`" part is optional and will usually be omitted entirely (depending on the nature of the specific project though). The parts between < > delimiters are always replaced by specific numbers. This looks more complicated and verbose than it actually is in practice. It's actually quite easy to use, and not at all open to subjective interpretation.

The "release number" is simply an incrementing count of the total number of publicly released versions of the software (i.e. not the internal version number, but the public-facing version number). Every time you make a new public release of the software you simply increment this number. You never attempt to determine if it is a "major", "minor", or "patch" change, since such a determination is nearly impossible to ever make rigorously within the constraints of human limitations.

The "support patch number" is only included when the developers have seen fit to continue adding patches to older versions of the software, for the sake of supporting customers who are still on those versions for whatever reason. This ideally shouldn't happen much, and thus the support patch number shouldn't need to be included very often. It's there though, if you need it. It basically represents a bug support branch that has broken off from the main line of development of the rest of the project.

The "change number" is simply the commit number of the developer's internal version control system corresponding to this version of the software. In a sense, it is the *real* version number of the software, since it is the most finely grained and accurate of all the numbers. However, the change number often becomes far too big for members of the public to find pleasing to refer to, and so it is thus placed in a more peripheral position in the version identifier text (i.e. in this case in the parentheses).

Finally, we also add in the date corresponding to the commit associated with the change number. The reason we do this is because even with all the other version numbers, it is often still unclear to the user how old a particular piece of software actually is in terms of real time. The timestamp for the commit thus fills this role, thereby helping the user to get a better sense for the real age of any given version of the software in a clear and easy to understand way.

That's the official full-length format. However, you are also free to write just "`r<release number>s<support patch number>`", as shorthand, when you don't want to bother with the extra parenthetical info. Furthermore, since the support patch number usually won't actually be used much, you can usually get away with just writing the "`r<release number>`" part, which is extremely concise and easy to use. You aren't expected to impose pointless tedium upon yourself. Simply use whichever of the longhand and shorthand versions seems best to you in whatever the given context is.

Notice also that decimal points are not used anywhere in this versioning scheme. That is very much intentional. By avoiding the use of decimal points, we prevent users from becoming confused as to whether or not the numbers represent fractions or else merely incremented numbers. Notice also that this scheme will tend to cause the release version number (i.e. the main public-facing version number) to become quite large, which improves public perception of quality, but not *too* large (unlike the change number, which does tend to become too large to use as the public-facing version number).

This rapidly growing version number is probably economically advantageous. It makes the public much more aware of the fact that numerous different versions of the software have been produced, and hence that great care has been taken in continuously improving the software quality and features over time. This creates a much better and much more attractive impression on the public. It is probably well worth doing. This alternative software versioning system merits a name I think. Here is the corresponding definition:

Definition 190. *Traditional software versioning of the form "major.minor.patch" tends to be both highly subjective and economically disadvantageous. It is helpful to have an alternative versioning system which is more foolproof, more rigorous, and more profitable. One such possible system has been explained above. Let's henceforth refer to this specific system as* **Simplified Rigorous Versioning***, or* **SRV** *for short. Having an explicit name for it makes it easier to talk about and to compare and contrast with other versioning approaches (e.g. traditional versioning, semantic versioning, etc).*

For example, using Simplified Rigorous Versioning, if we wrote "r7 (c341, 2018-04-14)" as our version identifier, then this would mean that it is the 7th publicly released version of the software, corresponding to version control commit number 341 internally, which was committed to version control on 2018-04-14. Notice how nothing here is subjective or arbitrary, unlike in the traditional "major.minor.patch" versioning system and similar variants.

It is also extremely easy to understand *exactly* which version this is (using the change number, if necessary) and to find precisely when it was originally created and committed to version control. Additionally, it is far easier to calculate distances between versions with this system and get a sensible result. For example, version r17 would clearly be 10 release versions newer than version r7, and there's nothing subjective about that. This system is uniformly conceptually consistent and reliable.

In contrast though, using traditional versioning, it would be completely unclear what the real distance between (for example) "1.3.7" and "2.0.0" actually would be in terms of how many changes happened between each respective version. For all we know, "2.0.0" might be the very next version after "1.3.7", but it could also just as easily be the 500th version after "1.3.7". Under the "major.minor.patch" versioning system, it would be impossible to know without scanning the entire release history of the software, which would surely be an extremely tedious endeavor.

As such, the sense of how much newer any given version is compared to any other version under the "major.minor.patch" system is thus mostly conceptually incoherent and meaningless. This arguably destroys much of the point of even having version numbers to begin with. This is yet another reason to instead prefer to use the Simplified Rigorous Versioning system that I have designed here, or at least something similar.

7.3 Society

7.3.1 A possible solution for out-of-control medical prices, one which will cost almost nothing and carry almost no risks

Prices in the healthcare industry are out of control. This is especially true in the United States, where I live. The costs of medical and healthcare services have ballooned to absurd proportions. Even worse still, few of these increased costs have been accompanied by proportional increases in healthcare quality. Claims by the industry that these high costs pay for "research" (and other things) are probably mostly just lies designed to create plausible deniability. The cost of healthcare has become vastly higher than the actual value of healthcare.

People can easily be forced into bankruptcy even for some of the most common and easily treated medical conditions. Much of the healthcare industry has seemingly become less about helping people and more about holding people's lives hostage for as large a sum of money as they can possibly extract from desperate sick and injured people. Much of it has become an industry of systematic extortion and exploitation. Corruption and greed, not humanitarianism and compassion, have come to be perhaps the most salient distinguishing characteristic of the healthcare industry as it currently stands (at least in the United States).

It doesn't have to be that way though. To understand some of the likely reasons why things have become the way they are though, we need to take a moment to understand *why* healthcare businesses are able to so easily set prices arbitrarily high. There are probably many contributing factors really (e.g. bribery of politicians through campaign contributions, local healthcare monopolies, etc), but one contributing factor in particular especially stands out to me as likely one of the biggest contributors: the complete lack of any real form of free-market price transparency.

Think about how the healthcare industry currently works. You make an appointment with a doctor or visit an urgent care center or hospital, and besides perhaps just a paltry standardized up-front copay fee, the actual costs of what your services will end up being are not specified until you are already forced to owe and to pay those costs. Even if you try calling the medical staff in advance to get up-front estimates for prices, the staff will most often just *pretend* to not have even the slightest idea of even a vague range of what the price could end up being.

This is despite the fact that these healthcare institutions literally spend all day sending out bills that clearly specify their exact prices every day. They must surely have some idea of what their own prices generally are, and yet they still nonetheless constantly pretend to be completely incapable of giving any information whatsoever about that. Do not be deceived. The real purpose of this behavior is clearly systematic economic exploitation, contrary to what the healthcare industry may try to convince the public to otherwise believe.

Free-market forces only operate in environments where the consumer can actually know the price in advance, so that they can therefore easily pick lower-cost providers whenever they need to. The modern healthcare industry though is designed from top to bottom to make price awareness as difficult as possible for consumers. This is probably done intentionally, in order to make free-market forces unable to function well within the healthcare industry, thereby allowing them to consequently charge whatever arbitrary amount they want, thus causing prices to balloon more and more out of control.

Imagine if your local supermarket or general store worked like this. Imagine if the next time you went grocery shopping, and arrived at the cash register, a giant metal cage suddenly fell down from the ceiling, locking you inside it. The cashier then told you that you now owe 1, 000 for the bag of carrots and the pack of bubble gum that you put onto the conveyor belt, and that by law you are now legally required to pay it, with no option to undo that choice, and that if you don't pay it they'll send collections after you and take away your car or house. This is exactly how the current healthcare system works. Disgusting, isn't it?

Isn't it obvious that free-market forces could never operate properly in such a ridiculous environment? Yet, the healthcare industry's constant propaganda and systematic bribery of a controlling majority of politicians has made it so that they can get away with it regardless. They've essentially brainwashed the public into thinking that this way of operating is perfectly normal and totally acceptable, and how dare you for ever "naively" suggesting otherwise.

Don't you care about their ability to provide "quality healthcare"? Apparently, according to them, "quality healthcare" is only possible if nobody is ever allowed to know the cost in advance. Imagine if a different industry tried to say the same thing. Imagine if groceries or software or furniture had unknown and arbitrarily high price tags. Would you believe them when they said that such pricing tactics are necessary in order to provide adequate service quality and funding? Does this sound like an ethical business practice to you?

Literally at least tens of thousands of people probably die because of this. Why then aren't the healthcare industry people responsible *charged for extortion, ransom, and negligent homicide* then, considering that that's essentially the actual net effect of what they are doing by structuring things the way they have? If they were treated according to their effect on the world, then many of these people would be punished as such.

Many of the CEOs of big medical companies probably actually deserve to spend the rest of their lives in prison, due to how many *thousands of people* they have *indirectly killed*, all just for the sake of their own petty greed, e.g. so they can have 7 vacation homes instead of only 3, or so that they can have 2 private jets instead of only one, etc. The magnitude of the immorality of it all is truly stunning.

Luckily though, there is an easy way of potentially solving a good chunk of this problem in a very inexpensive and low-risk way. It's probably not a perfect solution and won't solve everything, but I am almost certain that it would at least help significantly. It could easily save thousands of lives, by gradually reducing prices enough for more people to afford healthcare and therefore preventing their deaths. It could save a lot of people from a lot of suffering.

There's zero legitimate reason I can see why a sane person would ever say no to implementing the idea I have in mind here (or at least a similar variant). You'll see why soon. Only political corruption or incredible stupidity could stop it, and even then it could probably be extremely embarrassing and damaging for whoever the lawmakers were who stopped it from happening. They'd literally be letting thousands of people potentially die by doing so, in a way that probably could be easily and cheaply prevented. They'd be effectively killing people all just for the sake of taking some bribes from the healthcare industry, in a very obvious and publicly visible way.

What's the idea? Well, it's quite simple really. Just force price statistics (in a very specific and carefully chosen form) to be posted by all healthcare providers, in such a way that it would be generally impossible for patients to miss it. Make it as easy as possible for people to instantly see the overall price difference between any two given healthcare providers they look at. To make an analogy, it would essentially force healthcare providers to attach *price tags* to themselves, just like you'd see in any other kind of shopping experience, thereby making it instantly apparent which providers are more expensive and which are cheaper, thereby allowing free-market forces to begin operating again in the healthcare industry at least somewhat.

The exact details of how to do this in an effective way are critical though. It might be too difficult to get healthcare providers to give price statistics for every specific individual procedure they offer, since unexpected things do happen sometimes. They'd have lots of plausible deniability to fallaciously use to defend against an idea like that. I mean, it would still probably be good to force them to provide that kind of per-procedure info too, ideally, but let's take this one step at a time. I have a much easier idea though.

Simply force all healthcare providers to collect data on the total cost statistics per patient per year, for all patients who have been billed at least once in that year in any context, as a rolling yearly window, updated each quarter, and then force them all by law to publish those statistics on giant billboards at every entrance, and to also recite those statistics over the phone automatically whenever someone calls. In this way, you guarantee, by federal mandate, that all healthcare providers are forced to always make their relative cost differences apparent to the public in a non-overwhelming yet still representative form. Here is the exact format I would use, if it were me:

> By mandate of the Healthcare Price Transparency Act of <year law was passed> all healthcare providers are required to provide the following information.
>
> From <quarter> of <year> to <quarter> of <year>[7], the total costs of healthcare per patient for the entire year at <this healthcare provider>, including both the out-of-pocket and insurance-paid portions of all payments, summed together, have the following statistics:
>
> - 95% of patients paid more than <5th percentile of yearly cost per patient>
> - 75% of patients paid more than <25th percentile of yearly cost per patient>
> - 50% of patients paid more than <50th percentile of yearly cost per patient>
> - 25% of patients paid more than <75th percentile of yearly cost per patient>
> - 5% of patients paid more than <95th percentile of yearly cost per patient>

Each piece of text within < > delimiters is replaced by the corresponding data. The beauty of this approach is that it makes it now extremely easy for patients to instantly get a rough estimate of what their actual potential costs and financial risks of doing business with a specific healthcare provider might be, in advance, in a format that makes it extremely difficult for the healthcare providers to obfuscate the data or to otherwise overwhelm the patient with too many different possible outcomes.

Creating this law as a federal mandate would take away healthcare providers' ability to constantly corruptly pretend that they don't even know their own basic price statistics (a completely absurd claim, by the way, considering that any sane business *always* knows their own price statistics). The only cost of doing this would literally be the cost of data collection, the cost of the billboards, and the cost of updating the phone systems to put the notice somewhere.

This would amount to almost zero cost and would also carry almost zero risk, and yet it would have a very high chance of potentially saving thousands of people's lives, via healthcare cost reduction, via free-market forces. It would also force healthcare providers to openly face the well-deserved shame of the absurd prices they currently charge patients per year, in a very embarrassing way, which may provide even more pressure for them to subsequently reduce their prices (besides just the free-market competition forces).

I also want to point out that I have chosen the wording and statistics in this mandate very carefully, with great care taken to consider human psychology and to account for how people tend to reason about things. The choice to use percentiles instead of merely an average is

---

[7] as a sliding one year window, updated per quarter

critically important. The system wouldn't work if you used an average. Averages can easily be tampered with in order to mislead people. For example, extrema can easily be deliberately inserted into a data set to make an average highly misleading.

Percentiles are a lot more reliable, especially if you give multiple of them. I also made sure to phrase everything in terms of what patients paid *more than* instead of *less than*, so that it would put people more in the mindset of risk analysis, thereby causing them to be more likely to carefully weigh their choices. Changing the wording or the stats in this suggestion could easily ruin it and make it not end up having any real impact.

Every little detail makes a difference. It will only work if it gives people a reliable understanding of what the cost risks are, and that will probably only happen if you provide them with multiple percentile data points to give them a proper sense for the real shape of the risk distribution. The information must also be worded very clearly so that anybody can understand its implications immediately. That's why I designed it the way I did.

Every day that we wait to do something about the out-of-control prices in the healthcare industry, the more people will needlessly suffer and die. This idea has almost no costs and almost no risks. Why wouldn't we do it? Isn't it worth at least trying it, just for any chance at all that it could save thousands of people's lives? Aren't those people's lives worth it? I'm tired of nothing ever being done to change things. Mindless social momentum carries way too much weight in our culture currently. That's why I came up with this idea. There's no sane non-corrupt reason why we wouldn't try something like this. Stop letting all these innocent people die just for the sake of petty greed or lazy adherence to the status quo. It is beyond ridiculous that this disgusting behavior in the healthcare industry has been allowed to continue for as long as it already has. It's time to finally start taking steps to put an end to that.

7.3.2 The pillars of human prosperity and poverty

To defeat your enemy you must know your enemy. And, similarly, to help your friend you must know your friend. The same is true for humanity as a whole, with respect to the primary forces that tend to either make our lives better or make our lives worse. As such, if we ever hope to one day attain an ideal society, where our well-being is generally maximized and our suffering is generally minimized, we must be able to identify what the primary sources of our well-being or suffering usually are, so that we can therefore respond accordingly.

As such, I have decided to make an attempt at such an identification here. I will divide the relevant forces into two separate categories: (1) the pillars of prosperity and (2) the pillars of poverty. I will then briefly explain some of my reasoning in making these choices. Let's get started. Here are the two sets of "pillars" that I identified as possibilities:

The pillars of prosperity

1. Objectivity

2. Creativity

3. Benevolence

The pillars of poverty

1. Exploitation

2. Self-Limiting Beliefs

3. Anti-Intellectualism

4. Crime

5. Violence

6. Scarcity

For the pillars of prosperity, I tried to identify the smallest and broadest possible set of forces operating in human nature that collectively tend to cause us to prosper. In this respect, it occurred to me that it seems like only when all three of objectivity, creativity, and benevolence are highly prevalent in society does society tend to consistently and predictably do well.

A little bit of thought shows that this seems to be true. Think about it: With only objectivity and creativity, but not benevolence, there would be nothing to stop us from harming each other, regardless of how intelligent and creative we would be. Similarly though, with only objectivity and benevolence, we would be compassionate and smart, but our lives would be stale and without deeper meaning and joy. And, lastly, with only creativity and benevolence, we would have good intentions, but would lack the reasoning skills to consistently make those good intentions into reality. Thus, a high prevalence of all three pillars of prosperity in society does indeed seem to be a necessary condition for maximizing prosperity.

On the other hand though, for the pillars of poverty, I tried to identify a minimal set of causes for human suffering which together tend to cover the majority of human suffering, and the above list is what I came up with. Of the six pillars of poverty, five of them are essentially man-made and exist only because of our own imperfect behavior as human beings. These are the first five: (1) exploitation, (2) self-limiting beliefs, (3) anti-intellectualism, (4) crime, and (5) violence.

The sixth pillar is a bit different though. The pillar of scarcity is sometimes caused by natural environmental factors that are beyond our control. However, nonetheless, the pillar of scarcity can still generally be strongly influenced by our choices as a society, such as (for example) by doing a better job of conserving resources and protecting the environment, etc.

Identifying these forces for good and ill in our society can help clarify what goals to strive for if we want to make life better for our fellow human beings. For example, sometimes it might be difficult to address one of the pillars, in some context, but you could still nonetheless try to address some of the other pillars instead, so that you could at least continue making at least some kind of progress in making things better for people. This is strategically useful in that respect. Basically, in summary, as much as possible, we want to maximize the strength of the pillars of prosperity while simultaneously minimizing the strength of the pillars of poverty. Doing so would clearly tend to create much more optimal social outcomes.

7.3.3 Some obvious but currently underappreciated ways of greatly improving the fundamental structure and stability of democracy

The current system of representative democracy is certainly a lot better than dictatorship and monarchy. However, be that as it may, and as treasured as the current system is by many people, the current system is nonetheless actually very deeply flawed in many ways. It isn't the concept of democracy itself that is the problem though, but rather it is actually the implementation details. Many of the existing design choices in the structure of current democracies (e.g. especially democracies designed to imitate the United States) are not just badly designed, but are just about the worst possible ways of designing democratic systems to function in several major respects.

The first major flaw is plurality voting. Plurality voting is just about the worst possible way of structuring vote counting that you could ever possibly pick. It is intuitively obvious, making it an easy thing for naive people to instinctively think of and to reach for, but nonetheless its properties and ability to accurately represent the will of the people are actually terrible. It's so bad that even literal random selection of candidates (instead of holding elections) often can vastly out-perform it at statistically representing the will of the people. Plurality voting is so bad that it sometimes (under the right conditions) can gravitate towards actually selecting the *opposite* of the best candidate.

What's plurality voting exactly though? Well, plurality voting works by only allowing each voter to select a single candidate they'd approve of for their vote for each elected position. This sounds perfectly fine at first glance of course, based on "common sense", but in practice it has disastrous consequences. Plurality voting is a virtually guaranteed recipe for a very nasty political effect called the *spoiler effect* to happen very easily.

The spoiler effect causes additional candidates to easily potentially throw an election in the opposite direction of what the majority of the population actually wants. And, the more candidates participate, i.e the more democratic diversity and passionate public support there is for a position, the more likely the spoiler effect is to occur and hence the more likely it is that the election will be thrown in a different direction than what the public actually wants.

Over time, this also probably has the net effect of causing all but one or two political parties to become extinct, since the spoiler effect massively disincentivizes voting for anything other than the two most popular parties, even if those parties don't correspond with what the majority of the voters actually even want. Plurality voting practically guarantees that a democracy will probably deteriorate until either (1) nothing but two irrationally vindictive and petty parties remain, each mostly a highly polarized mirror image of the other, in a state of constant interference with each other, or else (2) a single party dictatorship.

Unfortunately, the two party state under plurality voting could easily be unstable. All it takes is a strong push in either direction by historical accident and a plurality based democracy could easily transform into a single party dictatorship, one with a fake election system that it now uses to legitimize itself for propaganda purposes. Nations that use plurality voting remain democracies only by sheer dumb luck.

This is probably a big part of why so many attempts to forcibly create plurality based democracies in other countries so often fail. Plurality voting naturally behaves very badly. You need two approximately equally sized and equally powerful polarized social subcultures in order for a plurality based democracy to not tend to degenerate into a dictatorship over time. Plurality based democracies are like gunpowder kegs, rigged to explode at any moment. All it needs is the right spark.

Let's do a quick example to make it more clear why plurality based democracies are so bad. Suppose, for example, that there are five candidates running for election one year in a plurality based democracy. Four of the candidates are very similar to each other and support approximately the same overall policies. Each has approximately 19.75% of the popular support. However, the fifth candidate is a violent lunatic, but has a vocal but minority base consisting of just 21% of the population.

In a plurality based democracy, *the violent lunatic would win*, even though he or she possesses only 21% of the popular will. This is despite the fact that nearly 80% of the population wants a candidate similar to the other four candidates. In contrast though, notice that if we instead just selected from the candidates completely randomly (with no election at all) with equally distributed probability, then there would be an 80% chance that the majority's preference would be successfully represented. Thus, exactly as I told you, plurality voting is *so bad* that often even literal random selection outperforms it greatly.

Luckily though, these problems are easily fixed. There are far better voting systems that you could use, ones which behave much better and are vastly more politically stable. The two alternatives that I personally would recommend choosing between are *approval voting* and *score voting*. They are both very simple and easy to understand. Either one would be a vast improvement over plurality voting.

Approval voting just means that you allow people to vote for as many people as they would approve of, instead of only restricting them to voting for just one candidate. Plurality based systems are easily modified to support this. Approval voting instantly eliminates the spoiler effect, and hence also the corresponding toxic effects that plurality voting ultimately has on society as a whole. The other good option is score voting, which has a similar effect but is just more fancy and precise.

Score voting works by allowing voters to give scores to all of the candidates (e.g. numbers between 0 and 10, or whatever) as a way of rating the perceived quality of the candidates. It is basically a more fine-grained version of approval voting. It means that your voting system for political candidates would now work similarly to how product review ratings (e.g. for online shopping, books, movies, video games, restaurants, etc) work. Voters would score all of the candidates and then the candidate with the highest overall average score would win the election.

Approval voting and score voting also help block the formation of one or two party monopolies (which typically exist mostly just because of the spoiler effect). Plurality based democracies seem to often degenerate until there are only one or two viable political parties remaining, but approval voting and score voting in contrast allow multiple other parties to much more easily come into existence and to proliferate.

This should greatly reduce political polarization and toxicity probably. It would make the political landscape vastly more diverse, nuanced, and interesting, and also vastly more likely to be capable of representing the real will of the people. Done properly, it could eventually wear down the black-and-white one-dimensional style of highly polarized and toxic discourse that typically characterizes plurality based democracies

Never underestimate the power of a seemingly simple change to how a system works. The resulting domino effects can often be vastly greater than what the average naive person would ever expect to happen. Be that as it may though, and as huge of an improvement as switching to approval voting or score voting would be, there would still be plenty of flaws remaining in the current system. Every incremental improvement helps a lot though. As always in life, we have to take things one step at a time. As such, let's take some time to talk about some of the other typical flaws in existing democracies too.

Another widespread systematic problem in representative democracies is the fact that elected representatives tend to highly favor the wealthy. This is partly just due to simple corruption, but it is also due in large part to the fact that often only people who are themselves wealthy and powerful (or else who court donations from the wealthy) can afford to successfully run for office. The more resources a person has the more likely they are to win elections. This favors wealthy and donation-dependent candidates greatly over poor candidates. It practically guarantees that almost all candidates will either be themselves wealthy or else have huge financial conflicts of interest.

This therefore further exacerbates the tendency for elected officials to create corrupt laws that inherently favor the wealthy and systematically exploit the poor. When the vast majority of elected officials *are* themselves wealthy (or else dependent on wealthy donations), why should we be surprised when they almost always vote in favor of policies that overwhelmingly help the wealthy at the expense of everyone else? That's obviously what would happen.

Additionally, the winner-takes-all nature of how elections work causes minorities of all kinds to often receive almost no real representation or power in the government, even when they constitute a large percentage of the electorate. Majority tyranny prevails far too often. This is an inherent problem in how representative democracies work. Minorities often don't have the voting power to elect anyone to office. The winner-takes-all approach to democracy is just fundamentally not fair.

Luckily though, there is yet again actually an easy (but often overlooked) solution to all of this, and its name is *sortition* (also known as *demarchy*). What's sortition? Well, it's quite simple really. It just means that for part (or all) of a representative body, instead of choosing representatives via elections, you choose them by randomly selecting people from the entire population with equal probability. If you select a large enough number of such representatives this way, then it nearly statistically guarantees that all subgroups of a population will usually be represented in approximately equal proportions to how many of them there actually are in society, regardless of how wealthy or able to run for office they are, or any other form of discrimination they face.

Done properly, sortition therefore has the effect of eliminating all biases against minorities in that representative body, thereby instantaneously ensuring proportional representation (instead of winner-takes-all power monopolies), as long as the selection process is done fairly and not tampered with. It works kind of like jury duty, in other words, except that you put the selected people into legislative or executive positions instead of merely using them as judicial observers. It is perhaps one of the only reliable ways of ensuring minorities are always given at least some representation in a government.

In theory, you could make the entire government work like this (i.e use only sortition and not elections), but sortition does have some pros and cons, and the selection process *could* become corrupted potentially, so a mixed approach would therefore seem wiser to me. Elected officials also often have more lawmaking skills than randomly selected people do. Therefore, my personal suggestion would be to create a new legislative branch which is composed entirely of a very large number of people selected via sortition, and to add that on to the existing elected legislatures as a check against their power. For example, this new body could have the power to veto decisions made by the elected representatives, as a way of blocking corrupt power grabs from the wealthy, among other things. If it was me, here's what I would call it:

Definition 191. *If a new representative body is added on to the legislature of a democratically elected government, such that the members of this new representative body are selected via sortition instead of via election, in order to ensure at least some fair statistical representation of all subgroups of a society, then I suggest letting such a legislative body be known as a **congressional jury**. This term was chosen by analogy to the concept of a judicial jury, since judicial juries use a roughly similar process of randomly selecting people from the general public in an unbiased way.*

If done properly, a congressional jury could serve as a very useful check against the power of elected officials. Elected legislative bodies tend to become corrupt surprisingly often. It appears that a large portion of this corruption originates from campaign finance and pre-existing wealth, which tend to cause many elected officials to overwhelmingly pass laws that disproportionately favor the wealthy and corporations, with no regard to the rest of society.

A congressional jury would probably be mostly immune to these effects. Members of a congressional jury have no up-front financial costs associated with being a representative, since they do not hold re-election campaigns of any kind, and hence have vastly fewer reasons to make financially corrupt decisions than elected officials have. They can vote on pure conscience, with no consideration of any re-election constraints and without any need to please donors.

This makes a congressional jury perhaps a perfect counterbalance to exactly the kinds of corruption that most tend to afflict elected officials. One could therefore make the argument that a congressional jury is actually an essential missing component of the checks and balance systems of many democratic governments. One could say that it is kind of a fourth (currently missing) branch of a properly balanced government.

Finally, in addition to these ideas to change to a better voting system (e.g. using approval voting or score voting) and to establish a congressional jury, it would also be wise to universally ban all gerrymandering (e.g. such as replacing human-made districting with a completely objective algorithmic system instead), to institute mandatory voting (so that election turnout problems don't randomly throw elections), and to remove any other systematic obstacles to a fair democratic system (e.g. by disallowing all of the anti-competitive strategies and policies that the existing parties currently use to block new parties from ever coming into existence).

If you ask me, an ideal democracy should actually be a hybrid of four different distinct systems of government: (1) direct democracy (e.g. via occasional referendums), (2) representative democracy (e.g. elected officials), (3) sampled democracy (i.e. sortition or demarchy), and (4) technocracy (i.e. rule by skilled and knowledgeable people). The current system of democracy (especially in the United States, where I live) unfortunately doesn't get this balance even nearly correct. There's lots more work to do. The existing democracies are far from perfect, but I think there are some really obvious ways that they could be vastly improved, very easily. This is important for the sake of the future of humanity. All it takes is one crazy person to be randomly elected to trigger a chain reaction of mass suffering on a global scale. We should prevent that future before it's too late.

7.3.4 Ban the use of "national security" as an excuse for reducing the freedoms and rights of innocent people

Throughout history, fearmongering has long been used as a favorite tool by corrupt politicians to acquire more power. When a democracy changes into a dictatorship, it is often because "national security" was somehow used as an excuse to do so. This has been true since at least the days of Julius Caesar, and probably long before that too. Indeed, it seems like such moves are one of the most prevalent methods for damaging or destroying a democracy and stripping basic human rights from its citizens. This pattern seems to have repeated itself many times throughout history.

People should stop buying into the excuses and plausible deniability that make these kinds of power grabs possible. Innocent people, for which there is no existing evidence of wrongdoing or danger, shouldn't have their human rights and basic freedoms taken away under any circumstances. Invasions of people's rights and privacy should require real evidence, and should otherwise nearly always be illegal.

Attempts to reduce citizens' freedoms or rights due to some kind of supposed "national security" event are very often just power grabs in disguise on the part of the government responsible.

The government will try to convince people that the move is purely intended to protect its citizens, when in reality the move is actually often really just intended to protect the government's own iron-fisted grip on power and dominance.

For this reason, it would be wise to actually add in a universal ban on all such power grabs at the constitutional level of the government. These protections against power grabs should be right in there with all the other most basic and most thoroughly entrenched human rights. Very firm boundaries should be set in place. It should be made extremely difficult for power grabs to be made on the basis of so called "national security" concerns. Such moves are historically very often grossly disproportionate to the actual magnitude of the supposed threat.

Also, in addition to setting hard boundaries that the government is not ever allowed to cross, the government's ability to pass laws that reduce freedom or privacy in the short and medium term aftermath of any "national security" event should be temporarily frozen. Such proposals should only be allowed to be considered during times of relative peace, when such an event hasn't occurred in a little while, so that any temporary public hysteria and fear thereby becomes more difficult to exploit. For example, a rule could be created that in the aftermath of a "national security" event no law whose effect is in any way to reduce the freedom, privacy, or human rights of ordinary citizens can be passed until at least 3 months have passed since the event. That's just one random idea among many possibilities. Use your imagination.

7.3.5 Expanding the foundational principles of a properly separated and sovereign government

The government of the United States (and of many other governments too) has the principle of the separation of church and state and of freedom of religion as two of its most important guiding ideas. This is certainly a good thing. However, we can actually take it even further than that. We can expand these kinds of principles even more to create even more benefit and justice for society.

Consider the separation of church and state for example. This principle was wisely included in the foundation of the constitution of the United States because the founders knew that when religion and government are allowed to mix it almost always results in widespread human suffering and corruption. It damages both the religion and the government involved. However, the church is certainly not the only institution about which this is true. Much the same also applies to the media and to business.

Indeed, mixing government with media and business often creates greatly increased corruption and tyranny, and yet there are currently not enough protections against this kind of mixing in our government. In other words, the foundation of our government is actually missing two additional critically important principles: (1) the separation of media and state and (2) the separation of business and state. Both of these are probably necessary in order to prevent many associated forms of corruption.

When the government and the media are allowed to mix too much, it often creates essentially state-run media of some kind, either of the one party variety (e.g. like in dictatorships) or of the multiple party variety (e.g. like on cable news in the United States right now, where almost all the news channels are basically just propaganda outlets working for either one political party or the other). To prevent this, it is necessary to put in place rules that ban political parties from owning any form of controlling influence over the media.

Similarly, when government and business are allowed to mix too much, financial corruption and widespread exploitation are the inevitable results. In this kind of scenario, once the corruption spreads deeply enough, businesses basically become able to purchase politicians at will, and the voting behavior of politicians will thereafter no longer reflect the will of the people and will instead only reflect the will of corporations. It turns the government into a corporate puppet. This kind of behavior must be blocked somehow, or else it is the only outcome one should ever expect.

Thus, as you can see, separation of media and state and separation of business and state are indeed additional principles that should probably be added to the government. The already existing principle of the separation of church and state (wonderful as it is) isn't really enough. In addition to these principles of separation though, it would also be wise to expand "the freedom of religion" so that it now takes the much broader form of "the freedom of thought" instead. It is actually *all* intellectual freedom, not merely just the freedom of religion, which should actually be universally protected.

It also occurs to me that the concept of "unbreakable rights" should also be introduced into the foundations as well. This is especially true with respect to legal contracts. Under the current legal system, it is far too easy for companies to strip people of far too many of their basic civil rights via the power of contracts. People are nominally guaranteed a whole bunch of civil rights under the constitution, but then contracts are just allowed to take away most of those rights anyway, thereby effectively rendering many of those rights meaningless in practice.

Abusing people's civil rights like this should not even be possible. That's where the idea of "unbreakable rights" comes in. Basically, the idea is that you could define a set of "unbreakable rights", such that no contract can ever take those rights away, not even voluntarily, thereby limiting the power of contracts to be exploitive. Paper has too much power in our current legal system. The continued willingness of participating parties to be a part of a contract should often carry far more weight than the paper.

In most circumstances, most average people actually have almost no leverage for negotiating in most contracts, or else are simply highly naive and uninformed about their rights, or else don't even read contracts, and are thus constantly exploited and taken advantage of. Oh, and finally, legal contracts that are too long or too cryptically worded for the average person to actually understand should not even be considered enforceable. Someone cannot really agree to something that they cannot even understand.

Also, as if that's not bad enough: Saying that a poor person should just consult a lawyer is generally very out-of-touch advice. Only wealthy people can afford to consult lawyers for contracts regularly. Losing even just $500 could very easily mean that a poor person will now have zero money left in the budget for the rest of the year for entertainment, relaxation, and hobbies. Would you be willing to give up *all your leisure money for an entire year* just to understand one contract a little bit better, on the off chance that it may be risky?

Even for an average person, such an expense would still usually be considered to be prohibitive. Only wealthy people can actually afford to consult lawyers regularly. Everyone else basically can't actually safely afford to make use of the legal system, and therefore very often cannot rightfully be expected to ever understand the contracts they sign. The wealthy basically own the legal system, for all practical purposes. Other people have hardly any realistic ability to use it.

7.3.6 A few simple ideas for making the power dynamics between businesses and consumers slightly more symmetrical

Businesses and consumers are not equal parties in agreements. The businesses almost always have far more power. Businesses can pretty much dictate the terms they want to consumers (and employees) and the consumers (and employees) often have little to no leverage with which to push back against this. Indeed, often the only real negotiation power consumers and employees have against a business is to choose not to do business with them, which is something that many consumers and employees can seldom afford, especially when monopolistic forces come into play.

One especially common way that businesses exert unequal and unjust power over consumers and employees is through the use of so called "arbitration clauses" in contracts. These arbitration clauses have the effect of taking away the consumer's or employee's right to pursue legal action against the business in the normal court system, and to instead force them to use an arbiter to decide the outcome of any case.

Arbiters work outside the normal legal system and depend on repeat business from corporations for the majority of their income, and thus tend to disproportionately side with businesses in cases, regardless of the details or merits of the legal case. In other words, arbitration is essentially a separate legal system that the corporate world has successfully carved out for themselves, one consisting mostly of corporate cronies. It is essentially a way for corporations to circumvent the normal legal system entirely and to instead just have their own separate legal system that is mostly under their own control and thus which usually rules in their favor. Evil, right?

So, since consumers and employees can generally never really expect fair treatment under this system, and are often forced into signing arbitration clauses in contracts anyway, it stands to reason that something should be done about this. As such, my suggestion is this: Arbitration agreements between parties of unequal power should be universally banned. In other words, more precisely stated, I'm saying that arbitration agreements should only ever be valid in business-to-business and person-to-person cases. In contrast though, arbitration agreements in business-to-person cases should be universally illegal, in order to prevent the systematic exploitation that such arrangements almost always cause.

Likewise, besides arbitration, another way that businesses like to systematically exploit the public is in making opt-out processes much more difficult, tedious, and labor-intensive than opt-in processes. For example, a business will often make it very easy for a person to sign up for their service and to agree to their terms, but will make the opposite process (opting out and leaving the service) as difficult as possible.

For instance, it might be trivially easy to sign up for a service online, but the company may require people to send paper mail to their corporate headquarters in order to opt-out and to delete all the data the company has accumulated. They'll also often artificially delay the entire process as much as they can at each step. Similarly, healthcare insurance companies seem to deliberately make the process of filing claims as painful as possible, in order to make it less likely that people will ever actually successfully file those claims for the benefits they are legally owed, thus saving the healthcare insurance boatloads of money in unfiled claims in a very unethical way.

It doesn't take much brains to see the insidious nature of what this kind of behavior is actually designed to do. Don't fall for these businesses' attempts to create an appearance of plausible deniability for why they have to do this. They definitely don't have to make things so hard. The real motivation is entirely greed. There's no real reason why opting-out has to be more difficult than opting-in. They could easily make it just as easy if they wanted to.

Instead though, these companies seem to deliberately make things difficult purely for the sake of their own profits, at the expense of the human rights and well-being of the public. This too should be banned, just like asymmetrical arbitration agreements. In particular, I suggest that a law should be made that requires that opting-in and opting-out of a service (or agreement) with a company must always be approximately equally easy. You can have a warning message or prompt asking if the user truly wants to opt-out and delete their stuff of course, to prevent accidents, but other than that it should be equally easy to opt-out as to opt-in. Also, deliberately making claim processes unnecessarily difficult, tedious, or slow (like insurance companies do) should likewise be illegal.

7.3.7 A theory on the evolutionary benefits of depression and why it probably exists, and therefore also a possible idea for how to naturally prevent, treat, or cure it

Depression is a state of mind that many people suffer from. Admittedly, some people may only feel depressed occasionally, whereas other people may feel depressed quite frequently, but regardless of whatever the specific case may be, it is clear that depression appears to be a fairly common emotion in society. Also, depression is generally quite an unpleasant emotion to experience, so it is therefore understandable that many people would instinctively view it as inherently a negative thing and hence would want to get rid of it and thus would feel inclined to treat it like some kind of disease or abnormality.

However, from an evolutionary perspective, this does not make sense. In fact, thinking about it logically, it seems overwhelmingly likely to me that depression actually serves a very important and frankly very obvious beneficial purpose. The likely evolutionary function of depression actually becomes quite easily apparent if you simply observe the typical causes and effects that are associated with it and then connect the dots together.

Let's take some time to think about this carefully, so that the point is more clear. Firstly, I want you to especially notice that just because an emotion *feels* negative doesn't make its actual *net effect* on human well-being and survivability negative. Fear is a fantastic example of this. Fear is an emotion that most certainly does *not* feel good. Yet, without fear, there would be absolutely nothing to stop people from starting wrestling matches with lions nor from running in front of the path of a speeding car. Fear keeps us alive. It is a necessity.

If negative emotions were all inherently bad, as some people seem to believe, then getting rid of all fear would be a good thing. However, this cannot possibly be true. Getting rid of all fear would be a death sentence for humanity. It would actually be one of the worst possible things you could ever do. It's a classic example of "be careful what you wish for" in action. Fear is simply an inherent part of the human experience. It is like a guardian angel watching over all of humanity. The fact that it just happens to feel unpleasant does nothing to diminish its essential and extremely evolutionarily beneficial nature.

The same is also true of depression. Depression is actually an absolutely essential and healthy emotion. You see, while it is more difficult to see why depression is useful than to see why fear is useful, it is nonetheless still true, as will be explained shortly. Like many other negative emotions, depression is actually a critical part of our defense and survival system. Silencing that defense system in any kind of unnatural way will often (depending on the specific circumstances) have potentially disastrous consequences and unintended side-effects.

A good analogy would be taking a medication that reduces the ability of your skin to detect temperature changes, and then putting your hand onto a burning-hot stove afterwards. The medication you took would reduce the unpleasant sensation you felt, but its overall net effect on your health would actually be negative, since it would result in increased injuries due to burns. Much the same can be said of how depression is currently treated in society (and probably many other conditions as well).

Indeed, just as with the temperature sensing example, suppressing the natural reaction that depression is trying to stimulate will very likely sometimes (depending on circumstances) set off a chain reaction of unintended counterproductive consequences. However, to understand what those negative consequences will be and why they will often be counterproductive, we must first understand what depression's evolutionary purpose actually is.

All emotions, no matter how negative and unpleasant, have beneficial underlying purposes. Every emotion has a role to fill. No emotion exists for no reason. Evolution is not that wasteful and not that stupid. For example, as we have already seen, fear is designed to protect us from direct harm. Likewise, happiness is designed to reward us for engaging in behaviors that our instincts crudely estimate will increase our survivability and future prosperity. What does depression do though?

The answer is simple: Depression is actually mostly an energy conservation and motivation balancing system. Besides just these two primary purposes though, which are relatively light-hearted and benign, depression also sometimes serves a much darker peripheral role of significantly speeding up natural selection processes in a very cruel yet very evolutionarily effective way: by culling what it estimates to be extremely poorly performing humans out of the gene pool by spurring them to commit suicide, thereby statistically reducing lower quality genes before those genes ever have a chance to replicate[8].

Luckily, the much darker side of depression generally only comes into play in extreme cases (e.g. in cases involving repeated, long-term, and/or intense depression). The usual role depression plays, in stark contrast, is actually very much designed to protect the individual life and future prosperity of the person it afflicts and does not at all have a spirit of cruelty underlying it. A more detailed explanation of the exact reason why depression is useful will make this clearer.

Here's how it works: Every action a person takes expends energy, time, and resources. No human being ever has an unlimited amount of any of these things. It is therefore necessary, from a natural selection standpoint, that a human being must carefully conserve these things, otherwise they would jeopardize their survival. How does a person decide where to invest their energy, time, and resources though?

The answer is through motivating forces, such as guiding emotions (e.g. hunger), tantalizing opportunities, and risk-reward mechanisms. Sometimes these motivating forces are too strong though, and thus may sometimes create a disproportionate and wasteful pattern of behavior in the human being who is experiencing them. Consider, for example, what would happen if someone is highly motivated to do something, but still continues to repeatedly fail to accomplish that goal every time they attempt it. They would lose a lot of energy, time, and resources because of that, wouldn't they?

Such waste can mean death from a natural selection standpoint. Too much motivation in the face of repeated failures is lethal. Therefore, in order to ensure high rates of survival, evolution requires an effective means by which to forcibly reduce how much energy, time, and resources a human is willing to exert in the face of repeated failures towards their goals. It therefore

---

[8]Mother nature is a cruel yet beautiful mistress. She does not care whether something causes increased suffering so long as it also causes increased survivability via a net gain in the overall gene quality of a species. If this upsets you, then that's very understandable. I too am repulsed and disgusted by this idea. However, the pursuit of truth is not about discerning what we like or dislike. It is simply about discerning what is real.

responds by causing that human to feel profoundly unmotivated, lethargic, and pessimistic. In other words, it *intentionally* makes them feel *depressed*, as a desperate attempt to make them stop engaging in what appears to be extremely wasteful (hence eventually lethal) behaviors. It would much rather save those resources for later, when conditions may have improved, or else for a different kind of opportunity entirely.

Depression is like a pressure valve for misdirected motivation. It responds to wasteful behavior in the short-term by rapidly deflating a person's sense of energy and reward incentives (as an emergency measure designed to prevent any devastating loses), and then also responds in the long-term by creating a very memorable unpleasant feeling (as an attempt to thereby discourage any additional similarly wasteful behavior in the future). In other words, depression is to energy management and motivation balancing what fear is to danger management and physical protection.

Thus, saying that you want to make depression cease to exist in a person is effectively equivalent to saying that you want to drop a nuclear bomb on the energy management and motivation balancing part of their brain. Be careful what you wish for indeed. This is yet another great example of how "common sense" thinking (e.g. "depression feels bad, therefore I should destroy it") can have extremely counterproductive effects. Depression is not a disease. Only the suicidal variant of it really merits medical treatment, if you ask me.

Oh, and by the way, if you're still wondering why depression also doubles as a suicide trigger, I'll tell you: It's because depression's energy management and motivation balancing capabilities require the ability to *detect and weigh the number, frequency, and magnitudes of failures to accomplish goals*. Depression is therefore the *perfect emotion* to also attach a suicide trigger to, since the whole point of a suicide trigger from a natural selection standpoint is that it should *only be triggered* when that member of the species has seemed (in a crudely instinctive way) to demonstrate evidence of consistently very poor performance, i.e. exactly the kind of thing that *only a facility with the ability to detect and weigh the number, frequency, and magnitudes of failures to accomplish goals could ever know*. Thus, seeing as mother nature is efficient and seldom wastes a chance to kill two birds with one stone, the suicide trigger attached itself to depression, even though depression normally wouldn't otherwise deal with such things. The suicidal side of depression is thus probably purely a coincidental evolutionary trick attached to an otherwise very benevolent and healthy emotion. That's my theory anyway.

Makes a lot more sense now, doesn't it? Depression very clearly fills a crucially important role in our natural energy management and motivation balancing instincts. It is an indispensable component of good mental health, just like fear is. In fact, someone who never experiences depression is very likely *not* mentally healthy. Yet, if you were to ask a psychiatrist about these things you'd probably walk away with the impression that depression is just as much a disease as the flu is. I wonder if that has anything to do with the fact that most of psychiatrists' money probably comes from over-diagnosing anti-depressants... ☺

Knowledge is power though, you know. There's a big difference between trying to solve something when you don't even understand the nature of the problem, versus trying to solve it when you *do* understand the nature of the problem. This brings me to my idea for an alternative prevention and treatment plan for normal (i.e. non-suicidal) cases of depression:

Definition 192. *It is the nature of depression to act as a deterrent against behavior that may waste energy, time, and resources. It does this by causing a person who engages in such behavior to subsequently feel profoundly unmotivated, lethargic, and pessimistic, as an ingenious way of automatically conserving valuable assets in response to repeated (or intense) failures or loses. In other words, failures to achieve goals or to avoid losses are frequently the root cause* *of depression. The more such failures or losses happen, the more intense the corresponding feelings of depression will tend to become in response.*

Therefore, it stands to reason that the opposite process should undo most of this effect. In other words, by ensuring that a person is continuously exposed to many repeated (or intense) successes, depression will likely be proportionally reduced. These repeated (or intense) successes will cause the energy management and motivation balancing mechanisms that form the basis of depression to relax, by convincing those mechanisms that not too much is actually being wasted, since successes do seem to be happening pretty often. For ease of reference, let us refer to such a method of preventing and treating cases of depression as **repetitive success therapy**, *for obvious reasons.*

Even the smallest of successes will probably help a bit, although there will likely also be a degree of proportionality to it, such that the *magnitude* of the successes will also impact how much they reduce any feelings of depression. As such, people with more severe depression may benefit from starting with some moderately small tasks, such as simple household chores like cleaning or gardening (etc), and then gradually branching out to more impactful and creatively fulfilling tasks from there, as they begin to feel better and more capable. The more successes are experienced, and the more it seems like those successes create a real impact, the better the treatment response will probably be.

Finally, one interesting connection I also noticed was that you can also see this same pattern of predictable depressive responses in people when you give them more free time than they know what to do with. Retired people are a good example. Many retired people actually may start to feel depressed from just sitting around doing nothing ("relaxing") too much, odd as it may sound, and may even start feeling nostalgic for their past work because it made them feel more accomplished and meaningful.

This is yet more evidence that the theory here about the nature of depression is correct. Depression really does seem to center around energy management and motivation balancing. Indeed, failing to set yourself enough actionable and attainable goals will *itself* begin to feel like a kind of failure, and will start to trigger corresponding increases in depression, as per usual. Human beings need to feel like they are being productive and are making an impact, just as much as beavers need to gnaw on wood or cats need to scratch things with their claws. Such behavior is critical for maintaining good health. Depression is a snowball effect. You just need to break out of it, one incremental step at a time, through repeated successes and focusing on what you can control. Not setting yourself up for frequent successes in your life is probably an almost guaranteed recipe for depression.

7.3.8 The underappreciated value of self-study and automated education resources

Traditional education, where students study in-person with the guidance of a local teacher, certainly works, but it is also in many ways actually highly inefficient and scales poorly. Self-study and automated education resources, in contrast, have vastly more potential, both in terms of education quality (for most subjects) and also in terms of economic efficiency and scalability.

In fact, I would go so far as to say that the only real competitive advantages of traditional education are (1) creating pressure and accountability to motivate learning even in cases where the student would otherwise not be motivated enough to learn something, (2) socialization, and (3) subject matter and projects that inherently require in-person hands-on instruction. In all other aspects though, self-study and automated education resources would probably be superior if done properly.

You see, the beauty of self-study and automated education resources is that they are (1) infinitely replicatable at little to no cost, (2) able to be easily pulled from the highest quality education sources (e.g. the best teachers in the nation or the world), instead of only from the (usually much lower quality) local sources (e.g. typical average teachers), (3) able to adapt to any student's schedule in any arbitrary way, (4) capable of supporting instantaneous feedback via automated testing systems, thereby vastly speeding up the associated learning cycle, (5) much more easily customizable to each student's specific needs and desired learning outcomes, (6) able to perfectly match each student's own specific learning pace, instead of always being forced to go slower or faster, and (7) much more capable of providing a very diverse set of educational options, such that students can immediately switch between different sources whenever one source is found to not be very effective or to otherwise not satisfy the student's preferences or goals (in stark contrast to being locked into the same crappy local source for a long span of time, as often happens in traditional education unfortunately, e.g. with bad teachers).

What exactly do I mean by self-study and automated education resources though? Well, I mean things like books, videos, websites, and educational software (e.g. automated testing and tutorial programs etc), combined with properly motivated study and participation in such things. The outcomes of such systems, done properly, can easily be much better than traditional education, both in terms of education quality and in terms of economic efficiency and scalability. Some progress has been made towards supporting more self-study and automated education resources in society (e.g. the existence of Khan Academy), but it still seems like the real potential of it is still vastly underappreciated and underutilized.

For example, huge swaths of the education system could be put under self-study and automated testing and such, and the cost reductions of this could be absolutely enormous. You could probably cut out 80% or more of society's existing educational costs simply by properly taking advantage of the value of self-study and automated education resources. This would also enable every person to learn from a diverse set of all of the nation's (or world's) best teachers.

Why make most students study under an average teacher, who often won't be very good, when you can instead find all of the best teachers for each subject and then have them create a bunch of automated education resources as a substitute for the traditional approach. This would cause the entire student population to now learn only from the best of the best, instead of being locked-in to comparatively lower quality local education institutions.

The key to making self-study and automated education resources work effectively though, is to provide sufficient motivation and accountability that students do not simply avoid doing the work and learning things. There are many things that could be done to address this, such as providing an incentivization and punishment system (e.g. real-life impactful consequences for succeeding or failing to meet basic educational goals), gamification, explaining to the student more clearly what the benefits of learning the material will be (thereby providing more inherent motivation), and so on.

The absolute ideal scenario is when the student ends up becoming self-motivated to study. There is nothing more powerful (in educational terms) than self-motivated study. That's the kind of thing that produces truly exceptionally well-performing and creative members of society. The value of this when it happens is enormous. Self-motivated study is often self-perpetuating and tends to cause educational and economic outcomes to be maximized as much as possible. It also instills a spirit of independence and free-minded thinking in students. It therefore strikes me as much wiser to teach students (1) how to be motivated on their own, (2) how to keep themselves accountable, and (3) how to manage their own time effectively, than to merely get them used to expecting that other people will always be around to forcibly prod them forward towards doing productive work.

Traditional education tends to train people to not be self-sufficient. This causes a lot of people to consequently become very bad at managing their own time and creative potential later in life. The school system often subconsciously trains students to only work on what they've been instructed to work on and to only do so when forced to by either rewards or punishments or both. This is not really ideal. It isn't intellectually or creatively healthy. It takes years to undo these habits[9].

It wouldn't have to be like this though if we tried harder as a society to emphasize, nurture, and support self-study and automated education resources properly. We need to prioritize teaching people to be self-motivated and capable of managing their own time and futures better. The traditional approach to education is massively economically wasteful in many different ways. The cost of the self-study and automated education resources approach though, by comparison, is absolutely tiny. Think about the difference between the cost of simply buying (or freely accessing) a bunch of books, videos, websites, and educational software (which is generally quite low) versus the cost of something like going to college (which is unfortunately extremely high, currently).

Indeed, even a poor person can fairly easily afford a wonderfully high quality education via self-study, by simply spending maybe a few hundred dollars on the necessary resources. I personally find this fact very inspiring. It means that with sufficient motivation and persistence, anybody can learn most subjects without going really far into debt, so long as they simply are capable of holding themselves accountable and managing their time well. However, even though self-motivated study very often produces people who are much more knowledgeable than people who come out of the traditional education system, self-study is not viewed with as much respect and credit as it truly deserves.

There should be more systems in place for people who learned by self-study to successfully prove that they are qualified for the positions they have studied for. The current system of requiring most people to go to college and get degrees before they can be considered adequately qualified is very inefficient and suboptimal. Society should invest much more strongly in creating a rock-solid system of self-study qualifications, and then instill those systems with real economic credibility. Doing so is probably the key to overcoming the problems with the ever-rising college tuition costs we are all currently facing. Self-study is far more economical and competitive (in principle) than colleges are.

Essentially, self-study and automated educational resources would allow you to completely cut out the middle man (i.e. colleges etc) for the majority of subjects. It would therefore allow people to get complete educations in whatever subjects they want at only a tiny fraction of what college would cost. Not only that, but the self-motivated study involved in this would likely produce students who actually know their material *better* than most college grads do. Indeed, in my own personal experience, all of the most valuable things I've ever learned in my profession and in my life in general have actually been through self-study. In stark contrast, shockingly little of what the traditional education approach has taught me has held as much value.

There's a lot more that can be done to support self-study and automated education resources

---

[9]Me writing this book is actually an example of this. It took me *years* to finally truly get a handle on holding myself accountable and on managing my own time well when nobody was around to potentially scold me for slacking, and this was true even despite the fact that I was extremely highly motivated. The bad habits that traditional education instills in us can be quite hard to kill, but we can still overcome them eventually though, if we are persistent enough. I know from experience.

better in society. The government could make it so that the associated books, videos, websites, and educational software are not subject to any taxes. They could fund massive content creation projects designed to create as much of this kind of educational material as possible. They could do more research into how to make educational software as effective as possible. They could establish stronger credibility for certification programs for self-study. They could integrate more automated education resources into the existing education infrastructure better.

The list goes on and on. Yet, it still seems like society hasn't really fully grasped the incredible potential economic power of this way of doing education. In fact, I'd go so far as to say that properly investing in self education and automated education resources could easily set off a new period of massive economic growth. Nations that don't invest more in it will probably experience large economic declines relative to other nations. It truly is an underutilized system, one with almost unimaginably huge potential if done properly. It is critical though, in implementing this system, that each piece of subject matter should be covered in multiple different ways, by creating a diverse set of multiple different sources and educational approaches to each piece of subject matter, rather than just creating one monolithic and homogeneous central authority.

7.3.9 Diversity and decentralization of control are essential to good education

Homogeneity is often dangerous. Take biology for example. When a population of organisms lacks genetic diversity, the risk of the entire population being wiped out is greatly amplified. This is bad. Too much similarity and centralization in a system will often make it fragile and incapable of adapting. This is just as true in education as it is in biology. Too much homogeneity in an education system is inevitably toxic and counterproductive.

Standardization can have value, to an extent, but if you take it too far then it actually begins to make things worse. This is a major problem that afflicts many education systems. You often encounter a scenario where a centralized power of some kind (e.g. the government, the principle of a school, etc) has mandated that a very specific curriculum be followed by all teachers. There is a huge danger that this kind of centralization of curriculums will actually make education outcomes worse instead of better.

Teachers are trained to be creative and to adapt to their students needs, and that is indeed the ideal case, but mandates from centralized education authorities often undermine that. Such mandates often inhibit the ability of teachers to properly adapt to their circumstances and to provide an adequately diverse and fulfilling educational experience to their students. Intellectual health requires a diversity of knowledge and a willingness to consider many different perspectives. However, centralized curriculums and educational mandates often severely reduce that diversity though. This in turn tends to create worse outcomes for students.

Therefore, in order to reduce the impact of these kinds of counterproductive and capricious whims from centralized authority figures in education, it is important that teachers be provided with at least a certain minimum amount of independent autonomy over their own classrooms. Central authorities should not be able to constantly micromanage and interfere with teachers' ability to choose how to educate their students. Only broad strokes kinds of controls should be used. As with many things in life, a light touch is usually best. Centralized authorities should stop treating teachers like pawns, and instead empower those teachers to adapt to their circumstances as they see fit, within reason.

7.3.10 Prisons should focus less on punishment and more on rehabilitation and personal growth

Society is too addicted to eye-for-an-eye thinking and vindictive self-righteousness. We see this kind of thinking embedded in so many different aspects of our culture. For example, novels and movies all too often choose to frame everything in terms of "good versus evil" in really contrived and comically exaggerated ways. Real life is far more nuanced than that though. I think a lot of this black-and-white thinking probably originates from people's subconscious tendency to try to rationalize themselves as always being "the good guy" and other people they've had conflicts with as always being "the bad guy", for egotistical reasons, but I suspect a large part of it is also just a cultural quirk of history. In either case though, whether through nature or nurture, I'm sure we could all do a lot better about this as human beings.

The prison system is a great example of this. It's certainly true that bad behavior in society needs to be deterred and penalized somehow, but doing it in a vindictive and needlessly cruel way (as often is unfortunately the case) strikes me as distinctly counterproductive. Correcting the bad behavior in as benevolent a way as possible should be the focus. However, despite that ethically imperative fact, it's almost as if society is far more interested in just rubbing it in and metaphorically poking prisoners in the eyes, for purely vindictive reasons, rather than actually rehabilitating them. This seems backwards to me.

There's more than one way to correct bad behavior. Cruel punishment is certainly not the only way. There's also the nurturing way, the approach that uses ethical, intellectual, creative, social, and personal growth to gradually make people into better versions of themselves, for the benefit both of the recipient and of society as a whole. The nurturing approach seems to me like it would be far more likely to stick and to create lasting positive change. In contrast though, merely applying vindictive and fruitless punishment seems far more likely to just create resentment and to train the prisoners (by example) to likewise themselves continue being vindictive and petty, since that's how society itself is treating them, thereby communicating (subconsciously) that such behavior must be good.

One of the most disgusting examples of how prisoners are mistreated is in how some prison systems don't allow their prisoners to read anything but a highly censored and limited list of books. This is intellectually and ethically stifling. Doing this often vastly reduces the prisoners' capacities for eventual personal growth, which only makes them even more likely to continue engaging in bad behavior even after their stay in prison is over.

The ability for prisoners to grow as people is one of the most important things to maximize in a properly functioning and ethical prison system. Prisoners should have access to almost any book, so long as the book is not about things like weapon making or anything else of that nature. Prisons should be designed as rehabilitation centers. That would be the most helpful way to structure the system. Petty one-dimensional thinking and needlessly cruel punishment, in contrast, doesn't really help anyone.

7.4 Productivity, creativity, and economics

7.4.1 Time pools: a method for ensuring productivity even for people who hate schedules

I've never really liked fixed schedules very much when it comes to creative endeavors. I like to be able to wake up every day and kind of just go with the flow. My sleep schedule also has a tendency to drift around randomly almost constantly, whenever I can get away with it (i.e. when not forced to adhere to someone else's schedule), and consequently my attempts to make myself adhere to fixed time frames in which I am supposed to start and stop working sometimes feel strained and uncomfortable. I like to have the flexibility to let my work time dynamically reshape itself to fit whatever ends up happening on any given day.

However, the thing about letting your work just shift around arbitrarily during the day is that it tends to make you a lot more likely to not get as much done, unless you are very careful about it. When your guiding principle for each day is just "get a whole lot of work done" it often becomes quite easy to slack off. In other words, free-form time management often creates accountability and discipline problems. I tried working like that for a while when I first began writing my book, but I ended up not getting nearly enough work done. I wasted a lot more time than I intended to.

Luckily though, during the course of writing this book, I came up with an interesting alternative way of managing my time each day, one which I found to actually be extremely effective and which allows me to have the best of both worlds: both a flexible and fluidly readjustable daily work schedule, with no fixed starting or stopping times whatsoever, and also strict accountability and consistently high productivity. I call this system **time pools**, to contrast it to the more typical "time schedules" system that most people use.

The key insight in devising this system for me was to start thinking of time like it was a physical expendable resource, like gasoline in a car or hit points in a video game. Imagine if instead of scheduling exact intervals of time (e.g. "9:00am to 11:00am") in which to get things done you instead simply decided (in advance) how many hours and minutes of time you want to invest in each of a set of activities each day. For example, suppose you ideally wanted to do 8 hours of work, 3 hours of eating, 1 hour of exercise, 3 hours of relaxation and fun, and 1 hour of miscellaneous stuff and chores each day. This totals to 16 hours, out of 24 possible hours per day. You should leave those remaining 8 hours reserved for sleeping though.

You could create a fixed time schedule for this, but (if you're like me) it would probably feel like a drag. It would feel very confining and rigid and you wouldn't be able to dynamically adjust to the random events of the day as easily. Rigid schedules can suck. However, all is not lost. You can still consistently achieve this productivity plan each day if you simply approach it the right way.

Instead of thinking in terms of rigidly scheduled intervals of time, try thinking of each of the different allocated durations of time you want to achieve each day as expendable resources that start counting down whenever you engage in those activities and stop when you stop. Thus, for example, when you're working, the corresponding amount of work you have remaining to do that day is counting down, whereas meanwhile all the other durations for the other activities you aren't currently doing remain unchanged.

However, once any of these durations is reduced to zero, you are no longer allowed to participate in that activity and must now switch to one of the others. I refer to each of these decreasing durations as a *time pool*. That's where the name I came up with for this method of keeping track of allocated time came from. Anyway though, in the event that you *do* go over your time limit for an activity, as does sometimes happen, you can simply divert the excess time into whatever currently remaining duration among the other activities has the lowest priority and least value to you. You can also leave a chunk of time reserved as "buffer time" or "overflow time" that you can switch to as a margin of error, if you want, so that when you do mess up it doesn't immediately start eating into your other time allocations.

You also ideally want to combine this system with some kind of advance periodic warning system, such as an automated voice system that verbalizes how much time is remaining in the currently active time pool every time a certain amount of time has passed. For example, you could download a timer app[10] of some kind onto your phone or computer and have it use a text-to-speech system to verbally recite the remaining time and current task type after every elapsed 30 minutes.

This gives you early warnings for each of the time pools you have, so that you therefore have time to react and thus to manage and adjust your activities as you go, instead of being suddenly caught off guard when one of the time pools depletes all the way to zero. This also helps you gradually build up a better sense of time and of how long the activities you do are really taking. You'll also want an alarm to go off for each time pool whenever one does indeed reach zero.

The net effect of this is that you now have a completely flexible system for managing how much of any arbitrary set of activities you want to get done each day, but without ever having to allocate any of your time to any rigidly preset intervals of time. This also makes it vastly easier to tweak your time allocations than a rigid schedule would. You simply change the corresponding durations and you're done.

I find that this system feels much more pleasant and also responds better to unexpected events than rigid time schedules do. You get both flexibility and accountability this way. It's the best of both worlds. As for me, I know that when I personally implemented this system in my own time management for writing this book it massively improved the consistency of my daily productivity. I really recommend you give it a try, especially if you're like me and find fixed time schedules uncomfortable.

7.4.2 Crucial components of achieving high productivity while simultaneously enhancing perception of quality

When working on a project, it is very tempting to just get lost in the details and to forget to see the big picture. It's easy to get stuck in a tarpit of focusing on the wrong things and managing your time and resources in highly inefficient or ineffective ways. People often get stuck in this mode where they continue working consistently, but do so in a kind of intellectually and creatively half-asleep way.

They start just slogging along, working on anything and everything they think to do, absent-mindedly following the path of least resistance and greatest mental inertia, regardless of whether or not their existing plans or assumptions are actually realistic. They continue following an existing production plan or idea of how they think the project will go, without really ever taking a step back to reconsider things or to actually test their assumptions. They fall in love with the *feeling* they imagine the end product will end up having, but all too often end up pursuing that end goal in a relatively aimless, poorly planned, and wasteful way.

To make matters even worse though, there is also a lack of awareness of even the basic contours of what it is that really causes one product to be perceived as higher or lower quality than another. It's not enough to just keep slapping on more changes in an absent-minded and complacent way. You need to think more purposefully and tastefully than that. The work methodology you use each day on the journey to your goals changes the substantive outcomes.

---

[10]The timer app I personally use for doing time pools is called "Watson Multi Stopwatch Timer", which is available in the app store on Android. It was actually surprisingly difficult to find an app that had enough features to support this way of working. So many of the timer apps on the app store are far too dumbed down and have far too many automated "helpful" features (e.g. always automatically resetting depleted timers) that get in the way of doing time pools properly.

Wishful thinking is not enough. You can't just suddenly make the outcome better even despite of working in ineffective ways every day up to that point. A bad work methodology won't generally produce a high quality product, except by sheer random luck. You need to lean *into* your challenges and highest value tasks instead of away from them. Failure to do this is a very common cause of ineffective and disappointing product releases. Too many products are designed with a lack of principled thinking and a lack of awareness of where the real value points in the product actually are.

It seems to me that much of this phenomenon is really just the result of not consciously orienting your frame of mind towards the right thing. Too many people just think of their work in terms of some generic amorphous notion of "just get a bunch of things done today" every day. Too few people have ever been taught what the real factors that determine success generally are. It's not enough to just think to your self "I need to add as much additional content onto the project as possible" each day. That's the wrong thing to try to maximize.

The real thing that you should probably be trying to maximize is not the sheer number of actions you invest in the creation process, but rather it is actually two far more nuanced and impactful quantities: (1) the *ratio* of value produced per unit of expenditure and (2) the *ratio* of value per unit of content that exists in the product. The sheer volume of junk that you add to the project is not usually the determining factor in success, generally, but rather it is these two ratios. You want both of these ratios to be as high as possible. Indeed, these ratios are important enough that they merit their own corresponding terms:

Definition 193. *The ratio of the value produced per unit of expenditure (whether that expenditure be time, money, or any other resource) invested in the process of working on something may be referred to as the* **value efficiency** *of that expenditure. Value efficiency is what determines whether the gains per unit of expenditure will be high enough to surpass the costs. In other words, value efficiency is what determines profitability.*

If value efficiency is too low, then the project will not be very economically effective. It is thus important to maximize value efficiency if you care about the profitability and/or impact of your projects. Ignoring value efficiency may cause success to become impossible, even if the end product gets lots of attention, because having low value efficiency may drive your costs higher than your gains.

Definition 194. *The ratio of the value per unit of content that exists in a product (or in part of a product) may be referred to as the* **value density** *of that product (or that part of that product). Value density is what determines how high-quality a product will be perceived as being, and also whether or not it will be seen as worthwhile to use or to purchase. It is a measure of the amount of value a person using the product would get out of it per unit of content from that product that they consume.*

If value density is too low, then a product will seem tedious, boring, weak, low quality, or mundane. When people evaluate their options on what to do or to buy, they often use value density (subconsciously) as a way of estimating what the best choices will be. If the value density of something seems too low, especially relative to other options that are currently available, then people will tend to feel demotivated to pursue that option. Thus, value density is one of the most important determining factors in how many people will find your product interesting and in how interested in it they will be.

Maximizing these ratios (preferably both at the same time) is essential to consistently achieving strong economic and creative outcomes in life and in business. Accounting for this dynamic is a necessary part of competing effectively against other people's products. Ignoring it makes you vulnerable to waste and complacency. Falling into the mode of "just getting more stuff done" in a mechanical and unthinking way is so easy to do, but it can also often be very lethal to your future economic and creative success. However, admittedly, accomplishing at least something is still far better than getting trapped in analysis paralysis and accomplishing nothing though, so keep that in mind as a point of nuance and counterpoint here during this discussion. Strive for a healthy and pragmatic balance.

Anyway though, notice that all you have to do in order to change these ratios for the better is to simply allocate your time, energy, and other resources in a more impactful and conscientious way. Everybody ultimately ends up only having a fixed (but initially unknown) amount of time and creative energy each day to invest in their work. Thus, an intelligent strategy for success is actually less about trying to cram as many tasks as you can into each day and more about trying to maximize the value efficiency and value density of every action that you take and every choice that you make. This approach is not only vastly more effective, but it is also often vastly more pleasant.

Probably the biggest dividing line between people who accomplish a lot with their lives and people who don't is how well they are able to identify high-value choices over low-value choices. Not only do they need to be able to identify the higher-value choices though, but they also need to actually be able to consistently follow through with the implementation of those choices. They need to be able to delay gratification in order to do this, and to also not allow lower-value things to constantly pull their attention away from higher-value things. Exceptional productivity is all about prioritization, pragmatism, and focus. Merely "working hard" is not at all sufficient.

Understanding value efficiency and value density on a deeper level also helps immensely in understanding and predicting consumer behavior and market outcomes. On the business side, if value efficiency is not past at least a certain threshold, then the endeavor may not be profitable, perhaps even regardless of how well it does. On the other hand though, on the consumer perception side, if value density is not likewise past at least a certain threshold, then people will not feel motivated or interested enough in the product to ever actually use or buy it. It strikes me that these critical thresholds also merit names of their own, just as the other terms did:

Definition 195. *If the value efficiency of a product is not greater than a certain threshold value, then the product will not be profitable and therefore will become unsustainable. We will call this threshold the **profitability threshold**. The profitability threshold is measured in the same units as the corresponding value efficiency measurement is (i.e. in value produced per unit of expenditure).*

Definition 196. *If the value density of a product is not greater than a certain threshold value, then the product will probably not be perceived as high quality by consumers, and therefore they will probably not be inclined to use or to buy it. They will view the product as not being worth their investment. This will manifest as feelings of boredom, tedium, disinterest, pointlessness, or disdain (etc). We will call this threshold the **effort threshold**. It is a measure of how much value payoff per unit effort a person would require before they would be willing to do something*[11].

This threshold can vary from person to person and from moment to moment, according to each person's random mood, state of mind, and personality. For example, a consistently high effort threshold may indicate that a person has a fundamentally lazy personality. The effort

---

[11] Alternatively, one could say that value density is the "diamond in the rough" factor involved in something. It tells you how much "diamond" there is per unit of "rough", and hence how worthwhile "digging around" would be. You don't want to wear people out too much by making them expend too much effort per unit of value they extract.

threshold is measured in the same units as the corresponding value density measurement is (i.e. in value per unit of content present in the product), but varies more widely and is more difficult to control. Still though, the effort threshold of most people will probably generally be relatively predictable and will probably usually have a fairly stable average value across a large population.

The effort threshold plays a huge role in what causes people to become bored. As people extract value from something (e.g. by reading a book, by playing a video game, etc), it is often the case that the remaining value left to be consumed from that thing will thereby be gradually reduced. This is because not all forms of value are repeatable. Sometimes, once a piece of content has been consumed or experienced, then there will be little or no additional value to be gained by repeating the process.

Thus, people's collections of items which they have previously extracted value from often tend to become less and less interesting to them over time. This is because the remaining value density for each of those items has decreased far enough that it is now *below* that person's effort threshold instead of above it like it originally was. This is what causes people to become bored with things they used to like. Willingness to do something is primarily determined by whether or not the value density of that activity seems greater than the effort threshold, but also combined with people's tendency to prefer activities with higher value density over those with lower value density whenever possible, even when multiple activities are above the effort threshold.

This also provides a bit of insight into how to undo some of this effect, so that you become bored less often and are more able to re-experience the things that you've previously enjoyed. One way is to lower your effort threshold, which means changing your attitude so that you are less lazy and also more attentive to seeing the value in subtle things. Another way is to change the way you approach the activity so that you create a more fresh experience, such as by changing what you focus on or otherwise doing things in a more unconventional or non-standard way.

For example, in a video game, instead of merely completing the game, you could try exploring the environment very slowly while savoring the artwork and atmosphere, or beating the levels as quickly as you can (i.e. speedrunning), or playing without using specific weapons/tools/abilities (i.e. handicapping), etc. This kind of approach enables you to seek out parts of the game's possibility space that you otherwise wouldn't have been naturally compelled to fully experience, thus temporarily revitalizing the freshness of the game. Similar ideas also apply to things besides games of course.

This point is fairly obvious in some ways, I suppose, but it does nonetheless provide good supporting evidence that this notion of value efficiency and value density (etc) is on the right track, and so is still worth saying regardless. Oh, and one more final note: Value efficiency, value density, profitability thresholds, and effort thresholds are not the only factors in play in these kinds of situations though. There are many other external factors and special cases that can potentially come into play and change the dynamics some.

Try not to think about this in too much of a one-dimensional or oversimplified way. Treat these concepts in broad strokes. Use these insights as a reminder of how to frame your goals in such a way as to tend to maximize your outcomes, but remember that real life is more nuanced than this. When in doubt, don't let these criteria cripple your ability to be productive and creative. Creating anything is almost always better than creating nothing. Let your creative spirit breathe freely and experiment frequently, regardless of any concerns over what choices may be "best". Do what needs to be done and be pragmatic. An inefficient and uncertain path is still better than no path.

7.4.3 The cult of minimalism: when simplification is taken too far

Lately, minimalism has become very trendy. Several major companies have become very successful with minimalistic approaches to design (e.g. Google, Twitter[12], etc), so of course a bunch of other companies and individuals have thoughtlessly decided to imitate them in dogmatic and overzealous ways, as per usual. This has resulted in minimalism being taken to a very destructive and counterproductive extreme by many companies and individuals, and thus has created a lot of economic waste and lost opportunities[13].

Groupthink, small-mindedness, complacency, poor organizational skills, lack of imagination, and conformity are probably largely to blame for this. Too many companies and individuals make their design decisions based on merely imitating successful examples of others, in an uninspired and uninventive way, rather than by actually thinking according to first principles or being genuinely consciously aware of what the underlying value proposition of their work really is.

Minimalism is sometimes perfectly fine. It does sometimes provide a very efficient and effective way of creating value. The problem with minimalism though, is that people tend to often take it to so far of an extreme that they actually end up destroying a lot of value by doing so. It's true that sometimes less is more, but it is also equally true that sometimes more is more. It really depends on the specific circumstance. The decision as to whether or not to be minimalistic in the design of something only makes sense in context.

Minimalism is not inherently virtuous and neither is complexity. Thus, adhering to minimalism as some kind of blanket philosophy, as if it is somehow inherently a better way of designing things, is a false and counterproductive worldview to hold. A true master designer only applies minimalism when it is actually merited, i.e. only in cases where it really does increase the overall value of the end product. Sometimes only a more complex solution offers the maximum amount of value to users though.

Dogmatic adherence to the cult of minimalism thus is inherently handicapping. A designer who is stuck in the mindset of only ever being minimalistic is a designer who can never really achieve true versatility in their craft, at least not for all possible cases. Some situations inherently require increasing complexity in order to maximize value, and therefore no strict minimalist can ever create an optimal solution for such cases. As usual, dogmatism and overly ideological thinking tend to have toxic effects.

Part of the problem here is that people often get so fixated on streamlining the most common uses cases and removing other noise and clutter that they start grossly disproportionately undervaluing the less common use cases. Just because a use case is less common doesn't make it less valuable. That's fallacious thinking. The value of a use case can often easily be much greater than how commonly used it is.

It is also true that people tend not to learn much unless you provide them with ample easy opportunities for exploring new things, and make sure that they can do those things with relatively little friction. Excessive minimalism destroys such opportunities and therefore causes the

---

[12] By the way, on a side note, if I was the one who got to choose Twitter's motto, it would be: "**Twitter: Where Twits Go To Communicate With Each Other**". I have a feeling Twitter's corporate managers wouldn't approve though. Alas... It captures the essence of their platform so well. Don't worry about them overhearing us talking about this though. Twitter users don't have a large enough attention span to understand anything over 140 characters long. ☺

[13] Even the most successful minimalistic companies seem to often be greatly hindering the true underlying expressive potential of their own products. Google's search system, for example, has lately become very prone to narrowly interpreting queries. It often refuses to interpret queries according to anything other than the most popular or most commercialized possible intents. The results are often too dumbed down and too difficult to precisely control, even when using all the "advanced" features. Thus, even "successful" examples of extreme minimalism are often actually being severely held back by overzealous adherence to it.

perspectives of users to needlessly stagnate and narrow. It makes users dumber and less capable of accomplishing things in more effective ways, in other words. It causes people's perspectives to shrink instead of to expand. Constantly catering to the least common denominator, i.e. the dumbest and most lazy portion of the human population, probably ultimately has some very toxic effects on the growth and future of society as a whole.

Also, most of the value produced by the use of well-designed tools is created by the small minority of the strongest users of those tools. The weakest users (in stark contrast) typically don't produce much of the value that those tools enable. It is therefore not really that reasonable, from an economic standpoint, to always cater to the weaker portion of the user base so much. Life is more nuanced than that. The sheer reach in number of users of a product is not the same thing as the total amount of value those users actually create.

All these things considered, it becomes clear that dogmatically catering to only the most simplistic use cases and the dumbest portion of users is thus a backwards way of thinking about many product designs. It does depend on the specific product though, and who that product's target audience really is. Fundamentally though, excessive minimalism is ultimately just an incorrect way of attempting to maximize the value density[14] (i.e. the value per unit of content) of a product, such that the person or company responsible takes the "less is more" mantra so far that they actually start *reducing* the net value density of the product instead of increasing it.

This is commonly caused by falsely believing (whether subconsciously or consciously) that less commonly used features are *so* low-value that they don't even merit being kept around anywhere, which is seldom actually true (i.e. the designers grossly overemphasize "noise reduction" in the design, at great expense to actual value creation). It seems like we should define a term for this common (and unfortunately currently trendy) pattern of making things *worse* through attempts at simplification instead of better. As such, here's a corresponding definition:

Definition 197. *When an attempt to simplify something actually causes that thing to become worse instead of better, let us refer to this as **value-reducing simplification**[15]. Value-reducing simplification is caused by having an incorrect understanding of what it really means to maximize the value density (i.e. the value per unit of content) of a product. One of value-reducing simplification's most common forms is removing "noise" and "clutter" from a system at the cost of completely removing or severely crippling previously useful components of that system, which is actually seldom a good trade-off in most contexts.*

Value-reducing simplification is extremely counterproductive, but is very easy to fall into for some people because it works based on a kind of mindless and dogmatic minimalism that requires far less careful thought than the correct way of optimizing things does, and many people don't like having to ever actually genuinely think about things and would instead much prefer to be given brain-dead easy (and often very wrong) "universal" answers instead.

Luckily though, not all forms of simplification and minimalism have these bad traits. There is a much more effective alternative. The key difference is that instead of just aggressively cutting out useful features for the sake of dogmatic adherence to minimalism, you instead think in terms of *refactoring* the existing components of the product in such a way as to simplify it *without* destroying any significant amount of the underlying value it provides. To do this properly

---

[14] See page 768 if you want to see where we originally defined value density.

[15] I also came up with a much more comical and colorful idea for what term to use for this, but the language involved was too coarse for my tastes. You might like it though. The idea was to use **simplif**kation** (i.e. "simplification" with the f-word inserted into it) to mean value-reducing simplification. This is a fun idea, but I ultimately decided against it. You can still use it though of course, if you like it more.

though, you need to actually take the time to adequately consider the existing use cases, so that you don't accidentally damage them. This alternative (and typically much better) approach to simplification should also (like value-reducing simplification) be given a term of its own:

Definition 198. *When an attempt to simplify something does not damage the overall value of that thing significantly, or sometimes even causes that thing's overall value to increase, let us refer to this as **value-conserving simplification**. Value-conserving simplification tends to consistently increase the value density of a product, therefore likewise increasing the perception of that product's quality and reducing the barrier to entry so that more people will find the product both more useful and more easily understandable.*

Value-conserving simplification really is a much better approach generally. However, unfortunately, recent trends have seemed to lean pretty strongly in the direction of value-reducing simplifications sometimes. For example, in the software world, I have come to often dread software updates on things like operating systems, web browsers, phone apps, and utility programs because of how often those updates have broken things and permanently removed useful features upon which I've depended, all just because the designers prioritized a bunch of arbitrary superficial factors (often decided on a random whim) at the expense of the core utility and versatility of the system, all in the name of a cult-like irrational adherence to minimalism. I'm tired of so many people dumbing things down in very damaging and counterproductive ways.

7.4.4 Value parasitism: an underappreciated source of economic instability

With the way some economic commentators talk, you'd think that all profits are inherently a good thing, as if everything that companies and individuals do to get money is automatically good for the economy as a whole (e.g. the whole "greed is good" mantra corporations love so much, etc). This is not true in reality though. Creating money is actually generally only a good thing for the economy if that money is actually tied to real value creation.

Money that isn't tied to a corresponding amount of real value creation though, in contrast, is actually generally damaging to the economy, even if it nominally looks like valid "wealth creation" on the surface. It's not enough for money to merely be legally obtained for it to have a positive economic impact. You need more than that to be true, if you care about money actually fulfilling its empowering role in society correctly. Yet, unfortunately, many people have become so focused on the superficial surface-level details of money production that they have forgotten that it is not money production itself that really forms the basis of the economy, but rather it is value creation and trade.

Money is just a proxy for value creation and trade. It is not itself value. It is merely a way of enabling trade to be conducted in a much more fluid, uninhibited, and efficient way. Without money, society would be forced to operate as a *barter economy*. A barter economy is an economy where there is no such thing as money, but instead all products and services are exchanged directly. For example, in a barter economy, if you wanted to buy a loaf of bread from a baker, you would have to offer something tangible in return. You could offer to trade the baker one of your knives or to mow their lawn (or whatever else) in exchange for the loaf of bread. That's bartering. No money is exchanged, only the products and services themselves are.

The problem with barter economies though is that they are very inefficient and limiting. It is difficult to barter in a perfectly fair way. The product or service that one person is offering the other is usually more valuable than the other, and it's hard to bridge that gap in any kind of precise way. A barter economy also only makes it possible to exchange products or services with people who actually *want* those products or services, which greatly reduces how often you will be able to find anyone who is interested in what you are offering as a trade. This causes all but the most commonly desirable products and services to become much less economically viable and thus greatly damages economies of scale and product diversity. Indeed, highly specialized products are often not even worth producing at all in barter economies, because they are so difficult to barter with. Society would suffer greatly for this.

That's why we have money. It is a proxy for the value of products and services, so that we can trade them *indirectly* with people in a much more efficient way. Done properly, and as long as all involved parties act in good faith, this system vastly improves economic outcomes. However, the catch with money based economies is that they are much more easily exploited and manipulated than barter economies are.

There is a much higher risk in a money based economy that money will be produced which actually does *not* correspond to any real value creation. Sometimes this is done in obvious ways, such as by counterfeiting, but it is equally true that it is also frequently done more indirectly, such as when social structures cause some people to be monetarily compensated for value that they did not themselves actually create. All money created in such cases, whether through obvious ways like counterfeiting or through social manipulation, is *unearned* money.

Here's the problem though: Money only works insofar as it is tied to real value, so the net effect of the creation of unearned money is always that it dilutes the value of all of the other money in circulation. It does this by essentially robbing other people in that economy of the face value of the unearned money. Thus, for example, if someone acquires $50 without ever actually earning it (and it isn't simply a gift or any other kind of intentional transfer), then $50 worth of value somewhere else in the economy will (in effect) be parasitically transferred away from the people who actually produced that $50 worth of value and will now be given to the person who did nothing whatsoever to earn it.

An economy cannot survive without *real* value creation though. It is only the *real* value creation that puts bread on people's tables. If only unearned money existed in circulation, it would have zero value and the entire economy would therefore immediately collapse. People would be forced to barter, because it would now be the only safe option. Every single time that unearned money comes into existence (in any quantity whatsoever) it automatically damages the ability of the *real* value producers in that economy to live sustainable lives.

And, if the amount of unearned money in circulation becomes large enough, the basic ability of producers to even so much as sustain the tools they need to produce enough products and services to keep everyone else in society alive may collapse as well. Thus, in the worst case scenario, this can result in millions of people dying from starvation and untreated diseases (etc). Unearned money is always accompanied by a roughly proportional amount of suffering and unfairness being afflicted upon someone else (a real producer) in the economy. That's the net effect, regardless of whether you can observe it directly. Every dollar of unearned income is a dollar's worth of real value that was stolen from someone else.

This principles applies at all levels of the economy, from rich to poor, from private citizen to government, from legal businesses to criminals, etc. It also applies across all industries. Almost all economic transactions are influenced by it. Wherever there is unearned money being received, there is always corresponding damage to the economy. The only exceptions are things like basic social safety nets, which are actually often extremely economically valuable because of the stability and upward mobility they provide, and gifts (i.e. intentional "unearned" transfers of wealth), and other kinds of similar special cases.

Also, it is important to realize that unearned money is not a black-and-white thing. Money can be partially earned and partially unearned at the same time. For example, if a healthcare provider charges $1,000 just to say hello to your doctor and to talk to them for 5 minutes, then probably at least 95% of that $1,000 (i.e. $950) is unearned income, and hence will correspondingly damage the economy by a proportional amount, but the remaining 5% (i.e. $50) *might* be earned income, but only if the doctor actually created some *real value* for the patient during that time. In no other case is economic compensation really merited. If money is not genuinely earned then it is actually economically destructive, not productive.

I know this may go against some of the existing views on economics, but I very much suspect it is true. In fact, I bet that most economic instability and recessions and depressions actually are caused by increases in circulation of unearned income from exactly these kinds of less obvious sources. *All* unearned, undercompensated, or overcompensated money contributes to economic instability. The worse a job money does of accurately tracking real value creation, the more unstable the economy will become. That's my theory. It makes perfect sense too, if you think about it in terms of what money is really designed to do, then compare it with what a barter economy would do, and then think about what that implies.

We can expand this kind of thinking far beyond just the overpriced doctor example. Another example of something that produces a huge amount of unearned income is things like massive bonuses being given to CEOs in the aftermath of scandals. This is unfortunately a very common occurrence in the current economic culture of society. A company gets caught doing something that has caused massive economic damage (i.e. destruction of value) to a large number of people, often through intentional systematic exploitation or deception of the public, and instead of being properly punished the CEOs often receive huge compensation packages totaling tens or hundreds of millions of dollars for their actions.

In such cases, not only does the initial act of exploitation itself damage other people's economic situations directly, but so too does the massive unearned bonus package the CEOs receive. Every single dollar of those multi-million dollar bonus packages ends up effectively being *stolen* from the pockets of the real value producers in society. The CEO can go and use that money to exchange for *real* goods and services, but the CEO never actually did anything (i.e. never created any real value) to merit that money. The CEO's use of that money therefore has the same net effect as theft. Understand?

Here's another way of thinking about it, a simple thought experiment designed to make the truth of this fact clearer and more impactful: Forget about the CEO for a moment. Imagine that a shoplifter raided a grocery store and stole a bag of apples. The shoplifter enters the store, provides nothing of value to trade for the apples, and then takes them and leaves. Clearly, this is economically damaging to other people who actually did the hard work to produce those apples. The apple business and grocery store business would not be sustainable if too much of such theft occurred.

In contrast though, imagine if the CEO who received the multi-million dollar bonus package for the corporate scam walked into the grocery store. The CEO enters the store, hands the cashier some of his dirty bonus package money, and then takes the apples and leaves. To a random onlooker, this may look like a fair transaction. Think about it more deeply though. What did the CEO provide to the rest of society to merit him being allocated that bag of apples from society's limited resources as a trade? Nothing. The CEO never provided anything of value to trade for the apples.

Money just makes that harder to see. In a barter economy, the CEO never could have even bought the apples, and would have been forced to shoplift them. All the CEO did was to damage society by scamming them, which is not something that has a tradable value. The CEO's bonus package money is therefore actually worth nothing in real terms. It is just meaningless paper. It has no correspondence whatsoever to any tangible good or service, and therefore is in some sense not even real money. The cashier just doesn't know that. Neither does the rest of society who sees the money. Nonetheless though, money is nothing but a proxy for value. That value doesn't just come from nowhere. *Someone* had to produce those apples. The CEO is providing nothing to society to compensate for that though. Therefore, the CEO's transaction to buy the apples is actually economically and logically *equivalent to theft*.

The only substantive difference between the shoplifter and the CEO in this scenario is that the CEO has paper in his hands and gives that paper to the cashier. Both the shoplifter and the CEO are stealing an equal amount of value from the rest of society by walking out of that grocery store with that bag of apples though. These two people are both thieves. One (the CEO) is just better at manipulating the structure of society than the other (the shoplifter) is.

Likewise, both of these people are performing an equally economically destructive action by taking the apples. If everyone behaved like this, then society would immediately collapse and everyone would subsequently die long and horrible deaths by starvation. The CEO and the shoplifter don't care about that though. That's the kind of people both the CEO and the shoplifter are.

Only when you think in terms of the real products and services being exchanged in an economy can you truly understand the health of that economy. Mere money, in contrast, can easily be very misleading. Yet, much of society has gotten so used to thinking in terms of the abstraction of money that they've lost sight of the fact that money doesn't actually mean anything at all except insofar as it is tied to real products and services being traded. Economies where the overwhelming majority of people act in good faith and where exploitation is rare will probably tend to be stable and to experience consistently strong growth.

In contrast though, economies where lots of unearned income is being parasitically extracted will tend to become destabilized and to more frequently risk collapse. I think this is probably necessary to understand if you want to actually fix economic problems. It's the underlying value exchange system that determines the strength of an economy, not merely the money. Don't get distracted by superficial factors. Fundamentally, allowing people to be systematically exploited always causes proportional damage to the economy. As long as governments keep allowing that exploitation to happen in abundance, maximal economic health and growth will be completely impossible to achieve.

The people who perpetuate money creation that isn't tied to real value creation are parasites. They are a drain upon society. This is just as true of corporations and governments as it is of criminals. They only make all of us poorer, including even themselves in the long term. They massively damage the structure and stability of society, all just in the name of their own short-sighted destructive greed. Value, not money, is the true beating heart of an economy. Money that isn't tied to value is just parasitism in disguise. Let's create a term for this:

Definition 199. *Because money is just a proxy for value, used to make trade more efficient and flexible, the creation of money that isn't tied to a corresponding amount of real value is therefore parasitic in nature. Unearned money thus has the same net effect as theft. We may refer to this effect as* **value parasitism**. *Value parasitism weakens any economy that it afflicts. Also, the more value parasitism exists in an economy, the more likely recessions or economic collapses probably become.*

Also, while it may be tempting to try to assign all of the blame for value parasitism to some specific group of people in society (e.g. wealthy people or poor people), doing so would be very wrong. Class warfare is the wrong way of thinking about it. Value parasitism afflicts *all* social and economic classes of society. Anywhere where income or resources are being taken without being properly earned (but not including basic social safety nets that help provide upward mobility, and hence which help grow the economy), that's value parasitism. It's something that both poor and rich people do.

Properly understood, it's less a matter of rich versus poor and more a matter of producers versus parasites. Some rich people are producers and some are parasites. Likewise, some poor people are producers and some are parasites. It depends on each specific person. A corrupt CEO and a poor shoplifter are both parasites, even though they come from very different economic classes. On the other hand though, a rich but honorable car salesperson and a diligent teenage burger flipper both have earned a good and livable compensation, regardless of any biases society may have against them. Likewise, there are actually lots of very honorable CEOs and businesspeople, but the bad ones seem to sometimes get more attention in the press because of scandals. Most rich people are probably very nice people, just like most poor people.

I think we should all learn as a society to spend less time pointing the finger at people merely for having more (or less) than us, and to instead focus our attention in a more united way against parasites in general, regardless of which economic classes they come from. We need to stop getting distracted by fruitless in-fighting, jealousy, condescension, and stereotyping and learn to instead think more carefully and fairly. There is both good and bad behavior in abundance in virtually all parts of society. If we want a truly healthy free market economy though, we need to make sure that we compensate everyone fairly, by neither paying them too much nor too little money, but without micromanaging and stifling things with regulation and interference too much either. Balance in all things. Proportionality and principled thinking are key. A free market is not truly free until it is free of value parasitism.

7.4.5 The counterproductive nature of worrying about things you can't control

As human beings, we often worry about a lot of different things. This can feel a bit uncomfortable, but properly moderated, worry and concern are essential components of a healthy mind and a healthy life. Worry helps us to protect ourselves from harm and to safeguard our opportunities and assets. This is critical for achieving good outcomes in life and for maximizing our well-being and minimizing our suffering. However, taken to too great an extreme or focused on the wrong things, worry can begin to have very counterproductive effects on our lives.

The key to determining whether or not worry is good or bad is how much power we actually have to influence the thing that we are concerned about. If something is completely beyond our power to influence in any way, then concerning ourselves with it will inherently be a waste of our time, energy, and resources. Think of it like a fork in the road. On one path, we have a problem that we spend a lot of time worrying about and still can't change. On the other path, we have the same problem but we don't bother thinking about it or wasting any energy or resources on it.

Which path results in a better life? The answer is clearly the second path. The only difference between the two paths is that the first path causes the person to suffer both from the problem they can't change and also from wastefully worrying about it, whereas the second path only causes the person to suffer from the problem but not from worrying about it. Worrying about things we cannot control does nothing but add additional suffering to what is already there.

Therefore, there is no point in ever doing so. It would only make our life worse. Thus, any time a person is tempted to worry about something that they can never change, they would be better off simply remembering this fact and deciding not to worry about it. This will result in a much more peaceful, stress-free, and enjoyable life.

However, this effect doesn't just apply to cases where you have no control whatsoever over an outcome. That's just the most extreme and most easily understood example. There's actually an entire spectrum of how much you should concern yourself over things relative to how much you can control them. The relationship is really quite simple. The less you can control something the less you should worry about it or invest time, energy, or resources in trying to resolve it. Let's create a new term corresponding to this more nuanced view of proportional concern, for ease of reference:

Definition 200. *Investing time, energy, and resources into changing things that are not under your power to change is inherently a waste of time and therefore only causes a person's life to become worse. However, this is true not just at the extreme of having no control at all over something, but also over the whole spectrum of how much control over any given concern a person has. Let's refer to this spectrum as the* **control-concern spectrum**, *and to the broader notion that you should only ever worry about things in proportion to how much you can actually control them as the* **control-concern proportionality principle**.

On the one extreme, we have zero control over a concern and therefore should devote zero time, energy, and resources to it. We shouldn't even so much as think about it in that case. On the other extreme though, we have nearly total control over what we want to change, and therefore worrying about it is actually likely to be very productive and helpful. We should definitely take whatever opportunities we can to improve our life in honorable and ethical ways.

Life is about making your circumstances as good as they possibly can be. However, the trick to doing that properly is to remember that getting yourself so worked up about making things better that you start worrying about things that you cannot control or which would be very costly (in terms of time, energy, and/or resources) to control will only make your life worse, by making you suffer more from trying to change things that are impossible to change or from otherwise wasting disproportionately high amounts of your very limited time, energy, and resources in life to change things that are difficult to control and that may not yield much benefit even if you do manage to change them.

A truly wise person knows to account for the difference between the benefits and the costs of something before they choose to invest a very large amount of their time, energy, or resources in that thing. Don't get pulled into fighting fights that you can't win without a huge cost, generally. Getting pulled into those kinds of low-value conflicts will tend to greatly damage your life outcomes. Be more balanced and nuanced than that. Don't let yourself get tied up in knots over a bunch of fruitless anxieties.

For example, avoid sources of unnecessary and toxic stress and distraction such as the vast majority of news[16], anything about your body or genetics that you can't change (or that would

---

[16] At most, you should only ever expose yourself to enough news to be aware of things that may directly impact your own life and well-being (e.g. the weather, major wars, etc) and perhaps also to know who you will vote for. Even then you should be extremely skeptical and distant. Political news (for example) is very often *designed* to keep people as artificially stressed out and worried as possible, in order to hook people's attention and to trigger repeated daily viewing. Almost nothing such news channels talk about is realistically controllable by the viewer though. News often severely interferes with a person's ability to apply the control-concern proportionality principle to their lives. Consuming a lot of news is one of the worst life choices any human being can ever make. You can waste an enormous amount of your precious little time in this world that way, and will have absolutely nothing to show for it once it's all over. It is almost always completely unproductive.

be too costly or risky or harmful to change), the arbitrary (and often capricious, biased, and irrational) opinions and commentary of other people, unexpected harmful life events, and so on. Simply make the best of every moment, using whatever resources are currently available to you. If you don't have the power to change something, or if you do but just aren't going to even try, then don't waste any of your attention on it. Stop letting it bother you. Recognize that worrying about such things only needlessly makes your life worse.

Focus your life on higher value things instead. Cut out waste and distraction. Nurture your creative spirit and your zest for life. Live in balance, with a strong sense of internal validation and persistence. The control-concern proportionality principle is a central and indispensable component of how to make the best of your live in a stable and reliable way. Used properly, control-concern proportionality will greatly improve your ability to successfully live a peaceful and wonderful life. It will liberate you from the tyranny of futile concerns and will teach you to find joy and prosperity in every moment of life, as best you can.

7.4.6 Creativity at will: how to be creative even when you don't feel inspired

Inspiration for creative work is not always easy to come by, or at least it feels that way for a lot of people. It can be very frustrating when you want to get creative work done but just can't seem to get into the right mood to do so. Many people have this problem, and it causes them to often produce much less creative work than they really want to. Lots of people have tons of motivation to work on creative projects, in spirit and in principle, but still can't seem to find the time, energy, or inspiration to do so.

However, this kind of attitude and experience is actually generally an illusion. It isn't really all that difficult to be creative at will once you know how to approach it, but a lot of people have never been properly taught how to do this. People who work in creative professions (e.g. art, game dev, film, etc) basically don't have a choice though, and eventually learn how to make at least some creative energy happen every day, otherwise they would not be able to be financially successful in a sustainable way. Many of these people seem to gradually learn how to be creative at will *subconsciously*, and may have trouble actually explaining it to other people in a way that really clicks.

Well, as it happens, I'm going to be making my own attempt at conveying how to be creative at will to you here, in a way that I hope you will understand. This method I will be describing will be based on my own creative experiences. There might be other ways that some creative people do it, but I can only speak to my own personal experiences and insights and what works for me. I do think that the method I will be describing here should work for most people though, as long as they understand and apply it properly.

There are multiple components to making creativity at will work. The first component though is simple: Stop making excuses about "not being inspired" to rationalize why you don't make time to sit down and do work. It's so easy to use this as a feel-good excuse to subconsciously justify laziness and avoidance. Don't do it. Even if you don't feel the slightest bit inspired, you need to make time (preferably every day) to sit down, remove distractions, and make an honest attempt to create at least something.

Creativity very much has a snowballing and momentum factor. You very often will have to forcibly get that ball rolling again each day if you want to be a successful creator, and that involves making a consistent habit of sitting down to create things each day. Sometimes you will just naturally be "in the zone" and this will come easily, but it very often won't, and you will never finish your projects if you don't learn to consistently make a habit of creating things even when you don't feel inspired.

Secondly, stop being so afraid of the judgment and accountability that comes with creating something real. Learn to be a more bold and dynamic person and to be less mentally fragile. Every creation feels perfect in its creators mind until they actually start having to make it real. Reality is much less forgiving than our imaginations are. The human brain can easily experience sensations of intense joy when fantasizing about an idea even when that idea would not be so wonderful if it actually existed in reality. Don't be deceived by this. Test your ideas and question your assumptions. Don't be a coward.

Running away from actually making your ideas real lets you avoid ever being held accountable for your ideas and being judged accordingly for them. This is a big part of why so many people have plans to create things "eventually" but never actually do. They subconsciously don't want to expose themselves to the real consequences and constraints that making the project real would imply. As long as they don't actually create the idea in reality, it can always remain flawless in their mind.

In contrast though, there is virtually no such thing as a perfect creation in reality. Stop being so afraid of this. Imperfection is no big deal. Every leaf on a tree is imperfect, but does that make the tree any less beautiful or valuable? No. The same is true of creative work. Stop letting perfectionism and fear cripple your ability to accomplish anything. Learn to appreciate the beauty of all the little details and nuances of everything around you. Remember that just as those random little flaws and quirks in those things don't detract from the beauty of those things in any real significant way, neither will they generally detract from your own creative work.

Learn to live in reality and to respect its arbitrary and beautifully imperfect nature, and stop trying so hard to fight that. Identify hidden sources of arrogance, cowardice, and disproportionate priorities within yourself that drive you to perfectionism and try not to let those sources have too much power over your. Don't be a slave to perfection and fear. Let yourself experience the wonderful peace and tranquility that comes with being able to truly appreciate everything exactly as it already is.

Try not to be too uptight, and don't worry so much about what other people think about every tiny little thing. Think more pragmatically than that and in broader strokes. Other people's opinions are just as arbitrary as the leaves on a tree are. They can provide value and insight, but they can also just be irrelevant noise that's trying to intrude upon your path in life, the meaningless product of nothing more than randomness and social momentum.

Thirdly though, I want to talk about the essence of creativity on a much deeper level. These previous two items (about stopping making excuses about "not being inspired" to rationalize why you don't make time to sit down and do work and about stopping being so afraid of the judgment and accountability that comes with creating something real) are just prerequisites to being able to get creative work done efficiently and effectively at all. This third item though is more about creative expression itself, and how to turn it on and off at will like a faucet or a light switch. It's about how to *force* yourself to be instilled with at least some creative power even when you can't otherwise find any natural inspiration.

Let's think about this and see if we can figure out what the trick is. Let's start by first framing the problem. Our problem is that we don't always have inspiration available to guide our creative work. We don't always know what choices we want to make when designing something or when trying to come up with a new idea. And, while it's true that we can force ourselves to sit down to work, and that's half the battle, that's not the same thing as actually successfully producing new and interesting ideas.

Yet, it is not as if we can just bang our head against the wall enough times that inspiration suddenly strikes us, at least not in any kind of consistent way. We cannot achieve fresh creative inspiration by sheer strength of frustration alone. We need something more than that. How then are we just supposed to conjure up inspiration out of thin air then? It seems impossible, doesn't it?

Indeed, in a sense, it is. Inspiration can't be conjured up at will. Creation isn't the same thing as inspiration though. That's the key. You actually don't even *need* inspiration in order to be highly creative and innovative. There is a much more powerful force for spontaneous creative energy operating in the universe than just inspiration. What is it though? The answer is all around us. You see it everyday. You see it everywhere. Even you yourself are a product of it.

The answer is evolution, and also other emergent effects more broadly[17]. Evolution and emergence are both mindless forces, nothing more than logical consequences of the way things interact with each other, and yet they have together created everything beautiful there is in our entire world and our entire universe. We are all of us children of evolution and emergence. Beauty itself, I would argue, is actually an emergent effect. Emergence is not just a contributor to beauty, but rather it is the fundamental essence of it[18].

Thus, the only thing we need to do in order to achieve creativity at will is to instill our actions with the power of evolution and emergence. We must simply act in harmony with the creative forces of the universe itself. When we express ourselves that way, such that our actions are structured in such a way that they cause evolutionary or emergent effects to occur, spontaneous creation of beauty and of other interesting structures is the inevitable result. This is true regardless of whether we even possess "inspiration" or not.

This may sound like an intimidating or hand-wavy thing to do, but it is actually quite easy and straightforward. Allow me to explain. Let's take evolution for example. Evolution is all about natural selection. It's all about allowing multiple instances of something to be made, randomly mutating a bunch of those instances, and then applying selection pressure to them to see which members of that group survive best, i.e. which ones have the highest overall fitness. There's no reason why we can't also apply this kind of process to the way we do creative work.

For example, suppose an artist is working on a digital painting and can't seem to find inspiration for what to do next. Many artists at this point would start metaphorically banging their heads against the wall in frustration or would start waddling around town making excuses about "not feeling inspired". In both cases this would cause the artist to probably not get much work done. Suppose though that the artist instead thought to themselves: Why don't I try approaching this problem in an evolutionary way?

Suppose the artist then, instead of engaging in fruitless frustrated groaning or other counterproductive behaviors, decided to simply make a bunch of extra copies of their digital painting file, then modify each of them in some small but appreciable way on a random whim (in a very relaxed way, without falling victim to perfectionism and analysis paralysis along the way), and then took a step back and asked which of these new versions they like best, selected one or more of those, and then continued the process for a while. That's evolution. Something doesn't need to be biological to still be capable of evolving.

---

[17] Evolution is itself an emergent effect. Emergence includes far more than just evolution though. The beauty of snowflakes and of crystals are both examples of emergent effects, for example. So is every random map generator ever put into any video game. Emergence is a very broad and extremely powerful phenomenon. It is also probably the bridge between abiogenesis and life.

[18] Unlike vaguer theories of beauty though, emergence is quantifiable and actionable. It can even be automated in the form of computer programs.

Evolution is more than just a biological phenomenon, contrary to what many people think. Properly understood, it is actually a *logical* phenomenon, one that just happens to also occur abundantly in nature. It is part of the structure of all logical systems, wherever sufficient forces exist, whether artificial or natural, and whether this universe or some other one. It can be performed manually, by hand, like in the digital painter example, but it can also be performed in automated ways such as on a computer simulation. Evolution is very versatile and can actually be applied to virtually any productive activity, often with great gains.

One of the most beautiful things about evolution is that its ability to gradually improve upon things and to innovate is essentially unstoppable. Think about it logically. If you have a bunch of random things, and a bunch of random mutations of those things, and then you apply selection pressure to those things and only allow the best of those things to survive, it is inevitable that the design and desirable qualities of those things will improve over time (even if all of the forces acting upon those things are completely mindless).

The weaker versions of things won't be allowed to survive. Thus, evolution provides a way to *force* creative changes to continue occurring, even when there is no conscious design or inspiration driving them in any particular specific direction. This is very empowering, if properly understood. It means that the sky is the limit for quality and creativity, as long as you are willing to continue the evolutionary process long enough and don't mess it up.

Besides just evolution though, there is also the broader phenomenon of emergence. Just like evolution, emergence also provides insight into how to better achieve creativity at will, although it does so in a more primal and elemental way than evolution does. Applying evolution to creativity is mostly about working in such a way that you emulate mutation and selection pressure. Applying emergence to creativity, in contrast, is more about exploiting the tension between the forces of order and chaos and about allowing your creative work to speak to you and to tell you where it naturally wants to grow. It's about allowing your work to take on a life of its own, as it usually will, if you let it.

To understand emergence better though, we need to take a bit of time to think about its nature. Emergence is all about the spontaneous creation of higher-level complexity from lower-level things. The reason why it works is because of how causality works, in that lots of little things interacting together can often create huge domino effects that cause more complex higher-level structures and properties to spontaneously emerge[19]. These chains of events are often so granular, subtle, and myriad that we humans have great difficult in intuitively predicting their outcomes, which is why so many people are so frequently tempted to attribute it all to magic and deities and such.

However, emergence typically only creates really interesting effects when chains of events somehow manage to get tangled up in non-trivial ways. For this to happen, there must usually be a strong mix of both order and chaos in play in the system. In the study of emergence, we sometimes refer to such a system as being on the *edge of chaos*. If a system is on the edge of chaos, then this means that in some sense it exists roughly on or near the boundary between chaotic forces (like heat) and orderly forces (like cold) in the universe. For example, the planet Earth exists in the edge of chaos of our solar system. It is neither too hot (chaotic) nor too cold (orderly). That's why it's capable of supporting life. Life is an emergent effect, one that seems to require a very strong mix of both order and chaos in order to occur.

---

[19] Indeed, in a sense, emergence can be thought of as being kind of like a lever effect, in that it amplifies the impact that even the tiniest of actions and events can potentially have, except that instead of a physical lever it uses causality itself to create the leverage it provides. In fact, the widely known "butterfly effect" is just one specific example of this much broader phenomenon of emergence.

Other things besides life work the same way though. Crystals and gemstones are an example. Crystallization only occurs in places where the balance between heat, cold, pressure, and the chemical properties of the participating molecules are all in just the right balance to cause the associated emergent properties necessary for crystallization to become possible. Once those properties are in the right balance though, crystals and gemstones will begin to spontaneously form over time.

This is just a less complicated form of the same kind of balance between order and chaos that enables biological life to exist. Biology and crystals are both simply examples of systems created by emergence. Different as they may be, they are in a certain sense like siblings. Both are children of emergence. Do you see the common pattern of stunning beauty involved in both of them though? The same will be true if we apply those same forces to our creative work. Spontaneous beauty creation will become inevitable, if done properly. All we need to do is to make sure to instill our actions with the power of emergence and it will happen.

How though does one do such a thing? How could we possibly instill our actions with the power of emergent effects? It sounds like too intimidating and hand-wavy of a task to actually accomplish, just as evolution did before we analyzed it in greater depth. However, here again some careful thought can be very illuminating. It is in fact not really very difficult to apply emergence to the way you do creative work.

The key is to understand what it means (by analogy) to apply chaos or order to your work. Chaos is analogous to things like uninhibited creation, brainstorming, sloppy drafts, and aggressive modification. Order is analogous to things like careful editing, cleanup, deletions, criticism, noise removal, caution, and polish. Creativity is maximized when you drift back and forth between chaos and order in a seamless, diverse, and efficient way. That's a big part of what it means to apply emergence to your work.

Think of your creative work like it is a crystal forming deep within the earth. Only through the proper balance of both chaos and order can the crystal become truly beautiful and continue growing. The same is true for our creative work as human beings. Another analogy I like to make is that applying emergence to your creative work is kind of like watching the clouds and trying to see shapes and objects in them. The clouds provide the chaos, your mind provides the order, and the result is pure beauty and wonderment.

For example, back when I did some digital sculpting work to create some art assets for a game I made during college, I applied this exact same "cloud watching" kind of thinking in order to do so. I needed to create a creature model to be used for both the player-controlled and the enemy-controlled combatants in the game. However, I didn't want to just reuse any pre-existing creatures or mythology or to imitate any other games. I wanted to create something truly unique and original, something unique to this game alone and not derived from any other creative source.

I therefore decided to apply this kind of emergent thinking idea to how I performed my sculpting. I literally did not even have a clue what I wanted the creature to look like when I already started sculpting it. I just started adding on digital clay however my random whims struck me, in a highly uninhibited way, until I had a shape that very vaguely seemed interesting to me. I was thinking in chaos mode.

After I was done with that though, I then switched to thinking in order mode. Instead of just randomly adding stuff in a chaotic and uninhibited way, I now started carefully refining, smoothing, and reshaping things. I cut out the stuff that didn't sit well with the critic inside me. A took that amorphous cloud-like blob of digital clay that the chaos phase had given me, and I trimmed and chiseled it down until it became much more of a clear and well-defined structure, one that left much less to the imagination. The shape became more intentional looking and more thought provoking, but remained generally crude and lacking in fine detail.

The order I imposed upon the creature did make it look better, but it still didn't seem good enough. No amount of additional reductive edits would have got it beyond that point. It would have simply stagnated. It needed more aggressive changes to reach the next level. I therefore switched back to chaos mode and made a bunch more uninhibited and whimsical changes, just doing whatever random things struck me in the moment. That caused the creature sculpture to start to look unrefined and ugly again though. So, naturally, I switched back to order mode and carefully edited the new changes I had made to trim them down to size.

I ping-ponged back and forth between chaos and order like this, over and over again, until at long last I was satisfied with the sculpture of the creature that sat before me. The creature hidden in the cloud had finally emerged. Not even I had any idea that this was what was going to be the end result. I simply allowed my creation to speak to me as I went along. I let it grow naturally, just like a plant in a garden would. The result was something truly unique and original, something I was happy with for my little college game project, something unconnected to any previously existing mythology, something that had evolved naturally instead, a spontaneous creation emerging from the primordial ooze with newborn life.

That's the other part of putting emergence into your creative work. It's not just the ping-ponging back and forth between chaos and order. It's also the part where you listen to what your creation is trying to tell you and let it grow wherever it naturally wants to grow. This is how to create truly original content. Let the power of emergence and evolution simply flow through your hands and your mind as you work, and infinite creative variety will simply spontaneously emerge from the mists for your enjoyment, just as it does in all the bountiful wonders of the natural world, which operates based on these very same two principles of emergence and evolution.

Just as a crystal cannot form without the proper balance of heating and cooling, so too will a creative work struggle to fully emerge without the proper balance of chaos and order. The loosening and tightening of constraints, the push and pull of wild whims against stringent critiques, the heating and cooling of the ground below us, and all the intricacies of life on Earth; these are the rising and falling tides of what it means to be creative, the heartbeat of beauty as it exists in our world. Simply work in harmony with that, with emergence and evolution at your fingertips, and you will never find yourself completely without creative power. Don't worry about depleting these resources either. The infinite canvas that emergence and evolution provide us know no bounds that any finite being could ever reach. Just ignite your creative spirit and enjoy the ride.

7.5 Addendums, reference, and technical

7.5.1 Some existing literature

After finishing writing this book, I became aware of another formal system which has a fairly close resemblance to transformative logic: production rules. Production rules are rules for generating new symbol sequences from old ones, and do so in a way that is similar to how transformative logic does it. The literature on production rules is mainly under the domain of computer science. A set of production rules specifies what is known as a "formal grammar", which is like the production rule analog of transformative logic's notion of a "transformative language".

Production rules are apparently mostly used in things like compiler theory, which is far outside of my own specialization within computer science. That's why I wasn't aware of it when I wrote the book. I wanted to inform you of the existence of this formal system though, both for the sake of giving proper credit to related pre-existing literature even if I wasn't aware of it, and also for the sake of enabling you to refer to the literature to extract any useful and relevant content from it. Naturally, since production rules have a close resemblance to transformative logic, it wouldn't be surprising if some theories from the literature of production rules could be carried over to transformative logic, and vice versa.

However, be that as it may, transformative logic and production rules still appear to have some very distinct and important differences. It is therefore probably not safe to just automatically assume that things that apply to one system will also automatically apply to the other. They are different beasts. Each was motivated to fill a different purpose and each reflects that in its design. Proceed with caution. Besides production rules though, L-systems are also closely related to transformative logic, but we've already discussed that and I was already aware of it when I wrote the book.

7.5.2 Sources / citations / references / bibliography / etc

You may notice that this book lacks a list of references and citations, even though such things are normally common in research. The reason for that is because, except for the basic overview information that I found in online encyclopedias and such (to refresh my memory and to find new broad concepts to consider), and for the included (often commonly known) review material and historical facts and such, most of the content in this book is original research.

The process I used to write this book was to simply scan through online encyclopedias and other basic summary websites in order to get a general idea of what the principles and motivations of things were, and then to just write the corresponding material for my book based on my own thoughts about it. In hindsight, I wish I had kept track of everything more carefully and followed a much stricter citation policy. However, I can't change the past, and I will still be providing an overview of what sources I used here as best I can.

The two sources I used the most, by far, were *Wikipedia* and *The Stanford Encyclopedia of Philosophy*. If I had created a list of references and citations when I first wrote the book, almost all of them would simply point to these two sites. Writing such a list would be highly redundant, and hence doing so felt wasteful and pointless at the time, but really I probably should have done it anyway. Perhaps I should have dug into the citations on some of those pages too, instead of ignoring them in favor of just reading summary info.

Regardless though, all that aside, if you want to get a close approximation of what I used, then simply look up any of the pre-existing terms in this book on the internet and whatever pages come up from these two sites (*Wikipedia* and *The Stanford Encyclopedia of Philosophy*) are probably what I read when I was doing my research. I also used the *Internet Encyclopedia of Philosophy* under the category of *Philosophy of Mathematics* occasionally. The time period during which I conducted such research was mostly roughly from sometime in 2014 to early 2018, and thus the archived versions of the pages from that time period would be the most representative of what I saw.

On the rare occasions where I used other sources, I generally mention that directly in the text. For example, there is a section (on page 124) where I quoted a big piece of text straight from George Boole's own work, with slight modifications for cleanup. I cited him directly within the free-flowing body of my text, in that case. Most of everything else (not including my original research) in the book is basically just common knowledge material that I gathered from summary websites like Wikipedia, or simply recalled from my own memory.

One possible exception might be some of the purely historical trivia that I included in a few parts, which may potentially have not all come from encyclopedias and summary pages, but I can no longer recall what any such sources may have been, if indeed I used them at all. In other words, for historical trivia (such as the history of how logic originally developed) there may be things that I forgot to cite that I really should have cited. If so though, I no longer have any clue how to fix that or to find the original sources, if indeed I used them at all. My memory has faded too much. Sorry if I accidentally missed a few citations.

All this being said though, despite most of this book being original research, it is definitely true that I never would have been able to write this book if the existing literature and research hadn't already laid a massive foundation upon which I would eventually discover the concepts that originally inspired my ideas. For example, the material in my book on the various non-classical logics and how they apply to unified logic could never have existed without all the corresponding literature also existing. Seeing those concepts summarized in online encyclopedias was absolutely critical to causing me to have the idea to write this book. I owe an immense and insurmountable debt of gratitude to anyone and everyone who has ever contributed to that literature, or to the literatures of any other subject I have ever discussed in this book for that matter.

It really is true what they say about standing on the shoulders of giants. None of this would be possible without that. There are probably hundreds upon hundreds of logicians, mathematicians, and programmers, both named and nameless, who I owe a heartfelt thank you to for their work in the literature. This work has created the amazing world we all now live in today and has made so many wonderful things possible for us humans in the modern age. It has made dreams into reality. So, thank you very much for that. The world we all live in is now a much better place for it. It is not just all researchers, but all of humanity itself, that owes this accumulated volume of wisdom a boundless debt of gratitude.

Oh, and on a more miscellaneous side note, the fancy gold and black cover of my book is based on a royalty-free vector-art background image that I bought non-exclusive rights to via a stock image website. After buying usage rights for it, I edited it myself to suit my needs. It has a nicely simple and elegant look. Technical books don't really need to get too crazy with the covers they use. Generic covers are often sufficient in such cases. If I was writing a novel though, I'd never put together the cover myself. Novels need custom professional artwork usually.

Anyway though, the image I used for the cover is available on Shutterstock and Colourbox (and maybe other places too). The artist who created the image goes by the name "Megin" on Shutterstock and Colourbox. I'm probably not required to credit her, but felt like doing it anyway. Feel free to look at her images and maybe buy some for your own covers (or other uses) if you feel like it. I always believe in giving credit where credit is due and supporting good creative work.

7.5.3 Some recommendations for good math resources

For those of you who still want more deep insights into mathematics: I would recommend that you try reading a website called *Math Better Explained* by Kalid Azad. It is one of my favorite mathematics websites. There are also some related books that have been published by Kalid Azad that are well worth reading. Kalid Azad has a special talent for explaining concepts in insightful and illuminating ways. His material has an unusually high degree of clarity compared to most sources on mathematics. In fact, the clarity of his own explanations of mathematical concepts was part of the inspiration for the clarity-oriented focus of this book. I mean, I already have always emphasized clarity and have always been passionate about it, but reading some of Kalid Azad's work helped to inspire me to do some writing of my own.

For those of you who want to review their understanding of standard mathematical material, or to learn it for the first time: Khan Academy is an extremely effective way to do so. It is an online free video website which hosts some of the best videos on learning mathematics (and other subjects) that currently exist on the internet. I have used them many times to review material that has faded from my memory. They were also a very valuable resource for times in college when some of my math professors were not doing an effective job of teaching the material clearly. Khan Academy is great for filling those kinds of gaps. It is truly a testament to the underappreciated value and immense untapped potential of automated education resources.

7.5.4 LaTeX typesetting code (packages and commands)

This book was typeset using an advanced typesetting program called LaTeX, which is the most commonly used system for writing documents about mathematics and is very heavily used in the world of science. The specific program I used to compile my LaTeX source code was TeXstudio (version 2.12.8), via MiKTeX (version 2.9). LaTeX is not a "what you see is what you get" approach to document editing. It works by compiling source code, just like compiling a computer program from source code, except for that it produces documents instead of executable code. This makes it powerful.

However, imitating someone else's document and symbols without access to their LaTeX code can be difficult and tedious. Therefore, as promised, I am providing most of the preamble of my source code file for this book here, so that you can see what packages I imported and what commands I defined, so that you can use them in your own work if ever you want to create documents that need to use the same symbols and formatting that I used. I'm not an expert LaTeX coder though, so I just did whatever was needed to get the job done. My code may have flaws in it or could be designed better. Not all of the commands and packages that I specified were necessarily actually used. Anyway though, here's the code:

```
\documentclass[12pt]{memoir}

\usepackage{
    verbatim,
    stix,
    amsmath,
    amssymb,
    amsthm,
    booktabs,
    longtable,
    siunitx,
    float,
    wrapfig,
    subcaption,
    placeins,
    accents,
    graphicx,
    nicefrac,
```

```
    tikz,
    dictsym,
    mathtools,
    fontawesome,
    enumitem,
    centernot,
    cancel,
    calculator,
    multirow,
    multicol,
    listings,
    etoolbox,
}

%Conflict: wasysym and bbding both define \Square (etc)
%Conflict Resolution:
\usepackage{savesym}
\usepackage{bbding}
\savesymbol{Square}
\savesymbol{sqsubset}
\savesymbol{sqsupset}
\usepackage{wasysym}
%Each use of \savesymbol{SymbolName} here renames the existing
    version of \SymbolName to \origSymbolName. This was done so
    that once wasysym is imported it will not generate name
    conflicts with the existing packages.

\usepackage[type1]{allrunes}
%The type1 option prevents rune pixelation.
%It forces allrunes to use vectors instead of bitmaps.



% == Command Definitions ==

\newtheorem{FormalDef}{Definition}

%Use \centernot{...} or \cancel{...} to draw a slash through
    something to indicate negation. The two commands behave a bit
    differently.

% -- Logic Ops --

%\lnot for NOT
%\lor for OR
%\land for AND
```

```latex
\newcommand{\Nor}{\mathbin{\downarrow}}
\newcommand{\Nand}{\mathbin{\uparrow}}

\newcommand{\SmallX}{\scalebox{0.5}{$\times$}}
\newcommand{\Xor}{\mathbin{%
    \begin{tikzpicture}[baseline=-0.5ex]%
    \node[inner sep=0pt] at (0ex,0ex) {$\lor$};%
    \node[inner sep=0pt] at (-0.025ex,0.5ex) {$\SmallX$};%
    \end{tikzpicture}%
}}

% -- Quantifiers --

\newcommand{\ThereExists}{\exists}
\newcommand{\ForAll}{\forall}

% -- Material Implication --
\newcommand{\MatImplR}{\rightarrow}
\newcommand{\MatImplL}{\leftarrow}
\newcommand{\MatEquiv}{\leftrightarrow}

\newcommand{\NotMatImplR}{\nrightarrow}
\newcommand{\NotMatImplL}{\nleftarrow}
\newcommand{\NotMatEquiv}{\nleftrightarrow}

% -- Formal Implication --
\newcommand{\FormalImplR}{\Rightarrow}
\newcommand{\FormalImplL}{\Leftarrow}
\newcommand{\FormalEquiv}{\Leftrightarrow}

\newcommand{\NotFormalImplR}{\nRightarrow}
\newcommand{\NotFormalImplL}{\nLeftarrow}
\newcommand{\NotFormalEquiv}{\nLeftrightarrow}

% -- Syntactic and Semantic Consequence --
\newcommand{\SynConseqR}{\vdash}
\newcommand{\SynConseqL}{\dashv}

\newcommand{\SemConseqR}{\vDash}
\newcommand{\SemConseqL}{\Dashv}

% -- Transformative Implication --
\newcommand{\TransImplR}{\mathbin{\rightarrowtriangle}}
\newcommand{\TransImplL}{\mathbin{\leftarrowtriangle}}
\newcommand{\TransEquiv}{\mathbin{\leftrightarrowtriangle}}

\newcommand{\STransImplR}[1]{\mathbin{\TransImplR\!_#1}}
```

```latex
\newcommand{\STransImplL}[1]{\mathbin{{_#1}\!\TransImplL}}

\newcommand{\NotTransImplR}{\centernot{\TransImplR}}
\newcommand{\NotTransImplL}{\centernot{\TransImplL}}
\newcommand{\NotTransEquiv}{\centernot{\TransEquiv}}

% -- Interpretation Injectors --

\newcommand{\InterpInject}[2]{\text{\textbf{#1}}(#2)}

\newcommand{\InterpSet}[1]{\InterpInject{Set}{#1}}
\newcommand{\InterpClassical}[1]{\InterpInject{Cla}{#1}}
\newcommand{\InterpArithmetic}[1]{\InterpInject{Arith}{#1}}
\newcommand{\InterpEquation}[1]{\InterpInject{Eq}{#1}}

\newcommand{\InterpRegex}[1]{\InterpInject{Regex}{\texttt{#1}}}
\newcommand{\ProgVarRegex}{[a-zA-Z0-9\_]+}
\newcommand{\ProgTextRegex}{[a-zA-Z0-9\_\textbackslash{}s]*}

% -- Transcience Delimiters --

\newcommand{\NarrowEmptyHalfCircleL}{\scalebox{0.8}[1.0]
    {$\Leftcircle$}}
\newcommand{\NarrowEmptyHalfCircleR}{\scalebox{0.8}[1.0]
    {$\Rightcircle$}}
\newcommand{\NarrowEmptyTriangleL}{\scalebox{0.6}[1.0]{$\lhd$}}
\newcommand{\NarrowEmptyTriangleR}{\scalebox{0.6}[1.0]{$\rhd$}}

\newcommand{\NarrowFilledHalfCircleL}{\scalebox{0.8}[1.0]
    {$\LEFTCIRCLE$}}
\newcommand{\NarrowFilledHalfCircleR}{\scalebox{0.8}[1.0]
    {$\RIGHTCIRCLE$}}
\newcommand{\NarrowFilledTriangleL}{\scalebox{0.6}[1.0]{$\LHD$}}
\newcommand{\NarrowFilledTriangleR}{\scalebox{0.6}[1.0]{$\RHD$}}

\newcommand{\dotvert}{\rotatebox[origin=c]{90}{$\dotminus$}}
\newcommand{\vertdot}{\rotatebox[origin=c]{-90}{$\dotminus$}}

\newcommand{\TransFreeL}{\NarrowEmptyHalfCircleL}
\newcommand{\TransFreeR}{\NarrowEmptyHalfCircleR}
\newcommand{\TransFree}[1]{\TransFreeL{}#1\TransFreeR{}}

\newcommand{\TransFixedL}{\NarrowEmptyTriangleL\:}
\newcommand{\TransFixedR}{\:\NarrowEmptyTriangleR}
\newcommand{\TransFixed}[1]{\TransFixedL{}#1\TransFixedR{}}

\newcommand{\TransClosedL}{\dotvert}
```

```latex
\newcommand{\TransClosedR}{\vertdot}
\newcommand{\TransClosed}[1]{\TransClosedL{}#1\TransClosedR{}}

\newcommand{\TransEqualL}{\NarrowFilledTriangleL\:}
\newcommand{\TransEqualR}{\:\NarrowFilledTriangleR}
\newcommand{\TransEqual}[1]{\TransEqualL{}#1\TransEqualR{}}

\newcommand{\TransOpenL}{\NarrowFilledHalfCircleL}
\newcommand{\TransOpenR}{\NarrowFilledHalfCircleR}
\newcommand{\TransOpen}[1]{\TransOpenL{}#1\TransOpenR{}}

% -- Reverse Transcience Delimiters --

\newcommand{\ReverseTransFreeL}{\NarrowEmptyHalfCircleR
    \hspace{2pt}}
\newcommand{\ReverseTransFreeR}{\hspace{2pt}
    \NarrowEmptyHalfCircleL}
\newcommand{\ReverseTransFree}[1]{\ReverseTransFreeL{}#1
    \ReverseTransFreeR{}}

\newcommand{\ReverseTransFixedL}{\NarrowEmptyTriangleR
    \hspace{1pt}}
\newcommand{\ReverseTransFixedR}{\hspace{1pt}
    \NarrowEmptyTriangleL}
\newcommand{\ReverseTransFixed}[1]{\ReverseTransFixedL{}#1
    \ReverseTransFixedR{}}

\newcommand{\ReverseTransClosedL}{\vertdot\hspace{2.5pt}}
\newcommand{\ReverseTransClosedR}{\hspace{2.5pt}\dotvert}
\newcommand{\ReverseTransClosed}[1]{\ReverseTransClosedL{}#1
    \ReverseTransClosedR{}}

\newcommand{\ReverseTransEqualL}{\NarrowFilledTriangleR
    \hspace{1pt}}
\newcommand{\ReverseTransEqualR}{\hspace{1pt}
    \NarrowFilledTriangleL}
\newcommand{\ReverseTransEqual}[1]{\ReverseTransEqualL{}#1
    \ReverseTransEqualR{}}

\newcommand{\ReverseTransOpenL}{\NarrowFilledHalfCircleR
    \hspace{2pt}}
\newcommand{\ReverseTransOpenR}{\hspace{2pt}
    \NarrowFilledHalfCircleL}
\newcommand{\ReverseTransOpen}[1]{\ReverseTransOpenL{}#1
    \ReverseTransOpenR{}}

% -- Transformative Misc. --
```

```
%\newcommand{\JapaneseQuoteL}{\lceil}
%\newcommand{\JapaneseQuoteR}{\rfloor}
\newcommand{\JapaneseQuoteL}{\mathopen{\ulcorner}}
\newcommand{\JapaneseQuoteR}{\mathclose{\lrcorner}}
\newcommand{\JapaneseQuotes}[1]{\JapaneseQuoteL#1\JapaneseQuoteR}

%\newcommand{\EmptyForm}{\scalebox{1.5}[1.0]{\textvisiblespace}}
\newcommand{\EmptyForm}{\JapaneseQuotes{}}
\newcommand{\UniversalForm}{\mathord{\wasylozenge}}


\newcommand{\RootForm}{\mathord{\plustrif}}

\newcommand{\ChestOfDrawers}{
    \begin{tikzpicture}[xscale=0.25, yscale=0.25, baseline=0.1ex]
        \coordinate (LowerLeftCorner) at (0,0);
        \coordinate (LowerRightCorner) at (1,0);
        \coordinate (UpperRightCorner) at (1,1);
        \coordinate (UpperLeftCorner) at (0,1);

        \coordinate (LeftMid) at (0, 0.5);
        \coordinate (RightMid) at (1, 0.5);

        \coordinate (LowerDrawerCenter) at (0.5, 0.25);
        \coordinate (UpperDrawerCenter) at (0.5, 0.75);

        \draw (LowerLeftCorner) -- (LowerRightCorner) --
            (UpperRightCorner) -- (UpperLeftCorner) -- cycle;
        \draw (LeftMid) -- (RightMid);
        \draw[fill] (LowerDrawerCenter) circle[radius=0.05];
        \draw[fill] (UpperDrawerCenter) circle[radius=0.05];
    \end{tikzpicture}
}
\newcommand{\NonEmptyForm}{\mathord{\makebox[1.5ex]{
    \ChestOfDrawers}}}

\newcommand{\SansSerifCapitalF}{\textnormal{\textsf{F}}}
\newcommand{\Finity}{\raisebox{-0.275ex}{\rotatebox[origin=c]{-90}
    {$\SansSerifCapitalF$}}}
\newcommand{\SubscriptFinity}{\scalebox{0.6}{\Finity}}
\newcommand{\Infinity}{\infty}

\newcommand{\MetaImplRSymbol}{
    \begin{tikzpicture}[scale=0.8, baseline=-0.6ex]
        \coordinate (Center) at (0ex, 0ex);

        \coordinate (RTop) at (2ex, 1ex);
```

```
            \coordinate (RMid) at (2ex, 0ex);
            \coordinate (RBottom) at (2ex, -1ex);

            \draw[thick, ->] (Center) to[out={0.0*360}, in={0.625*360}]
                (RTop);
            \draw[thick, ->] (Center) to[out={0.0*360}, in={0.375*360}]
                (RBottom);
        \end{tikzpicture}
}

\newcommand{\MetaImplLSymbol}{
    \begin{tikzpicture}[scale=0.8, baseline=-0.6ex]
        \coordinate (Center) at (0ex, 0ex);

        \coordinate (LTop) at (-2ex, 1ex);
        \coordinate (LMid) at (-2ex, 0ex);
        \coordinate (LBottom) at (-2ex, -1ex);

        \draw[thick, ->] (Center) to[out={0.5*360}, in={0.875*360}]
            (LTop);
        \draw[thick, ->] (Center) to[out={0.5*360}, in={0.125*360}]
            (LBottom);
    \end{tikzpicture}
}

\newcommand{\MetaEquivSymbol}{
    \begin{tikzpicture}[scale=0.8, baseline=-0.6ex]
        \coordinate (Center) at (0ex, 0ex);

        \coordinate (RTop) at (2ex, 1ex);
        \coordinate (RMid) at (2ex, 0ex);
        \coordinate (RBottom) at (2ex, -1ex);

        \draw[thick, ->] (Center) to[out={0.0*360}, in={0.625*360}]
            (RTop);
        \draw[thick, ->] (Center) to[out={0.0*360}, in={0.375*360}]
            (RBottom);

        \coordinate (LTop) at (-2ex, 1ex);
        \coordinate (LMid) at (-2ex, 0ex);
        \coordinate (LBottom) at (-2ex, -1ex);

        \draw[thick, ->] (Center) to[out={0.5*360}, in={0.875*360}]
            (LTop);
        \draw[thick, ->] (Center) to[out={0.5*360}, in={0.125*360}]
            (LBottom);
    \end{tikzpicture}
```

```
}

\newcommand{\MetaImplR}{\mathbin{\MetaImplRSymbol\,}}
\newcommand{\MetaImplL}{\mathbin{\,\MetaImplLSymbol}}
\newcommand{\MetaEquiv}{\mathbin{\MetaEquivSymbol}}

\newcommand{\IdentitySet}[1]{\InterpInject{id}{#1}}

\newcommand{\Validate}[1]{\InterpInject{val}{#1}}

\newcommand{\FormBoundaryL}{\rvzigzag}
\newcommand{\FormBoundaryR}{\lvzigzag}
\newcommand{\FormBoundary}[1]{\FormBoundaryL{}#1\FormBoundaryR{}}

\newcommand{\InclusionDirective}{\oplusrhrim\;}

\newcommand{\AdjustedFivePointStar}{\raisebox{-0.2ex}{
    \scalebox{0.7}{\mbox{\FiveStarLines}}}}
\newcommand{\CrossHair}{
    \begin{tikzpicture}[xscale=0.25, yscale=0.25, baseline=0.1ex]
        \coordinate (Center) at (0.5, 0.5);

        \coordinate (LeftCenter) at (0.0, 0.5);
        \coordinate (RightCenter) at (1.0, 0.5);
        \coordinate (DownCenter) at (0.5, 0.0);
        \coordinate (UpCenter) at (0.5, 1.0);

        \draw (Center) circle[radius=0.35];

        \draw[thick] (Center) ++ (-0.15,0.0) -- (LeftCenter);
        \draw[thick] (Center) ++ (+0.15,0.0) -- (RightCenter);
        \draw[thick] (Center) ++ (0.0,-0.15) -- (DownCenter);
        \draw[thick] (Center) ++ (0.0,+0.15) -- (UpCenter);
    \end{tikzpicture}
}

\newcommand{\AssumptionMarkSymbol}{\AdjustedFivePointStar}
\newcommand{\InferenceMarkSymbol}{\therefore}
\newcommand{\TargetMarkSymbol}{\CrossHair}

\newcommand{\AssumptionMark}{\AssumptionMarkSymbol\;\;}
\newcommand{\InferenceMark}{\InferenceMarkSymbol\;\;}
\newcommand{\TargetMark}{\TargetMarkSymbol\;\;}

% -- Set Operations Etc --

\newcommand{\SetL}{\{}
```

```latex
\newcommand{\SetR}{\}}
\newcommand{\Set}[1]{\SetL #1 \SetR}
\newcommand{\SetAutoSize}[1]{\left\SetL #1 \right\SetR}

\newcommand{\SetEmpty}{\varnothing}
\newcommand{\SetEmptyOpaque}{\widehat{\SetEmpty}}
\newcommand{\SetEmptyTransparent}{\SetEmpty}
\newcommand{\SetUniversal}{\mathscr{U}}

\newcommand{\SetNot}{\lnot}
\newcommand{\SetOr}{\mathbin{\cup}}
\newcommand{\SetAnd}{\mathbin{\cap}}
\newcommand{\SetDiff}{-}
\newcommand{\SetXor}{\mathbin{%
    \begin{tikzpicture}[baseline=-0.5ex]%
    \node[inner sep=0pt] at (0ex,0ex) {$\SetOr$};%
    \node[inner sep=0pt] at (-0.025ex,0.3ex) {$\SmallX$};%
    \end{tikzpicture}%
}}
\newcommand{\SetNor}{\mathbin{\forksnot}}
\newcommand{\SetNand}{\mathbin{\rotatebox[origin=c]{180}
    {$\forksnot$}}}
\newcommand{\SetCartesian}{\mathbin{\times}}
\newcommand{\SetBinaryPairing}{\mathbin{\resizebox{1.5ex}{1.5ex}
    {$\between$}}}
\newcommand{\SetConcat}{\mathbin{\dualmap}}
\newcommand{\SetPowerSet}{\mathrm{PowerSet}}

\newcommand{\SetElemOf}{\in}

\newcommand{\SetSub}{\subseteq}
\newcommand{\SetSuper}{\supseteq}
\newcommand{\SetProperSub}{\subset}
\newcommand{\SetProperSuper}{\supset}

\newcommand{\SetNotSub}{\nsubseteq}
\newcommand{\SetNotSuper}{\nsupseteq}
\newcommand{\SetNotProperSub}{\mathrel{\centernot{\subset}}}
\newcommand{\SetNotProperSuper}{\mathrel{\centernot{\supset}}}

\newcommand{\OpaqueSetL}{\lBrace}
\newcommand{\OpaqueSetR}{\rBrace}
\newcommand{\OpaqueSet}[1]{\OpaqueSetL #1 \OpaqueSetR}
\newcommand{\OpaqueSetAutoSize}[1]{\left\OpaqueL #1 \right\OpaqueR}

\newcommand{\TransparentSetL}{\SetL}
\newcommand{\TransparentSetR}{\SetR}
```

```latex
\newcommand{\TransparentSet}[1]{\TransparentSetL #1
    \TransparentSetR}
\newcommand{\TransparentSetAutoSize}[1]{\left\TransparentSetL #1
    \right\TransparentSetR}

% -- Existential Truth/Falsehood Operators --

\newcommand{\SetExis}{\ThereExists}
\newcommand{\SetNotExis}{\cancel{\SetExis}}

\newcommand{\SetExisSub}{\mathrel{\sqsubseteq}}
\newcommand{\SetExisSuper}{\mathrel{\sqsupseteq}}
\newcommand{\SetExisProperSub}{\mathrel{\origsqsubset}}
\newcommand{\SetExisProperSuper}{\mathrel{\origsqsupset}}

\newcommand{\SetExisEquiv}{\mathrel{\gleichstark}}

\newcommand{\EmptyDiamond}{\raisebox{0.25ex}{\scalebox{0.8}
    {$\mdlgwhtdiamond$}}}
\newcommand{\RHalfFilledDiamond}{\raisebox{0.25ex}{\scalebox{0.8}
    {$\diamondrightblack$}}}
\newcommand{\LHalfFilledDiamond}{\raisebox{0.25ex}{\scalebox{0.8}
    {$\diamondleftblack$}}}
\newcommand{\FilledDiamond}{\raisebox{0.25ex}{\scalebox{0.8}
    {$\mdlgblkdiamond$}}}

\newcommand{\SetPossibly}{\mathrel{\EmptyDiamond}}
\newcommand{\SetPossiblyLNSR}{\mathrel{\LHalfFilledDiamond}}
    %-LNSR: Left Not Subset of Right
\newcommand{\SetPossiblyRNSL}{\mathrel{\RHalfFilledDiamond}}
    %-LNSR: Right Not Subset of Left
\newcommand{\SetPossiblyNS}{\mathrel{\FilledDiamond}} %-NS: Neither
    is a Subset of the other

\newcommand{\SetExisAnd}{\mathrel{\sqcap}} %conceptual synonym for
    "\SetPossibly", helps show relation
\newcommand{\SetExisOr}{\mathrel{\sqcup}}

\newcommand{\SetMutex}{\mathrel{\raisebox{0.25ex}{\scalebox{0.8}
    {$\boxtimes$}}}}
\newcommand{\SetMutexAltA}{\mathrel{\cancel{\SetExisAnd}}}
\newcommand{\SetMutexAltB}{\mathrel{\cancel{\SetPossibly}}} %a.k.a.
    "impossibility"
\newcommand{\SetNotPossibly}{\SetMutexAltB}
\newcommand{\SetNotExisAnd}{\SetMutexAltA}

\newcommand{\SetMunex}{\mathrel{\raisebox{0.25ex}{\scalebox{0.8}
```

```latex
    {$\boxplus$}}}}
\newcommand{\SetMunexAlt}{\mathrel{\cancel{\SetExisOr}}}
\newcommand{\SetNotExisOr}{\SetMunexAlt}

\newcommand{\SetSurvival}{\mathrel{\raisebox{0.25ex}{\scalebox{0.8}
    {$\boxminus$}}}}

\newcommand{\SetExisXor}{\mathbin{
    \begin{tikzpicture}[baseline=-0.5ex]
    \coordinate (CupTopLeft) at (-0.5ex, 0.7ex);
    \coordinate (CupTopRight) at (0.5ex, 0.7ex);

    \coordinate (CupBaseLeft) at (-0.5ex, -0.5ex);
    \coordinate (CupBaseRight) at (0.5ex, -0.5ex);

    \draw[thick] (CupTopLeft) -- (CupBaseLeft) -- (CupBaseRight) --
        (CupTopRight);

    \coordinate (XCenter) at (0ex,0.3ex);
    \node[inner sep=0pt] at (XCenter) {$\SmallX$};
    \end{tikzpicture}
}}

\newcommand{\SetExisNor}{\mathbin{
    \begin{tikzpicture}[baseline=-0.5ex]
    \coordinate (NorTopLeft) at (-0.5ex, 0.3ex);
    \coordinate (NorTopMid) at (0ex, 0.9ex);
    \coordinate (NorTopRight) at (0.5ex, 0.3ex);

    \coordinate (NorBaseLeft) at (-0.5ex, -0.4ex);
    \coordinate (NorBaseMid) at (0ex, -0.4ex);
    \coordinate (NorBaseRight) at (0.5ex, -0.4ex);

    \draw[thick] (NorTopLeft) -- (NorBaseLeft) -- (NorBaseRight) --
        (NorTopRight);
    \draw[thick] (NorBaseMid) -- (NorTopMid);
    \end{tikzpicture}
}}

\newcommand{\SetExisNand}{\mathbin{\rotatebox[origin=c]{180}
    {$\SetExisNor$}}}

\newcommand{\SetNotExisSub}{\mathbin{\cancel{\SetExisSub}}}
\newcommand{\SetNotExisSuper}{\mathbin{\cancel{\SetExisSuper}}}
\newcommand{\SetNotExisProperSub}{\mathbin{\cancel
    {\SetExisProperSub}}}
\newcommand{\SetNotExisProperSuper}{\mathbin{\cancel
```

```latex
        {\SetExisProperSuper}}}

\newcommand{\SetNotExisEquiv}{\mathbin{\centernot{\SetExisEquiv}}}

% -- Unified Implication --

%(sub implication)
\newcommand{\SetSubImpl}{\mathbin{\rightharpoonup}}
\newcommand{\SetSubImplR}{\mathbin{\barrightharpoonup}}
\newcommand{\SetSubImplL}{\mathbin{\leftharpoondownbar}}

%(super implication)
\newcommand{\SetSuperImpl}{\mathbin{\leftharpoondown}}
\newcommand{\SetSuperImplR}{\mathbin{\rightharpoonupbar}}
\newcommand{\SetSuperImplL}{\mathbin{\barleftharpoondown}}

%(unified equivalence)
\newcommand{\SetUnifiedEquiv}{\mathbin{\leftrightharpoondownup}}
\newcommand{\SetUnifiedEquivAlt}{\mathbin{\rightleftharpoons}}

% -- Over/Under Brackets --

\newcommand{\ThinOverBracket}[1]{\overbracket[0.5pt][2pt]{#1}}
\newcommand{\ThinUnderBracket}[1]{\underbracket[0.5pt][2pt]{#1}}

% -- Interval & Range Notations --

\newcommand{\IntRange}[2]{#1{..}#2}

% -- Relational Operations Etc --

\newcommand{\RelPlaceAtSymbol}{@}
\newcommand{\RelPlaceAt}[2]{\RelPlaceAtSymbol{}(#1, #2)}


\newcommand{\RelExplicitExtractorSymbol}{\seovnearrow}
\newcommand{\RelExplicitExtractor}[1]
    {\RelExplicitExtractorSymbol(#1)}
\newcommand{\RelInsertionMark}[1]{\ThinOverBracket{#1}}

%\newcommand{\RelImplicitExtractor}{\resizebox{2ex}{0.8ex}
    {$\linefeed$}}
\newcommand{\RelImplicitExtractor}[1]{\ThinUnderBracket{#1}}

\newcommand{\RelRenameSymbol}{\vbrtri}
\newcommand{\RelRename}[1]{\RelRenameSymbol(#1)}
\newcommand{\RelRenameToSymbol}{\curvearrowright}
```

```latex
\newcommand{\RelRenameTo}{\RelRenameToSymbol}

\newcommand{\RelRecursion}{\scalebox{1}{$\circlearrowright$}}

\newcommand{\FormLengthSet}[1]{\mathrm{FoLenSet}(#1)}
\newcommand{\FormLengthFilter}[2]{\mathrm{FoLenFilter}(#1, #2)}

% -- Blueprint Expressions --

\newcommand{\BlueprintL}{\lBrack}
\newcommand{\BlueprintR}{\rBrack}
\newcommand{\BlueprintExpr}[1]{\BlueprintL #1 \BlueprintR}
\newcommand{\BlueprintExprAutoSize}[1]{\left\BlueprintL #1
    \right\BlueprintR}

% -- Set Size Operators --

\newcommand{\CountOf}[1]{\mathrm{Count}(#1)}

\newcommand{\SubsizeOf}[1]{\mathrm{Subsize}(#1)}
\newcommand{\Incomparable}{\mathrel{\glj}}

\newcommand{\MeasureOf}{\mathrm{Measure}}

\newcommand{\TerminacyCount}[1]{\mathrm{Ter}(#1)}

\newcommand{\DirAdjCount}[1]{\mathrm{DirAdjCount}(#1)}
\newcommand{\DirAdjSet}[1]{\mathrm{DirAdjSet}(#1)}

\newcommand{\BiAdjCount}[1]{\mathrm{BiAdjCount}(#1)}
\newcommand{\BiAdjSet}[1]{\mathrm{BiAdjSet}(#1)}

\newcommand{\AdjCount}[1]{\mathrm{AdjCount}(#1)}
\newcommand{\AdjSet}[1]{\mathrm{AdjSet}(#1)}

\newcommand{\DirAdjVar}{d\!j}
\newcommand{\BiAdjVar}{b\!j}
\newcommand{\AdjVar}{j}

% -- Common Sets (Old Names) --

\newcommand{\SetTradNat}{\mathbb{N}}
\newcommand{\SetTradInt}{\mathbb{Z}}
\newcommand{\SetTradRat}{\mathbb{Q}}
\newcommand{\SetTradReal}{\mathbb{R}}
\newcommand{\SetTradComp}{\mathbb{C}}
```

```latex
% -- Common Sets (New Names) --

\newcommand{\SetNumbers}{\mathtt{Num}}
\newcommand{\SetCounting}{\mathtt{Coun}}
\newcommand{\SetIntegers}{\mathtt{Int}}
\newcommand{\SetContinuum}{\mathtt{Cont}}
\newcommand{\SetFractional}{\mathtt{Frac}}
\newcommand{\SetOrientational}{\mathtt{Ori}}
\newcommand{\SetDefinite}{\mathtt{Def}}
\newcommand{\SetIndefinite}{\mathtt{Ind}}

\newcommand{\SetConvergent}{\mathtt{Conv}}
\newcommand{\SetDivergent}{\mathtt{Div}}
\newcommand{\SetRepeating}{\mathtt{Rep}}
\newcommand{\SetPositive}{\mathtt{Pos}}
\newcommand{\SetNegative}{\mathtt{Neg}}
\newcommand{\SetEven}{\mathtt{Even}}
\newcommand{\SetOdd}{\mathtt{Odd}}
\newcommand{\SetPrime}{\mathtt{Pri}}

% -- Statue Scenario --

\newcommand{\SkullVectorArt}{\raisebox{-0.5ex}{\includegraphics
    [width=2.3ex, height=2.3ex]{skull_cross_publicdomain.pdf}}}

\newcommand{\Statue}{\raisebox{-0.20ex}{\scalebox{1.1}
    {\dsjuridical}}}
\newcommand{\StatueEmptyHand}{\raisebox{-0.20ex}{\scalebox{1.0}
    {\dstechnical}}}
\newcommand{\StatueBook}{\raisebox{-0.20ex}{\scalebox{1.0}
    {\dsliterary}}}
\newcommand{\StatueSword}{\raisebox{-0.20ex}{\scalebox{1.0}
    {\dsmilitary}}}
\newcommand{\StatueShield}{\raisebox{-0.10ex}{\scalebox{1.0}
    {\dsheraldical}}}
\newcommand{\StatueFailure}{\SkullVectorArt{}}
\newcommand{\StatueSuccess}{\raisebox{-0.20ex}{\scalebox{1.0}
    {\dscommercial}}}

\newcommand{\StatueNotSword}{%
    \begin{tikzpicture}[baseline=-0.5ex]%
    \node[inner sep=0pt] at (0ex,0ex) {$\dottedcircle$};%
    \node[inner sep=0pt] at (0ex,0.03ex)
        {$\scalebox{0.65}{\StatueSword}$};%
    \end{tikzpicture}%
}
```

```
\newcommand{\StatueNotShield}{%
    \begin{tikzpicture}[baseline=-0.5ex]%
    \node[inner sep=0pt] at (0ex,0ex) {$\dottedcircle$};%
    \node[inner sep=0pt] at (0ex,-0.05ex)
        {$\scalebox{0.7}{\StatueShield}$};%
    \end{tikzpicture}%
}

\newcommand{\StatueAnything}{%
    \begin{tikzpicture}[baseline=-0.5ex]%
    \node[inner sep=0pt] at (0ex,0ex) {$\dottedcircle$};%
    \end{tikzpicture}%
}

% -- Rune Scenario --

% Some random runes listed in no particular order and with no
    regard for meaning:
\newcommand{\RuneA}{\textara{I}}
\newcommand{\RuneB}{\textara{p}}
\newcommand{\RuneC}{\textara{\ea{}}}
\newcommand{\RuneD}{\textara{\rex{}}}
\newcommand{\RuneE}{\textara{\oe{}}}
\newcommand{\RuneF}{\textara{\ng{}}}
\newcommand{\RuneG}{\textara{\stan{}}}
\newcommand{\RuneH}{\textara{q}}
\newcommand{\RuneI}{\textara{k}}
\newcommand{\RuneJ}{\textara{x}}
\newcommand{\RuneX}{\textara{g}}

% -- Transformation Path Tracing Lines --

%Use \TightUnderline instead of \underline if boxing creates an
    abnormal gap when underlining something:
\newcommand{\TightUnderline}[1]{\underline{\smash{#1}}}

%Some additional miscellaneous lining commands I needed for part of
    the book:
%(due to the fact that underlines and overlines can't be improperly
    nested)
\newcommand{\TightOverline}[1]{\overline{#1}} %Smashed version of
    this seems to have wrong effect.
\newcommand{\TightUnderOverLine}[1]{\underline{\smash{
    \overline{#1}}}}

% -- Better (Non-Surprising) Names For The "Alignment" Table
    Environments --
```

```latex
\newenvironment{AlignColTrioWithExtraCenterSpacing}
    {\begin{eqnarray}}{\end{eqnarray}}
\newenvironment{AlignColTrioWithExtraCenterSpacing*}
    {\begin{eqnarray*}}{\end{eqnarray*}}

\newenvironment{AlignColPairs}{\align}{\endalign}
\newenvironment{AlignColPairs*}{\csname align*\endcsname}{\csname
    endalign*\endcsname}

\newenvironment{AlignColPairsFlush}{\flalign}{\endflalign}
\newenvironment{AlignColPairsFlush*}{\csname
    flalign*\endcsname}{\csname endflalign*\endcsname}

% -- Miscellanous --

\newcommand{\EmptySquare}{\raisebox{0.25ex}{\scalebox{0.8}
    {$\square{}$}}}

\newcommand{\GenericOpSquare}{\mathbin{\EmptySquare}}
\newcommand{\GenericOpMulA}{\mathbin{\circledcirc}}
\newcommand{\GenericOpMulB}{\mathbin{\circledast}}
\newcommand{\GenericOpS}{\mathbin{\invlazys}}

\newcommand{\WFF}{\mathscr{w}}

\newcommand{\cliche}{clich\'{e}}
\newcommand{\cliches}{\cliche{}s}

\newcommand{\ModalNecessity}{\EmptySquare}
\newcommand{\ModalPossibility}{\EmptyDiamond}

\newcommand{\Radians}{\:\,\mathrm{rad}}
\newcommand{\Degrees}{^{\circ}}
\newcommand{\Cycles}{\:\,\mathrm{cyc}}

\newcommand{\CStyleComment}[1]{\texttt{/*}\;#1\;\texttt{*/}}

\newcommand{\Luk}{\L{}ukasiewicz}

\newcommand{\Min}[1]{\mathrm{min}(#1)}
\newcommand{\Max}[1]{\mathrm{max}(#1)}

\newcommand{\Bezier}{B\'{e}zier}

\newcommand{\Vector}[1]{\langle #1 \rangle}
\newcommand{\VectorAutoSize}[1]{\left\langle #1 \right\rangle}
```

```
\newcommand{\ForceBlankPage}{
    \newpage
    \thispagestyle{empty}
    \mbox{}
    \newpage
}

\newcommand{\ForceSinglePage}[2]{
    \newpage
    \thispagestyle{plain}
    \begin{minipage}{#2\textwidth}
        #1
    \end{minipage}
    \newpage
}
%\ForceSinglePage is used for the copyright page.
%The text given to it must be manually formatted though.
%Use the second parameter to constrain how much of the page
%the text is allowed to span, e.g. 0.5 for 50% of the page.

% == End Of Command Definitions ==



% == Options/Setup ==

\captionsetup{justification=centering}

\usepackage[yyyymmdd]{datetime}
\renewcommand{\dateseparator}{-}
```

7.5.5 Contact info

If you would like to contact me, whether it be to comment on my book, notify me of typos, or for any other reason, simply email me at my proxy email address, which has "**emailproxyj**" as its identifier and "**gmail.com**" as its host domain. Combine those two previous quoted pieces of text together to form an email address. Be warned however that I am sometimes slow to respond to my emails (especially on my proxy account) and often have a very busy schedule. I am constantly trying to learn new things and to expand my horizons and to pursue creative work, in addition to attending to many other things, so answering emails is often given a lower priority.

I'm trying to make the best of my life, just like all people. Don't take it personally if I don't respond soon. When I become fixated on learning something or creating something, I tend to become less attentive to emails and such and may not check for a while. I always have more on my to-do list than I can possibly ever do. My combined interests in game development, mathematics, logic, programming, hiking, art, music, gaming, and many other miscellaneous things together provide me with much more than a lifetime's worth of things to invest my time in. Completion of everything I want to eventually do has become utterly impossible, realistically speaking, so my life's all about prioritization now.

Anyway though, by the way, in case you can't guess why I told you my email address in such an obtuse way, it's to prevent spam bots from automatically harvesting it from the text of this book. Spam bots constantly scan the internet for text that's formatted like an email address. Any time a spam bot finds text that matches that format, they add it to their spam list. In contrast though, by writing my email address conversationally in prose, I make it so that only spam bots that can understand natural language would be able to detect it, which would be astronomically more difficult and computationally expensive to do.

And, with that, dear reader, our time together has come to an end. Thank you for coming along on this journey with me. I hope that at least something I have said in this book has added value to your life, whether through expanding your perspective on logic, providing random bits of life advice, or whatever else the case may be. It has truly been a joy to write this book. Keep fighting the good fight. Remember to question everything. Keep searching for new insights and better ways of living. Good luck. Oh, and remember that you can also rate and review my book online if you feel so inclined. That's the other channel of communication available to you, besides my proxy email address.

Also, I have a personal website, on which I will sometimes post various things of interest to me, which may also sometimes be of interest to you as well. Here's the address:

JesseBollinger.com

7.5.6 Other books by me

In the time since when I first wrote and published this book, I have also written and published a second book. In contrast to my first book though, which is very long at 800+ pages and is about logic and mathematics, my second book is very short at just under 50 pages and is about game development. My second book is a collection of various different useful game dev tips, each explained concisely to fill exactly one page each. I wanted to experiment with writing a book on the opposite extreme of the length spectrum relative to *Unified Logic*. My second book is much cheaper and much faster to read than *Unified Logic*, and every section inside it is completely self-contained and modular.

The next page contains a product description for that book, in case you're interested.

40 Underappreciated Game Dev Tips

Feature Overview:

- a collection of carefully selected game dev tips, designed to provide general advice on a broad range of subjects to help you avoid common pitfalls and to increase your chances of success

- a very concise and easy to read format, with little to no unnecessary fluff, designed so that the entire book can be read in less than a day with very little effort and with very little time investment, while still providing a lot of value regardless

- a mix of both high-level principles about how to work effectively and powerful special techniques, many of which are not widely known or not really adequately appreciated or fully understood, which will give you an edge over your competitors

Here's just a few examples of some included tips:

- how to be creative at will, regardless of whether you are feeling inspired or not

    (PS: This is just a shorter version of the same discussion we had in this book in the "Creativity at will" section on page 779. I included it for people who don't own *Unified Logic*. The *Unified Logic* version is better though.)

- why and how players become bored and how to prevent it using "game arcs"

- how to keep multiplayer games alive longer

- how to balance your games more effectively

- how to create puzzles much more easily

Made in the USA
Coppell, TX
16 November 2020